Inside
SolidWorks®

4th Edition

David Murray

THOMSON

DELMAR LEARNING™

Australia   Canada   Mexico   Singapore   Spain   United Kingdom   United States

Inside SolidWorks®
4th Edition

David Murray

Vice President, Technology and Trades SBU:
Alar Elken

Editorial Director:
Sandy Clark

Senior Acquisitions Editor:
James DeVoe

Senior Development Editor:
John Fisher

Marketing Director:
Dave Garza

Channel Manager:
Dennis Williams

Marketing Coordinator:
Stacey Wiktorek

Production Director:
Mary Ellen Black

Production Manager:
Andrew Crouth

Production Editor:
Jennifer Hanley

Technology Project Manager:
Kevin Smith

Technology Project Specialist:
Linda Verde

Technology Project Specialist:
Linda Verde

Editorial Assistant
Tom Best

For more information, contact:
Thomson Delmar Learning
Executive Woods,
5 Maxwell Drive, PO Box 8007
Clifton Park, NY 12065-8007.
Or find us on the World Wide Web at
http://www.delmarlearning.com

For permission to use material from the text or product, contact us by
Tel : 1-800-730-2214
Fax: 1-800-730-2215
www.thomsonrights.com

ISBN: 1-4180-2085-0

NOTICE TO THE READER

About the Author

David Murray is Training Manager for CADimensions, a computer-aided design software, hardware, and networking solutions provider based in East Syracuse, NY. Murray is the author of the Inside SolidWorks series, now in its fourth edition. He has used SolidWorks since its initial beta release in 1995, is a Certified SolidWorks Professional, and has been a Certified SolidWorks Instructor since 1997.

Acknowledgments

I would like to thank Daril Bentley and Carol Leyba for their ongoing professionalism in creating the *Inside SolidWorks* books. Many thanks also go out to all people involved at Thomson Delmar Learning for their efforts and for putting up with me when the book goes over budget. Thanks to SolidWorks Corporation for continually striving to make even better what is already the best solid modeling program in the world, and to the CADimensions team for being the number one SolidWorks reseller in New York. Finally, thank you to my wonderful wife who has never once stopped believing in me.

To Joe Smith, a true friend and constant source of inspiration.

Contents

How-Tos and Exercises

How-Tos

Exercises

Introduction

SOLIDWORKS IS PRODUCTION SOLID MODELING for the desktop in the Microsoft Windows environment. SolidWorks Corporation, the developer of the software, is focused on providing mechanical design solutions and depth of modeling for mechanical assemblies for desktop platforms. It is hoped that this book will allow you to learn and get the most out of the SolidWorks product.

Readership and Intent

Inside SolidWorks is written for the beginning to advanced SolidWorks user who wants to learn SolidWorks basics or expand existing knowledge of the software. This book takes the user through beginner to increasingly more advanced SolidWorks functionality and thinking by describing and showing through examples and step-by-step guides how to apply the software to a multitude of widely divergent manufacturing processes.

Students and those with little or no prior CAD experience will welcome the book's clear and easy-to-comprehend approach to solid modeling basics. In addition, each chapter ends with a Summary, which is intended to reinforce the fundamentals and key points that will help new users reach higher levels of expertise and experience in a very short time.

In short, *Inside SolidWorks* is intended to allow both the student and new user reach a highly productive level of SolidWorks program use. The book is also meant to guide current users of the software through specific commands and processes needed to model a wide range of parts from many manufacturing disciplines, and to serve as a reference for performing operations and solving problems.

Approach and Book Features

This book takes an easy-to-understand, step-by-step approach to Solid-Works functionality and its command procedures and combinations in a way that allows the reader to create real-world parts using a variety of manufacturing techniques. This approach is also intended to allow readers to extrapolate what they have learned to new situations and new types of parts.

At the same time, the book imparts "how-to" functionality. *Inside Solid-Works* discusses the "why" and "when" of applying functionality to projects. The thinking involved and the mastery of functionality are reinforced through examples and exercises. Because of the wide range of processes covered, many SolidWorks users will find the book an invaluable resource and reference guide to SolidWorks command functionality.

Hands-on examples are consecutively numbered and are called How-Tos. These are functions and processes you need to know how to perform in operating the software. Exercises bring several functions together to provide practice in building and manipulating parts and in performing more complicated processes. The text is also supplemented with consecutively numbered reference tables, with lists, and with notes, tips, and warnings. Notes highlight important information, tips provide time- and work-saving suggestions, and warnings steer you clear of actions that potentially cause problems.

Throughout the book are steps that point the way to a particular command or option. These steps are typically menu picks, but may also require clicking on a tab or button. In cases where these picks are listed, they will appear separated by greater than (>) symbols. For example, access to the Line command would appear as Tools > Sketch Entities > Line. This means that the reader should click on the Tools menu, followed by the Sketch Entities menu, followed by the Line command.

With regard to mouse functionality, the reader will often see the term drag used throughout this book. This is a common term in the computer industry and refers to holding the left mouse button down while moving the mouse. What will not be seen is the phrase "drag with the left mouse button," as the left mouse button is already implied with the term drag.

Version Specificity

The SolidWorks software is always evolving to bring to the end user an ever more powerful and full-featured solid modeling program. SolidWorks Corporation has periodically released major upgrades to its very popular software. The fourth edition of *Inside SolidWorks* is based on SolidWorks 2005, but includes special notes that highlight enhancements or new functionality in SolidWorks 2006. References to SolidWorks 2006 are highlighted, typically denoted with the label *SW 2006*.

This book contains all of the information you need to learn and productively operate the SolidWorks software, whatever version of the software you are using. However, future versions of SolidWorks may have enhancements that could not be foreseen or included in this edition.

Prerequisites to Using This Book

Readers of *Inside SolidWorks* should be familiar with the Windows operating system; preferably Windows 2000 or Windows XP. It would also benefit the reader to be somewhat familiar with Microsoft Excel. This only holds true if the reader wants to take advantage of SolidWorks Design Table (family of parts table) functionality or the Excel-based automated bill of material (BOM) routine.

Prior experience with mechanical design, drafting, or computer-aided design is helpful, but not a requirement. Prior experience with SolidWorks is not a requirement, as *Inside SolidWorks* will start new users on the right path from the beginning.

Images

Although every effort has been made to make *Inside SolidWorks* as accurate as possible, some of the screen shots or graphics may be slightly different than what you see on your computer screen. This is due to the nature of constantly evolving software and differences in operating system interfaces. Even so, we are highly confident this book will provide you with an in-depth and insightful educational experience with what has become one of the industry leaders in the solid modeling market.

Happy modeling!

CHAPTER 1

A SolidWorks Overview

THE ACRONYM *CAD* ORIGINALLY REFERRED TO COMPUTER-AIDED DRAFT-
ING. At that time, most of the companies using CAD software were using it
to create their 2D standard engineering layouts. Employees of those companies did not start out by creating elaborate 3D solid models. The technology
had not been introduced. These days, CAD is more often taken to mean
computer-aided design. When both technologies are referred to, CADD is
used (computer-aided design and drafting).

The world evolves, industry evolves, and with such changes CAD
evolves. Where once CAD was a better way of creating design drawings,
now it is a better way of creating a representation of an actual 3D object.
Early solid modeling software was extremely cumbersome and was not well
suited for designers. This is no longer the case. Solid modeling software has
become fairly easy to use, and gives designers many tools with which to alter the model and see those changes take effect immediately.

Though drawing boards have become nearly extinct, most offices still
revolve around detailed drawings. Some offices are paperless, sending 3D
models directly to rapid prototype machines from their computers. Stereolithography rapid prototype files (polygonal representations of a model)
can be used to create prototypes, sometimes in a matter of hours, allowing
designers and engineers to hold an actual part in their hands. Nonetheless,
most companies still need a hardcopy printed or plotted drawing, rather
than or in addition to an electronic solid model.

Where to Begin

Now that solid modeling has emerged as a standard process for mechanical design, the need to begin with a 2D drawing is no longer necessary. It is not that a 2D drawing is no longer needed. Rather, it is just not the starting point. The place to start the design process with a solid modeler is the solid model itself. The 2D drawing can then be generated automatically from the model. This process is much more efficient and allows designers to work in a mode more in tune with the human mind. It is much easier to visualize a 3D object than it is to visualize orthogonal 2D views of that same object.

Being able to generate a drawing from the solid model is only the beginning. There are huge downstream benefits to having an electronic 3D representation of an object that will hopefully be manufactured some day. 3D models of designs can be tested and analyzed on the computer before ever actually being built, thereby reducing errors and shortening a product's time to market. Assemblies can be assembled on the computer, and interference checks can be made. Mass properties can be obtained, and the amount of material for a specific part can be calculated—all automatically by SolidWorks.

Add-on programs can test a model for stress, strain, and thermal conditions; can find the resonant frequency of a part; can show air flow through or around a model; and much more. Animations can be created to show the dynamic movement of parts in an assembly, or to show assembly sequences for those putting the product together. Kinematics packages help people visualize dynamic motion and apply conditions such as gravity or spring effects. Photorealistic rendering utilities can produce beautiful images for use in such material as advertising brochures.

The benefits of solid modeling can easily be seen. Even so, there are many companies that continue to use drafting boards or use computers strictly for their 2D layouts. How can a mechanical design firm compete with another company employing solid modeling software, such as SolidWorks? The bottom line is that they cannot. This is the point at which CAD software ceases to become a luxury. To remain competitive in the computer age, one must learn to evolve with technology. It is not as much a question of *whether* your company should join the computer age or move out of the 2D environment as it is of *when*.

Basic Terms and Concepts

SolidWorks is much more than a simple CAD program. If it had to be summed up in one phrase, one could call SolidWorks a feature-based, parametric, solid-modeling design program. First, let's examine the terms associated with this statement, and then we can explore a few basic concepts that will help you learn the software.

Feature-based

The term *feature-based* is used to describe the various component properties of a model. For example, a part can consist of various types of features, such as extrusions, holes, fillets, and chamfers. These features constitute the overall part, just as parts constitute an overall assembly. The mechanics behind some simple features are discussed in Chapter 3. Figure 1–1 shows typical features on a simple part.

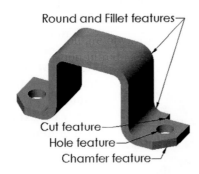

Round and Fillet features

Cut feature
Hole feature
Chamfer feature

Fig. 1–1. Basic feature types on a simple part.

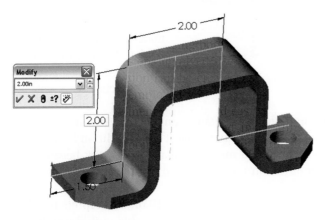

2.00

Modify
2.00in

2.00

1.50

Fig. 1–2. Example of parametrics: modifying a dimension to alter the geometry.

You will hear the word *features* used constantly throughout this book. Understand that any feature has a definition, which includes the parameters that define that feature. As an example, a hole may have a certain depth, or may contain a draft angle. Existing features can be edited, thereby changing the design of a model.

Parametric

Parametric is a term used to describe a dimension's ability to change the shape of model geometry if the dimension value is modified. This is a different way of thinking for most people accustomed to nonparametric CAD software. For instance, when lines or arcs are being created in a 2D drafting layout, accuracy is extremely important. Machine shop personnel might be taking measurements from the drawing. In many CAD programs, if the geometry is not of the correct size the dimensions themselves will not be accurate.

This is not the case with Solid-Works. Because SolidWorks is parametric, the dimensions drive the size and shape of the geometry, instead of the reverse. Figure 1–2 shows an example of what you might see when modifying a dimension (you will learn this technique in Chapter 3). Due to parametric capabilities, where dimensions are placed makes a big difference. This is known as design cussed in material that follows. It is one of the most important fundamental aspects of SolidWorks.

Solid Model

A solid model implies that there is actually a 3D solid model (with density and mass) that you can hold in your hands. These properties are true of the computer-generated solid model except for the ability to hold it in your hands. However, for all intents and purposes a computer model might as well be real. The model on the computer screen can be assigned material properties or a specific density, depending on what type of material it is to be made of. It has a center of gravity, otherwise known as its centroid. It also has weight and volume, at least as far as the computer is concerned.

You can rotate a solid model on the screen just as if it were actually sitting in the palm of your hand. You can measure it, and can extract a great deal of information from it (see Figure 1–3) as easily and perhaps more easily than if it were the genuine article.

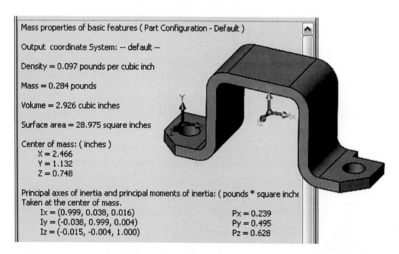

Fig. 1–3. Extracting mass property data from a solid model is only a few mouse clicks away.

There are a number of ways in which we observe material objects in our everyday lives. Most of us see the world as a series of 3D objects we must act upon or interact with. These objects may be organic or inorganic, animate or inanimate. If we pick up a rock from the ground we know that it has weight and that it has solidity. We know these things and much more about the rock because our senses tell us so. A computer, on the other hand, interprets data much differently.

Understanding the question of just how SolidWorks interprets the data it is given can sometimes help you feel more comfortable with the software. In the case of SolidWorks, the software understands all of the surfaces that make up the outer boundary conditions of the model. This is known as the model's *topology*. SolidWorks also understands what side of each face is the inside versus the outside. The various faces of the model are "knitted" by the software and in that way the software can understand that the model is an enclosed volume, and therefore "solid."

Sketch

We all know what the word *sketch* means, but in the context of working with SolidWorks it has special connotations. SolidWorks is a feature-based modeler, but in order to create most features, a sketch is needed. A sketch is typically a 2D profile that describes the shape of the feature you wish to create.

Creating a sketch that encompasses your design intent is a key to developing a good working foundation for using the SolidWorks software. A sketch is the starting point from which features are formed, and features are the objects that in turn make up the solid model itself. Not all features require a sketch, as you will see as you progress through this book. Some features, such as rounds or chamfers, can be placed directly on the model. These are known as applied features, as opposed to sketched features, both of which you will learn about in upcoming material.

Design Intent

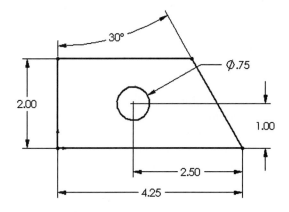

Fig. 1–4. Design intent, example 1.

What may be more important than any other aspect of the software is the fact that SolidWorks is a design tool. There is more to this simple statement than meets the eye. The ability to go back to some earlier stage in the design process and make changes by editing a sketch or changing some dimensions is extremely important to a designer.

Design intent refers to your intentions of how your model should behave when changes are made. The changes may be dimensional, or they may be of some other nature, such as components moving within an assembly. Because dimensions drive model geometry, design intent becomes very significant. Examine Figure 1–4. The person that placed the dimensions on this sketch determined that the angle of the line on the right should be at an angle of 30 degrees. Additionally, the hole in the sketch should be located from the bottom right-hand corner of the sketch. These parameters are what the designer deemed most important for this particular model.

Remember that the placement of dimensions is very important, because they are being used to drive the shape of the geometry. Take the horizontal 4.25-inch dimension in example 1 of Figure 1-4. If this value is increased, the overall length of the part will be increased and the 30-degree angle will be maintained. The hole will maintain its position from the bottom right-hand corner. If the locations of any of the dimensions are changed, how the geometry is driven will be changed.

In Figure 1–5 (example 2), you see that what is important to the individual designing this part is the horizontal dimension at the top of the part, rather than the angle of the line on the right. In this second example, changing the 4.25-inch horizontal dimension will alter the length of the sketch, but the angle of the line will change. The line's angle will not be maintained, as it was in example 1.

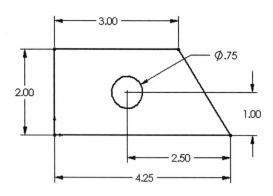

Fig. 1–5. Design intent, example 2.

◆◆ **NOTE**: *Keep in mind that the dimensioning scheme can be changed at any time. You are not locked into a specific design. You can also design without dimensions, rough out a sketch, and return to the sketch at a later time to dimension and fully define it.*

Take a look at one last example (example 3), shown in Figure 1–6. Perhaps the design calls for the hole to be a particular distance from the top left-hand corner, as opposed to the bottom right. In this case, the dimension arrangement would be best. But then again, it all depends on your *design intent*.

It is obvious that there can be quite a few variations on dimension placement. If there are so many choices, how can you be sure you will make the right decisions? The answer is very simple. Add dimensions on any part of the sketch geometry you want to control, and do not worry if you make a mistake. The ability to go back and add, remove, and change dimensions is never more than a few mouse clicks away. Chapter 3 covers the commands for adding, deleting, and modifying dimensions.

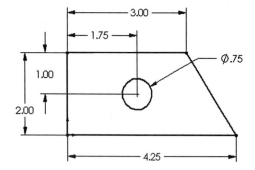

Fig. 1–6. Design intent, example 3.

◆◆ **NOTE**: *Do not be overly concerned with dimensioning to datum points or stacked tolerances in the part. These issues can be addressed in the drawing layout. Be more concerned with your design intent at this stage.*

Then and Now

Design intent is an important enough concept in SolidWorks that it cannot be overstressed. Primitive modeling software and nonparametric modelers did not incorporate the concept of design intent and were all but worthless from a design perspective. The reason for this is that primitive features lacked any editable definitions. In actuality, they were not features at all, at least not in the way we understand features from a SolidWorks standpoint.

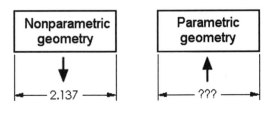

Fig. 1–7. Parametric dimensions drive geometry, such as in the example on the right.

Early modelers (and even some present-day modelers) lacked parametric dimensions. The main difference between a nonparametric modeler and a parametric one can be summarized in one simple concept. In a nonparametric modeler, the geometry drives the dimension; in a parametric modeler, the dimension drives the geometry. This simple concept is illustrated in Figure 1–7.

In the example on the right in Figure 1–7, there are question marks in place of the dimension value. This is to help illustrate that any value can be entered. Upon entering a value, the geometry then updates to reflect that value. This is the essence of parametric dimensions, but the simplicity of the concept belies its significance.

Without parametric dimensions, it would be extremely important to make certain the geometry in the model is created with the utmost accuracy. If it were not, it would probably be necessary to start over. When working with a parametric modeler, it is possible to create the geometry in an almost carefree way, because the work gets done when the dimensions get placed on the model. In essence, SolidWorks will ask "What would you like the value of the dimension to be?" You would type in a value, and the geometry would conform to that value.

Predictable Behavior

The benefits of a parametric modeler are far-reaching. Dimension values can be changed at any time, and the model will update accordingly. Instead of reworking the model, just tweak a few dimensions. This assumes that your design intent has been correctly established. If your intentions are accurately represented by the way in which the model is created, you should be able to predict how it will react if a change is made.

Examine Figure 1–8 to understand how parametric dimensions play a role in predictable behavior. For clarity, the sketch for the slot is being shown along with the dimensions. There is also a shaded view of the model shown as an inset. Note in particular the 2-inch dimension driving the length of the slot, and the 3.25-inch dimension that ties the center of the slot to the center of the part. If these dimensions are modified, we could easily predict how this feature will behave.

For the sake of argument, let's change the 3.25-inch dimension to a value of 2.25, and the slot length from 2 inches to 1.25 inches. This will shorten the slot, and at the same time draw its center in closer to the center of the model. Upon rebuilding the model, the results would look as they do in Figure 1–9, which shows the model in a shaded view.

Please don't be concerned with the technical aspect of editing dimensions or geometry at this time. These topics are covered in their entirety in chapters to follow. At this time, it is important that you first understand the basic concepts of working with SolidWorks.

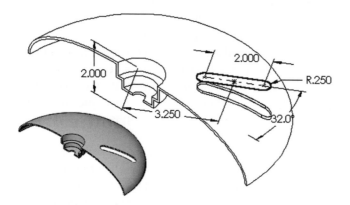

Fig. 1–8. Model with slot showing underlying dimensions.

Fig. 1–9. Results of changing two dimension values.

Benefits of Solid Modeling

Hopefully, some of the benefits of working with a parametric feature-based solid modeler have already become apparent. Another simple and immediate benefit would be the ability to see the part as a dynamic 3D image as it is being created, rather than a static 2D representation. It may sound odd, but the ability to visualize the model on screen allows you to make your mistakes faster. There are always going to be design issues. Working in 3D helps resolve such issues much more quickly.

Another benefit of solid modeling is the ability to determine material properties, as previously mentioned. Because the computer is constantly keeping track of the model and its structure and characteristics, mass property data can be extracted in a matter of seconds. Moments of inertia, section properties, and other data can be obtained at any stage of the design process.

Many CAD programs allow for some degree of shaded rendering. Because SolidWorks already keeps track of the topology information of the model, it requires no extra skills to render the model. Apply a texture or material to the model, add a few lights if desired, and when you are happy with the view save the image. This can be done through a screen capture or by saving an image directly within SolidWorks. If true photorealistic images are required, add-on programs such as PhotoWorks will accomplish this for you.

Third-party programs that completely integrate with SolidWorks let the user accomplish many tasks only dreamt of 10 years ago. Such programs include fluid dynamics, wire harnessing, and piping layouts, to name only a few. Questions regarding add-on programs should be directed to your local value-added reseller or SolidWorks salesperson. They are there to answer your questions and help you find the best products to suit your needs.

Why SolidWorks?

There are a number of reasons SolidWorks is an attractive software. Of primary importance is that SolidWorks was created with ease of use and powerful functionality at the top of the priority list. SolidWorks has a very intuitive interface and has the look and feel of other Windows-based programs. Many functions performed in SolidWorks mimic those you are already familiar with in Windows, such as renaming files or utilizing Windows Explorer.

SolidWorks was written exclusively to take advantage of the Windows operating system, and as such will run only on a Windows operating system or its future incarnations. Because SolidWorks was written for Windows from the ground up, it has many of the benefits of Windows-based programs. For example, cut-and-paste functionality exists throughout Solid-Works. The ability to drag a feature from one model into another for the purpose of duplicating the feature instead of recreating it is another great advantage. Drag-and-drop functionality exists throughout the software, further enhancing its ease of use.

SolidWorks also allows having numerous files open at one time. Toolbars can be resized and customized, much like toolbars in Microsoft Word or Excel. Common command sequences can be recorded and played back, thereby simplifying repetitive tasks. Excel-based spreadsheets can be used to create family part tables that drive part geometry or assembly design requirements. Objects, such as logos, can be embedded or linked to 2D drawings. Hypertext links to specification sheets can be added to SolidWorks documents, or other documents can be embedded directly in a SolidWorks file. These are just a small fraction of the capabilities you will have available to you while using SolidWorks.

SolidWorks Corporation has succeeded in creating world-class solid modeling design software with what could arguably be called unmatched functionality and ease of use, at a price point that puts it in reach of many companies and individual designers and engineers. Solid modeling may not be for the architectural or mapping professions. However, for most designers and engineers creating mechanical elements—whether toys or trains, springs or sprinkler systems—solid modeling is the way to make the most productive use of your time and increase your company's overall efficiency.

Running the Software

Obviously, you will require a computer. But what type of hardware is best? The minimum system requirements for running SolidWorks are always climbing, and individual needs differ. For these reasons, it is somewhat tough to pin down precise minimum requirements. Keep in mind that *minimum requirements* is a term that should fit loosely. Your minimum requirements will almost certainly be much more than those SolidWorks Corporation recommends.

To find out what the current minimum system requirements are as far as SolidWorks Corporation is concerned, access *www.solidworks.com* and view the data sheet on SolidWorks. It is actually quite vague. What follows is a more realistic approach to what is required to run the software. If you wish to discuss hardware requirements at a more detailed level, or have further questions, contact your nearest SolidWorks reseller for more information.

Processor

An Intel Pentium 4 or AMD Athlon processor will run SolidWorks just fine. The faster the processor the better, due to the large amount of mathematical computations constantly being performed. It is also best to run on a hardware platform with the fastest front-side bus speed available.

Is there a benefit from dual processors? If there is a question of purchasing a dual processor system over a faster single processor, you would be better off with the single processor system. However, SolidWorks is constantly evolving to take advantage of hardware developments. Multicore processors are imminent, and will enable multithreaded applications to process more quickly. Technologies must develop in unison, and thus expect to see SolidWorks take more advantage of multithreading technology as multicore processors become prevalent.

Should you obtain one of the latest 64-bit processors? This is a tougher question to answer. Do your homework and try to dig up some information on benchmarks. Until SolidWorks is compiled to run on a 64-bit platform, it will not be able to take advantage of a 64-bit processor. However, late-generation 64-bit processors have very efficient 32-bit pipelines and can run 32-bit code faster than their straight 32-bit predecessors.

Memory

The amount of memory you purchase depends on what you want to accomplish with the SolidWorks software. Even if you are not creating complex parts or assemblies, your SolidWorks workstation should have 512 MB. There is just no reason to have less. Let's face it, SolidWorks is cutting-edge, resource-intensive software.

For those who want to create assemblies of hundreds and even thousands of components, or if complex parts with hundreds of features or complex geometry are more your style, consider 1 GB or more of RAM. If large assemblies are the norm for you, 2 GB of RAM would not be overkill. Modern motherboards can carry as many as four modules of 1-GB memory, thereby raising the bar to 4 GB of total system RAM. Most users will not require this much memory, but at least the option is there for those who do.

Hard Drive

SolidWorks by itself takes a bare minimum of perhaps 600 MB of hard drive space. However, you will need more than that. Consider any add-on software you might purchase, where you will be keeping your work files, library parts, related documents, and so on. Realistically, if purchasing a new computer your minimum starting point should be a 60-GB hard drive. An SCSI drive is not necessary for a standalone SolidWorks workstation, but a rotational speed of 7,200 rpm minimum is recommended.

Graphics Card

There are traditionally two types of graphics cards sold on the market: gaming cards and workstation-class graphics cards. Workstation-class cards have a tendency to develop much more refined drivers that have a slant on quality and accuracy over fast frame rates. Graphics cards designed for workstation applications will often tweak their drivers for specific applications as well, including SolidWorks. Figure 1–10 shows the OpenGL driver settings for an Nvidia Quadro FX graphics card. Note that SolidWorks can be selected from a drop-down list.

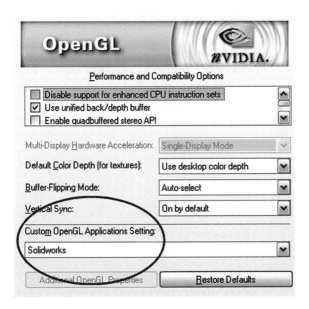

Fig. 1–10. SolidWorks can be selected from a list of applications.

Another aspect to consider when purchasing a graphics card is how important realism is when modeling. Solid-Works can apply very realistic materials to objects and display them in real time while manipulating the geometry on screen. This is known as RealView graphics, which requires a certain type of graphics card chip set.

Check the SolidWorks web site at *www.solidworks.com* for the current list of video cards that have been tested. The list includes information as to whether or not an individual card passed or failed, whether or not there is a setting specifically for the SolidWorks software, and whether RealView graphics are supported by the card. The list even includes what driver version was used when a particular card was tested.

Monitor

If you will be staring at a computer monitor a good portion of the day, you will want to see what it is you are doing without suffering eye fatigue. Selecting a monitor size is a major factor. There is not much of a decision to be made here, however. When doing CAD work, make sure the monitor has a minimum of 17 inches of diagonal viewing area. A 19- to 21-inch monitor might be better, assuming you have the budget and available desk space.

If you are considering a flat panel display, make certain your graphics card has the proper output connector and that the flat panel will actually be able to connect to it. Though flat panel displays are more expensive than their CRT counterparts, they use less energy, emit almost no heat, are much easier on the eyes, take up much less desktop space, and don't have the annoying flicker commonly associated with tube-based monitors.

Other Hardware

A CD-ROM drive is required for installing the software. However, if purchasing a new computer consider a CD-RW (rewritable) drive for backing up data onto your own CDs. For greater storage capacity, a DVD-RW drive can be used. DVD disks can hold about seven times the data a standard CD can hold. Dual-function units are also available that offer the ability to write to both CDs and DVDs, including rewritable versions of each, meaning that you can reuse the same disks over and over.

When purchasing a mouse, consider a three-button mouse, or more preferably a wheel mouse. Wheel mice are especially nice when running SolidWorks due to the ease with which a model can be rotated on screen, or zoomed in or out.

Additional Requirements

A Windows-based operating system is required. Currently supported operating systems are:

- Windows 2000 (Service Pack 4 or better)

- Windows XP (Service Pack 1 or better) – recommended

Additional software you should have is Internet Explorer 6 or later. Internet Explorer allows for taking advantage of certain online functionality associated with the SolidWorks web site. It is also recommended (but not required) that Microsoft Excel 2000, XP, or 2003 be installed. This is to gain advantage of bill of material (BOM) and design table functionality. There is BOM functionality available through built-in SolidWorks tables as well, but design tables will require Microsoft Excel.

The SolidWorks Interface

As previously mentioned, SolidWorks is strictly a Windows-based program. As such, it uses Windows conventions. An example would be shortcut keys (hot keys) used for standard Windows functions. Ctrl-C and Ctrl-V are the hot-key combinations used to copy and paste objects to and from the clipboard, respectively. These standard hot keys work for most Windows programs, not just SolidWorks. Considering that SolidWorks takes advantage of these standard Windows functions, typical Windows hot keys will function in SolidWorks the same way you would expect them to in any other Windows program.

Menus

There are a few basic ways of accessing commands in SolidWorks. The primary two methods are to utilize the pull-down menu structure (herein referred to simply as "menus") or to click on an icon located on a particular toolbar. There are keyboard shortcuts for some common commands, but for the most part SolidWorks commands are accessed either via toolbars or menus.

Menus are easily customizable. This can have both its good and bad side. On the good side, it is an absolute breeze to customize menus so that they only contain selections you prefer. The down side is that somebody not familiar with menu customization could be hunting for a command for a long time and never find it.

As is the case with any program, clicking on a menu will cause the menu to drop down and present you with a list of commands. At the bottom of every menu in SolidWorks is an option named Customize Menu, shown in Figure 1–11. Clicking on this option will present you with the same menu, but with checkboxes before each command. Unchecking a menu item removes it from the menu, and checking it brings it back. It's as simple as that.

What you should learn from this is that if a particular command does not appear where you think it should be try clicking on the Customize Menu option at the bottom of the menu. The command you are looking for may just be unchecked, making that command temporarily inaccessible from the menu structure.

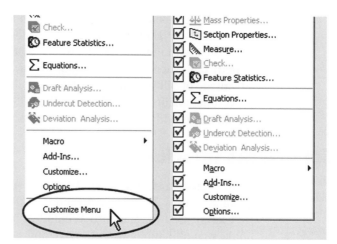

Fig. 1–11.
Customizing a menu.

Toolbars

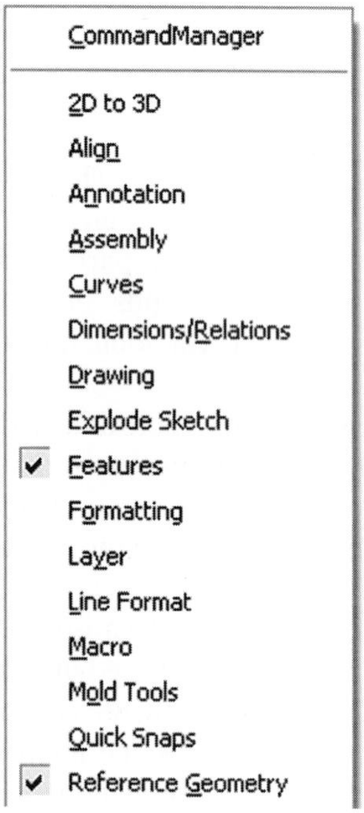

CommandManager

2D to 3D

Align

Annotation

Assembly

Curves

Dimensions/Relations

Drawing

Explode Sketch

✔ Features

Formatting

Layer

Line Format

Macro

Mold Tools

Quick Snaps

✔ Reference Geometry

Fig. 1–12. The Toolbars submenu, found in the View menu.

Many of the icons used in the SolidWorks toolbars are icons commonly used by many programs. The New, Open, and Save icons are the de facto standard icons found in many programs. This holds true for other icons. Of course, there are other functions particular to SolidWorks or solid modeling software in general for which icons had to be invented. Nonetheless, you will probably find that the icons employed by SolidWorks are easy to use and visually represent their functions quite well.

In the View menu, you will find the Toolbars submenu that lists all of the toolbars available to the SolidWorks user. The various toolbars can be toggled on and off by clicking on the toolbar's name. The upper portion of the View > Toolbar submenu is shown in Figure 1–12.

SolidWorks contains a number of standard and specialized toolbars. Included are toolbars you will not be able to do without, and those you may only occasionally use. Most of the available toolbars are customizable, and a few are not (such as the Font and Web toolbars). You should not worry about customizing the toolbars at this time. Become familiar with the program first. Once you can find your way around SolidWorks, then perhaps try customizing the toolbars.

✓ **TIP**: *An easy way to turn a toolbar on or off is to right-click on any docked toolbar.*

When the SolidWorks software goes through a new release, the toolbars sometimes change. This is due to functionality and enhancements being added to the program. Additionally, because most toolbars can be customized the toolbar images pictured in this book may look slightly different from yours. This is nothing to be concerned about, but is something you should be aware of in case you notice discrepancies between what is on your screen and what you see in this book.

All of the icons normally associated with file operations or with the Windows clipboard can be found on the Standard toolbar, shown in Figure 1–13. Users can also print or perform a print preview from this toolbar. Other commands include rebuilding a part, redrawing the screen's graphics, and accessing the help function. If you do not understand the exact function of some of the icons mentioned here, rest assured that their meanings will be explained over the course of this book.

Fig. 1–13. Standard toolbar.

*Fig. 1–14.
Drag a toolbar
to reposition it.*

As is the case with any Windows-compatible program, toolbars can be repositioned anywhere on screen, or docked to any side of the screen. To move a toolbar, drag the toolbar with your left mouse button. Place the cursor over the title bar if the toolbar is floating, or you can grab the toolbar by its drag "handle" if it is currently docked. This is illustrated in Figure 1–14.

FeatureManager

FeatureManager is a part of SolidWorks you should become familiar with right away. It is the key to editing your model and to understanding how a model was built. Think of FeatureManager as a chronological history of events, or a "feature tree." The feature you created first will be near the top of FeatureManager, and the feature you created last will be at the bottom. An example of FeatureManager is shown in Figure 1–15.

The work area needs little explanation. It is where you will see the model you are working on. It is where you will sketch and where your SolidWorks model will come to life. The triad, which can also be seen in Figure 1–15, represents the x, y, and z axes. The triad rotates as the model is rotated, giving the user an idea as to where the coordinate system axes are in relation to the model.

➥ **NOTE**: *It really is not necessary to display the triad for basic modeling. You can turn it on or off by selecting Tools > Options > Display/Selection section and changing the* Display reference triad *option.*

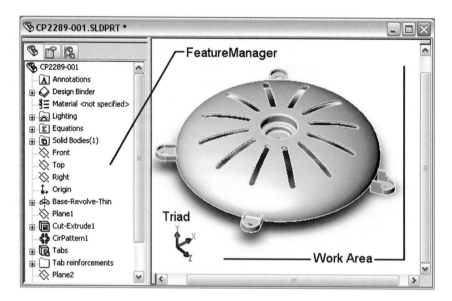

Fig. 1–15. SolidWorks' FeatureManager and work area.

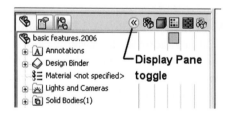

Fig. 1–16. FeatureManager tabs.

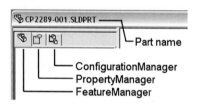

Fig. 1–17. Display Pane toggle switch.

FeatureManager Tabs

Tabs at the top of FeatureManager gain access to other "managers" within SolidWorks. There are a total of three tabs, shown in Figure 1–16. Sometimes there are more tabs, depending on the third-party software you may have purchased in addition to SolidWorks. The name of the SolidWorks file is displayed at the top of the window, which is also visible in Figure 1–16.

By clicking on any of these tabs, you can navigate to the various managers. ConfigurationManager is discussed in more detail in Chapter 11. PropertyManager's main function is to display the various options and input parameters when executing a command. Options and parameters vary depending on the command being executed. It is not typically necessary to utilize the PropertyManager tab to gain access to it, as PropertyManager automatically appears when accessing commands.

SW 2006: The Display Pane toggle switch hides or shows the Display pane, shown in Figure 1–17. The Display pane is discussed in more detail in Chapter 16.

✓ **TIP:** *PropertyManager buttons change appearance depending on the scheme (or "skin") currently in use. The skin can be changed by selecting Tools > Options > System Options > Colors and then changing the setting for PropertyManager Skin.*

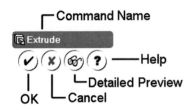

Fig. 1–18. PropertyManager buttons.

To complete commands, use the buttons at the top of the PropertyManager display. These buttons, shown in Figure 1–18, are self-explanatory. Once the applicable parameters have been input into PropertyManager, click on OK to accept the parameters and carry out the command, or on Cancel to exit without completing the command. The Help button is context sensitive, and will display help for whatever command you are currently engaged in. Next and Back buttons will not always be present. They only appear when PropertyManager requires multiple steps to complete a command.

The Detailed Preview button is only occasionally available for certain features. It gives a more refined preview of whatever feature is being created. Most of the time, it is not necessary to view a detailed preview, as the standard previews are enough to visualize what is taking place.

There are a few commands that require multiple steps. In such a case, Next and Back buttons (not shown) will appear in PropertyManager. These are simple arrow buttons that can be selected to move forward or backward through the command options, very similar to the forward or back buttons in most web browsers.

➥ **NOTE**: *PropertyManager should appear automatically when carrying out a command. If it does not, select Tools > Options > General section and then make sure the Auto-show PropertyManager option is checked.*

Fig. 1–19. Examples of pushpins.

Pushpins

Another button that makes an occasional appearance in PropertyManager is the pushpin, shown in Figure 1–19. They will also appear in other locations throughout the software. No matter where pushpins appear, their function is always the same, which is to make a command or window "sticky." That is, when a pushpin appears to be sticking into the screen, whatever command happens to be currently active will stay active so that you can use it multiple times.

Separator Bar

The separator bar is a vertical bar that separates FeatureManager and the work area. This vertical bar can be repositioned by placing the cursor over the bar and dragging the bar to a different position (as mentioned earlier, "drag" implies using the left mouse button). When the cursor is over the bar, the cursor turns into a double-sided arrow, enlarged for clarity in Figure 1–20. When this symbol appears, repositioning the bar is possible.

From time to time the separator bar needs to be repositioned so that the names of the features in FeatureManager can be seen in their entirety, or so that more of the work area can be used. If this vertical bar is repositioned and the file is saved, the next time the file is opened SolidWorks remembers where the bar was placed.

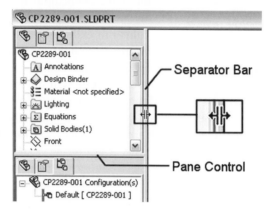

*Fig. 1–20.
Separator bar
and pane control.*

When repositioning the vertical bar, there is one position where the bar will want to "snap" into position. This is the ideal position for the vertical bar in most cases. It will allow you to see options in PropertyManager in their entirety and still maintain a maximum work area.

Pane Control

Another aspect of FeatureManager not quite as apparent as the vertical bar is a horizontal bar that allows for multiple panes. This horizontal bar is found at the very top of FeatureManager. As with the vertical bar, the cursor will change into a double-sided arrow when positioned over the horizontal bar. When this happens, drag the horizontal bar down to a new position, thereby creating a new pane in FeatureManager. An example of this can be seen in Figure 1–20.

Once FeatureManager is split into two panes, the FeatureManager tabs can be used in any combination for each pane. As discussed earlier, these tabs are for PropertyManager and ConfigurationManager. To change back to having one pane, simply drag the horizontal bar back to its original position at the top of FeatureManager.

Flyout FeatureManager

When in a command, it is often necessary to have access to FeatureManager. This is usually due to the need to select objects contained within FeatureManager. For this reason, FeatureManager will appear with a transparent background in the graphics area while in a command. An example of this is shown in Figure 1–21.

✓   **TIP**: *Clicking on the command name at the top of PropertyManager will expand or collapse the flyout FeatureManager.*

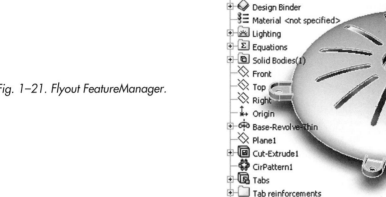

Fig. 1–21. Flyout FeatureManager.

Task Pane

Fig. 1–22. SolidWorks
Resources tab in the Task pane.

The Task pane allows for conveniently accessing resources or files that exist outside currently open or active SolidWorks documents. These resources range from links to SolidWorks documentation to accessing files located on your computer that you may wish to open in SolidWorks. An example of the Task pane displaying the SolidWorks Resources tab is shown in Figure 1–22.

There are three main categories within the Task pane. These categories are the aforementioned SolidWorks Resources (shown in Figure 1–22), the Design Library, and File Explorer. The various components of the Task pane can be accessed via tabs typically near the right-hand side of the SolidWorks interface (shown in Figure 1–23). There is also a button that allows for expanding or collapsing the Task pane.

Fig. 1–23.
Task pane tabs.

The Design Library option is an important aspect of the Task pane, as it allows for reusing geometry you find need for on a regular basis. The Design library is fully customizable (discussed in Chapter 9). File Explorer is exactly what it sounds like—a convenient way of accessing files on your computer or network.

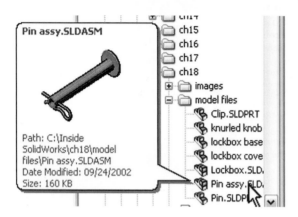

File Explorer is basically a miniature version of Windows Explorer, and functions in the same manner. Right-clicking on files in File Explorer will gain access to the familiar-looking menu you are accustomed to when right-clicking on files in Windows Explorer. There are also enhancements befitting the SolidWorks File Explorer, such as a preview image that appears when the cursor lingers over a file or an object in the Design library. An example of this is shown in Figure 1–24.

The entire Task pane can be docked to the left- or right-hand side of the Solid-Works interface, or can be left floating in space. You may decide to position the Task pane in a docked position on the left-hand side of the screen. This way, it can share the screen real estate used by FeatureManager, thereby maximizing your work area.

Fig. 1–24. File preview in File Explorer.

✓   **TIP:** *For those familiar with HTML code and web page creation, the SolidWorks Resources tab is driven by an HTML web page named* wel-com.html *located in your* SolidWorks\data\welcomepage *folder.*

Document Associativity

There are many different file types associated with SolidWorks, but there are three documents you will be primarily working with. These documents are parts, assemblies, and drawings. For those accustomed to working on a CAD system with a solitary file type, realizing that SolidWorks has three file types might come as a surprise. For those coming from an AutoCAD background, it is common to refer to a CAD document as a drawing. A single drawing might in actuality be a genuine 2D design drawing, or it might be a 3D model, an assembly, or something else.

SolidWorks part files are typically a single chunk of geometry. They may be simple or complex, but they are usually a single contiguous object. Multiple bodies can exist in a part, but this is more for design flexibility. Part files have an *SLDPRT* file extension.

Assemblies contain part files. An assembly may contain only a few components, or it may contain 10,000 components or more. Each component is a part file in its own right. When a part is inserted as a component in an assembly, the assembly references the part file on the computer hard drive. Assemblies have an *SLDASM* file extension.

Finally, there are drawings, the third SolidWorks document type. The term *drawing* will be used in this book to refer to an actual 2D design drawing. A drawing is a 2D representation of a part or assembly. Drawings reference part and assembly documents on the hard drive. Drawings have an *SLDDRW* file extension.

Now that you understand what the three main document types are in SolidWorks, let's get to the interesting aspect of this discussion, which is associativity. What this means is that when a part is altered, and that part is part of an assembly, the part also changes within the assembly. If there is a drawing of that same part, the drawing automatically updates to reflect the changes made to the part.

If you are working in an assembly and decide that a particular part isn't going to work, the part can be redesigned within the assembly. The changes made to the component will be reflected in the part file automatically. Additionally, changes made to dimensions in a drawing will be reflected in the part or assembly file.

The fifty-cent term used to describe the behavior we have been discussing is *bidirectional associativity*, illustrated in Figure 1–25. All three SolidWorks document types are inexorably linked to one another. This behavior causes issues, but nothing that cannot be resolved. For instance, sending an assembly to a co-worker means they must also have the part files being referenced by the assembly, or they will not be able to fully utilize the assembly file.

Yes, having three separate file types associated with one another does pose some challenges and may raise some interesting questions. However, this is just one of the many topics you will learn about as you make your way through this book. After the next chapter, you will be ready to start getting your hands dirty.

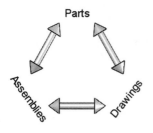

Fig. 1–25. Bidirectional associativity.

Summary

In this chapter, you have discovered that SolidWorks is a feature-based, parametric, solid modeling design tool. Creating solid models is the proper starting point when working with SolidWorks, and design drawings are then generated from the model. You also learned some very important terms associated with the core of SolidWorks functionality.

Parametric dimensions within SolidWorks help to build intelligence into the models you create, thereby making design changes a simple task. The fact that SolidWorks incorporates parametric dimensions makes it a sign tool. Design intent is a very important concept in SolidWorks. Because dimensions drive geometry, it is important where those dimensions are placed. Ideally, you should be able to predict how the model will behave if dimension values are changed.

You have also learned in this chapter that there are a great many benefits from working with solid modeling software. These benefits include the ability to see and manipulate the shaded model on screen. Other benefits of solid modeling include the wealth of information you can obtain from solid geometry (such as mass properties) before a part is ever actually built. There are also the numerous third-party applications, such as finite element analysis or photorealistic rendering software, which let you extend the functionality of SolidWorks far beyond its initial scope.

An important aspect of this chapter was the required hardware needed to run the software. If ordering or building a new computer to run SolidWorks, you should now be able to make a more intelligent decision as to how the computer should be configured.

This chapter introduced you to the SolidWorks interface, and how the program's look and feel is similar to many other programs designed to run on the Windows operating system. You should be familiar with basic toolbar and menu operation, and FeatureManager should be a new addition to your vocabulary. FeatureManager is a history tree, a chronological timeline of the features you have created. Those same features comprise the model shown in the work area.

The Task pane allows access to various SolidWorks resources, including web sites, documentation, the Design library, and files and folders located on your computer or obtained over a network. The Design library contains features and models that can be added to open SolidWorks files. File Explorer represents a very convenient way of accessing data within SolidWorks, without having to open Windows Explorer.

Finally, you have learned that there are primarily three file types you will use when working with SolidWorks. These files are parts, assemblies, and drawings. Drawings are 2D representations of parts and assemblies. Assemblies are collections of parts. All three documents are interrelated and exhibit bidirectional associativity. In other words, changing dimensions in one file (such as a part file) will automatically propagate to the other files (such as an assembly or drawing).

CHAPTER 2

Getting Started

BEFORE YOU CAN WORK WITH SOLIDWORKS, you have to know how to think like SolidWorks. This is especially true if certain habits have been developed from working with other software programs. Certain programs are easier than others from which to make the transition. Those who have worked with other parametric solid modelers usually find the transition to be quite smooth. Those who have worked strictly in 2D usually have the most difficult time, because of the old habits and mind-set developed when working with such programs.

For the CAD population accustomed to working in a 2D environment, the jump to 3D solid modeling is not something you should let intimidate you. SolidWorks is easy to learn, and there is actually a fair amount of entertainment value that can be gained from the software. Once you have created a few models, you will probably find that being able to view the shaded model on screen is something of a rewarding experience.

A Sketch Overview

A sketch is nothing more than a collection of lines, arcs, or other similar geometry, whose purpose is to define a feature. SolidWorks is in fact a sketch-based modeler. A sketch is the 2D geometry foundation that precedes the 3D solid model. As with any foundation, whether it be for a house or for CAD software, the way in which the foundation is laid is extremely important.

When speaking of a parametric modeler, the ability to edit dimensions and change the shape of a model is the most obvious aspect of parametrics. This functionality presents itself within the first sketch drawn, and is sometimes unexpected by those who are used to working in a nonparametric 2D drawing environment.

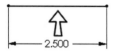

Fig. 2–1.
Dimensions drive
geometry.

When creating sketch geometry in SolidWorks, it is not necessary to create geometry with 100-percent accuracy. At first this may seem counter-productive, but in reality it is not. The sketch geometry becomes accurate when the dimensions are added, because the dimensions drive the geometry (illustrated in Figure 2–1).

Try to take the meaning of the word *sketch* literally. It need only be the approximate size and shape of the feature (or part) you are trying to create. When dimensions are added to a sketch, the geometry will change size and shape, depending on the values of the dimensions. This is where the work gets done, and is the essence of parametric modeling. This way of thinking goes hand in hand with the term *design intent*, as mentioned in the previous chapter. Where dimensions are placed will make a difference as to how your model behaves when a dimension is altered. It is very important to remember this fact.

The first feature built is considered the base feature. The sketch used to create the base feature should describe the basic overall shape of the part, but should not contain every bit of information needed to describe every feature. It is usually best to keep things simple. This facilitates creating a part with a greater number of features, rather than creating elaborate sketch geometry and fewer features. Following this guideline will give you more control over the part and make editing the part an easier task. Keeping sketch geometry simple also aids in troubleshooting if and when things go wrong.

Sketching is done not on a sheet of paper, but rather on a geometric plane. It is also possible to sketch on any planar (flat) face of the model. Planes can be defined in a number of ways (discussed in Chapter 4). However, SolidWorks gives you three planes to start out with, so getting started is easy.

Starting a New Part

Because you will be exploring the SolidWorks interface in the material that follows, it would be best if you knew how to start a new part. This way, you can follow along on your computer. How-To 2–1, which follows, shows you how to start any new SolidWorks document, whether part, assembly, or drawing.

How-To 2-1:   Starting a New SolidWorks Document

To begin a new SolidWorks document, perform the following steps.

1. Select New from the File menu, or click on the New icon on the Standard toolbar.

2. Click on the Advanced button, if present (the button toggles between Advanced and Novice new document interface modes).

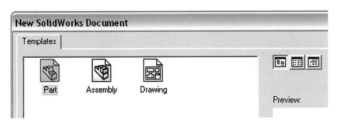

Fig. 2-2. New SolidWorks Document window.

3. Select the type of new file (Part, Drawing, or Assembly) from the New SolidWorks Document window, shown in Figure 2-2 (your tabs may look different).

4. Click on OK to start the new file.

Usually when starting a new document, you will want to specify a particular template to begin with. This can be done while in the Advanced new document interface mode. The more generic Novice mode allows for specifying the document type only. Default settings will be used for the new document. Feel free to use the Novice mode for now, but you will likely want to switch to Advanced mode once you have learned how to create your own templates.

When using the Advanced mode, the icons present in the New Solid-Works Document window are the actual templates, which contain predetermined settings, such as dimensioning standards or working units. In the next chapter, you will learn how to create templates and customize the tabs found in the New SolidWorks Document window. For now, let's continue exploring the basics and show you how to save your work.

✓ **TIP:** *Double clicking on a template will start a new document using that template.*

HOW-TO 2-2: Saving Your Work

It's a good habit to save your work right away. The first time a document is saved, it will need to be named. Always bear in mind the golden rule of computing, which is to save regularly and often. To save your work, perform the following steps.

1. Select Save from the File menu, or click on the Save icon found on the Standard toolbar.

2. Enter a name for the file. Navigate to an appropriate directory, if desired.

3. Click on the Save button to save the file.

If a change is made to a SolidWorks document and it has not been saved, an asterisk will appear to the right of the file name in the title bar. This is to let you know the file has not been saved since a change was made. Incidentally, it is never necessary to enter the file extension when saving a SolidWorks document. The software does that automatically.

10-second Topic: Tooltips

Fig. 2–3. Using tooltips.

Tooltips are the small hints that appear next to an icon when the cursor hovers over it for a second. These tooltips can be small, or they can be somewhat larger, with an expanded description of what a particular icon does. It would be beneficial at this stage to turn on the larger tooltips, as they will help to teach you the software.

With a SolidWorks document open, you can enable large tooltips. To do so, select Tools > Customize, and then select the Toolbars tab (which should already be selected). Look near the bottom left-hand corner of the window and make sure the *Use tooltips* and *Show large tooltips* options are both checked (as shown in Figure 2–3), and then click on OK.

Deciding on a Sketch Plane

One of the first decisions you need to make is which plane to begin sketching on. It is important to visualize the part you are about to create before actually beginning a sketch. For example, what is going to be the top of the part, what will be the front, and so on. It is possible to change a model's orientation after the design process has been started, but it is much easier if you make up your mind first and get the part's orientation correct to begin with. Once an initial sketch plane has been decided on, the first sketch can be created and the first feature can be built.

There are really only two reasons for wanting to start on the proper plane. The first is aesthetics. You want the part to look right when viewing its top, for example. If you were designing a wine glass, you wouldn't want it to appear upside down, would you?

The second reason involves generating a 2D design drawing (covered in Chapter 12). Standard top, front, and right side views are generated automatically. If the part is oriented correctly to begin with, less manipulation is required to generate the proper views.

Fig. 2–4. Determining what should be the top of the part.

Not all parts have an obvious choice for what should be top or front. In such a case, make a judgment call. The part shown in Figure 2–4 is a good example. The part is a guard for an engine. When the actual model gets manufactured, the guard will be placed at the top of the engine, and the central axis of the guard will point toward the sky. Knowing this fact, we could then make a determination as to which plane to begin sketching on.

The engine guard's base feature is a revolved feature (discussed in Chapter 5). Therefore, the front plane would suffice for the first sketch plane. After the sketch was revolved, changing to a top view would put the viewer's line of site directly down the axis of revolution.

A question that often gets asked is "Can the orientation of the part be changed after it has been created?" The answer is yes, it can, but this is not the time to discuss that topic. You will discover a method used to reorient a model in the next chapter (see "Updating Views").

Using the Mouse Buttons

SolidWorks does not require a three-button mouse, but it would definitely be to your advantage if your SolidWorks workstation had one. If Solid-Works is your primary tool on the computer, it would be beneficial to invest a little extra cash to obtain a three-button mouse. Even better would be a three-button mouse that has a wheel for the middle button. The reason for this is explained in the material that follows.

Selecting Objects

The left mouse button's primary purpose is to select objects and initiate commands. Objects can mean anything, from sketch lines to planes, features, or any other geometric entity. Commands might be selected via the menus, or perhaps through a toolbar. This is no different than any other Windows program.

There are many pick-and-drag operations used with the left mouse button that can be performed when sketching, and at other times, as you will discover shortly. The left mouse button can also be used to edit dimensions through double clicking, or for drag-and-drop functionality. These functions will seem like second nature to you after you have received a little hands-on experience with SolidWorks.

When an object is selected in the work area, it becomes highlighted with a different color (usually green). It is also possible to select objects in FeatureManager. To select multiple items, you must hold down the Control (Ctrl) key. This rule is lifted if in a command, and you can simply click on one or more objects to select them.

To deselect objects, simply click anywhere in a blank (empty) portion of the work area. If in a command, it is possible to deselect objects by clicking on them a second time.

The Context-sensitive Menu

The right mouse button is used for bringing up what is known as a context-sensitive menu. By clicking the right mouse button (i.e., right-clicking), you can bring up a menu containing various options. The options that appear are dependent on what is right-clicked on. Hence the term *context sensitive.*

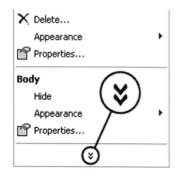

Fig. 2–5. Click to expand the menu.

Right-clicking while working in sketch mode may bring up a number of shortcuts to often-used commands. Right-clicking on a feature or sketch in FeatureManager will result in a menu containing various editing options. As you begin to learn Solid-Works, you will see that the right mouse button menus are very predictable. The context-sensitive menus are not filled with needless commands, but instead have a very logical and well thought out selection of choices that directly pertain to the task at hand.

Occasionally the context-sensitive menu will be expandable. This occurs when there are more options than can comfortably fit in the menu, in which case a truncated version of the menu is shown. To expand the menu, click on the "double V" symbol at the bottom of the menu, and you will be presented with the full menu (Figure 2–5).

Middle Mouse Button Functionality

If you have access to a three-button mouse, you will find the middle mouse button very useful in rotating, zooming, and panning. By holding down the middle button and moving the mouse, the model on the screen can be rotated, identically to using the Rotate command. Rotating a model can be a little cumbersome to the uninitiated, but will soon become a function any SolidWorks user will not want to live without.

To zoom in or out with the middle mouse button, hold down the Shift key while depressing the middle mouse button and moving the mouse. To pan, use the Ctrl key. Panning, zooming, and rotating model geometry are discussed in detail in Chapter 3. If you feel like experimenting, by all means open a sample part and try out the middle mouse button.

Some three-button mouse drivers will program the middle mouse button for a special function, such as an automatic double click. This can be very useful in Windows or other programs, but will keep SolidWorks from assigning its own functionality to your mouse's middle button. If your middle button is not functioning properly in SolidWorks, remove any special functionality assigned to the middle button by your mouse driver software. Typically, this can be determined via the Windows control panel by double clicking on the Mouse icon. Contact your company's system administrator or your hardware vendor if you have further questions.

If the middle mouse button of your mouse is a wheel, zooming just got easier. Simply rotating the wheel will allow for zooming in and out of a model. All other middle mouse button functions described previously will still function as advertised.

✓   **TIP:** *When zooming in to a model using the wheel, position the cursor over the area you would like to zoom in to. This welcome functionality is known as a scrolling zoom.*

Toggling Plane Display

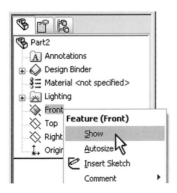

Fig. 2–6. The Show command.

The right mouse button can be used to toggle planes on or off (in SolidWorks terminology *show* or *hide*, respectively). The Show and Hide commands are accessed via the right mouse button's context-sensitive menu, and act as toggle switches. That is, a plane is either being shown or hidden at any one time. The three default planes that exist in any new part document are listed in FeatureManager and are named Front, Top, and Right. The act of showing the front plane is illustrated in Figure 2–6.

Show all three planes to see what they look like. If you are unaccustomed to right mouse button menus, try to become more familiar with this functionality. Right-clicking on objects is a crucial aspect of utilizing the SolidWorks software.

✓ **TIP:** *If planes do not display in the work area after showing them, make sure they are globally enabled by checking the Planes option under the View menu.*

Planes being shown will appear as a gray outline or translucent (read on to discover the option that controls plane appearance). If the plane border appears with small square "handles," it is because it is still selected. Deselect it by clicking in a blank area of the screen. Be aware that it is possible in the work area to right-click on the plane itself to hide it. You must click directly on the plane's border for this to work. When a plane is hidden, its icon in FeatureManager appears ghosted, as is the case with any hidden object.

It should be noted that planes actually extend infinitely in all directions, and that the gray border is there for display purposes only. When sketching on a plane, you are not limited to sketching within the visual borders of the plane.

A very nice display option in SolidWorks is the ability to display planes using a translucency effect that shows one side of a plane in one color and the opposite side of a plane in another color. It is suggested that you turn this option on. How-To 2–3 shows you how to change various settings with regard to plane display.

HOW-TO 2-3: Plane Display Settings

It is assumed you have already shown at least one plane prior to performing the following steps. First, you will want to make certain the global setting for displaying planes is enabled. Note that this setting is also a nice way of hiding every plane in the part file with one click.

1. Click on the View menu and make sure Planes is checked. If not, click on Planes.

   To display planes with transparency, rather than simple gray borders, continue with the following steps.

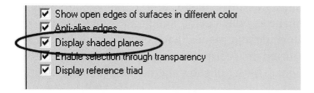

Fig. 2–7. The Display shaded planes *option.*

2. Select Options from the Tools menu.

3. Click on the System Options tab and select the Display/Selection section.

4. Check the *Display shaded planes* option (Figure 2–7), found near the bottom of the window.

✓ **TIP:** *The triad display can be turned on and off via the* Display reference triad *option, found below the* Display shaded planes *option.*

To control the transparency and color of planes, continue with the following steps. If running through this chapter for the first time, it probably is not prudent to change the color and transparency of the planes at this time. That information is only here for your future reference.

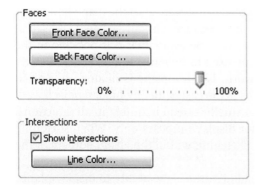

Fig. 2–8. Plane display options.

5. Click on the Document Properties tab found to the right of the System Options tab. Note that the Document Properties tab is only available when a document is open.

6. Select the Plane Display section. This will display the options shown in Figure 2–8.

7. Make the desired changes to the front and back face colors of planes and change the transparency. A setting of 90 to 95% works well.

8. Click on OK when finished.

For your reference, the *Show intersections* option (shown in Figure 2–8) will show the intersections of planes as dashed lines. The color of these lines defaults to black, but can be modified by the user. Try showing all three planes and rotating the planes using your middle mouse button. This will give you a feel for the 3D Cartesian coordinate system you will be working with.

✓ **TIP:** *If unaccustomed to working in a 3D Cartesian coordinate system, it is advisable to show only the plane you wish to sketch on. This will help keep you spatially oriented to your current sketch plane.*

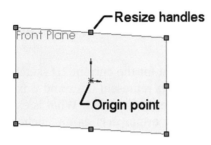

Resize handles

Front Plane

Origin point

Fig. 2–9. Plane resize handles.

Planes can be resized by dragging the small, square resize handles (shown in Figure 2–9). As previously mentioned, planes actually extend infinitely, but the visual representation of those planes can be any size. The reasons for resizing plane borders have to do with aesthetics and convenience. Plane border size does not physically affect the model geometry in any way.

A method of keeping a plane's border at a respectable size is to use the Autosize option. Autosize resizes a plane automatically to keep it the same size as the surrounding geometry. When the model becomes larger or smaller, so does the plane.

The Autosize option is automatically turned off if a plane is manually resized. Likewise, the Autosize option will not be available if Autosize has already been implemented. To experiment, try resizing a plane using the resize handles mentioned in the previous material. Once you have done that, right-click on the plane in FeatureManager and the Autosize option will be present in the menu.

The Origin Point

Fig. 2–10. The origin point.

The origin point is where the three default planes intersect and is indicated by two arrows in the center of the SolidWorks screen. Think of the origin (as it is commonly referred to) as the World 0,0,0 reference point. 0,0,0 refers to the *x-y-z* coordinates of the Cartesian coordinate system. The small arrow represents the *x* axis, and the long arrow represents the *y* axis. The origin is shown in Figure 2–10.

If there is no origin displayed, it may be because its display has been turned off. To see if this is the situation, click on the View pull-down menu and make sure there is a check in front of the Origins option. If there is not, select it to turn origin display on.

There are two types of origins: the World origin and the sketch origin. The World origin is represented by the arrows you see when starting a new part (or assembly) document. The sketch origin, the more important of the two origin types, will appear whenever you begin a new sketch. Sketch origins will appear a different color (typically red or blue), to differentiate them from the World origin. (Sketch origin arrow color is dependent on your version of SolidWorks. This book assumes that the arrows are red.)

The red sketch origin arrows are not always in the same position as the World origin point. If in a sketch, look for the red arrows. You may have to alter your view to see them, but they will definitely be there. The red sketch origin arrows are always aligned with whatever plane you happen to be sketching on.

✓ **TIP:** *If the view is rotated, the origin will also rotate. This helps to give an indication of your current sketch plane and can aid in establishing your bearings.*

The sketch origin acts as a 0,0 reference point for the current 2D sketch plane. The short and long red sketch origin arrows represent the *x* and *y* axes, respectively. However, these details are not very significant. What is extremely important, however, is that the sketch origin acts as an anchor point. Do not let this seemingly trivial fact escape you!

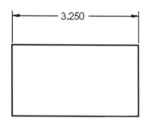

Why is this so important? The importance may be summed up in two words: *predictable behavior.* If a rectangle were sketched in space and not anchored in position, and a horizontal dimension were added to the rectangle (see Figure 2–11), which side of the rectangle would move if the dimension were changed? Try to answer this question prior to moving on to the next paragraph.

Fig. 2–11. What side moves if the dimension is changed?

Did you come up with an answer to the question? Admittedly, this was something of a trick question. The answer should have been "it is impossible to tell." The rectangle does not exhibit predictable behavior because it has no point of reference. If as a designer you cannot predict how your model is going to behave, you are certainly not going to be able to incorporate your design intent into the model.

Using the same scenario, if the rectangle's bottom left-hand corner were anchored at the sketch origin, you could say beyond a shadow of a doubt that it would be the *right* side of the rectangle that moves if the dimension is changed. In this case, predictable behavior is exhibited by the geometry, and your design intent can be maintained. This is a simple example, but the theory of predictable behavior encompasses nearly every aspect of SolidWorks.

In summary, the sketch origin serves as a point of reference for your first sketch. This is because of the origin's ability to anchor geometry. Subsequent sketches will typically reference existing geometry rather than the sketch origin.

Starting a New Sketch

Sketch

Fig. 2–12. Sketch icon.

Once a plane has been chosen, it is time to enter sketch mode. This is done by clicking on a plane to select it, and then clicking on the Sketch icon (shown in Figure 2–12). The order is not important. In other words, you can click on the Sketch icon first and then select a plane. When starting the first sketch, SolidWorks will ask which plane you would like to begin sketching on if one has not already been selected. SolidWorks will also temporarily show the three planes for you in the work area.

SolidWorks is a sketch-based modeler. What this means is that most of the features created in SolidWorks require a sketch. These feature types can be referred to as *sketch-based features.* The other type of feature that can be created is known as an *applied feature.* An example of an applied feature would be a fillet or chamfer applied directly to the model, which would not require a sketch.

Because SolidWorks is a sketch-based modeler, it is imperative that you understand whether or not you are in sketch mode. So how can one tell if they are in a sketch? There are a number of things that indicate you are currently editing a sketch, but the most obvious indicator is the following.

- The Sketch icon will appear pushed in.

This is the first sign you should look for if you are not sure you are in sketch mode. Following this obvious indicator, some other signs are:

- The red sketch origin arrows will be visible (though the view may need to be changed).

- The title bar will indicate precisely which sketch is being edited.

- The status bar will indicate that a sketch is being edited.

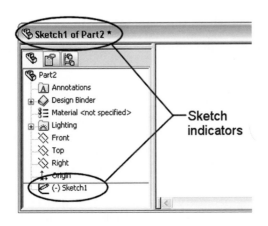

Fig. 2–13. Title bar while sketching.

Figure 2–13 shows what the title bar might look like when a sketch is being edited. If a sketch is not being edited, only the name of the model will be shown. What could be easier? With all of these indicators, there is no excuse for not knowing if you are in sketch mode or not.

When a new sketch is begun, the sketch will appear listed at the bottom of FeatureManager. This can be seen in Figure 2–13 as well. Do not be concerned with the number following the sketch. SolidWorks automatically increments the number of every new sketch. If the current sketch is exited and another sketch begun, it would be named *Sketch2*. This is SolidWorks' way of keeping track of things. This holds true for a number of objects in SolidWorks. Part file names are another perfect example. When a new part is started, its default name is *Part1*. Subsequent new parts would be named *Part2*, *Part3*, and so on. Obviously, you can rename the part files anything you want.

✓   **TIP:** *To exit a sketch, click on the Sketch icon a second time. If you exit a sketch by accident, right-click on the sketch in FeatureManager and select the Edit Sketch option.*

Creating Geometry

At this point, you have learned many of the fundamental aspects of the software. You have also learned how to start a new part and how to start a new sketch. Now you are ready to create some geometry.

There are two methods that can be employed when sketching any type of sketch entity. These methods are known as click-drag and click-click. Each has its benefits, which are explored in the following section. Sketching lines is the most basic, so let's start there. Both methods of sketching can be used with other entity types (such as circles, arcs, and so on), but for the sake of describing both sketch methods the Line command will be used. How-To 2–4 takes you through the process of sketching a line using the click-drag method.

How-To 2-4:   Sketching a Line Using the Click-Drag Method

To sketch a line using the click-drag method, perform the following steps. It is assumed you have entered sketch mode at this time.

1. Click on the Line icon, found on the Sketch toolbar, or select Tools > Sketch Entity > Line.

2. Hold down the left mouse button where you want the line to start.

3. Move the mouse, and then let go of the mouse button to position the second endpoint.

The first thing you will notice upon clicking on the Line icon is how the cursor changes appearance. Where once there was an arrow there is now what appears to be a marker. Along with the marker there is a symbol of a line. If you had been sketching an arc, there would have been an arc symbol. If you had been sketching a circle, there would have been a circle symbol, and so on.

The second aspect of sketching geometry you should have noticed is what is referred to as *system feedback*. The system feedback is there to help you create geometry, and will appear as dynamically changing symbols attached to the cursor, dashed lines that flash on screen, and a numerical readout. If you did not notice any system feedback the first time around, sketch more lines. We will discuss system feedback in more detail in material to follow. For now, let's examine another method of sketching.

How-To 2-5:   Sketching a Line Using the Click-Click Method

To sketch a line using the click-click method, perform the following steps.

1. Click on the Line icon, found on the Sketch toolbar, or select Tools > Sketch Entity > Line.

2. Click once where you want the line to start. (Do not hold down the left mouse button.)

3. Click once where you want the line to end.

4. Click to establish subsequent endpoints, and continue adding line segments.

5. To break off the command, press the Escape (Esc) key.

Note that to create a single line segment the click-drag method is most fitting. To create more than one segment, use the click-click method. Either method can be applied to all types of sketch entities. Which method is used depends on what you are trying to accomplish and on personal preference. Additionally, it is not necessary to break off a command prior to accessing the next command. Simply click on the icon of whatever sketch command you wish to begin next.

✗   **WARNING:** *An option titled* Single command per pick *can impede the ability to draw connected line segments. This option will be turned off throughout this book. To turn the option off, select* Tools > Options, *and in the* General section *make sure* Single command per pick *is unchecked.*

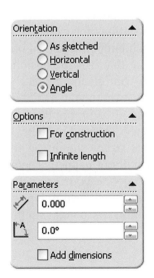

Fig. 2–14. Insert Line panel.

The Insert Line Panel

When you first click on the Line command, PropertyManager will display the Insert Line panel (shown in Figure 2–14). This happens prior to sketching the line. Use the *As sketched* option to get a feel for simple line sketching prior to trying out the other options.

The *As sketched* option is what will be used throughout this book. The Horizontal and Vertical options allow for drawing horizontal and vertical lines of the length you specify. The Angle option allows for entering an angular value along with the line length. Dimensions can be added automatically provided the *Add dimensions* option is checked. The length and angle values can be set to zero, in which case the values are determined by where the mouse is clicked in the work area.

The ability to enter dimension values or angle values prior to sketching a line is not necessarily an advantage. It is simply a different way of entering data. Traditionally, lines (and other sketch entities) are created first, and then the dimensions are added. Understand that lines can be made horizontal or vertical without using the options of the same name in the Insert Line panel. You will understand how this is done after reading the section on system feedback.

Construction geometry (and the *For construction* option) are discussed in the section "Construction Entities." The *Infinite length* option does just what it sounds like it should do it creates lines of an infinite length. This is occasionally useful, but probably not an option most people will use on a daily basis.

Sketch Entity PropertyManager

While sketching any entity (such as a line or circle), or even selecting a sketch entity, PropertyManager will take the place of FeatureManager (see "FeatureManager Tabs" in Chapter 1). This is normal behavior. It is a function that can be turned off, though you wouldn't really want to. Property-Manager displays panels containing useful data, such as what geometric relations are present on a selected object.

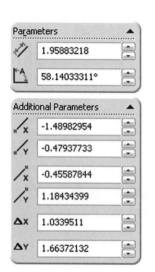

If PropertyManager is not appearing when you select an object, it means the Auto-show PropertyManager option is turned off. You can turn this option back on by selecting Tools > Options > System Options tab, and then checking the Auto-show PropertyManager option.

Try selecting one of the lines that were sketched a moment ago. Note that the Line Properties panel appears. In this panel you will find a section titled Parameters and another titled Additional Parameters. The problem with modifying parameters is that it doesn't accomplish anything. The parameters panel for a line is shown in Figure 2–15. This panel contains settings for changing a line's angle, its length, *x-y* coordinates of each endpoint, and so on. But don't be fooled by this. Make no mistake that it is dimensions and geometric relationships that will define your sketch geometry, not the Parameters panel.

Fig. 2–15. Parameters panel.

Many new users have fallen into the trap of using the Parameters panel for positioning their sketch geometry. This is a mistake. It is this author's opinion that the Parameters panel should be removed from the software, but that probably will not happen. Therefore, consider this your warning. Use dimensions to define your geometry (which you will learn about in the next chapter). And if you need to reposition some sketch object, don't bother with the Parameters panel. Simply drag the geometry to the desired position.

Exiting Sketch Commands

It is not usually necessary to exit out of a sketch command, most of the time. Typically, if it is necessary to initiate another command simply click on the applicable icon. In this respect, it is more a matter of switching commands than of having to exit out of one command and start another. This will obviously reduce the number of required mouse clicks.

If it does become necessary to exit a sketch command, there are a few choices. Take note that we are not discussing exiting a sketch. Rather, we are discussing exiting a sketch command. When not in any particular sketch command, you are in what is known as *select* mode. This simply means that you have the ability to select geometry.

Fig. 2–16.
Select icon.

One way to exit a sketch command is to click on the Select icon, typically found on the Sketch toolbar (see Figure 2–16). Assuming your cursor looks like the traditional arrow cursor, the Select icon will look just like your cursor. A second method of exiting a sketch command is to click on the command's icon a second time. Finally, if you prefer a keyboard shortcut, press the Esc key. Any of these methods accomplishes the same task, and it is a matter of personal preference which method you use.

NOTE: *The Select command is also available from the right mouse button menu and the Tools menu.*

Selection Options

Selecting geometry is required for various reasons, such as for deleting geometry or for having some command act upon it. Whatever the reason, selecting objects can be done via the left mouse button, as was discussed earlier in this chapter. When a sketch object is selected, it turns green. What if it becomes necessary to select multiple objects?

To select multiple objects, hold down the Ctrl key while selecting. This holds true for selecting any objects in SolidWorks, including sketch geometry. The only time it is not necessary to hold down the Ctrl key is while in a command. This refers to any command that causes the PropertyManager or a window to appear.

Multiple sketch entities can be selected at one time by dragging a box (rectangle) around the geometry. In other words, pick where one corner of the box should be located, and then drag the cursor to position where the opposite corner of the box should be placed. Whatever is enclosed in the box will be selected if dragging from left to right. Dragging from right to left creates a crossing box, which selects not just the objects enclosed in the box but those crossing over the edges of the box.

Objects can be deselected in the same manner they can be selected. Clicking on an object a second time will deselect it, as long as the Ctrl key is held down. To deselect everything, click anywhere in an empty portion of the work area.

Deleting Objects

Deleting unwanted objects is accomplished in the same manner for nearly everything in SolidWorks, including sketch entities. To delete an object, perform one of the following steps.

- Select Delete from the Edit pull-down menu.

- Right-click on an object and select Delete.

- Click on the Delete icon, found on the Standard toolbar.

- Press the Delete key.

The Delete key might be the quickest option for some people, but the choice comes down to user preference. Features, drawing views, parts in an assembly, and just about anything else can be deleted using the previous options. For significant items, such as features, you will get a confirmation window warning you that something is about to be deleted if you proceed.

Command Persistence

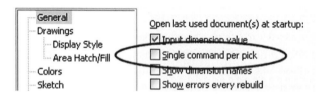

Fig. 2–17. Single command per pick *option.*

Whether or not you remain in a command after carrying out that command depends on an option named *Single command per pick*, shown in Figure 2–17. If this option is enabled, the icons will work a little differently. Specifically, clicking on an icon will allow you to use that command once, with the command then exiting automatically. If *Single command per pick* is not checked, commands persist until canceled or until a new command is entered. How-To 2–6 takes you through the process of changing command persistence.

How-To 2-6: Changing Command Persistence

To modify the *Single command per pick* option, thereby altering whether or not a command persists when used, perform the following steps.

1. Select Options from the Tools pull-down menu.

2. Select the General section in the System Options tab.

3. Modify the *Single command per pick* option as desired.

It should be noted that even with *Single command per pick* enabled, double clicking on an icon will allow for repeated use of that icon. If *Single command per pick* is not checked, one click on an icon is all it takes to use that command indefinitely. This is another of those options that depend on the preference of the user. This book will assume you do not have *Single command per pick* enabled, so you may want to leave it unchecked for now.

System Feedback

Your sketch will serve in large part to determine whether the part being built is a well-constructed part or a sloppy part. A well-constructed part makes good use of existing model geometry, fully defined sketch geometry, well-placed dimensions, and proper geometric relations. A well-constructed part will also incorporate your design intent, so that dimensions can be modified without fear of introducing rebuild errors in the model. These are all aspects of modeling you will learn about in good time.

During the sketch process, SolidWorks will supply a steady stream of information. This information is known as *system feedback*, which is dependent on where the cursor happens to be positioned at any given time. Symbols will appear next to the cursor. These symbols will change depending on what the cursor is pointing at, or to relay other important information. System feedback is SolidWorks' way of communicating with you, and plays a large role in helping you add the proper geometric relations to the sketch, depending on your design intent.

Geometric relations, otherwise known as *constraints*, are added automatically as you sketch. System feedback will inform you of this fact by displaying a symbol indicating what type of geometric relation will be added. As long as you are paying attention and keeping your focus on the cursor as you sketch, very little can go wrong. If a constraint is accidentally added, it is possible to remove it at any time. The process of manually adding and removing constraints is discussed in the next chapter.

➻ **NOTE:** *It cannot be stressed enough how important system feedback is. Watch the cursor!*

Inference lines are another aspect of system feedback, as are numerical readouts. The dashed inference lines help to align sketch entities with other endpoints or center points. They also indicate where a geometric relationship will be added should you decide to position the cursor at that location.

Numerical readouts are used when sketching to help determine a sketch object's approximate size. But remember, this is only a sketch. Numerical values do not have to be exact, because the dimensions will control, or drive, the entity's size. (Dimensions are examined in the next chapter.)

Constraints

Because constraints are added automatically, it is important to understand what constraints are before going much further. A constraint is a geometric condition or relationship set on an object or between objects. For instance, a line may be constrained vertically or horizontally. Once the line is drawn horizontally, it will remain horizontal unless the relation is removed. The line can be repositioned or resized, but must remain horizontal in order to satisfy the condition placed on it.

There are other constraints that can be placed on sketch objects, many of which have cursor symbols associated with them that are displayed as the sketch is created. If a line is about to be drawn parallel to another line, the cursor displays the symbol for parallelism. If an arc is about to be drawn tangent to a line, the tangent symbol is displayed. All in all, there are over a dozen types of constraints. Not all can be added automatically as the sketch is being created. Some constraints must be added manually by the user, and others may need to be removed. Manually adding and removing constraints is discussed in the next chapter.

✓ **TIP:** *There may be additional geometric relation symbols visible on any sketch geometry you create, above and beyond the system feedback. These are persistent symbols that will stay visible until you decide to turn them off. Their purpose is to show what geometric relations have been added to the sketch. To toggle the display of these symbols, check or un-check Display Relations in the View menu.*

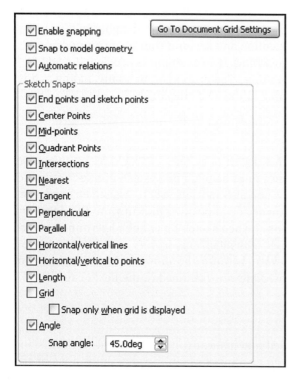

Sketch Snaps

The ability to automatically add geometric relations depends on settings known as sketch snaps (or quick snaps). As a matter of fact, you should think of sketch snaps as the automatic relations themselves. To explain the reason for this, consider that a perpendicular geometric relationship to another line can only be added automatically if the perpendicular sketch snap is enabled.

Sketch snap settings can be accessed by selecting Tools > Options, and then selecting the Relations/Snaps section in the System Options tab. These settings are shown in Figure 2–18. The *Enable snapping* option is what turns on sketch snap functionality. There is no reason you would want to turn this function off, so leave it checked.

Fig. 2–18. Relations and snap settings.

Snap to model geometry should normally be left on, because it is what controls the ability to reference geometry outside the sketch. It also automatically adds relations to that geometry. Likewise, *Automatic relations* should remain on. Otherwise, no geometric relations would be added automatically whatsoever, and you would be left adding them all manually.

✓ **TIP:** *If for some reason you do not want a relation to be added automatically while sketching, simply hold down the Ctrl key. This will temporarily disable automatic relations.*

The general rule with establishing the proper settings in the Relations/Snaps section (shown in Figure 2–18) is to leave everything turned on, with the possible exception of the Grid option. A grid snap is just not necessary in SolidWorks because dimensions will drive the geometry size anyway. Matching your settings to those shown in Figure 2–18 is a safe approach.

Sketch Color Codes

There are a number of colors SolidWorks uses to tell you that certain conditions exist in a sketch. The most common color codes are blue, black, and red. Green simply means that something has been selected, and is not considered a sketch color code. SolidWorks allows you to work with underdefined geometry, though having underdefined geometry in a production model is not recommended. Underdefined geometry indicates that there are not enough dimensions on the model to manufacture it.

If sketch geometry is underdefined, it will be color-coded blue. This is the way sketch geometry usually appears when first drawn. As more and more constraints or dimensions are added to a sketch, the geometry will turn black. This means that the geometry has become fully defined. When the entire sketch is black, the sketch has become completely defined. This should be your goal.

It will always take some combination of constraints and dimensions to fully define a sketch. What type of constraints and where dimensions are placed or the type of dimension used (i.e., angular or linear) depends on your design intent. As we learned in the first chapter, design intent is king. It is how your model will behave if you make a change to it. If a sketch is not fully defined, it is not possible to predict how it will behave, and your design intent will not be imparted to the model geometry.

Once a sketch is fully defined, adding any other dimension would serve to overdefine the sketch. Overdefined sketch geometry turns red, and should always be cleared up as soon as possible. If you choose not to correct overdefined geometry, you will likely create more problems for yourself down the road. Table 2–1 outlines the color codes used by SolidWorks.

Table 2-1: SolidWorks Color Codes

Color	Significance
Black	Fully defined
Red	Overdefined
Blue	Underdefined
Pink	No solution found
Yellow	Invalid solution found
Brown	Dangling relation

Occasionally there are situations in which a valid solution simply cannot exist within a particular sketch. This happens because dimensions or constraints have resulted in a sketch that cannot be drawn due to its geometric nature. This turns the sketch geometry pink, and the sketch geometry is not solved. Another rare occurrence is that the sketch is actually solved but has invalid geometry, such as a zero-length line, in which case the offending geometry will turn yellow.

The last color code, brown, is more common. Brown means that there is either a Odangling dimension or constraint in the sketch. An example of a dangling dimension would be if a circle were dimensioned to another feature and at some point in the future the feature were deleted. SolidWorks would not delete the dimension. Instead, it would assign the dangling attribute to it. It would then be up to you to edit the sketch and fix the dangling dimension, which you could easily spot due to its brown color. You will learn more about dimensioning in Chapter 3.

Dragging Geometry

Underdefined geometry can be moved and reshaped, assuming you are in sketch mode and not currently in a command (in other words, the cursor looks like an arrow). Any geometric relations placed on sketch geometry must be maintained during the drag process. Because of this fact, dragging can allow you to quickly and easily determine what constraints might need to be added to the geometry, or what constraints already exist. This is valuable information to a SolidWorks user.

What part of a sketch is dragged will make a difference as to how the sketch will react. This can most easily be demonstrated using a simple line. Draw a line at an angle, exit the command, and then place the cursor over the line and drag the geometry to a new position by holding down the left mouse button. Note that the entire line moves but remains parallel to its original location.

Next, try dragging a single endpoint of the same line. Note that it is possible to reposition just that one endpoint being dragged. You will notice that other geometry reacts differently as well. A circle will drag differently from its center point versus its perimeter, and so on.

As more dimensions and relations are added to a sketch, less and less geometry remains blue. Sometimes it is difficult to figure out why geometry is not fully defined. This is precisely why it is so beneficial to be able to drag geometry. When the question "Why isn't the sketch defined" arises, drag any remaining blue geometry and you will almost always find the answer you are looking for.

Basic Sketch Entities

You are obviously going to need to sketch objects other than lines. This section introduces you to the most common types of sketch entities and how to create them. Any special properties individual sketch entities may have are noted, along with any quirks or noteworthy behavior they may exhibit.

➥ **NOTE:** *It is very likely that not every icon mentioned in this section will be present on your Sketch Tools toolbar. It is the default nature of Solid-Works to not include all icons on all toolbars. SolidWorks toolbars can be customized to include any missing icons, or to remove those not frequently used. Customization is discussed in Chapter 25. If an icon is not present on your toolbar, it is suggested that you access the command via the Tools > Sketch Entities pull-down menu.*

It would be redundant to list the steps for creating each type of sketch entity using both the click-drag and click-click methods. This book lists the steps involved using the click-click method. Any sketch object in this section can be created using either method. Click-dragging is most commonly employed while sketching lines for the reason of creating a single line segment quickly and easily, yet remaining in the command.

Steps involved in creating the basic sketch entities specify the "click points." In the case of a line, for instance, you would see the following.

1. Click to establish the line's start point.

2. Click to establish the line's endpoint.

It is recommended that you try creating the various sketch entities on your own. This will give you a chance to get your hands dirty, and hopefully will allow you to feel more comfortable with the software. Feel free to experiment, and remember to watch the cursor for any system feedback Solid-Works is relaying to you.

➥ **NOTE:** *In each of the sections that follow, it is assumed that you have already initiated the applicable command, either via the Sketch toolbar or the Tools > Sketch Entities menu.*

Circle

Creating circles requires very little effort from the user and is a very simple operation. The Circle and Perimeter Circle command icons are very similar, so pay attention to your tooltips. To create a circle, perform the following steps.

1. Click to establish the circle's center point.

2. Click to establish the circle's radius.

Dragging the completed circle will modify the circle in different ways, depending on the point being dragged. Dragging the circle's perimeter will allow you to resize the circle. Dragging the circle's center point will allow you to reposition it.

Perimeter Circle

The Perimeter Circle command allows for defining a circle by picking three points. To create a perimeter circle, perform the following steps.

1. Click to establish the first point on the circle's perimeter.

2. Click to establish the second point on the circle's perimeter.

3. Click to establish the final point on the circle's perimeter (as well as to establish the circle's diameter).

Tangent Arc

Sketching a tangent arc is quite easy, but there is one stipulation. You must select an existing entity endpoint before sketching the arc. The following steps take you through the process of creating a tangent arc.

1. Click an existing endpoint where the arc is to start.

2. Click to establish the tangent arc's endpoint.

Tangent arcs can be drawn tangent to lines, other arcs, or splines. The system feedback must be used so that you will know when you are over an endpoint. The endpoint your cursor is positioned over will highlight, and that is when you know it is safe to create the arc. If you do not begin the tangent arc on an endpoint, you will receive an error message stating so.

When a tangent arc has been created, a tangent constraint is added between the arc and the tangent entity. This relationship will remain if any of the associated geometry is moved or dragged to a new position.

Tangent Arc Using Auto-transitioning

When sketching lines, it is possible to automatically transition to sketching arcs with little effort. This is accomplished in a number of very simple ways. The only requirement is to use the click-click method of sketching. You are encouraged to experiment and try this on your own. To begin, sketch a line using the click-click method. Then position the cursor as if you were about to establish the endpoint of a second line segment. You will see the typical thin line, which acts as a preview. Next, perform one of the following steps.

- Press the A key.

- Move the cursor back to the endpoint of the first line, and then away from it.

At this point, an arc preview should be visible in place of the line initially present. Clicking will position the endpoint of the arc, and then you can proceed from there, either drawing another line or auto-transitioning to another arc. In this fashion, a series of lines and arcs can easily be created, all of which will be tangent to the previous entity.

✓ **TIP:** *If you auto-transition to an arc and then change your mind, press the A key to transition back to a line.*

Fig. 2–19. Possible arcs created by auto-transitioning.

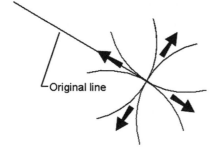
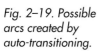

By moving the cursor back to the previously sketched object, and then away from it, it is possible to sketch an arc in one of four directions. The arc will either be tangent to the line or tangent to a perpendicular vector of the line. In this way, up to eight different tangent arcs can be created. All eight possible arcs are shown in Figure 2–19.

3 Point Arc

The 3 Point Arc command is commonly used when the locations for the arc's endpoints are known. This, however, is not a requirement. Three-point arcs can be created anywhere. To create a three-point arc, perform the following steps.

1. Click to establish the start point of the arc.

2. Click to establish the endpoint of the arc.

3. Click to establish the arc's radius.

Fig. 2–20.
System feedback
when creating a
three-point arc.

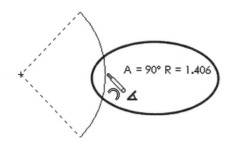

A = 90° R = 1.406

Bear in mind that when performing step 3 the pick point establishes where the arc will pass through, which in turn defines the arc's radius. Watch the system feedback prior to completing step 3 and it will give an indication of what the included angle and radius of the arc are going to be. This, dependent on the Angle quick snap being enabled, is shown in Figure 2–20.

Centerpoint Arc

A center-point arc (using the Centerpoint Arc command) can be created by performing the following steps.

1. Click to establish the arc's center point.

2. Click to establish where the arc will start.

3. Click to establish where the arc will end.

As you may have noticed already, the second pick point will also define the radius of the arc. Once again, as is always the case, let the system feedback give you an "in-the-ballpark" estimate of the arc's radius and included angle. In the end, dimensions or geometric relations will control the geometry.

Rectangle

Rectangles are constrained to be rectangles, meaning that the top and bottom lines of the rectangle are horizontal and its sides vertical. To create a basic rectangle, perform the following steps.

1. Click to establish one corner of the rectangle.

2. Click to establish the opposite corner of the rectangle.

What if it becomes necessary to create a rectangle at an angle? There are two solutions. One solution would be to manually draw the rectangle using the Line command, and then add any additional constraints that might be required to maintain the rectangles shape. An easier solution is to use the Parallelogram command, discussed in the next section.

Parallelogram

The Parallelogram command has a dual identity. It can be used to create both a rectangle at an angle and a parallelogram. Different relations are automatically added when using the Parallelogram command, depending on whether you are creating a parallelogram or an angled rectangle. An angled rectangle will have two perpendicular relations and one parallel relation associated with it. In the case of a parallelogram, the opposite sides of the parallelogram are geometrically constrained to be parallel, just as you would imagine them to be. To create a parallelogram, perform the following steps.

1. Click to establish one corner of the parallelogram.

2. Click to establish the length of the first side of the parallelogram.

3. Press the Ctrl key, and then click to establish the length of the adjacent side of the parallelogram.

To create a rectangle at an angle, do not press the Ctrl key when performing step 3. As you can see, it is actually easier to create an angled rectangle rather than a parallelogram, because there is one less key press to worry about. Incidentally, if the Ctrl key is pressed by accident, and it was originally your intention to create an angled rectangle, it will be necessary to exit the Parallelogram command and begin again.

Polygon

The easiest way to create polygons is through the use of the Polygon command. When a polygon is sketched, SolidWorks patterns one side of the polygon in a circular fashion. There is also a construction circle that helps control the polygon geometry. One side of the polygon is made tangent to the construction circle prior to patterning. This is all done automatically by SolidWorks, and is transparent to the user. To create a polygon, perform the following steps. Note that the first two steps will require making use of the Polygon panel displayed in PropertyManager, a portion of which is shown in Figure 2-21.

1. Specify the number of sides the polygon should have.

2. Specify whether the polygon should be inscribed or circumscribed.

3. Click to define the center of the polygon.

4. Click to define the diameter of the polygon.

Fig. 2–21. Polygon panel.

Other than to dictate the number of sides the polygon will have, and whether it is inscribed or circumscribed, the additional parameters in the Polygon panel do not serve much purpose (see Figure 2–21). You will want to define the polygon through the use of dimensions and constraints anyway, which the Parameters section in the Polygon panel cannot accomplish. This is true for all sketched entities and is not specific to polygons.

Specifying the number of sides and the inscribed/circumscribed option can be performed immediately after a polygon has been created, as well as prior to its creation. To create a second polygon with a different number of sides than the first, click on the New Polygon button. To modify the number of sides a polygon has after exiting the Polygon command, right-click on one side of the polygon and select Edit Polygon. This will open the Polygon panel and allow you to change the number of sides the polygon has.

Construction Entities

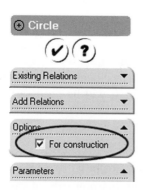

Fig. 2–22. Creating a construction entity.

Fig. 2–23. Construction Geometry icon.

Construction entities do not contribute to solid geometry. For this reason, construction entities can be used for a variety of tasks, such as reference geometry or to aid in adding dimensions. Construction lines are often used to help construct symmetrical geometry, or to control tangency conditions at the end of spline curves. A construction circle could be used to define a bolt hole circle diameter. These are just a few situations in which construction entities would prove useful.

Any sketch entity can be turned into a construction entity. Doing so is just a matter of checking the *For construction* option found in PropertyManager (see Figure 2–22). This is most easily done immediately after sketching an entity, because the entity will still be selected and its PropertyManager displayed on screen. However, a sketch entity can be selected at any time to bring up its PropertyManager to gain access to the *For construction* option.

Another method of turning sketch geometry into construction geometry is to use the Construction Geometry icon, found on the Sketch toolbar (see Figure 2–23). Clicking on this icon when a sketch entity is selected accomplishes the same thing as checking the *For construction* option.

The Construction Geometry icon can have an alternate way of functioning that might prove useful. If nothing is selected prior to clicking on the Construction Geometry icon, it enters what might best be described as "toggle mode." When in this mode of operation, sketch objects can be selected one after the other. Any sketch entity will convert to construction geometry, and vice versa. In other words, construction geometry can also be converted to regular sketch entities with the Construction Geometry icon, or by unchecking the *For construction* option in PropertyManager.

Centerline

Fig. 2–24. Centerline icon.

Centerlines are also known as construction lines. There is no difference between the two as far as SolidWorks is concerned. Centerlines are often used when mirroring sketch geometry or adding symmetrical relationships between sketch entities, which you will learn about in Chapter 4.

Centerlines are common enough that they have their own icon, shown in Figure 2–24. The process used to create a centerline is identical to that used to create lines. All of the functionality is also the same, including the ability to auto-transition to an arc.

Point

Fig. 2–25. Point icon

Points can be created by clicking on the Point icon (shown in Figure 2–25) and simply picking where you want to place the point entity. Sometimes points can be beneficial when trying to dimension or constrain to other entities. Points also play a useful role in the creation of advanced feature types, such as when constraining guide curves for the sake of lofted or swept features, which you will learn about at a later time.

Virtual Sharp

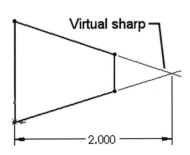

Fig. 2–26. A virtual sharp.

The alter-ego of the Point command is its ability to create virtual sharps. A virtual sharp can be thought of as the virtual intersection point of two lines or edges. It might be necessary to create a virtual sharp for the sake of adding a dimension to the projected intersection of two lines (as an example).

The key to creating a virtual sharp is to select the lines or model edges whose projected intersection the virtual sharp will be added to. Do this first, and then click on the Point icon. The virtual sharp should appear at the proper intersection. An example of a virtual sharp is shown in Figure 2–26.

✓ **TIP:** *To control the appearance of virtual sharps, access the Virtual Sharps section of the Document Properties window (Tools > Options > Document Properties tab).*

Advanced Sketch Entities

Eventually you may find the need to draw sketch entities that delve beyond simple lines, arcs, and polygons. Sketch geometry of a more complex nature, such as ellipses or splines, can also be created. If you do not feel as though you will have a need for these more complex shapes, it is possible to skip over this section without losing the flow of the material in this book. This section has been added for reference purposes and to complete the readers understanding of all 2D sketch entities that can be created in SolidWorks.

Ellipse

An ellipse is normally defined by the length of each of its two axes. The major axis is the long axis; the minor axis is the shorter of the two. To create an ellipse, perform the following steps.

1. Click to establish the center of the ellipse.

2. Click to establish the first axis length.

3. Click to establish the second axis length.

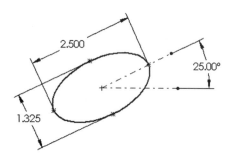

Fig. 2–27. Controlling an ellipse through dimensions.

It makes no difference whether the first axis is the major or minor axis. Once the ellipse has been created, you can drag its shape or rotate it by dragging its quadrant points. Reposition the ellipse by dragging the ellipse itself (from some point other than a quadrant point). An ellipse will have four quadrant points that can be dimensioned, as shown in Figure 2–27. This is normally an appropriate way of controlling the size and shape of an ellipse. It would also be common to add construction lines that can be dimensioned in order to control the rotation angle of the ellipse. Adding dimensions is covered in the next chapter.

✓ **TIP:** *To turn the display of quadrant points (or spline control points) on or off, select Tools > Options > System Options tab > Sketch section, and change the* Display entity points in part/assembly sketches *setting.*

Partial Ellipse

Partial ellipse is another term for an elliptical arc. It is one of the most laborious of all sketch commands because creating this sketch entity involves four steps. Creating a partial ellipse begins exactly the same as using the Ellipse command. However, it involves an additional operation that determines the elliptical arc's start points and endpoints. To create a partial ellipse, perform the following steps.

1. Click to establish the elliptical arc's center point.

2. Click to establish the first axis length.

3. Click to establish the second axis length (this pick point will also determine where the elliptical arc begins).

4. Click to establish the elliptical arc's endpoint.

The Partial Ellipse command can be a little tricky until you have gone through it a few times. Do not worry about being exact. You can always drag the sketch geometry after the arc is completed. Note that an elliptical arc has four quadrant points, just like would be seen with a complete ellipse. If you have not drawn a partial ellipse on your screen, Figure 2–28 shows an example of what a partial ellipse might look like. The arc's endpoints differ from the quadrant points and are noted in the figure.

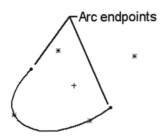

Fig. 2–28. An example of a partial ellipse.

Elliptical arcs can be difficult to manage. Dimensioning the arc is fairly easy. Attach dimensions to the quadrant points, just like with a complete ellipse. Where the difficulty comes in has to do with the endpoints of the arc and how they are related to the quadrant points. The arc's endpoints may or may not be coincident with the quadrant points, so coincident relations may need to be added or removed (manually adding relations is covered in the next chapter).

The included angle of the arc may need to be defined with a dimension as well. Typically, the arc will be attached to other geometry, which would help to define the arc. If tangent relationships are required, they can be added between the arc and some other adjacent object.

Parabolas

A parabola is one of those sketch entities most designers do not need very often, unless they are in a particular manufacturing discipline. Perhaps you fall into this category. The Parabola icon may not be present on your Sketch toolbar. If that is the case, use the Tools > Sketch Entities pull-down menu.

A parabola consists of a focal point, a directrix, and a parabolic curve. The curve's endpoints can be positioned to the user's liking. The steps that follow take you through the process of creating a parabola. If you are unfamiliar with the terms used in the steps, see Figure 2–29 for guidance.

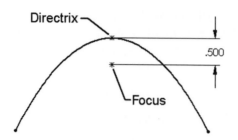

Fig. 2–29. Terms used to define a parabola.

1. Click to establish the focus.

2. Click to establish the length of the directrix.

3. Click to establish where the parabola will begin.

4. Click to establish where the parabola will end.

Splines

Spline entity types are best used for creating free-form curves. Splines take the place of French curve drawing guides in the CAD world. Splines have many peculiarities associated with them, and there are many things you need to be aware of when working with spline curves, which this section will teach you.

SolidWorks employs two types of splines: proportional and nonproportional. Nonproportional splines are created by default unless you specify otherwise. Proportional (fixed shape) splines retain their shape if their endpoints are moved. Both proportional and nonproportional splines can be reshaped if you drag their control points. The steps that follow take you through the process of creating a spline, and are the same whether creating a proportional or nonproportional spline.

1. Click to establish the start point of the spline.

2. Click to establish subsequent control points of the spline.

3. Double click to establish the endpoint of the spline.

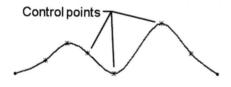

Fig. 2–30. An example of a spline.

In place of step 3, it is also possible to press the Esc key on your keyboard after establishing the endpoint of the spline. This has the effect of ending the spline and simultaneously exiting the Spline command. An example of a completed spline is shown in Figure 2–30. Some of the spline's control points are noted for reference.

Control points are used to change the shape of the spline via dragging. Dragging the endpoints of a spline will also change its shape if the spline is a nonproportional spline. Dragging on the spline itself (some location other than a control point) will allow you to reposition the spline.

Proportional Versus Nonproportional Splines

Because nonproportional splines have a larger degree of freedom of movement and are not as tightly regulated as proportional splines, all of their control points can easily be dimensioned. Figure 2–31 shows an example of a dimensioned nonproportional spline. Ordinate dimensions were used in the example, a dimension type explored later in the book.

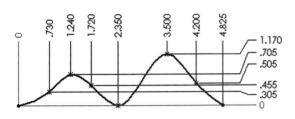

Fig. 2–31. Dimensioning a nonproportional spline.

Fig. 2–32. Making a spline proportional.

Proportional splines start out as nonproportional splines. What makes the spline proportional is an option in PropertyManager, shown in Figure 2–32. You may decide a proportional spline would serve you better because its size can be scaled up or down proportionally. If this is the case, simply select the spline and check the Proportional option.

Proportional splines look just like nonproportional splines. They just react differently when their endpoints are moved. Try it for yourself and you will see the difference. You will be able to scale a proportional spline by dragging one of its endpoints. By their very nature, proportional splines do not require as many dimensions to fully define them.

Splines can be open or closed. To create a closed spline, place the spline's endpoint on its start point. It is also possible to drag the endpoint of a previously created spline to its start point for the sake of closing the spline.

Adding and Deleting Control Points

Once a spline has been created, it may be necessary to add control points, thereby allowing for a greater degree of control over the spline's shape. Such is the case when it is discovered that there are not enough control points to achieve the desired contour. More control points would remedy this situation, and it is easier to add control points than to delete the spline and start from scratch. How-To 2–7 takes you through the process of adding control points to a spline.

How-To 2-7: Adding Control Points to a Spline

To add control points to a spline, perform the following steps.

1. Right-click on the spline and select Insert Spline Point.

2. Using the left mouse button, click on the spline to add a control point at the desired location. Repeat as necessary.

3. Click on OK or press the Esc key when finished.

Fig. 2–33. Removing a control point from a spline.

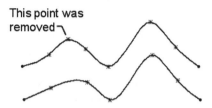

This point was removed ⟍

Once the new control points have been added, they can be dragged to reshape the spline, or dimensioned as required. Deleting spline control points is even easier than adding them. Just click on the spline's control point you want to delete and press the Delete key. SolidWorks will recalculate the new spline for the remaining control points. See Figure 2–33 for a before-and-after depiction of a spline with a control point removed.

Simplifying a Spline

There is another way of removing spline control points, and that is to simplify the spline. This method is performed by specifying a lesser tolerance level for the spline. This is sometimes referred to as "loosening" the tolerance, and is directly related to the degree of control one has over the shape of the spline. By specifying a lesser tolerance level, fewer control points are required, thereby reducing the number of control points on the spline.

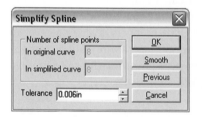

Fig. 2–34. Simplify Spline window.

The difference between deleting control points and simplifying a spline is that during the simplification process SolidWorks attempts to retain the original shape of the spline as much as possible. However, with a looser tolerance setting the spline becomes straighter, and loses its ability to retain its original curvature. How-To 2–8 takes you through the process of simplifying a spline, rather than simply deleting control points. The Simplify Spline window is shown in Figure 2–34.

How-To 2-8: Simplifying a Spline

To simplify a spline, perform the following steps.

1. Right-click on the spline and select Simplify Spline.

2. Enter a new tolerance for the spline.

3. Click on the OK button.

Figure 2–35 shows a spline that has been simplified. Note that the simplified version retains much of the original's basic shape. The simplified spline just has fewer control points. In the spline pictured, a tolerance value of .4 inch was used, which resulted in the spline using five control points, as opposed to the original eight.

Fig. 2–35. A spline before and after it has been simplified.

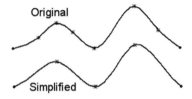

Original

Simplified

Another method of simplifying the spline would be to use the Smooth button found on the Simplify Spline window. This loosens the tolerance in incremental steps, thereby reducing the number of control points. The Previous button reverses this effect, similar to an undo function.

✓ **TIP:** *When entering a new tolerance, press Enter to see an updated preview prior to clicking on the OK button.*

Spline Handles and Tangency Conditions

Splines can be made tangent to lines, centerlines, arcs, elliptical arcs, and parabolas. If the entity the spline is tangent to moves, the spline will reform itself to accommodate the tangent relationship. Read the section "Adding Relations" in Chapter 3 to understand how to add geometric relations.

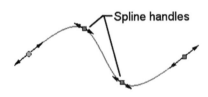

Spline handles

Fig. 2–36. Spline handles.

Spline handles, shown in Figure 2–36, are another way of manipulating a spline. They can be used to control the tangency conditions at each spline control point as well as the endpoints of the spline. Selecting the spline serves to display the spline handles, whereby the handles can then be manipulated by dragging the arrows at either end of each handle. Changing the orientation of a handle changes the tangency direction vector, and lengthening the handle will increase the force the tangency exerts on that particular spline control point. All of this serves to change the shape of the spline, and the easiest way to understand how the handles work is to simply try them out for yourself.

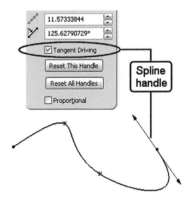

Fig. 2–37. Spline handle tangency control.

↝ **NOTE:** *If spline handles are not visible, even when the spline is selected their visibility may be disabled. Select Tools > Options, and make sure the* Enable Spline Tangency and Curvature handles *option is checked in the Sketch section of the System Options.*

Originally, spline handles will appear in gray, assuming you have selected the spline. If a handle is moved, the Tangent Driving option (shown in Figure 2–37) will appear checked, and the associated handle will appear blue. The handle will also stay visible to serve as a further indicator that it is a controlling force in the shape of the spline. This can also be seen in Figure 2–37. Clicking on the Reset This Handle button will remove any tangency control being exerted by that particular spline handle (as will unchecking the Tangent Driving option), and will return the handle to its original state.

↩ **NOTE:** *Only nonproportional splines have handles.*

Adding Tangency and Curvature Control

Spline handles can be added to a spline by right-clicking on the spline and selecting Add Tangency Control. It will then be necessary to position the handle at the desired location along the spline. Moving the mouse will "slide" the handle along the spline. Clicking the left mouse button locks the handle at its current position.

Adding a spline handle is really the same as adding a control point and then turning on the Tangent Driving option for that handle (discussed in the previous section). The Add Tangency Control command just makes the process a little easier.

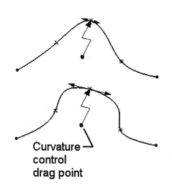

Curvature control drag point

Fig. 2–38. Adding a curvature control point.

Adding curvature control is done the same way as adding tangency control: by right-clicking on the spline and selecting Add Curvature Control. The similarity does not end there. Move the mouse to slide the curvature control arrow to the appropriate position on the spline, and then click the left mouse button to lock the curvature control point to that location. When curvature control is added to a spline, the control point will be displayed as shown in Figure 2–38. The jogged arrow differentiates the curvature control point from a standard tangency handle.

By dragging the handle at the end of the arrow (indicated in Figure 2–38), the curvature at that spline control point can be controlled. The process can be repeated to increase or decrease the curvature to a greater degree. Figure 2–38 shows the same spline with the curvature control adjusted to varying degrees.

Spline Minimum Radius and Inflection Points

Two diagnostic utilities available for splines are the Show Minimum Radius and Show Inflection Points options. Both options can be accessed via the context-sensitive menu when right-clicking on a spline. Both options are toggle switches and can be turned on or off by simply checking or unchecking the applicable option.

Minimum radius is self-explanatory. There would only be one portion of a spline that is determined to have a minimum radius. Inflection points are the points along a spline that momentarily become flat. In other words, curvature is zero at the inflection points. A spline doesn't necessarily have to have inflection points, but if it does the inflection points are represented as small "bow ties." Figure 2–39 shows an example of a spline whose minimum radius and inflection points are being displayed.

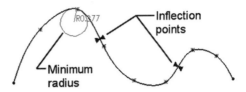

Fig. 2–39. Displaying a spline's minimum radius and inflection points.

Control Polygons and Relaxing Splines (SW 2006 Only)

Control polygons represent another method of controlling the shape of a spline, and can be thought of as the framework for a spline. A control polygon, which appears as a series of dashed lines, is shown in Figure 2–40. The endpoints of the dashed lines can be dragged, thereby reshaping the spline. To toggle the display of control polygons, use the Display Control Polygon icon found on the Spline Tools toolbar.

Fig. 2–40. Spline control polygon.

There is a setting in the Sketch section of the System Options (Tools > Options) that will affect the display of control polygons. The option is titled *Show spline control polygon by default*, and will show the control polygon for any newly created spline by default when checked.

Relaxing a spline is done by selecting a spline and clicking on the Relax Spline button found in the Spline PropertyManager. A relaxed spline's control points will behave differently than a spline that has not been relaxed. Dragging a control point of a spline along the spline itself can have a drastic effect on the spline shape. When the control point of a relaxed spline is moved along the length of the spline, the spline handle remains tangent to the spline. This allows the spline to retain more of its original shape.

Also new to SolidWorks 2006 is the ability to add dimensions to spline handles. An example of this is shown in Figure 2–40. The dimension in the example controls the magnitude of the tangency force exerted on the spline. However, angular dimensions can be added as well. Make it a point to first reposition (drag) a spline handle so that it will remain visible, and then add dimensions as desired.

✓ **TIP:** *If spline control polygons are too difficult to see due to the color used to display them, change the color of Temporary Graphics in the Colors section of your System Options (Tools > Options).*

Summary

This chapter has taught you how to begin a new SolidWorks document, as well as the general steps involved in creating a new part model. The Solid-Works interface makes use of typical standard Windows functionality, such as pull-down menus and toolbars. SolidWorks also makes intelligent use of the mouse. The left mouse button is primarily used for selecting entities or commands, and the right mouse button brings up a context-sensitive menu. Use the middle mouse button for panning, zooming, or rotating the model.

Selecting the correct plane to sketch on is an important first step in sketching. Try to decide what side of a new model is going to be the top, the front, and so on. Use the right mouse button to click on planes in FeatureM-anager to toggle plane display on and off. Click on the Sketch icon to enter sketch mode.

There is a great deal of system feedback SolidWorks is constantly show-ing you during the sketch process or while selecting entities. It is extremely important to pay close attention to the cursor, as SolidWorks will add con-straints to sketch geometry during the sketch process. Which constraints are added will be displayed by the cursor. This system feedback is an integral part of SolidWorks.

Fully defined sketch geometry will be color-coded black. Underdefined geometry will be blue. It is okay to work with underdefined geometry, but it is best to fully define all sketch geometry. If geometry is fully defined, its be-havior can be predicted if dimensions are changed. Red geometry is overde-fined and should be corrected as soon as possible.

Finally, you learned how to create a wide variety of sketch entities in SolidWorks. Any sketch entity can be created using either the click-click method of sketching or the click-drag method. It is okay to use centerlines and points as reference geometry, if the need arises. Centerlines and con-struction lines are one in the same as far as SolidWorks is concerned. Con-struction geometry does not contribute to solid geometry when creating features.

Now that you know how to create various sketch entities that can be used to form a shape, the next step is to learn how to turn that shape into a solid feature. This you will learn how to do in the next chapter.

CHAPTER 3

Basic Features

NOW THAT YOU HAVE LEARNED MOST OF THE FUNDAMENTAL ASPECTS OF SOLIDWORKS, it is time to begin putting that knowledge to use. You have learned how to create sketch geometry, but there are guidelines that should be followed when creating sketch geometry that will eventually become a feature. These guidelines are explored following a quick review on how to start a sketch, a short summary of the steps used to create a feature, and a look at some preliminary settings you should know about prior to sketching.

Sketching: A Quick Review

You may recall from the section "Starting a New Sketch" in Chapter 2 that a plane must be selected in order to begin a sketch. This is accomplished by simply clicking on the desired plane in FeatureManager. If the plane is currently being shown in the work area, you could also just as easily select it from there by clicking on the plane's border.

Now enter sketch mode. This will begin a brand new sketch on the plane you have selected. Enter sketch mode by clicking on the Sketch icon, shown in Figure 3–1. In the illustration, note that the Sketch icon appears pushed in. This indicates that sketch mode has already been activated. Looking at the Sketch icon is the best way to tell if you are in a sketch.

It is not a requirement, but it is also good technique to show the plane being sketched on. This generally helps new users orient themselves. It also serves as a reminder as to what plane is being sketched on. To reiterate, right-click on the applicable plane in FeatureManager and select Show.

All told, there are generally five steps you will have to perform over and over in order to create most of the features that constitute the model. These five steps are summarized as follows.

Sketch icon

Fig. 3–1. Entering sketch mode.

1. *Select a plane or planar face:* You must always have a plane on which to sketch. This can be one of SolidWorks' default planes or one that you create. It can also be a planar (flat) face of existing geometry.

2. *Enter sketch mode:* Enter sketch mode by simply clicking on the Sketch icon (Figure 3–1). One click with the left mouse button is all it takes.

3. *Create a sketch:* Most of the mechanics involved in creating a sketch were covered in the previous chapter. The following section discusses what constitutes valid sketch geometry.

4. *Add dimensions and constraints:* Adding dimensions and constraints is not a requirement, but is good practice and highly recommended. You will learn the technique for adding constraints and dimensions in this chapter. Adding constraints helps achieve your design intent on the part. Adding the dimensions is good technique simply because you cannot fully control the model without them. Additionally, a sketch that is not fully defined may inadvertently change, producing unwanted results.

5. *Create the feature:* Creating features is what you will be spending a lot of time learning from this point forward. There are many options, and many features that can be created in SolidWorks. This book will show you a large majority of those feature types and how to create them.

These five steps to creating a sketched feature are a generality. You would not perform all of these steps when creating, for example, an applied feature, because no sketch is required. This should become self-evident to you over time.

Sketch Guidelines

Fig. 3–2. Self-intersecting geometry.

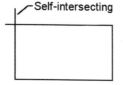

Self-intersecting

There are a few guidelines you should keep in mind when creating a SolidWorks sketch that will eventually become a feature. These are simple rules to follow, but must be obeyed. SolidWorks allows for not having to worry about exact size and shape when sketching, as dimensions and geometric relations will drive the geometry. However, it would not be wise to be sloppy. For instance, sketch profiles should be non self-intersecting. Figure 3–2 shows an example of self-intersecting sketch geometry.

Sketch rule 1: Avoid self-intersecting geometry.

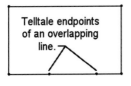

Fig. 3–3. Overlapping geometry.

If a sketch contains self-intersecting geometry, clean up the corners and get rid of any dead-end geometry. It is easy to create a sketch whose corners all meet perfectly in the first place. Just make it a point to pay attention to the system feedback.

Another type of problem is overlapping geometry. In Figure 3–3, there is a line overlying the top of an existing line. These errors are sometimes difficult to find because the overlapping geometry is difficult to see. In the figure, you can make out two of the overlapping line's endpoints. The endpoints of the extra line are visible due to an option that makes this possible.

Sketch rule 2: Avoid overlapping geometry.

There are two options in the System Options window you will almost certainly want to enable. These options are titled *Display arc centerpoints in part/assembly sketches* and *Display entity points in part/assembly sketches*. The first option allows for seeing the centerpoints of arcs and circles. The second option displays endpoints as small dots. This second option will aid you when creating sketch geometry, and will aid in troubleshooting. Both options, found in the Sketch section of the System Options window (Tools > Options), is shown in Figure 3–4. Make sure you have these options checked.

Fig. 3–4. Activating two important system options.

One last rule to follow would have to do with closing a profile. In this case, the "rule" is not really a steadfast rule, but more of a guideline for certain feature types. Leaving gaps in sketch geometry will result in a feature type that is probably not intended. Open profiles are legal for certain functions and for thin features, but not when trying to create typical solid geometry. As shown in Figures 3–5 and 3–6, the SolidWorks user was attempting to create a simple extruded part. However, the individual did not realize that a small portion of the sketch was open (Figure 3–5). Figure 3–6 shows the result of extruding an open profile, and the reason for the result.

Sometimes you might intend to sketch an open profile in order to create what is known as a thin feature. You may also use open profiles for more advanced feature types. Swept features, for instance, typically use open profile curves as a sweep path, or trajectory. All of these feature types are covered in upcoming chapters.

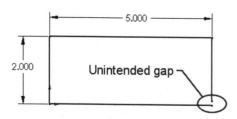

Fig. 3–5. Note the small gap in the profile.

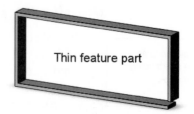

Fig. 3–6. The unintended result of extruding an open profile.

One last sketch guideline has to do with nested profiles, or sketch profiles within other sketch profiles. Profiles within profiles are acceptable to a point, but keep in mind that your editing options will be limited. Consider the scenario depicted in Figure 3–7. Here is a case in which the circles are nested within the larger profile of the rectangle. Because the circles and rectangle are part of the same sketch, they will have the same feature definition. What exactly does this mean to the user?

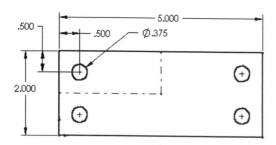

Fig. 3–7. Nested profiles.

If the holes in the plate are to always go all the way through the plate, the sketch with the nested profiles would suffice. However, if at some point in the future the designer changes their mind, a fair amount of editing would be required. If the circles were part of a separate sketch, they would have a separate definition as well. The depth of the holes could be independently changed with respect to the depth of the plate, because the holes and the plate would be independent features.

Sketches should be separated into editable chunks of geometry. Do not try to cram everything into a few complex sketches. It makes for a part that is difficult to manage, difficult to troubleshoot, and very inflexible.

Check Sketch For Feature

Fig. 3–8. Check Sketch For Feature window.

The Check Sketch For Feature function is very useful for anybody learning the SolidWorks software. You have the ability to make SolidWorks check a sketch and tell you if it is a valid sketch for any particular feature type. This function is found in the Tools > Sketch Tools menu. It is a very straightforward and easy-to-use command. The window for this function is shown in Figure 3–8. How-To 3–1 takes you through the process of using the Check Sketch For Feature command.

HOW-TO 3-1: Using Check Sketch For Feature

To use the Check Sketch For Feature command, perform the following steps. It is assumed a sketch has already been created and sketch mode is active.

1. Select Check Sketch For Feature from the bottom of the Tools > Sketch Tools menu.

2. Using the drop-down list, select the feature type you want to create.

3. Click on the Check button.

4. Click on Close when finished checking your sketch.

SolidWorks will display a message informing you as to whether or not you have a valid sketch for the feature type you chose. If there is a problem with your sketch, SolidWorks will tell you what is wrong. If there are no problems with the sketch, a message will inform you of this as well.

The Check Sketch For Feature command is commonly used when troubleshooting existing geometry. When an existing sketch goes bad due to some sort of design change, Check Sketch For Feature will tell you what is wrong. When used for troubleshooting a sketch for an existing feature, the correct feature will automatically be selected from the Feature Usage drop-down list. All the user has to do after calling up the command is click on the Check button (previous step 3).

Fully Define Your Sketch

SolidWorks does not require you to completely define sketch geometry. This means that if you did not want to add any constraints or dimensions you would not have to. This flexibility can be nice, but it is not necessarily good practice. It is good practice to fully dimension and constrain all of your sketch geometry. This allows you greater control over your sketches because the sketch must follow the set of conditions you have placed on it. If you do not have full control over a sketch, the sketch will not behave in a predictable fashion.

Use the origin point to your advantage when creating the first sketch of a part. Remember that the origin acts as an anchor point. This means that it will lock your sketch in place, thereby allowing it to be fully defined. Sketch geometry created later can be anchored or dimensioned to features that come before them. This creates a predictable model—one that is not prone to errors if design changes are needed later on.

Remember that your sketch geometry will turn black when it is fully defined. Consider this your goal. If you would like SolidWorks to force you to fully define sketch geometry (leaving you with one less detail to remember), there is an option that will accomplish this. Select Tools > Options, and in the Sketch section check the option *Use fully defined sketches*. Once this option is enabled, it will not be possible to create a feature unless the sketch is fully defined, which will help you develop the proper habits with the software.

Units of Measurement

One of the options usually set up before beginning a new part is the working units. Typical working units are inches and millimeters, but other options are available. How-To 3–2 takes you through the process of changing the working units of measurement. Note that the units can be changed at any time, even after a model is completed.

HOW-TO 3-2:   Changing Working Units

To change working units, perform the following steps.

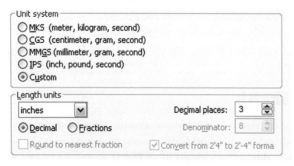

Fig. 3–9. Changing the units of the model.

1. Select Options from the Tools menu.

2. Select the Units category from the Document Properties tab.

3. Select the desired unit system, or use custom settings (see Figure 3–9).

4. Select units for use with dual dimensioning and angular dimensions, or just leave these values at their default settings.

5. If custom settings are desired, make selections for additional values, such as density and force units. These settings will not be available unless the Custom option is selected.

6. Click on OK when finished.

When the units of a part are modified, the change will only affect the current document. To have the units of measurement already set when starting a new part (or assembly or drawing), the document must be saved as a template with the desired units already specified. That way, when a new document is begun the template with the proper units can be selected. See the section "Creating Templates" later in this chapter for more information on this matter.

Grid and Snap Settings

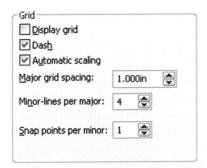

Fig. 3–10. Grid/Snap options.

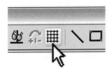

Fig. 3–11. Grid icon.

The Document Properties Grid/Snap section is shown in Figure 3–10. Placing a check in the *Display grid* setting will turn on the grid. The grid is strictly a user preference. If you find it useful, by all means turn the grid on. Some find the look of a clean, unencumbered screen more desirable. Others like the grid turned on because it helps orient the user as to what plane they are sketching on. Note that the grid is only displayed while in sketch mode.

If you decide to use the grid, the other optional settings in the Grid/Snap section will allow you to customize the grid's appearance. *Major grid spacing* allows for changing the spacing between the primary grid lines. *Minor-lines per major* sets how many secondary grid lines are placed between the major grid lines. In other words, if the *Major grid spacing* is set to 1 inch and *Minor-lines per major* is set to 4, there will be a grid line every quarter inch.

Incidentally, the Grid icon (shown in Figure 3–11) is a shortcut to the Grid/Snap section of the Document Properties window. However, this does not mean that you can only adjust the Grid or Snap settings once the window is open. Feel free to change any of the System Options or Document Properties.

✓ **TIP:** *The Grid sketch snap (see previous chapter) causes sketched objects to snap to the grid. However, this function has limited usefulness, considering that dimensions will drive the geometry.*

Sketch Plane Indicators

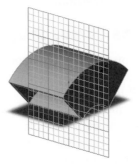

Fig. 3–12. Display grid.

As mentioned previously, the display grid can be useful because it serves as an indicator as to what plane is currently being sketched on. If this is the reason you find the grid useful, you should be aware of other options that serve as sketch plane indicators. Figure 3–12 shows an example of what the display grid might look like.

Fig. 3–13. Sketch display plane.

The grid only displays when in a sketch, so it can also be useful to new users in that it serves to indicate when a sketch is being edited. If the grid is not your cup of tea, another alternative is to use the *Display plane when shaded* option. An example of the display plane is shown in Figure 3–13. You will find the *Display plane when shaded* option in the Sketch section of the System Options (Tools > Options). Similar to the display grid, the display plane only appears when editing a sketch.

Some users prefer to have full control over what planes are shown, and when. This is typical of those who have more experience with the software. Chapter 2 discussed how to manually display a plane. There is something to be said for working with an uncluttered screen without grids and shaded planes appearing every time a sketch is started. However, these options are there to help, and you should use them if they benefit you.

Snap Behavior

There are two types of snaps in SolidWorks. There is the ability to snap to the grid, and there is the ability to snap to geometric locations, such as endpoints or midpoints. Snapping to a grid is not recommended, as it can actually impede the user's ability to sketch. Consider the fact that SolidWorks is a parametric program. This means that dimensions and constraints are controlling the sketch. Why would one need to snap a line to a precise length when a dimension is required to drive the length of the line anyway? Snapping to a grid would not be necessary, and that is precisely the point.

You learned about sketch snaps in the previous chapter. Not all sketch snaps add relations automatically. An example of one such snap would be the *Horizontal/vertical to points* snap. This particular snap setting will make it easier to line up endpoints of lines horizontally or vertically with other points in your sketch (via dashed inference lines), but no horizontal or vertical relations will be automatically added between those points. How can one tell if a relation is being added automatically or not? Read the next section to find the answer to this question.

Automatic Relations Versus Snaps

System feedback, which you first learned about in the previous chapter, is critical to determining if geometric relations are being added automatically. Various snap symbols will be displayed next to the cursor while sketching. Some examples of snap symbols are shown in Figure 3–14. These happen to be the angle and horizontal snap symbols. The horizontal symbol is displayed because the endpoint of the line currently being sketched is aligned horizontally with another line's endpoint.

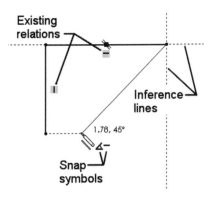

Fig. 3–14. Snap symbols and other system feedback.

The main question is in understanding whether or not a geometric relation is being added during the sketch process. This fact can be discerned by the color of the snap symbol being displayed. If the symbol is yellow, a relation is being added automatically. If it is white, no relation is added. Existing relations will be shown as blue symbols attached to various sketch entities. These symbols can also be seen in Figure 3–14.

✓ **TIP:** *To turn sketch relation symbols on or off, check or uncheck Sketch Relations in the View menu. This affects existing relation symbols only. It does not affect system feedback or the model geometry in any way.*

Inference lines are just another convenience factor when sketching. They facilitate positioning the cursor correctly in order to add relations while sketching. In addition to the relations added automatically to sketch geometry, it will also be necessary to add some relations manually to fully incorporate your design intent. This is discussed in the next section.

Adding Geometric Relations

To this point, you have learned that constraints are added automatically while sketching. This is why paying attention to system feedback is so important. It is important to watch for the feedback displayed by the cursor, so that you will know what constraints are being added to the sketch geometry.

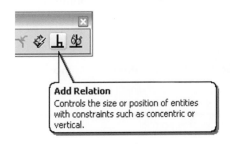

Add Relation
Controls the size or position of entities with constraints such as concentric or vertical.

Fig. 3–15. Add Relation icon.

↦ **NOTE:** *The terms* constraints *and* geometric relations *have identical meaning and are often used interchangeably.*

Another method of adding constraints is to use the Add Relation icon, shown in Figure 3–15. This icon is found on the Dimensions/Relations toolbar, and allows for manually adding constraints after sketch geometry has already been created. How-To 3–3 takes you through the process of manually adding geometric relations to sketch geometry.

HOW-TO 3-3:   Adding Relations

To add a geometric relation, perform the following steps.

1. Select Relations > Add from the Tools menu, or click on the Add Relation icon.

2. Select the entity (or entities) to add a relation to. You may also select entities you wish to add a relation between (such as two lines you wish to make perpendicular).

3. In PropertyManager, click on the relation you want to add. Be aware that the relation is added *as soon as* you click on the icon.

4. Repeat steps 2 and 3 as necessary.

5. Click on the green check when finished.

✓ **TIP:** *It is not actually necessary to click on the Add Relation icon. Simply selecting an entity (or entities) will suffice. All the Add Relation icon really does is to keep the Add Relations panel visible so that multiple relations can be added one after the other without the need to hold down the Ctrl key.*

When adding relations, the software tries to guess which relation you want to add by making the text for that relation bold in PropertyManager. This often makes finding the relation you want a little easier. Other than this fact, the bold text does not serve any purpose. SolidWorks does not always guess correctly, but is correct more often than not.

Sketch geometry will move position or change shape to accommodate the relations being added. Examine Table 3–1 to better understand what adding various constraints will accomplish. The term *arc* in this table refers to circles or arcs. The term *point* refers to point entities, endpoints of lines and arcs, centerpoints of arcs and ellipses, and so on. This table is meant to give a wide range of examples, but does not show every combination of possibilities.

Table 3-1: Geometric Relations and Examples of Use

Relation	Sketch Entities Used	Changes Sketch Entities Undergo
Horizontal or Vertical	One or more lines, or two or more points	Lines become horizontal or vertical; points are aligned horizontally or vertically.
Parallel	Two or more lines	Lines become parallel.
Perpendicular	Two lines	Lines become perpendicular.
Collinear	Two or more lines	Lines lie on the same theoretically infinite line.
Coincident	A point and a line, arc, parabola, or ellipse	Point lies on the line, arc, parabola, or ellipse.
Midpoint	A point and a line	Point remains at the midpoint of the line.
Intersection	Two lines and one point	Point remains at the intersection of the lines. This can be a projected intersection.
Coradial	Two or more arcs	Arcs share the same centerpoint and have an equal radius.

Relation	Sketch Entities Used	Changes Sketch Entities Undergo
Tangent	An arc, ellipse, or spline, and a line, arc, parabola, ellipse, or spline	Items remain tangent.
Concentric	Two or more arcs, or a point and an arc	Arcs share the same centerpoint, or a point remains at the arc's centerpoint.
Equal	Two or more lines, or two or more arcs	Line lengths or arc radii remain equal.
Symmetric	A centerline and two points, lines, arcs, ellipses, splines, or parabolas	Items remain symmetrical about the centerline.
Pierce	A sketch point and an axis, edge, line, arc, or spline	Point is coincident to where the axis, edge, line, arc, or spline pierces the sketch plane.
Merge	Two entity endpoints (such as line or arc endpoints)	Points are merged into a single point (similar to coincident).
Fix	Any entity	Entity is locked in position.

There are a few additional aspects of the last two constraint options, Merge and Fix, you should be aware of. When two points are merged, the two points essentially become coincident. However, because the points are actually merged into one point it is not possible to remove a Merge constraint. Any other relation can be removed, which you will learn shortly.

Regarding the Fix constraint, it is one you should typically avoid. Fixing items locks them in position. It is rare this action will be necessary, and almost always results in overdefining the geometry. Points (such as a line's endpoints) and entities (such as a line itself) can be fixed independently of each other.

It is not intended that you memorize this table. Its purpose is for reference. Adding relations is normally a fairly straightforward procedure. The descriptions of the various relations are meant to help the reader who is unfamiliar with some of the terminology being used by SolidWorks.

Viewing Existing Relations

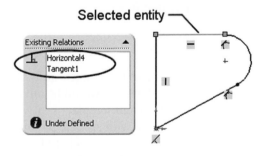

Fig. 3–16. Relation information in PropertyManager.

PropertyManager displays relations that have already been added to a selected object. It will also show relations between objects if more than one object is selected. This information is displayed in the area labeled Existing Relations, shown in Figure 3–16. In the case of Figure 3–16, PropertyManager is telling us that there is already a horizontal and tangent relationship associated with the selected line. Also visible are the words *Under Defined*, informing us of the current status of the selected sketch entity.

Aside from selecting a sketch entity to see what relations are attached to it, the small blue symbols that appear next to the geometry also inform the user of existing relations. The color of the geometry itself serves to indicate the state of the geometry, such as underdefined or fully defined.

✓ **TIP:** *If geometry is underdefined, try dragging it to see how it reacts. This will often give you clues as to what relations might be required to fully define the sketch.*

Deleting Relations

Fig. 3–17. Display/ Delete Relations icon.

The easiest way to delete a geometric relation is to delete it from PropertyManager or delete its associated symbol from the work area. Clicking on a symbol will highlight it in a different color. Once that has been done, pressing the Delete key is all it takes to remove the geometric relation from the sketch entity. To delete a relation via PropertyManager, select the relation from the Existing Relations list box and press the Delete key.

One last method used for displaying or deleting geometric relations is via the Display/Delete Relations command, whose icon is shown in Figure 3–17. It really is not necessary to use the Display/Delete Relations command to show or remove relations from just a few sketch entities. Where this command becomes beneficial is when it is necessary to see all relations associated with an entire sketch. The command is also helpful when troubleshooting.

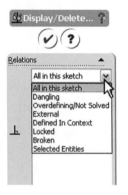

Fig. 3–18. Using the Display/Delete Relations command.

The best part of the Display/Delete Relations command is the dropdown list shown in Figure 3–18. Using this list, it is possible to view all relations in the sketch, or only relations with certain errors, such as overdefining relations. This often helps when troubleshooting a sketch. There are other options listed as well, such as locked or broken relations, which are discussed later in the book. For now, it is enough to understand that relations can in fact be filtered out.

You probably will not need the Display/Delete Relations command on a daily basis, but it is good to know it is available. If your sketch exhibits errors of one sort or another, you may find this command useful.

Using Templates

Rather than have to set up units of measure, grid/snap settings, and any other number of settings every time a new part is started, it would be advisable to create what are known as *templates*. A document template is a way of saving a SolidWorks part, assembly, or drawing file with specific parameters set up by the user. There is almost no limit to the variations of templates that can be created. Templates can be created for different drawing sheet sizes, for part files created in English or metric units, or for different standards, such as ANSI or ISO dimensioning standards. These are the most common reasons for saving a template. You may have others.

Template files have a different file extension than the regular three SolidWorks document types. What are file extensions? They are the characters at the end of a file name the Windows operating system uses to identify a file type. This has more to do with Windows than it does with SolidWorks.

The file extensions used by templates differ from those used by the standard document types so that SolidWorks (and Windows) can tell them apart. The naming convention used for the template file extensions is very logical, as are most things in SolidWorks. The following are the file extensions used by SolidWorks templates.

- Part templates: *.prtdot*

- Assembly templates: *.asmdot*

- Drawing templates: *.drwdot*

If you are fairly new to computers and do not fully understand the reasons for file extensions or what their significance is, it is not that important. The computer adds file extensions to your file names automatically, so it is nothing the computer user has to worry about.

Modifying Document Properties

You have already discovered how to modify some settings that can be saved within a template. Those settings are the units of measure and the various grid and snap settings. Accessing these settings is a matter of clicking on the Options command in the Tools menu. There are many other optional parameters that can be adjusted. It would be overwhelming to go through all of these various parameters now. Many of these parameters are discussed throughout the book.

Be wary of altering too many settings prior to saving a template, because you may change something you should not. Basically, stick with what you know, such as modifying the working units. If you are changing properties of the document with the intention of creating a new template, there is a general guideline you should follow. Start a new part file (or assembly or drawing) using the generic part template included with the software. You can then make changes to the document's properties and save the template with a file name that is meaningful to you. How-To 3–4 takes you through this process.

How-To 3-4: Preparing a Template

To modify a document's properties prior to saving the document as a template, perform the following steps.

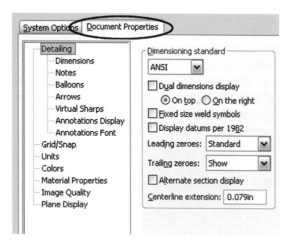

Fig. 3–19. Document Properties tab and various option categories.

1. Start a new part (or assembly or drawing) using the generic part (or assembly or drawing) template included with the SolidWorks software. You learned how to do this in Chapter 2.

2. Select Options from the Tools menu.

3. Click on the Document Properties tab. Anything changed within this tab will be saved with the document.

4. Select the desired category from the pane on the left (see Figure 3–19).

5. Make the desired changes.

6. Click on OK when finished.

Of primary importance in understanding how to create a template is that changes made via Document Properties only affect the current document. Changes made via System Options globally affect every SolidWorks document. Therefore, and settings changed in the Document Properties will become permanent changes once the template is saved. How-To 3–5 takes you through the process of saving the template.

HOW-TO 3-5: Saving a Template

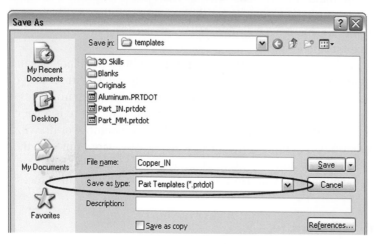

Fig. 3–20. Save as type *drop-down list.*

To save a file as a template, perform the following steps.

1. Select Save As from the File menu.

2. From the *Save as type* drop-down list (see Figure 3–20), select Part Templates as the type of file to be saved. Obviously, if starting with an assembly you should select Assembly Template, and so on.

3. Navigate to the appropriate directory where you would like the file to be saved. By default, this is the *Templates* directory. It is suggested that you save the template there for now.

4. Type in a name for the file. Use a logical name that makes sense to you.

5. Click on the Save button.

Now you can use the newly saved template when beginning a new SolidWorks project. Templates for all three SolidWorks document types should be created, unless you feel content using the three standard templates SolidWorks provides. It is very common to create templates that already have the units of measurement established, or for different dimensioning standards. Your reasons for creating templates may vary.

File Locations

By default, the place where SolidWorks stores template files is the *SolidWorks\data\templates* folder. This is also the same place where SolidWorks looks for templates when starting a new document. The location where SolidWorks looks for templates can be changed, or other locations can be added to the path listing, thereby forcing SolidWorks to look in multiple locations.

There are many different types of files that are used by SolidWorks, above and beyond document templates. The paths that point to the location of these various files can be specified using a command found in the System Options. How-To 3–6 takes you through the process of modifying or adding search paths to various SolidWorks files or folders. In future chapters you will learn about some of the other files and folders used by SolidWorks. For now, you can use the following procedure to set up your template locations.

HOW-TO 3-6: **Modifying Folder Location Paths**

To modify or add paths SolidWorks uses to find document templates (or other SolidWorks files or folders), perform the following steps. Note that file directories (folders) should be set up prior to pointing SolidWorks to a particular location. Creating file directories is a function of the Windows operating system and is outside the scope of this book.

1. Click on Options in the Tools menu.

2. Select the File Locations category in the System Options tab.

3. Using the *Show folders for* drop-down list, select the item to set the path for. In Figure 3–21, Document Templates has been selected from the drop-down list.

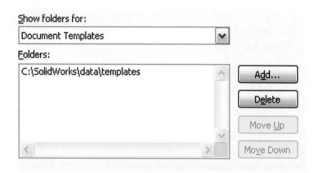

Fig. 3–21. Setting a path for Document Templates.

4. Click on the Add button to add a new or additional file location path.

5. Specify a new file or folder path and click on OK.

6. To remove a currently listed path, select the path in the Folders listing and click on the Delete button.

7. Click on OK when finished to close the System Options window.

When a path is selected from the Folders list, it is sometimes desirable to move that path higher or lower in the path listing. This is really not necessary when setting up locations for templates, but in other cases where SolidWorks must search through a list of paths the search order may be a priority due to reasons of speed. If an externally referenced file can be found more quickly, for instance, it may accelerate how quickly other files can be opened.

If you took a moment to examine the list of files and objects in the *Show folders for* drop-down list, you will have found that the list is quite extensive. For example, there is a listing for BOM Templates, Bend Tables, Textures, and Macros, just to name a few. You will learn about many of these items as you read through this book.

Template Tabs

After adding a path to the list of locations SolidWorks uses to search for template files, a new tab will appear in the New SolidWorks Document window the next time a new document is begun. This fact is illustrated in Figure 3–22. For tabs to appear, at least one template must reside in the folder specified in the path. The names of the tabs will match the folder names listed in the paths created in How-To 3–6.

Three of the tabs shown in Figure 3–22 were created a different way. Specifically, any folders created within the *Templates* folder itself will automatically appear as tabs in the New SolidWorks Document window. As mentioned previously, there must be at least one template in a folder before it will appear as a tab.

Fig. 3–22.
New tabs
appear.

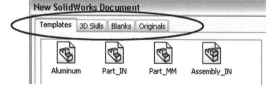

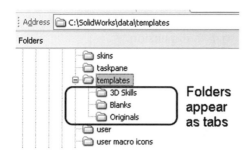

Fig. 3–23. Folders that appear as tabs when beginning a new document.

A portion of the Windows Explorer window is shown in Figure 3–23. There are three folders that reside in the path *C:\SolidWorks\data\templates*. This path is the same one specified earlier in Figure 3–21, and specifies the location of the Solid-Works templates. Note that the names of the folders in Figure 3–23 match the names of the tabs shown in Figure 3–22. This is no coincidence. This is intended behavior that makes it very easy to add a bit of customization to the software and control what appears as tabs when beginning a new document.

Dimensioning

Placing dimensions on a sketch is a very straightforward process. Most of the dimensions placed on a sketch (and on a part in general) transfer to the 2D drawing later on. This is another reason it is important to fully define sketch geometry. If all sketches are fully defined, theoretically the 2D drawing will have all required dimensions for the manufacturing department to build the part. Do not worry about 2D drawings just yet, as they are covered in Chapter 12.

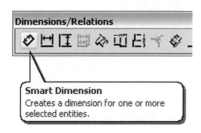

Fig. 3–24. Smart Dimension icon.

The main dimensioning tool in SolidWorks is aptly named the Smart Dimension command (shown in Figure 3–24), which is found on the Dimensions/Relations toolbar. This one icon does it all, from aligned dimensions to diameter dimensions and everything in between. So much so that the other dimension commands on this same toolbar are typically not even needed. The dimension type created largely depends on what is selected to place the dimension on. If a circle is selected, a diameter dimension is created. If a line is selected, a linear dimension is created, and so on.

This section is intended to get you started with the basic dimensioning skills needed to get by. Dimensioning requires picking the objects to be dimensioned, and then picking a location to place the dimension. Many specific dimension tips and tricks are covered throughout this book. For now, How-To 3–7 will take you through the process of adding generic dimensions to a sketch.

HOW-TO 3-7: Adding Dimensions

To add a dimension to a sketch, perform the following steps.

1. Select the Smart Dimension icon.

2. Click on the item (or items) you wish to dimension.

3. Click to position the dimension.

4. Type in a value for the dimension if prompted.

5. Press Enter or click on the green check to accept the dimension value.

Step 4 states that you should type in a value for the dimension *if prompted*. Whether or not you are prompted for a value depends on a particular setting in the System Options (Options > Tools). This option is named *Input dimension value,* shown in Figure 3–25. Make sure the option is checked. This will make SolidWorks ask for a value each time a dimension is added to the sketch.

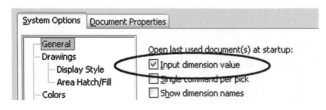

Fig. 3–25. Input dimension value option.

Fig. 3–26. Modify window.

Sometimes it is necessary to add a dimension between two objects, such as when dimensioning the angle between two lines. What throws most new users off is the dimension's preview. The point to remember is "don't stop short"! Picking a second object will automatically update the preview to show the correct dimension.

The rest is a no-brainer. Pick where you would like to see the dimension value be positioned and the Modify window, shown in Figure 3–26, will be displayed. The easiest way to finish up this process of adding a dimension is to simply type in a number. Do not click in the highlighted area first, where the current value is displayed. Just type. Once you have typed in a value, press Enter and move to the next task.

Users from other CAD programs may have a tendency to pick endpoints as opposed to lines when adding a linear dimension. This is not necessary. Just select the line. It is one click less and will do the job just as well.

↝   **NOTE:** *SolidWorks contains an automatic dimensioning tool named Autodimension. It is more appropriate for imported geometry (discussed in Chapter 24).*

10-second Topic: Locking In Previews

The Smart Dimension command is found on the Relations/Dimensions toolbar. Although there are a number of other dimension commands on the same toolbar, such as for creating vertical or horizontal dimensions, you will not necessarily need them. The reason for this is that you can force SolidWorks to create certain dimensions by locking in the dimension preview.

If you were to click on an angled line for the purpose of dimensioning it, you would find that moving the mouse causes the dimension preview to change. You would see an aligned dimension preview, along with a vertical and horizontal dimension preview, depending on where the mouse is moved. Anytime you are presented with multiple preview options when adding a dimension, simply right-click to lock in the desired preview. It is then possible to position the dimension value anywhere, and the dimension type will remain consistent.

Modifying Dimensions

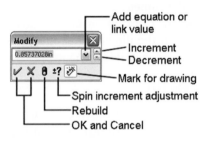

Fig. 3–27. Commands available from the Modify window.

This section describes the various functions of the Modify window and other general topics of interest related to dimensions. For example, how would you change the value of a dimension? The answer is simple. Double click on the dimension, at which point a new value can be entered. Press Enter or click the green check to accept the change, just like you would when adding a dimension for the first time. Deleting a dimension is accomplished by selecting the dimension and pressing the Delete key.

There are other commands on the Modify window besides the green check. Figure 3–27 depicts what functions clicking on the various icons would perform. Following the image, Table 3–2 provides a short description of each of these functions.

Table 3-2: Modify Window Commands

Command	Function
OK and Cancel	Accepts or rejects any changes made to the dimension value.
Rebuild	Rebuilds the model (or sketch) using the new dimension value.
Spin Increment Adjustment	Allows for entering a new value for Increment and Decrement arrows (see material following).
Mark For Drawing	This option, toggled on by default, sets whether a dimension has been marked for importation into a drawing. Toggle this setting off to keep a dimension from appearing on a drawing.
Increment and Decrement arrows	Increases or decreases the displayed value by an incremental amount.
Add equation or link value	Shortcut for linking the dimension to another dimension, or for creating a new equation containing the dimension. Linking dimensions and equations are discussed in Chapter 18.

Clicking on the Spin Increment Adjustment icon opens another window that allows for typing in an increment value. Pressing the Enter key will accept the new value, and the Increment and Decrement arrows will then change by that particular amount. This only affects the Modify window for the dimension you are modifying. However, it is possible to make the new spin increment value the default setting for all dimensions. To do this, check the Default option present after clicking on the Spin Increment Adjustment icon.

As one might expect, there is a location in the System Options (Tools > Options) where the spin increment can also be adjusted. It is found in the section of the System Options appropriately titled Spin Box Increments. There are settings for English and metric units, as well as for angular values, as shown in Figure 3–28.

Clicking on the Rebuild icon found in the Modify window will let you see how the dimensional change affects the sketch before you accept it. Clicking on Cancel closes the Modify window, whereupon the dimensional change is rejected and the sketch returns to its original shape.

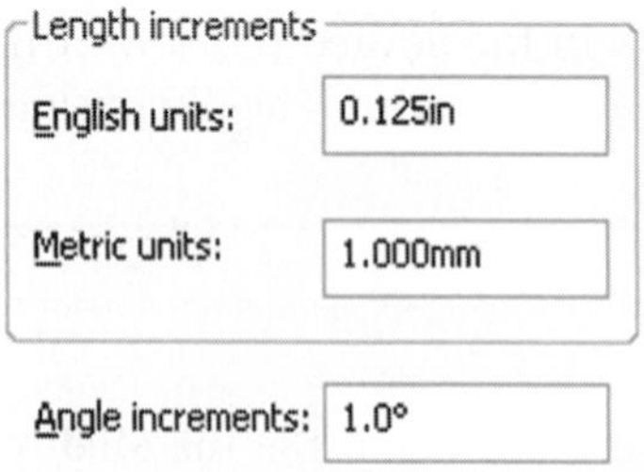

Fig. 3–28. Spin Box Increment adjustments.

Resolving Conflicts (SW 2006 Only)

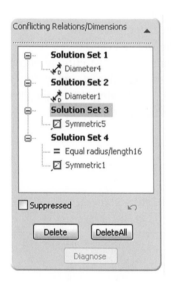

Fig. 3–29. Resolving conflicts in an overdefined sketch.

When a sketch becomes accidentally overdefined, understanding what is causing the overdefined situation can be challenging. An overdefined sketch can occur if there are too many dimensions or geometric relations. A tool that will help diagnose possible solutions is the Resolve Conflict command, which can be accessed from the Tools > Sketch Tools menu. This same command can also be accessed by clicking on the overdefined sketch warning, which will appear on the status bar of the SolidWorks window.

If you decide to take advantage of the Resolve Conflict command, the Resolve Conflicts PropertyManager will appear, a portion of which is shown in Figure 3–29. Initially, all of the conflicting dimensions and relations will be shown. Clicking on the Diagnose button (grayed out in the figure) will separate the conflicting dimensions and relations into solution sets. This is precisely what has been done in Figure 3–29.

By examining the solution sets provided by the diagnosis, it is easier to make an intelligent decision as to what should be deleted. Deleting one of the solution sets should serve to remove the overdefining situation. Another option is to suppress a solution set by checking the Suppressed option. Suppressing objects is akin to temporarily turning an object off (this is discussed in more detail in Chapter 11).

⊷ **NOTE:** *When you suppress a solution set containing a dimension, the dimension is not actually suppressed. Rather, it becomes a driven (reference) dimension.*

Base Features

The very first feature created in a part file is called the base feature, and nearly every SolidWorks model starts with one. The base feature typically represents the general overall shape of the part. All features fall into two categories: sketched and applied. Sketched features require a sketch. This is the feature type for which you will need the most practice. Applied features are applied directly to the model and do not require a sketch. Therefore, there is typically much less manual labor involved in creating an applied feature. All base features are sketched features, for the simple reason that an applied feature cannot be created if there is nothing to apply it to!

Sketched features can be created by extruding a sketch, revolving a sketch, sweeping a sketch along a path, or lofting (blending) two or more sketches. The names for these commands are Extrude, Revolve, Sweep, and Loft, respectively. Any one of these sketched features can be a boss or a cut; that is, they can either add or remove material. However, the base feature must add material to the model (for what should be an obvious reason). Base features can be any of the aforementioned four sketched feature types, or may be some other advanced feature type, many of which you will learn about in this book.

Once the first base feature has been created, building up the rest of the model is primarily a matter of adding more features to the model. These features may be sketched or applied, and may add material or remove material. The earlier chapters in this book deal primarily with extrusions and revolved features. Sweeps and lofts are covered in more detail later in the book, as they are generally more complex feature types.

Pivot Arm Project: An Overview

There's nothing like getting your hands dirty in order to learn how to do something. That is the reason for this project. One of the main points of focus in this project is basic extruded features. Other topics you will be introduced to in the next chapter include cut extrusions, renaming features, the Hole wizard, mirroring, and trimming and extending sketch geometry. Various end condition types in the Extrude panel are also explored.

At the end of the project (which will span multiple chapters) you will also see how easy it is to make design changes on a completed part. This will include modifying dimension values and editing definitions of features. The pivot arm you will begin in this chapter is shown in Figure 3–30.

Fig. 3–30. Pivot arm part.

This part will start like any other. A new file must be created, and the working units should be established. The pivot arm will be created using inches as the unit of measurement. If you have forgotten how to do this, Exercise 3–1 will guide you through this process. As with any sketched feature, you must begin by selecting the sketch plane, entering sketch mode, and creating the sketch that will be used to create the feature.

✓ **TIP:** *Command sequence is not important, as SolidWorks is very forgiving. For example, if the Rectangle command is selected prior to starting a sketch, the software prompts the user to select a sketch plane.*

The pivot arm's base feature and a likely plane on which to create its sketch need to be determined. Looking at this part, it is fairly safe to say that the main body of the part would be a good candidate for the base feature. Remember to keep the sketch geometry simple. Try to break the part up into its individual feature components.

EXERCISE 3-1:   Beginning the Pivot Arm

This exercise steps you through the process of creating the pivot arm. You will begin by starting a new part and selecting the appropriate units to work with. You will then create the sketch and extrude it using a Blind end condition. In the next chapter, you will learn how to modify the features you create in this chapter.

If you experience problems in this first exercise, skip ahead to the section "Common Mistakes." It will give you hints as to what problems you may be having. Following this exercise, you will learn the pan and zoom commands necessary to change your view of the model.

1. Select File > New, or click on the New icon, to begin a new SolidWorks document. Pay attention to the tooltips or refer to the Toolbar Quick Reference at the beginning of this book if you need help distinguishing icons.

2. Select the Part template and click on OK (or double click on the Part template).

3. To check the current working units, select Options from the Tools menu.

4. In the Document Properties tab, select the Units section. Set the unit system to IPS (inch, pound, second).

5. Click on OK.

6. Right-click on the Front plane and select Show. Note that this action also serves to select the Front plane.

7. Click on the Sketch icon (see Figure 3–1).

8. Create the sketch shown in Figure 3–31. Add dimensions and constraints as needed to fully define the sketch geometry (i.e., turn it black).

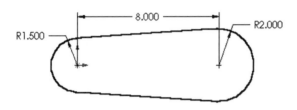

Fig. 3–31. Sketch profile for the base feature.

It is not necessary to show the sketch plane in order to sketch on it, but sometimes it helps to maintain your bearing. If you do not feel as though you need to see the plane you are sketching on, do not feel obligated to show it. It was also not necessary to select the Front plane prior to clicking on the Sketch icon. If a plane is not selected first, you will be prompted to pick one.

It will be necessary to attach the sketch to the origin point before it becomes fully defined. You may have to drag the center of the left arc to the origin. This will establish a coincident relation to the origin, thereby anchoring the sketch. If the sketch is black (fully defined), you are ready to create the base feature. The sketch you just completed should still be active, meaning that the Sketch icon should still appear depressed. There should also be only one sketch listed in FeatureManager. If this is not the case, start a new part and try again. Otherwise, continue with the following steps.

9. Select Insert > Boss/Base > Extrude, or click on the Extruded Boss/Base icon found on the Features toolbar.

10. In PropertyManager, set the depth to .75 inches. Figure 3–32 shows the PropertyManager settings and a preview similar to what should appear on your screen.

11. Click on OK (the green check) to create the extrusion.

Fig. 3–32. PropertyManager settings for the base feature.

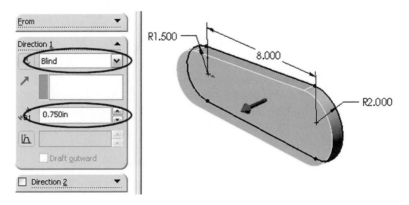

Were you able to create the sketch and feature without too much difficulty? The sketch is easiest to create if a line is drawn first. You can then use auto-transition (discussed in the previous chapter) to create an arc, sketch another line, and auto-transition to the second arc. Don't draw more geometry than necessary with the idea of trimming away what you don't need. This creates more work than is necessary (and besides, we have not discussed the Trim command yet!).

If you are running into some stumbling blocks, try not to get discouraged. The section "Common Mistakes" following this exercise will help you along. If you are having some difficulties, skip ahead to that section and read through it. Otherwise, continue with the exercise and try a slightly more challenging sketch.

This next sketch will be for the slotted angled arm of the pivot arm part. The sketch is a bit more complex and contains construction geometry. There are also constrains that you will need to add to help position the sketch. To complete the next feature, you will first need to select a plane to sketch on, and then create the sketch for the slotted arm. To complete this process, continue with the following steps.

12. Select the Front plane.

13. Click on the Sketch icon.

14. Change to a Front view. To do this, press the space bar on your keyboard and double click on the word *Front*. (You will learn more about changing views shortly.)

15. Complete the sketch as shown in Figure 3–33.

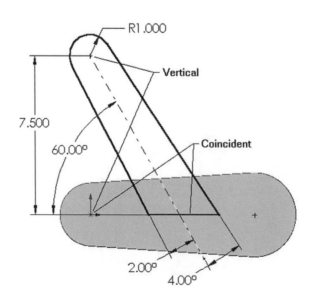

Fig. 3–33. Sketch for the slotted arm feature.

Figure 3–33 shows almost everything you need to know to complete the sketch. You may have to add some relations on your own. Two of these relations are pointed out, along with the entities the relations should apply to. Be careful when selecting entities, as it is possible to select either a line or its endpoints, and it makes a difference.

Once you have the sketch completed, you are ready to create the boss extrusion. Make sure your sketch geometry is fully defined. If your sketch still contains geometry that is blue, fully define the sketch before proceeding. To create the slotted arm feature, continue with the following steps.

16. Select Insert > Boss/Base > Extrude, or click on the Extruded Boss/Base icon.

17. Change the view orientation by holding down your middle mouse button. You need to be able to see the preview of the extrusion to determine its direction.

18. Specify a Blind end condition and a Depth of .375 inches. The extrusion should be going away from the first feature. Click on the Reverse Direction option if necessary, which is to the left of the end condition dropdown list (which states Blind).

19. Click on OK to accept the settings.

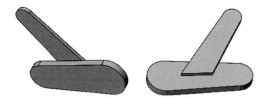

Fig. 3–34. Pivot arm after creating the first two features.

At this point, the pivot arm should look as shown in Figure 3–34 (as seen from two vantage points). To be honest, the two features just created need to be modified, but this was the intention. The next chapter explores basic editing techniques, and you will learn just how easy it is to make changes to geometry in SolidWorks. Therefore, make sure to save your work.

20. Select File > Save, or click on the Save icon.

21. Navigate to the directory of your choice and type *Pivot Arm* as the name for the file.

22. Click on the Save button to save your work.

If you have gotten this far, you are doing well. The most difficult aspects of working with SolidWorks are understanding how geometric relations work and coming to the realization that there is actual intelligence built into a sketch. Geometry reacts the way you want it to react and isn't just a bunch of lines and arcs on the screen. In SolidWorks, design intent is king.

Now that you have some geometry on the screen, it's time to explore how the pan and zoom functions work. If you are having some trouble with sketching, read over the next section. Otherwise, feel free to skip ahead to the "Pan and Zoom Commands" section.

Common Mistakes

There are some common mistakes new users have a tendency to make. One such mistake is the tendency to accidentally exit out of the sketch before the feature is created. To complicate things, a second mistake is to accidentally start a new sketch when in reality it is the original sketch the user should be editing. Read on, and chances are you will understand why you are having difficulty completing the pivot arm base feature.

If you cannot even seem to get started, keep in mind that you must be in sketch mode before you can create a sketch. This is done by clicking on the Sketch icon and selecting a plane. The order in which you perform these two acts is not important.

If your model does not look the same as the model pictured, it could be that you are simply not seeing the same view. This is not really important to the model; it just alters your perspective of the model. You can use your middle mouse button to rotate the model on screen. Move your mouse while holding down the middle mouse button.

Next, ask yourself the following questions. Is some or all of the sketch geometry gray instead of blue or black? Do you see more than one sketch in FeatureManager? If so, perform the steps outlined in How-To 3–8.

How-To 3-8:   Reentering a Sketch

To reenter a sketch, perform the following steps.

1. Click on the Sketch icon to exit out of the current sketch, if you have not already done so. If the Sketch icon does not look "pushed in," you are not currently in a sketch.

2. You need to edit the sketch you were originally in (typically the first sketch listed in FeatureManager). Right-click on the sketch in Feature-Manager and select Edit Sketch. Now the sketch is active and you can edit it.

If you have somehow managed to create more sketches than you need (there are multiple sketches listed in FeatureManager), go ahead and delete what you don't want. Click on the sketch in FeatureManager and press the Delete key. You will not be able to delete a sketch if you are currently editing that sketch, as that would be like pulling the rug out from beneath your own feet.

Pan and Zoom Commands

Fig. 3–35. Pan and zoom icons of the View toolbar.

Now that you actually have some geometry on screen that you can see, it's a good time to discuss in detail how to manipulate the view of the model. The ability to zoom in close, pan the model from side to side, or even rotate the model is critical to getting any work done. This section will teach you how to accomplish all of those functions and more.

The zoom command icons are located on the View toolbar and are easy to use. Most of the zoom or pan commands require holding down the left mouse button and moving the mouse. Figure 3–35 shows only the left-hand side of the View toolbar. The right-hand side of the toolbar contains display options, which are discussed in material to follow.

On the View toolbar there are a variety of icons that allow you to pan or zoom the view, or change it in a number of ways. They are all easy to learn, with the possible exception of the Rotate icon, which sometimes takes a little practice. In the section that follows, the full selection of zoom icons is described, and the corresponding hot keys, if any, are listed. You are encouraged to experiment with these various options to gain familiarity with them.

View Orientation
hot key: space bar

View Orientation

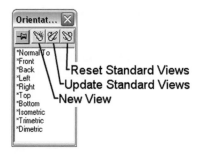

Fig. 3–36. Orientation window.

The Orientation window, shown in Figure 3–36, lists predefined views created by SolidWorks, along with any user-defined views. Once the Orientation window is open, double click on any of the listed views to display that view. Use the Pushpin icon to keep the Orientation window on top of the work area. The icons at the top of the Orientation window allow for saving user-defined views or resetting the system views. This functionality is explained in the following material.

Rather than use the Orientation window to change the current view, many users prefer to use the Standard Views toolbar, shown in Figure 3–37. Using the toolbar means that it only takes a single click to change views, rather than a double click. The Standard Views toolbar can also be easily tucked away in a convenient corner of the screen and takes up less screen real estate than the Orientation window. It really comes down to personal preference, so use whatever technique works best for you.

Fig. 3–37. Standard Views toolbar.

Normal To

The Normal To view can be thought of as a plan view. It exists in the Orientation window and on the Standard Views toolbar (icon furthest to the right). If you have never heard the term *normal* used in the context of CAD software, think of it as meaning perpendicular. It is a term common to the CAD industry.

If in a sketch, Normal To will orient the view perpendicular to the current sketch plane. If not in sketch mode, a plane or planar face must be selected. If you hold down the Ctrl key and select a second face, SolidWorks will attempt to orient the second face toward the top of the screen. In this way, you have more control over the final orientation of the model.

System Views

Any view with an asterisk before it is a system view created by SolidWorks. System views cannot be deleted. Parts and assemblies will have a number of system views. 2D drawings will only have one system view (Full Sheet). Drawings are covered in Chapter 12.

New Views

The New View icon is for adding your own views. This can be done in parts, assemblies, or drawings. User-defined views are saved with the file, so they will be available the next time the document is opened. How-To 3–9 takes you through the process of adding a user-defined view.

How-To 3-9: Adding a User-defined View

To add a user-defined view to the listing in the Orientation window, perform the following steps.

1. Establish the desired view using the pan, zoom, and rotate commands.

2. In the Orientation window, click on the New View icon.

3. Type in a name for the new view.

4. Click on OK.

The new view name will be added to the list, and will stay there forever unless deleted. User-defined views will not have an asterisk before them. It is possible to delete user-defined views by selecting the view and pressing the Delete key.

Not only is the orientation of the part remembered when saving a user-defined view, but so is its size and positioning on screen. Make sure you have sufficiently zoomed or panned the object on screen before saving your view. User-defined views cannot be redefined, so if you decide you do not like a particular view delete it and try again.

Updating Views

Update Standard Views and Reset Standard Views go hand in hand. Update Standard Views is used for changing the default orientation of the views. For instance, if you want the Right view to actually be the Front view, Update Standard Views would allow you to change that. Think of updating the views as redefining all system views simultaneously. How-To 3–10 takes you through the process of using the Update Standard Views command.

HOW-TO 3-10: Updating the Standard Views

To update (redefine) the standard system views, perform the following steps.

1. Using any method you wish, change to a view you would like to redefine.

2. Select (one click only) the view in the Orientation window you would like to set the current view to.

3. Click on the Update Standard Views icon. A warning message will appear, explained in material to follow.

4. Click on Yes to update the standard system views, or click on No to keep the default system view arrangement.

The reason for the warning message, and what it actually means, is that any drawing of the model whose views you are updating will be affected. For example, imagine a top/front/right-side view of a part in a 2D drawing. These views are directly linked to the solid model. If the system views of the model are redefined, the drawing is directly affected. Specifically, the drawing views will update to accommodate the new view definitions. If you have not yet created a drawing of the model, the warning can be safely ignored.

One reason users often update the system views is that the model was not created on the proper plane to begin with. It is always best to try to begin sketching on the correct plane in the first place. That is, if you are creating the first sketch for the part and it is a profile of the part as seen from the top, you should be sketching on the Top plane. This is always the best practice, because the part will be oriented correctly when finished, and the system views can stay as they are.

Resetting Views

The Reset Standard Views command is needed if you change your mind after updating the standard views. It allows for resetting the system views to their default state. Clicking on the Reset Standard Views icon causes a query message to appear asking if you are sure you want to reset the views. As with updating system views, resetting views will directly affect any drawing of the model. Click on Yes and you will be back to the default view arrangement. Any changes made using the Update Standard Views icon will be negated.

Previous View

The Previous View icon reverts back through the last ten previous views. You will probably never need to go back through more than ten views, so that number should suffice.

> Zoom To Fit
> hot key: F

Zoom To Fit

Zoom To Fit is analogous to an auto-scale function. The model will be resized to fit on screen, with a little room left to maneuver around the outside of the part. This is the easiest way to get an overall view of the part.

Zoom To Area

Zoom To Area requires a small amount of user intervention. If Zoom To Area is selected, pick and drag opposite corners of a window that define the area you want to zoom in on. The smaller the window the closer you will zoom in.

> Zoom In/Out hot
> keys: Z , Shift + Z

Zoom In/Out

Zoom In/Out also requires some mouse work on the part of the user. Once the icon has been selected, hold the left mouse button down and move the mouse either forward or backward. Moving the mouse forward is like walking toward the screen, with the object on screen increasing in size. Moving the mouse backward is like moving away from the screen.

If the mouse you are using has a middle mouse button wheel, scrolling the wheel will zoom the model in and out. Using the mouse wheel behaves slightly differently than using the Zoom In/Out icon, however. In the case of the wheel, zooming in will be accomplished with respect to where the cursor happens to be positioned. This is known as a scrolling zoom, and is extremely convenient. Using the icon zooms in or out with respect to the screen's center and does not depend on cursor location.

One last option is to hold the middle mouse button and Shift key down simultaneously while moving the mouse. This can be quicker than clicking on the icon if your hand is already near the keyboard, and is preferred by some people. Note that this technique is identical to using the icon, and not to using the wheel.

Zoom To Selection

Zoom To Selection allows you to quickly and effortlessly zoom to a particular entity. Usually a face is selected that you want to zoom in on, and then the Zoom To Selection icon is clicked. This will zoom in on the selected entity until it fills the screen.

Rotate View

Rotate View is probably the most difficult of all zoom commands to get accustomed to. New users sometimes find it somewhat awkward. However, with a little practice using Rotate View will seem like second nature to you. Just hold the left mouse button down while gently moving the mouse to get a feel for how this command works.

It is not necessary to click on the Rotate View icon if your mouse has a middle mouse button or wheel. Simply hold the middle button or wheel down while moving the mouse.

Rotate View has another capability if you desire a higher degree of control while rotating the part. Specifically, it is possible to rotate about an edge, about a selected position on a plane, or about a vertex point. All you need to do is select an edge, plane, or vertex point while in the Rotate View command. This helps to precisely control the point you want to rotate about, making it easier to get the exact view desired.

To gain a higher degree of control without using the Rotate View command, click on a model edge, face, or vertex point with your middle mouse button, and then hold your middle mouse button down and rotate the model. This may take a little practice, as you must hold the mouse perfectly still while middle clicking. With a wheel mouse, this can be a bit tricky!

Pan hot keys: Ctrl + Arrow keys

Pan

Pan is the last of the zoom icons and can be used to pan the model right, left, up, or down. Panning will always be parallel to the screen. After clicking on the Pan icon, hold the left mouse button down and move the mouse to slide the model about the screen. This function is quite easy to perform and does not require much practice, if any.

Another pan option involves the middle mouse button again. Hold down the middle mouse button (or wheel) and the Ctrl key simultaneously while moving the mouse. This function is identical to using the Pan icon.

Additional Zoom Options

In addition to the previously listed zoom and pan commands, there are a few extra hot key combinations that do not have any counterpart icons. These additional hot keys, along with those previously mentioned, are outlined in Table 3–3 for your convenience. Additionally, middle mouse button shortcuts are listed in Table 3–4 for easy reference.

Table 3-3: Zoom, Pan, and Rotate Hot Keys

Function	Hot Key(s)
Rotate view	Arrow keys
Rotate view in 90-degree increments	Shift + Arrow keys
Rotate model parallel to the screen	Alt + Left or Right Arrow keys
Pan the view	Ctrl + Arrow keys
Zoom in	Shift + Z
Zoom out	Z
Zoom to Fit	F
Front view	Ctrl + 1
Back view	Ctrl + 2
Left view	Ctrl + 3
Right view	Ctrl + 4
Top view	Ctrl + 5
Bottom view	Ctrl + 6
Isometric view	Ctrl + 7

Table 3-4: Middle Mouse Button Shortcuts

Procedure	Function
Middle mouse button	Same as Rotate View command (see Rotate View)
Middle mouse wheel (scrolling)	Similar to Zoom In/Out, but acts as a scrolling zoom (see Zoom In/Out)
Middle mouse button + Shift	Same as Zoom In/Out
Middle mouse button + Ctrl	Same as Pan

Display Options

Most of the main display options are fairly self-explanatory. However, there are a lot of little tweaks and adjustments that control how a model can be displayed in conjunction with the display options. This section should help clarify what all of your options are. The main display icons are located on the right-hand side of the View toolbar, shown in Figure 3–38. Icons not shown in Figure 3–38 (which may exist on your toolbar) are those for viewing curvature and zebra stripes. These are advanced viewing functions discussed in Chapter 20.

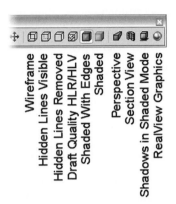

Fig. 3–38. View toolbar display icons.

Wireframe

In wireframe mode, all model edges are displayed, which is not recommended for most tasks. Keep in mind that the term *wireframe* in the context of display options has to do with display only. The model itself is still a solid model, with all of the characteristics of a solid model, regardless of how it is displayed.

Hidden Lines Visible and Hidden Lines Removed

With Hidden Lines Removed, as the name implies, any model edges hidden behind the model will not be displayed. All edges hidden behind the model will be displayed as dashed or gray lines with Hidden Lines Visible turned on. Whether gray or dashed depends on a setting in the Display/Selection section of your System Options (Tools > Options). Figure 3–39 shows the setting for changing this option.

Fig. 3–39. Setting how hidden edges should appear.

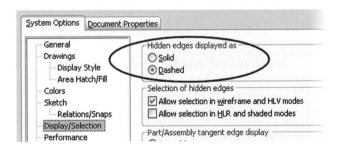

Whether or not hidden edges can be selected is controlled by another option. The option, found directly below the *Hidden edges displayed as* setting (shown in Figure 3–39), is titled *Selection of hidden edges*. You can specify whether or not it is possible to select hidden edges when using the Wireframe and Hidden Lines Visible options, and whether or not it is possible to select hidden edges when using the Hidden Lines Removed and Shaded display options. These are both independent toggle switches, and really depend on the user's preference.

✓   **TIP:** *Check only the* Allow selection in wireframe and HLV modes *option. This allows you to select hidden edges only when switching to Hidden Lines Visible display mode. In this way, you can then follow the rule of "if you can see it, you can select it."*

Draft Quality HLR/HLV

What this partial acronym stands for is "Draft Quality Hidden Lines Removed/Hidden Lines Visible." What this means in English requires further elaboration. This option is a toggle switch that affects Wireframe, Hidden Lines Visible, and Hidden Lines Removed display modes, but its effect is most apparent with hidden lines removed. When turned on, the model is treated like a shaded model, even though it is being displayed in a wireframe-style display mode. This is also known as a facetted display mode, and typically will result in faster graphics display, but at the cost of a less accurate graphics display.

When rotating a model that is being displayed in wireframe, "hidden lines removed," or "hidden lines visible" display mode, SolidWorks degrades the display while the model is being rotated when Draft Quality HLR/HLV is toggled off. To be specific, the model is temporarily displayed in wireframe mode without horizon lines (such as those along the length of a cylinder). When the model stops rotating, the hidden lines are calculated and the model is correctly displayed on screen.

With the Draft Quality HLR/HLV option toggled on, SolidWorks displays the model in the correct state during the rotation process. In other words, you do not have to wait to let go of the mouse button before the model's hidden lines are displayed correctly. As mentioned previously, the price of this nicety is a small degradation in model appearance.

Shaded and Shaded With Edges

Using the Shaded option, the model will be displayed in shaded mode. Shaded With Edges highlights the edges of the model. The latter setting makes it easy to differentiate between parallel faces that are facing you while viewing the model.

How you decide to view the model as you work is strictly a user preference. It is recommended that you use whatever is easier for you. Most people will probably find switching between one of the shaded modes and Hidden Lines Visible options the most beneficial.

✓ **TIP:** *Models saved while shaded make for better previews when opening files.*

When using the Shaded With Edges display setting, there are a few other optional settings that will further change the way the model looks. For easy reference, these settings and their end result are outlined in Table 3–5. The full menu picks to change these settings are included as well. Feel free to experiment with these settings to see how they will change the appearance of your model. Just make sure to click on the Shaded With Edges icon first.

Table 3-5: Settings That Affect Shaded With Edges Mode

Optional Setting with Menu Picks	*Options Function*
Tools > Options > System Options > Display/ Selection section > Edge display in Shaded With Edges mode	Highlights only visible edges when set to HLR (preferred). Highlights all edges when set to Wireframe.
Tools > Options > System Options > Display/ Selection section > Anti-alias edges	Smoothes highlighted edges to reduce the jagged appearance.
Tools > Options > System Options > Colors section > Use specified color for Shaded With Edges mode	When off, highlighted edges are displayed in the color of the model. When on, highlighted edges are displayed in the color set in the Colors section as set by the Edges in Shaded With Edges mode system color setting (see Figure 3–40). The default color is black.

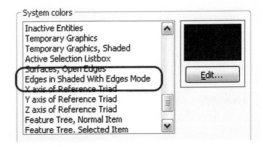

Fig. 3–40. Edges in Shaded With Edges mode setting.

Perspective

When perspective is toggled on, a vanishing point will be used to display the model. This gives the model a more lifelike appearance. If Perspective view is enabled, it is possible to alter the vanishing point distance. In other words, the relative distance the model is from the user's eye can be increased or decreased. How-To 3–11 takes you through the process of modifying the degree of perspective.

How-To 3-11: Modifying the Perspective Vanishing Point

Fig. 3-41. Modifying perspective.

To modify the vanishing point distance when in Perspective view, perform the following steps.

1. Select Modify from the View pull-down menu.

2. Select Perspective.

3. Specify a value for Observer Position, as shown in Figure 3-41.

4. Click on OK.

When changing the perspective Observer Position value, a smaller value increases the degree of perspective. The default value is 3. It is possible to type in a value less than 1. This greatly increases the degree of perspective. Increasing the value of the setting has a much more subtle effect.

With regard to perspective, it is not always a good idea to edit the part with Perspective enabled. Depending on the degree of perspective, certain objects may look odd, such as planes or notes and dimensions. Usually, perspective views are used as an enhancement when saving snapshots of the model or creating rendered images.

✓ **TIP:** *The easiest way to create a snapshot of the model is to click on the Print Screen key on your keyboard, and then paste the image into a paint program of your choice.*

Section View

Clicking on the Section View icon opens the Section View PropertyManager, shown in Figure 3-42. Any of the default planes (top, front, or right) can be used to section the model. It is also possible to select any planar face on the model as the section plane. The section plane offset can be adjusted by entering a value, or by dragging the arrows attached to the plane itself. The section plane angle can also be adjusted, or the edge of the plane can be dragged to alter the section plane angle.

By experimenting with the Section View command, you will probably find that the interface is very user friendly. The fact that the view dynamically updates as the section plane is repositioned is fun to see in action. The section view can be reversed by clicking on the Reverse Section Direction icon, found on the Section 1 panel. It is also possible to have multiple section planes. Clicking on the Section 2 option, for example, allows for selecting a second section plane and introduces yet another panel labeled Section 3.

The "section cap" is the blue cross section where the section plane crosses geometry. Turning the section cap off removes this cross section and makes the model geometry almost appear as though it is hollow or made out of paper. The section cap color can be altered using the Edit Color button.

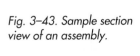

✓ **TIP:** *Selecting a planar face on the model allows for using that face as the section plane.*

Clicking on the Save View button causes a small window to appear, which prompts you for a view name. You can accept the default *SectionView1* name or type in something else. Clicking on OK will then save the view with the name you specify and bring up the Orientation window to show you that the new view has indeed been added to the list. Clicking on Reset will discard any offset or angle adjustments you have made to the section plane and cause the Front plane to become the section plane, which is the default setting.

If you wish to retain the currently displayed section view without necessarily saving a named view in the Orientation window, click on OK at the top of PropertyManager. This will exit the Section View command while retaining the section view of the model and allow you to continue working. Figure 3–43 shows an example of a section view. The cross section areas of the model geometry will be the same color as the original model geometry, rather than the cap color shown while in the Section View command.

Do not confuse the Section View command with creating section views in a drawing. The procedure for creating design drawing section views is much different and is covered later in the book. If you need to read up on creating drawing section views at this time, see Chapter 12.

Fig. 3–42. Section View PropertyManager.

Fig. 3–43. Sample section view of an assembly.

Shadows in Shaded Mode

When toggled on, shadows are displayed as if a light were shining onto the model from above. The lowermost physical point of the model determines the shadow plane, or "floor." When shadows are displayed, the orientation of the model determines what direction the "light" is pointing from. By turning shadows off, rotating the model, and turning shadows back on, this direction can be reset.

Shadows are a nice effect, but they decrease graphics performance. If you notice graphics performance degradation, try turning shadows off.

RealView Graphics

RealView graphics represent a way of viewing a model in a very realistic manner. Steel parts look like steel, wrought copper looks like wrought copper, and so on. Not all graphics cards support RealView graphics. The only way to know for sure if a particular card is supported is to check the Solid-Works web site at *www.solidworks.com*. Look for the Support area and follow the links to the graphics card test center.

As impressive as RealView graphics are the first time you see them, it is not something you would use on a day-to-day basis. The high-end real-time graphics display takes a toll on performance. This will become less of an issue as computers get faster, which is always the case in this age.

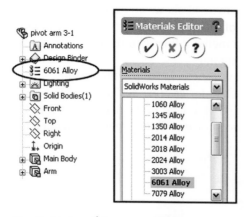

Fig. 3–44. Specifying a material for a part.

Although there is no real functional benefit from using RealView graphics, it is a step in the direction of realism. Showing models to clients or customers with RealView graphics turned on can often help a sale due to the realistic display properties.

Turning on RealView graphics by clicking on its icon will not do much good unless a material is specified for the model. This is done by right-clicking on the Material object in FeatureManager, (shown in Figure 3–44) and selecting Edit Material. A material can then be selected from a list of common alloys and other nonmetallic materials, such as various woods and plastics, to name a few. A portion of the Materials Editor panel is also shown in Figure 3–44.

Because materials play a large role in a part file's life, they will be discussed in more detail later. Specifically, materials have physical properties that carry over when investigating a part or assembly's mass properties, discussed in Chapter 8. Materials also play a role in finite element analysis, which is outside the scope of this book.

10-second Topic: Viewports

Depending on personal preference, SolidWorks users may find the ability to have multiple viewports a convenience. Others may decide viewports clutter up the screen too much. Use the viewport sizing handles, shown in Figure 3–45, to create additional viewports and then decide for yourself whether or not viewports are beneficial. Dragging these handles away from the edge of the current single viewport serves to create additional viewports if none exist. One, two, or four viewports can exist at any one time.

Once multiple viewports exist, as shown in Figure 3–46, each viewport can have its display settings changed independently. Drag the horizontal or vertical bars separating viewports to resize viewports. Drag the bars to any side of the screen (top, left, bottom, or right) to eliminate viewports.

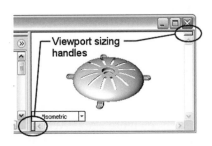

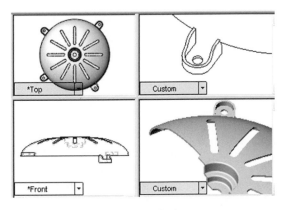

Fig. 3–45. Viewport sizing handles.   *Fig. 3–46. Utilizing multiple vxiewports.*

SW 2006: The small menu at the bottom left-hand corner of the viewports shown in Figure 3–46 can be used to establish multiple viewports or change views in a specific viewport. Use the Link Views option found in this menu to link all orthogonal views when multiple viewports are in use.

Image Quality

The quality of the model on screen can be controlled for both shaded and wireframe display modes independently. To put it simply, the higher you set the display quality for the model the more facets it will have (with regard to shaded image quality). The image quality itself is controlled through slider bars, shown in Figure 3–47.

To make adjustments to the image quality, click on the Image Quality section in Document Properties (Tools > Options menu). Moving the slider to the left decreases quality, and moving it to the right increases quality.

✗   **WARNING:** *Increasing image quality may degrade graphics performance and increase rebuild times due to a higher polygon count. Find a setting that is a good compromise between quality and performance.*

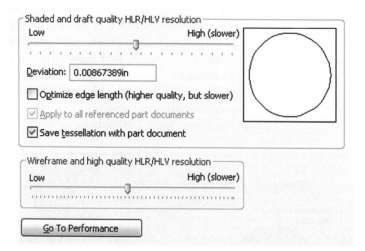

Fig. 3–47. Image Quality settings.

When the Draft Quality HLR/HLV display option is turned on, the image quality is controlled via the upper slider bar (see Figure 3–47). When turned off, the lower slider bar controls the quality of the display for wireframe display modes. Note that when Shaded display mode is used the model is always facetted and the Wireframe slider bar has no effect.

Increasing the Shaded quality will increase the number of facets used to display the model. (Actually, it decreases the chordal deviation, but increased facet number is a close enough approximation.) This results in curved surfaces appearing smoother. The Wireframe slider increases the accuracy of the wireframe display, as opposed to the number of facets used, but the end result is much the same.

✓ **TIP:** *When working with large assemblies, lowering the shaded image quality while the* Apply to all referenced part documents *option is checked will decrease the number of polygons used by all components in the assembly, thereby increasing performance.*

Saving Tessellation Data

Tessellation data is the polygonal information used to display the model. If the *Save tessellation with part document* option is turned off, the model will not be displayed if opened in View-Only mode, which is a checkbox option available when opening an existing SolidWorks document. The model will also not be viewable if it is opened in the SolidWorks Viewer. View-Only mode and the SolidWorks Viewer are both described in the following sections.

The only advantage to not saving tessellation data with a part file is that the model will take up less space on the computer's hard drive. Sometimes this space reduction can be significant, especially if the file has many curvy surfaces, as in the case of a spring. You can safely turn the *Save tessellation with part document* option off, but only if you are sure nobody will be using the SolidWorks Viewer to view your files.

View-Only Mode

View-Only mode is convenient for a few reasons. If it is necessary to view a large assembly or complex part without actually making any changes to it, View-Only mode allows for opening the SW file without loading any of the file's data. FeatureManager will be empty, and editing the part or assembly will not be possible. The file, however, will open extremely quickly. View-Only mode also allows for opening an assembly or drawing for viewing purposes even if the associated components (in the case of an assembly) or model files (in the case of a drawing) are not available.

SolidWorks Viewer

What is the SolidWorks Viewer? It is a free viewing program available to anyone who wishes to view SolidWorks files but does not have SolidWorks. It can be installed from the original SolidWorks installation disks, and is available on the SolidWorks web site at *www.solidworks.com.*

Documents opened in the SolidWorks Viewer can be zoomed or panned. They can also be printed, which makes the free viewer an excellent way for non-engineer or designer types (such as administrative personnel or print department employees) to print SolidWorks documents. Those using the Viewer cannot change the original SolidWorks document in any way.

Summary

You have learned the basics of feature creation in this chapter. To begin with, you should be feeling more comfortable with how to work in sketch mode. Understanding what system feedback is and getting a feel for automatic relations and sketch snaps should start becoming more familiar to you.

You have seen firsthand how to add and remove constraints and dimensions. Click on sketch entities to view the constraints, or turn on the relation symbols by checking View Relations in the View menu. Use the SolidWorks color codes to determine if your sketch is underdefined or fully defined. Fully constrained geometry will appear black, and underdefined will appear blue. Always fully define sketch geometry, as it gives tighter control over what is happening in the sketch and results in predictable behavior. If a sketch is not fully defined, its behavior is not predictable and your design intent has not been incorporated into the sketch.

Keep sketch geometry simple and include only the geometry necessary to create the feature at hand. The sketch will be easier to maintain and control that way. It is better to have a greater number of features and less complex sketch geometry. This makes for a flexible model that is more editable. Due to the nature of geometric relationships, an overly complex sketch can be very difficult to maintain. It can also make troubleshooting a difficult task if something goes wrong while making design changes.

Use hot keys and mouse shortcuts to be more productive. For example, the F key will fit the model to the screen, and the middle mouse button will rotate the model. Work in any view you find convenient. Save user-defined views using the Orientation window to make it easy to return to a particular view.

Use a display mode that suits your needs yet makes it easy to select geometry being edited. Hidden Lines Visible allows for seeing the part clearly and makes it easy to pick hidden edges on the other side of the model. Shaded mode is a good choice as well, and by adjusting the image quality it is possible to find a good setting that allows for reasonable graphics performance while maintaining a decent image quality on screen.

CHAPTER 4

Castings

CAST PARTS CAN BE CREATED EASILY USING THE SOLIDWORKS EXTRUDE COMMAND, which you were introduced to in the previous chapter. The Extrude command also happens to be one of the most basic SolidWorks feature types, and thus cast parts make an excellent starting point for learning SolidWorks. Throughout this book, you will be applying what you have learned in previous chapters and expanding on that information. Portions of this book are dedicated to applying the knowledge you have gained directly to parts you will create in exercises within each chapter.

The layout of this book is designed with the student in mind. Much more can be absorbed and retained when the student is trying firsthand the topics covered in the chapter. People have a tendency to wool gather or even fall asleep when an instructor does nothing but lecture. Any person who has attended college knows this to be true, unless the instructor is speaking on an interesting topic or is simply an outgoing and energetic person. Lab exercises are much more interesting. Therefore, let's start with a quick review, briefly introduce you to some new features, and then continue with the pivot arm exercise.

Creating Sketched Features: A Quick Review

It would be beneficial to reiterate a few of the important topics covered in the previous chapters with regard to creating sketched features. Most of these topics apply to all sketched features, not just extrusions, but now is still a good time to reinforce these important issues.

Before a feature can be created, you should have some idea of what that feature will look like. If creating the very first feature of the part, the sketch profile should describe the basic shape of the model. In short, it is important to *determine the best profile* for the feature you wish to create.

When it comes to sketch geometry, keep it simple. Simple sketch geometry is easier to manage and will provide more control when performing edits later on, because there will be fewer relationships between sketch geometry that can conflict with each other.

Examine Figure 4–1. Note that even though there are a number of features that make up this part the sketch used to create the base feature is actually quite simple. Incidentally, the sketch for the base feature was created on the Right plane. This makes perfect sense if the final orientation of the heat sink should appear as it does in the isometric view shown.

Before creating a sketch for the base feature, decide which side of the part would be best suited for representing the part's top, front, and right sides. Once you have determined the best profile to start with, decide which side of the part the profile represents, and *select an appropriate sketch plane*. This will aid you in the creation process later on, because you will have already determined in your mind's eye the correct orientation of the part. The part will also be oriented correctly when it comes time to generate a 2D drawing of the model, which is probably the most important reason for starting off on the proper plane.

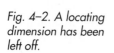
An appropriate first sketch

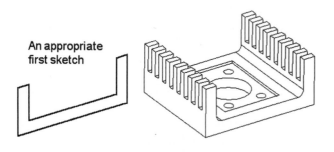

Fig. 4–1. A logical first sketch for the heat sink part.

Fig. 4–2. A locating dimension has been left off.

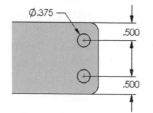

Note that part orientation does not play a significant role in how an assembly will be created. Components in an assembly are placed in position with absolutely no respect to how the part model was originally oriented in space.

SolidWorks does not require that you add dimensions and constraints, but it is extremely prudent to do so. Without enough dimensions or constraints to fully define geometry, your model will not behave in a predictable fashion. Examine the simple example shown in Figure 4–2. A designer creates a piece of plate steel that requires two holes to be positioned on the right-hand side of the part. The holes have been constrained to have equal diameters, and their centerpoints are vertical. A locating dimension has been left off, as the designer was not sure how far to place the holes from the right-hand side of the part.

Neglecting to add the final locating dimension in this case would not have been an error in and of itself, as long as the designer went back and added the dimension at some point in the future. However, in this scenario let's assume the designer forgot to add the dimension. Let's also imagine that the overall length of the finished part required a design change, and needed to be shortened by half an inch. The result is shown in Figure 4–3.

Fig. 4–3. The result of not fully defining a sketch.

It would be a safe bet to state that the holes cutting through the right-hand edge of the part was not the original design intent the designer had in mind. The designer may not have known how far the holes should have been from the right-hand side of the part, but should have added a dimension anyway. Don't rely on memory to save you from design flaws, and always follow the most important rule of sketching: *Fully define your sketch geometry.*

Sketch color codes tell when a sketch is fully defined (black), so it is easy enough to tell when enough dimensions or constraints have been added. What type of dimension or constraint to add can sometimes be confusing, though, so keep in mind that underdefined (blue) geometry can be dragged via the left mouse button. Much information can be deduced from dragging entity endpoints.

✓ **TIP:** *SolidWorks system feedback symbols show what constraints are added while sketching. Always watch the cursor!*

Cut Features

Removing material from a model is accomplished in almost exactly the same way material is added. The only difference is the command used. Rather than creating a base or boss feature, you will be creating a cut feature. With regard to the menu picks, it would be a simple matter of selecting Insert > Cut, instead of Insert > Boss/Base. Let's continue with the pivot arm exercise to see firsthand how cuts are created.

EXERCISE 4-1:   Cutting Holes in the Pivot Arm

Begin this exercise by opening the pivot arm part file, if it is not already open. Change to a front view of the part using any of the techniques you have learned to this point. The front face of the main body of the part is the face you should select for the next sketch plane.

In this exercise, you will first create one of the holes on the main body, followed by the slot on the arm feature. Use the steps that follow if you need guidance accomplishing this task.

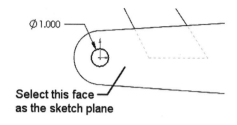

Ø1.000

Select this face
as the sketch plane

Fig. 4–4. Creating a new sketch for the hole.

1. Select the face on which to begin sketching. This is shown in Figure 4–4.

2. Click on the Sketch icon.

3. Create the sketch for the hole, as shown in Figure 4–4.

4. Fully define the sketch by adding a 1-inch dimension to the circle and a concentric relation between the circle and the arc edge of the existing geometry, if necessary.

10-second Topic: Waking Up Centerpoints

When creating the base feature of the pivot arm, the location of the origin point on your model may be different than the model displayed in this book. This is acceptable, but the circle for the hole would have to be located using a concentric constraint rather than using the sketch origin to anchor it in position. Another method of locating the circle's center would be to "wake up" the centerpoint of the arc edge so that it could be inferred to.

Waking up a centerpoint of an edge can be accomplished by pausing over any arc or circular edge with the cursor. This can be accomplished while in a sketch command (e.g., Line or Circle), or even by simply dragging an entity endpoint or centerpoint, such as the center of the circle. It would then be possible to place the center of the new circle over the newly displayed centerpoint of the arc edge, thereby inferring to it. This adds a coincident relation between the arc's centerpoint and the center of the circle, making them concentric. Use this method if necessary to complete the sketch, and then continue with the following steps.

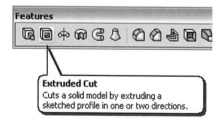

Features

Extruded Cut
Cuts a solid model by extruding a
sketched profile in one or two directions.

Fig. 4–5. Extruded Cut icon.

5. Select Insert > Cut > Extrude, or click on the Extruded Cut icon (shown in Figure 4–5).

6. Set the end condition to Through All, as shown in Figure 4–6.

7. Rotate your view using the middle mouse button and check the preview to determine if the cut is going in the proper direction. Click on the Reverse Direction button if necessary (see Figure 4–6).

8. Click on OK to accept the settings and make the cut.

The next cut will take a little more skill because the sketch is a bit more complicated than a simple circle. Adding the necessary constraints may be difficult for new users, but it will be good practice. Pay special attention to what is selected before adding constraints. If you do not have the correct items selected, the correct constraint will not be available. The Add Relations display panel in PropertyManager will show you what is selected. Pay close attention to the Selected Entities portion of this panel, shown in Figure 4–7.

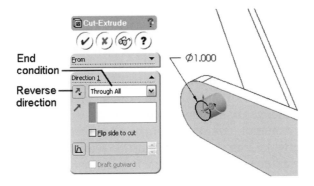

End condition

Reverse direction

Ø1.000

Fig. 4–6. Setting the end condition and creating the hole.

Add Relations

Selected Entities

Arc4
Line5

Existing Relations

Fig. 4–7. The Selected Entities list box will list what has been selected.

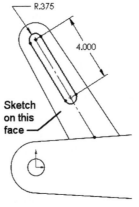

R.375

4.000

Fig. 4–8. Creating the sketch for the slot.

Sketch on this face

9. Change to a front view.

10. Select the face of the arm shown in Figure 4–8 and begin a new sketch.

11. Create the sketch shown in Figure 4–8. (Hint: start with the centerline drawn from the centerpoint of the arc edge to the midpoint of the edge at the base of the arm.)

12. Add dimensions and additional coincident or tangent relations necessary to fully define the sketch.

13. Create an extruded cut feature using the same technique used to create the first hole feature (Hint: Insert > Cut > Extrude, or Extruded Cut icon.)

14. Save your work.

A Special Note on Midpoint Constraints

One of the constraints that causes new users a particular amount of grief is the Midpoint constraint. When adding a Midpoint constraint between a line and a point, select the line, not the line's midpoint. SolidWorks will automatically find the midpoint for you. With regard to the Midpoint constraint added in the exercise, you will need to select the endpoint of the centerline and the edge at the base of the arm.

When sketching something, such as a line, it is possible to automatically infer to an existing midpoint of a line or edge. Sometimes it is necessary to manually add a midpoint constraint. With that said, is it possible to select the midpoint of a line or edge when not actively sketching an entity? Yes, it is, and the way this is done is through the right mouse button. In Select mode (the cursor looks like an arrow), try right-clicking on a linear edge or sketch line. You will see an option named Select Midpoint.

If it is possible to force SolidWorks to select a midpoint, why not just select that midpoint first and add a relation to that? You can do precisely that, but the constraint would no longer be a midpoint constraint; it would be coincident. It is really just a matter of terminology. A point can be made coincident to another point, in which case two points must be selected. A point can be made coincident to a line's midpoint, in which case a point and line must be selected.

Extrusion End Conditions

When creating an extruded feature, it is important to specify how you want to terminate the extrusion. In other words, you need to specify the condition that will determine how long the extrusion will be. In its simplest form, you would specify a distance for the extrusion and the sketch will be extruded that distance. This is known as a *Blind* end condition, which you were introduced to in Exercise 3–1. You were also recently introduced to the *Through All* end condition, in Exercise 4–1.

Over the course of the last few exercises, you may have noticed an arrow in the work area whenever performing an extrusion. This arrow is a handle that can be dragged with the left mouse button in order to dynamically alter the extrusion distance. It can be a useful visualization tool, but in the long run it is better to type in a precise value for the sheer reason of maintaining accuracy.

Blind and Through All are probably the two most common end conditions when performing extrusions. There are quite a few other end conditions you should become familiar with, as you will probably need to use nearly all of them over time. Not all end conditions are always available. We will discuss

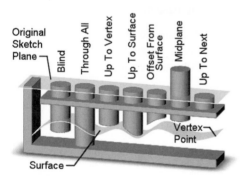

Fig. 4–9. Various end condition types and their results.

this in a moment. First, let's examine Table 4–1 to get a better understanding of the various end conditions. Refer to Figure 4–9 while reading through the table to better understand the end condition descriptions.

Table 4-1: End Conditions

End Condition Name	Results of Using the End Condition
Blind	Extrusion distance is specified by the user.
Through All	Extrusion continues through the entire part.
Up To Next	Extrusion terminates at the next face encountered.
Up To Vertex	Selected vertex or point dictates extrusion distance.
Up To Surface	Selected surface, face, or plane dictates extrusion distance.
Offset From Surface	Extrusion terminates the specified distance either before or after the selected surface, face, or plane.
Up To Body	Selected body dictates the extrusion distance.
Mid Plane	Extrudes equal amounts in opposite directions. Distance specified is the total distance of the extrusion.

10-second Topic: Solid Modeling Terms

Throughout this book you will hear them again and again; terms such as *faces* or *bodies*. What does it all mean, exactly? It is really all quite simple. A surface, for instance, is nothing more than an object with no thickness, volume, or mass. A surface has surface area, but not much else. Surfaces are the underlying objects that form a solid.

Solids have volume and mass properties. Solids are a set of surfaces that form an enclosed boundary. In this way, the software can "see" the surfaces as a solid model. When referring to the surfaces that make up a solid, those surfaces are referred to as faces. There really is no difference between faces and surfaces, except that faces form an enclosed volume that represents a solid.

Finally, there are bodies. Bodies exist in two forms: solids and surfaces. That's all there is to it. Now you can converse like a professional!

Certain end conditions have additional parameters that will require elaboration. Such is the case when using the Up To Vertex, Up To Body, or Up To Surface end condition. Each of these three end conditions requires that some geometry be selected. In the case of Up To Vertex, any endpoint of a model edge or any sketch entity point can be used. Up To Surface can be taken to mean not only surfaces but any model faces or any user-defined plane. This holds true for the Offset From Surface end condition as well.

If using any of the aforementioned end condition types, make sure you actually select the appropriate vertex, body, or surface. Otherwise, Solid-Works will keep prompting you to select something. (The creation of planes is discussed later in this chapter. Bodies and surface creation will follow in later chapters, as they are typically related to advanced modeling techniques.)

The Through All end condition is most often used with cut features. However, it can also be used when creating a boss. An example of this can be seen in Figure 4–9. Up To Next is a common end condition and convenient to use because it can be a "set it and forget it" end condition. Use this end condition when you want the extrusion to go up to the very next face it encounters, no matter what distance that face (or faces) may be from the original sketch plane.

✓ **TIP:** *If the Up To Next end condition is not available, try using the Reverse Direction button to the left of the end condition list in PropertyManager.*

Fig. 4–10. PropertyManager when using Offset From Surface.

Offset From Surface is an end condition with some extra parameters that must be considered. PropertyManager changes when using Offset From Surface, as shown in Figure 4–10. There is the offset distance, of course, which dictates how far the extrusion will be offset from the selected surface. There is the *Reverse offset* option, which moves the offset from one side of the selected surface to the other. Finally, there is the *Translate surface* option, which requires elaboration.

The *Translate surface* option is most easily explained by example. Figure 4–11 shows two extrusions, each the same except for the *Translate surface* option. A cylinder has been extruded using Offset From Surface. The spherical surface of the lower bodies in the image are the surfaces the extruded cylinders were offset from. The cylinders have been cut in half for clarity.

On the right in Figure 4–11, the translate setting was turned on. Note that the lower portion of the cylinder takes on the shape of the spherical surface below it. It's as if the spherical surface of the lower body has been translated upward by the specified offset distance, and that is precisely where the extrusion terminated. On the left, the translate setting was turned off, which caused the spherical surface to be truly offset. Leaving translate surface unchecked results in an end condition more representative of the name of the end condition itself.

The last end condition we have not discussed in detail is the Mid Plane end condition, which is very well suited to cast parts. Cast parts are created (cast) from a mold. The mold itself can be of a wide variety of types. Many cast parts will contain draft so that they will come out of the mold more easily. If you are not familiar with the term *draft*, the following simple example should help. In Figure 4–12, one part contains draft and the other does not. Which do you think would come out of the mold more easily?

Five degrees of draft have been added to the base feature of the part on the right, and also to the cut extrusion at the top of the part on the right. The line where the draft begins (on the main body of the part) is known as the parting line. This is where the mold splits in two and pulls apart. Creating an actual mold is covered in detail in Chapter 14.

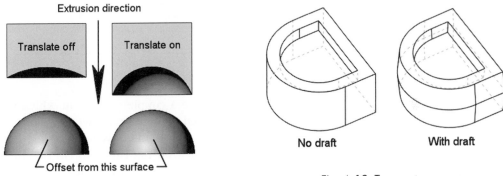

Fig. 4–11. Understanding the Translate *surface option.*

Fig. 4–12. Two parts, one with draft and one without.

The Mid Plane end condition type allows a sketch to be extruded in opposite equidistant directions simultaneously. Combine this with draft during the extrusion process to get a good start on the creation of a cast part. Using the Mid Plane end condition and adding draft simultaneously creates the parting line for you.

Start Conditions

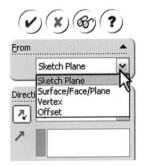

Fig. 4–13. From panel.

The From panel, shown in Figure 4–13, allows for specifying where an extrusion starts. Typically, this is the plane or planar face on which the sketch was created. Sometimes an extrusion needs to begin at a position where there is no plane or planar face. In a case such as this, the choices are to either create a plane (discussed later in this chapter) or modify the start condition.

Modifying the start condition of an extrusion is convenient because it no longer becomes necessary to create an additional plane for sketch purposes. There are three start condition options above and beyond the default setting of Sketch Plane. These options, along with the default setting, are outlined in Table 4–2 for your reference.

Table 4-2: Start Conditions

Start Condition Name	Results of Using the Start Condition
Sketch Plane	Sketch is extruded from the plane or planar face on which it was originally created (this is the default setting).
Surface/Face/Plane	Sketch is extruded from selected surface, face, or plane. If a surface or face is selected, the surface or face does not have to be planar, but it does have to fall within the path of the entire sketch if the sketch were projected perpendicular from its original sketch plane.
Vertex	Sketch is extruded from selected vertex or sketch point.
Offset	Sketch is extruded from the original sketch plane, but offset a user-specified distance in one direction or the other perpendicular from the original sketch plane.

Extruding with Draft

Fig. 4–14. Draft settings while extruding.

There are a couple of ways in which draft can be added, but for now we will focus on the most basic method. When creating an extrusion, click on the Draft On/Off button (shown in Figure 4–14). Once turned on, a draft value can be entered. The *Draft outward* option reverses the draft angle and causes the draft to be applied in the opposite direction. If working in shaded display mode, a preview of the draft will be shown. Thus, it is easy to know for certain if you need to reverse the draft.

Extrude Direction

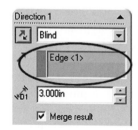

Fig. 4–15. Specifying an extrude direction.

By default, extrusions are perpendicular to the sketch plane, but that doesn't have to be the case. Another direction can be specified by clicking in the *Direction of extrusion* list box (shown in Figure 4–15). Linear model edges, or even sketch lines, can be used to dictate what direction the extrusion should take. If a sketch line is used, the geometry does not have to belong to the sketch you are extruding. A before-and-after preview of this option is shown in Figure 4–16.

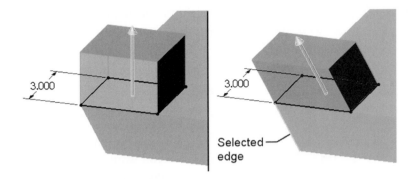

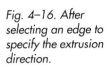

Fig. 4–16. After selecting an edge to specify the extrusion direction.

Bidirectional Extrusions

Not to be confused with the Mid Plane end condition, extrusions can be in two directions simultaneously. End conditions for each direction can be specified, and they do not have to be the same. Even draft can be specified for each direction independently.

To enable a bidirectional extrusion, select the Direction 2 option found in PropertyManager when creating the extrusion. The option is visible in Figure 4–14. Once Direction 2 is checked, make your selections for end condition, draft, and so on, just as you would for Direction 1. As previously mentioned, these settings are independent from Direction 1, and as such can be completely different. There is one exception. If an extrude direction is applied to Direction 1 (as discussed in the previous section), that direction will also be applied to Direction 2.

↬   **NOTE:** *If Mid Plane is selected for Direction 1, the Direction 2 option will not be available because it would be redundant.*

Editing Techniques

The following material is critical to being a productive SolidWorks user. One of SolidWorks' main strong points is the ease with which you can edit existing material. This section shows you how to implement a few of Solid-Works' most basic editing commands. Make it a point to learn this material well.

Editing a Sketch

There are times when you will find it necessary to get back to the sketch used to create a feature in the first place. Sketch geometry is hidden and absorbed by the feature it was used to create, but that geometry is never far from reach. The reasons for editing a sketch will vary, but the process is always the same. The right mouse button comes into play for this procedure, as it does quite often while working with SolidWorks. To edit an existing sketch, perform the steps outlined in How-To 4–1.

How-To 4-1:   Editing an Existing Sketch

To edit an existing sketch, perform the following steps.

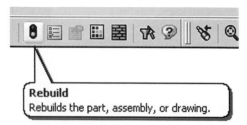

Fig. 4–17. Rebuild icon.

1. In FeatureManager, right-click on the sketch you want to edit. Right-clicking on the feature created from the sketch will also suffice.

2. Select Edit Sketch.

3. Make the desired changes.

4. Select Edit > Rebuild, or click on the Rebuild icon (shown in Figure 4–17).

Be aware that if you exit out of a sketch, whether on purpose or by accident, the sketch geometry will turn gray. When preparing to create an extruded or revolved feature, it is not necessary to exit out of the sketch. When the feature is created, the sketch geometry is absorbed by the feature and the user is automatically taken out of sketch mode. If you see gray sketch geometry, it could mean you have accidentally exited the sketch and you should edit it in order to gain access to that same sketch.

Model rebuild hot keys: Ctrl + B

Rebuilding a Model

Anytime a dimensional change is made to your SolidWorks model (a technique discussed momentarily), or if a sketch has been edited, the model must be rebuilt. Rebuilding the part is how you tell SolidWorks to take into consideration the changes you have made. This does not necessarily mean you have to rebuild after every change. For instance, it is perfectly acceptable to modify multiple dimensions before rebuilding the part.

If you are in the process of editing a sketch, there are a number of ways in which you can exit that sketch and rebuild the part simultaneously. Each of the following methods will suffice, so use whatever method is easiest for you.

- Click on the Sketch icon (so that it no longer appears depressed).

- Right-click in an empty area of the work area and select Exit Sketch.

- Select Exit Sketch from the Insert menu.

- Select Rebuild from the Edit pull-down menu.

- Press Ctrl + B (the hot key combination for the Rebuild command).

- Click on the Rebuild icon.

If you are not currently editing a sketch, rebuilding the part will do simply that. If you have made any dimensional changes, they will be taken into account when the part is rebuilt. This includes dimensions associated with a feature, sketch, plane, or anything else. It is acceptable to postpone rebuilds if there is more than one dimensional change to make. When you are finished making those changes, however, rebuild the model.

If you make the mistake of forgetting to rebuild the model after making dimensional alterations, SolidWorks will eventually prompt you to rebuild when you attempt to save the file. This warning should be taken seriously. Not rebuilding the model before saving is asking for trouble. If your modifications caused problems, you will not know it until you open the part at a later date. Certain operations will automatically rebuild the part, such as when editing a feature. There is also an option for regenerating a model, as opposed to rebuilding it. Editing features is covered in material to follow. For now, read on to understand the difference between a rebuild and a full regeneration.

Model regeneration hot keys: Ctrl + Q

Regenerating a Model

When performing one of the previously mentioned steps to rebuild a part, the model is not really completely rebuilt. In other words, the part file's entire database is not completely recalculated. Only the features marked as needing a rebuild by the software are recalculated. In contrast, it is possible to fully regenerate the entire part database by forcing a full regeneration. A full regeneration takes longer than a rebuild, so be forewarned: if the model is complex, you will have to wait a little longer.

It is usually not necessary to do a full regeneration. Sometimes, however, it can serve a useful purpose. If for some reason your model does not seem to rebuild correctly, try performing a regeneration. This happens rarely and is really nothing to be concerned with at this time.

Editing a Feature

The mechanics of editing a feature are exactly the same as for editing a sketch. All it requires is the right mouse button. All feature types can be edited, and the beauty of it is that editing a feature takes you to the same PropertyManager panel used to create the feature in the first place. There are no cascading menus, no complicated commands to navigate—just a simple right-click. If you have learned how to create a feature, you already know how to edit it. How-To 4–2 takes you through the process of editing a feature's definition.

How-To 4-2: Editing a Feature

To edit a feature, perform the following steps. Note that it doesn't matter what type of feature you wish to edit; they can all be edited in the same fashion. It is this type of consistency in the software that makes SolidWorks so easy to use.

1. In FeatureManager or the work area, right-click on the feature to be edited.

2. Select Edit Feature.

3. Make the desired changes.

4. Click on OK to accept the changes.

When you click on OK to accept the changes you made to a feature, the part is automatically rebuilt. Of course, you can always click on Cancel in PropertyManager if you change your mind about making any changes. This holds true for nearly all SolidWorks functions, not just when editing a feature.

Editing Dimension Values

Chapter 3 taught you how to modify a dimension. Double clicking on a dimension brings up the Modify window, in which you can enter a new value. Clicking on the green check in the Modify window or pressing Enter on the keyboard will accept the change. But how can you access those dimensions in the first place?

If you double click on a feature or sketch (or any other object that has a dimension associated with it) the dimensions associated with that object will be shown. Then you can double click on the dimension you wish to change and enter a new value. Just don't forget to rebuild the model when finished, so that you can see the changes take effect. It's really that easy.

✓ **TIP:** *You can modify more than one dimension on more than one sketch or feature before rebuilding the model. This can save time waiting on rebuilds with complex models.*

Click on the plus sign (+) located to the left of any sketched feature to see the sketch that belongs to that feature (see Figure 4–18). This is known as expanding the feature tree. The same can be done with folders (directories) in any Windows operating system to see the files that folder contains. Collapsing the feature tree is done the same way, except that there will be a minus sign (–) instead of a plus sign.

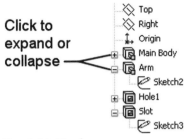

Click to expand or collapse ———

- ◇ Top
- ◇ Right
- ↳ Origin
- ⊞ 🗐 Main Body
- ⊟ 🗐 Arm
  - 🖉 Sketch2
- ⊞ 🗐 Hole1
- ⊟ 🗐 Slot
  - 🖉 Sketch3

Fig. 4–18. The feature tree can be expanded and collapsed.

If you double click on a sketch, you will see the dimensions associated with that sketch. Double clicking on a feature shows the sketch dimensions as well as the dimensions associated with the feature. Feature dimensions are added by SolidWorks, not the user. Take an extruded sketch as an example. If a sketch were extruded 3 inches with 5 degrees of draft, those 3-inch and 5-degree dimensions would be added to the model automatically.

The reason this is being pointed out to you is to make it easier to find a particular dimension when the need arises. Dimensions added by you during the sketch process will typically be color coded black, assuming your sketch geometry is fully defined. Any dimensions added by SolidWorks automatically during the feature creation process will be color coded blue. This makes it easier to find certain dimensions when modifications need to be made.

10-second Topic: Renaming Features

Feature renaming hot key: F2

It will be easier to talk about the pivot arm if we rename the features. Renaming features is a simple process and is no different than renaming files in Microsoft Windows. Click on an item in FeatureManager, wait a moment, and then click again. This is known as a slow double click. Type in a new name for the object, preferably something meaningful, and then press Enter. Nearly anything in FeatureManager can be renamed.

When the list of features in FeatureManager grows long, as it has been known to do, it is easier to find things later on if they have meaningful names. This makes life easier for you, as well as others who may need to work with the model. Common practice is to not worry about renaming every item, just major features you may need to refer back to at a later time. Exercise 4–2 incorporates some of the new editing commands you have learned in this chapter and takes you through the process of using them to make changes to the pivot arm.

EXERCISE 4-2: Editing the Pivot Arm

The pivot arm is supposed to be a cast part, but none of the features currently contain any draft. The arm should also be flush with the base feature. These modifications should be easy to make, so let's get right to work. Start by opening the pivot arm if you have not already done so, and then perform the steps that follow.

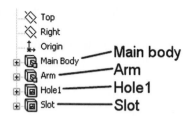

Fig. 4–19. Renaming the first four features of the pivot arm.

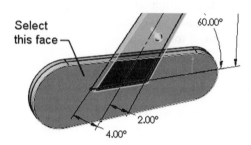

Fig. 4–20. Selecting a surface to terminate the end condition.

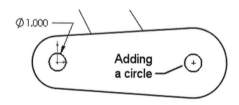

Fig. 4–21. Adding a circle to the Hole1 sketch.

1. Rename the features to something more descriptive. Use the names shown in Figure 4–19.

2. Edit the Main body feature by right-clicking on it and selecting Edit Feature.

3. Change the end condition to Mid Plane.

4. Add 5 degrees of draft.

5. Click on OK to accept the changes.

6. Edit the Arm feature.

7. Add 5 degrees of draft to the arm.

8. Change the end condition to Up To Surface and select the surface indicated in Figure 4–20. The reason for this action is explained in a moment.

9. Click on OK to accept the changes.

Why was it necessary to change the end condition of the arm feature? It wasn't if the extrusion depth of the Main body feature never changed. However, if the depth does change, the outside face of the arm would no longer be flush with the face of the Main body feature. Once the end condition of the arm is changed to Up To Surface, we no longer have to be concerned with the first two features being flush. They will always be flush, regardless of the thickness of the first feature.

10. Edit the sketch for *Hole1* by right-clicking on *Hole1* or its sketch in FeatureManager and selecting Edit Sketch.

11. Add another circle, as shown in Figure 4–21.

12. Make sure the circle is centered with respect to existing geometry. Add a constraint if necessary.

13. Add an equal constraint between the two circles to keep them the same size.

14. Rebuild the model.

15. Double click on the Arm feature to access its dimensions.

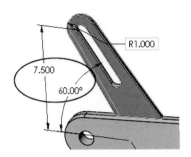

Fig. 4-22. Changing dimension values.

16. Change the two dimensions circled in Figure 4–22 to 6.5 inches and 50 degrees by double clicking on each dimension and entering the new values.

17. Rebuild the model.

18. Save your work.

As you can see from this exercise, making changes to a model is a fairly simple task. If you have managed to follow along with the steps in this exercise, your pivot arm should look similar to that shown in Figure 4–23. In the following section we will continue expanding your sketching capabilities by introducing you to some common sketch tools.

Fig. 4-23. Redesigned pivot arm.

Mirroring Sketch Geometry

Symmetrical sketch geometry can be created in a number of ways. Sketch geometry can be mirrored, in which case SolidWorks adds symmetrical constraints to the sketch geometry. Symmetrical relationships can be added manually with the Add Relation command. Geometry can also be mirrored dynamically as it is sketched.

Creating symmetrical sketch geometry is most easily done using the dynamic mirroring procedure. For that reason, it is the most common procedure and the procedure we will discuss first. The steps for mirroring sketch geometry dynamically are outlined in How-To 4–3.

HOW-TO 4-3: Dynamically Mirroring Sketch Geometry

It is assumed in the following series of steps that you are already in sketch mode. No other preparations are necessary. To dynamically mirror sketch geometry, perform the following steps.

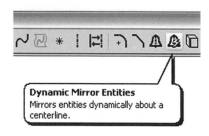

Dynamic Mirror Entities
Mirrors entities dynamically about a
centerline.

Fig. 4–24. Dynamic Mirror icon.

1. Select Tools > Sketch Tools > Dynamic Mirror, or click on the Dynamic Mirror icon (shown in Figure 4–24). (Don't confuse the Dynamic Mirror icon with the standard Mirror Entities icon. The Dynamic Mirror icon has a lightning bolt on it. The standard Mirror Entities icon can be seen to the left of Dynamic Mirror in Figure 4–24.)

2. Select a line you wish to use as the mirror line. This is commonly a centerline, but can be any sketch line or linear model edge.

3. Create the desired sketch geometry. Stay on one side of the mirror line or the other and the entities will be dynamically mirrored to the opposite side as you sketch them.

4. Click on the Dynamic Mirror icon again to turn dynamic mirroring off.

When dynamically mirroring, the mirror line contains small hash marks on either end of the line indicating that dynamic mirroring is turned on. Once dynamic mirroring is turned off, the hash marks disappear.

Dynamic mirroring is certainly the easiest way in which to create symmetrical sketch geometry. One point you should be aware of, however, is to not cross over the mirror line while sketching. This has a tendency to create overlapping sketch geometry, which can cause problems when attempting to create the feature, assuming the sketch will be used for this purpose.

It is possible to mirror sketch geometry after it has been created as well as dynamically. Any type of sketch geometry can be mirrored, including construction geometry and centerlines. How-To 4–4 takes you through the process of mirroring existing sketch geometry.

How-To 4-4:   Mirroring Existing Sketch Geometry

Fig. 4–25. Mirror PropertyManager

To mirror existing sketch geometry, perform the following steps.

1. Select Tools > Sketch Tools > Mirror, or click on the Mirror Entities icon (shown in Figure 4–24 to the left of the Dynamic Mirror icon).

2. Select the geometry to be mirrored.

3. Click in the area labeled *Mirror about* in the Mirror Property-Manager (shown in Figure 4–25) and select a line to mirror about.

4. Click on OK to finish creating the mirrored geometry.

✓ **TIP:** *Linear model edges can be used as mirror lines.*

As mentioned previously, SolidWorks adds symmetrical relationships automatically when entities are mirrored. However, you can add this constraint manually if need be, as long as there is a centerline in the sketch. To add a symmetrical relationship, use the Add Relation command, discussed in Chapter 3 (see How-To 3–3). As long as two similar sketch entities are selected, along with a centerline, the Symmetric relation will be available.

The next features to be added to the pivot arm form a symmetrical recessed area on the front of the part. Exercise 4–3 takes you through this operation.

Exercise 4-3: Adding a Feature to the Pivot Arm

In this portion of the ongoing pivot arm project, a recessed area is added to the model. Make sure the pivot arm file is open. Make use of the following steps if you feel you need them. Some of the more confident readers may decide to use the illustrations and just skim over the steps. Do not do this unless you feel you are ready.

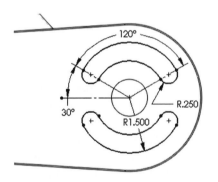

Fig. 4-26. Sketch for the recess feature.

1. Change to a Front view of the pivot arm.

2. Start a new sketch on the front face of the Main body feature.

3. Create the sketch shown in Figure 4–26.

Having trouble creating the sketch geometry? Often there are multiple ways to achieve the same geometry, but in this case there is definitely a preferred method to creating this sketch. Draw the horizontal centerline first, and turn on dynamic mirroring. This will make creating the sketch much easier. Use the Centerpoint Arc command to create the 1.5-inch arc and its smaller concentric counterpart. You will only have to worry about half the sketch, in that the other half will be dynamically mirrored for you.

Use the Tangent Arc command to add the rounded ends, and then add the centerlines to control the included angle of the slot. You can turn off dynamic mirroring before putting the centerlines in, which exist strictly for dimensional purposes. See Figure 4–27 for guidance.

Once the basic sketch geometry is in place, adding the dimensions is easy. There may be tangent relations that will have to be added manually as well. This is material that has already been covered, so the details will not be covered again here. Finish the recess feature by continuing with the following steps.

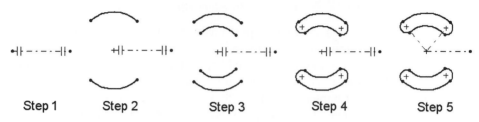

Step 1   Step 2   Step 3   Step 4   Step 5

Fig. 4–27. Recommended steps for sketching the recess feature.

4. Change your view so you can properly observe the preview that will be shown when performing the next step.

5. Cut the sketch into the part at a depth of .25 inch.

6. Rename the feature *Recess*. The completed recess feature is shown in Figure 4–28.

Fig. 4–28. Completed recess feature.

7. Save your work.

Creating Planes

Often you will need to create a new plane for one reason or another. A common reason is simply because you need a plane to sketch on, and the Front, Top, or Right planes are not satisfactory for the purpose. Creating planes is easy, but the method used is dependent on the geometry you have to work with.

For some of us, it has been a while since geometry class in high school. However, most of the plane creation options are very logical, and often a matter of common sense. For instance, it is possible to define a plane with three noncollinear points. This is very basic geometry, and is an example of one of the options you have at your disposal for defining reference planes in SolidWorks.

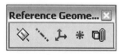

Fig. 4–29. Reference Geometry toolbar.

As most SolidWorks commands go, the command for creating reference planes can be accessed via the menu structure or a toolbar. The toolbar in this case is the Reference Geometry toolbar, shown in Figure 4–29. There is also a Reference Geometry menu under the main Insert menu. The toolbar is very handy, and is one you will probably want to keep available, largely for the sake of the Plane icon. Planes are just one of those objects that get created quite frequently, so the toolbar will save you time.

The steps used to create a plane are provided in material to follow. First, however, take a look at the various methods available. There are six methods in all, and each method uses a different combination of geometry to define a plane. Use Table 4–3 to choose the method best suited to your needs. Once again, which method is used depends on what geometry you have to work with.

➥ **NOTE:** *In regard to Table 4–3, the term* edge *refers to a model edge.* Point *can mean sketch point, endpoint, or midpoint, whereas* vertex *is an endpoint on a model edge.* Line *refers to a sketch line. The term* curve *means any sketch line, centerline, arc, spline, parabola, axis, or edge. The creation of axes is discussed in Chapter 7.*

Table 4-3: Plane Creation Methods

Method	Description	Required Geometry
Through Lines/Points	Plane passes through three points or a line and point	A combination of three points or vertices, or an axis, edge, or line and a point or vertex
Parallel Plane at Point	Parallel to a plane and passing through a specified point	Plane or face and a point or vertex
At Angle	At an angle from another plane and passing through a line	Plane or face and an axis, edge, or line
Offset Distance	Offset from another plane	Plane or planar face
Normal to Curve	Perpendicular to a curve and passing through a specified point	Any curve and a point or vertex
On Surface	Tangent to a surface, typically cylindrical or conical, where an axis, edge, line, or plane intersects said surface	Surface and an axis, edge, line, or plane

Whatever method you decide to use, creating a reference plane always starts in the same fashion. That is, you click on the Plane icon or select Insert > Reference Geometry > Plane in the pull-down menus. The Plane icon is the first icon on the Reference Geometry toolbar, shown in Figure 4–29. Once that has been done, the Plane PropertyManager will appear as shown in Figure 4–30.

Some plane creation methods are easier than others. There are even a few shortcuts that can be used to create planes. One such shortcut creates an offset plane. Another creates a plane normal to a curve. These shortcuts are explored in material to follow, and are clearly marked for easy reference.

The order in which the plane creation methods are presented in this section follows the order in which the options appear in the Plane Property-Manager. To create a plane using any of the methods available in SolidWorks, complete the steps in one of the How-To sections that follow. Let's begin with creating a plane using Through Lines/Points, outlined in How-To 4–5.

Fig. 4–30. Plane PropertyManager.

→   **NOTE:** *How-Tos 4–5 through 4–10 assume that the Plane Property-Manager is already open.*

How-To 4-5:   Creating a Plane Using Through Lines/Points

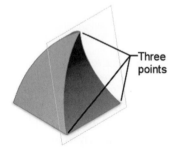

This method allows for defining a plane by specifying three (3) points through which the new plane will pass. A single point and a line can also be used. Perform the following steps to create a plane using the Through Lines/Points method. Figure 4–31 shows an example of a plane created using three points.

1. Click on the Through Lines/Points icon in the Plane Proper-tyManager.

2. Select any combination of three (3) sketch points, mid-points, or vertex points. Alternatively, select any combination of a single sketch point, midpoint, or vertex point and a single line, linear edge, or axis.

3. Click on OK when finished.

Fig. 4–31. Example of a plane defined with three points.

✓   **TIP:** *In any of the plane creation methods used, it is not necessary to click on the applicable icon first. For example, in this How-To skipping step 1 and simply selecting three points would have sufficed.*

Note in Figure 4–31 that the horizontal edge at the bottom of the part could have been used in place of the bottom two points to create the same plane. The end result would be the same.

HOW-TO 4-6: Creating a Plane Using Parallel Plane at Point

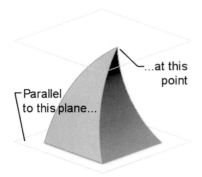

This method allows for defining a plane by specifying a plane or planar face the new plane is to be parallel to, and a point through which the new plane will pass. The point can be a sketch point, endpoint of a sketched object, or endpoint of a model edge. Perform the following steps to create a plane using the Parallel Plane at Point method. An example of a plane created using the Parallel Plane at Point method is shown in Figure 4–32.

Fig. 4–32. Example of a plane created using Parallel Plane at Point.

1. Click on the Parallel Plane at Point icon in the Plane PropertyManager.

2. Select a plane or planar face.

3. Select a sketch point, endpoint of a sketched object, or endpoint of a model edge.

4. Click on OK when finished.

As you can see, the Parallel Plane at Point method is quite simple. This holds true for the At Angle method as well, described in How-To 4–7.

HOW-TO 4-7: Creating a Plane Using At Angle

This method allows for creating an angled plane. Occasionally it is necessary to create some construction geometry prior to using this option, in order to get the new plane positioned correctly. Of course, this depends on the situation and model requirements. Any sketch line or construction line (centerline) can be used as a basis for determining the "hinge" location for the angled plane. Alternatively, any linear model edge can be used as well. A plane or planar face must also be specified, which will be the plane from which the angle is measured. Perform the following steps to create an angled plane.

1. Click on the At Angle icon in the Plane PropertyManager.

2. Select a plane or planar face.

3. Select a sketch line or linear model edge.

4. Enter an angle for the new angled plane in PropertyManager.

5. Click on the *Reverse direction* option if necessary to flip the plane to the opposite side of the plane selected in step 2.

6. Click on OK when finished.

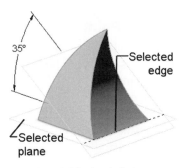

Fig. 4–33. Angled plane.

An example of an angled plane is shown in Figure 4–33. The plane's angular dimension is shown to help illustrate how the angle value is being applied, but this should be self-evident. As is the case with any driving dimension, the angle value can be altered after the plane has been created. Like any other object with an associated dimension, double clicking on the plane gains access to the dimension associated with the plane.

Another method of creating a plane is to use the Offset Distance method. How-To 4–8 takes you through this process.

How-To 4-8: Creating a Plane Using Offset Distance

Fig. 4–34. Specifying an offset distance.

This method allows for creating a plane offset from an existing plane or planar face. An offset distance must be supplied, similar to supplying an angle value for an angled plane. To create a plane using the Offset Distance command, perform the following steps.

1. Click on the Offset Distance icon in the Plane PropertyManager.

2. Select a plane or planar face the new plane will be offset from.

3. Specify an offset distance for the new plane (see Figure 4–34).

4. Click on the *Reverse direction* option if necessary (shown in Figure 4–34) to flip the plane to the opposite side of the plane selected in step 2.

5. Click on OK when finished.

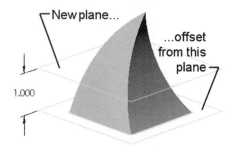

Fig. 4–35. Offset plane.

Both the At Angle and Offset Distance plane creation methods allow for optionally specifying the number of planes to be created. This setting, shown in Figure 4–34, defaults to a value of 1. Increasing the number of planes creates a series of planes with the spacing determined by the offset distance or angle. Figure 4–35 shows an example of an offset plane.

Offset planes are extremely common. You will probably find yourself creating them on a regular basis. For this reason, it's nice that SolidWorks has a shortcut for their creation. This shortcut is explained in the following section.

Shortcut: Offset Distance Planes

There is an alternative to creating an Offset Distance plane that is very convenient. It is a shortcut that involves holding down the Ctrl key. This shortcut works only if you want to create a plane offset from another plane, not a planar face.

To create an offset plane using the shortcut method, hold down the Ctrl key, and then hold the left mouse button down while the cursor is positioned over the plane's border. Drag the cursor away from the plane and you will see a new plane spring into view. You will also see the Plane PropertyManager appear. Release the mouse button, and then the Ctrl key, and type in a value for the offset distance as you normally would when creating an offset plane. Click on OK or press the Enter key and you are finished. Let's continue with the plane creation methods by examining the Normal To Curve option.

HOW-TO 4-9: Creating a Plane Using Normal To Curve

The Normal To Curve plane creation method is sometimes referred to as the Perpendicular To Curve method. Both terms mean exactly the same thing. A "normal" in CAD lingo refers to a perpendicular vector. If the term *normal* is unfamiliar to you, just replace it with the word *perpendicular* anytime you see it used by the SolidWorks software.

To create a plane using the Normal To Curve method, a curve must be selected, along with a point the new plane will pass through. The resulting plane will be perpendicular to the selected curve and will pass through the selected point. This may sound confusing, but hopefully it will not be after you see an example.

Be aware that a curve can be any number of things. For example, a sketch entity such as a line, arc, or spline is considered a curve. Model edges are considered curves as well. Even axes can be selected when using this plane creation method. Points can be sketch points, endpoints, or vertex points. The following steps take you through the process of creating a Normal To Curve plane. Figure 4–36 shows an example of a plane created using Normal To Curve.

1. Click on the Normal To Curve icon in the Plane PropertyManager.

2. Select the curve the new plane is to be perpendicular to.

3. Select a point the new plane will pass through.

4. Click on OK when finished.

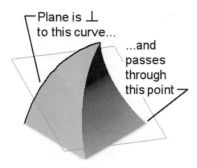

Fig. 4–36. Example of a Normal To Curve plane.

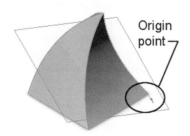

Fig. 4-37. Set origin on curve option not checked.

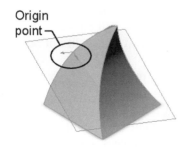

Fig. 4-38. Set origin on curve option has been turned on.

If you examine Figure 4–36, you might think there should be numerous solutions for the positioning of the new plane, given the selected geometry. However, consider that the selected edge does not have a consistent radius.

It is spline-like in nature. Even if it were an arc with a consistent radius, the selected point (a vertex point) would not be the centerpoint of the arc. In light of these facts, there can be only one solution for the placement of the new plane.

When creating a Normal To Curve plane, there is an option called *Set origin on curve*. This option is irrelevant if the point being selected to define the plane is already set on the selected curve. Such would be the case if a spline were used to define the new plane, along with one of its control points. In the case of the example shown in Figure 4–36, the vertex point exists at some distance from the curve.

The *Set origin on curve* option is sometimes significant when creating a sketch on the new Normal To Curve plane. In Figure 4–37, the *Set origin on curve* option has not been checked. As you can see, the origin point is at the same location as the point selected to define the plane.

In Figure 4–38, the *Set origin on curve* option has been checked. When a sketch is begun on the plane, the origin point appears attached to the curve instead of the point used in defining the plane. This often does not make any difference when sketching on the Normal To Curve plane unless it is necessary to use the origin for adding geometric relations. For most work, you can safely ignore this option.

Shortcut: Normal To Curve Planes

When it is necessary to create a plane normal to a sketch curve (a sketched entity, in other words), the manual method of creating the plane outlined in How-To 4–9 must be used. However, if the new plane must be normal to a model edge, there is an excellent shortcut that will save you some time. In this case you select near the endpoint of the model edge where the new plane should be and then click on the Sketch icon. A new plane will be created normal to the selected edge and passing through the edge's endpoint.

You really should try this to see how it works. It is as easy as it sounds. Typically, a plane or planar face is selected prior to sketching. When implementing this shortcut, do not worry about selecting a plane. SolidWorks creates the plane for you. Open up any existing SolidWorks model and give this shortcut a try.

✓ **TIP:** *This shortcut works on curve types other than model edges. For example, helical curves and 3D sketch splines, to name two, will also accept this shortcut. These curve types are discussed later in the book.*

Let's continue with the examination of plane creation methods by working through the process of creating a plane via the On Surface option, outlined in How-To 4–10.

HOW-TO 4-10: Creating a Plane Using On Surface

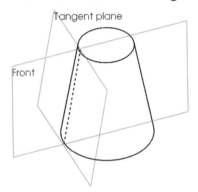

Fig. 4–39. On Surface plane, example 1.

Of all plane creation methods, creating a plane on a surface is often the trickiest to implement. The On Surface plane creation method has a variety of options you will be exploring in this section. In this first example, shown in Figure 4–39, the only two items selected were the conical surface of the cone and the Front plane. As can be seen from the illustration, the new tangent plane is tangent to the conical face where the Front plane intersects the face.

The following steps will vary slightly in the examples, depending on exactly how the On Surface plane is created. In regard to the first example, the process is fairly simple.

1. Click on the On Surface icon in the Plane PropertyManager.

2. Select a cylindrical or conical surface.

3. Select a plane that passes through the surface selected in step 2.

4. Use the Other Solutions option if necessary to flip the new plane's tangency from one side of the cylindrical or conical surface to the other.

5. Click on OK when finished.

It should be noted that this first example assumes that the plane selected in step 3 passes through the *center* of the cylindrical or conical surface. If this is not the case, a different scenario is necessary. Let's assume there is a plane offset some distance from the center of the cone. In this second example, the resultant plane will pass through the center of the cone and not be tangent to the conical surface at all. Furthermore, there will be an option for adjusting the angle of this new plane with respect to the selected plane (which would be the offset plane in this example).

Figure 4–40 shows what you might see using the scenario described in this second example. Superimposed on the image is a view of PropertyManager that shows the setting for adjusting the new plane's angle.

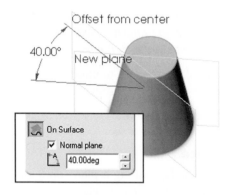

Fig. 4–40. On Surface plane, example 2.

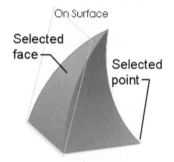

Fig. 4–41. On Surface plane, example 3.

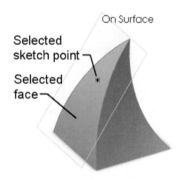

Fig. 4–42. On Surface plane, example 4.

To this point, you have been using faces and planes in the creation of On Surface planes. However, points can be used as well. In Figure 4–41, the On Surface plane creation method is implemented in one of its most basic ways. Note that the items selected consist of a nonplanar face and a vertex point.

The combination of selected objects works in example 3 because of the normal vector described by the point and the face. If a line were drawn from the selected point perpendicular to the selected face, the line would represent the *normal* used to define the new On Surface plane. The new plane is tangent to the face where the normal vector is defined. Considering that an imaginary line drawn from the point to the face is perpendicular to the face at only one position, this combination of entities works for defining the On Surface plane.

For this third example, simply substitute a point for the plane specified in step 3 of How-To 4–10 and you have the required process. If there is only one solution available for the entities selected, there will be no Other Solutions option displayed in PropertyManager.

Sticking with the theme of selecting a point, let's use a sketch point instead of a vertex point. In example 4, shown in Figure 4–42, a point was sketched on one of the nonplanar faces of the model. This was done using the 3D Sketcher, discussed in Chapter 10. Suffice it to say for now that the sketched point can be constrained to a particular face of the model. By controlling the position of the point on the face, the tangency point of the new On Surface plane is also controlled.

Simple 2D sketch points can also be used to define an On Surface plane. Sometimes the sketch point does not reside on the surface where the new plane must be tangent. This is often the case with a standard 2D sketch point. The point resides in space somewhere, sketched on a plane or planar face.

Because the sketch point lies at some location away from the nonplanar face where the On Surface plane must be tangent, there is a choice to be made as to how the point is projected to the nonplanar face. It can be projected perpendicular to its sketch plane or simply to the closest location found on the surface. Figure 4–43 should help you visualize this.

Note the sketch point in Figure 4–43. This point was sketched on a plane some distance from the model. When seen from the side, it is easy to see how the point can be projected in two possible directions—either perpendicular to the sketch plane or closest to the surface—whichever you choose. The options used to control this functionality (shown in Figure 4–44) will appear in PropertyManager. The Reverse option can be used if the projection direction needs to point the opposite way. This option is only available if *Project onto surface along sketch normal* is used.

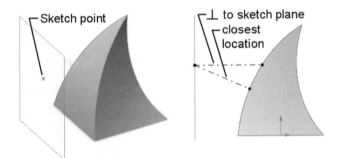

Fig. 4–43. A sketch point can be projected in two ways.

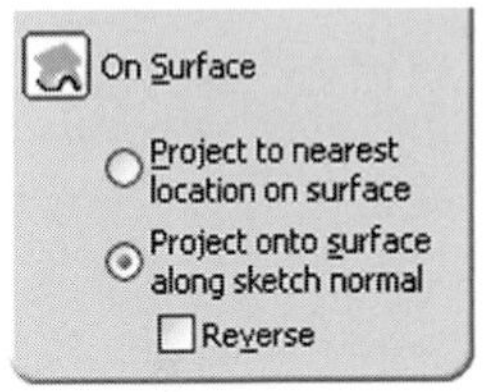

Fig. 4–44. Specifying how the point should be projected.

There is still one other combination of entities that can be used for defining an On Surface plane. This combination consists of a surface, an edge, and a plane. This is one of those cases for which a picture speaks a thousand words, so let's start there. Figure 4–45 shows two views of a vase. The view on the left shows the curve highlighted for purposes of reference. Note the point of tangency that defines the location of the new On Surface plane. But what defines the point of tangency?

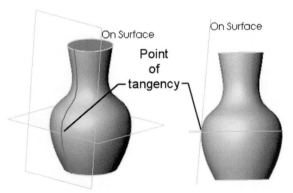

Fig. 4–45. Using a plane, surface, and curve to define a point of tangency.

The point of tangency is defined by selecting the surface on which the new plane is to be created, along with a curve and a plane. The curve should reside on the surface, but should not lie on the plane. Where the curve intersects the plane defines the point of tangency.

Sometimes the On Surface method does not seem to work properly. This is usually a case of operator error. You just have to know what to select. Where you can get into trouble is when selecting the curve. For example, sketch curves will not work. Neither will projected curves.

On the other hand, split lines and model edges work just fine. (Chapter 10 deals with curves in depth.)

This section has taught you how to create planes using a variety of methods. The main reason for creating planes is so that you will have something to sketch on. With the wide variety of plane creation techniques at your disposal, you should be able to place a plane just about anywhere you desire.

Summary

In this chapter you have learned a great deal about creating extruded features and the various end conditions that can be used to specify how the extrusion should terminate. You have also seen how easy it can be to add draft to a part during the extrusion process. The Mid Plane end condition lends itself very nicely to drafted parts because the draft can be added simultaneously for both extrusion directions. Additionally, adding draft during a mid-plane extrusion creates a parting line automatically by its very nature.

Very important editing commands were covered in this chapter. Specifically, you learned how to edit a sketch or feature by right-clicking over the sketch or feature in FeatureManager. You also learned how to modify dimensions by double clicking on a sketch or feature in FeatureManager to access its dimensions, and then double clicking on the dimension value.

You should now be more familiar with mirroring sketch geometry, and the various methods used to create symmetrical geometry. Sketch geometry can be mirrored either after it has been sketched or dynamically as it is being sketched. You should also have greater facility and confidence in adding relations.

You have also learned some good SolidWorks work habits in this chapter. One such habit is to rename features as you create them, to help make those features more easily recognizable. This is helpful if you need to go back at some later time and make design changes, or if somebody unfamiliar with the part is trying to understand the model.

The final section on creating planes has given you the ability to position planes anywhere you might find necessary. Sometimes a little additional construction geometry may be needed to define a plane, such as sketch points, but this is perfectly acceptable. Understanding how to create planes is important because they are often required prior to beginning a sketch. Being able to create a variety of planes for sketching will also greatly aid you in creating more complex feature types, covered in subsequent chapters.

CHAPTER 5

Turned Parts

AS FAR AS SOLIDWORKS IS CONCERNED, the main difference between cast parts and turned parts is the base feature used to create such parts. Turned parts are generally created by a lathe or similar manufacturing process. This chapter deals with parts that contain radial symmetry, at least at some point in the design process.

Many chapters of this book describe how parts manufactured by a specific process can be modeled in SolidWorks. That does not mean to say, however, that a part with radial symmetry cannot be a cast part. It may very well be that you have a cast part that could most easily be modeled with the Revolve command, or a turned part with the Extrude command.

The point is that this book covers a number of manufacturing processes and the most common means of modeling those types of parts with Solid-Works. With that said, always bear in mind that the first and foremost deciding factor of how you actually model the part is the basic shape of the part, not how the part will be manufactured.

Revolved Features

Turned parts are typically very well suited for creation via the Revolve command. The Revolve command parameters, which you will be working with in material to follow, are very simple. There are not as many options associated with the Revolve command as with the Extrude command. Usually when a sketch is being revolved, you must decide the number of degrees to revolve the sketch, and in which direction, clockwise or counterclockwise.

There are a few simple guidelines that must be followed when creating a revolved feature. These guidelines are easy to remember and should not pose a problem. First, a revolved feature must have something to define the axis of revolution. Centerlines are typically used with revolved features, but this is not a requirement. Any sketch line is suitable for dictating the axis of revolution.

Centerlines are preferred over regular sketch lines when creating a sketch for a revolved feature for two reasons. One reason is the added capability of dimensioning, which we will explore in a moment. The second reason has to do with the assumption SolidWorks makes. If there is a centerline in the sketch, SolidWorks will assume you will want to use it to revolve the sketch about. This is nearly always a correct assumption.

✓   **TIP:** *Model edges or axes can be used to revolve about, but only if co-planar with the sketch plane.*

Good strategy when creating a revolved base feature sketch is to position the sketch so that the origin point will be at the center of the part. This can be beneficial in the long run because two of the system planes will pass through the center of the model. This can aid in the creation of features added later in time.

If there is more than one centerline in your sketch, make sure you select one of them to represent the axis of revolution. Likewise, if there are no centerlines in the sketch, it will be necessary to select a line in the sketch to represent the axis of revolution. Otherwise, SolidWorks will not know what to revolve the sketch about to create the revolved feature.

Rules Governing Revolved Sketch Geometry

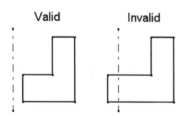

Fig. 5–1. Valid and invalid sketch geometry for a revolved feature.

When you are creating a sketch for a revolved feature, under no circumstances should you allow any sketch geometry to cross the centerline. Figure 5–1 shows valid and invalid geometry for a revolved feature.

Another steadfast rule to follow when creating a revolved feature is to never let the sketch geometry touch the centerline at an isolated point. During the revolve process, zero-thickness geometry would be created, and that is not allowed. Solid modeling software has a difficult time mathematically defining such geometry. Figure 5–2 shows examples of geometry not allowed with the Revolve command.

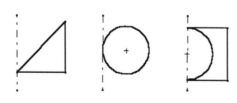

Fig. 5–2. Touching the centerline at an isolated point: examples of invalid sketch geometry.

You may think that the second and third examples in Figure 5–2 are legal operations, but they are not. In the second operation, the circle is tangent to the centerline. Mathematically speaking, the point of tangency is a single point. In the last example, the sketch touches the centerline at two points. However, each point can be defined as an isolated point, which rules that scenario out as well.

Allowing the sketch to touch the centerline is perfectly acceptable, as long as the sketch follows the previous rules. Any of the examples shown in Figure 5–3 would make for an acceptable revolved feature sketch. In the second and third examples, a centerline is not actually necessary. There are sketch lines, collinear with the centerlines, that could have been selected for the axis of revolution in each case.

Fig. 5–3. Examples of valid sketch geometry for a revolved feature.

The first example in Figure 5–3 would create a feature with a hole in its center (assuming the centerline is used as the axis of rotation). The second example in Figure 5–3 would create a sphere.

To summarize, do not cross the axis of revolution and do not touch the axis of revolution at an isolated point. Other than that, the guidelines governing basic sketch geometry in general still apply (such as when creating a sketch for an extruded feature). If you need to refresh your memory as to what those guidelines are, see the "Sketch Guidelines" section in Chapter 3.

Revolve Command Panel

When employing the Revolve command, the Revolve Parameters panel is displayed in PropertyManager. This panel is shown in Figure 5–4. The mechanics behind creating a revolved feature are very straightforward. There are only a few parameters, which amount to specifying the number of degrees to revolve the sketch and in what direction, clockwise or counterclockwise (CW or CCW).

Fig. 5–4. Revolve Parameters panel.

The drop-down list for the revolve types consists of the options One-Direction (visible in Figure 5-4), Mid-Plane, and Two-Direction. In other words, you may want to revolve the sketch in one direction only (CW or CCW), in both directions equal amounts, or in both directions different amounts. More often than not, use of the One-Direction option is all that is necessary. If it becomes necessary to use the Two-Direction option, you will need to plug in angular values for each revolve direction (both CW and CCW).

Exercise 5–1 takes you through the creation of a revolved feature. You will more than likely find the process quite simple. During the exercise, concentrate on your sketching technique. Pay close attention to the cursor and system feedback, and take care to connect lines from endpoint to endpoint.

EXERCISE 5-1: Creating a Revolved Feature

Begin by starting a new part. Make sure your units are set to inches, with precision set to three decimal places. Save the part right away and give it the name *Valve Stem*. Figure 5–5 shows the valve stem only partially completed, with features created in this chapter. In chapters to follow, you will complete the valve stem.

You will start by sketching the profile of the valve stem on the Right plane and revolving it 360 degrees. To create the valve stem base feature, perform the following steps.

1. Right-click on the Right plane and select Show.

2. Make sure the Right plane is still selected, and then click on the Sketch icon.

3. You will be placed in a Right view automatically. Create the sketch for the base feature, as shown in Figure 5–6.

4. Add collinear constraints, if needed, to fully define the geometry.

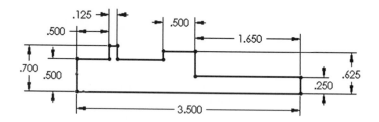

Fig. 5–5. Valve stem.

Fig. 5–6. Base feature sketch for the valve stem.

Make it a point to fully define the sketch by adding any necessary constraints. Remember, your goal should be to fully define the sketch (turn it black) so that it will exhibit predictable behavior. This is always in your best interest. If you feel you have all of the necessary dimensions on the sketch but it is still underdefined, it is probably not locked to the origin. In the sketch shown in Figure 5–6, the lower left-hand corner of the sketch was made coincident to the origin, thereby locking the sketch in position. Also keep in mind that dragging blue geometry will aid in identifying what additional constraints may be necessary.

5. Select Revolve from the Insert > Boss/Base menu, or click on the Revolved Boss/Base icon (shown in Figure 5–7).

6. Select the bottom horizontal line as the line you wish to revolve about.

7. Make sure the Angle setting is set to 360 degrees.

8. Click on OK. Figure 5–8 shows the valve stem at this stage in the process.

Save your work, and read on to see how you can change the appearance of dimension arrows, which side the leader is on, and other aspects of a dimension's appearance.

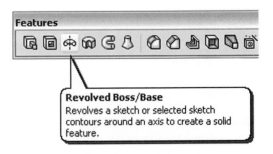

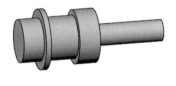

Fig. 5–7. Revolved Boss/Base icon.

Fig. 5–8. Valve stem so far.

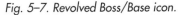

Dimension Properties

Almost all objects in SolidWorks have what are known as properties. What types of properties an object has depend on the type of object it is. Dimensions, for instance, have properties that determine whether arrows are inside or outside the extension lines of the dimension, whether the dimension is a diameter or radial dimension, the dimension's precision, dual dimensioning settings, tolerance values, and a host of other options.

In the following material you will take a look at how you can modify many of these properties, and what some of your options are. Consider the fact that the dimensions you add to a sketch will be used later on, in the 2D detail drawing. In SolidWorks, it is not necessary to recreate dimensions because they have been added to the model already.

Now consider how you might want those dimensions to look in the drawing. In many cases, the fine-tuning of a dimension's appearance is done in the drawing and not in the part. A perfect example of this is controlling where the ends of the extension lines are positioned. As a matter of fact, controlling that particular aspect of a dimension *must* be done in the drawing, and not in a part. However, some aspects of a dimension's appearance can be modified most conveniently while creating them in the model (part) file.

Figure 5–9 is for those not familiar with the terminology used in regard to dimensions. The extension lines shown in the illustration are sometimes referred to as witness lines. This book will always refer to them as extension lines.

SolidWorks refers to dimension lines as leader lines. However, when this book refers to a leader it is referring to the dimension leader line typically associated with radial dimensions, notes, or balloons, to name a few. Leaders, incidentally, can either be bent or straight, such as those shown in Figure 5–10.

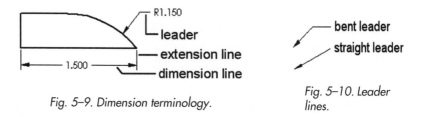

Fig. 5–9. Dimension terminology.

Fig. 5–10. Leader lines.

Bear in mind that this section discusses changing the appearance of individual dimensions. To change settings to dimension styles globally, the document's properties must be changed (Tools > Options > Document Properties tab). Specifically, the Detailing section and its various subsections of the Document Properties will on a global level control the appearance of all dimensions and annotations in a drawing.

Modifying Properties via PropertyManager

Most basic elements of a dimension's appearance can be changed in PropertyManager, which will appear after doing nothing more than selecting a dimension. If PropertyManager does not automatically appear after selecting a dimension, you should enable it. Turn on the Auto-show PropertyManager option found in the General section of your System Options (Tools > Options).

The following can be changed in PropertyManager. If what needs to be changed is not among the following, in all likelihood it will be necessary to access the dimension's full properties (discussed later in this section).

- Utilize dimension favorites

- Add tolerances such as basic, bilateral, and symmetric

- Modify tolerance values, such as in the case of bilateral or limit tolerance types

- Change the decimal place precision

- Add text to the dimension value

- Change the dimension line's arrow position

- Modify the arrow type

- Add text to a dimension value

- Add various symbols to the dimension value, and so on

The list is fairly extensive, and most of the items do not require further explanation. For example, if you need to add a tolerance to a particular dimension, simply select the tolerance type from the drop-down list. However, there are a few details that will prove valuable for a few of the properties, which are discussed in the material that follows.

Note that Dimension Favorites is not discussed in the following material. The topic is better suited to Chapter 12, which discusses detailed design drawings. See the section "Dimension Favorites" in that chapter for further information.

Dimension Tolerance and Precision

Fig. 5–11. Dimension Tolerance/Precision panel.

As should be evident from Figure 5–11, adding tolerance values or changing dimension or tolerance value precision can be done in PropertyManager. It is just a matter of choosing the desired settings for the selected dimension or dimensions. The Dimension Tolerance/Precision panel will change appearance depending on the tolerance selected from the drop-down list. Figure 5–11 shows the settings when choosing a bilateral tolerance.

To change the precision of either the dimension value or tolerance value, select the desired decimal place precision from the drop-down listings. One selection will be followed by the word *Document*. This indicates the default precision setting for the document as specified in the Units section of the Document Properties (Tools > Options). Document precision can also be altered by clicking on the Precision option found in the Dimensions section of the Document Properties.

10-second Topic: The Dimension Properties Window

PropertyManager only shows the properties that most frequently need to be changed. However, there are other properties that can be made available that are not present in the Dimension PropertyManager. The Dimension Properties window allows for modifying all of a dimension's properties, above and beyond those listed in PropertyManager.

Gaining access to the Dimension Properties window can be achieved in two ways. The most direct method would be to right-click on a dimension and select Properties from the menu. Another method would be to click on the More Properties option found at the very bottom of every dimension's PropertyManager. After selecting a dimension, you will probably need to scroll down to the bottom of PropertyManager, as the More Properties option will likely not be visible otherwise.

✓   **TIP:** *Properties for multiple annotations can be modified simultaneously. By selecting multiple dimensions or annotations and accessing the properties of any one of the selected objects, the common properties for all of them will be shown.*

To change the scale of tolerance values with regard to the dimension value, access the Dimension Properties window for that dimension. How-To 5–1 takes you through the process of changing the scale of the text used for the tolerance value.

How-To 5-1:   Changing Tolerance Text Scale

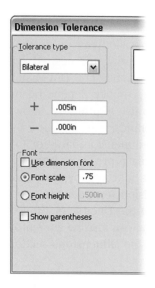

Fig. 5–12. Dimension Tolerance window.

To change the scale of the text used for tolerance values associated with a dimension, perform the following steps.

1. Right-click on a dimension and select Properties.

2. Click on the Tolerance button to open the Dimension Tolerance window, shown in part in Figure 5–12.

3. If a tolerance has not already been established for the dimension, one can be added at this time using the Dimension Tolerance window.

4. Uncheck the *Use dimension's font* option.

5. Select either *Font scale* (recommended) or *Font height*, and enter the desired value.

6. Click on OK to close the Dimension Tolerance window.

7. Click on OK to close the Dimension Properties window.

When changing the tolerance size, using the *Font scale* option is easiest. To modify the tolerance text size via the *Font height* option, you must know what the current height of the dimension font is ahead of time or the tolerance values will not look right. There is a preview area in the Dimension Tolerance window (not visible in Figure 5–12) that will help you gauge tolerance text size. To change tolerance scale globally, click on the Tolerance button found in the Dimensions section of the Document Properties (Tools > Options).

Dimension Text

Anytime text, symbols, or tolerances are added to a dimension value that information is carried over to the drawing when the dimensions are inserted. This can be useful, especially when two different individuals are doing the designing and the detailed drawings. If the data is already present in the model dimensions, less communication is required between the designer and drafter.

To add text to a dimension value, click in the text field shown in Figure 5–13. The dimension code will appear as the value < DIM >. You should not touch the less-than (<) or greater-than (>) symbols, or anything between them. This is the code that translates to the actual dimension value shown on the SolidWorks model or drawing. Mess with the code and you break the link.

Place the flashing cursor before the <DIM> code to add text before the dimension value, or after the <DIM> code to place text after the dimension value. The Enter key serves to add a line break, identical in operation to any basic text-editing program.

Symbols can also be added by clicking on the applicable icons, also shown in Figure 5–13. The More button will open up another window, which contains an expanded list of symbols. Use the additional icons on the Dimension Text panel for justifying all of the text and symbols vertically or horizontally.

Fig. 5–13. Dimension Text panel.

Primary Value Override

The Primary Value panel, shown in Figure 5–14, can be used to override the value of a dimension. This is not recommended, as the dimension will no longer display what the actual value of the dimension is. Nonetheless, if you discover a need to override a dimension's value the proper procedure is to check the *Override value* option and specify a value to be displayed in place of the true value.

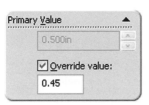

Another method of overriding a dimension's value is to simply delete the <DIM> code and enter a different value in the dimension text box, discussed in the previous section. This has the effect of disabling tolerance values, which is why it is preferable to use the *Override value* option. Under certain circumstances, such as when creating tables in a drawing, you may wish to delete the <DIM> code and type in text instead. An example might be when you want the dimension value to read the word *LENGTH*, as opposed to the actual dimension value or a character such as the letter A. In this way a tabulated drawing could be created.

Fig. 5–14. Primary Value panel.

Dimension Arrows

There are three options listed in PropertyManager associated with dimension arrow placement. These options are Outside, Inside, and Smart. The Smart option (button on far right in Figure 5–15) will place the arrows outside the extension lines if there is insufficient room between the extension lines.

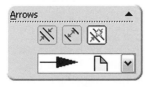

Fig. 5–15. Arrows panel.

The drop-down list below the buttons allows for changing the dimension arrow type. This is straightforward, but one item worth mentioning is the sheet of paper (or "document") pictured to the right of the arrow. This signifies that the arrow style will follow whatever is set as the default in the Document Properties window. This is typically what you would want.

✓ **TIP:** *Arrow positioning and style can be set for the entire document in the Dimensions section of the Document Properties window.*

10-second Topic: Flipping Arrows and Leaders

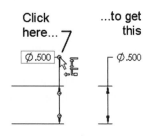

Fig. 5–16. Flipping dimension leaders.

An alternative method of modifying a dimension's arrow placement via PropertyManager is to use the green points present on the dimension arrow when the dimension is selected. First select the dimension by clicking on the dimension value. Next, click on one of the green dots that appear on either arrow. This action serves to flip the arrow direction and is easier than using PropertyManager.

On a related note, some dimensions have leaders that end in a small, horizontal line, an example of which is shown in Figure 5–16. When this is the case, clicking on the green dot at the end of the horizontal line will flip the leader and value to the other side. Of course, the dimension must be selected first for the green dot to appear.

Diameter Dimensions from Centerlines

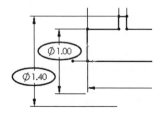

Fig. 5–17. Diameter dimensions from centerlines.

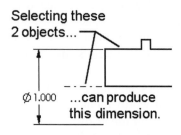

Fig. 5–18. Creating a diameter dimension using a centerline.

Eventually, a model such as the valve stem is going to get placed in a detailed drawing, complete with dimensions and various annotations. Considering that model dimensions can be used in drawings, it is often best practice to make sure the dimensions look right in the model. Such is the case with the diameter dimensions shown in Figure 5–17. Originally, these dimensions terminated on the horizontal line at the bottom of the sketch. This would have resulted in inappropriate dimensions in the drawing, as it would not be acceptable to have dimensions terminate in the center of the model. Keep in mind that this sketch is being used for a revolved feature.

The diameter dimensions shown in Figure 5–17 can be created quite easily, but you have to know the proper technique. To add a diameter dimension, make sure the centerline is one of the two entities selected. SolidWorks assumes you will be creating a revolved feature about the centerline, and allows for creating diameter dimensions. Figure 5–18 should help explain this task.

The most common mistake made when trying to add a diameter dimension is to select the endpoint of the centerline rather than the centerline itself. You must select the centerline, or the diameter dimension will not be created. Additionally, make sure to move the cursor to the opposite side of the centerline in order to see the correct diameter dimension preview.

✓ **TIP:** *Right-clicking locks in a dimension preview.*

Usually SolidWorks will automatically add a diameter symbol for you if the sketch is revolved about the centerline from which the diameter dimension was added. If it is necessary to manually add a diameter symbol, use the Dimension Text panel (discussed previously in this chapter in the "Dimension Text" section).

Dimensioning to a Tangency Point

Often when dimensioning to an arc or circle the extension line is attached to the arc's centerpoint. This default behavior is not always desirable, such as in the case of dimensioning an obround slot. Figure 5–19 shows what is meant by dimensioning to the tangency points of arcs. Note where the extension lines are attached to the sketch geometry.

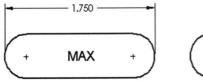
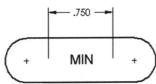

Fig. 5–19. Dimensioning to tangency points.

In Figure 5–19, the minimum and maximum arc conditions are used as extension line attachment points, rather than the arc's centerpoint. *Arc condition* is SolidWorks' terminology for the minimum and maximum tangency points. As you can see in Figure 5–19, SolidWorks can dimension to an arc's minimum tangency points even though no geometry is there. Of course, in this particular example you would not want to add a dimension to the minimum arc tangency points, but in other situations it may be necessary.

To dimension to tangency points, while holding down the Shift key pick near the tangency point where you would like the dimension extension line to terminate. Pay attention to the dimension preview and you should have no trouble with this procedure. If an extension line of an existing dimension needs to be repositioned, such as from an arc's centerpoint to its tangency point, try simply dragging the green handle at the end of the dimension's extension line. You will need to select the dimension first in order to perform this maneuver.

✗ **WARNING:** *The need to hold down the Shift key was added in a software patch for SolidWorks 2005 to address a certain issue. Specifically, drafters adding dimensions to very small holes could create dimensions to tangency points, rather than centerpoints, without realizing it. This could cause parts to be manufactured incorrectly if the mistake were not caught in time.*

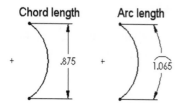

First arc condition :

○ Center   ○ Min   ⊙ Max

Second arc condition :

○ Center   ○ Min   ⊙ Max

Fig. 5–20. Arc condition options.

Occasionally, dragging extension line endpoints to tangency points on arcs does not work. If you run into this behavior, access the properties of the dimension in question. Assuming you picked an arc (or arcs) rather than its centerpoint (or centerpoints), you should see the arc condition options shown in Figure 5–20.

✓ **TIP:** *Any dimension can have its extension line repositioned if you first select the dimension and then drag the end of the extension line to some other object.*

Additional Dimension Types

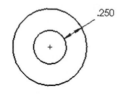

Fig. 5–21. Chord length versus arc length.

This section has been reserved for dimension types you will probably not use every day, and dimensioning techniques that are not quite as common. A good example of such a dimension would be an arc length dimension. This would be a dimension that shows the length of the arc as measured around its circumference, as opposed to the chord length (which would be measured linearly between the arc's endpoints). Examples of both dimension type are shown in Figure 5–21.

Selecting the arc itself results in a typical radial dimension. To create the chord length dimension, select both arc endpoints. To create the arc length dimension, select both arc endpoints and the arc. For none of the three aforementioned dimension types should you pick the centerpoint of the arc.

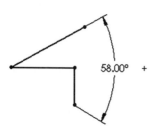

Fig. 5–22. Dimensioning concentric circles.

The distance between two concentric arcs or circles can be shown with a dimension. Such is the case in Figure 5–22, where the distance between the two circles is .25 inch. The circles or arcs must be concentric (share the same centerpoint), or it will not be possible to add this type of dimension. Selecting the two circles or arcs (or any combination thereof) will allow for adding this dimension type.

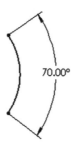

Fig. 5–23. Angular dimensions.

Angular dimensions can be added in two different ways. The most straightforward is to select two lines that form an angle. (Don't try adding an angular dimension between two parallel lines, as it just isn't possible.) Another method is to select three points, where the first point selected is the angle's vertex. The points can be sketch points, endpoints, arc centerpoints, and so on. Examples of angular dimensions are shown in Figure 5–23. Note that in each case there are not enough lines to add angular dimensions to, and thus being able to select points becomes essential.

Dual Dimensioning

Dual dimensioning refers to dimensioning in both English and metric units. Usually the extra dimension is placed in brackets. A typical example of a dual dimension value follows.

```
1.00 [25.4]
```

Dual dimensions will not be used in this book. However, your company or business may require you to use them. Normally you would probably want dual dimension display to be a global setting so that all dimensions in the drawing will be affected. To turn on dual dimensioning, perform the steps outlined in How-To 5–2.

How-To 5-2: Turning on Dual Dimension Display

To turn on dual dimension display for the entire drawing, perform the following steps.

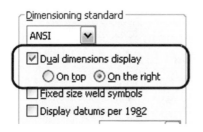

Fig. 5–24. Dual dimension display settings.

1. Select Options from the Tools menu.

2. Select the Detailing section in the Document Properties tab.

3. Place a check in front of the *Dual dimension display* option, shown in the Figure 5–24.

4. Specify whether you want the dual dimension value to be *On top* or *On the right* of the primary dimension value.

5. Click on OK when finished to accept your changes.

This setting is retroactive and will affect dimensions already present in the model. If there are only a select few dimensions you want to show dual dimensioning on, check the *Dual dimension* option in the PropertyManager of the dimensions in question.

It should be noted that dual dimension display can be turned on in a 2D detail drawing quite easily. Therefore, it really is not necessary to enable this option in a part model. This is especially true considering that dual dimension display will clutter things up quite a bit.

There are quite a few options when it comes to making changes to the way a dimension looks, as you can see. Chapter 12 covers many of the other options available. Many of these options are more important with regard to detail drawings, but do not play a role when modeling a part.

The following section takes you through some new sketch topics, including adding sketch chamfers and fillets, trimming and extending sketch entities, extracting sketch geometry from model edges, and offsetting sketch entities or model edges. As mentioned earlier in the book, it is very important to develop good sketch practices and skills in SolidWorks, as that is the foundation of becoming a good modeler. Practice these skills well.

⤖   **NOTE:** *Your toolbars may not contain all of the icons shown in this book. Customizing toolbars is discussed in Chapter 25, but you can also access any of the sketch tools discussed in the following material via the menu structure (Tools > Sketch Tools).*

Sketch Fillets

If you want to add fillets directly into your sketch, you can use the Sketch Fillet icon (shown in Figure 5–25). When you click on the Sketch Fillet icon, PropertyManager displays the appropriate panel (shown in Figure 5–26). How-To 5–3 takes you through the process of creating a fillet in a sketch.

Fig. 5–25. Sketch Fillet icon.

Fig. 5–26. Sketch Fillet PropertyManager.

How-To 5-3:   Creating a Sketch Fillet

It is suggested that you create a rectangle or two on your computer at this point, adding dimensions as well. This will give you something to experiment with when performing the steps that follow for creating a sketched fillet.

1. Click on the Sketch Fillet icon.

2. Specify a radius for the fillet.

3. Select two sketch entities to be filleted. These can be any combination of lines, arcs, splines, or even parabolas.

4. Click on OK when finished.

Be aware that sketch entities being filleted do not have to form a perfect corner. If there is a gap between the two entities, or if the entities overlap, SolidWorks will extend or trim them as necessary. If the entities overlap, click on the portion of the entities you want to keep when the fillet is added. If two entities form a perfect corner, it is possible to select the corner to be filleted. This is one less click than having to select both entities.

There is an option for keeping corners constrained. This option is necessary when there are existing dimensions (and sometimes geometric relations) on the items being filleted. If *Keep constrained corners* is not checked, any dimension terminating at the corner being filleted will usually be deleted, though this ultimately depends on what geometry was selected when the dimension was initially added.

To put it succinctly, dimensions cannot exist if the original geometry the extension lines attach to is deleted. As a general rule, leave *Keep constrained corners* checked all the time. SolidWorks will add what are known as *virtual sharps* where necessary to maintain any previously created dimensions (see Chapter 2, "Virtual Sharps"). Existing dimensions will be retained while the sketch fillets are added.

Sketch Chamfers

Chamfers can be added to a sketch in much the same way that fillets can. The process is almost identical. After clicking on the Sketch Chamfer icon (shown in Figure 5–27), the Sketch Chamfer PropertyManager will be displayed (shown in Figure 5–28). How-To 5–4 takes you through the process of creating a sketch chamfer.

*Fig. 5–27.
Sketch Chamfer
icon.*

*Fig. 5–28. Sketch
Chamfer
PropertyManager.*

How-To 5-4:   Creating a Sketch Chamfer

To create a chamfer in a sketch, perform the following steps.

1. Click on the Sketch Chamfer icon.

2. Specify how you would like to define the chamfer's dimensions by selecting either Angle-distance or Distance-distance.

3. Type in the applicable dimensional values.

4. Select the two sketch entities where the chamfer is to be applied.

5. Continue as required, and then click on OK when finished.

There is an *Equal distance* option that can be applied if the Distance-distance method of defining the chamfer is used. This option will not be available if the Angle-distance method is selected.

Similar to creating a sketch fillet, sketch chamfers can be created by selecting a corner instead of the two lines being chamfered. This is not recommended, because you have no control over which line has the first distance applied to it, and which line has the second distance. When selecting lines, as opposed to the corner, the first distance is applied to the first line and the second distance is applied to the second line. Of course, this is irrelevant if the distances are equal.

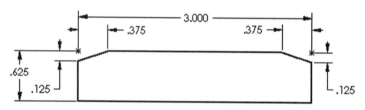

Fig. 5–29. An example of sketch chamfers.

For whatever reason, there is no *Keep constrained corners* option, as there is with the Sketch Fillet command. Dimensions may be deleted, with no warning, when adding sketch chamfers. For this reason, it is best to add sketch chamfers prior to adding dimensions to the sketch. Figure 5–29 shows an example of a sketch with two chamfers added.

Trimming, Extending, and Splitting

Fig. 5–30.
The Trim icon.

Because SolidWorks requires a non self-intersecting sketch, it is sometimes necessary to trim away some unwanted sketch geometry. The Trim icon, shown in Figure 5–30, can be found on the Sketch Tools toolbar. Clicking on the Trim icon brings up the Trim PropertyManager, shown in Figure 5–31.

There are five choices for how you would like to trim away unwanted sketch geometry. For those who have used versions of SolidWorks prior to SolidWorks 2005, the final option (*Trim to closest*) is the method you would be accustomed to. In previous versions, the Trim command did not present you with five choices in PropertyManager, but it never seemed to be an issue. With *Trim to closest* selected, simply click on what you would like to get rid of (trim). If you prefer a particular trim method, SolidWorks will default to that selection the next time you use the Trim command.

The Corner option allows for clicking on two objects that intersect and thereby trim them to meet at a perfect corner. With this option, click on the segments of each object you want to keep. Figure 5–32 shows a before-and-after example, including mouse picks where geometry was selected.

Trim away inside and *Trim away outside* work in a similar manner. Both trim options first require that you pick two objects to be the trimming objects. The third and subsequent picks should be between the first two objects and on the entities to be trimmed. Figure 5–33 shows a before-and-after example of the *Trim away outside* trim method, including mouse picks. If the *Trim away inside* method had been used, the picks would be identical. However, the outer portions of the horizontal lines would remain, and the inside segments would have been discarded.

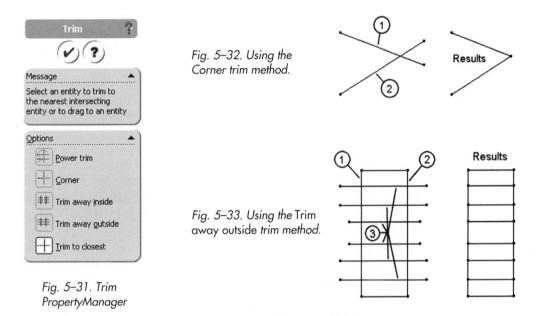

Fig. 5–32. Using the Corner trim method.

Results

Fig. 5–33. Using the Trim away outside trim method.

Results

Fig. 5–31. Trim PropertyManager

If we were saving the best for last, the final trim method we would explore would be the first method in the list, which is *Power trim*. To use this method, hold your left mouse button down and drag the cursor across anything to be trimmed. Careful, because it's easy to trim away more than you really want to! Objects can be completely deleted while trimming using this method (or while using the *Trim to closest* method).

Having five different ways to trim sketch entities is nice, but in the end *Trim to closest* is probably the best all-around trim method. It is very straightforward and very flexible. For example, it is possible with *Trim to closest* to trim to a specific object or to the projected intersection of two objects. To try this, hold the left mouse button down with the cursor positioned over what you want to trim, and then drag the cursor to the item you want to trim to. See Figure 5–34 for an example of how to perform a trim in this manner.

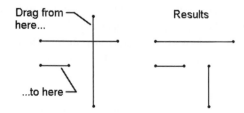

Fig. 5–34. Trimming to projected intersections.

Fig. 5–35. Extend Entities icon.

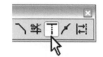

If you were able to get the hang of trimming, using Extend Entities should pose no difficulty. The two commands are very similar. The Extend Entities icon is shown in Figure 5–35. Similar to trimming, extending will give you a preview of what the extended entity will look like before you actually click the left mouse button. Extending is identical to trimming in all other aspects as well. The only difference between the two commands is that trimming takes portions of sketch geometry away and extending adds to geometry.

The Extend Entities command is not really necessary, because you can use the Trim command as an extend command. Use the same drag technique used when trimming to a specific entity in order to extend to a specific entity. In other words, using the *Trim to closest* method position the cursor over the entity to be extended and then drag the cursor with your left mouse button so that the cursor points to what the entity should extend to.

✓ **TIP:** *You can extend sketch entities using the* Power trim *method while in the Trim command if the Shift key is held down.*

Keep in mind that if geometry is blue it can be dragged and reshaped. If a line isn't long enough to intersect some other object, just drag its endpoint and drop it on the other object. For this reason alone, an extend command is often not necessary.

A command that can be used to split sketch entities into multiple segments is aptly named the Split Entities command. Think of the Split Entities command as a "break" command. Sketch entities can be broken at any location. For example, a line can be broken in two so that it is two line segments instead of one. Any sketch objects can be split, including parabolas, circles, and splines. To practice using the Split Entities command, try How-To 5–5.

How-To 5-5: Splitting Sketch Geometry

To split sketch geometry into multiple segments, perform the following steps. The Split Entities icon is visible to the right of the Extend icon in Figure 5–35.

1. Click on the Split Entities icon on the Sketch toolbar, or select Split Entities from the Tools > Sketch Tools menu.

2. Click on a sketch entity where you would like a break (or split) to occur.

That is all there is to it. If you want the split to be at a particular spot, add any additional relations or dimensions required. When a line is split, the resultant segments are collinear. When a circle is split, the resultant arcs are co-radial. If a spline is split, the remaining spline segments will be tangent to each other.

Converting Entities

Fig. 5–36. Convert Entities icon.

The ability to convert existing edges of a feature is an extremely important function. There is more to this than meets the eye. Therefore, the following material takes an in-depth look at this command. The Convert Entities icon is shown in Figure 5–36. When using either the Convert Entities command or the Offset Entities command (discussed in the next section), you must be in a sketch.

It has been mentioned repeatedly that you must have a sketch in order to create features such as extruded or revolved features. Considering that sketch geometry is a requirement of creating sketched features, it would occasionally be beneficial to directly convert existing edges of feature geometry into sketch geometry. This capability does exist, and it is basically a shortcut for obtaining the necessary sketch geometry. The Convert Entities command essentially "extracts" sketch data from the existing model edges.

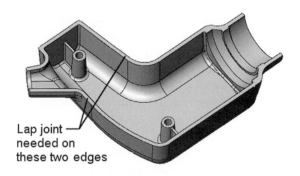

Lap joint needed on these two edges

Fig. 5–37. Model in need of a lap joint.

Let's consider a simple example in order to better understand this principle. Figure 5–37 shows half of a plastic molded part. The other half (not shown) would typically be held in place with screws or rivets. It is common for the seam between the two parts to fit together via some type of lap joint that would run along the edges of the part. One half neatly fits into place on the other half of the model.

To get a better grasp of exactly what is meant by a lap joint, examine Figure 5–38. Considering Figure 5–37, try to understand what type of sketch would be needed to create such a joint along the edges of the model. This is not as difficult as you might think, because Convert Entities allows us to extract the sketch geometry from the model edges and then use that sketch geometry to cut the part with.

It is easy enough to begin a sketch on the thin strip of either face near the edge of this model. We can then right-click on the outer edge of where the seam would be and select Select Tangency in order to select all outer edges along one side. Following this, clicking on Convert

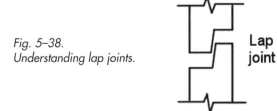

Fig. 5–38.
Understanding lap joints.

Lap joint

Entities is all it takes to create all of the sketch geometry shown in Figure 5–39.

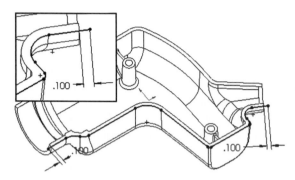

Fig. 5–39. Converting model edges into sketch entities.

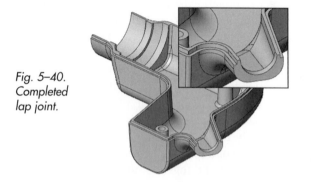

Fig. 5–40. Completed lap joint.

The inset in Figure 5–39 shows more clearly how one end of the converted geometry was pulled away a tenth of an inch from the model. This was done by dragging the endpoint and then dimensioning it from the outside edge of the model. This technique helps to remove all geometry when the cut is made, ensuring the removal of any leftover slivers of material. The cut itself was created using a thin feature, described in the next chapter. The model with the completed lap joint feature (one on each side of the part) is shown in Figure 5–40.

The steps for using Convert Entities are very simple. Select the model edge or edges to be converted, and then click on the Convert Entities icon. Because there are other choices with regard to what may be selected when using the Convert Entities function, Table 5–1 is included to outline other scenarios you may encounter. Use this table as a reference in terms of how the Convert Entities command functions.

Table 5-1: Convert Entities Functionality

Selection	Result of Clicking on Convert Entities Icon
Model edges on the current sketch plane	Sketch geometry is extracted from the model edges.
Model edges not on the current sketch plane	Sketch geometry is extracted from the projection of the model edges perpendicular to the current sketch plane.
A model face	All edges of the face are extracted. If the face is not on the current sketch plane, edges are projected perpendicular to the sketch plane.
Sketch geometry from an existing sketch	Sketch geometry is extracted from the projection of the existing sketch onto the current sketch plane.

One very important aspect of the Convert Entities command has to do with relationships. Any sketch geometry that has been converted from a model edge (or from another sketch) has a relationship to the originating geometry. This means that if the original geometry changes size or shape the sketch containing the converted geometry will change also. In fact, there is a direct dependency leading back to the original model edges.

The relationship created when converting geometry is known as an *on edge* relationship. When a model edge is converted into sketch geometry, its endpoints will appear to be fully defined. This is intended behavior. This behavior is convenient because you do not have to worry about fully defining converted sketch geometry if you don't want to. However, dragging the endpoints of a converted entity will "loosen" the endpoints, thereby rendering them underdefined. It would then be possible to dimension the endpoints of a line without overdefining it.

Offsetting Entities

Offset entities display the same characteristics as converted entities. They are both dependent on the underlying geometry and can "extract" sketch entities from model geometry. Endpoints of offset entities will appear fully defined unless they are dragged, similar to converted entities. Like converted entities, offset entities will change shape if the dimensions of the underlying geometry are modified. The main difference is that offset entities allow you to supply an offset distance from the π original model edges or sketch geometry.

The Offset Entities command has more optional parameters than the Convert Entities command. This being the case, How-To 5–6 takes you through the process of using the Offset Entities command.

How-To 5-6: Command

Using the Offset Entities

Fig. 5–42. Offset Entities PropertyManager.

To use the Offset Entities command, shown in Figure 5–41, perform the following steps.

1. Select the model edges (or sketch geometry) to be offset.

Fig. 5–41. Offset Entities icon.

2. Click on the Offset Entities icon located on the Sketch toolbar.

3. Enter an offset distance in the Offset Entities PropertyManager (shown in Figure 5–42).

4. Select the Reverse option to reverse the offset if necessary.

5. Click on OK when finished.

In the case of converting entities, you must select the entities first. That is not the case here. Select the entities to be offset either before or after clicking on the Offset Entities icon.

Offset entities will have an associated dimension added automatically by SolidWorks if offsetting model edges. This offset dimension is fully parametric and can be changed just like any other driving dimension. If offsetting sketch geometry, the choice to add a dimension is up to you. Usually it would be desirable to leave the *Add dimensions* option checked.

The *Bi-directional* option, like the name implies, offsets in two directions at the same time. If *Bi-directional* is used, the *Cap ends* option will also be available, in which case the ends of the offset geometry can be capped with either lines or arcs.

The *Select chain* option applies to offsetting sketch geometry only, and not model edges. If *Select chain* is checked and a sketch entity is selected, any sketch geometry connected to the selected segment will also be offset. This makes selecting sketch geometry much easier.

In the example shown in Figure 5–43, sketch geometry was offset bidirectionally. The *Cap ends* option was used, with arcs used for the caps. *Select chain* was turned on to make the selection process easier, and *Add dimensions* was checked, resulting in the dimensions seen in Figure 5–43. Finally, the *Make base construction* option was also checked, which turns the original geometry into construction geometry. Attempting to create this curvy geometry would have been much more tedious had it not been for the Offset Entities command.

SW 2006: Parabolas, elliptical arcs, and ellipsis are capable of being offset.

Continuing with out ongoing project, in Exercise 5–2 you will edit the sketch used to create the revolved feature and modify some of the properties of the dimensions. You will also create a groove feature that will later be used to create a circular pattern. This pattern will represent the portion of the valve stem used to grip the handle when the faucet is assembled. Some other processes you will get some practice with in this exercise are working with sketched chamfers, converting entities, and trimming.

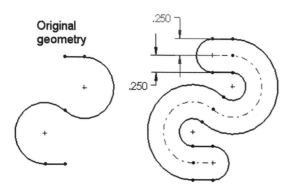

Original geometry

Fig. 5–43. Example of offset entities.

Exercise 5-2: Editing the Revolved Feature Sketch

Begin this exercise by opening the valve stem part you created in Exercise 5–1. Use some of the techniques learned in the "Dimension Properties" section to modify the way the dimensions appear in the sketch for the revolved feature.

1. Right-click on the Base-Revolve feature in FeatureManager and select Edit Sketch.

2. Delete the original .700-, .500-, .250-, and .625-inch vertical dimensions created in Exercise 5–1.

3. Add a centerline over the top of the bottom horizontal line. The centerline should extend perhaps a quarter inch or so beyond each end of the original line.

4. Add four (4) diameter dimensions to replace the four dimensions just deleted. (Hint: make sure one of the items you select is the centerline when adding the dimensions.)

5. By changing the properties of some of the existing dimensions, try your best to make your sketch appear like that shown in Figure 5–44.

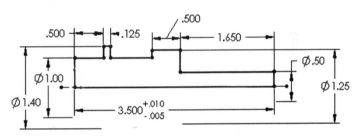

Fig. 5–44. Adding tolerance values and modifying dimension properties.

Were you able to get your sketch to look like that shown in Figure 5–44? You may have had trouble with the .500-inch offset text dimension near the middle of the sketch. Some dimension options hide in the right mouse button menu. Such is the case with offset text. Right-click on the dimension and look for a submenu titled Display Options. This submenu contains common options that can be toggled on and off. Toggle on the Offset Text option to achieve the effect shown in the figure.

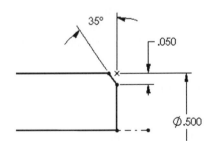

Fig. 5–45. Adding a sketched chamfer.

6. When you have finished modifying the dimensions, add a chamfer to the sketch, such as that shown in Figure 5–45 (image enhanced for clarity).

7. Click on the Rebuild icon when finished modifying the sketch.

8. Start a sketch on the small, round end of the stem where the chamfer was just added.

Next, you will be creating the sketch profile used to cut the groove into the small end of the stem. This will most easily be accomplished by starting out with some construction geometry. You will also use Convert Entities, the Trim command, and the Construction Geometry command to finish the sketch.

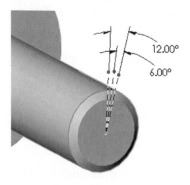

Fig. 5–46. Adding construction lines and dimensions.

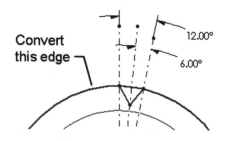

Fig. 5–47. Converting an edge and adding two lines.

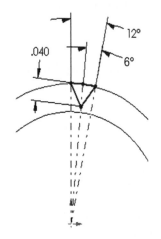

Fig. 5–48.
Finished sketch.

9. Add some construction geometry (center-lines), as shown in Figure 5–46, to help control the sketch used to cut the groove. Add angular dimensions as well. Ensure that one of the lines is vertical.

10. Convert the outer edge of the chamfer into sketch geometry. Add two lines, as shown in Figure 5–47. Pay attention to system feedback!

11. Trim back the portion of the converted circular edge that is not needed, as well as the ragged ends of the construction lines (just for the sake of neatness).

12. Add a .040-inch dimension to control the depth of the groove. (Hint: if you did not trim the center construction line, this will be difficult.)

13. Modify the properties of any dimensions as desired. Figure 5–48 shows the finished sketch. Note that the .040-inch dimension is an aligned dimension, not a vertical.

14. Create an extruded cut .50 inch deep.

15. Rename the new feature *V-groove*. The completed groove feature is shown in Figure 5–49.

Next, you will add a revolved cut, which will allow for the placement of an O-ring washer. First, let's resize the Right plane so that it more closely matches the size of the part.

16. Show the Right plane.

17. Change to a Right view.

18. Right-click on the Right plane in FeatureManager and select Autosize. The plane should shrink to match the size of the model.

19. Start a sketch on the Right plane.

20. Create the sketch shown in Figure 5–50. You will need to add a tangent relation between the circle and vertical edge of the model (which is actually a circular edge that appears as a linear edge in our current sketch plane).

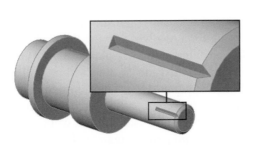

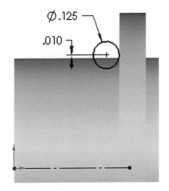

Fig. 5–49. Finished groove feature.

Fig. 5–50. Sketch for the O-ring groove.

21. Add the .010-inch dimension by selecting the centerpoint of the circle and the silhouette edge of the model.

22. Make sure to add a centerline along the axis of the model for the revolved feature.

Fig. 5–51.
Revolved Cut icon.

23. Click on the Revolved Cut icon, shown in Figure 5–51, and create a revolved cut 360 degrees around the model.

24. Rename the feature *O-ring groove.*

25. Save your work.

Fig. 5–52. Silhouette edge system feedback symbols.

Step 21 had you add a dimension to a silhouette edge of the model. Did you happen to notice the system feedback while performing this action? You should have seen a symbol attached to the cursor. This symbol represents a silhouette edge. In case you missed it, this symbol is shown in Figure 5–52 (enlarged for clarity).

Fig. 5–53.
Valve stem
with a few
new features.

If you managed to get through the exercise correctly, the valve stem should look similar to that shown in Figure 5–53. You will finish this part in subsequent chapters, so make sure you tuck it away for safe keeping.

Summary

Now that you have finished Chapter 5, you can add another feature type to your list of SolidWorks tools. Revolved features can be used to create turned or lathed parts, or any part that is radially symmetrical. Centerlines are often used in the sketch for a revolved feature, but are not required. If a centerline is not used, or if more than one centerline is used, make it a point when creating the feature to select a line or centerline you wish to revolve about.

When creating a sketch for a revolved feature, the sketch must not cross the centerline or touch the centerline at an isolated point. With regard to dimensioning, it is possible to dimension to a centerline and thereby create the appearance of a diameter dimension. When brought into a 2D layout, the diameter dimension will conform to typical drawing standards.

Modifying the properties of a dimension allows you to change many aspects of the dimension. These properties include changing the direction of the arrows, modifying decimal precision, changing between radial or diameter dimension types, extending a linear dimension to the tangency point of an arc or circle, and so on. Quite often, these dimensional properties are fine-tuned in the layout, but altering dimension appearance can be made directly in the part file as well. Use the green handles associated with dimension lines to flip the arrows from one side to the other.

Sketch chamfers and fillets can be created quickly and easily within a sketch. If sketch geometry forms a perfect corner, clicking on the corner is enough to add a sketch fillet. SolidWorks will add a virtual sharp if there are dimensions attached to the corner being filleted.

Other important sketch tools were covered in this chapter. You learned the importance of the Convert and Offset Entities commands, and that any converted or offset sketch entity is dependent on and associated with the edge it was converted or offset from. Offset Entities gives you the added ability to specify an offset distance from the edge it was converted from. When offsetting sketch geometry (as opposed to model geometry), additional options are available that aid in the creation of a sketch profile.

You also learned how to use the Trim and Extend commands. SolidWorks will give you a preview of the entity to be trimmed or extended. Clicking on an entity will trim or extend that entity to the next object it encounters. If you want to be more precise, you can pick an entity and drag it with the left mouse button to the specific entity you want to trim or extend to. Specific trim options are available, such as *Power trim*, which allows for dragging the cursor across objects to trim them or even to completely delete them.

CHAPTER 6

Molded Parts

IT WOULD BE IMPOSSIBLE TO COVER EVERY TYPE of molded part in a single book, much less in one chapter. Therefore, the component this book deals with is primarily the thin-walled type of molded part (usually plastic) you might find in products throughout your own household.

The types of parts this chapter deals with are typical of many household appliances. For example, the casing of your telephone or answering machine is probably a plastic molded part. How about your electric shaver, or even your disposable razor handle? If you are working near your computer (assuming the reader of this book would own one), take a look at your mouse, or maybe the faceplate of the computer itself.

Many of these components would be considered thin-walled parts in SolidWorks terminology. The parts generally have a uniform wall thickness, probably with a few ribs added here and there for structural support. They usually also have a number of bosses (often known as stand-offs) that can be drilled through. These bosses would be used for putting screws through, in most cases, to attach the part to another component in the assembly.

This should give you a pretty good idea of where this chapter is heading. The main topics covered are shelling, adding draft as a feature, and the Hole wizard. You will also learn how to add fillets and chamfers directly to a model (as opposed to in a sketch), and how to use the Rib command to easily create ribs. Last, you will learn how to analyze model draft, thereby allowing you to see if the part will come out of the mold.

The Shell Command

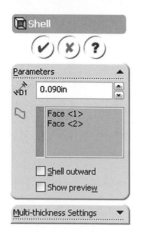

Fig. 6–1. Shell PropertyManager.

In its simplest form, the Shell feature is very easy to implement. In theory, the inside of the part is removed, leaving behind a thin wall. You have the ability to define the thickness of this wall. Typically at least one face is selected to be removed during the shell process, though this is not required. If no face is selected, a hollow part is created.

An option available to you is to add a wall thickness to the outside of a part. This is something like dropping the part into a vat of molten wax and letting the wax cool and harden. The outer layer of wax would be all that is left, and the original part would be removed (assuming that this feat could actually be accomplished).

The Shell feature panel is shown in Figure 6–1. In particular, note the Parameter area. It is this area where a wall thickness must be specified. Faces can be selected in the Parameter area to be removed during the shell process. The *Shell outward* checkbox option allows you to create the type of shell analogous to the "hot wax" scenario previously described. How-To 6–1 takes you through the process of creating a simple shelled part.

HOW-TO 6-1: Creating a Shelled Part

To create a shelled part with a consistent wall thickness, perform the following steps.

1. Click on the Shell icon found on the Features toolbar, or select Shell from the Insert > Features menu.

2. Specify a wall thickness for the shell.

3. Select the face or faces to be removed during the shell operation.

4. Specify *Shell outward* if necessary.

5. Click on OK to complete the shell operation.

Fig. 6–2. Before and after performing a shell operation.

If you take a look at Figure 6–2, you will see a simple example of the shell operation. On the left is the part before the shell was completed. On the right is the same part after the Shell command. In this example, a shell thickness of .060 inch was used, and four faces were removed during the shell operation: three tangent faces on the top of the part and one face from an inset area beneath the part (not visible prior to shelling).

Fig. 6–3. Removing tangent faces during the shell operation.

Face selection is an important aspect of shelling. Common practice is to select faces that do not have tangent faces. If there are tangent faces, the tangent faces are usually selected as well. This is not a requirement, just common practice. Figure 6–3 shows the same model with an additional two faces removed during the shell. The additional faces were on the perimeter of the model, and had been tangent to other perimeter faces prior to their removal.

An important aspect of shelling is the built-in intelligence of the command. Features too small to be shelled, such as small bosses or rounds, can usually be ignored by SolidWorks. Ideally speaking, it is best to try to perform a shell operation early in the design process. Due to design requirements, this may not always be possible, but it is good technique to perform shelling as early as the design permits.

✓ **TIP:** *If it is necessary to model a container with a specific volume, create the inside of the container and then shell outward. Use the Mass Properties tool (discussed later in this book) to find the volume.*

Multi-thickness Shell

Another aspect of the shell command is its ability to create a multi-thickness shell. What this means is that during the shell operation it is possible to specify varying degrees of thickness for specific faces.

Fig. 6–4. Multi-thickness settings.

In the Shell PropertyManager, you will see an area titled Multi-thickness Settings, shown in Figure 6–4. If you click inside the list box within this area, it is possible to select faces on the model that are to have varying degrees of thickness. You can specify whatever thickness you want for individual faces, within reason. The same rules apply with multi-thickness shells as with standard shell features. The shell feature has to be geometrically possible or SolidWorks will refuse to process it. How-To 6–2 takes you through the process of creating a multi-thickness shell.

HOW-TO 6-2: Creating a Multi-thickness Shell

To create a multi-thickness shell, perform the following steps. Note that the first three steps are exactly the same for creating a shell of constant wall thickness. The same option for shelling outward is still available, but is not present in the following steps because it is not applicable.

1. Click on the Shell icon found on the Features toolbar, or select Shell from the Insert > Features menu.

2. Specify a wall thickness for the shell.

3. Select the face or faces to be removed during the shell operation.

4. Click in the list box area of the Multi-thickness Settings section to highlight it, and then select any faces that are to have a different thickness than the overall part.

5. Select each face in the *Multi-thickness faces* list box and assign a thickness value to each. Make sure to assign a thickness value to every face listed if there is more than one. Values can be different if required.

6. Click on OK to complete the shell.

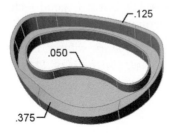

Fig. 6–5. Another multi-thickness shell example.

Figure 6–5 shows an example of what a multi-thickness shell might look like. Even though different thickness values were specified for tangent faces, the part is still shelled. SolidWorks overcomes the inherent difficulties of this scenario by tapering the thicker areas into the thinner areas. Note that the front of the model is .375 inch thick, whereas the rest of the perimeter faces are only .125 inch thick. The inner faces surrounding the cutout are even thinner, at .050 inch.

When creating a multi-thickness shell feature, it is not usually good practice to specify differing values for tangent faces. The results can be potentially undesirable. If one face is selected that does have tangent faces, it is typical to select the tangent faces as well.

When shell operations fail, it is usually because there is geometry that cannot be shelled and SolidWorks does not have the ability to simply ignore the geometry. Sometimes thinner values will work. Other times it is necessary to perform the shell operation at an earlier stage of development. If a shell does fail, pay attention to the screen. You will be shown a grid pattern on the area where the shell feature is having difficulty.

Thin Features

Thin-walled parts are created using the Shell command. Thin features differ in that they do not require the Shell command and are usually created with an open profile sketch. Similar to creating a shell feature, thin features do require that a wall thickness be specified.

Just because thin features are called thin does not mean that they are not solid model parts. Thin features are still considered solid geometry and have volume just like any other solid part or feature. Also, thin features do not necessarily have to be thin. They can be a foot thick. The term *thin feature* is really used to denote how the feature was created.

Fig. 6–6. Open profile sketch.

Fig. 6–7. Thin Feature option enabled.

Fig. 6–8. Thin-feature part.

Let's examine how a thin-feature part could be created. What would happen if an open profile sketch were extruded? Figure 6–6 shows a very simple open profile sketch.

Even though the profile is open, the sketch can be extruded. Feel free to create a sketch and try this out for yourself. When the Extrude panel is displayed, note that the Thin Feature option is checked (as shown in Figure 6–7).

Because the sketch is an open profile, SolidWorks forces you to create a thin-feature part (which is why the checkbox option is grayed out in Figure 6–7). You will see an option for specifying how thick you want the material to be. This wall thickness can be added to one side of the sketch or the other by clicking on the Reverse Direction button. This is typically dependent on which side of the material the dimensions are critical for, either the inside or outside. Make it a point to keep an eye on the preview so that the wall thickness is applied in the proper direction.

The Type drop-down list contains three selections for specifying how the wall thickness will be applied. These options are One-Direction, Mid-Plane, and Two-Direction. The Mid-Plane selection will apply a wall thickness equally in both directions (the total of which is user defined). Two-Direction allows for entering precise values for the wall thickness on either side of the profile sketch.

When creating a thin feature, there are still the end condition parameters that must be entered. This aspect of the Extrude panel has not changed, and thus you specify the end condition type and distance as you always have. It is even possible to add draft while extruding a thin feature. Figure 6–8 shows the possible outcome of extruding the sketch shown in Figure 6–6.

10-second Topic: Auto-fillet Option

How did the bends get placed in the model shown in Figure 6–8? There were no fillets added in the sketch. Rather, the fillets (which appear as bend regions in the model), were added by turning on the *Auto-fillet corners* option. This option can be seen in Figure 6–7.

When using *Auto-fillet corners*, the fillet radius parameter is applied to the inside of all sharp corners. This action creates the "fillets," though they are not really fillets in the true sense of the word. The fillets more closely resemble bends. This setting is an "all or nothing" setting, and is not suitable if creating sheet metal parts. For that matter, thin features in general should not be used for developing sheet metal parts at all. A different modeling process, discussed in its entirety in Chapter 9, is employed for sheet metal parts.

Revolving an Open Profile

It is possible to revolve an open profile, just as it is possible to extrude one. The revolve process functions slightly differently than the extrusion process when it comes to open profiles. Specifically, the user is given a warning message regarding non-thin features. The material that follows presents an example that explains the message.

The same sketch that was used in the previous example will be used for creating a revolved thin feature. A centerline will be added for the center of rotation. When the Revolve command is initiated, a window appears warning you that a non-thin feature requires a closed profile. This warning is shown in Figure 6–9.

Fig. 6–9. Warning seen while trying to revolve an open profile.

You have two choices at this point. You could click on Yes and let Solid-Works close the profile, or you could click on No and revolve the open profile as a thin feature. Usually this message is an annoyance, because if you wanted the sketch to be closed you could have drawn it that way.

When you let SolidWorks attempt to close a sketch automatically, it will add a sketch line between the two open endpoints of the sketch. In some cases this is fine. In other cases it is not. If adding a line results in a self-intersecting sketch, SolidWorks will inform you of this fact and will refuse to add the line. It will, however, continue to attempt to create the revolved feature.

Fig. 6–10. Creating a revolved thin feature.

Once in the Revolve PropertyManager, the Thin Feature parameters section will work exactly the same way it did in the Extrude PropertyManager. You simply specify the direction Type and the Wall Thickness, and use Reverse if necessary. There is no auto-fillet option when creating a revolved thin feature. Figure 6–10 shows what revolving an open profile sketch might look like. The same sketch geometry used earlier for the extruded thin feature was used here. The sketch was revolved 270 degrees (rather than 360 degrees) so that a cross section of the geometry would be visible.

Closed-profile Thin Features

Yes, it is possible to create thin features from closed profiles. When extruding or revolving open profiles, you do not have a choice. The feature must be a thin feature because there is no alternative. When starting with a closed-profile sketch, the choice is yours. If a thin feature is required from a closed profile, simply check the Thin Feature option and supply the desired parameters for the wall thickness.

If a thin feature is created and you change your mind, you will find that the Extrude or Revolve definitions no longer make the Thin Feature option available. The Thin Feature option will be there, but it will be grayed out. This is one of the very few options SolidWorks does not let you go back and edit. The workaround is simple enough, though. If you delete the feature, you will find that the sketch remains in FeatureManager. You can then reuse the sketch as you see fit.

Cap Ends Option

Fig, 6-11. Using the Cap ends *option.*

When creating a thin feature from a closed profile, there is an option to cap the ends of the thin-feature part. This creates a hollow, airtight part. If you do use the *Cap ends* option, you must also specify the wall thickness for the caps (as shown in Figure 6–11). This can be different from the overall wall thickness of the model. Also note that only extruded thin features offer the *Cap ends* option (i.e., revolved thin features do not).

→ **NOTE:** *The* Cap ends *option is only present when creating a thin feature as the first base feature.*

✓ **TIP:** *Any feature type can be created as a thin feature, including a swept or lofted feature.*

Fillets and Chamfers

In Chapter 5 you learned how to add fillets and chamfers in a sketch. In this chapter you will learn how to apply fillets and chamfers directly to the model. Let's begin by exploring fillets and rounds, which can be added simultaneously without difficulty. SolidWorks does not really differentiate between the two. From a terminology standpoint, fillets are applied to inside edges and rounds to outside edges. Some simple examples are shown in Figure 6–12.

→ **NOTE:** *For the sake of simplicity, both fillets and rounds are referred to as fillets throughout this book.*

Adding fillets is a very simple process, and in its simplest form you need only specify a radius for the edge being filleted. Whether you add a fillet as part of a sketch or as a feature is really up to you; the end result will be the same for simple fillets. More complex fillets need to be added as features, because the command parameters are much more flexible.

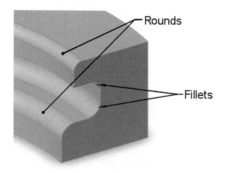

Fig. 6–12. Fillets and rounds.

As a general rule, it is usually best to add cosmetic fillets near the end of the design process. Where fillets affect the form or function of the model, they should be added early in the design process rather than toward the end. Cosmetic fillets complicate the model graphically and can slow down the model display. More polygons are needed to render curved surfaces than flat surfaces. There is no sense in overly complicating the model if there is still design work to be done.

When adding a sketch fillet, two lines can be selected. However, when applying a fillet to a model the edge being filleted must be selected. In addition, objects other than edges can be filleted. For example, you can select a face on the model, in which case every edge on that face would be filleted. Features can also be selected from FeatureManager, in which case every edge on the feature is filleted. What you select depends entirely on your design intent. Figure 6–13 shows an example of what would happen if an edge or face were selected on a particular part and a fillet applied.

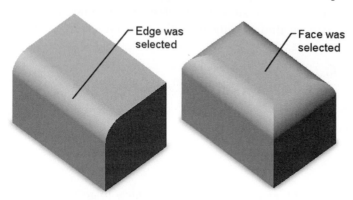

Fig. 6–13. Adding fillets to a model.

✓ **TIP:** It is generally preferable to select edges over faces when applying fillets. Edge selection offers more control, and selecting faces can sometimes cause more fillets to be applied than is desirable.

Control-selecting: A Reminder

It is normally preferable to select objects after initiating a command, because you do not have to hold down the Control (Ctrl) key. As a reminder, hold the Ctrl key down anytime you want to select more than one entity prior to entering a command. This is standard operating procedure throughout SolidWorks.

Fillets

Large fillets are usually applied to a model first. It is generally desirable to have smaller fillets wrap around the tangent faces of a larger fillet. However, it is possible in SolidWorks to make a larger-radius fillet wrap around the faces of a smaller-radius fillet. This may not be a desirable outcome, however. Take a look at Figure 6–14 for a better understanding of what is actually being accomplished in this type of situation. The edges are shown in black for clarity.

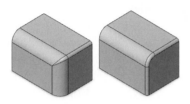

In Figure 6–14, the image on the left had a small .125-inch-radius fillet added first. The larger .5-inch-radius fillet was added later. As you can see, the larger-radius fillet propagates around the tangent faces created by the first fillet. A more desirable situation appears on the right of Figure 6–14. In this second example, the large-radius fillet was added first. It is easier to see how the smaller-radius fillet wraps around the larger fillet automatically in the right-hand image.

Ultimately, the order in which the fillets are added will depend on how you want the model to look. Just be aware that ordering is important and that it is best to construct a mental plan as to what that order should be. To add a fillet as a feature, perform the steps outlined in How-To 6–3.

Fig. 6–14. Wrapping a large-radius fillet around a small-radius fillet.

HOW-TO 6-3: Adding a Fillet Feature

First, bear in mind that these steps show how to apply a fillet directly to the model as a feature. The process is very simple, but you must not be editing a sketch when performing these steps. To add a fillet as a feature, perform the following steps.

1. Click on the Fillet icon found on the Features toolbar, or select Fillet/Round from the Insert > Features menu.

2. Specify a radius for the fillets in the Fillet PropertyManager, shown in Figure 6–15.

3. Select the items to be filleted (typically edges).

4. Click on OK to create the fillets.

Fig. 6–15. Fillet PropertyManager.

The previous steps show how to add a fillet, in its simplest form, using the Fillet command. Obviously, there is much more to the Fillet command than How-To 6–3 describes. Variable radius, full round, and face fillets are discussed in depth in Chapter 20. For now, though, some of the more important basic options you should be aware of are described in the material that follows.

Tangent Propagation

Selected edge   Propagation off   Propagation on

Fig. 6–16. Effects of tangent propagation.

By default, SolidWorks will continue filleting any edges tangent to the original edge selected for the fillet. This is known as the *Tangent propagation* option, which can be turned off if desired. By default, *Tangent propagation* is always turned on, even if it was turned off the last time the Fillet command was used. When this option is turned off, SolidWorks miters any corners instead of flowing around tangent faces. Figure 6–16 shows two examples of a fillet, the same edge of which was selected in both cases. The first example shows tangent propagation turned off, whereas the last example shows the same part with tangent propagation turned on.

Multiple-radius Fillets

The *Multiple radius fillet* option (found above the *Tangent propagation* option) is used when it is desirable to fillet more than one edge using different radial values in the same fillet command. Most of the time it is usually best to create similar-radius fillets as the same feature. This just makes good sense, but sometimes adding multiple fillets of varying radii simultaneously is very convenient.

When creating fillets on edges that will converge on a single vertex point, the *Multiple radius fillet* option can be beneficial if preparing to create what are known as *setback fillets*. You will learn about setback fillets in Chapter 20. How-To 6–4 takes you through the process of adding multiple-radius fillets.

HOW-TO 6-4: Adding Multiple-radius Fillets

To create multiple-radius fillets as a single feature, perform the following steps.

1. Click on the Fillet icon, or select Fillet/Round from the Insert > Features menu.

2. Select the *Multiple radius fillet* option.

3. Specify a radius for the fillets (this can be changed later).

4. Select the edges to be filleted (faces can be selected, but they must not have common edges when utilizing the *Multiple radius fillet* option).

5. Select any object from the *Items to fillet* list box and specify the desired radius for that particular object. Use the Ctrl key to select more than one object at a time.

6. Click on OK to create the fillets.

Fig. 6–17. Three fillets created simultaneously.

An example of fillets created with the *Multiple radius fillet* option is shown in Figure 6–17. The fillets use radial values of .25, .50, and .75 inch. Note in particular the corner at which the fillets meet. It would be possible to achieve this same effect without the *Multiple radius fillet* option, but it would be more time consuming and would require three separate fillet features. To duplicate this same look, one would have to add the fillets independently, with the *Tangent propagation* option turned off. This could certainly be accomplished, but would not be nearly as efficient.

Fillet Previews

Fillet previews come in two varieties: full and partial. If the partial preview is used, only one edge will show a preview, even if a face is selected. The full preview shows previews for all edges that wind up being filleted, regardless of what was selected for filleting. It is also possible to select *No preview* if you decide it just is not necessary to see the preview.

Whether or not you decide to use previews of the fillets you are about to create is totally up to you. Your decision will not affect the model geometry in any way. Occasionally, you may find it necessary to turn previews off strictly because it is often easier to select edges if the preview is not cluttering up the work area.

Filleting Features

We have been discussing adding fillets as features. But features themselves, such as bosses and cuts, can be selected when adding fillets. Features consist of faces and model edges. Because of this, selecting a feature to be filleted must be done carefully. Otherwise, a face or edge may get filleted instead of the entire feature.

To select the feature correctly, do not simply click on the feature in the work area. Rather, right-click on the feature and use the *Select feature* option found in the menu while in the Fillet command. Another option would be to use the fly-out FeatureManager, mentioned in Chapter 1. Figure 6–18 shows examples of what happens when a feature is selected for filleting (the reason for the difference between these examples is discussed in the following material).

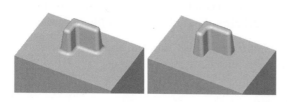

Fig. 6–18. Result of filleting a feature.

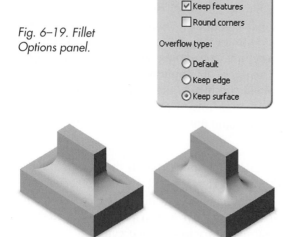

Fig. 6–19. Fillet Options panel.

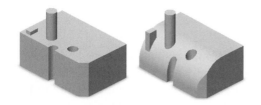

Fig. 6–20. Note the difference with Round corners enabled (at right).

Fig. 6–21. Features remain after fillet is applied.

When features are selected for filleting, there is an option that plays an important role. This option is known as *Omit attach edges*, which is only available if a feature has been selected. It is found in the panel of the Fillet PropertyManager labeled *Fillet options*, shown in Figure 6–19. (Material to follow explores the remainder of the fillet options.)

When *Omit attach edges* is checked, none of the edges that form a boundary between the existing model and the selected feature are filleted. The example on the right in Figure 6–18 shows the results of checking *Omit attach edges*.

Round Corners

Only certain types of geometric situations can take advantage of the *Round corners* option. There are no special instructions required for using this option. Simply enable it by checking the option. Figure 6–20 shows an example of a part with *Round corners* disabled (left) and with the same option enabled (right).

Keep Features

Simply put, when the *Keep features* option is checked features affected by the fillet are retained. This refers to features that are not being filleted themselves but are positioned on faces being directly affected by the fillet. Figure 6–21 shows this option in action better than can be explained.

The *Keep features* option is checked by default, which is more than likely the way you will want to leave it. In a perfect world, a SolidWorks user would have a good idea of the order in which the features are to be added to the model. As many readers are aware, however, this is often not the case. Models change and revisions are almost inevitable.

Because foresight is not always as accurate as we might like, the *Keep features* option gives you a little more flexibility in creating the model. The option can be turned off, but in doing so any fillet created will absorb features affected by the fillet. In other words, the features will disappear from the model.

➤ **NOTE:** *Features absorbed by a fillet are not deleted from the model, they just disappear. This is due to a fillet's definition and geometric necessity. The features are still present, and if the fillet is deleted the affected features will reappear.*

Overflow Control

When a fillet extends beyond an area that can physically accommodate it, something has to give. This is known as "overflow" in SolidWorks terms, and how the geometry surrounding the fillet is affected depends on the overflow setting. Either the surface of the fillet can change to accommodate the surrounding geometry or the surrounding geometry can change to accommodate the fillet.

The two settings for overflow control are Keep Edge and Keep Surface. There is also a setting of Default, but that just means that SolidWorks will use the Keep Edge setting. An example of each setting is shown in Figure 6–22. A description of these settings follows the illustration.

Fig. 6–22. Overflow control for fillets.

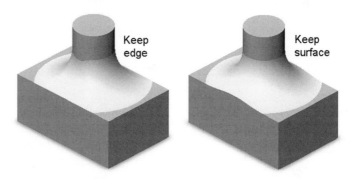

Keep Edge

When Keep Edge is used, the edge where the fillet overflows the geometry is kept the same. In other words, the edge does not change to accommodate the fillet. Instead, the fillet changes to accommodate the edge. Note in Figure 6–22 that the top of the block remains straight, whereas the top of the fillet dips down. The radius of the fillet is not changing. Rather, picture a rolling ball positioned between the block and cylinder. The ball would move downward along the axis of the cylinder in order to maintain contact with the edge of the block. This accounts for the dip you see where the fillet meets the cylinder.

Keep Surface

When Keep Surface is used, the surface of the fillet remains constant and the edges where the overflow takes place alter to accommodate the fillet. Note in Figure 6–22 that the fillet appears very consistent, with the top of the fillet appearing as a perfect circle around the cylinder. The top edge of the block, on the other hand, is pulling upward.

Chamfers

Chamfers are applied to a model in much the same way as fillets. Start the command, pick the objects to be chamfered, plug in some values to establish the size, and click on OK. There are a few other options that should be expanded upon, though. The sections that follow explore the three main options for creating chamfers.

The Chamfer PropertyManager is shown in Figure 6–23. Like fillets, it is possible to select either an edge or a face when adding chamfers. Just remember that if a face is selected chamfers will be applied to every edge on that face, so use discretion. Features cannot be selected for chamfering as they can be when adding fillets. It is highly unlikely you would want to do this anyway for any model machined in the real world.

Note that there are options in the Chamfer PropertyManager that are identical to those found in the Fillet PropertyManager. These options include *Keep features*, *Tangent propagation*, and the various preview options. Refer back to the "Fillets" section if you have questions about the functionality of these options.

Fig. 6–23. Chamfer PropertyManager.

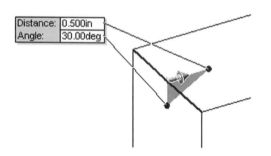

Fig. 6–24. Distance and angle parameters can be used to define a chamfer.

Angle-Distance

Angle-distance chamfers require you to specify a distance and an angle to define the chamfer dimensions. The preview arrow points in the direction the distance will be applied. Depending on the display mode, a cross-section representation of the chamfer should be visible as well. There will also be a callout with the current dimensions displayed. These graphical cues can all be seen in Figure 6–24.

When the angle-distance chamfer parameters are used, there is an option that allows for flipping the direction of the preview arrow. The option is appropriately named *Flip direction*. Likewise, clicking on the arrow (which indicates the direction the distance value will be applied) will also flip the chamfer. The preview will update accordingly.

Distance-Distance

When defining a chamfer using the distance-distance parameters, there is no preview arrow or option to "flip" the chamfer. Use the callout and cross-section preview to determine how the distance values will be applied.

If creating a chamfer using the distance-distance method, there will be an option for *Equal distance*. This disables the setting for the second distance parameter and automatically sets that distance to whatever the first distance happens to be.

✓ **TIP:** *Be aware that parameters can be changed via the callouts. This holds true for any command that utilizes callouts, not just for chamfers. Click on the dimension to be changed, type in the desired value, and press Enter.*

Vertex Chamfer

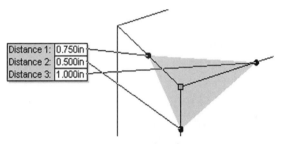

Distance 1: 0.750in
Distance 2: 0.500in
Distance 3: 1.000in

Fig. 6–25. Creating a vertex chamfer.

Vertex chamfers are a little different than their counterparts. Only vertices can be selected for this type of chamfer, and three dimension values must be specified. There will be a cross-section preview, which is critical in determining which dimension values are being applied to each of the three edges that meet at the selected vertex point. Figure 6–25 shows an example of a vertex chamfer and preview.

If this chamfer type is used, there will be an *Equal distance* option, equivalent to the *Equal distance* option in the Distance-Distance chamfer type. In the case of the vertex chamfer, equal distances are automatically assigned to all distance parameters. Vertex chamfers in general can only be created on corners with three edges, though the adjacent faces certainly do not need to be orthogonal. Geometry containing vertices where more than three edges meet is not all that common anyway.

Selection Techniques

In Chapter 2 you learned that the left mouse button was used to select objects. Now let's examine additional selection techniques that can be employed to select groups of sketch entities, model edges, and features. Some of these techniques will prove invaluable when it is necessary to select model edges for filleting. These selection techniques will also prove very useful elsewhere, so remember them well.

Select Midpoint

While sketching, it is possible to snap to a midpoint of existing edges or lines. While in certain commands, such as when creating a plane, it is possible to select midpoints directly. Occasionally, you may find it necessary to select midpoints for other reasons. If this is the case, right-clicking on a line or edge will make available the Select Midpoint option, thereby forcing SolidWorks to select the midpoint of that line or edge.

It is not possible to force SolidWorks to select a midpoint when in the Add Relation command. The reason has to do with how a midpoint relation is added. It is a matter of terminology. Consider this: It is possible to add a midpoint relation between a point and a line. It is possible to add a coincident relation between a point and a point. It is *not* possible to add a midpoint relation between a point and a point (a midpoint is considered a point).

Select Chain

A chain is nothing more than a series of sketch entities connected end to end. When right-clicking on a sketch entity, the Select Chain option is presented, allowing for the selection of all entities in the chain.

Select Loop

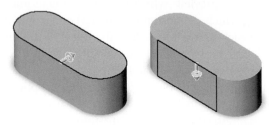

Loops are similar to chains, but typically refer to model edges as opposed to sketch entities. Loops usually, but not always, have one or more edges common to a pair of loops. To put this another way, right-clicking on a model edge allows for selecting all edges that wrap around either of the edges' adjacent faces. Examine Figure 6–26 to get a clearer picture of this.

Fig. 6–26. Using Select Loop offers two choices.

When the Select Loop option presents you with two choices, using the left mouse button click on the arrow to toggle to the other loop. Once the arrow is pointing in the proper direction, simply continue with whatever command you happen to be working with.

Select Partial Loop

What do you do if it becomes necessary to select a series of connected edges on a face, but not the entire loop? In this situation, use the *Select partial loop* technique. It is somewhat tricky the first time through, but easy once you have tried the process a couple of times.

Imagine you want to select a series of edges on a face. You know the starting edge you would like to select, and you know the final edge you want to select. Every edge between these two edges should also be selected. Keeping with this example, it is important that the edges between the first and last edges on the *right-hand* side of the model, not the left-hand, be included in the selection process. Note the darker sketch lines and arcs in Figure 6–27 to better understand this scenario.

This process works as follows. Select the first edge in the partial loop (see Figure 6–27). Next, hold down the Ctrl key and right-click (to bring up the menu) on the last edge in the partial loop. When right-clicking on this second edge, which side of the midpoint you right-click on is significant. The edges can flow around the model from one direction or the other relative to the first and last edges selected. It is the "click point" of the last edge that dictates this direction, and therefore which series of edges is selected. Select *Select partial loop* from the menu and you are ready to move on.

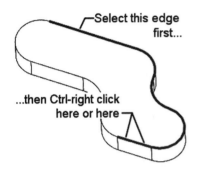

    Looking back at our original example, it was necessary to Ctrl/right-click near the right-hand side of the second edge in Figure 6–27. *Select partial loop* was then selected from the menu, and the desired edges were selected. What action is carried out on the edges at that point depends on what the user is attempting to do. In our example, Convert Entities was used to convert the edges into sketch entities.

Fig. 6–27. Using the Select partial loop *technique.*

Select Tangency

Like the Select Loop option, this option is available when right-clicking on model edges. As the name implies, all edges tangent to the original edge will be selected. Occasionally this option is necessary, even for commands such as the Fillet command. Although the Fillet command contains an automatic tangent propagation function, using Select Tangent often gives SolidWorks an extra boost in difficult situations.

Invert Selection (SW 2006 Only)

Selection sets can be inverted in SolidWorks 2006. As an example, if a face on a model is selected, inverting the selection will cause every other face on the model to be selected except for the originally selected face. To invert a selection, select one or more objects and then right-click in the work area and select Invert Selection. Objects originally selected should be of similar types (i.e., one or more sketch entities, one or more faces, and so on).

Select Other

This is one of the most useful of the various selection options. Use Select Other to select edges or faces that prove difficult or impossible to pick any other way. Also use it to pick faces that are on the other side of the model, facing away from you, or that are located deep within components of an assembly. It is often much more convenient to use Select Other than to rotate a model to get at a particular model face. How-To 6–5 takes you through the process of the Select Other technique.

How-To 6-5:   Using Select Other

When model faces are on the side of the model facing away from you, they are impossible to select using solely the left mouse button. Use the Select Other selection technique, outlined in the steps that follow, to select such faces without rotating the model. Note that shaded display mode should be used for this command to work correctly.

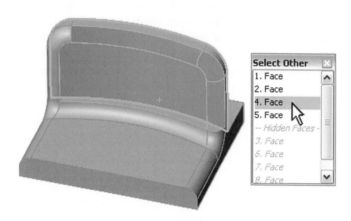

1. Right-click over the area where you imagine the face you are trying to select is located. Because the face is not visible, it is okay to guess its location.

2. Select Select Other from the menu.

3. Right-click on faces to hide them if they are in the way. Hidden faces will appear in the list box below the title *Hidden faces*, shown in Figure 6–28.

4. Left-click on the desired face either in the work area or in the list box.

Fig. 6–28. Using Select Other.

Fig. 6–29. Completed shaver housing.

All that is really happening when using Select Other is that various faces are being hidden (via the right mouse button) so you can get at the face you wish to select. It is really nothing more than that.

Now it is time for you to put to practice some of the topics you have learned so far in this chapter, along with some of the other knowledge you have gained. In Exercise 6–1 you will begin creating a shaver housing, shown in Figure 6–29. It is a fairly simple plastic part that makes up part of an electric razor. The features you will use to begin this part are features you learned previously. These features are an extrusion and a revolved feature. Afterward, you will add some fillets and a shell feature.

➡ **NOTE:** *You will begin to notice that fewer and fewer steps are spelled out for you as you continue through this book. Some things you will be expected to remember from previous chapters and/or current chapter material. The exercises become slightly more advanced without offering quite as much attention to detail. If you get stumped, return to earlier portions of the book so that you can refresh your memory on topics as necessary.*

Exercise 6-1: Creating the Shaver Housing

Begin a new part and save it as *Shaver Housing*. Start by sketching on the Front plane. You will begin the shaver housing here, by performing the following steps, and finish it in Chapter 7.

1. Create the sketch shown in Figure 6–30, with dimensions.

2. Perform a Base Extrusion.

3. Specify a Blind end condition of 3.5 inches. Do not add any draft.

4. Click on OK to complete the extrusion.

5. Change to an Isometric view.

Fig. 6–30. Complete this sketch on the Front plane.

6. Start a new sketch on the front end of the part, as shown in Figure 6–31.

7. Use Convert Entities to extract a new sketch (see Figure 6–31).

8. Create a revolved feature that revolves 90 degrees about the lower horizontal line. Flip the direction of revolution if required (using the preview as guidance).

9. Add .375-inch fillets to either side of the housing, as shown in Figure 6–32. You will need to select one edge on either side of the part.

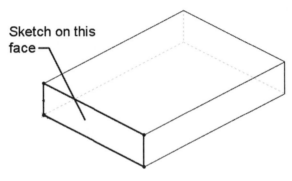

Fig. 6–31. Preparing to create a revolved feature.

Next, you will perform a shell operation. Figure 6–33 shows the correct two faces to select for removal during shelling. The model has been flipped upside down for illustration purposes. If your part does not look like that shown in the figure, try repeating the exercise and figure out where you went wrong. Otherwise, continue with the following steps.

Fig. 6–32. After adding the fillets.

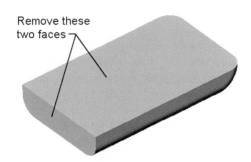

Remove these
two faces

*Fig. 6–33. Selecting faces to be
removed during the shell operation.*

10. Create a shelled part by selecting Shell from the Insert > Features menu, or by clicking on the Shell icon on the Features toolbar.

11. Select the two faces shown in Figure 6–33.

12. Type in a wall thickness of *1/16* (inch) for the shell feature. Try actually typing in the text *1/16*. You will find that it converts automatically to .0625 inch (rounded to .063 if using three-place precision).

13. Click on OK to complete the shell.

14. Save your work.

It is possible to type in mathematical operators when being asked for dimensions. This essentially works as a built-in simplified calculator. It is possible to use mathematical operators such as the plus sign (+), minus sign (–), asterisk (*), and forward slash (/). These symbols function as plus, minus, multiply, and divide operations, respectively. You can also use parentheses, the carat symbol (^) for exponents, and trigonometric functions. Table 6–1 outlines mathematical operators available when entering dimension values.

Table 6-1: Mathematical Operators

Symbol	Meaning	Example	Outcome
+	Add values	.0625 + 1.375	1.4375
-	Subtract values	.325 - .03125	.29375
*	Multiply values	3*.625	1.875
/	Divide values	1/16	.0625
^	Raise to the power of	2^3	8.000
( )	Parentheses	(1/16) + (.125*2)	.3125
sin	Sine of (value)	sin(1)	.8415
cos	Cosine of (value)	cos(1)	.5403
tan	Tangent of (value)	tan(1)	1.557
pi	Use the value of pi	pi	3.14159265

If you need to use the trigonometric functions, note that SolidWorks likes to work in radians, not degrees. If you are not used to working in radians, you can work in degrees and convert to radians. There are 2π radians in 360 degrees. Therefore, to convert to radians divide the number of degrees by 180 and then multiply that value by π (pi). (Trigonometric functions are best left to a mathematics class and are beyond the scope of this book.)

Fig. 6–34. After completing the shell feature.

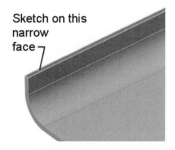

Sketch on this narrow face

Fig. 6–35. Beginning a new sketch.

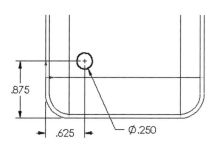

.875

.625 Ø.250

The shaver housing should now look like that shown in Figure 6–34. The next portion of this exercise will have you create a boss. This boss will create material that can be drilled through for connecting one side of the housing to the other. Later on, these features will be patterned, which you will learn how to do in Chapter 7. For now, finish Exercise 6–1 by continuing with the following steps.

15. Start a new sketch on the narrow face created on the bottom of the part during the shell operation. See Figure 6–35 for reference. Zoom in if necessary!

16. Change to a plan view (using the Normal To view option).

17. Create the sketch shown in Figure 6–36. Optionally, create a virtual sharp to dimension to.

18. Use the Up To Next end condition.

19. Create a boss extrusion.

20. Add 3 degrees of draft, and make sure draft is applied outward.

21. Click on OK to complete the operation. Figure 6–37 shows the boss added with draft, both before and after completing the extrusion.

22. Rename the boss-extrude feature *Boss1*.

23. Save your work. You will need this part later.

Fig. 6–36. Creating a sketch for the boss.

Fig. 6–37. Creating the extruded boss from a circle.

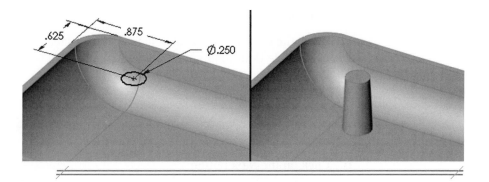

.625 .875 Ø.250

The Hole Wizard

The Hole wizard is an easy and convenient means of creating a wide variety of hole types. Holes can be counterbored or countersunk, drilled or tapered, and many combinations of these options. The Hole wizard steps you through creating a hole. The hole creation process is not difficult, but there are a lot of parameters that need to be set carefully to obtain the proper hole size.

When you create a hole using the Hole wizard, SolidWorks remembers the hole callout data. Later, when a drawing layout is created, you have the opportunity to use that callout information in the drawing. You will explore hole callouts in Chapter 12, along with other annotations.

SolidWorks has a special way of handling holes created with the Hole wizard. Specifically, two sets of sketch geometry are created. The first sketch consists of nothing more than point entities, also known as sketch points. There may be a single point, or there may be many. The point helps to locate the center of the hole via constraints or dimensions.

The second sketch is used to create the hole itself. The hole is created from a revolved feature. A sketch profile of half the hole's center cross section is automatically created and then revolved about a centerline, with the centerline passing through the locating point. If there is more than one point added by the user, SolidWorks creates a pattern. Most of this is transparent to the person creating the hole. All you do is fill in the desired dimensions and parameters in the Hole wizard window and specify some locations, and SolidWorks does the rest. To create a hole using the Hole wizard, perform the steps outlined in How-To 6–6.

How-To 6-6:   Creating a Hole with the Hole Wizard

To create a hole with the Hole wizard, perform the following steps.

1. Select a face on the part on which to place the hole. You should not be in a sketch when you do this.

2. Select Insert > Features > Hole > Wizard, or click on the Hole wizard icon found on the Features toolbar.

3. Select the tab for the type of hole you want to create. The upper portion of the Hole Definition window is shown in Figure 6–38.

4. Specify the parameters to define the size of the hole. You may have to scroll through the list.

5. Make sure to adjust the end condition parameter as required; for example, using Through All, Blind, and so on.

6. Click on the Next button.

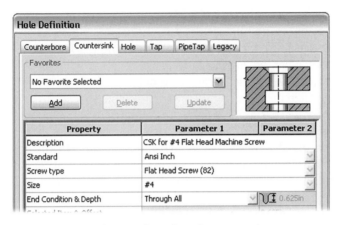

Fig. 6–38. The Hole wizard's Hole Definition window.

At this stage you will be asked to create points. This often confuses those new to the Hole wizard. Although in the midst of the Hole wizard command, you will essentially be in sketch mode with the Point command active. Be aware that you do not have to stay in the Point command. Your choices are to add points where copies of the hole will go or to add dimensions or relations to constrain the points already there. Actually, both of these tasks should be performed, but if you only want the one point (and subsequently the one hole) simply dimension or constrain the single point to the proper location.

7. Add points wherever copies of the hole should be located, or go to step 8 if only a single hole is needed.

8. Click on the Smart Dimension and/or Add Relation icon to locate any points you created.

9. Click on Finish.

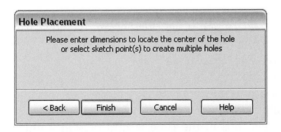

Fig. 6–39. Hole Placement message.

A confusing aspect of the Hole wizard is the Hole Placement message window (shown in Figure 6–39) that appears immediately after clicking on the Next button (step 6). The message states that you should "…select sketch point(s) to create multiple holes." This is very misleading because it implies that the sketch points already exist. In reality, the sketch points should *not* already exist. Rather, you should create them while in the Hole wizard. The message would read much more accurately if it read "…create additional sketch points to create multiple holes."

A common mistake when using the Hole wizard is to create a separate sketch containing sketch points prior to using the Hole wizard. While in the Hole wizard command, those sketch points are then selected to represent the hole positions. This places points on top of points, which is redundant and bad practice.

Fig. 6–40. Hole wizard
PropertyManager.

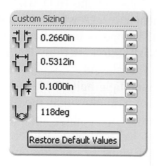

Fig. 6–41. Custom
Sizing panel.

If you expand the Hole wizard feature in FeatureManager (click on the plus sign to the left of the feature's icon), you will see the two sketches used to define the hole(s). Either sketch can be edited. The sketch containing the locating points (the first sketch) can have points added or removed from it, thereby changing the number of holes.

SW 2006: The Hole wizard interface received an overhaul in SolidWorks 2006 and was moved to the PropertyManager. The interface is much more pleasing to the eye and is easier to understand where data must be input for the desired hole parameters. The upper portion of the Hole wizard PropertyManager is shown in Figure 6–40.

The functionality of the Hole wizard remains unchanged for the most part. After supplying parameters for the desired hole, click on the Positions tab (visible in Figure 6–40) and add points as described in step 7 of How-To 6–6.

If values are changed in the Custom Sizing panel (shown in Figure 6–41), the background of that value is changed to the color yellow. This informs the user that a custom value is in use for that parameter (which is behavior that also occurs in SolidWorks 2005). If custom settings are used, a button titled Restore Default Values will appear, which makes it very easy to remove any custom values previously set.

Legacy Tab

Just what is the Legacy tab in the Hole wizard window? It is something you will probably want to avoid unless editing older SolidWorks models. The Legacy tab is what the old Hole wizard at one time looked like. It remains in the new Hole wizard for those times when an older model needs editing. For anything created in more current versions of the software, the other tabs in the Hole wizard present a much wider array of options.

Holes on Nonplanar Faces

The Hole wizard will indeed allow for placing holes on nonplanar faces. Faces can be cylindrical, conical, spherical, or any type of free-form shape. SolidWorks handles this by creating the locating points for the holes in what is known as a 3D sketch, explored in Chapter 10. An example of holes created on a nonplanar face is shown in Figure 6–42. A partial cutaway view is used for clarity.

Fig. 6–42. Holes on a nonplanar face.

What is important to understand from a user stand-point is the order in which a face is selected and the Hole wizard command initiated. If a planar face is selected prior to the command being initiated, a 2D sketch will be used for the hole-locating points. It is much easier to locate points in a 2D sketch than in a 3D sketch, so there is no sense in allowing SolidWorks to use 3D sketch points if it is not necessary. If the command is started prior to picking a face on which to place the hole (or holes), a 3D sketch will be used for the locating points. *This is true even if the selected face is planar!*

✓ **TIP:** *Although it is not required, you should always select a face prior to starting the Hole wizard. This allows SolidWorks to create the proper locating sketch (2D or 3D) according to the face selected.*

Adding Favorites

Each tab, with the exception of the Legacy tab, has a drop-down list called Favorites. This Favorites list is used to save Hole wizard parameter settings for hole types you use frequently. You can add as many favorites as you wish. Each tab will have its own Favorites listing, so do not worry that they will get mixed up between tabs or hole types.

Fig. 6–43. Naming favorite hole parameters in the Hole wizard.

To add a favorite hole size, simply click on the Add button. All of the parameters currently set for the hole will be saved in the Favorites list. You will be asked what to name your favorite, and you can accept the default name, add to it, or type in something completely different. This makes recalling your preferred settings for particular hole types very easy. Figure 6–43 shows an example of what you might see after clicking on the Add button.

The Update and Delete buttons are fairly self-explanatory. Clicking on the Delete button will delete the currently displayed favorite from the drop-down list. The Update button will only be active if a change to one of your favorites is made. You can then decide on whether or not to update the current settings so that the favorite will always reflect those changes.

Simple Holes

When simple cylindrical holes are needed, you basically have three options. One option is to use the Hole wizard to create a simple hole by specifying the Hole tab and plugging in the desired parameters. One of the benefits of using the Hole wizard for such basic holes is that the parameters for numeric and lettered hole sizes are already spelled out. This includes, from smallest to largest, #97 to #1 (followed by A-to-Z hole sizes).

Other benefits of using the Hole wizard for simple holes are the ability to enter near- or far-side countersink diameters and angles, to enter drill point angles (assuming blind end conditions), and to add multiple holes simultaneously. The ability to add holes to nonplanar surfaces is also a benefit.

A second method of creating holes is to simply sketch a circle and cut it through the model. Yet a third option is to use the Simple Hole command, found in the same menu that contains the Hole wizard. This is not to be confused with the Hole tab mentioned previously. The interface for the Simple Hole command is much different than the Hole wizard and very bare bones. How-To 6–7 takes you through the process of creating a simple hole using this method.

HOW-TO 6-7:   Using the Simple Hole Command

Fig. 6–44. Creating a simple hole.

To create a hole using the Simple Hole command, perform the following steps.

1. Select Simple from the Insert > Features > Hole menu, or click on the Simple Hole icon found on the Features toolbar.

2. Select a planar face on which to place the hole (these first two steps can be reversed).

3. Enter values for end condition, diameter, and so on (see Figure 6–44).

4. Click on OK to create the hole.

As you may have noticed, the Hole panel (shown in Figure 6–44) is essentially a stripped-down version of the panel displayed when creating an extruded cut. You can specify an end condition, add draft, and do nearly anything you would normally be able to do when performing a basic extrusion. The only additional option is for the diameter of the hole.

On the downside, the user gets little opportunity to tell SolidWorks just where it is the hole should be located. None of the sketch tools are available, and the opportunity to add points or position the hole when using the Hole wizard is missing in action when creating a simple hole.

It is possible to drag and drop the center of the sketched circle to an approximate location. It is also possible to wake up a centerpoint (see Chapter 4) while dragging and dropping the circle's centerpoint. In most cases, though, you must edit the sketch automatically created by SolidWorks when using the Simple Hole command in order to position the hole. Some may decide that using the Simple Hole command does not really have as many benefits as one might initially think, and may very well decide to simply sketch a circle and cut it through the model the old-fashioned way.

Split Lines

One command frequently associated with molded parts is the Split Line command. The Split Line command is often used for creating parting lines. As a matter of fact, the Split Line command might have been called the "Parting Line" command. Technically, though, a split line physically splits a face or set of faces, which is why it carries the name it does.

In a case where a parting line is being created, a sketch (often just a simple line) is projected toward the faces to be split. Any sketch profile can be used to split a face, even a closed profile. The Split Line command is used to prepare for creating a number of other feature types, many of which are explored in this book. In this chapter, we will use it to prepare for creating draft.

Adding draft using a parting line is discussed in material to follow. To illustrate what the Split Line command can do for you, let's take a look at the early design stages of a molded plastic part (shown in Figure 6–45).

Figure 6–45 shows the part in an isometric view. It also shows the side of the part. The dark line represents where the design specifications require the parting line to be. The parting line is represented by a sketch and is dimensioned, typical of any sketch. One aspect of this sketch worth mentioning is that its endpoints have been constrained to the silhouette edges of the model. This has been done in case the part changes size. If the part does change size, the parting line will shrink or grow with the model.

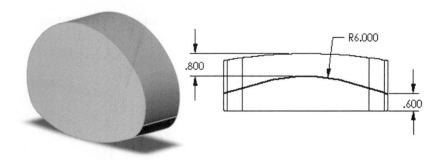

Fig. 6–45. Molded part in need of a parting line.

Next, the Split Line command will be used to project the sketch onto the faces containing the parting line. The steps involved in this process are outlined in How-To 6–8. The model will be assessed after How-To 6–8 to see how it fared. You will also discover how to add draft to the same model using the newly created parting line. The steps outlined in How-To 6–8 can also be used as a precursor to many other feature types in SolidWorks, not just parting lines. The Split Line command is extremely versatile, especially with advanced fillets.

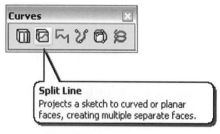

HOW-TO 6-8: Creating a Split Line

To utilize the Split Line command, perform the following steps. The Split Line command is found on the Curves toolbar, shown in Figure 6–46.

1. Create a sketch that will be used to define the split line. When projected perpendicular to the face or faces to be split, the sketch must extend up to the edges of the faces to be split (or beyond) but must not fall short.

2. Select Split Line from the Insert > Curve menu, or click on the Split Line icon. The Split Line PropertyManager will appear, which is shown in Figure 6–47.

3. Select Projection as the method for defining the split line. (Silhouette and Intersection are discussed in material to follow.)

4. Click in the Faces to Split list box and select the faces to be split.

5. Click on OK to complete the process and create the split line.

Fig. 6–46. Split Line command and Curves toolbar.

Fig. 6–47. Split Line PropertyManager.

Figure 6–48 shows a partially shaded view (for clarity's sake) of the faceplate after the split line has been created. It wraps all the way around the part, and for all intents and purposes can be considered the parting line. This example shows that the parting line does not have to be straight, if that is what your design requirements call for.

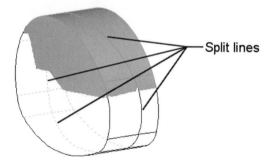

Fig. 6–48. After creating the split line.

Split lines

When working through the Split Line command, it is not necessary to select a sketch to project, as long as the sketch was being edited when the Split

Line command was initiated. Otherwise, a sketch must be selected. The *Single direction* option is useful if it is not necessary to project the sketch in both directions at once. There is no sense in making SolidWorks work any harder than it has to. As a general rule, use *Single direction* if feasible, as it is more efficient. Obviously, the *Reverse direction* option can be used to flip which direction the sketch is projected in if *Single direction* is enabled.

Silhouette Method

The silhouette split type mentioned earlier is used for creating split lines on parts without the need for a sketch. Instead of projecting a sketch, the user must supply a pull direction. The pull direction is the direction the mold would be pulled away from the part. This can be supplied by selecting an edge, an axis, two vertex points, or a planar face. In the case of the face, the pull direction will be perpendicular to the face.

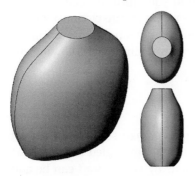

The silhouette method of adding a split line is very convenient and can save a lot of time. However, it is only meant to be used on particular parts. The faceplate would not have been a good candidate for the silhouette method because all of the faces that would have required the parting line were completely perpendicular to the mold's pull direction. Figure 6–49 shows a model that would be a good candidate. It has a free-form body with flat faces at the top of the neck and at the base only.

In the case of the example shown in Figure 6–49, the Front plane was used to define the pull direction, and the surface of the model was selected on which to define the parting line. Note that the parting line flows perfectly up either side of the model. The Silhouette option, in essence, extracts the silhouette edges from the model relative to the planar face you select.

Fig. 6–49. Split line created with the silhouette method.

Intersection Method

Creating a split line using the intersection method requires surfaces or faces. Specifically, a surface or face is selected that will be used to split some other surface or face. Unlike using the projection method, wherein a sketch must completely extend across the face being split, surfaces do not need to meet this requirement. See Figure 6–50 for an example.

The example shown in Figure 6–50 uses a surface to split the top face of the shaver housing. Note that the arced surface does not fully extend across the face being split in any of the examples shown. Rather, the surface is extended automatically in order to split the selected face. (In Figure 6–50, the resultant faces created by the Split Line command have been given different shades for the sake of clarity.)

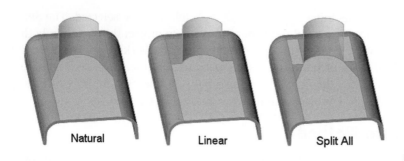

Fig. 6–50. Different outcomes of the intersection method.

When the intersection method is used, SolidWorks can extend the selected surface in either a natural or linear fashion. Options appropriately named Natural and Linear (not shown) will be available in the Split Line PropertyManager, along with a checkbox option labeled *Split all*. Depending on the original surface geometry being used for the split, the Natural and Linear options may behave differently than the example shown. Simply put, experiment with the options until the result is suitable to you. Turn on *Split all* if you find the need to break up a face into multiple pieces (a rare circumstance).

Now that you know how to add a parting line to a part, you can take it one step further and apply draft using that parting line. The following section shows you how to do just that.

Adding Draft

Draft is what often gets incorporated into a molded part so that the part can come out of the mold. Perimeter faces of a model are angled inward, starting at the parting line. There are three methods by which draft can be added to a model. One method is to incorporate the draft directly into a sketch. This first method was not explicitly laid out for you, but you should have been able to interpolate this process from material previously covered. Figure 6–51 serves as an example.

It can be observed in Figure 6–51 that the draft in the extruded feature appears on either side face but is missing from the front (and presumably the back) face. A second method of adding draft is through the extrusion process itself. Figure 6–52 shows the Draft option in the Extrude PropertyManager. Also shown is an example of extruding a simple rectangle using the Mid Plane end condition with draft enabled.

Adding draft while extruding works fine if the parting line is planar. Unfortunately, parting lines are often much more involved. It often becomes necessary to add draft to specific faces, to parts with curved or stepped parting lines, and so on. The Draft command allows for these types of scenarios. In addition, adding draft as a feature is not limited to using only parting lines. Draft can be specified using a neutral plane as well. First, however, let's examine how to add draft with a parting line.

Fig. 6–51. Incorporatin g draft into a sketch.

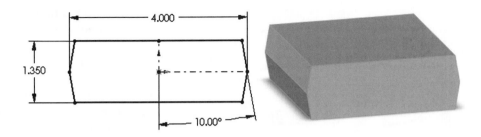

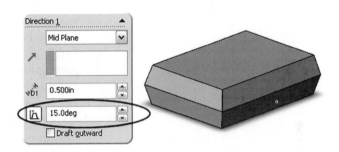
Fig. 6–52. Adding draft while extruding.

How-To 6–9 takes you through the process of adding draft as a feature. The draft will be applied directly to the model, eliminating the need for a sketch. A split line is not required to create parting line draft, but we will employ one in the following How-To. The illustrations that follow help guide you in the proper direction, and aid you in selecting the correct geometry.

HOW-TO 6-9: Adding Draft Using a Parting Line

To add draft as a feature using the parting line method, perform the following steps.

1. Select Draft from the Insert > Features menu, or click on the Draft icon found on the Features toolbar.

2. Select the Type of draft (in this case, Parting Line). Figure 6–53 shows the Draft PropertyManager.

3. Specify the Draft angle.

4. Click in the Direction of Pull list box and select a linear edge or planar face to indicate the Direction of Pull. This will be the direction the mold would be pulled away from the model. If a planar face is selected, the direction of pull will be perpendicular to the face.

5. Click on the *Reverse direction* icon to flip the direction of pull if necessary.

Fig. 6–53. Draft PropertyManager.

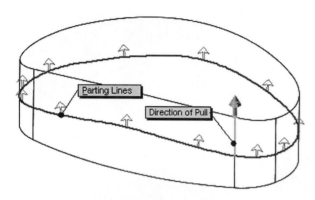

Fig. 6–54. In the process of adding draft.

There are a number of things going on at this point you should be aware of. The direction of pull will be represented with a callout and a large arrow (see Figure 6–54). Clicking on this arrow is enough to flip the direction of pull. The next step is to select the parting lines. There will almost certainly be more than just a couple, and you will need to select them all. Use any of the various selection methods at your disposal (such as Select Tangency). Once the parting lines are selected (step 6), many more arrows will appear. These smaller arrows point toward the faces being drafted. This can be reversed, but typically you will want all of the smaller arrows pointing in the same direction.

6.   Click in the Parting Line list box and select the parting lines.

7.   Click on OK to complete the draft.

Fig. 6–55. Model with draft applied to both sides of the parting line.

Figure 6–55 shows what the faceplate looks like after the draft has been added to both sides of the parting line. The drafting process was run through twice to achieve this effect. Because the direction of pull is different for each side of the parting line, all perimeter faces cannot be drafted at the same time.

If for whatever reason you wish to reverse one of the smaller arrows, which point to the face being drafted on either side of the parting line, select the appropriate edge from the Parting Line list box and click on the Other Face button. You will rarely want to do this, however, because it will often cause the draft to fail due to geometric conditions. Use discretion with this option.

Neutral Plane Draft

Whether you decide to add draft using a parting line or a neutral plane is really up to you, but this also depends on the model on which draft is being applied. Obviously, if the Parting Line method is used the part must contain a parting line. That is not the case if the Neutral Plane method is used.

What exactly is a neutral plane? The neutral plane dictates where the draft will be measured from. The neutral plane does not necessarily have to be adjacent to any of the faces being drafted. However, it may make face selection easier (see the section "Face Propagation Options" in material to follow). How-To 6–10 takes you through the process of adding draft using a neutral plane.

HOW-TO 6-10: Adding Draft Using a Neutral Plane

The following steps take you through the process of adding draft using a neutral plane (as shown in Figure 6–56). The process is similar to adding draft using a parting line, but with fewer necessary parameters. Therefore, it is a bit more straightforward.

Fig. 6–56. Adding neutral plane draft.

1. Select Draft from the Insert > Features menu, or click on the Draft icon.

2. Select Neutral Plane as the *Type of draft*.

3. Specify the Draft angle.

4. Click in the Neutral Plane list box and select a plane or planar face to indicate the neutral plane. This will determine where the draft angle is measured from.

5. Click on the Reverse Direction button (if necessary) to flip which side of the neutral plane the mold will be pulled away from. The preview arrow will indicate the direction of pull.

6. Click in the Faces to Draft list box and select the faces to which draft should be applied.

7. Click on OK to add the draft.

As is always the case, bear in mind which list box is highlighted (typically a salmon color), as this indicates which list box is currently active. This is important when working with commands such as Draft, which contain more than one area to supply information for.

Step Draft

Step draft is used in a situation in which the parting line does not consist of a simple line or tangent entities. Instead, the parting line consists of angled line segments. This is more easily described visually than verbally. Take the simple example shown in Figure 6–57. Note the stepped parting line created on the top face of the part. (This example will use a split line projected only onto one face, to clearly illustrate the effects of step draft rather than confusing the issue and wrapping the parting line around the entire part.)

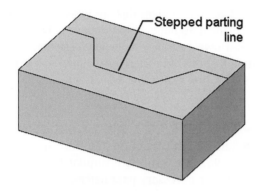

Fig. 6–57. A stepped parting line on the model prior to adding draft.

Next, let's add parting line draft to the model and see what results are achieved. If draft were added to this part without using the Step Draft option, the results would be as shown in Figure 6–58. SolidWorks creates oblique faces in order to accommodate the stepped parting line. A large draft value of 25 degrees was used to exaggerate the effect. The back face (behind the model in the image) was used to define the direction of pull.

When Step Draft is used, the results vary dramatically. When using Step Draft, the direction of pull must be specified in a distinct way. Specifically, a plane or planar face must dictate the direction of pull, whereas a model edge will not suffice. Direction of pull will be perpendicular to the plane, as usual. In Figure 6–59, Step Draft was used, and once again the back face of the model was used to define the direction of pull.

There is one last option that should be mentioned. An option for setting the step-draft "steps" to either tapered or perpendicular will change the appearance of the draft. It should be noted that the step draft depicted in Figure 6–59 is using the *Tapered steps* option. Figure 6–60 shows use of the *Perpendicular steps* option.

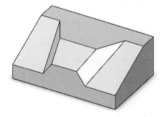

Fig. 6–58. Model with parting line draft.

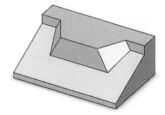

Fig. 6–59. Model with step draft.

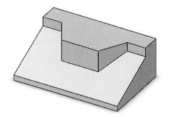

Fig. 6–60. Step draft with perpendicular steps.

Face Propagation Options

Face Propagation is no more than a convenient option that makes selecting parting lines or faces to be drafted a little easier. For example, if you are adding draft to a part that has perimeter faces that are all tangent, selecting *Along tangent* from the drop-down list will enable you to only have to select one parting line segment or one face, depending on the type of draft being created. SolidWorks will select all other segments of the parting line for you, or all perimeter tangent faces.

✓ **TIP:** *The* Along tangent *option is much easier to use with parting line draft. If you attempt to use this with neutral plane draft, the selected face must share a common edge with other faces to be drafted, and the edge must reside on a neutral or base plane. Results can be unexpected.*

When using the Neutral Plane method of defining draft, there are other options added to the Face Propagation drop-down list in addition to *Along tangent*. Table 6–2 outlines the various options associated with the Face Propagation feature. Figure 6–61 shows examples of some of the face propagation options. In all cases, the neutral plane is the face on the first "step."

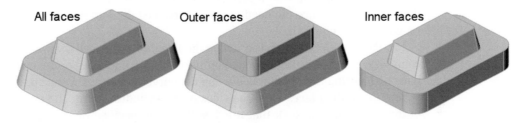

Fig. 6–61. Examples of face propagation options.

Table 6-2: Face Propagation Options

Face Propagation Option	Description
None	This setting is the default. No face propagation will be used for the selection process. Faces must be individually selected.
Along tangent	All edges tangent to the selected parting line segment will be drafted.
All faces	All faces extruded from the neutral plane will be drafted.
Inner faces	All faces extruded from within the boundaries of the neutral plane will be drafted.
Outer faces	All faces extruded from the perimeter of the neutral plane will be drafted.

Using any of the last three options in Table 6–2 means that no face selection is necessary. You do not need any faces listed in the Faces to Draft list box. The last three options in the table are only available when using neutral plane draft.

Parting Line Reduced Angle Option

The *Allow reduced angle* option is only available when creating draft using a parting line. It is not an option you will need to use on a regular basis. To understand exactly what this option does, it is best to start with an image. Let's imagine that a molded part had a protrusion that was cylindrical in nature. There are faces at the end of the cylinder that should be drafted. Such is the case in Figure 6–62.

It would be quite easy to add draft to the two end faces found at the end of the part. This is exactly what has happened in the image on the right in Figure 6–62. The outcome is exactly the same whether using the neutral plane or parting line method of defining the draft.

Just to make things interesting, imagine that it is necessary to define the draft from the curved edges of the flat faces, rather than the linear edges pointed out as the parting lines in Figure 6–62. It might be difficult to even imagine what the end result would look like. The resultant face could not possibly be planar. It would also not be possible to maintain a consistent draft angle, and hence the need to allow a reduced draft angle when geometric conditions require it.

In Figure 6–63, we can see where the need for a reduced draft angle becomes a requirement. In the image on the left, both end faces of the cylinder have had parting line draft added using the curved edges as the parting lines. Where the drafted faces meet the sides of the cylindrical part, the curvature of the drafted face is such that it is less than the specified draft angle. If the *Allow reduced angle* option had been turned off while creating the draft, the feature would not have been created and an error message would result.

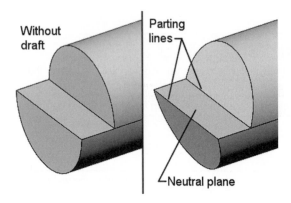

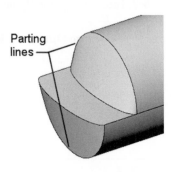

Fig. 6–62. Drafting faces at the end of a cylinder.

Fig. 6–63. Using the curved edges as parting lines.

Rib Tool

The Rib tool allows you to add ribs to a model with minimal effort. A sketch is required, but the sketch can be very simple in nature. This is the major benefit of the Rib tool. It allows for quickly and easily placing ribs in a model.

If a simple line is used as an example, we can examine what the Rib tool will accomplish. The length of the line is for the most part irrelevant. Solid-Works will extend the line to the next face it encounters. Good practice would be to sketch the line within the boundaries of where the rib should be positioned. Second, the line is "thickened," in a manner similar to that of a thin feature. This wall thickness is determined by the user when defining the rib. Third, the rib is extruded up to the first body it encounters. Let's look at an example.

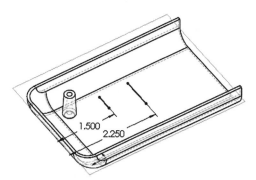

Fig. 6–64. Preparing a sketch for a rib.

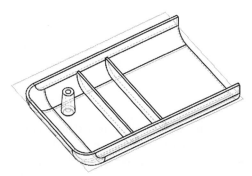

Fig. 6–65. After creating the ribs.

In Figure 6–64, a plane was created that is situated .200 inch below the top edge of the shaver cover. The two lines were then sketched on this plane. Note that the length common to the two lines has not been established. It does not need to be. This is the beauty of the Rib command. Plug in just enough information to establish your design intent, and then let SolidWorks do the rest.

In any other situation in which an attempt were made to extrude two open profiles, it would result in errors. However, with the Rib command the results are two new ribs added to the model. Draft can even be added during the creation process. Using the sketch shown in Figure 6–64, the end result might look like that shown in Figure 6–65.

The Rib command itself is fairly straightforward. How-To 6–11 takes you through the process of creating ribs with the Rib command. It is assumed that an appropriate sketch has already been created. What constitutes an appropriate sketch can be just about anything. This includes closed or open profiles or a combination of the two. Splines can be used as well. The important thing to remember is this: any open profile's projection must be bounded by a face of the model. In other words, if a line (for example) is extended the line must run into a wall.

How-To 6-11: Using the Rib Command

To create a rib using the Rib command, perform the following steps.

1. Click on the Rib icon found on the Features toolbar, or select Rib from the Insert > Features menu.

2. In the Rib PropertyManager, a portion of which is shown in Figure 6–66, enter a value for the thickness of the rib.

3. Establish to which side of the model the rib wall should be applied with respect to the sketch geometry. This will be one side or the other, or mid-plane (both sides), which is the default.

4. Using the *Extrusion direction* buttons and *Flip material side* option, specify a direction for the rib's extrusion. Use the preview arrow as a guide.

5. Add draft if required.

6. Click on OK to create the rib feature.

*Fig. 6–66. Rib
PropertyManager.*

Linear Versus Natural

An interesting aspect of the Rib command is its ability to extrude either parallel or perpendicular to the original sketch plane. This makes the command very flexible. Another interesting option gives you the ability to control how certain objects, such as arcs, are projected to the walls of the model. These are, specifically, the Linear and Natural options located at the bottom of the Rib panel.

An arc can be extended by projecting tangent lines from either endpoint (linear), or extended naturally by projecting the ends of the arc using the radius of the original arc. This is true of parabolic and elliptical arcs as well. The mathematical definition of the original parabolic or elliptical arc will be used to elongate the arc until it runs into a wall of the part. In the case of circular or elliptical arcs, the original arc segment will form a circle or ellipse if no boundary face is encountered.

➦ **NOTE:** *Splines are not affected by the Linear or Natural option.*

Figure 6–67 shows a simple sketched arc (left-hand image) that will be used for creating a rib. The endpoints of the arc have been constrained to control their locations, as you would want them to be if the Linear option were used. The reason for this will become clear in a moment.

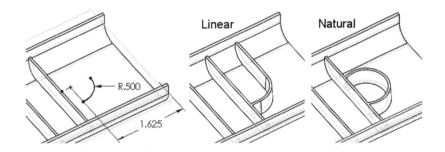

Fig. 6–67.
Linear versus
Natural options
used on the
same rib.

The center image shows the arc extending linearly until it connects with the walls of the model. The example on the right shows the same arc, only this time it has been extended naturally. The arc continues on its original path until it encounters either itself or a wall or walls of the model.

Obviously, if the arc is extended linearly it will make a big difference as to the included angle and positioning of the arc's endpoints. Care must be taken to add dimensions or relations to the arc. Otherwise, the arc will extend in what will probably be an undesirable direction.

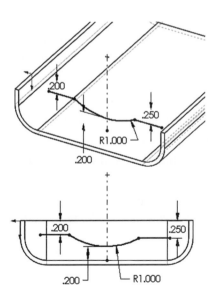

Fig. 6–68. Where would the draft be measured from?

Draft Reference for Ribs

Not all ribs are created equal. Ribs are not all linear on top. Some are stepped, have cutouts, or some other design characteristic. Take, for example, the sketch shown in Figure 6–68. Note the arc in the center, and the fact that the lines on either side of the arc are at different heights. Under such conditions, how would draft on the rib be defined, assuming it were added to the resultant rib?

The answer to the previous question is that you have a choice. When a sketch for a rib has multiple segments, the user can choose where the draft is measured (referenced) from. This is done via a button titled Next Reference. The preview arrow used to distinguish the direction of the extrusion also distinguishes what segment the draft will reference. Clicking on the Next Reference button serves to move the arrow from one sketch entity to the next.

Because the thickness of the rib to be created in this example is going to be .050 inch at the uppermost line segment, the draft should reference that specific line. Figure 6–69 shows the preview prior to creating the rib.

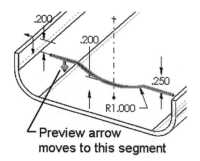

Fig. 6–69. Referencing the
upper line segment for draft.

Draft Analysis

An important tool that goes hand in hand with molded parts is the ability to analyze the draft of a model. The Draft Analysis tool, found in the Tools menu, allows for checking if faces of the model exhibit positive or negative draft, among other things. Draft analysis does not take much effort from a user standpoint. In a nutshell, select a face to indicate the direction of pull, enter a draft value, and click on a button. How-To 6–12 takes you through this process.

HOW-TO 6-12: Analyzing Draft

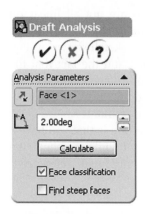

Fig. 6–70. Draft Analysis
PropertyManager.

To analyze the draft of a model, perform the following steps. Note that the model should be displayed in shaded mode in order to use this function.

1. Select Draft Analysis from the Tools menu. The Draft Analysis panel will appear, a portion of which is shown in Figure 6–70.

2. Select a face or plane that will determine the direction of pull. Direction of pull refers to the direction the mold will be pulled away from the model.

3. Click on the Reverse Direction icon if necessary to reverse the direction of pull. Use the preview arrow as a guide, which will be visible in the work area.

4. Enter a value for the draft angle.

5. Click on the Calculate button.

Once the Calculate button is selected, the model will be displayed using the colors designated in the bottom half of the Draft Analysis panel (shown in Figure 6–71). If the *Face classification* option is *not* used, the body of the model is checked for positive and negative draft, regardless of faces. Colors representing draft angle are displayed as a gradient over the entire model. In contrast to this, let's assume the *Face classification* option *is* used. With faces being classified as having positive or negative draft, entire faces are either one color or another. There are no color gradations within individual faces.

With *Face classification* turned off, an option for using a gradual color transition becomes available. This option uses the entire visible spectrum of colors, and is the most accurate way of checking the amount of draft over the entire model. It is also more than most people require. Once the analysis calculation is complete, moving the cursor over the part will result in the draft for that location being displayed on the cursor. This "cursor readout" functionality will work whether or not the gradual color transition is used.

Fig. 6–71. Color settings for a draft analysis.

Face Classification

The *Face classification* option does have some benefits. First, the graphics usually appear cleaner. That is, the draft analysis result is a little easier on the eyes. Second, the model can be saved with the colors used to display the draft analysis. This could be useful if extensive rework is needed.

There are some terms that might not be familiar to all readers with regard to how SolidWorks classifies certain faces. For this reason, Table 6–3 is provided, which outlines face classification designations. Note that the "straddle" and "steep" face designations are used only when the *Face classification* option is enabled. These designations are not otherwise necessary. The term *neutral plane* can be thought of as where the draft is measured from. It is typically where the mold would split. The term *draft specification* refers to the value typed in by the user in the Draft Analysis panel.

Face Classification	Description
Positive draft	Faces have draft equal to or greater than the current draft specification on the side of the neutral plane that signifies the current direction of pull.
Requires draft	Faces on either side of the neutral plane that have less draft than required by the current draft specification.
Negative draft	Faces have draft equal to or greater than the current draft specification on the side of the neutral plane opposite the current direction of pull.
Straddle faces	Faces that have both positive and negative draft. Only curved faces will exhibit this condition, as planar faces will fall into either one category or the other.
Positive steep faces	Faces that have both positive draft and areas that will require draft. Only curved faces can exhibit this condition.
Negative steep faces	Faces that have both negative draft and areas that will require draft. Only curved faces can exhibit this condition.

When *Face classification* is used and you are done with the draft analysis, clicking on the OK button will result in a message being displayed. This message asks "Do you want to keep face colors?" Clicking on OK will change the colors of all faces on the model.

✓ **TIP:** *If you accept face colors when exiting the Draft Analysis command and later want to change face colors back to their original color, change to Hidden Lines Visible display mode and then window-select the entire model while in the Edit Color command (discussed in Chapter 8). This will select every face on the model, and the face colors can then be removed.*

Draft Analysis Color Settings

The Color Settings section of the Draft Analysis panel, a portion of which is shown in Figure 6–71, allows for changing the colors used to indicate the various face classifications. Feel free to change these colors as you see fit, but understand that any color alterations made will be saved as system settings and not saved with the part. Color changes will take effect every time you use the Draft Analysis command. There is no easy way to reset them to their default values.

The small light bulb buttons will hide all faces with that particular face classification. This is good for seeing into a part to further examine other interior or difficult-to-see faces. The number present on the swatch of color is the number of faces that exhibit that face classification. For example, a number 3 in the yellow swatch indicates that there are three faces that require draft.

In Exercise 6–2, the *Shaver Housing* project is modified and a number of features are added. These features include a hole created with the Hole wizard and some rib features. In Chapter 7, you will pattern some of the features created in this exercise.

Exercise 6-2: Adding Features to the Shaver Housing

Open the *Shaver Housing* model started earlier in this chapter. You should have *Boss1* created, per Exercise 6–1. Next, you will need to add a hole to *Boss1* so that a screw can be inserted through the hole and connected to the other half of the housing. Note that this is an exercise, not a How-To description. Steps you should be expected to know at this point in the book will not be spelled out in detail (though hints may be given from time to time).

To create the countersunk hole, you will use the Hole wizard. This is the most economical method of creating a hole of this type. Remember that it is best to first select a face on which to place the hole. This will ensure that the Hole wizard creates a 2D sketch for the locating points. That is where you will begin. The face you should select is depicted in Figure 6–72. To create the hole through the boss, perform the following steps.

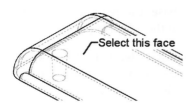

Fig. 6–72. Selecting a face for the new hole.

Fig. 6–73. After adding the countersunk hole.

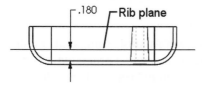

Fig. 6–74. Creating the rib plane.

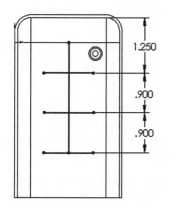

Fig. 6–75. Sketch for the ribs.

1. Select a face on which to place the hole, as shown in Figure 6–72.

2. Access the Hole wizard command.

3. Select the Countersink tab, and then plug in the following parameters.

   • ANSI inch, flat head (82), size #4

   • End Condition & Depth = Through All

   • Hole Fit & Diameter = Normal

   • Head Clearance & Type = Added C-bore at .020 inch

4. Leave all other parameters at their default values. Click on Next.

5. Click on the Add Relations icon, and then add a concentric relation between the locating point and the end of *Boss1*. This will center the hole properly. (Hint: change to Hidden Lines Visible display mode to make the selection process easier.)

6. Click on Finish to create the hole.

Figure 6–73 shows the newly created countersunk hole as seen from two different vantage points. Next, you will add some ribs to strengthen the model. If you added ribs by following along with the How-To sections earlier, delete them now to make room for the new ribs.

7. Create a new plane that is .180 inch above the inside face of the shaver housing, as shown in Figure 6–74. Rename this new plane *Rib Plane*.

8. Create the sketch shown in Figure 6–75. This sketch will be used to create the ribs.

9. Using the Rib command, create ribs with a wall thickness of .050 inch using the *Both sides* option (which is the middle icon) and 3 degrees of draft. The extrusion direction should be obvious, as there is only one possible choice. The final result is shown in Figure 6–76.

Performing a draft analysis will show that the part requires draft if the entire part is to have a draft of 3 degrees minimum. This last portion of the exercise involves performing a draft analysis, and then editing the base feature sketch to change the draft on the sides of the part. Continue with the following steps.

Fig. 6–76. Completed ribs.

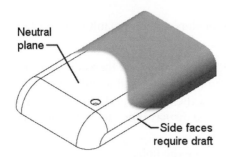

Fig. 6–77. Showing an Isometric view (with additional information).

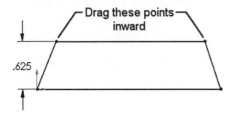

Fig. 6–78. Modifying the base feature sketch.

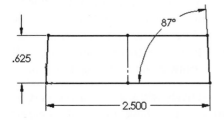

Fig. 6–79. Adding dimensions after dragging the endpoints.

10. Change to an Isometric view. If you have been following along correctly, the view on your screen should look like that shown in Figure 6–77.

11. Select Draft Analysis from the Tools menu.

12. Using the neutral plane shown in Figure 6–77, analyze the draft using a 3-degree draft angle specification. Use the *Face classification* option.

At this point you should be able to determine faces of the model that do not meet the current design requirements of 3 degrees of draft. It is necessary to alter the part in some way to establish 3 degrees of draft on the sides of the model. This can be accomplished by editing the base sketch of the part. Do not be afraid of altering the base sketch, as long as you do not change anything in a drastic way.

13. Close out of the Draft Analysis command and do not keep the face colors when prompted. (You will only be prompted to keep the face colors if you click on OK. It is acceptable to click on the Cancel button.)

14. Edit the sketch for the base feature and change to a Front view.

15. Remove the vertical relationships on each of the lines on the ends of the rectangle.

16. Remove the horizontal 2.500-inch dimension, as it may control the sketch in a way that is unsatisfactory.

17. Drag the endpoints of the top line inward (see Figure 6–78). This must be done prior to adding an angular dimension, as it is not possible to add an angular dimension between parallel lines.

18. Add a centerline between the midpoints of each horizontal line, as shown in Figure 6–79. Make sure the centerline is constrained to be vertical.

19. Add an 87-degree dimension to control the draft (shown in Figure 6–79).

20. Add the 2.500-inch dimension to the bottom line. If the geometry is not black, add any missing relations to fully define the geometry (such as symmetry). Alternatively, an 87-degree dimension could be added to the left-hand side of the sketch.

21. Rebuild the model.

22. Perform another Draft Analysis using the same parameters as in step 12.

23. Save and close the *Shaver Housing* project.

After performing the final draft analysis, it is found that the inner faces of the countersunk hole show up as requiring draft. Ideally, a small amount of draft should be added to the hole if this is to be a molded feature. Feel free to try this, if you wish, using the Draft command. In addition, the small thin edge at the end of the part shows up as requiring draft, but this can be safely ignored.

This concludes the exercise on the shaver housing in this chapter. You will pick it up again in Chapter 7, so keep the shaver housing file handy.

Summary

In this chapter, you first learned about thin-walled parts. You also discovered that a thin feature is one usually created with an open profile. The term *thin-walled* refers to shelled parts, whereas *thin feature* is the term reserved for extruded or revolved features given a wall thickness by the user.

The Shell command can be useful when creating a thin-walled part because it allows removal of the inside of a part, leaving a wall behind whose thickness you specify. Alternatively, it is possible to add a wall thickness to the outside of the part, akin to dipping the part in a vat of hot wax. This essentially removes the entire original geometry, leaving behind the newly created shell. Multi-thickness shelled parts can be created with one command.

A fillet or round can be created on a part, thereby rounding the sharp corner of an outside or inside edge. *Fillet* is the term used to describe a rounded inside edge, but is generally used to describe a fillet or a round. It is possible to fillet an entire face at once, if you want to fillet every edge on that face. Fillets and rounds can be added simultaneously, all within the same feature.

When fillets are added, the *Tangent propagation* option is selected by default. This function can be turned off if desired. When selecting items to be filleted, it is possible to select edges using a variety of options, including selecting chains, loops, or tangencies. This selection process works for other commands as well. When filleting, features can also be selected, in which case every edge associated with that feature is filleted.

Use the Select Other option from the right mouse button menu when trying to select a face on the opposite side of the part. While using the Select Other option, right-clicking on faces will serve to hide them temporarily, thereby giving the user the ability to visibly see and therefore click on a previously inaccessible face.

The Hole wizard allows for the creation of a wide variety of hole types, including countersunk, counterbored, and tapered. It is good technique to first select the face on which to place the hole. Holes can be placed on non-planar faces using the Hole wizard. You should not be in a sketch when using this function. The Hole wizard function allows you to position the hole's centerpoint before completing the command. Multiple holes can be added in one operation.

Other important topics learned in this chapter were how to create a parting line using the Split Line command and how to add draft as a feature. Draft can be added either by defining a neutral plane or through the use of parting lines. The neutral plane defines where the draft angle is measured from. Ribs can easily be added to a model with the Rib tool. Benefits of the Rib command include simplicity of sketch geometry and extrusion direction flexibility. Last, you learned how to analyze the faces of a model to see if those faces require draft.

CHAPTER 7

Patterns

MECHANICAL PARTS OFTEN CONTAIN PATTERNS. Many items you use in your everyday life also contain patterns of one type or another. The peanut butter jar in your cupboard may have some sort of ridge pattern around its outside perimeter near the lid. The razor you use to shave with in the morning probably has some type of rib or groove pattern so that it is easier to hold onto.

If you were to walk around your house and look at things the way a designer would, you might wonder how you would design such things in SolidWorks. Do you own a television with a remote? Chances are the remote control unit has buttons that form a pattern.

The point of this discussion is that patterns are a very common aspect of the design process. This chapter is devoted to patterns, both feature patterns and sketch patterns. A much simpler form of patterning is to mirror feature geometry, a process also covered in this chapter.

Copying and Pasting Features

Once you have taken the time to create a feature, it is often much more convenient to pattern that feature rather than to recreate the feature at another location. Depending on the situation and how many copies of a feature you require, it may be easier to perform a simple copy and paste.

When it is only one or two additional features that are required, or if additional features do not conform to any type of pattern, copying and pasting features can be the best method to use. This functionality makes use of the Windows clipboard, and can be used to copy a feature from one location to another, or even from one part to another.

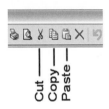

Fig. 7–1. Cut, Copy, and Paste icons.

When copying and pasting features, the Edit menu can be used, or the "hot key" combination specifically designed for the task. There are also icons on the Standard toolbar that can be used, as shown in Figure 7–1. No matter what Windows program you are using, Ctrl-C will copy the selected object to the clipboard, and Ctrl-V will paste the content of the clipboard into your document. This is standard operating procedure in any Windows program, and not limited to SolidWorks.

Dangling Relationships

When copying and pasting features, there are normally locating dimensions or constraints attached to the original feature's sketch. If this is the case (as it usually is), you will see a window asking you how you want to handle these relationships after performing the paste operation. This window, shown in Figure 7–2, offers the choices of either deleting the locating dimensions or constraints or leaving them dangling.

Fig. 7–2. Deciding what to do with locating dimensions or relations.

What does it mean to leave something "dangling"? In the case of copying and pasting features, a dimension or constraint is considered dangling if it originally located the feature but is now left "hanging" because the copied feature is now in a different location, or perhaps even on a different part. SolidWorks gives you the ability (via the window shown in Figure 7–2) to delete the meaningless relationships or to leave them dangling so that they can be repaired.

Dangling relations can be repaired by clicking on the object containing the dangling relation and then dragging the red "handle" to a new location. Dangling dimensions can be repaired by selecting the dimension and dragging the red handle at the end of one of the extension lines to some other object in the model. Dangling dimensions will appear brown, and dangling constraints will make the associated sketch geometry appear brown.

If there are no locating relationships attached to the feature being copied, you will not even see the Copy Confirmation window (shown in Figure 7–2). The feature will be directly pasted in wherever you tell it to be. In summary, if relations are deleted new ones should be added. If they are left dangling, repair them prior to continuing.

To copy and paste a feature from one location to another (or from one part to another), perform the steps outlined in How-To 7–1. Make certain you are not editing a sketch, or copying and pasting features will not work.

How-To 7-1: Copying and Pasting

To copy and paste a feature, perform the following steps. Note that any method (not just the Ctrl-C or Ctrl-V hot keys) can be used for copying and pasting, such as accessing the Edit menu or utilizing the Standard toolbar.

1. Select the feature from FeatureManager you want to copy to the clipboard.

2. Copy the item to the clipboard (Ctrl-C).

3. Select a planar face on the part to which you want to copy the feature.

4. Paste the item from the clipboard (Ctrl-V).

5. If there are dangling relationships, specify whether you want to delete them or leave them dangling.

6. Edit the sketch of the newly pasted-in feature in order to repair the dangling relations or add new relations to properly locate the feature.

7. Rebuild the model.

Copied features will not have an association to the original geometry. In other words, the newly pasted-in feature will be independent of the original. With regard to steps 6 and 7, you should already be familiar with editing a sketch, which was discussed in detail in Chapter 4.

Copying and pasting has its limitations. Copying more than one feature at a time, for instance, often does not work. Do not attempt to copy multiple features that exist on different planes. This simply will not be allowed. When pasting a feature onto a part, make sure a planar face is selected. Nonplanar faces cannot be used for pasting features; nor can construction planes.

For those not familiar with "cutting" objects to the clipboard, it works exactly the same as copying. The only difference is that the object being cut is deleted from the model, and then pasted in at a different location. Obviously, you will need to be careful with this option.

The Control-Drag Technique

Another technique that can be used to add copies of features to a part is the control-drag technique. The result is exactly the same as if you were to copy and paste features. This is just a different way of accomplishing the same thing.

Control-dragging amounts to nothing more than holding down the Ctrl key and dragging a feature from FeatureManager to a planar face on the model in the work area. It should be noted that control-dragging a feature from the work area also works. Whether you use copy-and-paste techniques or the control-drag method is up to you. Use the method that works easiest for your situation.

➥ **NOTE:** *If control-dragging a copy of a feature from the work area (as opposed to FeatureManager) to another location, the cursor must be positioned over a face on that feature for the procedure to work.*

If using the control-drag technique, make sure you are not actively editing a sketch. Otherwise, this method does not work. As the feature to be copied is being dragged, a preview will be displayed. The preview will appear to align itself with whatever planar face the cursor is over. Just release the mouse button and the feature will be created at that position. Make it a point to edit the sketch of the new feature afterward in order to repair the sketch or position it in the desired location.

✓ **TIP:** *Use Ctrl-drag to copy, and use Shift-drag to move.*

Copying and Pasting Sketch Geometry

Keep in mind that to this point we have been discussing features. Copying and pasting copies a feature's definition as well as its underlying sketch geometry. It is also possible to copy and paste sketch geometry by itself, possibly for the sake of reusing that sketch in another part.

Sketch geometry can be reused using the same basic principles of copying and pasting. For example, select the geometry, copy it to the Windows clipboard, and paste it in at the desired location. Clicking at some location with the mouse serves to specify the general area where the item will be pasted.

An entire sketch can be copied and pasted, or entities within a sketch can be copied and pasted. The main point to keep in mind is consistency. If you wish to copy an entire sketch, sketch mode should not be active when the sketch is copied, nor should it be active when the sketch is pasted. A plane or planar face should be selected prior to pasting the sketch. This can be in the same part file or in a completely different part.

If copying and pasting entities within a sketch, select sketch entities within the active sketch, and then paste those same entities at a different location either in the same sketch or some other sketch. If pasting the entities into a different sketch, you must be editing that sketch first prior to pasting the entities into it.

Understanding Error Symbology

Errors can occur in SolidWorks for any number of reasons. For instance, an error can occur if too many dimensions or relations have been added to a sketch, thereby overdefining the geometry. As you learned in the previous section, errors known as dangling dimensions or relations can arise when copying and pasting features. Trying to create a cut through empty space would result in an error.

Fig. 7–3. Error symbology.

The point of the matter is that errors will arise from time to time whether you want them to or not. FeatureManager will display various symbols and change the color of text to inform you of these errors. Errors are also coded by level, which helps you to assess the problem. Warning symbols inform you that a minor problem exists but geometry can still be created. More serious problems will result in SolidWorks' inability to build geometry, and these are known as rebuild errors. All SolidWorks error symbols are shown in Figure 7–3. The symbols containing the arrows are there to let you know that an error or warning is close by.

Overdefined Geometry

Another type of error that commonly occurs in a sketch is overdefined geometry. When sketch geometry has been overdefined, SolidWorks alerts you to this condition a number of ways. One way the software warns the user is to turn the color of all conflicting entities the color red, which you learned in Chapter 2. You may also notice warning symbols in FeatureManager telling you of the problem, and a plus (+) sign preceding the sketch. These error symbols only appear after a rebuild (or similarly, when exiting a sketch).

You typically will not see error symbols attached to a sketch in FeatureManager (at least not resulting from overdefined sketch geometry), as long as you make it a point to repair any overdefined conditions prior to building a feature or exiting a sketch. Red sketch geometry will clue you in to the problem before it ever gets to the point of FeatureManager error symbols appearing.

Fig. 7–4. Sketch states and symbols.

In contrast to the plus sign, you may also see a minus sign (–) preceding a sketch. This means that the sketch is underdefined. Ideally, nothing should appear before the sketch name in FeatureManager, which means the sketch is fully defined. Figure 7–4 shows examples of all of these scenarios (*Sketch1* is overdefined, *Sketch2* is underdefined, and *Sketch3* is just right!)

What's Wrong?

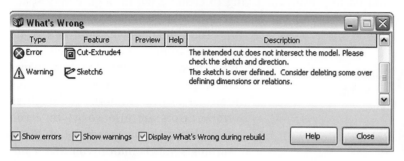

The What's Wrong function goes hand in hand with the error symbols shown in Figure 7–3. By right-clicking on any object in FeatureManager containing an error symbol, the What's Wrong menu item will be available.

Fig. 7–5. What's Wrong error window.

If the What's Wrong command is selected, a window with an explanation of the problem is shown. An example of this is shown in Figure 7–5. While learning the SolidWorks software, you will invariably experience error messages from time to time. This is fine, and is all part of the learning process. Do not be afraid to experiment, because errors can be an educational experience.

Both errors and warnings can be listed together or separately in the Rebuild Errors window. You will also see an option to *Display What's Wrong during rebuild*. If this option is checked, the What's Wrong window will appear every time the model is rebuilt or a feature is added. This is a good setting to leave on for new users, as it will be impossible to miss the What's Wrong window should you happen to do something that causes an error.

➥   **NOTE:** *Selecting Tools > Options and checking the Show errors every rebuild option will also display the What's Wrong window after every rebuild.*

Whenever a problem arises, whether from overdefined sketch geometry or something else, every effort should be made to correct the situation. If the problem is not remedied, more problems are likely to occur further down the road. At various stages throughout the book, you will look at common errors, how they occur, and how to fix them.

Linear Patterns

When you need to make a number of copies of one or more features, patterning is often the best method to use. Linear patterns are covered here. Other pattern types and the Mirror command (for mirroring features) are covered later in this chapter. Patterns in general can be employed more broadly than may be apparent. Do not underestimate the power of patterns in SolidWorks, as they may surprise you. A simple linear pattern is shown in Figure 7–6.

Fig. 7–6. Linear pattern.

You have the option of creating a linear pattern in either one or two directions. A two-directional pattern will create a grid-type pattern. Often, the pattern directions are orthogonal, but that is certainly not a requirement. The options associated with pattern commands will be something you learn about in this chapter.

One requirement when creating a pattern is to specify a direction for the pattern. This means you must select either an edge or a dimension that will signify the direction in which the patterned copies are created. If selecting a linear edge, that edge determines the pattern direction. If selecting a dimension, it must be a linear dimension, and the dimension line's arrows will point in the direction the pattern will take place.

An important aspect of patterning is that if the original patterned feature is modified the changes will propagate to the other features in the pattern. Because of this, it is very easy to maintain control over a large number of features on the part with very little effort. In actuality, the pattern feature is considered a single feature within itself, even if it contains a thousand instances. How-to 7–2 takes you through the process of creating a linear pattern.

How-To 7-2: Creating a Linear Pattern

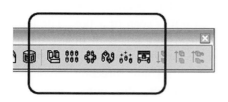

Fig. 7–7. Various mirror and pattern icons.

To create a linear pattern, perform the following steps. Note that all of the icons for various pattern types (shown in Figure 7–7) can typically be found on the Features toolbar. These command icons are, from left to right, Mirror, Linear Pattern, Circular Pattern, Curve Driven Patter, Sketch Driven Pattern, and Table Driven Pattern.

1. Select Insert > Pattern/Mirror > Linear Pattern, or click on the Linear Pattern icon. This will open the Linear Pattern PropertyManager, a portion of which is shown in Figure 7–8.

2. Click in the Features to Pattern list box and select a feature or features to be patterned.

3. Click in the *Pattern direction* list box for Direction 1, and select a linear edge or dimension to define the pattern direction.

At this point you should see a preview take shape on screen. For this reason alone it is best to specify the objects to be patterned, as well as a pattern direction, prior to plugging in any of the other parameters. Technically, though, these steps can be done in nearly any order.

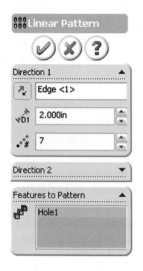

4. Click on the *Reverse direction* button if necessary to flip the pattern direction.

5. Specify the Spacing (distance) from one instance to the next.

6. Specify the Number of Instances to create in the pattern. This number includes the original.

When typing in values for spacing or number of instances, press the Enter key to update the preview. Using the spin box arrows updates the preview automatically.

7. If patterning in two directions, repeat steps 3 through 6 for the Direction 2 parameters. Otherwise, click on OK to accept the settings and create the pattern.

Fig. 7–8. A portion of the Linear Pattern PropertyManager.

There are a number of options available when creating patterns in general. Some of these options are common to all patterns, and some are specific to linear patterns. Let's look at some of these options.

Pattern Seed Only

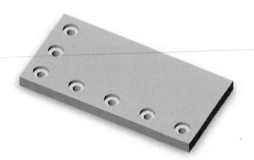

The Pattern Seed Only option is very easy to understand once it is seen in action. The term *seed* simply means the object or objects being patterned. The option is only available when patterning in a second direction. Typically when patterning in two directions, the second direction patterns the seed and every instance created from patterning in the first direction. If only the seed is patterned, however, none of the instances created from the first direction is patterned in the second direction (as seen in Figure 7–9).

Fig. 7–9. Results of using the Pattern Seed Only option.

Vary Sketch

The Vary Sketch option can be a very powerful feature when used correctly. This option is active only if you are working with sketch geometry that contains entities either converted or offset from existing feature edges, or that is constrained to existing geometry. The sketch must also be fully defined. One last criterion is that a dimension be selected, as opposed to an edge, to establish the pattern direction. If all of these conditions are met, the Vary Sketch option will be enabled.

What does Vary Sketch accomplish? This is easiest to explain through example. In Figure 7–10, a simple metal plate has been created that contains a single curved slot. The curved slot is rounded on both ends. The metal plate is pie-shaped.

Using this example, it would be convenient if the rounded slot could be patterned outward to create more slots. The dilemma occurs when creating the pattern. The design intent of the proposed slot pattern is that the distance between each rounded end of the slots and the edges of the pie-shaped plate be maintained during the pattern. However, during a typical pattern these distances would not be maintained. The distance would be maintained relative to one edge of the pie shape but not the other. This situation is depicted in Figure 7–11.

What would be ideal is if the slot could be made to somehow maintain a constant distance from the left edge of the plate to the left side of the slot, just as the right side of the slot is doing. It is possible to accomplish this feat using Vary Sketch.

To take advantage of the Vary Sketch option, some forethought must be used when creating the sketch for the feature to be patterned. The sketch must be fully defined and must contain some sort of constraint or dimension to existing feature edges. Converted or offset entities would fall into this category, but these are not a requirement. Figure 7–12 shows the sketch used for the slot feature. Such a sketch would allow the Vary Sketch option to become enabled.

In this sketch, the .125-inch dimensions control the distance between the arcs and the plate edges. The endpoints of the construction arc are coincident with the model edges. By dimensioning and constraining the sketch the way it is, the slot can be moved upward and away from the origin by altering the 1.000-inch dimension, and the rounded ends on either side of the slot will remain at constant distances from each edge of the pie-shaped plate. The 1.000-inch dimension plays an important role in the linear pattern. It specifies the direction of the pattern, and is required to enable the Vary Sketch option. Vary Sketch will become active once this dimension is selected to indicate the pattern direction.

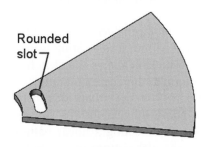

Fig. 7–10. A slot cut into the plate.

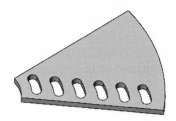

Fig. 7–11. Typical linear pattern of the slot.

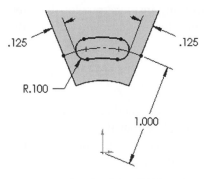

Fig. 7–12. This sketch would allow for using the Vary Sketch option.

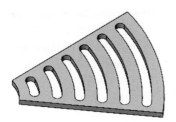

Fig. 7–13. Patterning the slot using the Vary Sketch option.

You saw what the patterned slot looked like (Figure 7–11) without the Vary Sketch option selected. Figure 7–13 shows the same pattern with the Vary Sketch option enabled. The difference is obvious.

When the Vary Sketch option is enabled, the relationships present in the sketch that constrain it to existing feature edges are taken into consideration. This allows the sketch to change shape as it is being patterned, thereby maintaining relationships to the rest of the part. It is not a bad idea to try this on your own. Make up some dimensions for the pie shape, and then model the slot. Try creating a linear pattern using Vary Sketch. Later you will perform a circular pattern on the entire model to create a round cover plate for a fan housing.

SW 2006: The Propagate Visual Properties option, when checked, will cause patterned instances to inherit the color or texture of the original feature or features being patterned. Cosmetic threads can also be propagated with this option. Propagation can occur if either faces or features have had their appearance changed. Model appearance is discussed in Chapter 8.

Deleting Pattern Instances

It is possible to remove instances from any type of pattern, but linear and circular patterns have a special option titled Instances to Skip. This option allows for easily choosing which instances to keep and which to get rid of. You can also change your mind and recover any instance previously deleted, without recreating the entire pattern. How-To 7–3 takes you through the process of deleting pattern instances.

How-To 7-3: Deleting Pattern Instances

To delete an instance from a pattern, regardless of whether it is a linear or circular pattern, perform the following steps. Note that these steps can be performed either during the pattern creation process or after the pattern has already been created (by editing the pattern feature).

⇥ **NOTE:** *Use these same steps for recovering skipped instances.*

1. When defining the pattern (see How-To 7–2), click inside the list box titled Instances to Skip.

2. A series of dots will appear on screen, representing the patterned instances. Click on the dots to add or remove instances from the pattern using the pattern preview as a guide. Any skipped instances will appear in the Instances to Skip list box, shown as an inset in the upper right-hand corner of Figure 7–14.

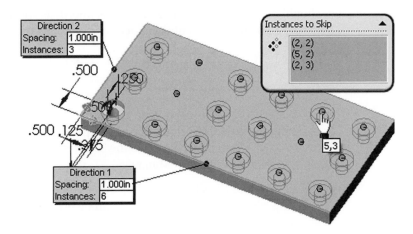

Fig. 7–14. Skipping instances in a pattern.

3. Click on OK to accept the changes or complete the pattern.

Any pattern instance can be skipped, as long as it is not the original feature the pattern was based on. Once an instance has been skipped, its instance position identifier is listed in the Instances to Skip list box. It is best not to even worry about the instance identifier. Simply use the preview to determine which instances will be skipped (deleted).

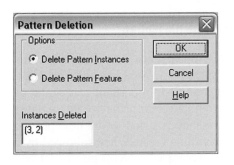

Fig. 7–15. Pattern Deletion window.

There is a shortcut for deleting pattern instances. If a face of a pattern instance is selected, pressing the Delete key will display the Pattern Deletion window, shown in Figure 7–15. This is nothing more than a confirmation window, asking whether or not you want to delete just the selected instance or the entire pattern.

It is essential to select a face, not an edge, in order to use this shortcut. In addition, more than one instance can be deleted simultaneously by holding down the Ctrl key and selecting faces from multiple instances. SolidWorks will even let you delete multiple instances from more than one pattern at a time. In this case, you will be presented with the Pattern Deletion window more than once.

Next, you will put some of the knowledge you have gained regarding linear patterns to the test. The shaver housing you started in Chapter 6 will be used in Exercise 7–1.

Exercise 7-1: Patterns on the Shaver Housing

To begin this exercise, open the *Shaver Housing* part you started in Chapter 6. Currently, there is only one boss containing a countersunk hole. There should be four such features. The boss and hole will be patterned simultaneously. To create the additional features, perform the following steps.

1. Add a .050-inch fillet to the base of *Boss1*.

2. Click on the Linear Pattern icon.

3. Click in the Features to Pattern list box and select *Boss1*, the countersunk hole, the fillet you added in step 1, and any additional draft you may have optionally added to the countersunk hole in Chapter 6. (Hint: use the fly-out FeatureManager.)

4. Click in the Direction 1 list box and establish a pattern direction. The long edge of the shaver housing can be used, as well as the .875-inch dimension. Click on the Reverse Direction button if necessary.

5. Specify two (2) instances for Direction 1.

6. Specify 2.75 inches for the spacing.

7. Click in the Direction 2 list box and establish a pattern direction. This time, use the .625-inch dimension. Once again, click on the Reverse Direction button if necessary.

8. Specify two (2) instances for Direction 2.

9. Specify 1.25 inches for the spacing.

10. Click on OK to accept the data and create the pattern.

Fig. 7–16. After creating the first pattern.

When finished with this first pattern, the shaver housing should look as it does in Figure 7–16. Next, an additional feature will be added, and then patterned. In particular, it would be beneficial to the design of the shaver housing to add grooves that would make gripping the shaver easier. In the next portion of this exercise, a simple cut will be created first, followed by a pattern of the cut. This will create a washboard effect that will keep the shaver from slipping in the hand of the person using it.

For this next feature, you will be sketching on the Right plane. This being the case, it is suggested that you show the Right plane. Resize the plane if you desire. Remember that you can resize a plane's border by dragging the green handles, or "autosize" a plane via the right mouse button. This does not affect the model and is for aesthetic purposes only.

The sketch that needs to be created is shown in Figure 7–17. This will be used to create the first traction groove. The feature will then be patterned. This should be fairly easy for you this time around. Afterward, the pattern itself will be patterned again.

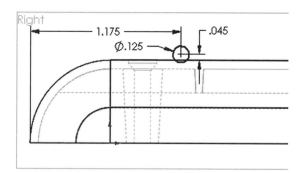

Fig. 7–17. Creating the sketch for the traction groove.

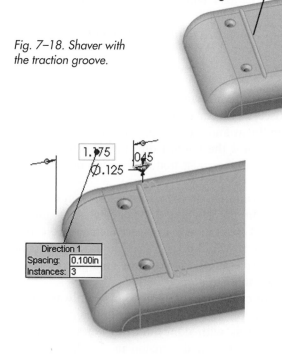

Fig. 7–18. Shaver with the traction groove.

Fig. 7–19. Creating the first groove pattern.

11. On the Right plane, create the sketch shown in Figure 7–17.

12. Perform a Cut-Extrude. Use the Through All end condition, and click on the Reverse Direction button if necessary.

13. Rename the cut-extrude *Traction Groove*. The feature is shown in Figure 7–18.

14. Click on the Linear Pattern icon.

15. Select the *Traction Groove* feature to be patterned.

16. Specify a pattern direction. In Figure 7–19, the groove's 1.175-inch dimension was used. Reverse the direction if necessary.

17. Set the number of instances to *3*.

18. Set the spacing to *.100* inch.

19. Click on OK to finish the pattern.

Next, you will pattern the pattern. Nothing unique or different needs to be accomplished in regard to this task. However, as a side note it is worth mentioning that when the pattern is patterned the original feature need not be selected, as it is automatically included as part of the first pattern. Continue with the following steps.

20. Click on the Linear Pattern icon again to create another pattern feature.

21. Select the last pattern you created as the feature to be patterned.

22. Specify a pattern direction. This time, use a linear edge that runs along the length of the shaver housing. Reverse the direction if necessary.

23. Set the number of instances to *5*.

24. Set the spacing to be *.4875* inch.

25. Click on OK to finish the pattern.

26. Rename this last pattern *Traction Pattern*.

27. Save your work.

Fig. 17–20. Select a different cell prior to double on clicking a feature.

The new groove patterns on the shaver housing should look like those shown in Figure 7–20. Optionally, add some .020-inch fillets to the ribs on the inside of the shaver housing. You can accomplish this most easily by selecting the rib feature from FeatureManager while in the Fillet command. This will fillet every edge on the rib feature with a minimal amount of effort.

Circular Patterns

Fig. 7–21. A portion of the Circular Pattern PropertyManager.

The Circular Pattern PropertyManager, shown in Figure 7–21, is very similar to that used for linear patterns. The only major difference between defining linear and circular patterns would be the requirements for what dictates the pattern direction. With regard to circular patterns, an axis is the most frequently used object to define the pattern direction.

Another way of determining circular pattern direction is to use an angular dimension. The point where the extension lines for the angular dimension would theoretically intersect dictates the center of rotation for the pattern. It would be necessary to use a dimension to dictate circular pattern direction, for example, on a part in which an axis was not readily available.

When working with cylindrical parts, it is almost always more convenient to use an axis or temporary axis for the center of rotation. In other cases, such as with noncylindrical parts, it may be easier to add some sort of angular dimension in the sketch of the feature being patterned. Even a reference dimension would suffice. Such would be the case if a hole were drilled through a square plate, for example.

Figure 7–22 shows typical dimensions for a hole on a square plate. The image also shows additional centerlines and an angular reference dimension. It is the angular dimension that will dictate the center of rotation for the circular pattern. The centerlines are only there for the sake of the dimension.

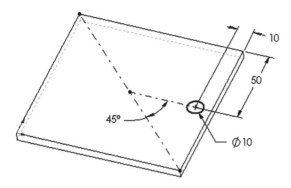

Fig. 7–22. This sketch makes a circular pattern easy.

How-To 7–4 takes you through the process of creating a circular feature pattern. Be aware that when it is necessary to select a dimension to define the pattern direction, whether it is a linear or circular pattern, it is helpful to first select the feature to be patterned. SolidWorks will then display the dimensions associated with that feature. (There are other ways of displaying dimensions, which are addressed in later chapters.)

How-To 7-4: Creating a Circular Pattern

To create the circular pattern, perform the following steps. A dimension will be used in this How-To, but an axis could be substituted for the dimension if one were available. The section titled "Axes" in material to follow explains how to create axes for a variety of uses.

1. Select Insert > Pattern/Mirror > Circular Pattern, or click on the Circular Pattern icon, found on the Features toolbar.

2. Click in the Features to Pattern list box and select a feature or features to be patterned.

3. Click in the *Pattern axis* list box and select an axis, linear edge, or angular dimension to represent the axis of rotation.

4. Specify the angle between instances.

5. Specify the number of instances.

6. Click on the Reverse Direction icon if necessary.

7. Click on OK to create the circular pattern.

Fig. 7–23. Multiple circular patterns.

Figure 7–23 shows an example of what circular patterns might look like. With regard to deleting patterned instances, linear and circular patterns follow the same set of rules. An *Equal spacing* option in the Circular Pattern panel acts as a built-in equation (explored in the following section).

Circular patterns can exist almost anywhere. This includes cylindrical faces or even spherical faces. Such is the case with the small slot pattern shown in Figure 7–23. The pattern resides on a spherical feature and was created using an axis to dictate the pattern's axis of revolution.

✓ **TIP:** *When creating patterns, specifying the objects to be patterned and defining an object to indicate the pattern direction first will result in a preview that will update as the number of instances or spacing is changed.*

Equal spacing Option

The *Equal spacing* option is only available with circular patterns. This option acts as a sort of built-in calculator. When checked, whatever value is typed into the Angle parameter is automatically divided by whatever value is specified for the Number of Instances parameter.

If *Equal spacing* is checked, the Angle parameter will be regarded as the Total Angle option and will automatically change to 360 degrees. The user can override the 360-degree setting, but to do so would usually be counterproductive. For example, if three small holes are "equally spaced" over an included angle of 90 degrees, it would be just as easy to set the Angle setting to 30 degrees and leave *Equal spacing* disabled. There is, however, one exception.

With *Equal spacing* enabled, a design change can be made quite easily regarding the number of instances in a circular pattern. The number of instances can easily be altered, and the spacing between the instances is automatically updated to fill the desired included angle. If the designated included angle happens to be less than 360 degrees, then so be it. *Equal spacing* still works.

✓ **TIP:** *The number of instances in a linear or circular pattern is treated as a dimension. When double clicking on a pattern in FeatureManager, the number of instances can be changed via a double click, just as any other parametric dimension.*

Geometry Patterns

An option available for all feature patterns is the *Geometry pattern* option. The *Geometry pattern* option is also available when mirroring feature geometry. A geometry pattern, simply put, disregards the definition of the original feature. It takes only the geometric information of the feature (i.e., size and shape) and patterns it. There are no special requirements when performing a geometry pattern. Simply check the *Geometry pattern* option and complete the pattern as you normally would.

To see what a geometry pattern will look like in comparison to a standard pattern, examine Figure 7–24. The cylindrical feature was extruded using the Up To Surface end condition. The angled plane was used as the terminating surface, which is the reason the cylindrical boss has an angled top. Each cylindrical boss is a different height because the patterned instances all take into consideration the end condition of the original boss.

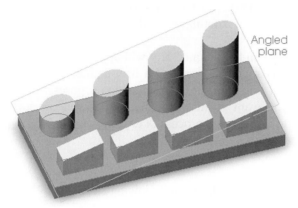

The rectangular boss was patterned using the *Geometry pattern* option. When the *Geometry pattern* option is turned on, SolidWorks ignores the end condition and patterns only the original geometry. The shape of the original is maintained because of this. In the example shown in Figure 7–24, note that the patterned rectangular bosses all look exactly the same. This is a direct result of creating a geometry pattern.

Fig. 7–24. Standard versus geometry pattern.

Axes

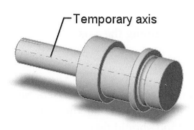

Fig. 7–25. Temporary axis.

In that circular patterns can benefit from having an axis present in the part, this is a good time to talk about axes. The first fact you should know about axes in SolidWorks is that there are two types: those created by SolidWorks and those created by the user. Axes created by SolidWorks are known as *temporary axes*, which are generated by cylindrical or conical faces. Temporary axes can be used for creating circular patterns, mating components in assemblies, and other functions. Figure 7–25 points out a temporary axis on a part created earlier in this book.

A temporary axis is "temporary" because theoretically another feature could completely remove the cylindrical or conical face being used to generate the axis. If the face is removed, so is the related axis, and hence the term *temporary axis*. From a cosmetic standpoint, there is not much you can do with temporary axes. You cannot rename them and cannot lengthen or shorten them; nor can you decide which temporary axes to display on an individual basis.

Displaying the temporary axes within a SolidWorks file is simple enough. Under the View menu, look for the Temporary Axes menu item, and make sure it has a check mark in front of it. Clicking on the Temporary Axes menu item will serve to toggle the display of temporary axes on and off.

↦   **NOTE:** *Certain faces, such as those created by adding a fillet feature, do not display temporary axes.*

User-defined Axes

Creating an axis is much the same as creating a plane. How you create the axis depends on the geometry you have to work with. You can define an axis by two vertex points, an existing edge, the intersection of two planes, and so on.

A user-defined axis can be renamed and will appear in FeatureManager as a feature. You can change the length of a user-defined axis (referred to simply as an axis from this point on) for aesthetic purposes. You can also control which axes you want to hide or show, just as you can for planes. How-To 7–5 takes you through the process of creating an axis.

How-To 7-5: Creating an Axis

To create a user-defined axis, perform the following steps.

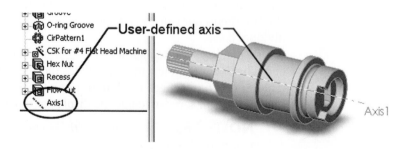

1. Select Insert > Reference Geometry > Axis, or click on the Axis icon, found on the Reference Geometry toolbar.

2. In the Axis PropertyManager (see Figure 7–26), select the method with which to define the axis. This will depend on the geometry you have to work with.

3. Select the applicable geometry for defining the axis.

4. Click on OK.

Fig. 7–26. Axis PropertyManager.

It is worth noting that step 2 can be completely bypassed. Simply select the geometry you wish to use to define the axis and SolidWorks will figure out which method should be used. Figure 7–27 shows a user-defined axis. It is possible to right-click on the axis either in the work area or in FeatureManager to hide or show it, using the same procedure you would for hiding or showing a plane. Select an axis and drag its endpoints to change the axis' length.

Fig. 7–27. A user-defined axis.

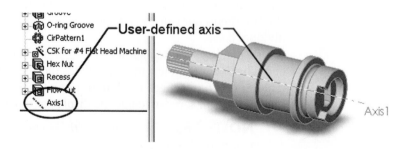

⟿ **NOTE:** *To globally toggle the display of all user-defined axes, make sure Axes is checked in the View menu. This is important! None of the axes you create will be visible if the ability to view them is not enabled.*

The part created in Chapter 5 was the Valve Stem. That same part is used in Exercise 7–2. Now that feature patterns have been discussed in detail, finishing up the *Valve Stem* project should not prove any difficulty. A number of other features will be added in addition to a pattern, so this exercise will give you some good practice.

Exercise 7-2: Adding Features to the Valve Stem

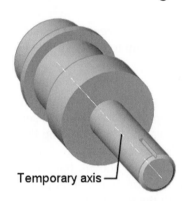

Fig. 7–28. Valve stem with temporary axis visible.

In this exercise, you will pattern the groove on the valve stem part originally created in Chapter 5. This feature will represent the area the faucet handle will grip onto. Some additional features, including a countersunk hole, will be added to the stem. Begin by opening the *Valve Stem* file. Use the illustrations that follow as guidance. Figure 7–28 shows the valve stem. Note that the temporary axis is being shown.

1. Turn on the display of temporary axes by selecting View > Temporary Axes. You will use a temporary axis for the pattern.

2. Click on the Circular Pattern icon, or select Insert > Pattern/Mirror > Circular Pattern.

3. Select the temporary axis that runs through the center of the part for the Pattern Axis.

4. Select the V-groove feature from FeatureManager as the Feature to Pattern. You should see a preview at this point.

5. Check the *Equal spacing* option.

6. Specify *24* for the Number of Instances. The Total Angle need not be supplied because it will default to 360 degrees automatically.

7. Click on OK to accept the settings and build the pattern.

8. Rename the pattern *Groove Pattern*.

9. Turn off the display of temporary axes, in that they are no longer needed at this point.

Figure 7–29 shows what the groove pattern should look like. Next, let's add a countersunk hole to the end of the stem. This should be old hat for you, in that the Hole wizard was covered in the last chapter. Continue with the following steps.

Fig. 7–29. After creating the circular groove pattern.

Fig. 7–30. Valve stem with a countersunk hole.

10. Select the face at the end of the stem on which to place the hole.

11. Click on the Hole Wizard icon.

12. Select the Countersink tab.

13. Specify an ANSI Inch, Flat Head (82), #4 screw. Use a blind end condition of .750 inch. Leave all other parameters at their default settings.

14. Click on Next.

15. Locate the single locating point at the center of the stem by adding a concentric relationship to the circular edge.

16. Click on Finish.

Figure 7–30 shows the valve stem with the countersunk hole. The last few features are simple features that will serve as good practice. Try to finish the valve stem by adding these last few features using the skills you have already learned. The sketch geometry and dimensions are shown, but you are on your own regarding adding the proper relations to fully define the sketches. Continue with the following steps.

17. Add the six-sided extrusion to the valve stem using the dimensions shown in Figure 7–31. Rename the feature *Hex Boss*. The extrusion depth is .500 inch.

18. Add a cut at the opposite end of the stem that is .150 inch deep. The sketch for the cut is shown in Figure 7–32, along with the completed feature.

19. Rename the feature *Recess*.

20. Create one final sketch at the bottom of the *Recess* feature. The sketch is shown in Figure 7–33.

21. Cut the sketch to a depth of 1 inch.

22. Save your work.

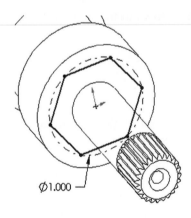

Ø1.000

Fig. 7–31. Adding the Hex Boss feature.

The valve stem could probably use a few more features before it is fully functional, but for the sake of brevity the exercise will end here. The point of the exercise was to give you some experience with a circular pattern and to practice sketching. Sketching is the foundation the rest of your SolidWorks knowledge will be built upon, so practice it well. The completed part should look as it does in Figure 7–34.

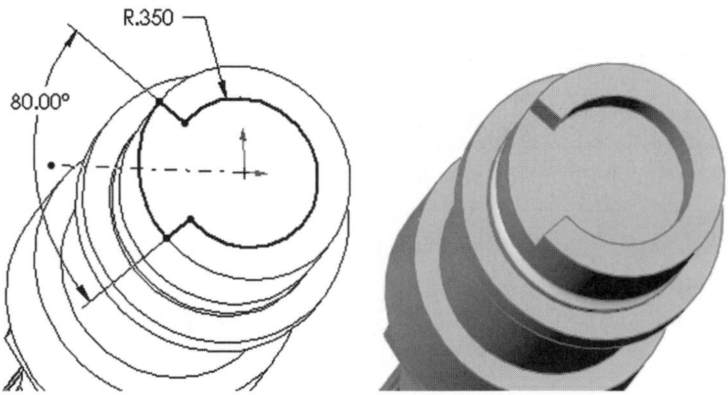

Fig. 7–32.
Creating the
Recess *feature.*

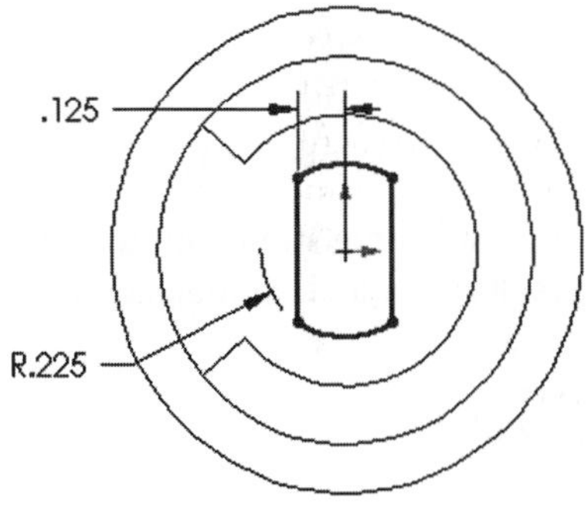

Fig. 7–33. Creating a sketch for one last cut.

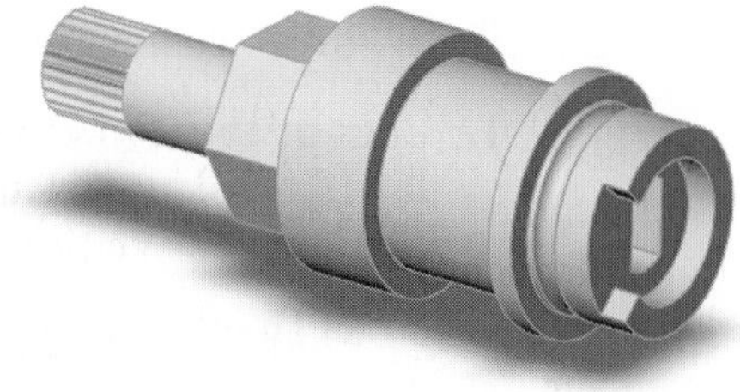

Fig. 7–34. Completed valve stem part.

Sketch-driven Patterns

Not every pattern is a perfect linear or circular pattern. Sometimes it is desirable to specify points where instances should be placed, as if telling SolidWorks to "drop a copy here." Sketch-driven patterns are very flexible, convenient, and easy to edit because they allow for placement of patterned instances wherever there is a sketch point.

The benefits of a sketch-driven pattern are that the pattern can be irregular or chaotic in nature. It is also extremely easy to create and modify. Because the pattern is based on sketch points, little effort is required on the user's part. Figure 7–35 shows a part with a group of features that need to be copied to various locations on a part. The image also shows a sketch containing points with ordinate dimensions added to fully constrain them.

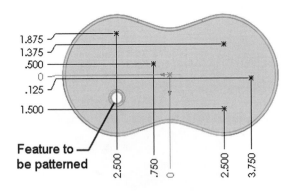

Fig. 7–35. Features about to be patterned.

To prepare for creating the pattern, the points were placed in their approximate positions, and then the dimensions were added to fully constrain them. In some cases, geometric relations were used, such as horizontal or vertical. The sketch shown in Figure 7–35 uses ordinate dimensions, but you can use any dimension type.

Because we have not explored ordinate dimensions yet, let's first learn how they can be added to a sketch. How-To 7–6 takes you through the process of adding ordinate dimensions.

HOW-TO 7-6: Adding Ordinate Dimensions

To add ordinate dimensions to a sketch, perform the following steps.

1. Click on the Horizontal or Vertical Ordinate Dimension icon found on the Dimensions/Relations toolbar, or select Tools > Dimensions > Horizontal or Vertical Ordinate.

2. Select an object to serve as the 0 datum point.

3. Pick to position the 0 datum point value.

4. Select additional points to add further dimensions.

In contrast to the Horizontal Ordinate or Vertical Ordinate dimension commands, the Ordinate Dimension option allows for creating ordinate dimensions at an angle. Because the dimension lines are not forced to be either horizontal or vertical, you should select a line or linear edge when establishing the 0 baseline location. The line or linear edge will determine the angle the ordinate dimension lines extend from the geometry.

To add a new ordinate dimension value to an already existing series, make sure you are not in any other command and then right-click on one of the ordinate dimension values. There you will find the Add To Ordinate option. You should also see a submenu titled Display Options, which will contain other functions such as Jog and Re-Jog. These options, along with the Break Alignment option (also found in the right mouse button menu), give you quite a bit of flexibility in regard to ordinate dimensioning.

Now that you can add ordinate dimensions, let's proceed to creating a sketch-driven pattern. How-To 7–7 takes you through this process. Because a sketch is a requirement with this type of pattern, for your reference that portion of the process is included in the steps.

How-To 7-7: Creating a Sketch-driven Pattern

To create a sketch-driven pattern, perform the following steps.

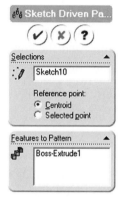

Fig. 7–36. Sketch Driven Pattern PropertyManager.

1. Create a new sketch and use the Point command to create points wherever copies of the original feature being patterned will be placed.

2. Add dimensions or geometric relations as required to locate the points.

3. Exit the sketch.

4. Select Insert > Pattern/Mirror > Sketch Driven Pattern, or click on the Sketch Driven Pattern icon. This accesses the Sketch Driven Pattern PropertyManager, shown in Figure 7–36.

5. Click in the Reference Sketch list box and select the sketch containing the points.

6. Click in the Features to Pattern list box and select the feature or features to be patterned.

7. Select whether to pattern by the selected object's centroid or some other reference point.

8. Click on OK to create the pattern.

The *Reference point* option allows for specifying a reference location on the parent feature, which will dictate where the patterned copies are positioned. By default, the centroid of the object being patterned is used. If there is more than one feature being patterned at a time, the centroid of the largest feature is used. If a reference point is used, select a sketch point, entity endpoint, or vertex point somewhere on the model. When patterning features such as holes, using the centroid is fine, but when patterning groups of features or features with irregular shapes you will probably need to use a reference point.

➠ **NOTE:** *A centroid is an object's center of gravity.*

If the *Reference point* option is still not making sense, just think of it as the "handle" by which the feature gets picked up. When it gets put back down on the sketch points, the "handle" is positioned on those sketch points. Figure 7–37 shows an example of a sketch-driven pattern.

If an instance in the pattern needs to be repositioned, modify the dimensions associated with that instance's respective sketch point. This is assuming the sketch points in the reference sketch were dimensioned in the first place, which they certainly should have been.

To take modifications one step further, edit the sketch that contains the sketch points. To delete an instance in the pattern, delete its respective sketch point. To add an instance, add a sketch point. It could not be any easier. The patterned instances are associated with the original feature, and therefore modifying the original will automatically update the copies.

Fig. 7–37. After completing the sketch-driven pattern.

✓ **TIP:** *Double clicking on a sketch-driven pattern feature will display all dimensions associated with the original sketch points.*

Table-driven Patterns

Table-driven patterns are different from sketch-driven patterns in one major fundamental way. Instead of the need to create sketch points that define the patterned instance locations, a table of points is used to describe those locations. In other words, a set of x- and y-axis coordinates determines where the patterned instances are positioned.

One additional item required when creating a table-driven pattern is a reference entity known as a coordinate system. Because of this, coordinate systems are discussed in the following section. Make sure to read the next section before attempting to create a table-driven pattern.

Coordinate Systems

A coordinate system consists of a set of x, y, and z axes, and a point that defines the coordinate's 0 reference point. Think of the SolidWorks origin point as a part or assembly's World coordinate system. It is the intersection of the three default planes and is the Cartesian coordinate system's zero reference point.

Certain functions of the SolidWorks software can alternatively employ user-defined coordinate systems (as opposed to the part's original Cartesian coordinate system). A perfect example of this would be when obtaining the mass properties of a part (discussed in more detail in an upcoming chapter). When obtaining the principal moments of inertia, for example, it might be preferable to reference a location other than the part's origin. In other cases, a coordinate system is simply a requirement, as in the case of a table-driven pattern.

Why is a new coordinate system required? Table-driven patterns are based on a set of *x-y* coordinates. Therefore, an *x-y* coordinate system must be established on the planar face where the pattern can be defined. How-To 7–8 takes you through the process of establishing a user-defined coordinate system.

HOW-TO 7-8: Creating a Coordinate System

To create a user-defined coordinate system, perform the following steps.

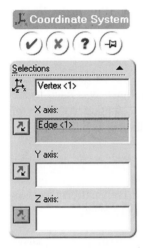

1. Select Insert > Reference Geometry > Coordinate System, or click on the Coordinate System icon, located on the Reference Geometry toolbar.

2. Select a point for the new coordinate system.

3. Optionally, click in the X Axis, Y Axis, or Z Axis box of the Coordinate System PropertyManager (see Figure 7–38) and select a plane, planar face, edge, or line to define the direction of that axis.

4. Repeat step 3 for an additional axis if required, and flip the axes if you wish.

5. Click on OK when finished.

Fig. 7–38. Coordinate System PropertyManager.

After creating a user-defined coordinate system, it will appear in FeatureManager as *Coordinate System1*. You will find it possible to rename the coordinate system, just like any other feature in Feature-Manager. If the coordinate system is not visible in the work area, turn on coordinate system visibility by selecting Coordinate Systems from the View menu. A user-defined coordinate system is shown in Figure 7–39.

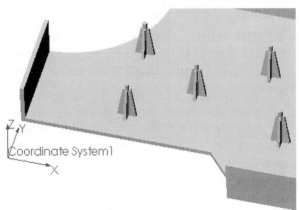

The coordinate system shown in Figure 7–39 was created at the project-ed intersection of two model edges. This serves to illustrate that a coordi-nate system can be defined anywhere, as long as there is existing geometry of one type or another from which to de-fine it.

✓ **TIP:** *Coordinate systems can be hidden or shown, just as planes can be, by right-clicking on the coordi-nate system in FeatureManager.*

Fig. 7–39. A user-defined coordinate system.

Even though coordinate systems have three axes, SolidWorks only makes use of the x and y axes with regard to a table-driven pattern. The x-y coordinates specified will dictate the placement of the patterned instances. Entering these coordinates can be accomplished in one of two ways, as follows.

- Create x-y coordinates on the fly by typing them in.

- Import a previously existing table of x-y coordinates.

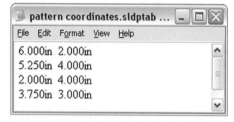

Fig. 7–40. A set of coordinates for a table-driven pattern.

If a table is to be used, the file extension of the coordinate table file must have either a *.txt* file extension or SolidWorks' own *.sldptab* file extension. Either is fine, and in either case the file itself is actually only a simple text file. Figure 7–40 shows an example of what one of these coordinate table files might look like. Windows Notepad was used to create the table shown. How-To 7–9 takes you through the process of creating a table-driven pattern.

HOW-TO 7-9: Creating a Table-driven Pattern

The following steps show you how to create a table-driven pattern. It is assumed that the feature or features to be patterned already exist.

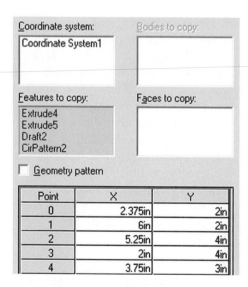

Fig. 7–41. A portion of the Table Driven Pattern window.

1. Select Insert > Pattern/Mirror > Table Driven Pattern, or click on the Table Driven Pattern icon, found on the Features toolbar.

2. In the Table Driven Pattern window (a portion of which is shown in Figure 7–41), click in the *Items to copy* list box and select the features to be patterned.

3. Click in the *Coordinate system* list box and select the previously defined coordinate system. The part's origin point will not suffice for this.

4. Specify the x-y locations for the patterned instances by double clicking on the appropriate cells in the Point listing, or optionally browse for a file that contains the point coordinates.

5. Click on OK to create the pattern.

The *Reference point* option (not shown in Figure 7–41), found in the Table Driven Pattern window, serves the exact same function it did with a sketch-driven pattern. A table-driven pattern was performed on the part shown in Figure 7–42.

Deleting an instance in a table-driven pattern can be done by selecting a face on any of the instances to be deleted. This is identical to deleting an instance in a standard linear or circular pattern. Another way to delete an instance in a table-driven pattern is to edit the definition of the pattern, select one of the instance rows from the table by clicking on its corresponding Point number, and press the Delete key.

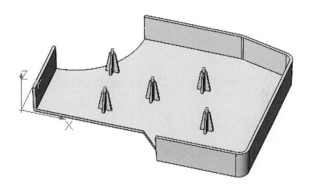

Fig. 7–42. A completed table-driven pattern.

✓ **TIP:** *Double clicking on a table-driven pattern feature in FeatureManager will display the dimension values from that feature's table.*

The other buttons found in the Table Driven Pattern window allow for saving the points as a SolidWorks *.sldptab* text file (Save and Save As). This assumes that the coordinates were manually entered in the first place. The Browse button, obviously, is for hunting down a text file containing the coordinate data.

Curve-driven Patterns

SolidWorks provides a method of patterning features along a curve. The curve can be a model edge, can be sketch geometry, and can even be non-planar. (Various methods of creating curves are discussed in Chapter 10.) The objects that constitute a valid curve are not unlimited, but there are really only a few stipulations. Only one curve can be used per direction, and all segments in the curve must be tangent.

Because the Pattern command contains no option for propagating along tangent edges, creating a sketch is common. Another option is to use the *Composite curve* command to add edges together (discussed in Chapter 10), but a sketch offers more flexibility. You will see why this is true in the forthcoming example. Figure 7–43 shows a sketch created preparatory to creating a curve-driven pattern.

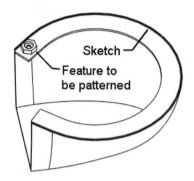

The sketch shown in Figure 7–43 was created using the Convert Entities sketch tool, which made creating the curve extremely easy. The Offset Entities sketch tool also works well in this situation. How-To 7–10 takes you through the process of creating a curve-driven pattern.

Fig. 7–43. Preparing for a curve-driven pattern.

HOW-TO 7-10: Creating a Curve-driven Pattern

To create a curve-driven pattern, perform the following steps. If using a sketch to drive the pattern, make it a point to exit the sketch prior to creating the pattern. The steps that follow assume that a sketch or curve of some sort has already been created.

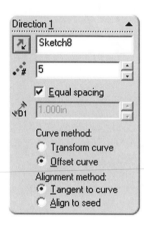

Fig. 7–44. Portion of the Curve Driven Pattern panel.

1. Select Insert > Pattern/Mirror > Curve Driven Pattern, or click on the Curve Driven Pattern icon, located on the Features toolbar.

2. Click in the Pattern Direction list box and select a sketch, model edge, or some other curve to drive the pattern for Direction 1. A portion of the Curve Driven Pattern panel (accessed in step 1) is shown in Figure 7–44. It is recommended that you select sketch or curve geometry from the fly-out FeatureManager (reasoning explained in material following).

3. Click in the Features to Pattern list box and select the features to be patterned.

4. Specify the number of instances to be included in the pattern. This is the total number of instances, as is typical for other pattern types.

5. Check the Equal Spacing option if desired.

6. If the Equal Spacing option is not used, enter a value for the spacing.

7. Click on the Reverse Direction button if necessary (use the preview as a guide).

8. Click on OK to create the pattern.

Fig. 7–45. Example of a curve-driven pattern.

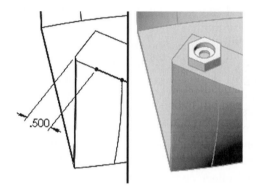

Fig. 7–46. Editing the sketch used by the curve-driven pattern.

Figure 7–45 shows an example of a curve-driven pattern. The hexagonal boss maintains the proper orientation as it continues around the curved perimeter of the part. The only problem is the last feature in the pattern, which overhangs the edge of the model. This problem can easily be solved by editing the sketch used as the curve to drive the pattern.

The original hexagonal boss was positioned .500 inch from the edge of the model. If the sketch used as the pattern curve is edited, the end of the last line segment in the sketch can be "pulled back" and dimensioned .500 inch from the opposite edge of the part, thereby spacing out all pattern instances appropriately. This is depicted in Figure 7–46.

There are two options that play a role in the outcome of the pattern. These options are *Curve method* and *Alignment method*. They are both easy to understand once seen in action. To better explain these options, we will look at a rectangular cut patterned along a sketched spline. In this first example, curve and alignment methods are set to *Offset curve* and *Tangent to curve*, respectively. This is shown in Figure 7–47.

The curve method controls the position of the patterned instances, whereas the alignment method controls the orientation of the patterned instances. This fact becomes clear once the curve or alignment methods are changed. In example 2, the alignment method has been changed to *Align to seed*. Note that the orientation of the feature remains constant, aligned with the original feature. This is depicted in Figure 7–48.

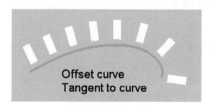

Fig. 7–47. Example 1.

Fig. 7–48. Example 2.

What happens if the curve method is altered? If the curve method is changed from offset curve to transform curve, the pattern changes dramatically. This can be observed in example 3. Transform curve in essence "moves" the curve and patterns along the new curve, whereas offset curve projects a new curve outward from the original, similar to how ripples flow outward when a pebble is dropped into a pool of water. This is shown in Figure 7–49.

Fig. 7–49. Example 3.

It is often difficult to understand how the patterned instances are relating back to the original curve. Without going into a lot of detail, let's just say that the centroid of the original geometry is related to the curve, and the patterned instance's position depends on the curve method used. If a second direction is used, a second curve can be selected, though this is not required. In the final example, shown in Figure 7–50, we see a slot patterned along a sketched arc for direction 1 and without a curve specified for direction 2.

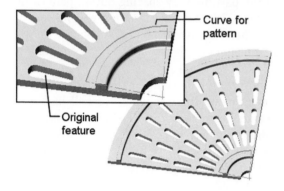

Fig. 7–50. A curve pattern in 2 directions.

When a second curve is not specified for the second direction, the final pattern is highly dependent on the curve method used and the position of the original feature relative to the curve used for direction 1. In Figure 7–50, the offset curve method was used for direction 1. Because a specific curve was not specified for direction 2, the "pond ripple" effect continues outward for a total of four ripples (to continue with the analogy), which is also the number of instances specified for direction 2. It is an interesting situation, and unique to curve-driven patterns.

SW 2006: Curve-driven patterns support 3D curve geometry.

Mirroring Feature Geometry

When sketch geometry is mirrored in a sketch, a centerline is needed. However, when feature geometry is mirrored a plane must be used. The Mirror Feature PropertyManager, shown in Figure 7–51, is extremely easy to use. To mirror a feature, simply select a plane or planar face to be the mirror plane, and select the items you want to mirror. You will see a preview of the mirrored objects. Click on OK to accept the new feature. This is all there is to it.

Parent/child relationships play a role when mirroring features (just as they do when patterning). For instance, you would not be able to mirror a hole in a boss without mirroring the boss also. In other words, features that are children cannot be mirrored without their parents. In addition, if a plane is used as the mirror plane you cannot delete the plane without also losing the mirrored features. In this case, the plane would be a parent of the mirrored features.

Fig. 7–51. Mirror Feature PropertyManager.

The Geometry Pattern option means the same thing when mirroring as it does when creating patterns. There is absolutely no difference. If Geometry Pattern is checked, no end condition information is taken into account. How-To 7–11 takes you through the process of mirroring feature geometry.

How-To 7-11: Mirroring Feature Geometry

To mirror features, perform the following steps.

1. Select Insert > Pattern/Mirror > Mirror, or click on the Mirror icon, located on the Features toolbar.

2. Select the plane or planar face to use as the mirror plane.

3. Select the features to be mirrored.

4. Click on OK to mirror the features.

Figure 7–52 shows a before-and-after image of a part that has features mirrored. Sometimes it is necessary to create a plane first that will serve as a mirror plane. In the case of the model shown in the figure, the Right plane (renamed to mirror plane in the figure) was convenient for this purpose because the original base feature sketch had been centered on the origin. Centering the first sketch on the origin point is common practice when creating symmetrical parts.

Fig. 7–52. Mirroring feature geometry.

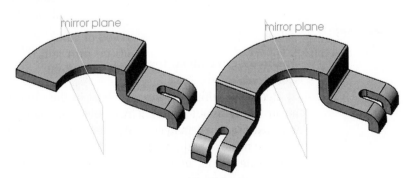

Mirroring and Patterning Faces and Bodies

Throughout the previous sections dedicated to various patterning commands and mirroring, you may have noticed areas in each of the command interfaces that allow for selecting faces or bodies. These areas can be seen in Figure 7–53, and in this case happen to be from the Mirror PropertyManager.

The ability to pattern or mirror faces is extremely important to those working with imported geometry. Because imported models rarely contain any feature data, there are no features that can be selected. In such a case, the ability to select faces becomes crucial. Faces can be mirrored or patterned with the same result as if features had been mirrored or patterned.

There are limitations to patterning or mirroring faces. If the faces being mirrored (for example) do not form a closed boundary condition, SolidWorks will fail to perform the operation. This is not necessarily true when mirroring faces for the sake of cutting away geometry.

On a somewhat similar note, surfaces can be patterned and mirrored, and no closed boundary condition is required. If you recall from Chapter 4, faces are nothing more than surfaces that have been used in the formation of a solid body. Surfaces are considered bodies as well, and thus can be patterned or mirrored just like a solid body would be. (Read more about surfaces in Chapter 21.)

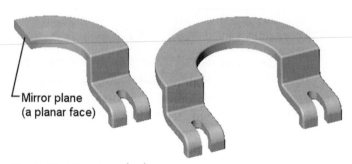

Mirror plane (a planar face)

Fig. 7–54. Mirroring a body.

In the case of symmetrical parts, it is often easier to create half of a part and then mirror that half to create the whole. This can be done with the same Mirror command discussed in the previous section. The main difference would be to click inside the Bodies to Mirror list box and select the model from the work area. Assuming the model is a single body, it can then be mirrored about the planar face of your choosing. Such an example is shown in Figure 7–54.

Because multiple bodies can exist in the same part file, it is possible to mirror bodies about planes. This could theoretically result in two (or more) detached pieces of material. Similarly, even if using a planar face on the model to mirror about, unchecking the *Merge solids* option would also result in separate pieces of geometry. Unless you have ulterior motives, leave *Merge solids* checked.

The *Knit surfaces* option only applies to mirroring surface bodies. Any seams that may exist between mirrored surfaces will be eliminated if this option is checked, assuming the surfaces do not intersect at a sharp corner (i.e., are tangent).

Mirrored Parts

Another option open to you is the ability to create a mirror image of a part. To clarify, a new part is created that is a mirror image of the original, as opposed to a single part (which is symmetrical). The new part will be a separate SolidWorks part file. In addition, a relationship (or "link") is created to the original part. Therefore, if the original part is modified the changes will appear in the mirrored part as well.

Fig. 7–55. Insert Part PropertyManager.

When a part is mirrored, SolidWorks uses the original part as the base feature for the mirrored part. Features added to the newly mirrored part will not affect the original, but changes made to the original will propagate to the mirrored part. The Mirror Part command is found in the Insert menu. A plane or planar face must be selected prior to accessing the command. Once the command has been accessed, the Insert Part PropertyManager (shown in Figure 7–55) is displayed.

The reason for the Insert Part PropertyManager is to allow for importing any additional reference geometry that might be relevant. Of significant note is the Cosmetic Thread option. If creating a left-hand version of a large part with numerous cosmetic threads, checking the Cosmetic Thread option keeps you from having to recreate all of those cosmetic threads in the mirrored part. Likewise, reference planes, axes, and surfaces can also be brought over from the original part file into the mirrored part.

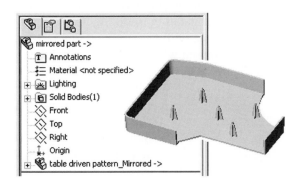

Fig. 7–56. After using the Mirror Part command.

In Figure 7–56, note the content of FeatureManager. The new part's base feature will use the name of the original part, with the text *_Mirrored* appended to it. Also note that an external reference has been created, which is symbolized by the arrow after the feature name (and after the name of the part at the top of FeatureManager). This reinforces the fact that this newly created part is indeed dependent on the original part. This also means that if the originating part file is deleted or renamed you will not be able to open the mirrored part. External references are very important and are explored in the section that follows.

External References

Anytime a part references another, an external reference is created. This is exactly what happened in the previous section. One part (call it part A) was used to create a second part (part B). Part B is now dependent on part A. If a change is made to part A, the change is propagated to part B. This is a unidirectional relationship only. Any changes made to part B will not propagate to part A.

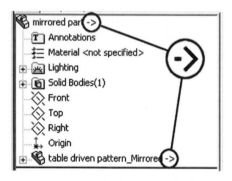

External references are denoted by a small arrow symbol after the name of the feature or part with the external reference. This is shown in Figure 7–57. The "arrow symbol" is actually a dash followed by a greater-than character (- >), but it serves the purpose. Any part containing an external reference will have this symbol attached directly to the part's name. This is to let you know that there is an external reference somewhere in the part. In addition, a symbol will appear on whatever feature actually contains the external reference.

Fig. 7–57. External reference symbols.

It is always possible to determine where an external reference is originating from. Right-clicking on any feature containing an external reference will gain access to the List External Refs command. Selecting this command will open the External References window, partially shown in Figure 7–58.

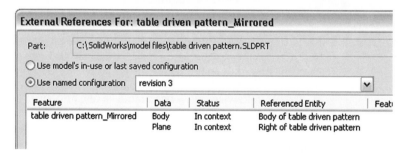

Fig. 7–58. Listing external references.

The path data at the top of the window will list the path to the part or assembly the feature is referencing. Other data is given as well, such as what type of entity is being referenced and whether or not the reference is *In context* or *Out of context*. If a reference is out of context, it means that the referenced geometry is unavailable because the file being referenced is not open at that time. Opening the referenced file will cause the status of the reference to be changed to *In context*.

Referencing Specific Configurations

If you refer back to Figure 7–58, you will see an option that allows you to either *Use model's in-use or last saved configuration* or *Use named configuration*. Configurations are explored in Chapter 11. Until then, know that it is possible to have different versions of the same part saved in a single part file. Since this is the case, the aforementioned options allow for specifying which, if any, configuration is referenced.

Using the last saved configuration will force the mirrored part to reference whatever configuration the original part was last saved in. This is usually undesirable. It is typically preferable to use a named configuration, which means that a specific configuration can be selected from a drop-down list present in the External References window. Visible in Figure 7–58, the configuration *revision 3* was selected from the drop-down list. This means that the mirrored part will be forced to reference the *revision 3* configuration of the original part.

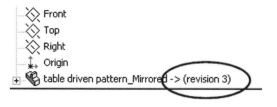

Fig. 7–59. Referenced configuration shown in FeatureManager.

When the referenced part has multiple configurations, the referenced configuration name will be displayed after the name of the base feature in FeatureManager. This is illustrated in Figure 7–59. The name of the base feature itself takes on the name of the original part.

Opening Referenced Files

It cannot be stressed enough that it is extremely important to realize when an external reference is out of context or not. *In context* means *in association with*. The meaning in SolidWorks is identical to that associated with speech. For instance, words can be taken *out of context*, which means that their original meaning is lost. When the same words are read *in context* to the original phrase, their true meaning is discovered. The same holds true with features that are *out of context* or *in context*.

Features that are out of context will appear with a question mark (- > ?). Be aware that when externally referenced files are not loaded into memory (opened, in other words) they are out of context. What exactly does this mean to you? It means that you may not be viewing the true nature of the model displayed on screen. Geometry may have changed. If the referenced file were altered in some way, you would not know it because it has not been opened yet to allow the referencing file to update.

To load the externally referenced file, simply right-click on the feature containing the reference and select Edit In Context. This will open the referenced file automatically. Performing a rebuild on the file with the external reference will cause the question mark to disappear. When that happens, you can rest assured the geometry being displayed on screen is up to date.

✓ **TIP:** *The Load Referenced Documents option found in the External References section of the System Options (Tools > Options) offers the ability to always load externally referenced files automatically. Selecting the All option enables this functionality.*

Locking References

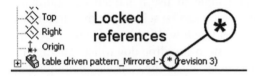

Fig. 7–60. Locked external references.

When references are locked, it means that any changes made to the referenced file will not show up in the file containing the reference. This can be useful if it is desirable to put a temporary "hold" on a model. If at a future time it becomes necessary to incorporate the changes made to the referenced file, the external references can be unlocked and the changes made to the referenced file will propagate forward to the model containing the reference. Figure 7–60 shows that an asterisk symbol (*) is used when external references have been locked. To lock (or unlock) external references, use the Lock All or Unlock All buttons found in the External References window.

Breaking External References

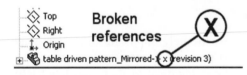

Fig. 7–61. Broken external references.

It is possible to remove external references on sketch geometry by editing a sketch and deleting any unwanted external references or relations (see "Deleting Relations" in Chapter 3). In the situation of a mirrored part, however, the external references are associated with a feature, not sketch objects. In this case, the external references can be broken using a different technique. Note that any external references can be broken, not just those created by mirroring a part. The symbol used to denote broken references is an x, shown in Figure 7–61.

To break external references, simply click on the Break All button found in the External References window. A warning message will appear. Clicking on OK will continue the process, and the external references will be broken. (You will be exposed to external references again later in the book.)

✗ **WARNING:** *Once references have been broken they cannot be reestablished. Heed the warning message displayed by SolidWorks and make sure the operation is what you want.*

Sketch Patterns

Sketch geometry can be patterned in a linear or circular fashion. The commands are known as *Linear Sketch Step and Repeat* and *Circular Sketch Step and Repeat*. Patterning feature geometry is recommended over patterning sketch geometry, as feature patterns are more flexible. However, there still may be times when the ability to pattern sketch geometry proves useful. Step-and-repeat commands are performed strictly on sketch geometry, never features, and are therefore considered sketch tools.

A notable difference step-and-repeat functions have over standard feature patterning is that there is no requirement to select an object that dictates the pattern direction. Instead, step-and-repeat functions reference the sketch origin. The *x* axis of the sketch origin, represented by the smaller of the two red arrows, would correlate to an angle of 0. The *y* axis would represent 90 degrees. Likewise, 180 degrees would represent the negative x direction, and 270 degrees would represent the negative *y* direction. This is exactly how a Cartesian coordinate system is arranged, and is otherwise known as a system of polar coordinates when combined with a distance value from the center of the coordinate system.

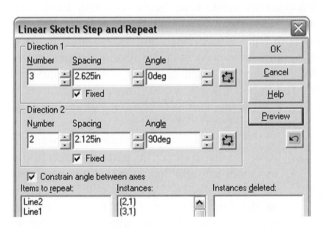

Fig. 7–62. Linear Sketch Step and Repeat *window.*

SW 2006: Linear and circular sketch-step-and-repeat commands have been renamed. The commands are now known simply as linear or circular sketch patterns. The Linear Pattern or Circular Pattern commands can be found in the Tools > Sketch Tools menu.

Like the name implies, the *Linear Sketch Step and Repeat* command will allow you to create copies of selected sketch geometry in a linear fashion. It is possible to create the linear pattern in a single direction, or in two directions at the same time. How-To 7–12 takes you through the process of using the *Linear Sketch Step and Repeat* command. The upper portion of the *Linear Sketch Step and Repeat* window is shown in Figure 7–62.

How-To 7-12: Performing a Linear Sketch Step and Repeat

To perform a *Linear Sketch Step and Repeat*, perform the following steps.

1. Click on the *Linear Sketch Step and Repeat* icon (shown in Figure 7–63), or select Linear Step and Repeat from the Tools > Sketch Tools menu.

Linear Sketch Step and Repeat

Fig. 7–63. Linear Sketch Step and Repeat icon.

2. Select the entities to be patterned. This can be one or more sketch entities.

3. Specify the Number of Instances (total) for Direction 1.

4. Specify the Spacing between instances.

5. Specify the Angle for Direction 1, where an angle of 0 is along the sketch origin's *x* axis.

6. Repeat steps 3 through 5 for Direction 2, if required.

7. Click on OK to create the pattern.

These steps cover the basics for creating a step-and-repeat pattern, but there are a few other options available in the window. For instance, select Fixed for either direction if you want a dimension added automatically for the Spacing distance. Select *Constrain angle between axes* if you want Solid-Works to add a dimension to control the angle between pattern directions. Both of these options are recommended. Do not confuse the Fixed option with the *Fix* geometric relation. They are not related in any way.

SW 2006: Sketch patterns (aside from being renamed) have been moved to PropertyManager, rather than being contained in their own window. This holds true for both types of sketch patterns (linear and circular). In addition, misleading options have been renamed to more clearly represent their functions. For instance, the Fixed option is more appropriately named Add Dimension. The option titled *Constrain angle between axes* is now labeled *Add angle dimension between axes*. Overall, you should find the sketch pattern interface easier to navigate.

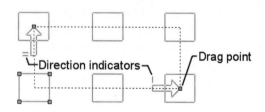

Fig. 7–64. Step-and-repeat preview.

If there are any instances you want to have removed from the pattern, select those instances from the Instances list box and press the Delete key. The instances will be displayed in the *Instances deleted* list box. Deleted instances can be retrieved (added back in) by removing their instance identifiers from the *Instances deleted* list box. Instance identifiers refer to the direction indicators in the preview, such as those shown in Figure 7–64.

Obviously, the arrow labeled 1 is indicative of Direction 1, and so on. Instance identifiers follow a grid pattern. That is, an instance identifier of (3,2) in the Instances list box refers to an instance positioned over 3 in direction 1 and across 2 in direction 2.

The dots at the arrow points can be dragged, thereby changing the spacing dynamically. The *Reverse direction* buttons will flip the pattern direction from one side to the other. In actuality, all that is happening when the *Reverse direction* buttons are used is that SolidWorks adds or subtracts 180 degrees from the Angle value. You could do this manually almost as easily.

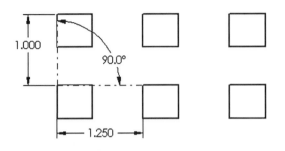

Figure 7–65 shows a completed sketch pattern. The dimensions were added by SolidWorks automatically, as were the construction lines. The linear dimensions were added because the Fixed option was checked, along with the *Constrain angle between axes* option. The 90-degree dimension is a result of the latter. Other dimensions and relations should still be added to the original sketch geometry to fully define it before using the sketch for a feature. As mentioned previously, fully defining your sketch geometry is always good practice.

Fig. 7–65. After completing a Linear Step and Repeat.

If you wish to pattern sketch entities in a circular fashion, use the *Circular Sketch Step and Repeat* command. The upper portion of the window that appears after initiating this command is shown in Figure 7–66. How-To 7–13takes you through the process of using the *Circular Sketch Step and Repeat* command.

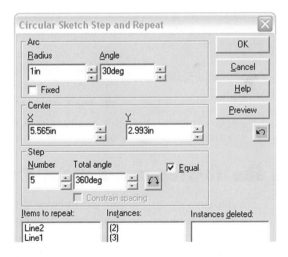

Fig. 7–66. Circular Sketch Step and Repeat window.

HOW-TO 7-13: Performing a Circular Sketch Step and Repeat

To perform a circular sketch-step-and-repeat, perform the following steps.

1. Click on the *Circular Sketch Step and Repeat* icon (shown in Figure 7–67), or select Circular Step and Repeat from the Tools > Sketch Tools menu.

Fig. 7–67. Circular Sketch Step and Repeat icon.

2. Select the entities to be patterned. This can be one or more sketch entities.

3. Specify where the center of the circular pattern should be positioned using either the Radius and Angle settings or the X and Y settings (explained in more detail shortly).

4. Specify the total number of instances.

5. Specify the spacing between instances, or the total angle if using the Equal option.

6. Click on OK to create the pattern.

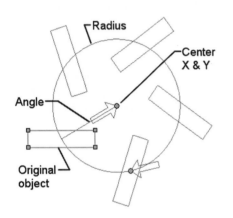

Fig. 7–68. Understanding the circular sketch pattern parameters.

If you have just tried to create a circular sketch pattern and found the process cumbersome, this is no surprise. This is not one of SolidWorks' more user friendly commands. To gain a better understanding of how the pattern is defined, examine Figure 7–68.

The Center X and Y parameters always reference the sketch origin. If you find it easy to dictate where the center of the pattern should be using Cartesian coordinates, enter values using the Center X and Y parameters.

The Arc Radius and Angle parameters use polar coordinates based on the centroid of the sketch objects being patterned. Referring to Figure 7–68, note the angled line emanating from the center of the original object. Assuming that the positive x axis is off to the right, the center of the pattern is currently at an angle of about 30 degrees. The radius value would be the distance from the center of the original object to the center of the pattern.

Editing a Sketch-Step-and-Repeat Pattern

Step-and-repeat pattern parameters can be edited after the command has been completed. To modify the spacing between instances, alter the dimension value associated with that distance. The dimension may have been added automatically, or it may be one that was placed by the user. Whatever the case, double click on the dimension to change its value. This holds true whether the dimension is for the spacing or for any other value related to the patterned geometry.

To change the number of instances within the pattern, right-click on any of the sketch entities in any of the pattern instances. You will see a command for Edit Linear (or Circular) Step and Repeat. Selecting this command will bring the original window back up, and the number of instances can be adjusted as required.

A quirk of the step-and-repeat commands occurs when editing the linear or circular step-and-repeat. The Spacing and Angle options will be grayed out. This is because the proper way of adjusting these values is to modify the dimensions associated with the geometry via double clicking. Nevertheless, the spin-box arrows still seem to work. Clicking on the spin-box arrows will even make the preview change. However, clicking on OK closes the window and none of the changes will take place. So again, to change the Spacing or Angle values just double click on the associated dimensions.

SW 2006: Sketch patterns no longer exhibit the odd behavior of being able to change parameter values that are grayed out (and which are now contained within PropertyManager).

Fill Patterns (SW 2006 Only)

As the name would imply, a fill pattern will fill an area with predefined shapes or user-defined features. The predefined shapes consist of circular, square, diamond, or polygonal cuts. User-defined features can be more complex shapes or bosses. Why would you use fill patterns? Typical uses would be for ventilation holes, aesthetics, tread plate or grip surfaces, and other applications.

Fill patterns must be applied to planar faces. However, the pattern can extend across multiple faces, as long as all faces are co-planar. Sketch geometry can be used to define the boundary of where the pattern will reside, or faces can be selected. How-To 7–14 takes you through the process of creating a fill pattern.

How-To 7-14: Creating a Fill Pattern

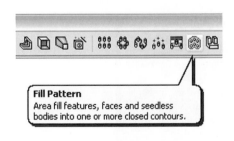

Fig. 7–69. Fill Pattern icon.

To create a fill pattern, perform the following steps. The Fill Pattern icon is shown in Figure 7–69.

1. Select Fill Pattern from the Insert > Pattern/Mirror menu, or click on the Fill Pattern icon, typically found on the Features toolbar.

2. In the Fill Pattern PropertyManager, a portion of which is shown in Figure 7–70, click in the Fill Boundary panel and select the boundary the pattern will fill. Allowable selections are co-planar faces, a single face, or closed-profile 2D sketch geometry.

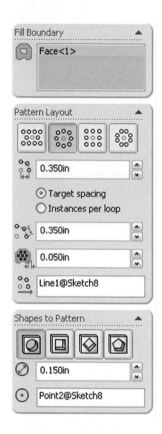

Fig. 7–70. Fill Pattern
PropertyManager.

3. Select the pattern shape from the Shapes to Pattern panel.

4. Specify the size parameters for the shape being patterned. Parameters will vary depending on the shape.

5. Optionally, specify a point where the fill pattern will be centered, such as a predefined sketch point or endpoint of a line.

6. Select the pattern layout type by clicking on one of the Pattern Layout buttons.

7. Depending on the pattern layout type, specify the parameters for the layout. This may include, but may not be limited to, spacing between instances, stagger angle, and so on. The Margin parameter is the spacing between the boundary and outer fill pattern instances. *Pattern direction* (optional) is a reference direction for the fill patterns' orientation, which can be dictated by selecting an edge or sketch line.

8. Click on OK to create the pattern.

Step 7 may take some experimentation, but it should only take a few minutes before the parameters begin to make sense. Depending on the pattern layout type, different parameters will be available, but these settings are not difficult to figure out. To pattern a user-defined feature rather than a predefined shape, expand the Features to Pattern panel (not shown) in the Fill Pattern PropertyManager and select a feature. This would replace steps 3 through 5 in the How-To.

Figure 7–71 shows a fill pattern on a sheet metal part. In the case of the part shown, the pattern was created to allow sound to travel through the part from an oval speaker. The predefined circle shape and a circular pattern layout were used. An ellipse was used to define the fill boundary.

Fig. 7–71. Fill Pattern
on a sheet metal part.

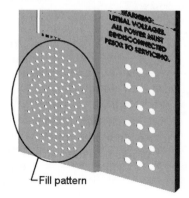

Summary

In this chapter you learned about patterns and how they can be useful, as well as the various forms of mirroring geometry. Patterns can be linear or circular in nature. In either case, an axis, dimension, or edge must be selected that determines the pattern direction or center of rotation. There are also patterns that can be defined via sketch points, *x-y* coordinates, or curves. These are sketch-driven, table-driven, and curve-driven patterns, respectively.

Like most anything else in SolidWorks, patterns are parametric. This includes the number of instances, which can be accessed by double clicking on a linear or circular pattern feature in FeatureManager. Instances in a pattern can be deleted by selecting a surface of that instance and pressing the Delete key. They can be brought back by editing the definition of the pattern feature. Linear patterns can be in one or two directions, and need not be orthogonal.

Once a direction has been specified for a pattern, and the objects to be patterned have been selected, a preview will be displayed. For circular patterns, the preview arrow points along the center of rotation for the pattern. Axes are the most common way of specifying the center of rotation for a circular pattern. Temporary axes are created automatically by SolidWorks. User-defined axes can be independently hidden or shown, and lengthened or shortened for reasons of aesthetics. User-defined axes also appear in FeatureManager.

Use copy and paste when only a few copies of the features are required but those instances do not necessarily constitute a pattern. The hot key combinations of Ctrl-C and Ctrl-V can be used to copy and paste, respectively. The hot keys are a Windows standard, and will work for any Windows program. Make sure to select a planar face on which to place the copied feature before pasting it in. Also bear in mind that you must not be editing a sketch for this functionality to exist

Patterned features will contain the original feature definition. If the Geometry Pattern option is enabled, feature definition information is not used. The patterned items will be duplicates of the original feature's geometry only. As is the case with any pattern, geometry pattern or otherwise, modifying the original feature will update all instances in the pattern. This is not the case when performing a simple copy and paste, as there is no link to the original.

Geometry can be mirrored by using a mirror plane or planar surface. It is also possible to mirror all of the geometry in an entire part. Use the Mirror Body list box in the Mirror command and select the model from the work area to accomplish this task. If a mirror image of a part is needed, you can employ the Mirror Part command, in which case a new part will be created that is an exact mirror image of the original. This latter function also creates an external reference to the original part file. If the original part is modified, the changes will propagate to the mirrored part. This effect is unidirectional only; if the mirror image part is modified, it will not affect the original.

Sketch-driven patterns are very useful when a pattern is going to be somewhat irregular in nature. A sketch is required that contains sketch points. Wherever a point is located, a pattern instance will appear in the completed pattern. Deleting a point in the reference sketch deletes its associated pattern instance.

Table-driven patterns can be useful when a set of *x-y* coordinates is available for positioning pattern instances. Coordinate text files can be imported into SolidWorks to dictate where the patterned instances will be located. Due to the nature of a table-driven pattern, a user-defined coordinate system must first be created. The coordinate system dictates the 0,0 reference point and *x-y* axes the coordinate data will reference.

Curve-driven patterns allow for patterning objects along a particular curve. The curve can be defined by a model edge or (more appropriately) a sketch. If a sketch is used, segments in the sketch geometry must be tangent for the curve-driven pattern to work.

Sketch step-and-repeat functions are another way of creating patterns. Step-and-repeat functions can only be used in a sketch to pattern sketch geometry. Feature patterns are typically easier to work with and provide more flexibility.

CHAPTER 8

Model Appearance

IF EVERY SOLIDWORKS MODEL HAD TO REMAIN GRAY, life as a CAD operator would certainly get boring. Not only is the ability to change part color an aesthetic nicety, it's also functional. For example, it would be nearly impossible to distinguish components from one another in an assembly if they were all the same color.

There are other functional reasons color and model appearance in general play an important role. Being able to distinguish faces that require draft during a draft analysis (which you learned in Chapter 6) is a perfect example. Sometimes it is beneficial for a model to simply look pleasing to the eye. A picture speaks a thousand words, and a customer has a better chance of being impressed by your product if it can be shown in a realistic or cosmetically appealing fashion.

This chapter will teach you not only how to change part, feature, or face colors but how to apply textures and materials. You will also learn how to manipulate lighting characteristics, such as specularity or transparency, and even how to add or modify light sources.

Modifying Part Color

Although there are different ways in which to alter the color of a model, one particular method stands out as the easiest. This method involves the Edit Color command, whose icon is shown in Figure 8–1, along with the Edit Material and Edit Texture icons. All of these commands can be found on the Standard toolbar. The Edit Color command allows for changing the color of an entire part, individual features, or even specific faces.

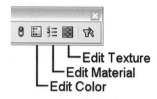

Fig. 8–1. Commands for changing part appearance.

There is a distinct hierarchy followed regarding colors of objects. Specifically, colors of faces override colors of features, colors of features override colors of bodies, and colors of bodies override colors of parts. Just remember the order (faces, features, bodies, parts) and the rest comes easy. For example, if the color of a feature has been changed it will retain its color if the color of the overall part is altered. How-To 8–1 takes you through the process of changing the color of faces, features, bodies, or parts.

How-To 8-1: Editing Colors

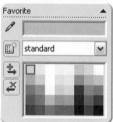

Fig. 8–2. Color and Optics PropertyManager.

To change the color of faces, features, bodies, or parts, perform the following steps.

1. Click on the Edit Color icon or select Color from the Edit > Appearance menu.

2. From the work area or the fly-out FeatureManager, select the object whose color is to be changed.

3. Select the desired color from the Color and Optics PropertyManager, shown in Figure 8–2.

4. Click on OK to accept the color changes.

In its most simple form, choosing a color is just a matter of selecting a swatch from the palette shown at the bottom of Figure 8–2. To change the color of the entire part, select the part from the top of the fly-out FeatureManager. Click on the pushpin if you wish to change the color of multiple items without restarting the Edit Color command.

There are filter buttons built into the Selection panel, as can be seen in Figure 8–2. It is important how objects are selected when applying color, which is why the filter buttons are there. They are not necessary most of the time if you are careful about how items are selected. For example, clicking on the model itself will serve to select a face on the model. Selecting a feature from FeatureManager will serve to select that feature. Bodies can be selected from the *Solid Bodies* folder, and so on.

The drop-down list in the Favorite panel (also visible in Figure 8–2) contains preset categories for dull, shiny, standard, and transparent colors. Actually, there are two sets of these lists, with the first four items consisting of system presets. The second set of four preset colors is user customizable. Feel free to add or remove swatches of color from the last four items in the list.

Selecting from the Favorite list changes the optical properties for that color palette. Selecting a swatch of color changes the color properties, such as the red, green, and blue (RGB) values. Both the color and optical properties can be adjusted manually as well. These panels are shown side by side in Figure 8–3.

The slider bars found in the Color properties panel can control either the RGB components of the color or what SolidWorks refers to as the hue, saturation, and value components. Most graphic artists think of the "value" as luminance. Checking the Numeric option allows for plugging in the actual RGB values, which allows for creating a color in a much more precise manner.

Fig. 8–3. Color and Optical Properties panels.

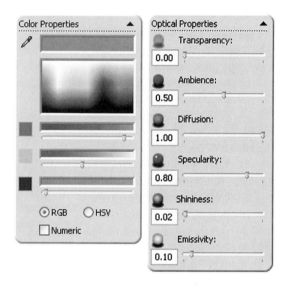

Optical Properties

The Optical properties panel (shown Figure 8–3) can be used to adjust everything from specularity to transparency. Not all of the six characteristics that can be modified are necessarily self-explanatory. Therefore, some further elaboration is in order. The sections that follow describe these various optical properties.

Transparency

Transparency is the one setting that everybody understands. Technically, it is the amount of light that can pass through an object. If the Transparency setting is turned up all the way, the part is invisible because it is passing 100% of the light being directed toward it.

Ambience

Ambient light is the light scattered throughout the room. Any light reflected from other objects that adds to the overall brightness of the scene is considered ambient light.

Diffusion

When light hits an object, a certain amount of light is reflected by the surface of the object equally in all directions. This is known as diffuse light.

Specularity

A surface's specularity is its ability to reflect light. Increasing the Specularity setting will increase the amount of light shining from a "hot spot" directly toward the user. This is similar to the shininess property (see following section), but where specularity changes the brightness of a hot spot shininess will change its size.

Shininess

If a part is very shiny, it will reflect light in a more concentrated beam. Increasing the Shininess setting will make the reflective hot spot smaller. Decreasing the Shininess setting all the way has the effect of making the part bright and washed-out looking because the reflective hot spot is no longer isolated to a narrow beam pointing toward the user.

Emissivity

Light being emitted from a part is essentially light being generated and projected outward from a surface. This setting is the least commonly used of all six settings. The Emission setting has its place in the grand scheme of things, but is not typically used on a day-to-day basis for most parts.

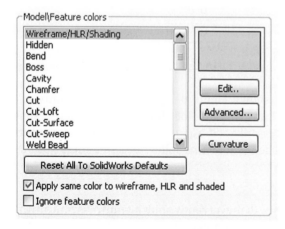

Fig. 8–4. Color section of the Document Properties window.

Feature Color Settings

In contrast to changing the color of features on a feature-by-feature basis using the Edit Color command, colors can also be set for specific feature types. For example, cuts could be blue, boss extrusions might be brown, and fillets and chamfers could be red.

Setting colors to specific features is done in the Color section of the Document Properties window, which is accessed by clicking on Options in the Tools menu. The Color section is shown in Figure 8–4. Changing a feature's color is simply a matter of selecting a feature from the list, clicking on the Edit button, and selecting the desired color.

The Advanced button (shown in Figure 8–4) allows for adjusting the six optical properties of the overall part. These properties are the same as were discussed in the previous section. The optical properties can only be adjusted for the overall part, and hence the Advanced button will not be available if an individual feature is selected.

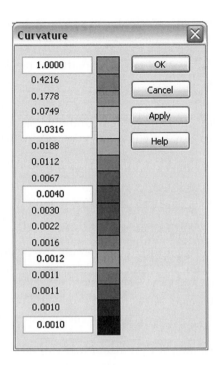

Fig. 8–5. Curvature color settings.

Finally, there is a button labeled Curvature. Clicking on this button opens the window shown in Figure 8–5. The colors associated with the curvature of the model cannot be edited directly. Rather, it is the curvature values associated with the colors that can be modified. Entering a different curvature value serves to modify all values above or below the modified value. Intermediate values are interpolated automatically by SolidWorks.

If the topic of curvature is new to you, it is best to leave the curvature settings at their default values. Curvature is discussed in detail in Chapter 20.

Textures

Adding textures to a model is another way to give the model a much more realistic appearance. There are a variety of textures, such as metal, stone, and wood. Textures are bitmap files that get mapped onto faces of the model. Users can create their own textures by creating bitmap images using a Windows paint program or photo-editing software.

Applying textures starts out in a manner nearly identical to changing the color of a model. However, due to the nature of textures there are additional steps involving positioning or scaling the texture. Wood grain may need to run in a particular direction, or tread plate may need to be a particular size, for example. How-To 8–2 takes you through the process of applying textures.

HOW-TO 8-2: Applying Textures

To apply textures to a model, perform the following steps.

1. Click on the Edit Texture icon found on the Standard toolbar, or select Edit > Appearance > Texture.

2. Select the faces, features, or bodies to which the textures will be applied. Optionally, select the entire part from the top of the fly-out FeatureManager.

pine.jpg

☐ Blend color

Fig. 8–6. Texture Selection panel.

3. Select the desired texture from the Texture Selection panel, shown in Figure 8–6.

4. Modify the texture properties as needed using the scale and angle slider bars, shown in Figure 8–6.

5. Select the *Blend color* option if the color of the selected object (i.e., the model) should be blended with the selected texture. Leaving this option off uses the texture as is.

6. Click on OK when finished to apply the textures.

A part that has had textures applied to it in a variety of ways is shown in Figure 8–7. Pine was used for the overall part, and a metal texture was used for the "inserts." This model is a part file, not an assembly, so how were the inserts made to look the way they are? The Split Line command, which you learned in Chapter 6, was used to create faces separate from the overall top surface of the model. The technique extends to the underside of the model as well. The black-and-white printing of this book does not do the image justice.

Another texture detail that can be seen if we look closer is the thread textures applied to the lower portion of the counterbored holes. An enlarged cutaway view is shown in Figure 8–8, which should give you a finer appreciation of what has been accomplished. Although the insert does not truly extend through the part, it could be made to appear that way should it become necessary to show the cutaway view in a brochure, for example.

The text on the model is a cut feature using sketched text, which you will learn about in Chapter 22. A special technique was required to display the engraved text, which appears bright red in the actual model file. Because textures take precedence over colors, a color applied to a feature will not appear in its true state, or may not appear at all.

The Blend color option is critical in forcing colors to appear. When used, the color of the selected feature (or entire model) is blended with the selected texture. If a light and plain texture is selected and the Blend color option turned on, the underlying color will appear very close to its true state. Better yet, create your own texture bitmap and make the entire texture white. This will allow the underlying color to appear perfectly natural. The section that follows teaches you how to do this.

Fig. 8–7. Model with textures applied.

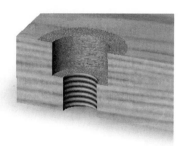

Fig. 8–8. Note the thread textures.

Creating Textures

SolidWorks has a fair number of textures it includes with the software, but you may want to create some of your own. Another software program will be needed to perform this task. It is outside the scope of this book to recommend what software program to use, but the basic guidelines can be given.

The standard bitmap images included with SolidWorks are found in the *data\images\textures* folder. Examining these images, we find that they have a *.jpg* (pronounced jay-peg) file extension (except for some of the special-use bitmaps) and are 600 by 600 pixels in size. Upon closer examination, these images have a 24-bit color depth (16.8 million colors) and a resolution of 72 pixels per inch (ppi). Textures you create should have these same characteristics.

It would be wise to create your own texture library, rather than add to the existing SolidWorks stock library. In this way, the chances of something becoming corrupted in the software is much less likely. Simply create your own folder at a location of your choosing, and then specify that path using the process outlined in How-To 3–6. As a reminder, click on Tools > Options and select File Locations from the System Options tab. You can then specify the path for Textures, as shown in Figure 8–9.

Fig. 8–9. Specifying a path for a custom texture library.

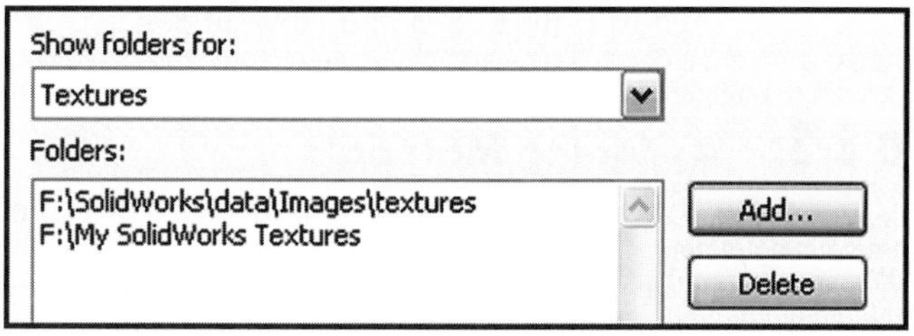

Show folders for:

Textures

Folders:

F:\SolidWorks\data\Images\textures
F:\My SolidWorks Textures

Add...

Delete

One added point should be made regarding custom textures. Most textures should be what are known as a *seamless pattern*. What this means is that if the textures were tiled one after the other in a quilt fashion it would be impossible to tell where one tile merged into the next, either side to side or top to bottom. Not all graphic art software programs have the capability to create a seamless pattern, so make sure you do some research before laying down your hard-earned cash.

Materials

Materials are different from textures in that they do not use bitmaps. Materials use standard colors and optical properties. If the computer graphics card is capable of displaying RealView graphics, the model will look very realistic. Materials also add physical characteristics to a model, such as density, yield strength, thermal conductivity, and others.

RealView graphics are more realistic than is typical. Chrome looks like chrome, complete with reflections from some unseen surroundings. (Reflective materials such as chrome would typically reflect what is in their environment. However, there is nothing "in" the computer-generated part file's environment and thus one is supplied by SolidWorks. The user sees this only in the reflection.) Wrought copper contains hammer marks, glass looks like glass, and plastic looks like plastic. RealView graphics really need to be seen on a computer monitor before they can be fully appreciated.

Because RealView graphics display realistic materials in real time, they require special graphics cards. Generally speaking, standard gaming cards will not suffice. Workstation-grade Open Graphics Language (OpenGL) cards are required. Check the SolidWorks web site for a list of graphics cards that support RealView graphics, at *www.solidworks.com*.

Applying a material to a model can have a detrimental side effect: it may reduce the performance of the model. This can be noticed in that rebuild times may become significantly longer and graphically the model may not rotate as smoothly. If you notice this occurring, consider removing the material assigned to the model. Applying and removing materials are covered in How-To 8–3 and How-To 8–4.

HOW-TO 8-3:   Assigning Materials

To assign a material to a model, perform the following steps.

1. Click on the Edit Material icon, or select Material from the Edit > Appearance menu. You can also right-click on the Material object in FeatureManager and select Edit.

2. Select a material from the Materials panel listing, shown in Figure 8–10.

Fig. 8–10. Selecting a material.

3. Click on OK to apply the material to the part.

If you have a RealView-graphics-capable video card, you can turn on the RealView graphics display. To do so, select View > Display > RealView.

When a material is applied to a part, the Material object in FeatureManager displays the assigned material. This is shown in Figure 8–11. Normally, the material object states "< not specified >" prior to assigning a material.

The Visual Properties panel of the Materials Editor PropertyManager is shown in Figure 8–12. If not using RealView graphics, there are settings that can be tweaked, similar to when positioning a texture. As a matter of fact, textures are used for many materials if RealView graphics is disabled or simply not available. There are the same scale and angle slider bars, along with the *Blend color* option. Additionally, there are *Use material color* and *Use material crosshatch* options.

Checking *Use material color* forces the color assigned to the selected material in the material database to be used for that material. Otherwise, the part will take on the natural appearance of the texture bitmap assigned to the material. The color of the part will be blended with the texture bitmap if *Blend color* is checked.

Fig. 8–12. Material visual properties.

Fig. 8–11. Material object in FeatureManager.

Checking *Use material crosshatch* forces the crosshatch assigned to the material (again, in the material database) to be used if a section view of the part is created in a drawing. Section views are discussed in Chapter 12. If *Use material crosshatch* is unchecked, the crosshatch assigned to the part is used. You will explore how to assign crosshatch to a part later in this chapter.

How-To 8-4: Removing Materials, Textures, or Colors

To remove a material from a part—or to remove textures or colors from parts, bodies, features, or faces—perform the following steps.

1. Enter the Edit Color, Edit Texture, or Edit Material command, whichever applies.

2. Select the part, bodies, features, or faces from which you wish to remove the material, texture, or color.

3. Click on the Remove button. The button will actually be titled Remove Material, Remove Texture, or Remove Color, depending on the command.

4. Click on OK to accept the changes.

Creating a Custom Material

Before a custom material can be created, a custom material database must be created. Both of these processes go hand in hand. The process is quite simple, but there are a fair number of steps which must be run through. How-To 8-5, which follows, will take you through the entire process.

How-To 8-5: Creating a Custom Material

To create a custom material, perform the following steps.

1. Access the Material Editor, such as by right-clicking on the material in FeatureManager (shown in Figure 8–11).

2. Select a material from the materials list that has properties similar to the material you wish to create.

3. Click on the Create/Edit Material button. This will open the window shown in Figure 8–13.

4. Click on the drop-down list (labeled 1 in Figure 8–13) and select < New Material Database > .

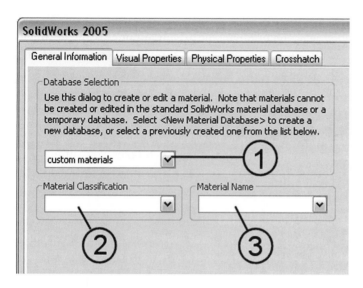

Fig. 8–13. Creating a new material.

5. Enter a name for the new database and click on the Save button. The name *My Custom Materials* might be a good choice, but you can name the database anything you wish.

6. Click in the area labeled 2 in Figure 8–13 and type in a Material Classification (i.e., metal, wood, plastic, and so on).

7. Click in the area labeled 3 in Figure 8–13 and type in a Material Name.

8. Click on the Visual Properties tab.

9. Select a color to represent the new material. Optionally, select a texture as well.

10. Select a material if your computer can display RealView graphics, and a PhotoWorks material if you have PhotoWorks (which is outside the scope of this book).

11. Click on the Physical Properties tab.

12. Double click on any of the physical properties listed and enter the desired value for the new material.

13. Click on the Crosshatch tab.

14. Specify how the crosshatch should appear for the new material.

15. Click on OK to create the new material database and material.

Fig. 8–14. Utilizing the new database.

Now that you have created a new material, as well as a custom database, you can add materials to that database. The database can be selected from the drop-down list shown in Figure 8–14, and the custom materials can then be chosen.

When is it necessary to apply a material? It is only necessary for two reasons. The first reason would be to take advantage of Real-

View graphics. The second reason is to perform a stress analysis on a model, which is outside the scope of this book. If you are not taking advantage of one of these functions, it isn't necessary to apply a material. In fact, it is not recommended. Because applying a material often contributes to a loss of performance and increased rebuild times, it is generally recommended that a part's material properties be changed instead. Modifying a part's material properties is the method by which a density and crosshatch can be associated with the part. This is much more efficient than assigning a material, and will not affect rebuild times. How-To 8–6 takes you through the process of modifying the material properties of a part.

How-To 8-6: Modifying Material Properties

To modify a part's material properties, and thereby change its associated density and crosshatch, perform the following steps. The Material Properties window is shown in Figure 8–15.

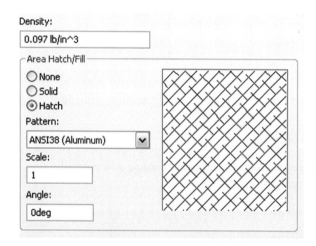

Fig. 8–15. Material Properties window.

❖ **NOTE:** *If a material has been assigned to the model, the Density value will be unavailable. Likewise, if the Use material crosshatch option is checked in the Materials Editor, the crosshatch settings will be unavailable.*

1. Select Options from the Tools menu.

2. Click on the Document Properties tab and select Material Properties.

3. Specify the density of the part.

4. Specify what crosshatch should be associated with the part in the event a section view is created of the model in a drawing.

5. Click on OK to accept the changes.

✓ **TIP:** *If working in one set of units, but the density is only known in another set of units, specifying the density in the known units will make SolidWorks convert the density to the current operating units. For instance, if working in inches and a density of .0024 gram/mm^3 is entered the value will be converted to .0867 lb/in^3.*

Mass Properties

Once the density of a part has been set, or a material has been specified for the part, the mass properties of the model can be accurately obtained. To do so, select Mass Properties from the Tools menu. There is also a Mass Properties icon available on the Tools toolbar. Accessing mass properties will bring up the window shown in Figure 8–16. Only the upper half of the Mass Properties window is shown, with a model overlaid on the window.

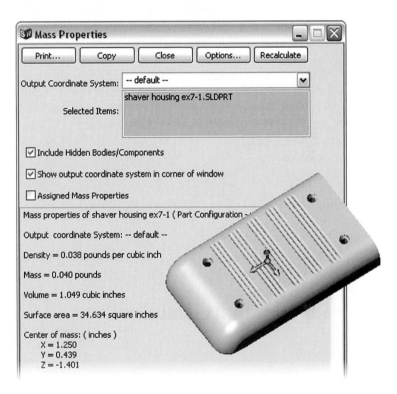

Fig. 8–16. Viewing a model's mass properties.

A good deal of relevant data can be obtained from the Mass Properties window, as you can see. Also included, but not visible in the figure, are the principal axes of inertia and moments of inertia. All mass properties will display in whatever the model's working units have been set to. Clicking on the Options button allows for displaying the mass properties in units other than those specified for the model. For example, if working in metric units the mass properties can still be displayed in inches and pounds.

Moments of inertia are displayed as "pounds * square inch." This reads as pounds per square inch, and would refer to pounds mass not pounds force. Pounds force is equal to pounds mass times acceleration, and there is no acceleration vector specifiable in a SolidWorks model. If all of this sounds like so much gibberish, consider this information as not pertinent to your line of work.

The center of mass (centroid) of the model is represented by a purple triad. This triad is visible in Figure 8–16 at the center of the part. Also visible are a few options that require explanation. The Include Hidden Bodies/ Components option is only relevant if working with parts containing multiple bodies or if working with assemblies. If a body or component (in an assembly) is hidden, the body or component can still be included in the mass properties calculation. Be forewarned that if a body or component is suppressed it will not be included in the calculations under any circumstances. Suppressed objects are removed from memory, whereas hidden objects are simply removed from view.

Show output coordinate system in corner of window refers to the reference triad, which is shown automatically when viewing mass properties. If this option is unchecked, the reference triad is displayed on the model rather than in the lower left-hand corner.

✓ **TIP:** *Use the Copy button to copy mass properties to the Windows clipboard, where they can then be pasted into another program, such as Microsoft Excel.*

When viewing mass properties of a model, a different coordinate system can be referenced other than the default coordinate system of the model. The user-defined coordinate system can be selected from the Output Coordinate System drop-down list in the Mass Properties window. To create a coordinate system, see How-To 7–8.

Mass Properties in Assemblies

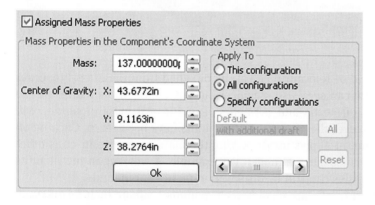

Fig. 8–17. Assigning custom mass properties.

Checking *Assigned mass properties* makes available an area in which custom mass properties can be added (see Figure 8–17). This capability would only be used under certain circumstances. Without going into all the details at this point, an assembly can be saved as a single part file. The part may be a simplified representation of the true assembly. This is done to show positional information of a subassembly, for example, without increasing the complexity of a top-level assembly. This is often referred to as a phantom assembly.

If a simplified representation of an assembly is used, the true assembly's mass properties can still be specified via the *Assigned mass properties* option. This allows the top-level assembly to still display the correct overall mass properties while benefiting from the use of phantom assemblies. You will learn a great deal about assemblies in Chapter 13.

The list box area labeled Selected Items in the Mass Properties window shows the items selected. In an assembly, this can be particularly important. The mass properties of individual components can be selected, in any combination, rather than being limited to viewing mass property data on the entire assembly.

In the case of an assembly, the mass property data is only going to be accurate if each component has been assigned the proper density first. In an assembly, the possibility of error is increased due to multiple components being present in the assembly. In that each component can have its own density setting, you will need to make it a point to set the density value for each component.

✓ **TIP:** *It is good practice to set the density and crosshatch properties of a part when it is being created so that you do not have to worry about setting these values later.*

10-second Topic: Section Properties

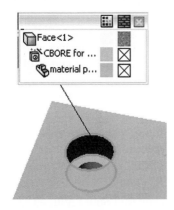

Fig. 8–18. Example of an appearance callout.

Another command that is very similar to Mass Properties is named Section Properties, which is also found in the Tools menu. Various bits of information related to a selected planar face can be determined via this command. Obviously, there would be no mass associated with a face, as it would have no thickness. However, moments of inertia and related data can be obtained.

Any planar face can have its section properties displayed. Although the selected face does not have to literally be a cross section of some portion of a model, it does have to be planar. The face's centroid will be shown, and the buttons found in the Section Properties window all perform exactly as they do in the Mass Properties window.

SW 2006: Appearance callouts, such as that shown in Figure 8–18, make it very easy to see what colors or textures have been assigned to a selected face. The color and texture applied to the associated feature, body (if relevant), and part are also displayed. Clicking on one of the color or texture swatches will open the Edit Color PropertyManager or the Texture Property-Manager, respectively. To turn on appearance callouts, check Appearance Callouts in the View menu.

Light Sources

There are three light sources that can be added by the user. They are directional lights, spot lights, and point light sources. This is in addition to the ambient light source that exists in every part and assembly. The user can control certain aspects of the default ambient light source, but cannot create additional ambient light sources. There would be no point in it.

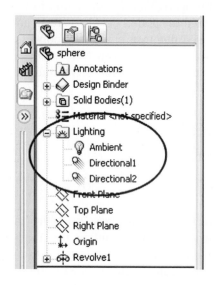

Fig. 8–19. Lighting folder in FeatureManager.

All lights are controlled through a special folder in FeatureManager named *Lighting*, shown in Figure 8–19. The *Lighting* folder usually contains three default light sources: Ambient, Directional1, and Directional2. New sources can be added via the *Lighting* folder, and existing light sources can be modified. Lights can be adjusted to your liking and made available for all future parts simply by saving a template (see Chapter 3).

SW 2006: The *Lighting* folder has been renamed *Lights and Cameras* to accommodate the inclusion of cameras, discussed at the end of this chapter.

Be forewarned that experimenting with lighting sources can be very entertaining and may decrease productivity! However, they can also be used to create some handsome effects if you are looking to spruce up a model. The following sections show you how to create and modify the various light sources. Following this, you will learn more about the individual light sources themselves. How-To 8–7 takes you through the process of adding and editing a light source.

How-To 8-7:   Adding and Editing a Light Source

To add a light source, perform the following steps.

1. Right-click on the *Lighting* folder.

2. Select Add Directional Light, Add Point Light, or Add Spot Light, as required.

3. The new light source will appear in the *Lighting* folder. Right-click on the new light source and select Properties.

4. Modify the light's properties as required. These properties are explained in detail in material to follow.

5. Click on OK to accept the changes.

After selecting the light source to be added, that particular light source automatically gets added to the *Lighting* folder. You do not have to modify the new light sources properties, but you will almost certainly want to. The difficult part is understanding all of the options available for the various light sources. This is covered in the following material, which starts with the easiest of all light sources, the ambient light source.

✓   **TIP:** *Light sources can be turned on or off by right-clicking on them.*

Fig. 8–20. Ambient light source PropertyManager.

Ambient

The one and only ambient light source is the one provided by SolidWorks. When the properties of the ambient light source are accessed, the Ambient light source PropertyManager appears (see Figure 8–20). There are not many adjustments that can be made, so this will be a short topic.

Adjusting the Intensity slider bar will increase or decrease the ambient light surrounding the model. A setting of .10 is a reasonable level. However, the setting can be adjusted anywhere from 0 to 1. Use the Edit button to control the color of the ambient light source. The small light bulb button serves as the on/off switch.

Directional

Directional lighting is added by default, but directional light sources can be added as required. As the name would imply, the direction of the light source can be altered to shine from any position surrounding the model.

Directional light sources are collimated, meaning that the light rays are parallel, which makes directional lights act as if they were from a very distant light source such as the sun. The target point of a directional light source (where the light points) is always at an *x-y-z* location of 0,0,0. The Directional light source PropertyManager is shown in Figure 8–21.

Directional lights have their own Ambient setting, along with slider bars for Brightness and Specularity. The Brightness setting controls the intensity of the directional light, whereas the Specularity setting controls the reflective properties of the light source.

If using colored light sources other than white, hot spots may appear dimmer. Specularity may need to be increased for these light sources to compensate for this. If you are unfamiliar with the term *hot spots*, Figure 8–22 points out what this term represents. Hot spots are typically smaller and brighter on smooth, glossy objects.

Although the position of a directional light can be controlled via the latitude and longitude slider bars (shown in Figure 8–21), it isn't really necessary. Directional lights can be dragged by the arrow on the directional light representation in the work area. Note the arrow on the directional light in Figure 8–22. To position the light, simply drag the arrow.

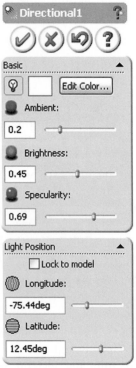

Fig. 8–21. Directional light source PropertyManager.

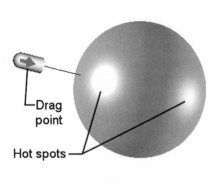

Fig. 8–22. Directional light drag point.

10-second Topic: Lock to Model Option

What is meant by Lock to Model? When Lock to Model is unchecked, the light source will remain in a constant position. The part can be rotated, but the light source stays where it was put. When Lock to Model is checked, the light source rotates with the model. Any light source (except for Ambient) can either be unlocked or locked to the model.

Whether or not the Lock to Model option is checked is dependent on your personal preferences and what effects you are trying to achieve. In the case of a light source meant to represent a light bulb in a lamp, the light should be locked to the model. If setting up light sources as though you were showcasing a model in a studio, leave Lock to Model off. With Lock to Model off, you can be the museum curator and the model your showpiece.

Point

Point lights act as very small pinpoint light sources that shine in all directions simultaneously. The location of the point light can be altered, but not its direction, because there is no specific direction in which the light is shining. The basic properties of a point light source are identical to those of a directional light source, so we won't bother rehashing that topic.

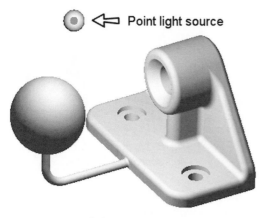

← Point light source

A point light source is graphically represented as a sphere. This serves to remind the user that light is being emitted from the point light in all directions simultaneously. An example of a point light source is shown in Figure 8–23. This graphical representation will only be shown when accessing the properties of the light source.

Accessing the properties of a point light source gains access to the Point light source PropertyManager, shown in Figure 8–24. Note that the light position can be specified using either Cartesian or spherical coordinates. As with directional lights, the point light source can also be dragged, thereby changing its position.

Fig. 8–23. Point light source.

Fig. 8–24. Point light source PropertyManager.

Being able to specify a precise *x-y-z* position is convenient. Imagine for a moment that you wish to position a point light source in the center of a light bulb in a lamp you just finished modeling. By entering the *x-y-z* coordinates for the light source, the light can be positioned exactly where it is needed.

✓ **TIP:** *The status bar at the bottom of the SolidWorks window will display x-y-z coordinates when a sketch point or vertex point is selected.*

In the example shown in Figure 8–25, the point light source was placed directly at the center of the spherical feature. Note that the shading of the model seems to reinforce this fact. The sphere itself, however, does not appear to be radiating any light. This is due to the way light is handled in a SolidWorks model.

Surfaces that face the point light source will be brighter, as though light is being shone upon them. This is true even for faces behind other objects (see Figure 8-26). Plain and simple, if faces point toward the light source the faces will be lit. If faces point away from the light source, they will be dark. Realistic shadows are not displayed in SolidWorks unless photo-rendering software is used (such as PhotoWorks). The lack of shadows is depicted in Figure 8-26.

Fig. 8-25. Point light positioned at the center of the sphere.

Fig. 8-26. Surfaces facing s light source are brighter.

⇥ **NOTE:** *Regarding shadows, the Display Shadows in Shaded Mode display option (View > Display) is in no way associated with light sources.*

Spot

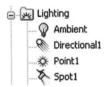

Fig. 8-27. Various light sources in the Lighting *folder.*

Spot light sources are extremely useful because they offer the unique capability of being able to have not just their position set but their target. Another useful function of spot lights is the ability to alter the cone angle at which the light beam is being projected. The cone can be narrowed to display a tight beam of light that shines in any direction the user desires. Spot lights (and the other light sources) are represented in FeatureManager as shown in Figure 8-27. The image has been enlarged for clarity.

✓ **TIP:** *Light sources can be renamed just like any other feature.*

Spot lights have the nicest graphical representation of all light sources. The light source appears as a sphere, similar to a point light, and the cone of the spot light's beam is depicted as well. The beam of the spot light will fall within the area encompassed by the cone. Any geometry that falls outside the cone will not be illuminated, at least not by the spot light source in question. Figure 8-28 shows how a spot light is represented in the work area.

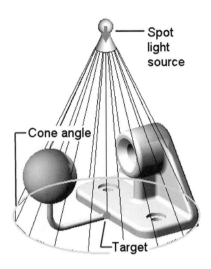

Fig. 8–28. Spot light in work area.

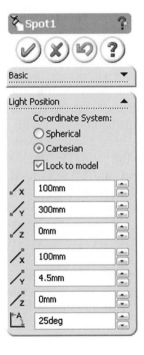

Fig. 8–29. Spot light PropertyManager.

Any of the elements that make up the definition of a spot light source can be edited by dragging those elements. This includes the spot light source location, the target (where the light is pointing), and even the cone angle. Dragging the circle at the base of the cone will adjust the cone angle. All of these elements can also be specified via the spot light source PropertyManager, a portion of which is shown in Figure 8–29.

The spot light PropertyManager contains slider bars for Ambient, Brightness, and Specularity settings (not shown). There is also the Light Position panel, visible in Figure 8–29, which allows for positioning both the light source and target via spherical or Cartesian coordinates. The Angle adjustment is how the cone angle can be controlled.

The location of the spot light in Figure 8–30 was made by adding a reference point in a sketch. The reference point is just a point entity constrained to be concentric to the circular edge of the hole. Once this was accomplished, the *x-y-z* location of the point could then be found by clicking on the point and reading the *x-y-z* coordinates from the status bar. The *x-y-z* coordinates can then be entered into the relevant fields of the spot light's PropertyManager. In a case such as this, we would want to make sure to check the Lock to Model option so that the light will not fall out of position if the model is rotated.

There is an additional panel in the spot light PropertyManager labeled Advanced, which is shown in Figure 8–31. This panel controls the "roll-off" aspect of the lights cone. The Attenuation setting is another way of controlling the brightness of the light source. Let's examine these settings in more detail so that you may more fully understand them.

In Figure 8–30, a spot light has been positioned at the center of the hole in the cylindrical boss. The beam of the spot light has been modified to point directly at the spherical feature. The cone angle in the example has been reduced to only 10 degrees.

If you visualize the cone of light created by a spot light, you discover that it is brighter at the center than it is near its edges. How quickly the light fades near the edges is known as the roll-off. The Exponent slider bar is what controls the light level's roll-off. To put it simply, it is how sharp the transition is between the circle of light and the surrounding area. The default value of 0 makes for a very sharp transition. In Figure 8–32, a spot light is being projected onto a "screen." The spot light source is positioned inside the hole in the cylindrical feature. The Exponent value has been set to 0.

Fig. 8–30. Positioning a spot light.

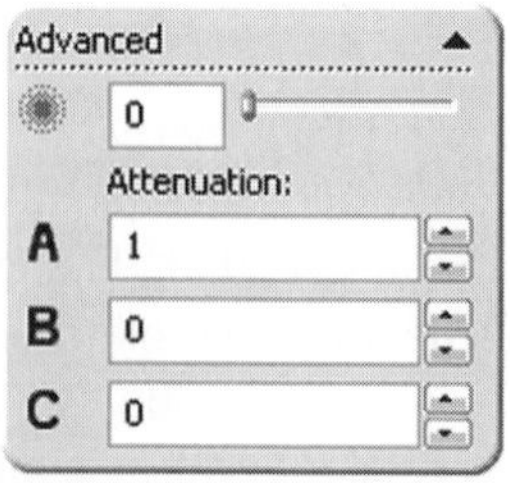

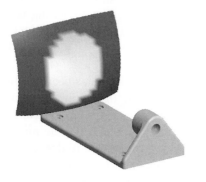

Fig. 8–31. Spot light PropertyManager Advanced panel.

Fig. 8–32. An exponent value of 0 (zero) produces a sharp transition.

Note that the circle of light is choppy-looking, not defining a true circle. This has to do with how the shading is handled internally. The polygons used to make up the model either fall within the light's beam or outside it. If a larger number of polygons is used, the circle of light will appear less choppy, but only to a certain extent. The number of polygons used to make up the model is controlled in the Image Quality section of Document Properties (Tools > Options).

Incidentally, the "screen" in the image is curved. This is necessary in order to force the software to use more polygons to create the screen. If a simple flat face were used, SolidWorks would create one or two polygons, as opposed to dozens or hundreds. The screen would either appear bright or dark, with nothing between. It would not be possible to focus the beam of light to form a circle in such a situation, even with the Exponent value set to 1, its highest setting.

Increasing the value of the Exponent setting will result in a circle of light that is smoother. The line of demarcation between light and shadow is not as sharply defined. The light falls off gradually as it moves away from the beam's axis. Figure 8–33 shows an example of this. In the image, the Exponent value has been set to .50. Because increasing the Exponent setting also decreases the size of the area of brightness, the cone angle was also increased a bit to compensate.

What is the term *exponent* referring to? Remember that light drops off exponentially with respect to distance. By moving the

Exponent slider bar, you are modifying the drop-off rate of the spot light source.

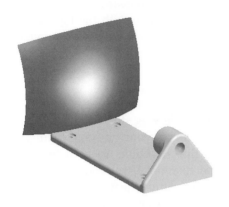

Fig. 8–33. A smoother transition with an increased exponent value.

The Attenuation setting is another way of adjusting the brightness of the light. Where the brightness slider bar controls the intensity of the light source, the Attenuation settings control how much the light is attenuated before reaching its target. The differences are subtle, so you will need to experiment. Decreasing the value in the box labeled A will increase the brightness. Adding values to the box labeled B or C will serve to decrease the brightness level of the light further. Valid values can be anywhere from 0 to 1.

Basically, the A setting can be thought of as a course adjustment. The B and C settings are medium and fine brightness level adjustments. This is the easiest way of looking at these settings. The actual formula used to calculate the brightness level is as follows, where D is the distance. It is much easier to simply adjust the brightness level in the Basic tab of the light source's properties.

```
Attenuation = 1 / (A + (B * D) + (C * D^2))
```

Cameras (SW 2006 Only)

Light sources will illuminate a model, but a camera will allow for looking at the model from a different perspective, literally. The model can be viewed from wherever the camera happens to be. The camera's field of view and perspective can be customized, or a standard "lens" can be used, such as a 35-mm wide-angle lens. Cameras can even be mounted to moveable components in an assembly.

Fig. 8–34. Solar system assembly.

Admittedly, cameras play a much more interesting role when add-in products such as SolidWorks Animator and PhotoWorks are available. Nonetheless, there is still a fair amount of usefulness that can be obtained from adding cameras to a model using SolidWorks without add-in software. To help understand how cameras can play a role in viewing a model, the solar system assembly shown in Figure 8–34 will be used.

The solar system assembly is not true to scale, and in fact is a toy model of the planets, as can be seen from the image. Turning the crank in the image causes the planets to orbit around the sun. The assembly encompasses topics this book has not explored yet, such as assemblies and gear mates. These topics are covered in Chapter 13. For now, let's continue with the topic at hand, which is adding a camera to this model. How-To 8–8 takes you through the process of adding a camera.

How-To 8-8:   Adding a Camera

Fig. 8–35. Upper portion of the Camera PropertyManager.

To add a camera, perform the following steps. Setting up a camera can be fairly involved. If this is your first time setting up a camera, it is recommended that selections be used to establish camera target, position, and rotation. If selections are not used for any of the aforementioned settings, leave the *Show numeric controls* option checked, which will give more flexibility in manipulating camera elements.

1. Right-click on the *Lights and Cameras* folder in FeatureManager and select Add Camera.

2. In the Camera PropertyManager, the upper portion of which is shown in Figure 8–35, select the Camera Type. Choices are *Aimed at target* (in which case a target can be selected) or *Floating*, in which case camera orientation must be established.

3. If *Aimed at target* was chosen, position the target either by dragging the target point on screen or by selecting a point somewhere on the model (which can be an actual point, a point on a face, and so on).

4. Specify the camera position. To attach the camera to an object, check *Position by selection* and select an object on the model. If *Position by selection* is cleared, check the *Show numeric controls* option in the *Camera type* panel to make manual positioning of the camera an easier task. Manual positioning of the camera can be done by either Cartesian or spherical coordinates.

5. Specify the camera rotation. If *Set roll by selection* is checked, select an object to indicate the camera's *up* direction. If *Set roll by selection* is cleared, it will be necessary to manually set the camera's Roll (twist). It will also be necessary to set the camera's Yaw and Pitch if a Floating camera type was initially selected (as opposed to Aimed).

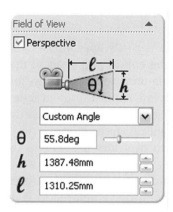

Field of View

☑ Perspective

Custom Angle

θ 55.8deg

h 1387.48mm

ℓ 1310.25mm

Fig. 8–36. Field of View panel.

6. Set the camera's field of view using the Field of View panel (shown in Figure 8–36), found at the bottom of the Camera PropertyManager. Leaving the Perspective option checked allows for selecting preset lens types from the drop-down menu shown in the image.

7. Click on OK to complete the camera insertion process.

While stepping through this How-To, you will have undoubtedly noticed how multiple viewports are used to aid in the camera setup. Viewports are nothing new to SolidWorks (see "10-second Topic: Viewports" in Chapter 3). When setting up a camera, one viewport (labeled Camera) shows what is seen through the camera. The other viewport will show the camera's position relative to the model.

When setting up the camera, the camera's position and target can be dragged with the mouse. This takes a bit of practice, and you may find it easier to use the slider bars and controls found in the Camera PropertyManager. This is the reason for checking the *Use numeric controls* option, as mentioned at the beginning of the How-To.

Figure 8–37 shows an example of what might be seen from the camera's perspective after setting up a camera. The sun was selected as the target, and the planet Neptune was selected for the camera's position. The top flat face of the model's base was used to establish the camera's rotation, which caused the camera's *up* direction to be perpendicular to the base.

As can be seen from the image, a fair amount of perspective has been applied. What cannot be appreciated from the image is how all of the planets move relative to their actual periods when the crank is turned, which is due to the gear mates added to the model. Additionally, a rotary motor has been added to the crank, causing the planets to move about their orbits. These are topics, covered in Chapter 13, that are not specific to SolidWorks 2006.

Fig. 8–37. Placing a camera on Neptune.

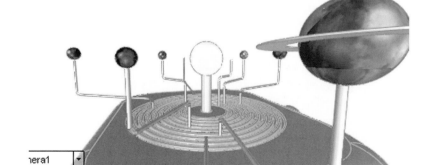

Summary

This chapter taught you many ways to modify a part's appearance. Not only is it possible to change a part's color, but the optical properties of the model can be adjusted as well. Parts can be made transparent or shiny, for instance. Individual faces or features can have colors applied to them independently. When applying color to individual objects rather than the entire part, bear in mind the order of precedence. Specifically, face color overrides feature color, which overrides body color, which overrides part color—in that order.

Textures can be applied to a model to give it a more realistic appearance. With a suitable graphics card, RealView graphics can be employed to give the model an even more realistic appearance. A material must be assigned to the part in order to use RealView graphics.

A number of light sources can be created in a part or assembly file. These include directional, point, and spot light sources. This can increase the realistic look of the model. All in all, adding and modifying light sources can enhance the look of your model, and can be fun. Experiment with light sources, textures, and materials to create models that appear more realistic.

CHAPTER 9

Sheet Metal

SHEET METAL PARTS COME IN A LARGE ARRAY of sizes and shapes, but they all have at least a few characteristics in common. They almost always have bends, there may be some sort of punched-out or cut-out shapes or holes, and often some sort of relief cuts may be needed so as to not stress the metal.

In this chapter you will learn how to create sheet metal parts, add bends, and perform numerous functions related to working with a sheet metal part. This includes showing the part in its flattened state and working on the part in either its flattened or formed state.

From a SolidWorks perspective, sheet metal is defined when extruding a base flange feature. Bends can also be added to a model, either created natively in SolidWorks or imported, in order to define it as a sheet metal part. When a model is defined as a sheet metal part in the SolidWorks software, it exhibits special properties, and varied additional functionality is available for editing and developing the model.

Defining a Sheet Metal Part

If you are going to be creating sheet metal parts, make things easy on yourself and first turn on the Sheet Metal toolbar, shown in Figure 9–1 (select View > Toolbars). Next, extrude the sketch (assuming one has already been created) using the Base Flange command, whose icon is also shown in Figure 9–1. SolidWorks will then create three features in FeatureManager that directly relate to the sheet metal model.

Typically, you want to start with an open-profile sketch prior to using the Base Flange command. The only time you would use a closed profile would be when developing a flat blank that would then be formed to take the desired shape. This would be known as *working in the flat*, as opposed to designing the formed model and letting SolidWorks calculate the appropriately sized flat pattern for you. How-To 9–1 takes you through the process of creating a sheet metal part using the Base Flange command.

Fig. 9–1. Sheet Metal toolbar.

How-To 9-1: Creating a Base Flange

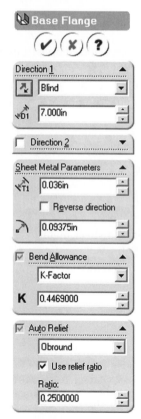

Fig. 9–2. Base Flange PropertyManager.

To create a sheet metal part in SolidWorks, perform the following steps. Note that either an open or closed sketch profile can be used, but if closed you will in essence be working with a flat piece of metal. This How-To will use an open profile, which is more common.

1. Select Insert > Sheet Metal > Base Flange, or click on the Base-Flange/Tab icon. This is a dual-function icon (thus the slash in its name).

2. In the Base Flange PropertyManager, shown in Figure 9–2, specify the end condition and depth, as for any simple extrusion.

3. In the Sheet Metal Parameters panel, specify a thickness for the material.

4. Check the Reverse Direction option if necessary to change the direction the wall thickness is applied.

5. Specify a value for the default Bend Radius.

6. Specify a value for the default Bend Allowance.

7. Specify the default values for relief cuts, if any, which should be made.

8. Click on OK to create the base flange. The part will be defined as a sheet metal part.

Options such as Auto Relief and Bend Allowance are discussed in material to follow. After creating a base flange, a number of features are added to FeatureManager. These are shown in Figure 9–3. The sections that follow explain these features in greater detail.

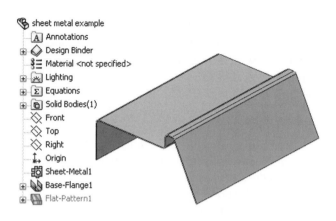

Fig. 9–3. The results of creating a base flange.

→ **NOTE:** *SolidWorks adds identification numbers to feature names so that it can tell similar features apart, such as* boss-extrude1, boss-extrude2, *and so on. This is also true for sheet metal features. This chapter will drop these numbers from feature names, as they are of no importance to the reader.*

SW 2006: A gauge table, in the form of an Excel spreadsheet, can be used to dictate K-factor, gauge thickness, and allowable bend radii. When gauge tables are used, the Base Flange PropertyManager changes appearance and limits choices to those values entered in the gauge table. Gauge tables provide an excellent way of reducing operator error by reducing choices to a limited set of values. Manual overrides are available in the PropertyManager that allow for specifying values outside the gauge table if this need should arise.

Sample gauge tables can be found in the *SolidWorks lang\English\Sheet Metal Gauge Tables* folder. Understanding how to create custom gauge tables is extremely easy, as can be seen by opening one of the sample tables. To set the path to where SolidWorks looks for gauge tables, use the method outlined in How-To 3–6.

Sheet-Metal

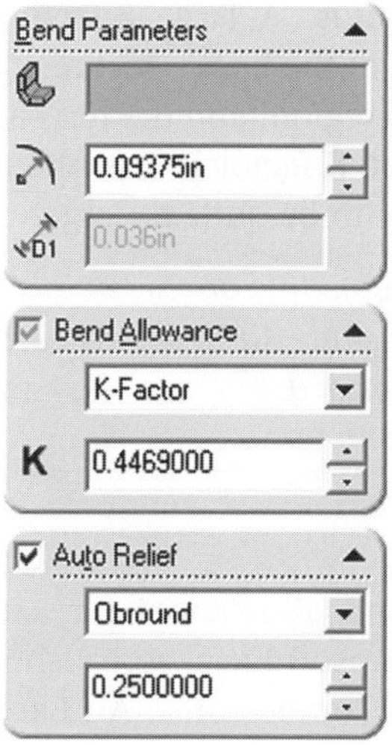

Fig. 9–4. Sheet-Metal PropertyManager.

Sheet-Metal is the name of the feature that defines the basic characteristics of the sheet metal part. The Sheet-Metal PropertyManager is shown in Figure 9–4. Its definition contains information such as the default bend radius, bend allowance, and information related to whether or not automatic relief cuts are made to the sheet metal part when the metal would otherwise tear. There are portions of this panel that are inaccessible at this time, such as the thickness setting. In this case, the thickness is part of the Base-Flange definition and can be changed there. Bend Allowance and Auto Relief are discussed in the following sections.

Bend Allowance

Fig. 9–5. Specifying a bend allowance table.

SolidWorks takes into consideration how much the sheet metal material is going to stretch, which is generally known as bend allowance. There are four options that can be used to account for material stretch. Using the pull-down menu in the Bend Allowance section, select Bend Table, K-Factor, Bend Allowance, or Bend Deduction. Keep in mind that the Bend Allowance setting in general will determine the default bend allowances used for all bends in the sheet metal part. If necessary, bend allowances can be defined on a bend-by-bend basis, which is described later in this chapter.

If Bend Allowance, Bend Deduction, or K-Factor is selected from the drop-down list, you will need to enter an appropriate numerical value. If Bend Table is selected, you will be required to click on the Browse button and select a bend table that already contains the bend allowance values (see Figure 9–5). Regardless of the method used in accounting for material stretch, somebody at some place and time will have had to do some calculations. If not, it may be up to you to break out the calculator.

10-second Topic: What Is K-Factor?

If you were to take a piece of metal and bend it, there would be some amount of compression on the inside of the bend area, and a bit of expansion on the outside. Somewhere in the middle of that piece of metal would be the neutral plane, where the metal is neither expanding nor compressing. The distance from the surface on the inside of the bend of the sheet metal part to the neutral plane, divided by the total thickness of the metal, is the K-factor.

The K-factor can be set anywhere from 0 to 1, assuming some value has been given for the bend radius. A large K-factor means that the neutral plane is closer to the outside of the bend. As a result, there is more compression of the metal in the bend region, and the flat pattern will need to be larger to accommodate this. If the K-factor is less, there is less compression and the flat pattern does not have to be quite as large.

✓ **TIP:** *If working with nonmetals or you simply want the formed and unformed model to be the same size (as measured along the outer face of the bends), use a K-factor of 1.*

Bend Table Helpful Hints

Bend tables contain the bend allowances for a certain type of material. They can be text files or Microsoft Excel spreadsheets. Text files are not recommended, as they are very cumbersome. An Excel spreadsheet is going to be much easier to manipulate than a text file. SolidWorks includes a number of preformatted tables for plugging in bend allowances. Some even have the bend allowances already plugged in, but they might not match the material you work with. Figure 9–6 shows a portion of an Excel bend table.

In the case of the bend table shown in Figure 9–6, the constant is the material thickness, and the bend radius and angle make up the horizontal and vertical axes of the table, respectively. The bend allowance values populate the table's body. If you decide to create your own bend tables, it is highly recommended you start with one of the preformatted tables SolidWorks pro-

Thickness:	1/32					
Angle	**Radius**					
	1/32	3/64	1/16	3/32	1/8	5/32
15	0.0118	0.0159	0.0200	0.0282	0.0364	0.0446
30	0.0237	0.0319	0.0400	0.0564	0.0728	0.0891
45	0.0355	0.0478	0.0601	0.0846	0.1091	0.1337
60	0.0473	0.0637	0.0801	0.1128	0.1455	0.1782
75	0.0592	0.0796	0.1001	0.1410	0.1819	0.2228
90	0.0710	0.0956	0.1201	0.1692	0.2183	0.2674

Fig. 9–6. Sample Excel bend table.

vides. You can then populate the table with the values appropriate to your manufacturing style, your existing empirical data, or your own calculations.

➥ **NOTE:** *By default, sample bend tables are found in the SolidWorks installation folder under* lang\English\Sheetmetal Bend Tables. *In addition, the English folder will be named after a different language if using a version of SolidWorks based on another language.*

Bend tables can be an excellent route to take with regard to sheet metal parts. The reason for this is that they are all-encompassing. In other words, a single bend table can include allowances for a wide range of bends. Differing bend angles and bend radius values can happily coexist in the same part. It may not be necessary to specify a bend allowance for a nonstandard bend that is added to the model.

The table shown in Figure 9–6 shows applies to material with a thickness of 1/32 inch. Other tables can exist in the same Excel worksheet that accommodate other thicknesses. This reduces the number of files a sheet metal fabrication house will have to keep track of.

How does SolidWorks find allowances for bend angles or radii not in the table? The answer to that question is interpolation. An on-the-fly calculation is made for bend angles and radii that fall *between existing values* listed in the table. It is only when a bend angle or radius of the model falls *outside* the table that SolidWorks will flag the model file with errors.

✓ **TIP:** *Create a table with bend angles up to (for example) 340 degrees and bend radii of 1/1,000 inch. This will allow for a variety of hems to be created, such as rolled or small gap (folded) hems without causing errors.*

Auto Relief

When a bend must be added to a part that would otherwise rip the metal, you need to use Auto Relief. If you do not have Auto Relief turned on and attempt to create such a bend, the bend will fail, giving you a warning message telling you of this fact. You then have the option of manually creating a cut in the area containing the bend or turning on Auto Relief.

To add relief cuts to the sheet metal part automatically, make sure Auto Relief is checked in the definition of the Sheet-Metal feature (see Figure 9–4). The Relief Ratio is the value that determines how deep the relief cut should be. If the material were .100 inch thick, and the Relief Ratio set to .5, the cut would be one-half the thickness of the material, or .050 inch deep. Additionally, this depth measurement is taken from the point where the bend ends.

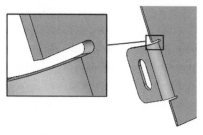

Fig. 9–7. Bend relief cuts on an edge flange.

Figure 9–7 shows a sample part with an edge flange (which you will learn about in material to follow). The relief cuts on either side of the flange are a direct result of turning on Auto Relief.

There are three types of relief: rectangular, obround, or tear. Use whichever type suits your purpose. *Obround* is a term combining the words *oblong* and *round* and is used to describe a slot or cut with rounded ends. Tear relief is similar to creating a cut with a pair of tin snips, rather than a saw blade, plasma cutter, or some other device. All three relief types are shown in Figure 9–8, with rectangular on the left, obround in the middle, and tear on the right.

*Fig. 9–8.
Examples of
available
relief cuts.*

Base-Flange

The Base-Flange definition contains parameters for the flange direction and end condition, along with the default thickness and bend radius values. Although the Bend Allowance and Auto Relief parameters are missing from this feature's definition, these parameters are present in the *Sheet-Metal definition* (as previously mentioned).

Flat-Pattern

A third feature added to FeatureManager is Flat-Pattern, and is the key to flattening the part. The Flat-Pattern PropertyManager, shown in Figure 9–9, will appear when editing this feature. The options contained in this panel are discussed in material to follow. First, however, let's find out how to flatten the part.

Fig. 9–9. Flat-Pattern PropertyManager.

Fig. 9–10. Flatten icon.

Flattening a Sheet Metal Part

Did you notice that the Flat-Pattern feature was suppressed in FeatureManager? The feature and text are grayed out (see Figure 9–3 for a glimpse of this). You may not have, in that we have not explored the suppression state of features. That can wait until Chapter 11. It is true that unsuppressing the Flat-Pattern feature will show the model in its flattened, or unformed, state. But all you really need to know at the moment is that clicking on the Flatten icon, located on the Sheet Metal toolbar, will flatten out the part. The Flatten icon is shown in Figure 9–10. Clicking on the Flatten icon a second time will form (or fold) the model back up.

The model can be flattened at any stage in the design process. Any tabs, hems, jogs, or flanges added to the model will flatten out as well. When the model is flattened, bend lines will be shown at each bend location (see Figure 9–11). These lines will automatically appear in a detailed drawing when the flat pattern view is inserted into the drawing. Detailed design drawings are discussed in Chapter 12.

SW 2006: Bend lines can have colors applied to them to help indicate bend direction (up or down), and other aspects of the sheet metal pattern. Colors only apply to flat-pattern views in a drawing, but can be changed through the Sheet Metal section of the Document Properties of the part file (Tools > Options).

➥ **NOTE:** *The options that follow are not crucial to understanding basic sheet metal functionality in SolidWorks. You can safely jump ahead to the "Edge Flange" section without losing the flow of this chapter.*

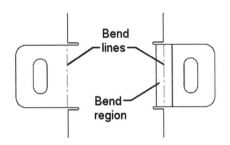

Fig. 9–11. Turning off Merge faces shows the bend regions.

Merge Faces

Checked by default, the *Merge faces option* treats all coplanar faces on the flattened model as one. Certain situations may require this option to be turned off, but typically it should be left alone. If turned off, the bend regions are shown in the flattened part, as indicated in Figure 9–11.

Note that *Merge faces* is an all-or-nothing setting, meaning that a single part cannot have just one bend showing a bend region. (The image shown in Figure 9–11 is a composite image.) Either all bend regions in the model are shown or none are shown.

Simplify Bends

The *Simplified bends* option has to do with how SolidWorks handles the edges of bend regions. The default method is to simplify bend edges, which can most easily be seen through example. Figure 9–12 shows a section of molding cut to fit around a 90-degree corner. An end view is also shown so that the cross-section shape of the molding can be discerned.

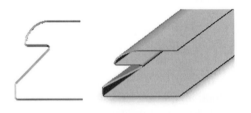

If this sheet metal molding is flattened out, we can see how the edges of the bend region appear straight. One area in particular is pointed out in Figure 9–13. The example on the left has the *Simplified bends* option checked. The example on the right has the *Simplified bends* option turned off. The edge in the example on the right is noticeably different. Whether or not you decide to turn this option off depends on what you want to send to the pattern makers who will be building this sheet metal part.

Fig. 9–12. Section of sheet metal molding.

Fig. 9–13. Molding in its flattened state.

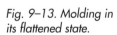 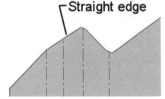 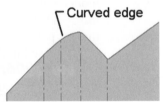

Corner Treatment

In the case where a feature such as a miter flange has been created, the flattened part can display the corners in one of two ways. Figure 9–14 shows a miter flange feature wrapping around the corner of a formed sheet metal part (left). The same part is also shown in its flattened state with corner treatment turned off (center), then again with corner treatment turned on (right).

Some designers choose to let the pattern maker create their own corner relief cuts, in which case the designer would check *Corner treatment* so that the relief cuts made by SolidWorks would not show up in the flat pattern.

Fig. 9–14. Corner treatment *option in action.*

Add Corner-trim

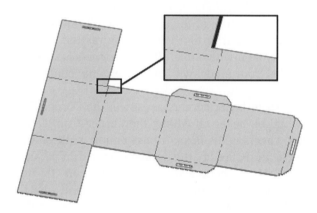

Fig. 9–15. *Prior to turning on Add Corner-trim.*

Fig. 9–16. Add Corner-trim *panel and related options.*

Corner trim refers to relief cuts for corners. The *Add Corner-trim* option functions somewhat differently from other options, in that turning it on causes another feature to be created and displayed in FeatureManager. *Corner trim* comes in a number of varieties. To demonstrate, a sheet metal container will be used, which is shown in its unfolded state in Figure 9–15.

If the *Add Corner-trim* option is checked, the panel expands to give the user additional options related to corner trim. An example of the panel you might see (depending on the options selected) is shown in Figure 9–16.

So what does this corner trim look like? There are three versions: circular, square, and bend waist. All three versions are shown in Figure 9–17. The Corner-Trim feature itself will appear after the Flat-Pattern feature in FeatureManager. Corner trim only appears in the flat pattern. If you wish to see relief cuts on corners in the formed state, use the *Trim side bends* option (described in the next section regarding edge flanges) or create relief cuts manually.

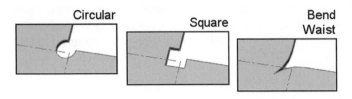

Fig. 9–17. Three versions of corner trim.

If you change your mind about applying corner trim to the flat pattern, simply delete the feature. To edit corner trim, edit the definition of the Corner-Trim feature. Specify the size of the corner trim either by supplying a distance or providing a ratio-to-thickness value.

The *Break corners* option becomes available after checking the *Add Corner-trim* option (see Figure 9–16). When breaking corners, the sharp edges of the sheet metal are removed using either fillets or chamfers. There is a separate Break Corner command (discussed in material to follow) accessible through the Sheet Metal toolbar.

The *Break corners* option is a bit tricky to understand. Although it only becomes available after checking the *Add Corner-trim* option, it is not possible to break corners and add corner trim to the same corner, in most cases. For example, adding circular corner trim to a corner removes the internal edge that could otherwise be broken with a fillet. If *Break corners* is checked after checking *Add Corner-trim*, your model will almost certainly contain errors.

Fig. 9–18. Corner-Trim PropertyManager.

If you do wish to break corners as part of the Corner-Trim feature, you would be better off editing the Corner-Trim feature itself rather than checking the *Break corners* option while editing the *Flat-Pattern1* feature. Specify the desired corner trim parameters, leave *Break corners* unchecked, and then click on OK. You can then right-click on the Corner-Trim feature found directly below the Flat-Pattern feature and select Edit Feature. At that time, you will be presented with the Corner-Trim PropertyManager, shown in Figure 9–18.

Once the Corner-Trim PropertyManager is displayed, the work area will show a preview of where corner trim is being applied. It is then much easier to deselect certain corners, by removing them from the list box in the Relief Options panel (see Figure 9–18), clicking in the upper list box in the Break Corner Options panel (see Figure 9–18), and selecting the corners to be broken.

Corners trimmed or broken using the aforementioned procedures only appear in the flattened state. They do not reflect the true nature of the model and should only be used to show the flat pattern in a particular way for internal process or pattern-making requirements. For broken corners that appear in both the formed and unformed states, refer to the section titled "Breaking Corners" later in this chapter.

Edge Flange

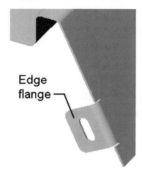

Fig. 9–19. An edge flange.

Edge flange

An edge flange is the first feature type we will be examining that is particular to sheet metal parts. In other words, the Edge Flange command can be used only on a sheet metal part, which you discovered could be created with the Base Flange command. Figure 9–19 shows an example of an edge flange feature.

The convenient aspect of the Edge Flange command is that it does not require the user to create a sketch first. The command will create a sketch for you, and then the sketch can be modified as required. How-To 9–2 takes you through the process of creating an edge flange.

How-To 9-2: Creating an Edge Flange

Fig. 9–20. Edge Flange PropertyManager.

To create an edge flange feature, perform the following steps. Note that you should not be in sketch mode at this point, and the model must be a sheet metal part.

1. Click on the Edge Flange icon found on the Sheet Metal toolbar, or select Edge Flange from the Insert > Sheet Metal menu.

2. Select a linear edge and drag it in the direction you want the flange to protrude. The angle will default to 90 degrees, but this can be changed later.

3. To use a radius other than the default (optional), in the Edge Flange PropertyManager (shown in Figure 9–20), uncheck *Use default radius* and enter a value.

4. Specify an Angle for the flange.

5. Specify the Flange Length. Accomplish this task by entering a length value, modifying the end condition if desired, and choosing where to measure the flange from (Inner Virtual Sharp or Outer Virtual Sharp button).

6. Specify a Flange Position (elaborated upon in material to follow).

7. Optionally specify a custom bend allowance or relief type.

8. Click on OK to create the flange.

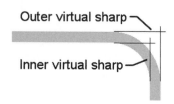

Fig. 9–21. Inner versus outer virtual sharps.

The flange length can be determined in a number of ways, the most basic of which is to enter a value using the Blind end condition. There is also the matter of where the flange length is measured from with regard to the bend. This can be specified using the Inner Virtual Sharp and Outer Virtual Sharp buttons mentioned in step 5. Figure 9–21 shows what is meant by "virtual sharp." The length of the flange will reference one of these virtual sharps, and this will definitely play a role in the final length of the flange, so make the choice accordingly.

Flange Position

Fig. 9–22. Flange position settings.

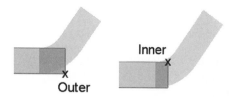

Fig. 9–23. Two possible outcomes using the Bend from Virtual Sharp option.

There is more to the Flange Position settings than meets the eye. First, it is very important to choose where the flange should be positioned with regard to the edge from which the flange originated. In its most basic form, it would be necessary to choose whether the flange should be positioned on the inside of the edge or the outside, or if the entire bend should be positioned on the outside of the original edge. This can most easily be seen in Figure 9–22, which has been annotated to show the related flange position settings.

There is a fourth setting, titled Bend from Virtual Sharp. This fourth flange position option relies on the flange length virtual sharp setting mentioned in How-To 9–2, step 5. Figure 9–23 shows two edge flange previews using the Bend from Virtual Sharp flange position option. The outcome of the flange is dependent on whether the Inner Virtual Sharp or Outer Virtual Sharp button is selected in the Flange Length panel.

Trim Side Bends Option

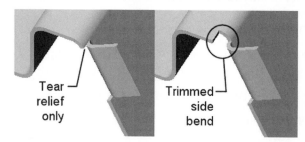

Fig. 9–24. Trim side bends option.

A side-by-side comparison example of the *Trim side bends* option is shown in Figure 9–24. On the left, the *Trim side bends* option was disabled. When the option is disabled, the area between the bend and flange has the appearance of being cut with a pair of tin snips, such as when using tear relief. This is the case even when rectangular or obround relief is used. On the right, *Trim side bends* was enabled.

Offset Option

The Offset option allows for performing some remarkable tasks. When checked, more options are presented, which allow for using various end conditions to define the offset distance. End conditions such as *Up to surface* make it possible to specify the flange position using some other surface in the part, or even some other externally referenced face of another component in an assembly. You will learn more about defining features in the context of an assembly in Chapter 13.

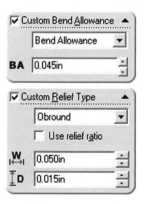

Fig. 9–25. Using the Offset option.

To use a simple example, let's understand what would happen if the Offset option were checked and the Blind end condition were used. In this case, a value could be entered for the offset distance. This serves to push the flange outward from the originating edge in one direction or the other. The direction in which the flange moves from the edge is dependent on the Reverse Direction button, located to the left of the end condition setting. Two examples of adding an offset value to offset the flange position are shown in Figure 9–25. The only difference between the two examples is that the Reverse Direction button was used for the example on the right. Relief cuts are added automatically when required (such as when the offset pushes the flange into the part).

Custom Relief and Bend Allowance

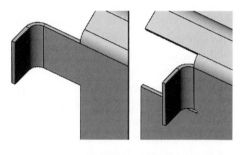

Fig. 9–26. Custom relief cut and bend allowance values.

Just because a default relief type, such as Rectangular, has already been established during the creation of a sheet metal part does not mean that this setting is required for every bend. This also holds true for the default bend allowance. By checking either the Custom Bend Allowance or Custom Relief Type option in the Edge Flange panel, the user is left with the ability to plug in any value desired. Figure 9–26 shows an example of this.

These optional custom parameters are present in every sheet metal feature that creates a bend (which is to say nearly every sheet metal feature). With regard to relief cuts, unchecking the *Use relief ratio* option allows for controlling the relief cut's width and depth. Keep these settings in mind as you progress through the rest of this chapter. You will encounter them frequently.

Editing the Flange Profile

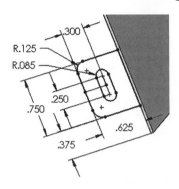

Fig. 9–27. Editing a flange profile.

One last topic must be discussed before wrapping up this section on edge flanges, and that is the flange sketch profile. If the default sketch created by SolidWorks specifically for the edge flange feature were always accepted at face value, it would always be the same width as the edge from which it emanated, and you would have control over its length only. That is hardly acceptable, and is certainly not the case.

When creating an edge flange, it is not necessary to define the flange length via the Flange Length panel (see step 5 in How-To 9–2). Instead, try clicking on the Edit Flange Profile button found on the same panel. A message window will appear, presumably informing you that the sketch is valid. Move the window out of the way. You will now have complete control over the sketch, an example of which is shown in Figure 9–27.

A flange profile sketch follows the same general rules as any sketch geometry. It should also be noted that the sketch will exhibit the same behavior found when using the Convert Entities command (see "Converting Entities" in Chapter 5). When first editing the flange profile sketch, the ends of the rectangular sketch must be "loosened" prior to dimensioning them. If this is not done, you risk overdefining the sketch. This is intended behavior.

Adding a dimension to control the length of the flange, such as the .625-inch dimension shown in Figure 9–27, will cause the Edge Flange panel to remove the Flange Length dimension parameter from the panel. This makes sense, in that the length is now being controlled by the dimension added to the sketch by the user.

Feel free to modify the sketch as necessary. That is, add or remove relations, add geometry for a hole, and so on. While the flange profile sketch is being edited, the small message window will remain on screen. This window will contain the buttons Back and Finish, as well as the usual Cancel and Help buttons. If the Back button is selected, you will be placed back in the Edge Flange PropertyManager. If Finish is selected, the flange will be created. It does not really matter which button is selected, as you can always edit the flange feature or the flange sketch anyway. Quite simply, if you have not finished specifying the parameters of the flange click on Back.

SW 2006: Multiple edges can be selected when adding an edge flange. If the edges happen to be adjacent, the gap distance between edge flanges can be specified via the *Gap distance* setting. The *Gap distance* setting can be found in the Flange Parameter panel, shown in Figure 9–28. If it becomes necessary to edit a specific flange profile, select the profile from the Flange Parameter list box prior to clicking on the Edit Flange Profile button.

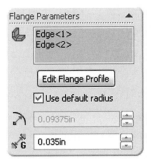

Fig. 9–28. Flange Parameter panel in SolidWorks 2006.

Miter Flange

Unlike an edge flange, a miter flange requires a sketch to be completed by the user. The sketch does not need to consist of much. In its simplest form, the miter flange sketch need be nothing more than a line. A miter flange is similar to an edge flange, but instead of emanating from a single edge a miter flange can wrap around corners. Those corners can either be sharp corners or rounded corners. If rounded, the miter flange can be made to automatically propagate around the tangent faces.

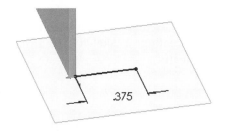

Fig. 9–29. Miter flange.

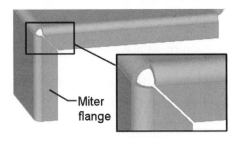

Fig. 9–30. Miter flange sketch.

Miter flanges are essentially edge flanges that fit together nicely where they meet at corners. Some functionality has been added that allows for controlling parameters such as the gap distance between flanges at each corner. Figure 9–29 shows an example of a miter flange with the corner shown in greater detail.

To create a miter flange, start with a sketch on a plane that is perpendicular to the edge the flange will run along. The easiest way to create the sketch plane is to pick near the endpoint of the edge where the flange should start, and then click on the Sketch icon. This neat little shortcut was mentioned in Chapter 4. Figure 9–30 shows the sketch plane used for the miter flange shown in Figure 9–29. It also shows the sketch used to create the flange.

As the figure shows, all that was needed was a single line segment. SolidWorks does the rest. How-To 9–3 takes you through the process of creating a miter flange feature on a sheet metal part. It is assumed the sketch has already been created.

How-To 9-3: Creating a Miter Flange

To create a miter flange, perform the following steps.

Fig. 9–31. Propagate icon.

1. Click on the Miter Flange icon found on the Sheet Metal toolbar, or select Miter Flange from the Insert > Sheet Metal menu.

2. The flange will flow along the first edge by default. To continue the flange around sharp corners, select additional edges. To force the flange to propagate around tangent faces, click on the Propagate icon, shown enlarged in Figure 9–31.

3. Specify where the flange position should be relative to the original edge. Like the Edge Flange command, this setting can be Material Inside, Material Outside, or Bend Outside.

4. Specify a Gap Distance, if desired, or just use the default value.

5. Click on OK to create the miter flange.

Fig. 9–32. Miter Flange PropertyManager.

A portion of the Miter Flange Property-Manager is shown in Figure 9–32. Occasionally, depending on the nature of the miter flange feature, the default gap distance may not be enough. Solid-Works will then supply an error message that states what the minimum gap distance should be. Using the minimum gap distance as a gauge, you can then specify a new gap distance.

With regard to adjusting the bend angle of a miter flange, there will be no angle option, as was present in the Edge Flange panel. This is because the angle of a miter flange is incorporated into the sketch. To edit the angle of the miter flange, edit the miter flange sketch. Lines and arcs can both be used when creating a sketch for a miter flange feature.

Offset Miter Flange

The Start/End Offset option (shown at the bottom of Figure 9–32) refers to the distance the miter flange is from either the beginning or end of the edges around which the flange is extending. To put it another way, the flange does not have to start at the end of the edge where the sketch plane was positioned.

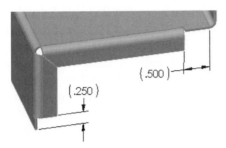

Fig. 9–33. Offsetting a miter flange.

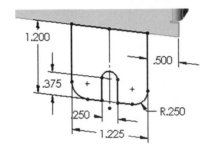

Fig. 9–34. A completed sketch that will become a tab.

Figure 9–33 shows an example of using a start and end offset value of .25 inch and .5 inch. Reference dimensions were added to the model for illustration purposes. If relief cuts become necessary, SolidWorks will present you with relief cut options in the Miter Flange PropertyManager. Either use the default auto-relief settings or enter the desired custom relief parameters (discussed earlier in this chapter.)

Tabs

Tabs are one of the easiest of all sheet metal features to create. The only manual labor involved is creating the sketch that will define the profile of the tab. An example of a sketch prior to creating the tab is shown in Figure 9–34. How-To 9–4 takes you through the process of creating a tab.

HOW-TO 9-4: Creating a Tab

To create a tab on a sheet metal part, perform the following steps.

1. Select a face on the model where a tab is to be located, and enter sketch mode.

2. Create a closed profile that describes the shape of the tab.

3. Click on the Base-Flange/Tab icon found on the Sheet Metal toolbar, or select Tab from the Insert > Sheet Metal menu.

Fig. 9–35. Example of a tab.

As you can see, there is very little to the creation of a tab. A large percentage of the work is creating the sketch. An example of a completed tab is shown in Figure 9–35.

Sketched Bends

If an additional bend were required on a sheet metal part, the bend could be defined by first placing a sketch line on a face where the bend is to occur. This bend type is known as a sketched bend. The principle behind sketched bend features is straightforward: place a bend wherever a line has been sketched.

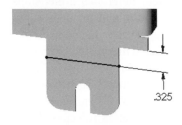

Neatness is a desirable trait when creating a sketch prior to defining a sketched bend feature. A good example of a well-drawn sketch line is shown in Figure 9–36. Note that the ends of the sketch line terminate on the edges of the tab, and in fact are coincident with the edges of the tab. The Sketched Bend command would still work regardless of whether the sketch line runs short or long, but it is good practice to terminate the sketch line on model edges.

Fig. 9–36. Preparing to add a sketched bend.

More than one line can be present in a sketch, but bear in mind that each bend will be driven by the same set of parameters. For the greatest amount of flexibility, limit the number of lines in each sketch used to create a sketched bend feature. Obviously, do not crisscross lines, as SolidWorks cannot place bends on top of bends. Placing a sketch line too close to a model edge may result in the bend not being created. A single line can span multiple tabs, but should not cross complex regions containing hems or existing bends. How-To 9–5 takes you through the process of creating a sketched bend.

HOW-TO 9-5: Creating a Sketched Bend

To create a sketched bend on a sheet metal part, perform the following steps.

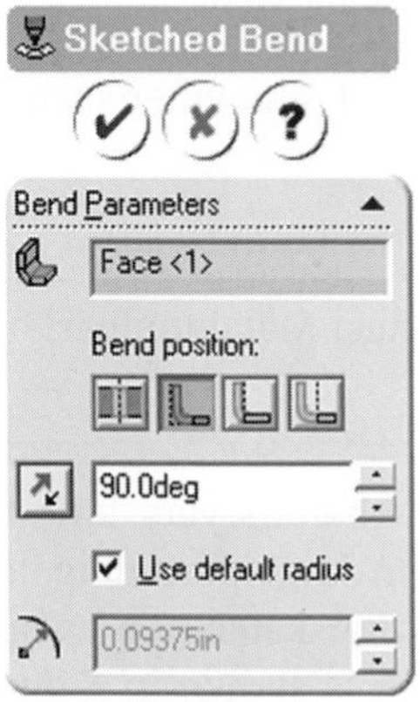

1. Create a sketch line on a face of the model where you would like to place a bend.

2. Click on the Sketched Bend icon found on the Sheet Metal toolbar, or select Sketched Bend from the Insert > Sheet Metal menu.

3. Select a face to be held stationary when the bend is created (i.e., click on one side of the line or the other). The face will be listed in the Sketched Bend PropertyManager, shown in Figure 9–37.

4. Specify where the new bend should be positioned with reference to the sketch line. Choices are Bend Centerline, Material Inside, Material Outside, and Bend Outside.

Fig. 9–37. Sketched Bend PropertyManager.

5. Specify a value for the Bend Angle.

6. Click on Reverse Direction if it is necessary to flip the bend direction. The direction will be indicated by a small preview arrow.

7. Click on OK to create the bend.

Fig. 9–38. Completed sketched bend feature.

Figure 9–38 shows an example of a completed sketched bend feature. Regarding the position of the bend with reference to the sketch line, as determined in step 4 there is a new choice that has not been discussed yet. That choice is Bend Centerline. The other three choices were discussed in the section "Flange Position" and are shown in Figure 9–22. The Bend Centerline option will ensure that the center of the bend region is at precisely the same position as the sketch line used to define the bend.

Jogs

The end result of a jog feature is only slightly more complex than a sketched bend, but starts out the exact same way. That is, a jog begins with a simple sketch line. Figure 9–39 shows the sketch line used to define the jog feature created in How-To 9–6.

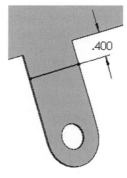

.400

Fig. 9–39. Preparing to create a Jog.

How-To 9-6: Creating a Jog

To create a jog feature on a sheet metal part, perform the following steps.

1. Create a sketch line on a face of the model where you would like to place a jog.

2. Click on the Jog icon found on the Sheet Metal toolbar, or select Jog from the Insert > Sheet Metal menu.

3. Select a face to be held stationary when the jog is created. The face will be listed in the Jog PropertyManager, shown in Figure 9–40.

4. Specify a value or end condition, which will indicate the Jog Offset distance. Also specify the Dimension Position for the jog offset (Outside Offset, Inside Offset, or Overall Dimension), which will indicate where the jog offset distance is measured from.

Fig. 9–40. Jog PropertyManager.

Fig. 9–41. Example of a jog.

5. Click on Reverse Direction if it is necessary to flip the jog direction. The direction will be indicated by a small arrow, and the preview itself should change accordingly.

6. Specify the Jog Position with reference to the sketch line.

7. Specify a Jog Angle. This value can be anything between 0 and 180.

8. Click on OK to create the jog.

Figure 9–41 shows an example of a jog feature. The jog is a result of one simple sketch line, so it is not difficult to see that this feature type could save a lot of time.

Step 6 mentions that the Jog Position should be specified with reference to the original sketch line. This procedure carries with it the same options present when creating a sketched bend. However, a jog contains not just one bend but two. So how is the position of the jog actually being defined? The answer is that the position is defined by the first bend in the jog. The second bend can safely be ignored with regard to the jog position.

When specifying the Jog Offset, there is an option titled *Fix projected length*. This option makes a difference in the final length of the material receiving the jog, so be careful. When *Fix projected length* is checked, SolidWorks extends the length of the material so that the end of the material extends to the same position prior to the jog. With the option off, the length of the material receiving the jog reacts as it would in the physical world. Figure 9–42 shows the same jog with the *Fix projected length* option turned off (left) and turned on, or checked (right).

Fig. 9–42. Fix projected length *option deactivated (left) and activated (right).*

Hems

Sheet metal hems, of which there are four types, are very easy to create. Hems can be either Closed, Open, Teardrop, or Rolled, examples of which are shown in Figure 9–43. No sketch is required prior to creating hems. It is simply a matter of selecting the appropriate edge or edges and then plugging in the hem parameters. How-To 9–7 takes you through the process of creating hems.

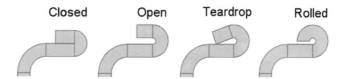

Fig. 9–43. Hems are of four types.

How-To 9-7: Creating a Hem

To create a hem on a sheet metal part, perform the following steps.

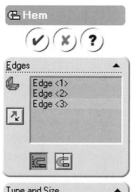

Fig. 9–44. Hem PropertyManager.

1. Click on the Hem icon found on the Sheet Metal toolbar, or select Hem from the Insert > Sheet Metal menu.

2. Select the edge or edges that are to contain hems. Any selected edges will appear in the Edges list box in the Hem PropertyManager, shown in Figure 9–44.

3. Using the icons on the Hem panel, specify where the hem should be positioned relative to the selected edge. This will be either Material Inside or Bend Outside.

4. Use the Reverse Direction option, if necessary, to flip the hem from one side to the other.

5. Specify the hem Type.

6. Specify the hem Size. This may be a combination of settings, depending on the hem Type, and may include Length, Gap Distance, Angle, and Radius.

7. Click on OK to create the hem.

The Miter Gap parameter, shown near the bottom of Figure 9–44, will not be available unless multiple edges are selected. Even then, the Miter Gap parameter may not be relevant. It depends on whether or not two edges are adjacent. An example of a closed hem with a miter gap is shown in Figure 9–45. If multiple edges are selected for hemming, and they do not share a common corner, ignore the Miter Gap setting.

When an edge is selected during the hem definition process, the hem will attempt to bend toward the selected edge. You can use this to your advantage, and therefore clicking on the Reverse Direction option will not be necessary. It is not possible to form hems in multiple directions at once (i.e., on different sides of the sheet metal part) when selecting multiple edges. The hems will need to be created as separate features in such a case.

Fig. 9–45. Example of mitered hems.

Special Note on Hems and Bend Allowance: Hems often require a custom bend allowance to be specified. The reasons for this have to do with the included angle of the hem, type of hem, and how the default bend allowance was specified for the overall sheet metal part.

Consider the following scenario. A bend table was used to specify the bend allowances for the bends in the sheet metal part. However, the bend table only includes bends up to 90 degrees.

When a hem is added to the model, SolidWorks suddenly presents you with an error. Common errors will inform you that "This part contains features that cannot be unbent" or that "The angle of this bend fell outside the bend allowance/deduction table."

The reason for errors in sheet metal parts vary widely, but almost always are due to the operator not inputting an appropriate bend allowance. If bend tables are used, the table must include values large enough to accommodate the hems in the part. These values might be bend angles large enough to accommodate a rolled hem with an included angle of 340 degrees or a bend radius small enough to accommodate an open bend with a gap of only .005 inch, in which case the bend radius in the table should be .0025 inch or smaller.

Fig. 9–46. Specifying a custom bend allowance for a hem.

If you do not want to take the time to modify your existing bend tables to include hems, simply plug in a custom bend allowance for the hem when creating it. You do this via the Custom Bend Allowance panel at the bottom of the Hem PropertyManager, shown in Figure 9–46. This is the same panel available in any sheet metal part containing a bend, and has been mentioned previously in this chapter.

Breaking Corners

A special sheet metal command known as Break Corner can be thought of as an enhanced fillet or chamfer command designed with sheet metal parts in mind. The primary benefit of the Break Corner command is that it makes selecting the corners to be broken extremely easy. By the way, breaking corners is sheet metal lingo for removing the sharp corners from a sheet metal part, usually for the sake of safety.

As you discovered in Chapter 6, adding a fillet or chamfer as a feature requires selecting the edge on which the fillet or chamfer is to be applied. With regard to sheet metal parts, that edge can be very small. For example, 19-gauge steel sheet metal is only .048 inch thick. That certainly does not present much of an edge to select, so it may require the user to zoom in close in order to pick it, or to use some other selection technique.

Break Corner uses a filtering technique to make selecting very small edges easy. Simply picking near the edge will select it without incorrectly selecting the longer edges adjacent to it. The Break Corner command works on sheet metal parts only, and therefore cannot be used on other thin-feature parts. How-To 9–8 takes you through the process of breaking corners.

HOW-TO 9-8: Breaking Corners

To add chamfers or rounds to (i.e., to break the corners of) a sheet metal part, perform the following steps.

1. Click on the Break-Corner/Corner-Trim icon found on the Sheet Metal toolbar, or select Break Corner from the Insert > Sheet Metal menu.

2. Select the edges to be broken. Faces can also be selected, but this will break every corner on that face (which may be more than you want), so use caution.

3. Specify the Break Type (chamfered or rounded).

4. Specify the break Distance or Radius.

5. Click on OK to finish breaking the corners.

The Break Corner PropertyManager is shown in Figure 9–47. It should be noted that when Break Type is set to Chamfer only one distance parameter is present. The chamfer that is created is limited to a simple 45-degree chamfer (assuming a right-angle corner, though this is not required). If something a little more specific is required, you will need to either use the Chamfer command or create a cut.

If the sheet metal part is in its flattened state, clicking on the Break-Corner/Corner-Trim icon will bring up the Corner-Trim PropertyManager (shown Figure 9–48). This is where the "Corner-Trim" portion of the icon name originates, making this a dual-use command. The Corner Trim PropertyManager is nearly identical to the Break Corner PropertyManager, except that it contains the Relief Options panel (shown in Figure 9–48).

The technical aspects of the Corner Trim command were discussed previously (see the section "Add Corner-trim"). In short, corners can be broken in the formed state, and broken or trimmed in the flattened state. If corners are broken or trimmed in the flattened state, a Corner-Trim feature will appear after the Flat-Pattern feature in FeatureManager. This also means that the Corner-Trim feature will only appear in the flattened state and not in the formed state. If corners are broken in the formed state, those broken corners appear in both the formed and flattened states.

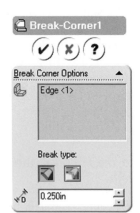

Fig. 9–47. Break Corner
PropertyManager

Fig. 9–48. Corner Trim
PropertyManager.

Closing Corners

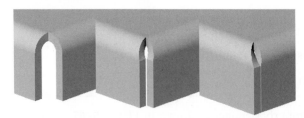

Fig. 9–49. A corner in need of closing, before and after.

The Closed Corner command operates on gaps, such as those a miter flange feature might leave behind. When Closed Corner is used, the corner is not truly closed. The Closed Corner command does not create an airtight "weld" between two adjacent flanges. Rather, it creates a butt (or overlapping region). Figure 9–49 shows the corner of a miter flange prior to using the Closed Corner command (left). Also pictured is the same corner closed using the Butt option (center) and again using the Overlap option (right). How-To 9–9 takes you through the process of closing corners.

HOW-TO 9-9: Closing Corners

To close a corner of a sheet metal part, perform the following steps. Note that only planar faces can be selected for closing, and that the corner faces must be perpendicular. Corners that can be closed are those typically resulting from a miter flange.

1. Click on the Closed Corner icon found on the Sheet Metal toolbar, or select Closed Corner from the Insert > Sheet Metal menu.

2. Select one face on the corner to be closed. You will not need to select the opposing face.

*Fig. 9–50.
Closed Corner
PropertyManager.*

3. Specify how to close the corner. Choices are Butt, Overlap, and Underlap. Use the icons in the Closed Corner PropertyManager, shown in Figure 9–50.

4. Click on OK to complete the command.

Note that more than one corner can be closed at a time. It should also be noted that Overlap and Underlap are exactly the same thing. The only difference is what face is selected for the corner to be closed. In other words, selecting one face and specifying Overlap is identical to selecting the opposing face and selecting Underlap.

SW 2006: Increased flexibility allows for closing a wider variety of corners and includes settings for gap distance and overlap/underlap ratio. For example, Figure 9–51 shows a corner where an edge flange and miter flange meet (left). The edge flange is at an angle of 75 degrees, yet the corner can still be closed (center). In some cases, the faces of the corner being closed do not even have to be perpendicular.

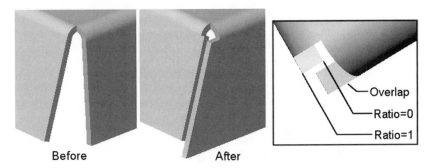

Fig. 9–51. Closing corners in SolidWorks 2006.

The inset in Figure 9–51 helps to show what the *Overlap/underlap ratio* setting refers to. When Overlap or Underlap is specified as the corner type, the ratio controls how far the overlapping face extends. The *Overlap/underlap ratio* value can be anywhere from 0 to 1, and the overlapping face will be positioned as shown in Figure 9–51.

Figure 9–52 shows the Faces to Extend panel of the Closed Corner PropertyManager in SolidWorks 2006 (left). The new *Gap distance* (currently set to a value of .030 inch) and *Overlap/underlap ratio* settings can be seen, along with an option labeled *Do not extend bends*. The image shows the results of checking the *Do not extend bends* option (center), and what happens when this option is left off (right).

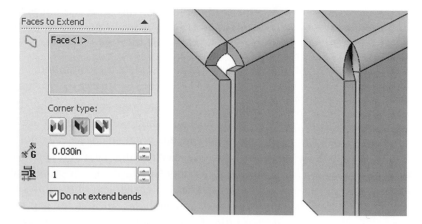

Fig. 9–52. The enhanced Faces to Extend panel.

Lofted Bends

A lofted bend feature allows for creating a free-form shape that can be flattened. Sheet metal duct work would be a good example, where one opening is rectangular and the opposing end is circular. However, there are some rules that must be followed if you wish to create a lofted bend.

Of primary importance is that each profile in the loft be open. Second, there can be two profiles only. Third, all segments of each profile must be tangent. Adding sketch fillets is an acceptable method of meeting this requirement. If all of these conditions are met, you can create a lofted bend. How-To 9–10 takes you through the process of creating a lofted bend.

How-To 9-10: Creating a Lofted Bend

Fig. 9–53. Lofted Bends PropertyManager.

To create a lofted bend, perform the following steps. It is assumed that each sketch profile has already been created, and that they meet the requirements dictated in the previous section. Sketch profiles do not have to reside on parallel planes, nor do they have to have the same number of sketch segments.

1. Click on the Lofted Bend icon found on the Sheet Metal toolbar, or select Lofted Bend from the Insert > Sheet Metal menu.

2. Select each sketch from the work area by clicking near common endpoints, which should match up when performing the loft.

3. In the Lofted Bends PropertyManager, shown in Figure 9–53, specify a thickness for the new lofted bend feature. You can also reverse the wall direction if necessary.

4. Click on OK to create the lofted bend.

Fig. 9–54. Lofted bend feature.

An example of a lofted bend feature is shown in Figure 9-54. An open rectangular shape and an arc were used in the example shown. Once a lofted bend feature is created, a Sheet-Metal and Flat-Pattern feature will appear in FeatureManager. This is similar to when creating a base flange feature. However, that is where much of the similarity ends.

You will probably find that there are limitations once the lofted bend feature has been created. Creating a miter flange, for example, will prove impossible. Edge flanges can usually be created if the transition is not round or elliptical. If a flange must be created on a round or elliptical transition, consider using swept features, discussed in the next chapter. Also, if the swept flange is a separate body (see Chapter 20), it will still be possible to flatten the lofted bend.

Bend Deviation

Lofted bends are what might be considered a unique feature type in Solid-Works. Once created, there is not a lot that can be done with them aside from creating cuts and a handful of other feature types. With any luck, once a lofted bend part has been completed it can still be flattened.

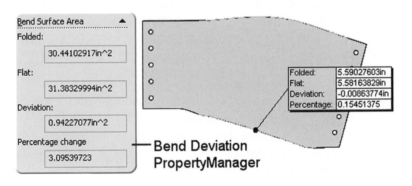

Bend Deviation PropertyManager

Fig. 9–55. Viewing Bend Deviation.

The lofted bend part shown in Figure 9–54 was designed as a guard to fit over existing components in an assembly. It has some holes running along either end so that it can be bolted in place. If the part is flattened, a tool unique to lofted bends can be accessed. This tool is known as Bend Deviation, shown in action in Figure 9–55. Only the maximum deviation is being shown on the model, and the left side of the image shows a portion of the Bend Deviation PropertyManager.

As can be ascertained from Figure 9–55, edge deviation as well as bend region surface deviation can be obtained. Accessing bend deviation data is done by right-clicking on the subfeatures associated with the Flat-Pattern feature. Right-clicking on the Flat-Pattern feature itself will not suffice. Figure 9–56 shows the proper features to right-click on in order to access the Bend Deviation menu item.

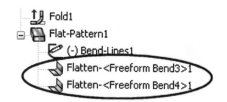

*Fig. 9–56. Accessing
Bend Deviation.*

Cutting Sheet Metal Parts

To this point you have learned about all of the basic sheet metal functionality present in SolidWorks. If you stopped reading at this point, you would probably be able to hold your own if asked to design a new sheet metal model. However, if you are to gain a more complete understanding of SolidWorks sheet metal commands and options, you should continue with this chapter.

Most sheet metal parts will not have bosses in the normal sense of the word. They will usually have at least a few cuts, though, and often more than just a few. There are some details regarding cut features that are specific to sheet metal parts. These are addressed in the sections that follow.

Link to Thickness

*Fig. 9–57. Link to thickness
and Normal cut options.*

When defining a sheet metal part, the thickness of the material is remembered by SolidWorks. The *Link to thickness* option is a direct result of creating a sheet metal part. When a cut (or a boss, for that matter) is added, the thickness of the cut can be linked to the original material thickness.

Figure 9–57 shows the *Link to thickness* option, which appears when creating an extrusion. Note that the Depth setting normally present when the end condition is set to Blind is not available. This is a direct result of checking the *Link to thickness* option. The Depth setting simply is not needed anymore.

The benefit of using *Link to thickness* is apparent if a design change is required at a later time. You may have spent hours developing a complex sheet metal part, and now your customer decides that the material should be of a heavier gauge. By double clicking on any feature utilizing the *Link to thickness* option, you can access the dimension for the thickness of the material and make the change. This change will be propagated, throughout the part, to any other feature utilizing the *Link to thickness* option.

Normal Cut

When the *Normal cut* option (also shown in Figure 9–57) is selected, the resultant cut will be perpendicular to the geometry it is cutting through. In contrast, cuts are usually perpendicular to the sketch plane. This can most easily be illustrated with a simple example.

Figure 9–58 shows what you would typically see when creating a cut through a part. A simple circle was sketched on the planar face shown in Figure 9–58. The circle was then cut through the model. The *Normal cut* option was not turned on. This allowed the cut to be normal to (perpendicular to) the original sketch plane. A cylinder could easily be slid through the cut shown in Figure 9–58. However, if the part were shown in its flattened state the cut would not represent the true nature of the cut had it been stamped out of the part in the physical world.

In Figure 9–59, *Normal cut* has been turned on. Note the shape of the hole on the angled geometry. With *Normal cut* turned on, the cut will be normal (perpendicular) to the geometry being cut. The extrusion direction is still projected in a perpendicular trajectory away from the sketch plane. A cylinder would still slide through the opening. However, the cut accurately represents a stamping or punching process of the sheet metal in its flattened state. The *Normal cut* option is available for sheet metal parts only.

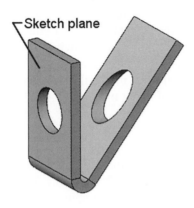

Fig. 9–58. Typical cut through a part.

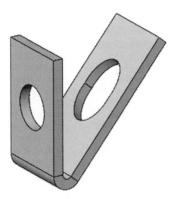

Fig. 9–59. Normal cut has been turned on.

Unfold and Fold Features

Yes, there actually are commands named, appropriately enough, Unfold and Fold. Unfold and Fold are features that appear in FeatureManager. Unfold allows for adding features to the sheet metal part in its flattened state, whereas Fold will form the part back up. The commands are similar and should always appear in pairs in FeatureManager.

Cuts in sheet metal parts can be added in either the formed or flattened state. Nothing special is required when creating cuts in the formed state, unless those cuts affect bend regions. A circular cut through a bend region, for example, would not look correct unless punched through the part in its flattened state.

The process for adding features in the flattened state is quite simple, as long as the proper technique is followed. First, use the Unfold feature to flatten out any bend regions that will be affected by a cut. Second, add cuts as needed. Third, use the Fold feature to form the part back up. Nearly any feature type can be added to a sheet metal part, but cuts are by far the most common, so that is what is shown in the following example. If adding features other than cuts (such as tabs), make sure the features do not intersect the bend regions. Otherwise, the part may fail to fold back up. How-To 9–11 takes you through the process of using the Unfold and Fold commands.

HOW-TO 9-11: Unfold and Fold Commands

To utilize either the Unfold or Fold command, perform the following steps. These commands should always be performed one following the other, in the proper order. Unfold the desired bend regions, edit the part as needed, and then fold those same bend regions back up.

1. Click on the Unfold icon or the Fold icon, found on the Sheet Metal toolbar, or select Unfold or Fold from the Insert > Sheet Metal menu.

2. Select a face to be held stationary.

3. Select the bend regions to be either folded or unfolded.

4. Click on OK to complete the operation.

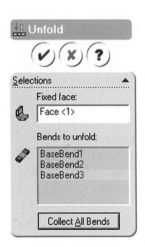

Fig. 9–60. Unfold PropertyManager.

The Unfold PropertyManager is shown in Figure 9–60. The Fold PropertyManager looks nearly identical. In step 2 you are required to select a face to be held stationary. This must happen so that SolidWorks will know what to fold or unfold relative to the rest of the part. The Collect All Bends button will select every bend in the model, which is convenient when there are a lot of cuts that must be added.

It is good practice to get all cuts out of the way prior to folding the model back up. That way, it will not be necessary to unfold portions of the part a second time. If you only have to perform an Unfold and Fold once, it makes for a more streamlined FeatureManager.

Let's see how the Unfold and Fold features are put into practice. Cuts will be added that would be very difficult to add without unfolding the model first. Figure 9-61 shows a portion of the part before performing an Unfold on it.

In Figure 9–62, the part has been unfolded and some sketch geometry has been added. Note that the sketch geometry will be cutting through various bend regions.

Next, the cuts are added and the part is folded back up (Figure 9-63). Would it have been possible to create the rounded slot without first unfolding the model? Yes, but it would have taken two features at a minimum. The holes would not have been possible at all.

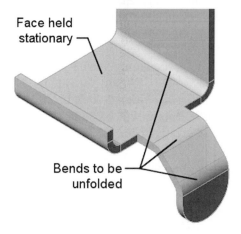

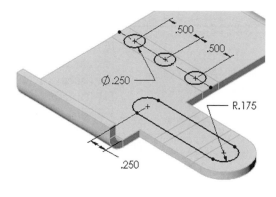

Fig. 9–61. Prior to unfolding.

Fig. 9–62. After unfolding and creating a sketch.

Fig. 9–63. After folding the model back up.

Working in the Flattened State

Sheet metal parts can be designed in the formed state, beginning with a base flange feature and then adding tabs, hems, jogs, and so on. You have been learning about these feature types throughout. A second alternative is to start out with flat stock (an extrusion representative of the flat pattern), and then add sketched bends, jogs, and various other sheet metal features. This is certainly possible, but is not recommended for most situations.

Usually when designing a sheet metal part, the final dimensions are known. Because SolidWorks figures in the bend allowances for you (which you specify), it is quite easy to design a sheet metal part with precise dimensions in the formed state. It would then be a simple matter of showing the flattened pattern in a design drawing and adding dimensions to it. (Detailed design drawings are explored in Chapter 12.)

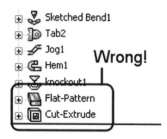

Fig. 9–64. Do not add cuts with the Flat-Pattern feature unsuppressed.

The real point of this section is how to create cuts in a sheet metal part while in the flattened state. You just learned how to correctly use the Unfold and Fold commands. You also learned that flattening a part can be done by clicking on the Flatten icon. When the Flatten icon is selected, the Flat-Pattern feature is unsuppressed, thereby showing the sheet metal part in its flattened state.

It is strongly stressed that you not create cuts with the Flat-Pattern feature unsuppressed! SolidWorks will add the new feature to the bottom of FeatureManager, as shown in Figure 9–64. Note in particular that the feature Cut-Extrude is below Flat-Pattern. This may seem trivial, but when the part is formed back up both the Flat-Pattern and the Cut-Extrude feature will be suppressed. You will *not* be able to see the cut except in the flattened state, and there will be no way around it!

The moral of this lesson is simple: Do not add features while the model is flattened. Rather, use the Unfold and Fold commands discussed in the previous section.

The Design Library

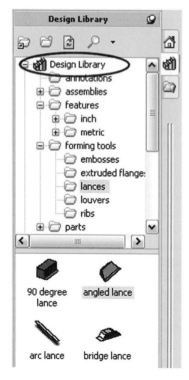

Fig. 9–65. Design Library.

A big part of working with sheet metal is the ability to perform the punching and deformation operations that typically accompany the manufacturing process. Punching refers not just to cuts but to creating features such as lances and louvers. Deformations include operations such as adding dimples or embosses. Forming tools will perform these operations, and more, on SolidWorks sheet metal parts. This section shows you how to use forming tools (as well as library features), and how to create your own.

Forming tools are part of a larger aspect of SolidWorks functionality known as the Design library, which consists of various categories that carry with them slightly different traits. The Design library, in general, allows for reusing geometry. It is a library of annotations, symbols, features, parts, and assemblies that can be used in other parts, assemblies, or drawings. The process of reusing these items operates via a simple drag-and-drop interface.

The Design library is accessed via the Task pane, first discussed in Chapter 1. The main category names found within the Design library are Annotations, Forming Tools, Features, Parts, and Assemblies. The table also points out the file types associated with each main category found in the Design library. Figure 9–65 shows the Design Library pane (which happens to be open to the *forming tools\lances* folder). Items are dragged from the lower portion of the task pane and dropped onto a model. Table 9–1 provides an overview of Design library functionality.

The intricacies of Design library parts, assemblies, and annotations are discussed in other chapters because they do not relate directly to sheet metal parts. If you would like to explore Design library annotations in more depth at this time, turn to Chapter 12 ("Favorites" section). Design library parts and assemblies are discussed in Chapter 13.

Table 9-1: Design Library Overview

Category	*What are they?*	*What are they used for?*
Annotations	"Favorite" files or blocks	Common annotations or symbols most typically added to drawings
Features	Library feature part files (*.sldlfp*)	Added to part files as a single sketch, feature, or groups of features, thereby eliminating the need to recreate commonly used items
Parts	Part files (*.sldprt*)	Added to assemblies where they become components in the assembly
Assemblies	Assembly files (*.sldasm*)	Added to assemblies where they become subassemblies
Forming Tools	Part files (*.sldprt*)	Added to sheet metal parts to create stamped, punched, or embossed features, dimples, and so on

Using the Design Library

Navigating through the folders of the Design library is accomplished in exactly the same fashion as navigating your computers folder structure in Windows Explorer, which anyone reading this book should be familiar with. Regardless of what is being used from the Design library, the process always begins in the same fashion. Hold the left mouse button down with the cursor over the item in the library you would like to use, drag the item into the SolidWorks document, and let go of the left mouse button. In other words, a simple drag-and-drop operation is all that is required.

After inserting a forming tool or a feature, the next step in the process is basically the same: position the object. The method used to position the object will vary depending on the type of object inserted (forming tool or feature). Because of this, the steps for inserting each of these objects are outlined separately. How-To 9–12 takes you through the process of using the Design library to insert a forming tool.

✗ **WARNING:** *Before forming tools can be used, a folder must be designated a Forming Tools folder. To do this, right-click on the desired folder in the Design library and make sure Forming Tools Folder is checked (as shown in Figure 9–66).*

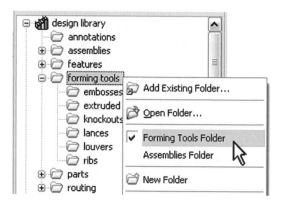

Fig. 9–66. Designating a Forming Tools folder.

How-To 9-12: Inserting Design Library Forming Tools

To insert forming tools via the Design library, perform the following steps. Inserting Design library annotations or symbols is discussed in Chapter 12. Inserting parts or assemblies is done with a simple drag and drop, though the finer details of this process are discussed in Chapter 13.

✓ **TIP:** *If your Task pane is not visible, it can be turned on by right-clicking on any docked toolbar.*

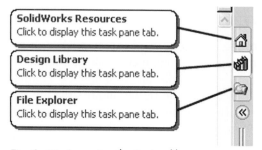

Fig. 9–67. Accessing the Design library and other areas of the Task pane.

1. Click on the Design Library tab, shown in Figure 9–67.

2. Under the *Forming Tools* folder, navigate to the folder that contains the desired forming tool.

3. Using the left mouse button, drag the item onto a planar face of the model. Use the Tab key on the keyboard to punch the feature through the model from one side or the other prior to releasing the mouse button.

4. Position the orientation sketch using dimensions or geometric relations. Use the Modify Sketch command (discussed in material following) if it is necessary to rotate the orientation sketch.

5. Click on Finish.

If inserting a feature, the process of positioning the feature works a bit differently. Rather than adding dimensions or relations, it is common that dimensions already exist. These dimensions must be made to reference some edge in the model the feature is being dropped into. You are also given the opportunity to modify dimensions associated with the library feature (or features, as there may be more than one). How-To 9–13 takes you through the process of inserting a feature via the Design library.

HOW-TO 9-13: Inserting Design Library Features

Fig. 9–68. Select the desired configuration.

1. In the Design library, under the *Features* folder, navigate to the folder that contains the desired feature.

2. Using the left mouse button, drag the item onto a planar face of the model.

3. If multiple configurations ("versions") are available, such as those shown in Figure 9–68, select the desired configuration.

4. If you wish the feature to be linked to the original library feature part file, check the *Link to library part* option, visible in Figure 9–68. Leaving this option unchecked results in a standalone feature that can be edited if necessary.

5. If presented with a preview window, such as that shown on the right-hand side of Figure 9–69, it will be necessary to select edges referenced by the locating dimensions in the library feature. Using the preview as a guide, pick edges (or sometimes other objects) from the model in the work area that will serve as references for the locating dimensions (i.e., what the dimension extension lines will attach to).

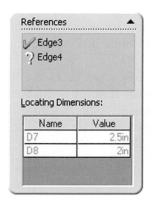
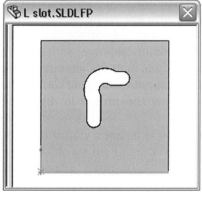

Fig. 9–69. Selecting reference edges.

Before continuing with the How-To, it may be helpful to expand on exactly what is happening at this point. The References panel displayed in the PropertyManager (see Figure 9–69, left) will list all of the references that must be selected. Each reference will correspond to a highlighted edge in the preview (see Figure 9–69, right). As various references are selected from the model, check marks will appear in place of the question marks in the References panel to indicate the references that have been satisfied.

It will appear in many cases that selecting certain edges on the model to correspond to the edges highlighted in the preview window should serve to rotate the feature being inserted. That is not always the case. Do not be concerned with the orientation of the feature at this time. Select the edges to satisfy the dimension references and move on.

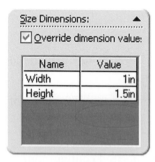

6. Once all references have been satisfied, click on the dimension values in the References panel and type in the desired values. Press Enter to accept each change.

7. If the inserted feature is not of a satisfactory size, click on the *Override dimensions* option (shown in Figure 9–70) and enter the desired dimensional values.

8. Click on OK to finish inserting the feature.

Fig. 9–70. Overriding dimensions.

As can be seen from How-To 9–13, the steps for inserting a Design library feature are quite extensive. However, you may not always be faced with all of these steps. For example, library features may not always have multiple configurations. In addition, selecting references for the locating dimensions can be ignored, or the feature may not even have locating dimensions in the first place.

If the locating dimension references are ignored, the resultant feature will have dangling relationships in the sketch. It will be necessary to repair these relations. If you need to review how this is done, see the section "Dangling Relationships" in Chapter 7.

When locating dimensions are not present in the original library feature, and there are no references of any type (such as geometric relations locating the library feature to geometry in the original library feature part file), there will be no References panel. Rather, there will be a button labeled Edit Sketch, which will allow for editing the underlying sketch for the library feature. In this way, the feature can be positioned exactly where it is needed during the insertion process.

Whether you add locating dimensions or relations when creating a library feature is a personal preference. Creating library features is explored in the following section. A common question when inserting features or forming tools is how to get either item oriented correctly. This task is accomplished via the Modify Sketch command. Because the Modify Sketch command is fairly extensive, a section is devoted to it later in this chapter. First, however, let's finish exploring library features.

Creating Library Features

A library feature part file is how a Design library feature begins its life. Technically, the Design library represents a method of implementing library features. A library feature is nothing more than some item that can be reused over and over, and the Design library is just a convenient point of access to those features.

the same thickness as the average stock from which your sheet metal parts are typically created.

The second stage in the process is to model the tool. The tool should be the same shape you would see on the inside of a piece of sheet metal if the tool were punched into it. This is very true to life, so creating the tool is easy to visualize. Step three would be to cut away the original base feature. Deleting the original base feature is not allowed, because every feature that comes after it would also be deleted. It must be cut away.

Last, an orientation sketch must be created. This orientation sketch is what will be seen when the forming tool is used. This is a necessary part of the process, even if you do not think you will need an orientation sketch. How-To 9–15 takes you through the process of creating a forming tool.

HOW-TO 9-15: Creating a Forming Tool

Fig. 9–77. Completed knockout.

To create a forming tool, perform the following steps. To help explain the process, an example is used. The example shown in the illustrations is of a knockout, but your forming tool may be different. Figure 9–77 shows the knockout forming tool in its completed state. (If using SolidWorks 2006, see the SW 2006 note at the end of this section, as the forming tool creation process has been significantly streamlined.)

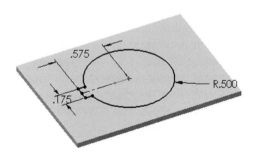

Fig. 9–78. A base feature "slab" and new sketch for the forming tool.

1. Create a base feature large enough to accommodate the forming tool. Good practice is to make the feature a similar thickness to the sheet metal parts the tool will be used on. Figure 9–78 shows the base feature "slab" with a sketch already created for the next feature, which will make up the main body of the forming tool.

2. Create the shape of the forming tool. Use extrusions, revolved features, fillets, and other techniques you have learned to this point. Figure 9–79 shows the sketch from Figure 9–78 extruded, with additional features added.

3. Cut away the geometry that makes up the main base feature. The easiest method for accomplishing this is to sketch on one side of the "slab," use the Convert Entities command, and then cut using a Through All end condition.

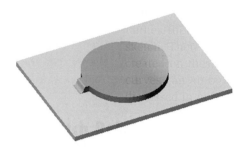

Fig. 9–79. The forming tool's geometry.

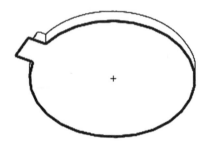

Fig. 9–80. Creating the orientation sketch.

4. Create an orientation sketch by sketching on the bottom of the forming tool and once again using the Convert Entities icon. Figure 9–80 shows the sketch on the underside of the knockout (enhanced for clarity).

5. Once the orientation sketch has been created, exit the sketch.

6. Hide the orientation sketch by right-clicking on the sketch in FeatureManager and selecting Hide. It will still appear when the forming tool is used.

7. Using the Edit Color command, apply the color red to any face that should be removed when the forming tool is used (explained in material to follow).

8. Change to a reasonable view orientation, zoom in, and display the model shaded. What you see on screen will represent the preview in the Design library.

9. Save the file.

Forming tools are typically stored in one of the folders in the path *data\design library\ forming tools*, found in your SolidWorks installation folder. As previously mentioned, the location of all design library items is covered in the section "Design Library Structure" to follow.

It is a good idea to centrally position a forming tool's main feature sketch on the sketch origin when it is being created. This will centrally locate the tool and make it easier to drag and drop onto a sheet metal model. Otherwise, the tool will be some distance from the cursor as it is being dragged onto the model. In addition, SolidWorks will add centerlines at the origin, for dimensional purposes, when you are using the tool.

✓ **TIP:** *Rectangles can be centered on the origin by adding a centerline from corner to corner and applying a midpoint relation between the origin point and centerline.*

In How-To 9–15, step 7 instructs you to apply the color red to faces to be removed when the forming tool is used. To better explain this, let's see what the knockout forming tool would look like if it were used on a sheet metal part. Figure 9–81 shows the resultant feature from two angles for clarification.

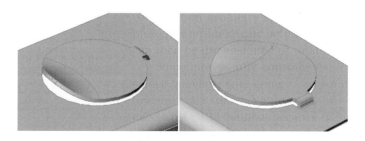

Fig. 9–81. After using the knockout forming tool.

In many cases, forming tools do not break or cut the sheet metal part whatsoever. Many forming tools do just that; they form, but nothing else. Other tools will literally stamp out a piece of material or leave open areas, such as with lancing operations.

How does the forming tool create open areas in a model? The secret is the color red. Specifically, any faces that have been given the color red will be removed when the forming tool is applied to the model. Red is the only color that has any modeling significance in SolidWorks, and this unique functionality only applies to forming tools. Only the true color red can be used, with an RGB (red, green, blue) value of 255,0,0 (the three comma-separated "numbers" representing the three color values, respectively).

Figure 9–82 shows the three perimeter faces that should be given the color red for the knockout to work properly. The bottom face should not be selected. Other features, or even the part itself, can be assigned other colors to no ill effect on the forming tool. Use the Edit Color command to assign colors to a model (see "Modifying Part Color" in Chapter 8).

Fig. 9–82. Faces that should be colored red.

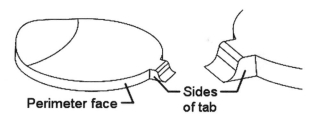

Perimeter face —

Sides — of tab

SW 2006: The steps for creating a forming tool have been simplified to a large extent. Create the geometry as you normally would per steps 1 and 2 of How-To 9–15, and then click on the Forming Tool icon (shown in Figure 9–83), found on the Sheet Metal toolbar (or in the Sheet metal menu). Using the panels found in PropertyManager (Figure 9–83 inset), select any faces to be removed (optional), along with the "stopping" face.

Think of the stopping face as the face that limits how far the tool can punch or stamp into the sheet metal part. Do not worry about removing the base material, creating the orientation sketch, or coloring the faces to be removed. That is all done automatically. All that remains is to save the forming tool at the desired location in the *Forming Tools* folder in the Design library.

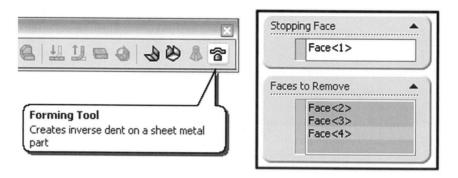

Fig. 9–83. Forming Tool icon and PropertyManager panels (inset).

Design Library Structure

SolidWorks stocks the Design library with some standard parts, shapes, and features, but you can customize the library to your heart's content. This is an extremely easy thing to do. If you know how to use Windows Explorer, you know how to customize the Design library. Figure 9–84 shows an expanded excerpt from Windows Explorer, which shows the directory structure of the Design library. Two of the branches have been expanded to show their underlying structure.

Fig. 9–84. Directory structure of the Design library.

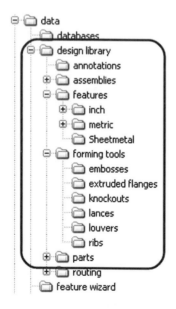

The folders in the Design Library window are directly linked to the actual names of the folders stored on the hard drive. This makes customizing the folder names in the Design library extremely easy, as most computer users should already be familiar with creating or renaming folders in Windows.

All SolidWorks installations will have a similar directory structure to start with. To specify different or additional paths where SolidWorks looks for Design library files, refer to How-To 3–6.

By default, SolidWorks will look in the locations specified in Table 9–2. Note that *< drive >* refers to the hard drive letter SolidWorks was installed on, and *< installation directory >* refers to the SolidWorks installation folder. By default, the installation drive and directory is *C:\Program Files* unless otherwise specified by the person installing the SolidWorks software.

Table 9-2: Design Library File Paths

Design Library Object	*Path*
Annotations	\<drive>:\\\<installation directory>\\SolidWorks\\data\\design library\\annotations
Features	\<drive>:\\\<installation directory>\\SolidWorks\\data\\design library\\features
Parts	\<drive>:\\\<installation directory>\\SolidWorks\\data\\design library\\parts
Assemblies	\<drive>:\\\<installation directory>\\SolidWorks\\data\\design library\\assemblies
Palette Forming Tools	\<drive>:\\\<installation directory>\\SolidWorks\\data\\design library\\forming tools

Renaming Dimensions

Renaming dimensions is a good idea in the case of Design library features because any dimension not designated as internal will appear in the Size Dimensions panel (see Figure 9–70) when inserting the feature into a model. Renaming dimensions is typically something that would be done while creating the library feature. Renaming features may be something you will also want to do. There are other reasons for renaming dimensions, such as when they are used in equations or design tables, discussed later in this book.

Dimensions have names that correspond to the sketch or feature with which they are associated. For instance, a dimension with the name D1 that is associated with Sketch 4 would have the full name of *D1@Sketch4*. The "at" (@) symbol is nothing more than a separator character SolidWorks understands. Dimension names such as *D1@Sketch4* do not mean much by themselves. A descriptive name such as *Diameter@Lower-holes* makes it much easier to understand what will be changed if the dimension value is altered.

To rename any dimension, access the dimension's properties by right-clicking on the dimension. The area where the dimension name would be changed is shown in Figure 9–85. Dimensions that belong to a library feature and appear in the *Dimensions* folder can be renamed in the same way as any other object in FeatureManager; that is, with a slow double click.

✓ **TIP:** *To display dimension names in the work area, check the Show dimension names option of the General section of System Options (Tools > Options).*

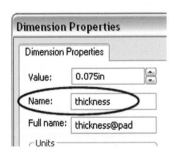

Fig. 9–85. Changing a dimension's name.

Modify Sketch Command

Sketch geometry can be mirrored, rotated, scaled, and translated using the Modify Sketch command. The Modify Sketch command consists of a small window, shown in Figure 9–86.

Fig. 9–86. Modify Sketch window.

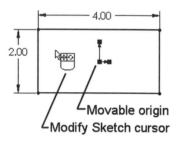

Fig. 9–87. Movable origin and Modify Sketch cursor.

Any sketch geometry can be modified, unless it has references to external model geometry. For instance, a sketch could not be moved (translated) if it were dimensioned to existing edges of a model. Scaling sketch geometry will work if there are dimensions associated with the sketch, but as with translating, it must not be fixed at one location.

Rotating a sketch is one of the more common uses of the Modify Sketch command, which will work even if the geometry has lines that are horizontally or vertically constrained. This is because the entire sketch, including the origin point, is rotated. Because of this, any entities constrained horizontal or vertical remain so, albeit relative to the movable origin.

When the Modify Sketch command is used, a symbol that looks like a black origin point with small squares (or "handles") attached to it will be visible. This is known as the movable origin. The handles can be used to mirror or flip the sketch geometry. In addition, the cursor changes and allows for dynamically translating or rotating the sketch with the left or right mouse button, respectively. The movable origin and Modify Sketch cursor are shown in Figure 9–87.

The section that follows covers the capabilities of the Modify Sketch command in a general way. The Modify Sketch command is very simple to use once you understand the basics. All of the functions contained within the Modify Sketch command are implemented in a very similar fashion. How-To 9–16 takes you through the process of using Modify Sketch to perform its basic functions. Following How-To 9–16, some additional functionality is pointed out.

How-To 9-16: Using the Modify Sketch Command

To use the Modify Sketch command to translate, rotate, or scale sketch geometry, perform the following steps. Note that you should be editing a sketch in order to use the Modify Sketch command. If not, the sketch to be modified must be selected in FeatureManager. Modify Sketch can also be used when a Design library object must be positioned.

1. Click on the Modify Sketch icon, shown in Figure 9–88, or select Modify from the Tools > Sketch Tools menu.

2. Click in the applicable area of the Modify Sketch window, depending on what you want to accomplish. For example, click in the Rotate area.

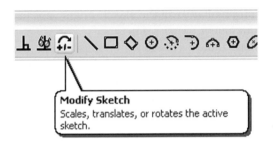

Modify Sketch
Scales, translates, or rotates the active sketch.

Fig. 9–88. Modify Sketch icon.

3. Enter a value. For example, enter *90* (degrees) for the Rotate value.

4. Press Enter.

5. Repeat steps 2 through 4 for other modifications as required and click on Close when finished.

As you can see, the Modify Sketch command in its basic form is extremely easy to use. However, there is some additional functionality, along with a few nice tips and tricks, you will want to be aware of. The following sections point out various points of interest and additional functionality of the Modify Sketch command.

✓ **TIP:** *The movable origin can be positioned or related to other geometry by dragging it with the left mouse button. This can play an important role when scaling and rotating.*

Rotating Using the Mouse

In addition to typing in values, you can use the mouse to rotate geometry dynamically. If you hold down the right mouse button, you will notice a line attached to the center of the movable origin. Move the mouse, and a small readout will be displayed on the cursor, informing you how much rotation you are placing on the geometry.

If using the mouse to dynamically rotate, note that the distance of the mouse from the movable origin affects the coarseness of the rotational increments. Closer to the origin will rotate in large increments; farther away will offer smaller incremental values and more precise control. To quickly rotate a sketch (say, 90 degrees), keep the cursor close to the movable origin.

Translating

Both x and y values can be changed at the same time if entering values. Pressing Enter will apply both values at once. To use the mouse to translate dynamically, pick anywhere on the screen and drag the sketch with the left mouse button. The short arrow (of both the movable origin and sketch origin) always relates to the positive x direction, and the longer arrow the y direction.

The *Position selected point* option allows for picking a specific sketch entity point after turning the option on, and then viewing the x-y coordinates of that point. Entering a value at that time and pressing Enter will move the selected vertex point to the x-y coordinates specified relative to the sketch origin. Note that the sketch will move relative to the sketch origin, not the movable origin.

Using *Position selected point* is a little confusing because when the sketch is translated the sketch origin moves as well. This is always the case, whether *Position selected point* is selected or not. This means that if in using this function you were to select the very same point it would have the same *x-y* coordinates it had prior to being translated.

Scaling

It probably goes without saying that a scale factor greater than 1 increases the size of the sketch, and anything less than 1 will decrease the size of the sketch. If a sketch contains dimensions, the values of those dimensions will change accordingly.

Make sure to specify what you would like to scale about: the sketch origin or movable origin. If the sketch origin is used and the sketch is some distance from the origin, the sketch will move away from or closer to the origin, depending on the scale factor. Positioning the movable origin near the center of the sketch and scaling about the movable origin provides good results.

Flipping

Flipping refers to mirroring geometry to the other side, but not retaining the original. Use the movable origin to flip the sketch left to right, top to bottom, or both. A simple right-click on the applicable black square will accomplish the flip. For example, clicking on the black square at the end of the *x* axis of the movable origin will flip the sketch across the *y* axis.

Move, Rotate, and Scale Sketch Tools

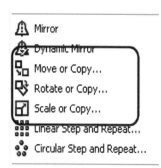

Fig. 9–89. Sketch tools to avoid.

This section has been added for the purpose of advising you not to use certain commands under most circumstances. The commands in question are the Move, Rotate, and Scale sketch tools, which are found in the Sketch Tools menu (shown in Figure 9–89). Each of these commands has the option of copying the original geometry, and hence the language "or Copy" shown following each command in the menu.

SW 2006: Copying sketch geometry can be performed as a separate operation. Commands have been renamed to reflect this fact and are now known as Move, Rotate, Scale, and Copy.

Why should you avoid these sketch tools? It is because these tools can damage your sketch relations. These tools are best suited for working with imported 2D geometry or for creating 2D geometry from scratch in a drawing (as opposed to a part). For example, if the Rotate command is used on a sketch with existing relations—such as vertical, horizontal, and so on—many of those relations will be automatically removed by SolidWorks in order to rotate the sketch. The Modify Sketch command does not cause these problems.

Let's look at a simple example so that we can understand the issue. Figure 9–90 shows a sketch with a fair number of small arcs and interesting geometry. There are a total of 40 dimensions and relations on this geometry. The sketch can be dragged from any point without losing its shape.

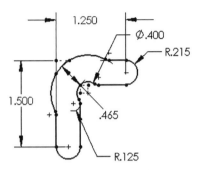

Fig. 9–90. Sample sketch.

Accessing the Rotate or Copy command from the Sketch Tools menu brings up the Move or Copy PropertyManager, shown in Figure 9–91. Actually, accessing any of the aforementioned three commands (shown in Figure 9–89) brings up the same PropertyManager. Move, Rotate, or Scale can then be selected from the Operation panel (visible in Figure 9–91). The sketch entities to be rotated were selected, the *Keep relations* option was checked, and a base point for the rotation operation was selected. Finally, a rotation angle of 90 degrees was entered, and the OK button was clicked on to complete the process.

After the sketch is rotated, the total number of relations drops from 40 to 34. All of the horizontal and vertical relations are dropped, along with a few others. Dragging the geometry at this point reshapes the geometry in an undesirable fashion, and it becomes difficult to maintain. Relations will have to be added back into the sketch to maintain the design intent.

In short, it is better to use the Modify Sketch command instead of the Move, Rotate, or Scale sketch tools. Use the latter tools if creating a 2D drawing without the benefit of a solid model, or if manipulating 2D imported drawing geometry. 2D drawings are discussed in Chapter 12.

SW 2006: Each of the Move, Rotate, Scale, and Copy sketch commands has its own PropertyManager. Although the interface for these commands has been altered slightly, the recommendations made in this section still hold true for SolidWorks 2006.

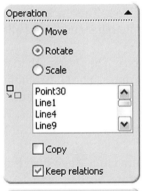

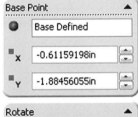

Fig. 9–91. Move or Copy PropertyManager.

Summary

At the beginning of this chapter, you learned that sheet metal parts are defined by creating a base flange feature. Once a base flange feature has been created, other sheet-metal-type features can be added. When creating an edge flange feature, SolidWorks automatically generates a sketch. This sketch can be modified by the user via the Edit Profile button found in the Edge Flange PropertyManager. Miter flanges require a sketch, though the sketch can be pretty basic. A simple line will suffice.

Other sheet metal features include hems, sketched bends, tabs, and jogs, to name a few. There are also fold and unfold features that should be used if there are cuts that need to be made to areas involving bend regions. Cuts can also be added to a sheet metal part without necessarily unfolding the model first. A simple rule to follow would be to create sheet metal features in the same way those features will be created in the real world. If a cut will be made to the formed model in the real world, create the cut on the formed model in SolidWorks.

Auto relief can be of three types: rectangular, obround, or tear. Bends can have their individual relief settings modified. This includes changing the type of relief, and the width and depth of the relief cuts. Bend allowance can also be altered on a bend-by-bend basis.

Forming tools are a part of the Design library. The Design library allows for the reuse of geometry. Features, forming tools, parts, assemblies, and even annotations make up the Design library. Design library features are library-feature part files, whereas forming tools are regular part files. Forming tools can be used on sheet metal parts only, but library features can be used on any part.

The Modify Sketch tool is an important command when inserting forming tools or library features. It allows for rotating, scaling, translating, and flipping sketch geometry. The fact that the Modify Sketch tool can rotate sketch geometry when inserting forming tools or library features is of utmost importance and is a definite necessity.

CHAPTER 10

Springs, Threads, and Curves

SPRINGS ARE A COMMON COMPONENT in many assemblies. They come in all sizes and shapes. Springs are quite easy to model in SolidWorks. Just about every spring begins with a helical curve. Threads also use a helix as a sweep path. Threads might be external threads on a bolt or internal threads on a nut. Helical curves, however, are not the only thing springs and threads have in common.

Both springs and threads must be created as swept features. Swept features can be used for many other purposes above and beyond creating springs and threads. You can also get into some very complex geometry with swept features. This chapter shows you some of the more common tasks you can accomplish with swept features.

Curves can play a large role in creating swept features. Curves can be used as the path for a swept feature, but curves can also be used for a number of other feature types. You will undoubtedly find many different uses for curves. Thus, even though springs and threads are two of the main topics in this chapter, you will find many applications for what is explored in the material that follows.

Swept Features

To this point, everything you have created in the way of sketched features has required one sketch per feature. For instance, an extruded boss or cut requires only one sketch. A revolved feature also requires just one sketch. With the introduction of swept features, you will find that one sketch is no longer going to be sufficient most of the time.

Why would you need two sketches to create a swept feature? It is because you must have a sketch you will be sweeping, as well as a sketch you can sweep along. These two sketches are called the profile and the path, shown in Figure 10–1. Sometimes the sketch geometry is referred to as the sweep section and trajectory.

There are a lot of options and physical parameters that can be used to obtain some very interesting geometry. Most of these options are covered in detail in this chapter. A swept feature, taken in its simplest form, is easy to understand. The profile is swept along a path, thereby creating a solid. Figure 10–2 shows the result of sweeping a rectangle along the U-shaped path shown in Figure 10–1.

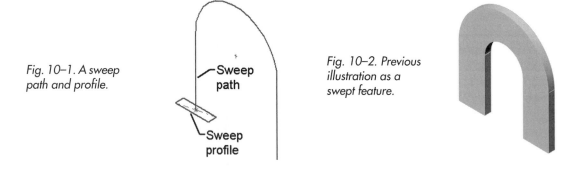

Fig. 10–1. A sweep path and profile.

Sweep path

Sweep profile

Fig. 10–2. Previous illustration as a swept feature.

Valid Profiles

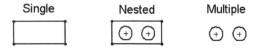

Single   Nested   Multiple

Fig. 10–3. Valid sweep profiles.

A sweep sketch can contain a single closed profile or multiple closed profiles. If the base feature sketch contains multiple profiles, understand that multiple bodies will be created. Sweep profiles can be separate, or nested two deep, as shown in Figure 10–3. *Nested* is a term that signifies how objects (in this case, profiles) can exist within other profiles. Creating a thin feature while sweeping is also an option. In other words, the general rules of sketching still apply, which were covered in Chapter 3.

Valid Paths

When creating a sweep path, you have quite a few options. However, there is one rule you absolutely must follow. This steadfast rule you must adhere to is that the path must never self-intersect. This means that if the path so much as touches itself at any point the sweep will not work. This does not mean to say that a sweep path must be open. For example, a circle can be used as a sweep path. Sweeping a circle about a circle is one method of creating a toroid (donut shape).

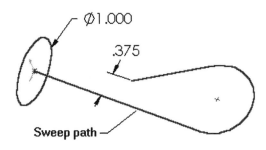

Fig. 10–4. This would result in a self-intersecting sweep.

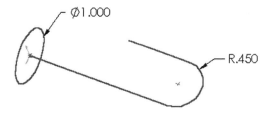

Fig. 10–5. More self-intersecting sweep sketch geometry.

Fig. 10–6. A candle holder.

An illegal situation that is very similar to a self-intersecting path is a self-intersecting sweep. Not only can the path not intersect itself, the resultant swept geometry must not self-intersect. This can occur in a number of ways. Figure 10–4 shows a simple example of this.

It should be obvious that by sweeping the circle along the path shown in Figure 10–4 the geometry would self-intersect. The .375-inch dimension could actually be a lot larger, say .750 inch, and the sweep would still self-intersect. This is because the distance between the two closest points of the sweep path should be greater than twice the radius of the circle, or 1 inch. Otherwise, the geometry created by the circle will touch. This is what is meant by self-intersecting geometry.

A situation that is not as obvious regarding self-intersecting geometry can also arise when sweeping along a path with arcs in it. In Figure 10–5, the circle's radius is greater than the radius of the arc it is sweeping along. As the profile circle moves along the arc in the path, it overlaps itself. If the arc on the path had a radial dimension of .501 inch instead of .450 inch, the sweep would work.

Just what can you get away with as valid path geometry? Quite a bit, really. The path does not have to be perpendicular where the sweep starts. The profile sketch plane does not have to intersect the start of the sweep path. The path does not have to be tangent along its entire length. As a matter of fact, the path can contain angled lines at less than 90 degrees. The path can even be a spline existing in 3D space. Figure 10–6 shows an example of what can be accomplished with a swept feature. The image is of a candle holder. It is a single swept feature, and you will learn how to create this model before this chapter is through.

Now that you are aware of what constitutes valid sketch geometry for a sweep path and profile, we can get down to the business of creating an actual swept feature. How-to 10–1 takes you through the process of creating a swept feature. Subsequent material explores some of the options available during the sweep process. It should be pointed out that the sweep can be a boss or cut. The process is identical except for the menu picks.

➤ **NOTE:** Swept features require that the final sketch be exited prior to the Sweep command becoming available. Exit out of any active sketch by clicking on the Sketch or Rebuild icon.

HOW-TO 10-1: Creating a Sweep

Fig. 10–7. Sweep PropertyManager.

To create a swept feature, perform the following steps. Remember that you must exit out of any sketch you may have created prior to accessing the Sweep command.

1. Select Insert > Boss/Base > Sweep, or click on the Swept Boss/Base icon found on the Features toolbar. This accesses the Sweep PropertyManager, shown in Figure 10–7.

2. Select the sweep Profile.

3. Select the sweep Path.

4. Click on OK to create the swept feature.

It is possible to select profile and path sketch geometry from either FeatureManager or the work area. This makes absolutely no difference in the outcome of the feature. It is okay to select either the profile or path first. Do whatever is easiest for you, but be aware that the list box that is active (salmon colored) will display whatever you select. It is best to activate a particular list box by clicking inside the box prior to selecting geometry.

Using Existing Edges as a Sweep Path

It is true that both a sweep profile and path are needed to complete a sweep. However, the sweep path does not have to be a sketch. The path can be an existing model edge, which simplifies the process of creating the swept feature because there is one less sketch that must be manually created. Figure 10–8 shows a part to which a swept cut will be applied. In this example, a small curvy cut will be added around the outside of the model. The sketch profile is shown in an enlarged inset for clarity.

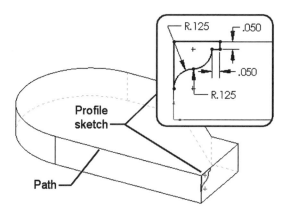

Fig. 10–8. Preparing for a swept cut.

Fig. 10–9. Outcome of the swept cut.

If the sweep is created without modifying any of the options in the Sweep PropertyManager, the outcome will appear as it does in Figure 10–9. As can be seen in the figure, the swept cut continues along the first edge, but does not continue along the tangent edges. To force the sweep to continue along the tangent edges, an option known as *Tangent propagation* must be enabled. The *Tangent propagation* option is found in the Options panel of the Sweep PropertyManager, and is only available when an edge is selected as the sweep path.

When *Tangent propagation* is enabled, the sweep will continue along the selected edge as long as there are tangent edges that come after it. As long as SolidWorks keeps finding tangent edges, the sweep will keep going. When it comes to the end of the last tangent edge, the sweep will terminate at that point. Figure 10–10 shows the completed swept cut. Note the material that remains at the end of the sweep. This is addressed in the following section.

Fig. 10–10. Completed swept cut.

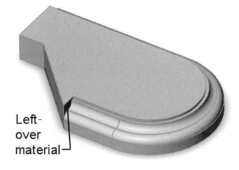

Left-
over
material

Sweep Options

The *Align with end faces* option (shown in Figure 10–7) is useful for situations in which the sweep path does not completely allow the profile to cut the desired material away during a swept cut (though this option can apply when creating bosses as well). Such is the case in Figure 10–10. However, if the *Align with end faces* option is enabled, the outcome looks as it does in Figure 10–11.

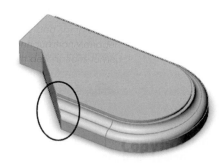

Fig. 10–11. After turning on Align with end faces.

Note that the sweep continues beyond the end of the sweep path, thereby cleaning up the remaining scrap of material on the part. The end faces are typically the next faces encountered on the part. Without the *Align with end faces* option, it would have been necessary to manually create another feature to perform this task for us.

The *Maintain tangency* option is a commonly misunderstood option. When *Maintain tangency* is checked, tangent segments in the sweep profile will result in the corresponding faces of the model being tangent. In other words, if a line and arc in a sweep profile are tangent before being swept along some path the corresponding faces that are created will also be tangent. This would not necessarily be so if *Maintain tangency* were not checked.

Just as it sounds, the *Show preview* option will force SolidWorks to show a preview of the swept feature. During complex sweeps, it may take some time to complete the sweep and the preview will also take a proportionately long time to generate. In such cases, once you become more familiar with the software you will probably want to leave the *Show preview* option off. For now, however, leave it on so that you can see what the feature will look like.

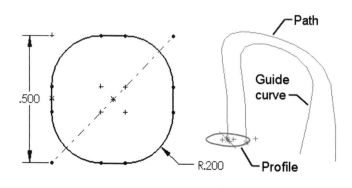

Fig. 10–12. Profile containing tangent segments.

Certain options do not appear until the proper conditions are met. Such was the case with Tangent propagation, which will not appear until an edge is specified for the sweep path. Another such option is *Merge tangent faces*, which appears only if the sketch profile contains tangent segments, such as arcs and lines. The profile shown in Figure 10–12 contains such tangent segments (shown enlarged for clarity), which will be used to explain the *Merge tangent faces* option.

You will also notice a guide curve in the figure, the usage of which is explained in the following section.

After the sweep is created (based on the geometry shown in Figure 10–12), note that the model consists of numerous faces. In this example, *Merge tangent faces* was left off. The edges in the model have been highlighted black in order to more easily see individual faces.

Fig. 10–13. Merge tangent faces disabled.

Fig. 10–14. Merge tangent faces enabled.

With *Merge tangent faces* turned on, the resultant geometry is the same from a volumetric standpoint, but faces that were previously tangent have been merged into a single face. This can be seen in Figure 10–14. The black edges that were seen in Figure 10–13 are no longer present. The entire outside face of the model is now a single face.

There will be times when it is desirable to merge tangent faces for cosmetic (or other) reasons. Sometimes you may feel that the option does not work properly. This might be the case if you expect to see cylindrical or planar faces merged into one. The fact is, only complex faces will be merged. Cylindrical and planar faces will be left as they are, and will not merge into a single face.

There are additional settings that will greatly affect the outcome of a swept feature. Before exploring what these additional settings accomplish, it is necessary for you to understand what role guide curves play when creating swept features. Guide curves are discussed in the following section.

Guide Curves

Guide curves can be employed when it is necessary to have a greater degree of control over swept or lofted features. (Lofted features are discussed in Chapter 20.) A guide curve is essentially an extra sketch that serves as a second path. You can use one guide curve or more then one guide curve simultaneously. Guide curves are typically open profile sketches, but do not have to be.

As an example of where a guide curve or two can play a very important role, let's use what might be a typical plastic bottle found in the home. To begin with, it would be beneficial to see what the underlying sketch geometry might look like. Figure 10–15 shows a total of four sketches used to create the bottle's basic shape.

In Figure 10–15, note the two main ingredients of a swept feature. These are the sweep profile

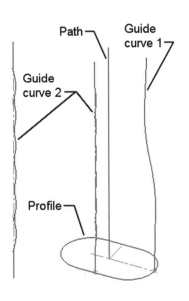

Fig. 10–15. Preliminary sketch geometry prior to creating the bottle.

and path. In this example, the path is a simple straight line, but this is not a requirement. The first guide curve consists of two lines and two arcs. The first guide curve will control the shape of the sides of the bottle. The second guide curve is more complex, and consists largely of arcs. It is shown in a plan view by itself in Figure 10–15 for clarity.

Fig. 10–16. Guide Curves panel.

Fig. 10–17. Resultant swept feature.

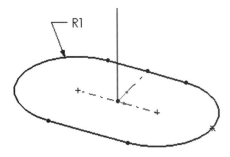

Fig. 10–18. Resultant swept feature.

In that we have already covered the basics of sweeping, let's focus directly on how to make use of the guide curves. In the Sweep PropertyManager you will find the Guide Curves panel, shown in Figure 10–16. Click inside this list box and select the guide curves.

As soon as the guide curves are added, the shape of the model changes dramatically. The profile changes shape, as dictated by the guide curves as the profile moves along the path. Figure 10–17 shows the resultant feature.

The simplicity of employing the guide curves belies the challenge of creating them. Creating the sketch geometry that constitutes a guide curve is not necessarily the challenging part. Rather, it is the technique one must follow.

One important fact to consider when creating guide curves is order of creation. Specifically, guide curves should be created prior to creating the profile geometry. Consider that the profile geometry will need to follow the guide curves. A special geometric relation known as a pierce constraint will help the profile geometry do exactly that: follow the guide curve. If the guide curves are created after the profile geometry, the sweep will be more prone to errors. The pierce constraint is discussed in material to follow.

Another important aspect of using guide curves is to ensure that the profile sketch is properly constrained. Figure 10–18 shows the profile used for the bottle with the guide curve hidden for clarity. Note the construction geometry added for the sake of positioning the profile and establishing symmetry. The profile is centered at the path's endpoint and is symmetrical about the path's endpoint. Without symmetry, the guide curve would simply pull or push the profile in one direction or the other, and the profile would not be able to change size during the sweep. The sketch point on the arc was added strictly for the sake of adding the pierce constraint to it.

Pierce Constraint

As mentioned previously, there is a special constraint in SolidWorks known as the "pierce" constraint. It is similar to a coincident constraint in that it typically constrains a point to some other object. The pierce constraint is a bit more restrictive than a coincident relation, however. A pierce constraint is very often used with regard to guide curves when creating swept or lofted features.

Technically, the pierce constraint might be explained as follows. The sketch point being pierced is coincident at the point where another object pierces the sketch plane. This description is a little difficult to understand without some illustrations. Let's try to visualize the pierce constraint in comparison to a coincident constraint. Examine Figure 10–19.

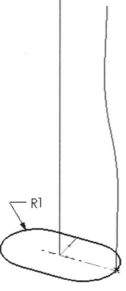

*Fig. 10–19.
Visualizing a
pierce constraint.*

Figure 10–19 shows the same sketch geometry used for the bottle minus the second guide curve. Imagine that the guide curve is a sharpened piece of stiff wire. It has been sharpened at the bottom, where it intersects the profile. It should be noted that a sketch point was added to the profile (specifically, one of the arcs) for the express reason of adding the pierce constraint. Now imagine that as the profile moves up the path the pierced point, as it were, must move to conform to the shape of the wire. Because the center of the profile will be locked onto the path, the pierce constraint causes the profile to change size.

How would the pierce constraint be different from a coincident constraint? Consider what would happen if a coincident relation were added between the guide curve and sketch point on the profile. The guide curve would be projected perpendicular to the sketch plane of the profile. This projection would actually be a line, and that is what the point on the profile would be coincident with. From previous experience, we know that "coincident" does not necessarily mean "touching," but rather "on." In other words, the point would be free to move anywhere along an infinite line, as defined by the projection of the guide curve to the profile's sketch plane.

This can all be rather confusing. To reiterate, a coincident relation to some other object in the model means that the object, if not on the current sketch plane, will be projected to the sketch plane. With the pierce relation, there is no projection. The point being pierced must intersect the object piercing it.

With that said, a point being pierced can be any point associated with a sketch, such as an endpoint, center of a circle, or sketched point. The object doing the piercing can be an axis, edge, line, arc, spline, or any number of other curves we will explore in this chapter. The pierce constraint is added the same way any other constraint is added. Use the Add Relation icon.

➥   **NOTE:** *It is not always a requirement to use the pierce constraint with guide curves. It depends on the situation. Good technique is to use a pierce constraint, as it will normally give you a higher degree of control.*

Orientation and Twist Control

*Fig. 10–20.
Selecting the
Orientation/
twist type option
in the Options
panel.*

The Orientation/twist Type option allows for controlling how the sweep profile is oriented as it is swept along the path. It also controls the twist exhibited by the swept feature if guide curves are used. The first two options in the Orientation/twist Type drop-down list will be options employed most often when sweeping, with *Follow path* being the most common. The Options panel in the Sweep PropertyManager is shown in Figure 10–20.

Follow Path

The *Follow path* option is the default when performing a sweep. It is the safest option to use, meaning that it is less likely to create an error in the geometry while sweeping. It is also by far the most commonly used Orientation/twist Type setting. An example of this option in use is shown in Figure 10–21. The original sketch geometry of the path and profile is shown in the same image.

*Fig. 10–21.
Follow path
option in action.*

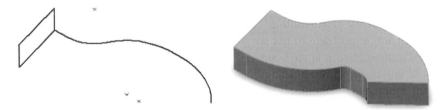

When *Follow path* is used, the sweep profile remains at a constant angle to the path along the entire length of the path. The *Follow path* option is used when creating springs or threads. It is also used in any situation in which a constant cross section is required.

Keep Normal Constant

When the *Keep normal constant* option is used, the sweep profile remains parallel to its original sketch plane as it is swept along the path. This option is not quite as common as the *Follow path* option, but it does have its uses. An example of a swept feature created with the *Keep normal constant* option is shown in Figure 10–22. The same sketch geometry used in the previous example (Figure 10–21) is being used here.

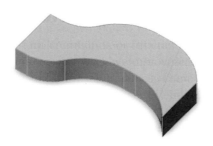

Fig. 10–22. Using the Keep normal constant *option.*

As is evident in Figure 10–22, the orientation of the profile plane never changes. The profile is always parallel to its original sketch plane along the entire path. This option sometimes makes it impossible to complete a sweep. For example, if a bend in the path were greater than 90 degrees the sweep would intersect itself. This is not allowed in SolidWorks.

Follow Path and 1st Guide Curve

Follow path and 1st guide curve requires the use of at least one guide curve. The sweep profile remains at a constant angle to the path as it moves along the length of the path, just as it does with the *Follow path* option. The twist of the sweep profile is based on a vector between the path and the first guide curve.

Until an example is shown, it is difficult to imagine what is happening when this option is employed. To help with the visualization process, Figure 10–23 shows what the sketch geometry looks like that will be used. A couple of different angles are used to help give an idea of how the three sketches are positioned in space in relation to one another.

Fig. 10–23. Sketch geometry for Follow path and 1st guide curve example.

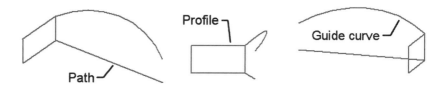

To best understand how the *Follow path and 1st guide curve* option affects the outcome, let's first look at the geometry if the default *Follow path* option were used. Figure 10–24 shows the resultant swept feature geometry from a few different angles. Note in particular how the guide curve changes the shape of the swept feature.

The geometric relations on the original sketch profile geometry will play an important role in the final shape. For example, there was a horizontal relation at the top of the rectangle, but there was no vertical relation on the right side. When the rectangle was swept, the guide curve forced the rectangle to change shape within the constraints applied to it. Another significant contributing factor is a pierce relation that was placed between the top right-hand corner of the rectangle and the guide curve, which forced that point on the rectangle to follow the guide curve.

Now let's examine what happens when the *Follow path and 1st guide curve* option is used. Figure 10–25 shows the same part as that of Figure 10–24, with the only change being the *Follow path and 1st guide curve* option used instead of *Follow path*. Once again, we will examine the model from a few different angles. The rectangle's left side in the example in Figure 10–25 is twisting upward. The changes are easily noticeable.

*Fig. 10–24.
Results with guide
curve using the
Follow path option.*

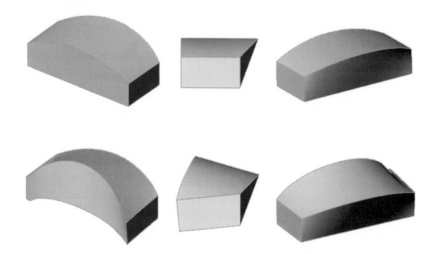

*Fig. 10–25.
Results using the
Follow path and
1st guide curve
option.*

Follow 1st and 2nd Guide Curves

The *Follow 1st and 2nd guide curves* option requires the use of two guide curves. Once again, the sweep profile remains at a constant angle to the path as it moves along the length of the path. The twist of the sweep profile is based on a vector between the first and second guide curves. Once again, let's use illustrations to help understand what is happening during the sweep process.

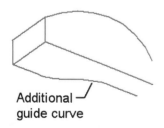

**Additional
guide curve**

*Fig. 10–26. Adding a second
guide curve.*

Because a second guide curve will help demonstrate this option, some new sketch geometry is in order. Figure 10–26 shows the additional guide curve. Although it is difficult to tell from the illustration, the second guide curve is actually coplanar with the path.

The three views in Figure 10–27 show the outcome of the sweep using the *Follow path* option. Figure 10–27 gives you a chance to see what effect multiple guide curves can have on a sweep. Keep in mind that the original sweep profile was nothing more than a simple rectangle. From this, you begin to appreciate what guide curves can accomplish.

In Figure 10–28, the *Follow 1st and 2nd guide curve* option has been used. In this example, the twist is determined via a vector between the first and second guide curves, as opposed to a vector between the path and the first guide curve (shown in Figure 10–25).

Fig. 10–27. Sweeping with two guide curves using the Follow *path option.*

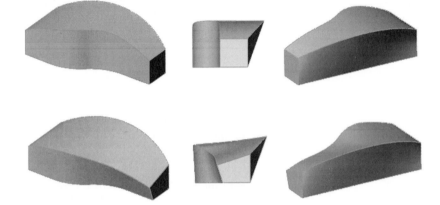

Fig. 10–28. Using the Follow *1st and 2nd guide curve option.*

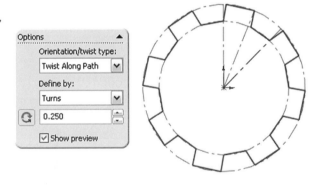

Fig. 10–29. Selecting Twist along path.

Fig. 10–30. Worm gear.

Twist Along Path

Adding twist during a sweep is a function that will make certain specific modeling tasks much easier. Creating gears is a perfect example. When the *Twist along path* option is chosen from the Orientation/twist Type list, the panel changes to appear as it does in Figure 10–29. The sketch shown in Figure 10–29 is the profile that will be swept.

The amount of twist can be determined by degrees, radians, or turns. If the sketch shown in Figure 10–29 is swept along a straight line, the worm gear model shown in Figure 10–30 is what can be achieved. Cosmetic fillets and a cut have been added to complete the model.

The only option that has not been discussed in the Orientation/twist Type listing is *Twist along path with normal constant*. Where *Twist along path* is akin to the *Follow path* option (but with the addition of twist), *Twist along path with normal constant* would be similar to the *Keep normal constant* option, but again with the addition of twist.

Fig. 10–31.
Path alignment.

Path Alignment

The path alignment settings shown in Figure 10–31 are only available when *Follow path* is selected from the Orientation/twist Type listing and there are no guide curves being used. Path alignment offers a bit more control when experiencing problems in swept feature creation, and attempts to align the profile sketch based on the option selected from the list.

The basic rule of thumb with path alignment settings is to not bother with it if you are not experiencing problems with the sweep. If the sweep is failing or there are odd surface fluctuations in the resultant swept feature, experiment with the settings. Table 10–1 outlines the path alignment settings and what they accomplish.

Table 10-1: Path Alignment Settings

Setting	Results
None	The default setting; no correction to the sweep profile is made.
Direction Vector	Attempts to align the profile perpendicular to the edge or line selected for the direction vector.
All Faces	Attempts to make the resultant sweep tangent to adjacent faces at the beginning and end of the sweep.
Minimum Twist	Attempts to reduce twisting of the profile during a sweep along a 3D path in order to eliminate self-intersecting geometry.

Helical Curves

A helical curve is a prerequisite to creating a spring or a thread, both of which require a sweep operation. Before you can create a helix in Solid-Works, you must first make a circle. After that, it is pretty much just a matter of plugging in the parameters that define the helix.

The circle defines the diameter of the helix and helps to determine the start point of the helical curve. Like everything else in SolidWorks, a helix is parametric. You can change the number of revolutions and the pitch and height of a helix by editing its definition or accessing its dimensional values by double clicking on the helix feature.

The Helix/Spiral PropertyManager, shown in Figure 10–32, is used to define a helix. Usually the first thing you would do after entering the Helix command is to determine how you want to define the helix. You have the following three options.

- Pitch and Revolution

- Height and Revolution

- Height and Pitch

Once you have determined how you want to define the helix, it becomes a matter of plugging in the appropriate information. The pitch is the distance between revolutions. Height is the overall height of the helix, and revolutions are the number of total turns the helix makes.

The Reverse Direction option specifies in what direction the helix is going away from the circle, such as up or down. The Clockwise and Counterclockwise options are self-explanatory. If creating threads from a helix, you might equate the clockwise or counterclockwise options with regular or reverse (left-hand) threads, respectively.

Fig. 10–32. Helix/Spiral PropertyManager.

SolidWorks also provides the ability to create a tapered helix, shown in Figure 10–33. This is very convenient for creating pipe threads. If you do create a tapered helix, make sure you specify an angle for the taper, and whether the taper is inward or outward.

The starting point of the helix is more important than it might seem. For example, if you were creating threads you would probably want the threads to begin at a particular point. Using the Starting Angle setting, you could rotate the helix so that it would coincide with a precise location. By clicking on the arrows to the right of the Starting Angle value, you can see the helix rotate clockwise or counterclockwise about its axis. It would be typical to simply type in a value from 0 to 360.

Fig. 10–33. Tapered helix.

When using a helix to create a simple spring, it is best to use a start angle of 0, 90, 180, or 360 degrees. Assuming the original circle defining the helix is centered on the origin point, the start point of the resultant helix will coincide with one of the default planes, such as the front or right planes. Although this is not required, it is good technique. How-To 10–2 takes you through the process of creating a helical curve.

✓ **TIP:** Selecting the helix near (but not on) an endpoint and then clicking on the Sketch icon will establish a plane perpendicular to the helix and passing through its endpoint.

How-To 10-2: Creating a Helical Curve

To create a helical curve, perform the following steps. The Helix icon is found on the Curve toolbar.

1. Create a sketch that contains a circle. The circle will define the diameter of the helix.

2. Select Curve > Helix/Spiral from the Insert menu, or click on the Helix icon.

3. In the *Defined by* section, specify how you would like to define the helix.

4. Specify the Height, Pitch, or Revolution parameters as required.

5. Specify a *Starting angle* of 0 to 360 degrees. If 0 degrees is required, the value must be typed in.

6. Select CW or CCW.

7. Check Reverse Direction if necessary.

8. Click on OK to create the helix.

If it is necessary to create a tapered helix, check the Taper Helix option. It will then be possible to enter an angle for the taper value and to specify whether or not the taper is outward or inward.

Spirals

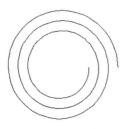

Fig. 10–34. Spiral curve.

To create a spiral curve, select Spiral from the drop-down list in the Defined By panel of the Helix/Spiral PropertyManager. Spiral curves can be used to create coil springs, such as those found in a carpenter's tape measure or a mechanical watch. A example of a spiral is shown in Figure 10–34.

Spirals are defined by pitch and revolution only, and have no height. The pitch, in the case of a spiral, is the distance between coils. The Reverse Direction option refers to whether the spiral winds its way outside the circle or inside the circle. If inside, you must use some common sense as to the number of revolutions and the length of the pitch. If the spiral winds its way in too far, it will have a negative radius. This is not allowed. This problem does not exist in the case of an outward-flowing spiral.

Variable Pitch Helix (SW 2006 Only)

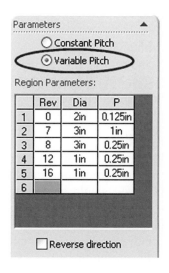

Parameters

○ Constant Pitch

◉ Variable Pitch

Region Parameters:

	Rev	Dia	P
1	0	2in	0.125in
2	7	3in	1in
3	8	3in	0.25in
4	12	1in	0.25in
5	16	1in	0.25in
6			

☐ Reverse direction

Fig. 10–35. Defining a variable pitch helix.

The Helix/Spiral PropertyManager contains an option for creating a variable pitch helical curve. A variable pitch helix can be created if either the Pitch and Revolution or Height and Pitch method is chosen to define the helix. The diameter of a variable pitch helix can also be adjusted, and does not have to adhere to the original circle diameter used to begin the helix creation process. Figure 10–35 shows the Parameters panel from the Helix/Spiral PropertyManager when defining a variable pitch helix.

In the example shown in Figure 10–35, the Pitch and Revolution method is used. The table shown in the figure contains values that define the helix. Values entered (reading the columns from left to right) are for revolutions (Rev), diameter (Dia), and pitch (P). The table reads such that at the beginning of the helix (row 1) the diameter will start at 2 inches, with a pitch of .125 inch. The diameter for row 1 cannot be changed, as it is defined by the original sketch circle.

Let's move on to the next row so that we can understand how to read the rest of the table. Row 2 states that from the beginning of the spring to the seventh revolution the diameter of the helix will increase from 2 to 3 inches, and the pitch will change from .125 to 1 inch. According to the table, the helix has a total of 16 revolutions that vary in diameter and pitch over its length. Figure 10–36 shows an example of a variable pitch helix after sweeping a small circle along its length, thereby creating a spring.

Entering or modifying values that define the variable pitch helix is done by clicking in the table's cells and typing in data. The remaining options that accompany a variable pitch helix are identical to a standard helix and have already been discussed.

Fig. 10–36. Variable pitch spring.

Springs

Once you know how to create a swept feature and define a helix, creating a spring is easy. Follow your fundamental rules regarding sweeping and you should do fine. Actually, you have already learned everything you need to know to create a simple spring. For the sake of convenience, the entire process is broken down into segments in the sections that follow, for easy reference so that you can try working through the process. If necessary, refer to previous sections of this chapter to gain a more complete understanding of each phase of the process. It is suggested that you follow along in the sections that follow.

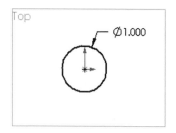

Fig. 10–37. Circle that will define the helix.

Phase 1: Start with a Circle

Start a new part and set the units to inches. Start a new sketch on the Top plane, and then sketch a circle with a diameter of 1 inch, as shown in Figure 10–37.

Phase 2: Define the Helix

Enter the Helix command, using either the Helix icon or the menus, and define the helix. Usually, you would be editing the sketch where you just created the circle prior to initiating the Helix command. If you have exited the sketch, just make it a point to select the sketch first from FeatureManager, which will make the Helix command accessible.

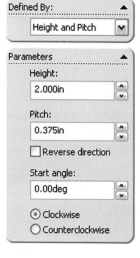

Fig. 10–38. Settings used to define the helix.

In the Helix Curve window, enter the parameters to define the helix. Let's use height and pitch settings, with a height of 2 inches and a pitch of .375 inch. Type in a starting angle of 0 degrees, and accept the default setting of Clockwise. Leave all other settings at their default values. The settings used in this example are shown in Figure 10–38.

Phase 3: Create a Sketch Plane

At this point, you should have a helix on your screen. It is necessary to establish a sketch plane next. If you have never actually tried the Normal to Curve plane creation shortcut we are about to use, you will enjoy this next sequence of steps. To create a new sketch plane for the sweep profile, select near the start point of the helical curve (shown in Figure 10–39) and then click on the Sketch icon.

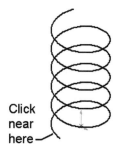

Fig. 10–39. Preparing to establish a sketch plane.

Click near here

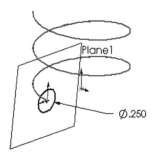

Fig. 10–40. Creating the sweep profile.

Fig. 10–41. Completed spring.

Fig. 10–42. Composite Curve panel.

Phase 4: Sketch a Sweep Profile

After clicking on the Sketch icon in phase 3, two things happened: a plane was created (though you might not see it) and sketch mode was entered. The plane will become visible at a later time. At this point, the sweep profile for the spring can be created. Figure 10–40 shows the new sketch plane and the completed profile. For our example, a circle with a diameter of .25 inch is used. Make sure to add a pierce constraint between the circle's centerpoint and the helix so that the circle will be properly positioned and fully defined.

Phase 5: Exit the Sketch and Create the Sweep

You are almost done at this point. Exit the sketch, as that is a prerequisite to creating a swept feature, and then enter the Sweep command. Pick the circle as the sweep profile, and the helix as the sweep path. Click on OK to complete the sweep. You should wind up with a spring on your screen, such as that shown in Figure 10–41.

By double clicking on the helix feature, access to the underlying dimensions is gained. This includes the helical curve's height, pitch, and number of revolutions. Even though only two of those parameters were used when defining the helix, any of the parameters can be modified. Because the helix has been absorbed by the newly created swept feature, expanding the swept feature will allow for gaining access to its constituent components. Use the small plus sign (+) to the left of the Base-Sweep feature.

Composite Curves

There are many occasions, especially when working with swept features, when the ability to join a number of sketches would prove beneficial. Being able to join existing feature edges to form a single sweep path would also prove useful. This capability exists within SolidWorks and is the domain of the Composite Curve command.

The Composite Curve command, found in the Curve menu, allows for joining feature edges, sketch geometry, or various curve types to form a single curve. This single curve can then be used for other SolidWorks functions. For example, a composite curve is commonly used as a sweep path, but could be used as a guide curve as well.

Implementing the Composite Curve command is extremely simple. The Composite Curve panel, shown in Figure 10–42, is very basic. When the items to be joined as a composite curve are selected, they appear in the list box area. To create a composite curve, follow the steps outlined in How-To 10–3.

How-To 10-3: Creating a Composite Curve

To create a composite curve, perform the following steps. The Composite Curve icon is found on the Curve toolbar.

1. Select Curve > Composite from the Insert menu, or click on the Composite Curve icon.

2. Select the items to be joined into a composite curve.

3. Click on OK to create the curve.

There is only one requirement when creating a composite curve, which is that all selected objects must form one continuous string. In other words, the objects must be connected end to end. There can be no openings between selected objects. The order in which items are selected is not a consideration and makes absolutely no difference in the outcome.

In Exercise 10–1 you will create the candle holder shown in Figure 10–6. This will require creating a helical curve and using the Composite Curve command to add sketch geometry to the helix. The resultant geometry will be used as a sweep path. This exercise will help strengthen your knowledge in the area of sweep paths and swept features.

Exercise 10-1: Creating a Candle Holder

Start a new part and save it with the name *Candle Holder*. Set the units to inches. You will begin by creating the individual sketches that make up the main base of the candle holder. Each sketch will be a simple 2D sketch. When joined using the Composite Curve command, the 2D sketch geometry and a helix will form a complex 3D path for the swept feature. Use the illustrations as guidance for completing this exercise.

1. Create the sketch shown in Figure 10–43. The sketch consists of a line and an arc. Make sure to fully define the geometry.

2. Exit the first sketch.

3. Create a new sketch on the Front plane. Sketch the geometry shown in Figure 10–44. Add tangent relationships to the arc to fully define it. The arc must be coincident to the end of the first sketch, which is also visible in Figure 10–44.

4. Exit the second sketch.

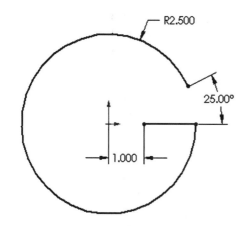

Fig. 10–43. Creating the first sketch for the candle holder.

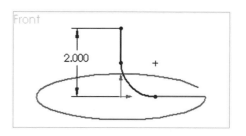

Fig. 10–44. Creating the second sketch on the Front plane.

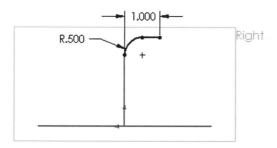

Fig. 10–45. Developing the third sketch.

5. Start a new sketch on the Right plane and create the simple sketch shown in Figure 10–45, which shows the completed sketch as seen from a Left view. Once again, tangent relationships between the arc and adjacent geometry are critical to the design intent of the model. Make sure the startpoint arc is coincident to the endpoint of the line in the second sketch.

6. Exit the third sketch.

You should be noticing a common theme by this point. The process is fairly repetitive, and the sketch geometry is simple, which makes the entire process very straightforward and easy to carry out. The procedure begins to get a little trickier at this point. A plane must be created for the next sketch, and the sketch geometry itself will be slightly more complex. A construction circle is used in this next sketch, which serves two functions. It defines the location of an arc, and it is used to define a circle for the helix. Continue with the following steps to see how this pans out.

7. Define a plane using the Parallel Plane at Point option. (Hint: if you forgot how to access the Reference Plane command, click on Insert > Reference Geometry.) Use the Top plane and point shown in Figure 10–46 to define the new plane.

8. Rename the new plane *helix plane.*

9. Start a sketch on the *helix plane.*

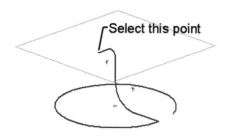

Fig. 10–46. Defining a new plane.

10. Create the sketch geometry shown in Figure 10–47. The sketch is shown from two angles to help you understand how the geometry is positioned. The existing geometry from the first three sketches is also shown, so do not confuse that with what is new. The new sketch should contain two (2) arcs and a construction circle. (Hint: to draw a construction circle, sketch a circle and then select *For construction* in the PropertyManager.) The end of the arc is horizontal to the origin point.

Fig. 10–47. Developing the fourth sketch.

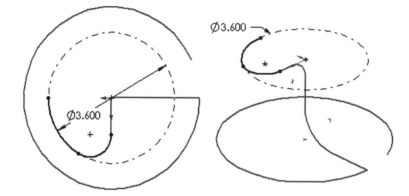

11. Exit the sketch.

12. Start a new sketch on the *helix plane* once again.

13. Select the construction circle from the fourth sketch and convert it to a sketch circle using the Convert Entities sketch tool. The construction circle will be converted into a regular circle at this point.

14. Enter the Helix command.

15. Use the following parameters to define the helix.

- Pitch = .75 inch
- Revolution = 3
- Starting angle = 270 degrees
- Clockwise is selected
- Reverse Direction and Taper Helix are both unchecked

16. Click on OK to create the helix once all parameters have been specified.

You should now have something similar to that shown in Figure 10–48. FeatureManager should list four separate sketches, along with a helix. The helix plane and other usual items will be listed also, but they do not concern us. You next need to join the four sketches and the helix to form a single curve. Continue with the following steps.

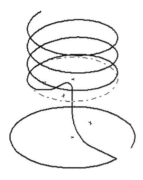

Fig. 10–48. Completed helix and sketch geometry.

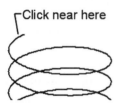

Fig. 10–49. Establishing a sketch plane for the sweep profile.

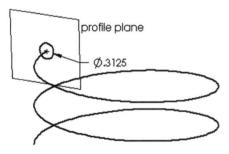

Fig. 10–50. Creating the sweep profile.

17. Enter the Composite Curve command.

18. Select the four sketches and the helix. This can be done via the work area or the flyout Feature-Manager.

19. Click on OK to complete the composite curve. The result will look similar to that shown in Figure 10–48, but will be a single curve.

20. Using the by-now-familiar shortcut to establish a sketch plane, click near the end of the composite curve shown in Figure 10–49 and then click on the Sketch icon.

21. Create a circle with a diameter of .3125 inch. The circle shown in Figure 10–50 will be used as the sweep profile. The sketch plane has been renamed *profile plane* in the illustration.

22. Add a pierce constraint between the center of the circle and the helix. It is important to select the helix near the end of the helix but not on the actual endpoint. The area where the helix is selected is also significant because theoretically there are different locations on the helix that could pierce the circle's centerpoint.

23. Exit the sketch.

24. Enter the Sweep command.

25. Select the circle as the profile, and the composite curve as the path.

26. Leave all default settings as they exist and click on OK to complete the sweep.

The final feature should look similar to that shown in Figure 10–51. Feel free to add any other finishing details or to make modifications to the model if you like. For example, you might decide to add a sketch fillet to the first sketch to remove the sharp corner from the sweep path. Just make sure you use a radius larger than the object being swept (the .3125-inch-diameter circle). You could also add a slight outside taper to the helix, or perhaps round off the upper end of the swept feature with a revolved feature. This would help keep wax from being scraped off as a candle is rotated into the holder.

Fig. 10–51. Final swept feature with a few embellishments.

Threads

The only real difference between springs and threads is the complexity of the sketch geometry for the profile. It does not take much to create a spring profile, especially when it is a round-wire spring. Threads, on the other hand, usually require a little more finesse.

The first question is: Do you really need to show actual threads? Most often the answer is no. There are alternatives to creating actual helical threads that require much less processing power and that render features that look like threads for all intents and purposes. Figure 10–52 shows the result of one such alternative.

The discriminating person can tell that the threads in this illustration are indeed simulated. There is no helical sweep being used. Real helical threads are only needed for very accurate visual renditions of a product, at least for parts such as fasteners. They are not usually needed for rapid prototype models, because threads are often machined into the prototype anyway.

Fig. 10–52. Threads: real or fake?

✓ **TIP:** Cosmetic threads can be added to a model by using the Cosmetic Threads command found in the Insert > Annotations menu.

Creating Threads

The threads shown in Figure 10–52 were created by sketching the profile of one thread near the end of the bolt. The profile was then revolved 360 degrees as a cut-revolve feature, and then patterned linearly up the bolt. In this case, the width of the thread and the distance between instances in the pattern were the same. This makes for a sharp thread, with no flats at the peak of each thread.

There are times, however, when nothing short of real threads will do the trick. Creating the threads on the neck of a soda bottle might be a good example. These are usually threads of a particular design, and cannot be called out with a simple note. Rather, they need to be modeled and shown in a design drawing. The procedure for creating this type of thread is similar to creating a spring. The mechanics are the same.

Assume you have a new soda bottle or something similar you are designing. You can use a simple cylinder for practice if you want to follow along with the book. Your dimensions and profile might be different, but the procedure is one you can use anytime you need to create threads.

You already know the beginning of the routine. You must create a circle, and then define a helix (the sweep path). After that, you can create the thread profile (sweep section), and then create the threads. If the threads are to start at a specific position from the top of the bottle, that is where you must sketch the circle to define the helix. Therefore, create an offset plane (per Chapter 4) where you want the threads to begin.

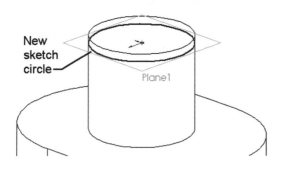

Fig. 10–53. Creating a circle at the proper height.

Figure 10–53 shows the offset sketch plane and the sketch circle. The plane was offset .075 inch below the top of the neck. The circle that will define the helix was created using the Convert Entities icon. This is by far the easiest method to use for creating the circle, and the threads will always be the right diameter, even if the diameter of the neck changes. The selected top edge of the neck was projected to the sketch plane, which is the nature of the Convert Entities command.

Now define the helix. You have been through this a couple of times if you have read through the previous sections in this chapter. Plug in the values necessary to define the helical curve that meets your requirements. Figure 10–54 shows a possible example. In the example shown, the helix was given enough revolutions so that any threads created will completely embed themselves into the rest of the model. This is certainly not a requirement, and the length of the threads you create will of course be dependent on your design requirements.

Next, it is simply a matter of defining a plane at the end of the helix and creating a profile for the threads. The profile should match the thread cross section, obviously. Figure 10–55 shows an example of what a thread cross-section profile might look like. What still needs to be accomplished in the example shown is to locate the profile in some way. A pierce relation is not required, but it works well for situations such as this.

A point that should be made has to do with the way in which the sketch profile was created as shown in Figure 10–55. Note in particular how construction geometry (centerlines) are being used. One centerline has been dimensioned 25 thousandths of an inch from the right side of the profile. The top endpoint of this centerline is where the helix will pierce the centerline, and the centerline has also been made parallel to the silhouette edge of the bottle's neck.

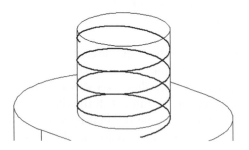

Fig. 10–54. Helix on the neck of a bottle.

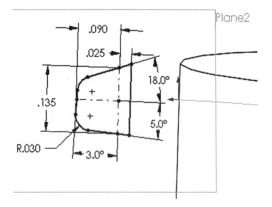

Fig. 10–55. Sketch profile for the threads.

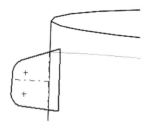

Fig. 10–56. After properly positioning the thread profile.

All of this effort to create the profile may seem unnecessary, but in fact it is very important. As the profile is swept along the helix and around the neck of the bottle, it is of the utmost importance that no gaps occur under the threads. Any miniscule twisting of the sweep profile that might occur during the sweep would cause gaps to appear. Because of the 25-thousandth-inch overlap built into the sketch, air gaps are not a concern.

What is the overlap area actually accomplishing? It is a built-in safety net. Software is not perfect, and there is a certain tolerance built into every CAD program. By establishing an overlap area, we allow for any built-in error factor within the software to be of no consequence. The thread feature will easily blend into the neck of the bottle, and the possibility of any gaps will be nonexistent. Figure 10–56 shows the sketch after adding a pierce constraint and exiting the sketch preparatory to performing the sweep command.

At this point the threads are all but done. Completing a swept feature should be familiar territory to you by this time. Complete the sweep, leaving all of the settings in their default states. The only setting that deserves some mention is the *Align with end faces* option. Turning this option on may cause the sweep to have undesirable results, usually because there are no end faces to align with. It depends on the situation, but a general rule of thumb is to leave *Align with end faces* off whenever creating threads or springs. Figure 10–57 shows what the completed thread feature would look like.

Typically, you would not want to leave the ends of the thread feature with flat faces. It would be best to round the faces off somehow, or have the ends of the threads gently taper into the neck. There is a very easy technique that can be used to accomplish this.

Fig. 10–57. Completed thread feature.

By creating a revolved feature at the end of the threads, you can gradually taper the threads instead of having them cut off abruptly. This is accomplished by sketching on the end of the thread and converting the feature edges into sketch geometry. A centerline can then be added, which the converted geometry can revolve about. Figure 10–58 shows what such a sketch might look like prior to creating the revolved feature.

A dimension was added between the centerline and the top right-hand corner of the converted geometry. A larger value for the dimension will make the thread taper off more gradually. You can experiment with this setting to obtain a look you prefer. An example of the revolved feature is shown in Figure 10–59.

It may be necessary to check the Reverse option if the feature revolves in the wrong direction. As far as the total angle of revolution is concerned, a value of 90 degrees or so is fine. The main concern is that you want to make sure the taper feature makes contact with the rest of the part.

Fig. 10–58. Preparing to taper the thread.

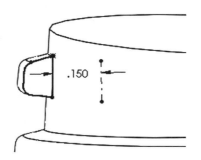

Fig. 10–59. Completed revolved feature for tapering the thread.

Springs and Threads: Performance Considerations

When creating springs, ask yourself if you really need to show them in the assembly. They usually make for a large part file. The simple spring previously shown had a file size of over 2 megabytes. This is large, especially considering there was only one feature in the part file! An alternative is to use a cylinder to represent a spring, or even just a helix itself, without creating the swept feature.

Threads will also have a tendency to make a part file quite large, and springs and threads are difficult to compute and to display because of their high polygon count. The technical reasons for this are beyond the scope of this book. The point of the matter is that you may notice rebuild times getting longer and model rotation speeds begin to slow after threads have been created, depending on the speed of your machine.

Springs and threads are fun to create, but consider whether you need true helical threads. Often you can get by with a patterned cut or even simple cosmetic threads. The decision should be made with regard to the level of performance you desire from your computer, and with regard to whether or not these threaded parts or springs will be placed in higher-level assemblies.

3D Sketcher

Fig. 10–60. 3D Sketch icon.

The 3D Sketch icon, shown in Figure 10–60, represents another means of creating a curve that exists in 3D space. It should not be confused with the Sketch icon used for 2D sketch geometry. The 3D Sketch tool is different from 2D sketching in implementation, operation, and functionality.

The first major difference between the 3D sketcher and the 2D sketcher is that with the 3D sketcher a plane or planar face does not have to be selected. Click on the 3D Sketch icon and you are ready to sketch in 3D space. When you click on the icon, the available sketch tools on the Sketch toolbar become active. The only entities that can be created while using the 3D sketcher are points, lines, centerlines, and splines. A few other tools, such as Convert Entities and Sketch Fillet, are also available.

SW 2006: Circles, arcs, and rectangles can be created in a 3D sketch.

The 3D sketcher is great for creating paths for wiring or piping. As a matter of fact, the 3D sketcher is the main tool used for routing in conjunction with the Routing module, which is an add-on program for pipe layouts and wire harnessing. Whereas the Routing module would be an extra investment, the 3D sketcher is part of the SolidWorks software.

When using the 3D sketcher, planes can still be selected to use as reference planes when sketching, but they are selected differently than with the 2D sketcher. When the actual sketch process begins, two large red arrows appear to help show what virtual plane is being sketched on. Figure 10–61 shows an example of these arrows, which can be thought of as an oversized floating sketch origin that changes position as the sketch is being created. (SolidWorks' on-line help refers to these arrows as "space handles.") The virtual plane being sketched on can be changed by pressing the Tab key while sketching a line, for example.

Another way of altering the sketch plane being referenced is to select other planes while in the 3D sketcher. This involves clicking on the plane to be referenced in FeatureManager. This can be done after beginning a sketch, and it does not matter if a sketch tool (such as the Line command) is active or not. The important thing to remember is that the plane must be selected in FeatureManager and not in the work area, due to the fact that you can be in the middle of a sketch command at the time you are switching reference planes.

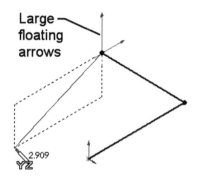

Large floating arrows

2.909

YZ

All of this probably sounds quite confusing. In all truthfulness, the 3D sketcher is a bit cumbersome to use the first few times. It takes practice, and until you get the hang of it you may find it a bit frustrating. Keep in mind that the 3D sketcher does not function at all like the 2D sketcher. Treat the 3D sketcher as a completely new command, because it might as well be for all intents and purposes. How-To 10–4 takes you through the process of creating a 3D sketch.

Fig. 10–61. What might be seen while using the 3D sketcher.

How-To 10-4: Using the 3D Sketch Command

Perform the following steps to gain a better understanding of how to use the 3D Sketch command. It is highly recommended that this command be experimented with to a large degree because there just is no other way to get a good feel for working with the 3D sketcher. You may find it necessary to run through the steps a few times, because the techniques for referencing different planes are not ruled by logic, and basically require memorization.

1. Select 3D Sketch from the Insert menu, or click on the 3D Sketch icon.

2. Select a plane to be referenced when sketching. The virtual sketch plane will align itself orthogonally to the reference plane.

3. Select the sketch tool to use, such as the Line icon.

4. Begin sketching using traditional techniques. For example, pick and drag the endpoints of a line, always paying attention to system feedback.

5. While sketching, press the Tab key to change virtual planes. For example, press the Tab key while dragging to position the endpoint of a line.

6. To change the referenced plane, select a plane from FeatureManager. This can be done while a command (e.g., Line) is active and you are between sketching line segments.

7. Add any dimensions or relations necessary. As always, it is good policy to fully define the sketch.

8. Exit the sketch when finished.

Figure 10–62 shows an example of what a 3D sketch might look like with all dimensions applied to it. Although it is difficult to appreciate from a 2D image in a book, the 3D sketch shown in Figure 10–62 does indeed weave its way back and forth through 3D space across the computer screen.

The dimensions shown are added in the same way they are added to a 2D sketch, so the topic of dimensioning does not need to be expanded on here.

Fig. 10–62.
A 3D sketch.

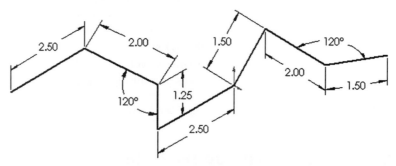

Odd as it may seem, a 3D sketch can be extruded. The sketch must form a closed profile, and it will be necessary to specify a direction for the extrusion. A plane can be used for dictating the extrusion direction, which will be perpendicular to the plane. Be forewarned, extruding a 3D sketch can result in some odd geometry.

Sketch fillets can be added in a 3D sketch in the exact manner they are added in a 2D sketch. Sketch chamfers can be added as well, and the Trim and Extend sketch tools will still function in the 3D sketcher. In that these commands have been discussed in previous chapters, they are not discussed here.

SW 2006: While in a 3D sketch, PropertyManager displays the two panels shown in Figure 10–63. The Visibility panel toggles the display of dimensions created in the 3D sketch, as well as sketch relation symbols that indicate the geometric relations within the sketch. Plane visibility can also be toggled on and off, but 3D sketch planes function somewhat differently than planes you are accustomed to working with (discussed in Chapter 4).

To understand 3D sketch planes, it might help to compare them to the reference planes you are already familiar with. Reference planes are usually defined for the sake of creating a sketch or some other function. 3D sketch planes are defined while within a 3D sketch. Reference planes can be defined from model geometry and update only after the model is rebuilt. 3D sketch planes update in real time within the 3D sketch.

How are 3D sketch planes beneficial? In a word: convenience. A 3D sketch might be developed for a number of reasons, such as for creating advanced lofted shapes and weldments. Rather than have to think out ahead of time where all of the planes need to reside, 3D sketch planes can be created on the fly as required.

Creating 3D sketch planes is implemented in a different fashion than reference planes, but the basic principle is the same. That is, geometry must be selected to base the new plane on. The geometry could be existing planes, sketch geometry, model geometry, and so on. The icon used in the creation of 3D sketch planes is shown in Figure 10–64. How-To 10–5 takes you through the process of creating a 3D sketch plane.

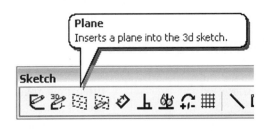

Fig. 10–63. 3D sketch PropertyManager.

Fig. 10–64. Icon used for creating 3D sketch planes.

HOW-TO 10-5: Creating 3D Sketch Planes

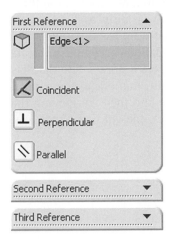

Fig. 10–65. Sketch Plane PropertyManager.

This How-To applies only to SolidWorks 2006. To create a 3D sketch plane, perform the following steps.

1. While in a 3D sketch, select Plane from the Tools > Sketch Entities menu, or click on the Plane icon (shown in Figure 10–64).

2. In the First Reference panel of the Sketch Plane Property-Manager, shown in Figure 10–65, select geometry (a point, edge, plane, and so on) that will be used to define the new plane.

3. Select a relation (see Figure 10–65) that will describe the relationship between the selected geometry and the new plane.

4. Optionally repeat steps 2 and 3 for a second and third reference, as needed.

5. Click on OK to create the 3D sketch plane.

Sketch planes created within a 3D sketch belong to the sketch. In other words, it would be necessary to edit the sketch in order to delete a plane. However, sketch planes can be used for other purposes. For instance, if it happens to be convenient to use a 3D sketch plane for a 2D sketch at some future point in time it is possible to do so.

3D sketch planes should not take the place of standard reference planes. Rather, they should be used when a plane is required while working in a 3D sketch. There are hidden benefits of 3D sketch planes. For instance, profiles created on these planes can be used for a lofted feature. Profiles and guide curves can all be created in a single 3D sketch. These advanced topics are discussed in upcoming chapters.

3D Sketch Relations

Fig. 10–66. Additional relations available in a 3D sketch.

If the Add Relations icon is depressed while creating a 3D sketch, three new relations will be made available, though you may not see them unless the correct combination of objects is selected. The three relations, shown in Figure 10–66, are only available when working in the 3D sketcher.

The three additional relations can be used along with any of the existing and more well-known relations. 3D sketch geometry can be overdefined, just like any other sketch, so use the same care when adding relations that would be employed when working in 2D. Table 10–2 outlines the three new relations available only when working in the 3D sketcher.

The Along Z relation is the most commonly used of the three relations particular to a 3D sketch, and is fairly easy to understand. If you are one of those people who want to fully understand what the Parallel ZX and Parallel YZ relations accomplish, create some angled planes, and then in a 3D sketch draw a line that has an endpoint locked onto the sketch origin. Add one of the parallel relations between the line and an angled plane, and then try moving the line or its endpoint. This should help give you an idea of how these relations can be used.

Table 10-2: 3D Sketch Relations

Relation	Description
Along Z	Two or more points can be aligned along the World Z axis. If lines are selected, the lines will align with the Z axis and be parallel to it.
Parallel ZX	A line and plane must be selected. The selected plane will represent the XY plane of a coordinate system whose ZX plane the line will be parallel to.
Parallel YZ	A line and plane must be selected. The selected plane will represent the XY plane of a coordinate system whose YZ plane the line will be parallel to.

Drawing Splines in 3D

Drawing a 3D spline using the 3D sketcher is definitely an interesting experience. It is fun, but in order to get any type of real control out of the spline construction geometry should certainly be used. A skeleton of construction lines (centerlines) can be created, and the control points of the 3D spline can then be dimensioned or related to the skeleton in some way.

Sometimes a fully controlled spline may not be what is needed. For instance, maybe a wire going from point A to point B needs to be shown and a free-form 3D spline would make a perfect sweep path. The path does not need to be fully defined, because the wire may just be routed through an assembly in some fairly loose fashion anyway.

If this is the case, use the 3D sketcher to roughly shape a route for the wire in your assembly and use it for a sweep path to show the wire. You may need to insert a new component into the assembly to do this, which is an advanced topic discussed in Chapter 19.

Sketching On Nonplanar Surfaces

A special command known as Spline On Surface can be found in the Tools > Sketch Entities menu. It is a somewhat unique command in that it is not necessary to select a sketch plane. After entering the command, pick points on a face of the model in order to create the spline curve. Such a curve is shown in Figure 10–67.

Although not completely discernible in the image, the curve shown in Figure 10–67 conforms perfectly to the face of the model. What can be done with such a curve? A number of things, actually. It can serve as a sweep path or it can be extruded, to name two possibilities. The points of the spline can be dimensioned to control the shape of the spline to a higher degree. If this is the case, dimensioning the spline's control points to planes (as opposed to edges) is the best practice, as this offers more control over the orientation of the dimensions.

Fig. 10–67. Spline on a surface.

Projected Curves

If it is necessary to create a sketch that contains something other than a spline curve, it is possible to first create a sketch on a planar face and then project the sketch onto the nonplanar face. The projected curve can then be used as a sweep path, or for a number of other functions. The command for performing this function is named, aptly enough, Projected Curve.

Another function of the Projected Curve command is its ability to take two existing 2D sketch profiles and project them together. The resultant curve is a combination of the original two curves. Projecting a sketch onto a face is a much more common action in SolidWorks than projecting a sketch onto another sketch. However, the section that follows presents an example of why it might be necessary to project a sketch onto a sketch, and how the functionality might be beneficial.

Projecting a Sketch onto a Sketch

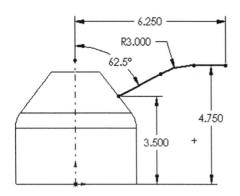

Fig. 10–68. Supply line path from the front.

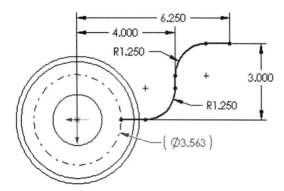

Fig. 10–69. Supply line path from the top.

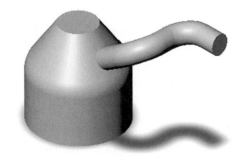

Fig. 10–70. A projected curve used in a swept feature.

This topic refers to two 2D sketches being joined mathematically. It is required that two sketches be created that are typically open profiles. The resultant curve will take on the shape of each original sketch as seen perpendicular to their original sketch planes. This is somewhat difficult to picture, even for those used to working in 3D. Following are some illustrated examples that should help to explain.

Let's use a hypothetical situation in which you have been given the task of designing a new vacuum fitting for an automobile. The size of the main feature has been determined, and the design intent for a supply line has been given. You have been given the dimensional requirements for the path the supply line must take with respect to the front and top views, but that is all you know at this point. Figure 10–68 shows the design requirements for the path as seen from the front.

What complicates matters is that there are other objects the supply line must route around. Those other objects are not shown here, as this is a hypothetical situation, but imagine that the supply line must take an odd path as seen from the top to avoid objects already in place. Figure 10–69 shows, from the top this time, the path the supply line must take.

Knowing what the path must be from both the front and top allows us to create the third curve. The resultant curve will have the same profile as the sketch shown in Figure 10–68 (as seen from the front), and will have the same profile as the sketch shown in Figure 10–69, as seen from the top. If you were to then use the resultant curve as a sweep path, the result might be a feature similar to that shown in Figure 10–70.

Although the name of the command is Projected Curve, what we have discovered from this section is that sketches are actually what are being projected. A sketch is projected onto another sketch to create a projected curve, but do not get bogged down in semantics. You will learn the steps required to carry out the Projected Curve command in the following section.

Projecting a Sketch onto a Face

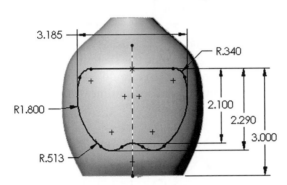

Fig. 10–71. Preparing to project a sketch onto a face.

To project a sketch onto a face, a sketch must first be created on a plane or planar face. The sketch can contain various geometry, such as lines, arcs, or whatever else is required. However, the sketch must contain one contour only. Once the sketch has been created, it can then be projected onto either a planar or nonplanar face.

A sketch to be projected onto a face will usually be a closed profile, though this is not a requirement. If an open profile is used, the sketch being projected must completely cross the geometry it is being projected onto. If an open-profile sketch falls short of completely crossing a face, it will fail to project onto that face.

Let's take a closed-profile sketch and project it onto the front face of a model for use as a label outline or cosmetic enhancement. Figure 10–71 shows the sketch that will be used. Now is also an ideal time to walk through the steps of the Projected Curve command. How-To 10–6 takes you through the process of creating a projected curve.

How-To 10-6: Creating a Projected Curve

To project a sketch onto another sketch or face, perform the following steps. It is assumed that at least one sketch has already been created. In the case of projecting a sketch onto a sketch, two sketches are required.

1. Exit out of the completed sketch if you have not already done so.

2. Select Curve > Projected from the Insert menu, or click on the Project Curve icon located on the Curves toolbar.

3. Select either the Sketch onto Sketch or Sketch onto Face option, as required.

4. Select the sketch to be projected. Select both sketches if Sketch onto Sketch was selected.

Fig. 10–72. Projecting
a sketch onto a face.

5. If the Sketch onto Face option was used, select the face or faces the sketch is to be projected on. The Projected Curve panel will appear (as shown in Figure 10–72) if the Sketch onto Face option was selected.

6. Click on the Reverse Direction option if required (pertains to the Sketch onto Face option only).

7. Click on OK to complete the command.

In Figure 10–73 we can see that the sketch has been projected onto the front face of the model. Note that the sketch conforms to the face, as you would expect it to. The curve could now be used as a sweep path to create, for example, a raised outline, which could serve as a border for the label.

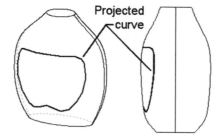

Fig. 10–73.
Completed
projected
curve.

Curves Through Points

The last couple of curve-related commands we will explore in this chapter sound very similar. These commands are named Curve Through Reference Points and Curve Through XYZ Points. The first allows for creating a curve by simply picking points on the screen, and the latter allows for typing in x-y-z coordinates. Let's start with the easier of the two commands to use, Curve Through Reference Points, found in the Insert > Curve menu or on the Curves toolbar.

Fig. 10–74.
Curve Through
Reference Points
panel.

The Curve Through Reference Points panel is shown in Figure 10–74. To use this command, simply pick points within the work area that will define a new curve. The order in which the points are selected is important, because the new curve will pass through the points in the order you pick them. The points can be anything, such as sketch points, endpoints of sketch geometry, or vertex points.

Where would you use this command? Use it anyplace you need to define a curve when you already have points established. A likely scenario is to establish construction geometry that dictates where the curve should travel. Once a "frame" has been established, the curve can be created.

Curve Through XYZ Points

If you need to create a curve based on precise *x-y-z* coordinates, Curve Through XYZ Points is the command to use. It allows for typing in *x-y-z* coordinates that dictate where the curve will pass through. The coordinates that are entered can be values you pick on the fly or coordinates generated through some other program. How-To 10–7 takes you through the process of creating a curve using the Curve Through XYZ Points command. Subsequently, the command is examined in a little more detail.

HOW-TO 10-7: Creating a Coordinate Based Curve

To create a curve through a set of *x-y-z* coordinates using the Curve Through XYZ Points command, perform the following steps.

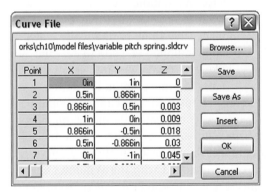

Fig. 10–75. Curve File window.

1. Select Curve Through XYZ Points from the Insert > Curve menu, or click on the Curve Through XYZ Points icon found on the Curves toolbar.

2. In the Curve File window that appears (shown with some coordinates already specified in Figure 10–75), double click on the first available button below the word *Point* (using the left mouse button) to add a set of *x-y-z* coordinates.

3. Type in a value for the *x* coordinate.

4. Press the Tab key.

5. Type in a value for the *y* coordinate.

6. Press the Tab key.

7. Type in a value for the *z* coordinate.

8. Double click on the button in the next available row.

9. Repeat steps 3 through 8 as required.

10. When you are finished entering coordinates, click on OK to create the curve.

Step 2 states that you should double click on the first available button below the word *Point*, but in reality double clicking anywhere in a blank row will suffice. Double clicking in any occupied cell will allow for altering that cell's value.

Clicking on a row number will highlight the entire row. If the Insert button is selected at this time, a new row will be inserted above the highlighted row. In this way, additional control points can be added anywhere along the curve. To delete a row, click on the button to the left of the row to highlight that row and then press the Delete key.

Curve Files

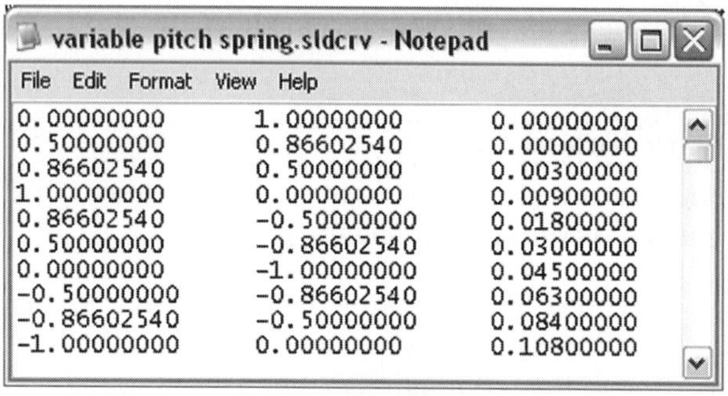

```
variable pitch spring.sldcrv - Notepad

File   Edit   Format   View   Help

0.00000000      1.00000000      0.00000000
0.50000000      0.86602540      0.00000000
0.86602540      0.50000000      0.00300000
1.00000000      0.00000000      0.00900000
0.86602540     -0.50000000      0.01800000
0.50000000     -0.86602540      0.03000000
0.00000000     -1.00000000      0.04500000
-0.50000000     -0.86602540      0.06300000
-0.86602540     -0.50000000      0.08400000
-1.00000000      0.00000000      0.10800000
```

Fig. 10–76. Curve file.

There is a special file type directly related to the Curve Through XYZ Points command and the Curve File window. This file type is known as a curve file, indicated by the file extension *.sldcrv*. Curve files can be saved directly from the Curve File window via the Save or Save As button. You may very well want to save a curve file if you have taken the time to manually enter a large number of coordinates and are likely to need the same set of coordinates at some point in the future.

A curve file is nothing more than a simple text file. The file should be a space-delimited file, and each set of *x-y-z* coordinates should be on a separate line. If more than one space separates each coordinate, that is fine, as it will not pose a problem. Figure 10–76 shows an example of a curve file opened in the basic Notepad text editor included with the Windows operating system. The file being displayed has the standard *.sldcrv* SolidWorks curve file extension, but a simple *.txt* text file extension is also compatible with SolidWorks.

Along with saving curve files, you can import them, as you would expect. Use the Browse button in the Curve File window to import either text or curve files with the *.txt* or *.sldcrv* file extensions, respectively.

How to generate the *x-y-z* coordinates is a problem that can be solved in a number of ways. One option is to use a spreadsheet to generate the coordinates using mathematical formulas. Those coordinates can then be saved as a space-delimited text file, or copied and pasted into a text file, for instance. The math and formulas behind such an endeavor can be intimidating, especially if your math or Excel skills are a little rusty. The point, however, is that it can be done.

Fit Spline Command

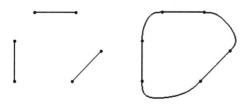

Fig. 10–77. Fitting a spline to three lines.

One method of creating a spline is to fit a spline to existing geometry. This is what the Fit Spline command accomplishes. Sketch geometry does not necessarily have to be connected, as is the case in Figure 10–77. The three lines shown in the figure have had a spline fitted to them, with the result shown on the right-hand side of the figure.

Reasons for wanting to fit a spline vary, but most reasons share a common theme. That is, a spline is typically fit to geometry where the resultant feature should not have any seams. When it is desirable to have a single contiguous face on the resultant feature, fitting a spline to the original sketch geometry is the perfect choice. Creating cams or belts would be a suitable application for the Fit Spline command. How-To 10–8 takes you through the process of fitting a spline to sketch geometry.

HOW-TO 10-8: Using the Fit Spline Command

To fit a spline to existing sketch geometry, perform the following steps.

1. Select Fit Spline from the Tools > Spline tools menu.

2. Select the sketch geometry the spline should be fitted to.

Fig. 10–78. Fit Spline PropertyManager.

3. In the Fit Spline PropertyManager, shown in Figure 10–78, select whether the resultant spline should be closed or open.

4. Choose whether to delete the original geometry.

5. If the original geometry is retained, choose whether the spline should be constrained to that geometry. If the spline is fixed, it will be locked in position.

6. Specify a tolerance for the spline. A smaller value fits the spline more tightly to the original geometry.

7. Click on OK to create the spline.

After fitting a spline to sketch geometry, the original geometry (if retained) will be converted into construction geometry. Using the Constrained option (visible in Figure 10–78) allows for altering the original sketch geometry and thereby reshaping the spline. Using the Fixed option is not recommended, as it removes the ability to edit the spline shape. Likewise, deleting the original geometry (*Delete geometry* option) also removes the ability to edit the spline.

Summary

This chapter taught you how to create some fairly impressive-looking geometry. You have seen how you can use the Sweep command to create springs and threads. A swept feature typically consists of two sketches. The sweep profile is guided along a sweep path to create the swept feature.

An important factor to keep in mind when creating a swept feature is that the geometry created by the sweep should not intersect itself. An important geometric relationship is the pierce constraint, which in essence locks a specified point onto an existing edge or curve. The point can move along the curve, but will not veer off the curve it is pierced by. The pierce constraint can be very useful for creating swept features because it can be used to help guide the sweep profile along a path or guide curve.

A guide curve will help control the shape of the profile as it is being swept. Guide curves should be created prior to the sweep profile, which then allows for piercing the profile geometry with the guide curves. Pierce relations are not required if using guide curves when sweeping, but employing the pierce constraint is good practice.

The Helix command can be used in the creation of springs and threads. You may want to gauge the relative importance of creating true threads with respect to system performance. It is often easier to create "fake" threads that look very much like the real thing but do not require as much computational power. Use revolved cuts and linear patterns to create the appearance of threads without the performance overhead.

Curves are a necessary part of SolidWorks and can be created in a variety of ways. The Composite Curve command allows for joining edges or sketch geometry to form a continuous curve. The Curve Through XYZ Points command generates a curve through a set of *x-y-z* coordinates that can either be specified by the user or imported via the Curve File window. Curve files can have either a *.txt* or *.sldcrv* file extension, and are nothing more than basic text files.

Projected curves can be used when it is necessary to establish a curve on a nonplanar face. The Projected Curve command can also be used to project one sketch onto another, thereby creating a new curve that takes on the appearance of the previous two sketches relative to each sketch's original sketch plane.

CHAPTER 11

Part Configurations

PARTS CONSIST OF FEATURES, AND ASSEMBLIES CONSIST OF PARTS. It is possible to turn individual features within parts on and off. It is also possible to turn components within an assembly on and off, which you will learn more about in Chapter 16. This ability to turn features or components off is known as suppression. When a feature is suppressed, it has been turned off. Likewise, unsuppressing a feature turns it back on.

It is possible to save a part with specific features suppressed. This is known as a configuration. You can have as many configurations in a part as you want. Each configuration can have different sets of features suppressed or unsuppressed, depending on your requirements. This chapter shows you how to create and manage these configurations within part files.

Another more powerful aspect of configurations has to do with dimensions. Size variations of a part can be stored as configurations, all within the same part file. An entire "family" of parts can peacefully coexist in the same file. Dimensions can even be driven via a table, to create a true family-of-parts table. If you have Microsoft Excel on your computer, you can create a table in which one part file can have literally hundreds of configurations, all driven by a single spreadsheet. Design tables (discussed in detail in Chapter 17) are useful, but it is mandatory that you have a good understanding of configurations first.

Areas of Configuration Use

Whatever area of manufacturing or design work you are in, configurations can be very valuable. You might need to take advantage of configurations for a variety of reasons. If you fall into any of the categories discussed in the sections that follow, you should seriously consider using configurations.

Assembly Performance

If you work with large assemblies, it is beneficial to create simplified versions of parts. These simplified versions can be configurations used in the assembly. This has a tendency to increase the performance of the assembly, in that the fine details of individual components do not have to be shown.

Sheet Metal Forming Operations

Using configurations, the entire forming process of a sheet metal part can be shown. Each bend-forming operation can be shown, step by step and in the proper sequence, until the forming process is completed.

Part Families

If your company designs components that have different size requirements, configurations can be used to represent the various sizes. Instead of maintaining and tracking 20 different part files representing the various sizes, only one part file containing 20 configurations is required. The logistics of tracking one file is undoubtedly easier than maintaining 20 files.

Application-specific Requirements

It may be necessary to have different versions of a part for use in different applications. One version may be used for finite element analysis, whereas another version may be used for kinematic studies.

Design-specific Requirements

Those working in the medical industry may have different design requirements than those working in the military, aeronautics, or civilian sectors. Configurations could be used to distinguish among these requirements.

The examples could go on and on. You may have other reasons for utilizing configurations. By the end of this chapter you will have a much better idea of what can be accomplished with configurations and why you may find them useful.

ConfigurationManager

There are three tabs at the top of FeatureManager. The second tab from the left is the PropertyManager tab, discussed in Chapter 1. The third tab will open ConfigurationManager. If you have access to SolidWorks, try to follow along on your own screen. Click on the ConfigurationManager tab (shown in Figure 11–1) to open ConfigurationManager, also shown in the figure.

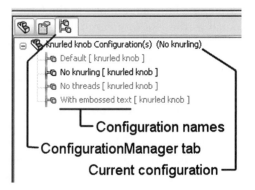

Fig. 11–1. ConfigurationManager.

Once in ConfigurationManager, the first thing you will notice is that there does not appear to be much going on. ConfigurationManager will have the name of the part at its top, followed by the word *Configurations*. Below this will be the names of all configurations. When you begin, you will see the Default configuration only, which is present in every part and assembly file. How-To 11–1 takes you through the process of adding configurations to a part or assembly file.

How-To 11-1: Adding a Configuration

To add a configuration to a part or assembly file, perform the following steps.

Fig. 11–2. A portion of the Configuration PropertyManager.

1. Click on the ConfigurationManager tab.

2. Right-click on the name of the model at the top of ConfigurationManager and select Add Configuration.

3. Type in a name for the configuration in the Configuration PropertyManager, partially shown in Figure 11–2.

4. Optionally add a description or comment if you wish, for your personal reference.

5. Click on OK to create the configuration.

Once you click on the OK button, the new configuration should appear in ConfigurationManager. When there are multiple configurations, SolidWorks automatically alphabetizes everything for you. There is no way around this, not that it really matters. To get back to the window that appeared in step 2, right-click on any configuration in ConfigurationManager and select Properties. The configuration does not have to be active in order to access its properties.

Only one configuration can be active, or current, at a time. The current configuration will have a yellow icon, whereas all other configurations will be gray. To change which configuration is currently being displayed, double click on the configuration you want to see. The upper limit of the number of configurations possible in SolidWorks is beyond any reasonable amount you would ever need. What can be altered and stored in configurations is discussed in material to follow.

Fig. 11–3. Advanced configuration options.

There are a couple of additional options in the Configuration PropertyManager, such as the *Use configuration specific color* option, which can be seen in Figure 11–3. Checking this option activates the Color button. You can then pick a color for the model and it will change the overall model color, but only in that configuration.

The *Suppress features* option refers to new features added in other configurations. That is, if *Suppress features* is checked features added to other configurations will be suppressed in the configuration in which the option was enabled (checked). This option is turned on by default, which means that features added to a model will only appear in the current configuration.

As a word of advice regarding the *Suppress features* option, you may want to make sure that all configurations have the option checked or that all have this option unchecked. The reason is that with numerous configurations it can be very difficult to keep track of which features are being added to which configurations, and which features are being suppressed. Temporarily turn *Suppress features* on or off as required, and then set it back so that all configurations are the same before closing the file and going home for the night.

Bill of Material Part Numbers

You will learn more about Bill of Material (BOM) creation in Chapter 13, but one aspect of BOM creation you should be made aware of now. In particular, the ability to control what is used for the part number in a BOM is controlled in a configuration's properties. The panel for controlling the part number is titled Bill of Materials Options (visible in Figure 11–3).

BOMs can be generated automatically in SolidWorks, saving you a great amount of time and typing. By default, it is the file name that is used in a BOM for the part number. Alternatively, the configuration name can be used. To make this happen, select Configuration Name from the drop-down list in the Bill of Materials Options panel.

Another choice is to select User Specified Name from the drop-down list. This will give you the ability to type in absolutely anything you wish. In this way, it is possible to control exactly what is shown in a BOM. It is an easy task to perform, but it is important that everybody remembers to do it. For example, if everyone on the engineering team is not on the same page, the person generating the BOM for an assembly will have a mishmash of part numbers in their BOM.

Obviously, the Bill of Materials Options setting is an important setting if the part being created will be placed in an assembly that will later have a BOM generated for that assembly. It is a setting everyone creating models in your company should be made aware of. Make a decision as to how parts should be represented in a BOM, and then make the appropriate selection in a configuration's properties. This should be done *even if configurations are not used!*

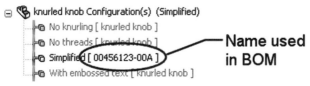

✓ **TIP:** *The name used as the part number in a BOM is displayed after the name of the configuration in ConfigurationManager, as shown in Figure 11–4.*

Fig. 11–4. Name displayed in a BOM.

Descriptions

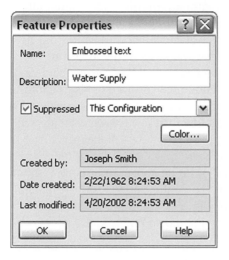

Fig. 11–5. Feature Properties window.

Descriptions can be assigned to both configuration names and features. These descriptions can then be displayed in FeatureManager, in ConfigurationManager, or in both. By default, the description of a feature is the same as its corresponding feature name. Likewise, configuration descriptions are the same as the corresponding configuration name.

To specify a description for a configuration or feature, right-click on the configuration or feature and select Properties. An example of the Feature Properties window is shown in Figure 11–5. The area where a description can be added to a configuration is aptly named *Description* (shown in Figure 11–2).

To display descriptions in FeatureManager or ConfigurationManager, right-click on the name of the model at the top of either manager and select Tree Display. You will see a set of options, the content of which depends on whether you are working with a part or assembly. In either case, the options basically amount to whether object names or descriptions are being displayed. The inset in Figure 11–6 shows what the Tree Display menu might look like. Also shown is a sample of what ConfigurationManager might look like after turning on configuration descriptions.

Because names of features or configurations can be basically whatever you want, and because descriptions can be added, FeatureManager and ConfigurationManager can wind up being quite informative. They can theoretically contain more information than most people require. If you do not feel the need to use descriptions, by all means do not bother with them. After all, configuration or feature names can be fairly "descriptive" in their own right.

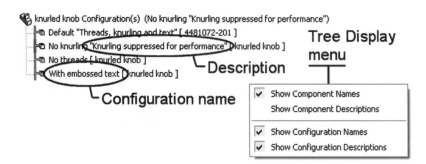

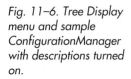

Fig. 11–6. Tree Display menu and sample ConfigurationManager with descriptions turned on.

Copying Configurations

Essentially, when a new configuration is added you are really just creating a copy of the current configuration. This fact can be used to your advantage. If an additional configuration is required, and the new configuration happens to be very similar to an existing configuration, it would make sense to make the existing configuration current before creating the new configuration. A little forethought can save you steps in the long run.

When a new configuration is created, all settings of the current configuration are also transferred to the new configuration. This even holds true for the *Suppress features* option. It also holds true for any configuration-specific color changes that were made, and of course for any features that were suppressed.

Configurations can be copied using the Windows clipboard. This has the advantage of making it possible to create many copies very quickly. To use this technique, select the configuration to be copied and then press the hot key combination of Ctrl-C to copy the configuration to the clipboard. You can then press Ctrl-V to paste a copy into ConfigurationManager. Pressing Ctrl-V again will paste a second copy in, and so on.

✓ **TIP:** *Configurations can be renamed, like any other feature in Solid-Works, by performing a slow double click on the configuration.*

Suppression States

What does it mean to suppress something in SolidWorks? It means to temporarily turn the display of that object off. To be more technically precise, a suppressed object is not so much hidden as it is completely ignored by the rebuild process. SolidWorks acts as though any suppressed items do not exist. You can unsuppress items at any time. Features can be suppressed in part files, and components can be suppressed in assembly files. You can read more about suppressing assembly components in Chapter 16.

✗ **WARNING:** Suppressing a feature is not the same as deleting a feature. When a feature is deleted, it is removed from all configurations and is no longer present in the model.

The whole idea of being able to suppress a feature raises an interesting issue. What happens to an object that is dependent on an object that has been suppressed? A simple example would be suppressing a block that contains a hole. With the block suppressed, what becomes of the hole?

Because of the dependencies involved with a feature-based modeler such as SolidWorks, relationships are established between features. These relationships can be described as parent/child relationships. When a feature is dependent on another feature, it is considered a child of that feature. When a feature has other features dependent on it to exist, it is known as a parent feature. If the block-and-hole example were used, the block would be a parent of the hole, and the hole would be a child of the block. With this simple block-and-hole example, it is easy to understand parent/child relationships. However, not all relationships are quite as straightforward.

Many seemingly innocent actions can create parent/child relationships. For example, adding a constraint to an object causes a relationship to exist to that object. If a circle were sketched on one face of a model, and the center of the circle made coincident with an edge on the other side of the part, the feature the edge belongs to would be considered a parent of the feature created by the circle.

Parent/child relationships are something a SolidWorks user grows accustomed to over time. At first, such relationships seem to be a cumbersome annoyance. In time, you will find relationships to be a very powerful way of imparting your design intent to a model.

With regard to suppressing features, parent/child relationships mean that suppressing a parent will also suppress any dependent, or child, features. There is no way around this, except for somehow removing the relationship back to the parent. Removing dependencies to other features is sometimes necessary for various reasons, and the removal process often involves the Display/Delete Relations command (Chapter 3) or some other editing process.

It is often necessary that the parent/child relationship information be available in some straightforward manner. The steps you perform to find out what the parent/child relationships are for a particular feature are simple and involve using the right mouse button. How-To 11–2 takes you through the process of accessing parent/child relationship information.

HOW-TO 11-2: Accessing Parent/Child Data

To access parent/child data, perform the following steps.

1. Right-click on the feature in question (usually via FeatureManager).

2. Select Parent/Child.

3. Click on Close when done viewing the data.

The Parent/Child window is shown in Figure 11–7. The left-hand portion of the Parent/Child window shows the parents of the feature in question, and the right-hand portion shows the children. The window can be resized if necessary. Right-clicking on any of the items listed in the window gains access to the same menu you would see if right-clicking on the same item in FeatureManager.

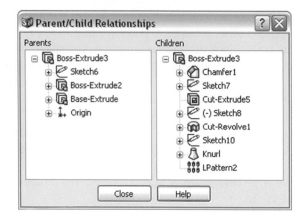

Fig. 11–7. Parent/Child window.

Now that you understand parent/child relationships, you know more of what can be expected when suppressing a feature. There are a number of simple methods that can be employed for changing the suppression state of features. One method is to use the Edit menu, which is the method described in How-To 11–3.

How-To 11-3: Changing Feature Suppression States

To suppress or unsuppress features, perform the following steps. Do not forget to hold down the Ctrl key if it is necessary to select more than one object.

1. Select the feature or features to be suppressed or unsuppressed.

2. From the Edit menu, select Suppress, Unsuppress, or Unsuppress With Dependents, as required.

3. Select This Configuration, All Configurations, or Specified Configurations, as required.

When features are suppressed, their icons appear grayed out in FeatureManager. As mentioned earlier in this chapter, if a feature is suppressed all of its children will be suppressed. Child features cannot exist without the parent feature, though it is possible to suppress a child without its parent.

As can be seen from step 2, it is possible to unsuppress a parent and its dependents simultaneously. The choice is up to you. Use the Unsuppress option to unsuppress the selected feature, or use Unsuppress With Dependents to bring back the selected feature, along with any dependent features.

Using the Edit menu to change the suppression state of features allows for being particular about exactly which configurations will be changed. Options include Current Configuration, All Configurations, and Specified Configurations. If there is only one configuration, the option This Configuration will be the only one available. Selecting Specified Configurations from the menu brings up the window shown in Figure 11–8.

Fig. 11–8.
Choosing
configurations
to modify.

From the window shown in Figure 11–8, it is possible to pick and choose exactly which configurations the selected feature or features are to be suppressed or unsuppressed in. This is extremely convenient, and can be something of a necessity when there are many configurations present in the model file. With numerous configurations, manageability becomes an issue.

Other Methods of Changing Suppression

Probably the easiest method for changing the suppression state of a feature is to use the right mouse button menu. Right-clicking on a feature will show the Suppress or the Unsuppress option in the menu, depending on the feature's current state. There is no Unsuppress With Dependents option available with this method.

Another choice is to use the icons usually found on the Features toolbar, shown in Figure 11–9. The icons may not be present on your SolidWorks workstation. If this is the case, they can be added with a little customization. Customizing toolbars is discussed in Chapter 25.

Finally, by right-clicking on a feature and selecting Properties from the menu the Feature Properties window appears (previously discussed). There will be a checkbox option for changing the suppression state of the feature in question, as shown in Figure 11–10. There will also be a drop-down list containing the options This Configuration, All Configurations, and Specified Configurations. These options have the same functions as their counterparts in the Edit menu (discussed in How-To 11–3).

The Feature Properties window has more to it than just the ability to suppress or unsuppress a feature. It also contains the information of who created the feature and when, along with when it was last modified. More accurately, the name in the *Created by* box is the log-in name (user name) of whoever logged onto the computer before the SolidWorks session was begun.

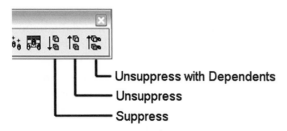

Fig. 11–9. Changing suppression via the Features toolbar.

Fig. 11–10. Suppressed option in the Feature Properties window.

Feel free to use the Feature Properties window to suppress or unsuppress multiple features at a time. If more than one feature is selected, accessing the properties of any one of those features will bring up the Feature Properties window. Only common properties will be displayed. None of the name, creation, or modified times will be displayed because those bits of data will likely be different for each feature.

Changing the color of a feature can also be accomplished through a feature's properties. However, the Edit Color command is more versatile, as it can be used to change the color of many objects, one after the other. See Chapter 8 for discussion of changing the color of parts and features.

Dimensional Configurations

Suppressing and unsuppressing features is just one aspect of configurations. Another capability has to do with being able to control dimension values within a configuration. There must be a minimum of two configurations before this functionality is present.

 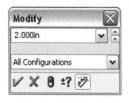

Fig. 11–11. The Modify window changes when there are multiple configurations.

If more than one configuration is present in the part file, and a dimension value is double clicked on with the left mouse button, the Modify window will appear. This is a basic editing technique, discussed in Chapter 3. However, the Modify window will look different when more than one configuration exists. With more than one configuration present, the Modify window will appear as it does on the right in Figure 11–11.

The implementation of this functionality is easy enough, and has been explained previously. Simply select This Configuration, All Configurations, or Specify Configurations for the dimensional change. If This Configuration is selected, the dimensional change will only affect the current configuration. When Specify Configurations is selected, the window shown in Figure 11–8 will appear.

This is an ideal way of creating multiple versions (or sizes) of a part. If there are many dimensions to be changed, or perhaps numerous configurations to be made, you might opt to create a family-of-parts table. This functionality requires what is known as a design table, discussed in Chapter 17.

Other Configurable Objects

Certain commands or functions can be made to affect particular configurations. So far, we have discovered this to be true for suppression states, colors, and dimension values. There are still other items that are configurable. These other items include end conditions, geometric relations, materials, and equations (discussed in Chapter 18), to name a few.

Many items that are configurable use the same process used to specify which configurations are affected when changing feature suppression states. The object being configured may be different, but the process is generally the same. The process usually involves editing some item in some way (such as an end condition in a feature's definition) and then specifying one of the three previously examined options: This Configuration, All Configurations, or Specify Configurations.

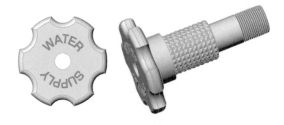

Fig. 11–12. Knurled handle.

Because it would be redundant to discuss the method for suppressing or unsuppressing every single configurable object in Solid-Works, let's pick one item and use that in the example. For instance, imagine that the knurled handle shown in Figure 11–12 will be used in an assembly. It is desirable to create a simplified version of this part, for reasons of improving computer performance. We will assume that a new configuration named *Simplified* has already been established. This part has knurling that has actually been modeled, along with embossed text. These features do not need to be shown at the top-level assembly, so they will be suppressed in the Simplified configuration.

Next, a texture will be applied to the face that previously contained the knurl feature. In that you learned how to apply textures in Chapter 8, we won't have to go into details as to how this is done. However, we only want the texture to be applied in the Simplified configuration. At the bottom of the Texture PropertyManager there is a panel with the by-now-familiar options pertaining to which configurations the texture should be applied in. This can be seen in Figure 11–13, along with the handle.

By specifying *This configuration*, we successfully place the texture on the model in the Simplified version only. All other configurations continue to have the knurl feature unsuppressed and will not display the texture.

As mentioned previously, there are many items that are configurable. Wherever you find the Configurations panel in a command, such as that

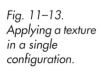

Fig. 11–13.
Applying a texture
in a single
configuration.

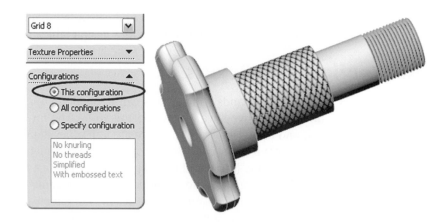

shown in Figure 11–13, that command is a configurable item. There are so many items that are configurable in SolidWorks it would almost be easier to specify what is *not* configurable rather than to try to list all items that are configurable. For example, section views (discussed in Chapter 3) are not configurable items. To rephrase, it is not possible to create a section view that is displayed in one configuration but not another.

If you did want to show a section view, or a "cutaway" view, how could this be accomplished? The answer is really quite simple. Add a configuration, and then create a feature that cuts away a portion of the model. The cut feature should be unsuppressed in the new configuration but suppressed in all others.

Just because many items are configurable does not necessarily mean that it is prudent to take advantage of this functionality. For instance, geometric relations between sketch entities are configurable. However, keeping track of configurations can often be complicated enough without worrying about what geometric relations are suppressed in this configuration, or unsuppressed in that configuration. Be forewarned: configurations are a powerful tool, but it is necessary to pay close attention to what is happening in the model document if you try to juggle too many of them at once.

Nested Configurations

Nested configurations do not require a lot of explanation. To put it quite simply, configurations can exist within other configurations. These are known as nested configurations, and the depth to which configurations are nested is up to the user.

To create a nested configuration, right-click on the configuration you want to add a nested configuration to and select Add Derived Configuration. Once that is done, the Add Configuration PropertyManager will appear, as it does when adding any configuration.

You know what to do at this point. Type in a name for the new configuration and click on OK. The short process is almost exactly what you learned in How-To 11–1. As far as the result is concerned, nested configurations establish parent/child relationships. The derived configuration is a child of the parent configuration. What exactly does this mean when you start changing dimensions? Let's use Figure 11–14 as an example, which contains configuration names for a spring.

Fig. 11–14. Configurations nested two deep.

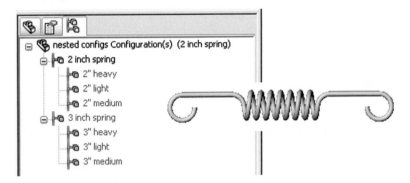

We can see from Figure 11–14 that there are light, medium, and heavy-duty springs for each spring size. The configurations named *2 inch spring* and *3 inch spring* are the parent configurations. The three additional configurations that come after each parent would be considered the dependent or derived configurations. Every configuration, regardless of its position in the "tree," must have a unique name. This is why the light, medium, and heavy-duty versions of each spring also contain a prefix indicating the spring length.

When a parent configuration dimension is altered, it affects the children as well. This means that if a dimension were changed with *2 inch spring* active each of the three dependent versions would also change. Essentially, dimensional changes roll downhill.

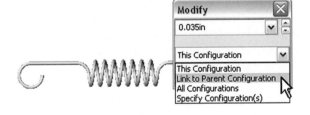

Fig. 11–15. Nested configuration dimension options.

When making a change to a dimension when a nested configuration is active, the Modify window appears slightly different. In particular, there is an additional selection in the drop-down list (shown in Figure 11–15).

If a dimension of a nested configuration has been changed and This Configuration is selected from the drop-down menu, the link to the parent is broken. This means that the parent can have the same dimension altered and will no longer affect the configuration derived from it.

If a nested configuration has had a dimension changed previously, Link to Parent Configuration will be one of the options in the drop-down menu if the dimension is changed again. As a matter of fact, the dimension value itself does not have to be modified by the user to link it back to the parent. By double clicking on a dimension and selecting Link to Parent Configuration, SolidWorks reassociates the value to the parent.

Using nested configurations for the sole sake of categorizing is a poor idea, because the only legitimate configurations would probably be the lowest-level configurations. All others would be categorical configurations that SolidWorks would see as separate configurations, which serves to complicate the model unnecessarily.

Summary

There are many reasons for adding configurations to a part. Most reasons boil down to the same situation, and that is one in which having multiple versions of a part within the same file would be a convenience. Parts can have different features suppressed in different configurations, and different dimensional values between configurations.

There can exist as many configurations as necessary within the same file, but only one configuration can be active at any one time. Double click on a configuration name to show that configuration (make it active). Suppress or unsuppress features while in an active configuration and the configuration will remember the suppression state of the features.

Features can be suppressed quite easily via the Edit menu or right mouse button menu. Right-clicking on a feature in FeatureManager or the work area will gain access to either the Suppress or Unsuppress command. Unsuppressing with dependents will unsuppress a parent, along with any child features.

Parent/child relationships exist between features when dependencies are created. Not all features are dependent on each other. When a feature is suppressed, its children will be suppressed as well. There is no way around this, as child features cannot exist without the parent feature.

Items other than features and dimensions are configurable. End conditions and geometric relations are two other examples of items that can be altered on a configuration-by-configuration basis. You must be careful to keep track of what is being altered in each configuration if taking configurations to this level.

As soon as there is more than one configuration in a part, care should be taken whenever making a dimensional change. Changes will take effect according to the options This Configuration, All Configurations, or Specified Configurations. The default is All Configurations, but SolidWorks remembers what was specified for each dimension. In short, use caution when changing dimension values.

CHAPTER 12

Design Drawings

NO MATTER HOW ADVANCED THE CAD INDUSTRY BECOMES, there is often still a need for paper drawings. Two-dimensional CAD systems have been around for some time, and many improvements have been made. Creating design drawings has gotten easier, with much of the process now automated. From a SolidWorks standpoint, adding a typical top, front, and right side view to a standard engineering drawing is literally a drag-and-drop operation. You will see how to do this in material to follow.

There are many view types that can be created in SolidWorks. These views include auxiliary views, section and detail views, broken views, and many others. Annotations of all sorts can be added to a design drawing to help explain your design intent. Notes, surface finish, and weld symbols are only a few examples.

Much of the tedium of adding dimensions has been eliminated from creating drawings. Most dimensions are added automatically for the user. It is basically up to the drafter to decide where to place the dimensions, and on which views. This is largely a matter of dragging the dimensions around to position them.

You will learn how to perform all of these tasks in this chapter. All of the SolidWorks view types, along with how to add dimensions and annotations, are explored. Many of the parts used in the examples throughout this chapter are parts created in earlier chapters, so they should look familiar to you.

File Associativity

Design drawings are a distinct file type within SolidWorks. Part files, which you have been learning about to this point, have an *.sldprt* extension. Drawing files have a different extension type, which is *.slddrw*. Assemblies use yet another extension, which is *.sldasm*. This is only important from the standpoint of knowing what file types to browse for when opening an existing file.

Bidirectional associativity is a complicated-sounding term that describes a simple concept. What it means is that all three basic SolidWorks file types (parts, drawings, and assemblies) are associative with one another. If a part is modified, the design drawing for that part will be updated the next time it is opened. Likewise, a part file can be altered from within a drawing file by modifying the dimensions in the design drawing. Design drawings of assemblies exhibit the same associativity.

Parts can be directly edited within an assembly. This can even be taken one step further. That is, parts can be built up from scratch completely within the context of an assembly, if you need to go to that extent. If a part has been placed into an assembly, the assembly will automatically update if the part file is altered in any way.

As you can see, associativity goes both ways between each pair of file types; that is, between parts and drawings, between drawings and assemblies, and between assemblies and parts. Hence, this functionality is described as *bidirectional associativity*.

✎ **NOTE:** *Existing CAD operators may take the term drawing to mean any CAD document. Regarding SolidWorks, the term drawing refers strictly to the 2D design drawing of a part or assembly, not part or assembly documents.*

New Drawings

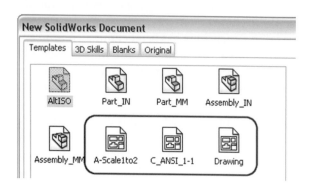

Fig. 12–1. Starting a new drawing.

You can begin a new drawing the same way you begin a new part; that is, by selecting New from the File menu. You then specify the drawing template you want to use for the new drawing (see How-To 2–1). Templates, which can exist for any of the three SolidWorks document types, were discussed in Chapter 3. Figure 12–1 shows three drawing templates available in the New SolidWorks Document window.

There may not be three drawing templates on your computer. As a matter of fact, there may only be one. If that is the case, you may wish to create additional templates based on sheet size, working units (i.e., English or metric), or perhaps a particular dimensioning standard, such as ANSI or ISO. If you wish to create additional templates, review the section "Using Templates" in Chapter 3.

Do not confuse the term *template* with the term *drawing format*. With regard to this book, *drawing format* refers to the title block and border data, whereas a template is the file used when opening a new SolidWorks document. How-To 12–1 takes you through the process of creating a new drawing. Subsequent material shows you how to create drawing formats, and addresses a few other relevant issues.

How-To 12-1: Starting a New Drawing

This How-To assumes the model you wish to generate a drawing from is already opened and visible on screen. To create a new drawing, including basic views of the model, perform the following steps.

1. With the model visible on screen, select Make Drawing From Part from the File menu. There is also an icon for this command on the Standard toolbar, or use the hotkey combination of Ctrl-D.

2. Select the template you want to use for the new drawing and click on OK.

3. If a generic template containing no format was selected in the previous step, you will be prompted to select a paper sheet size from a list. It will also be possible to select a format, though SolidWorks will try to select one for you. Click on OK when finished making your selections.

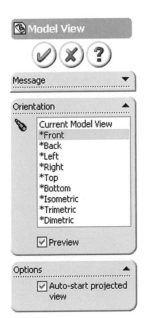

4. You should be presented with the Model View PropertyManager, shown in Figure 12–2. For the sake of this How-To, make sure the Preview and Auto-start Projected View options (see Figure 12–2) are checked.

5. Click on the view in the Orientation panel listing that you wish to place in the drawing, or leave it set to Front, which is the default selection.

6. Click on the drawing sheet where you would like to place the view. Note that views can be moved later, so it is not necessary to be precise.

7. Click on the drawing sheet to add views as required. Additional views are automatically projected from the first view placed in the drawing. Pay attention to the dynamic previews, as they indicate what views will be added.

8. Click on OK when finished.

Fig. 12–2. Model View PropertyManager.

If you have followed along with these steps on your own computer, you have just discovered how easy it is to create a drawing in SolidWorks. The Auto-start Projected View option is what enabled you to place the additional views in the drawing in a very smooth and automated fashion. You may have noticed that even isometric-style views can be added in this manner.

SW 2006: The Model View PropertyManager has been enhanced to make adding multiple views easier. If the Single View option is used (see Figure 12–3), the desired view can be selected from a graphical interface as opposed to a list. Views can then be added as described in the previous How-To, assuming the Auto-start Projected View option is checked.

When the Multiple Views option is chosen, the graphical interface allows for selecting multiple standard views. The views will be automatically positioned and spaced appropriately on the drawing sheet. A preview will show where the views will be placed, though the views can be moved at a later time (discussed shortly) if necessary.

Fig. 12–3. Model View PropertyManager in SolidWorks 2006.

Drawing Interface

After starting a new drawing, you will notice that the interface is a bit different from that of a part. Assemblies (discussed in the next chapter) and parts have essentially the same interface, aside from what is contained in the pull-down menus. Drawings, being in the 2D realm, are made to appear as if you were working on a sheet of paper. Figure 12–4 shows an example of a drawing in progress.

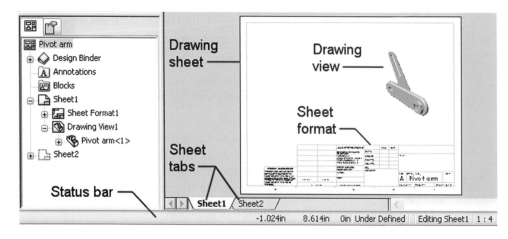

Fig. 12–4. A drawing in its beginning stages.

The drawing FeatureManager structure (visible in Figure 12–4) lists items pertinent to a drawing, rather than a part or assembly. Drawing sheets, which can be thought of as pieces of paper, are at the top level of the "tree." Sheets, which you will learn how to add in this chapter, will appear as tabs at the bottom of the screen.

Drawing views will appear in FeatureManager below the sheets on which they reside. Each view then lists whatever model is pictured in that view. The hierarchy you have grown accustomed to in a parts FeatureManager is exactly what you will see listed below the view name in the drawing FeatureManager. If you were to expand *Pivot arm 1* (shown in Figure 12–4) you would find that it contains every feature present in the *Pivot arm* part file, right down to the sketch geometry.

The model hierarchy plays an important roll when it comes time to add dimensions to a drawing. Features in certain views can be selected in order to have their associated dimensions imported into the drawing. Assembly components can be selected so that they may be displayed in a different line style. It should suffice to say that although the reasons vary FeatureManager still plays an important roll in drawings, as it did when creating the solid model.

The sheet tabs allow for easy switching between sheets in the drawing. The status bar—available for every document type—relays tidbits of information, some of which are more important than others. For instance, the sheet scale can readily be determined from the status bar. Sheet formats—which you will learn about in more detail shortly—contain the title block geometry, drawing border, zone information, and other standard drawing information.

There are different toolbars that will be needed when working in a drawing as opposed to a part (or assembly) document. These include the Annotation, Drawing, and Layer toolbars, to name a few. SolidWorks will remember the toolbar layouts for each of the three SolidWorks file types. In this way, you can "customize" the interface to contain the toolbars you choose for each of the three document types. How-To 12–2 takes you through the process of setting and saving toolbar layouts.

How-To 12-2: Setting and Saving Toolbar Layouts

Actual toolbar locations are saved based on document *type*, rather than on an individual document basis. That is, certain toolbars will be turned on or off when in a part, and different toolbars can be turned on or off when working in a drawing. This is also true for assembly documents.

Knowing this, there is a simple procedure you can use to set up your toolbars so that they always appear in the same place, depending on what type of SolidWorks file is open. This procedure involves opening one of each type of SolidWorks file. To establish toolbar layouts for all three SolidWorks file types, perform the following steps.

➥ **NOTE:** *Although we have not explored assembly files yet, you can still perform this procedure. A typical assembly toolbar layout would be the same as for a part, with the addition of the Assembly toolbar.*

1. With no SolidWorks files open, turn off any extraneous toolbars you do not care to see. You could turn them all off, or leave just those common to all three document types, such as the Standard and View toolbars.

2. Start a new SolidWorks part, assembly, and drawing using templates of your choice.

3. Turn toolbars on or off, positioning them as desired. (Hint: right-click on a docked toolbar.) Do this for each of the three documents you have open. Repositioning toolbars common to all three document types will reposition that toolbar in every document type.

4. Press Ctrl-Tab to cycle between the open part, drawing, and assembly files. Fine-tune toolbar placement as necessary while switching between document types.

5. Close all open documents. You need not save them.

6. Close SolidWorks. This will ensure that toolbar positions and layout are remembered and written to the Windows registry.

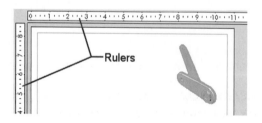

Fig. 12–5. Showing rulers in a drawing.

From this point forward, the toolbars associated with the various SolidWorks document types will appear when that document type is opened or activated.

✓ **TIP:** *Rulers, shown in Figure 12–5, can be turned on or off via the View menu. This holds true for the status bar as well. Grid settings (discussed in Chapter 3) affect the appearance of the rulers, and the grid itself can be displayed while in a drawing.*

Drawing Sheet Formats

First, just as a reminder, sheet formats and templates are two different objects. Formats contain title block and border geometry; templates contain document settings. Drawing templates may or may not contain sheet formats. It all depends on whether or not a format was present when saving the template (discussed in Chapter 3).

There are two types of sheet formats: standard and custom. Standard formats are those SolidWorks loads onto your computer hard drive when you install the software. Custom formats are any formats you create or import into SolidWorks. If a new drawing is begun using a template that does not contain a sheet format, you will be given the opportunity to select one. This is done through the Sheet Format/Size window (shown in Figure 12–6), which will appear automatically.

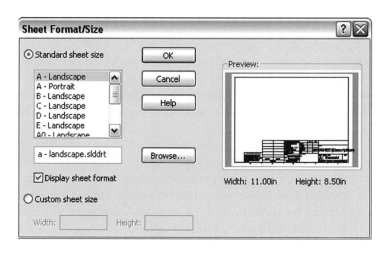

Fig. 12–6. Sheet Format/Size window.

Picking a standard sheet size from the listing shown in Figure 12–6 will also cause a standard format (which fits the sheet size chosen) to be selected automatically. Use the Browse button to specify a custom format, or uncheck *Display sheet format* if you wish to use a blank sheet of paper with no format. Select *Custom sheet size* if it is necessary to enter a nonstandard size for the drawing sheet. A nonstandard size, as an example, might be a 36-inch by 100-inch drawing that could be printed out on a roll plotter.

✓ **TIP:** *The format displayed on a drawing can be changed at any time by right-clicking over the drawing sheet and selecting Properties.*

The standard formats included in the software are actually a very good place to start if you want to customize and create your own formats. They provide a basic starting point for the most common sizes of sheet formats, with title block and additional information already added.

To edit a sheet format, you must first tell SolidWorks you wish to do so. To create a new sheet format from scratch, you must in essence "edit" a blank sheet format. Format geometry and drawing geometry are kept separate. This is a good thing, as it helps reduce the risk of human error. How-To 12–3 takes you through the process of editing a drawing sheet format.

HOW-TO 12-3: Editing a Drawing Sheet Format

To edit a sheet format, perform the following steps. Use these steps as well to create a new sheet format from scratch.

1. Select Sheet Format from the Edit menu, or right-click anywhere in a blank area of the drawing sheet and select Edit Sheet Format.

2. Edit the format as needed. If creating a new format, use various sketch tools learned in Chapter 2 to create the geometry that will constitute the format.

3. When finished, select Sheet from the Edit menu, or right-click anywhere in a blank area of the drawing sheet and select Edit Sheet.

When editing a format, any drawing views present on the sheet will temporarily disappear. Do not be alarmed, as this is meant to happen. Drawing views will reappear when editing the sheet. Format geometry will appear gray when editing the sheet, and blue or black when editing the sheet format. You will always be in either one mode or the other; editing either the sheet or sheet format.

As mentioned in step 2, use any of the tools on the Sketch toolbar to edit the format. Use the Add Relations command to constrain geometry in the format. Add dimensions if you want to control the exact size and shape of title blocks or anything else in the format. If you do add dimensions, you can hide them by placing the dimensions on a layer and turning off the layers visibility. Layers are discussed later in this chapter.

✓ **TIP:** *The Linear Sketch Step and Repeat command, discussed in Chapter 7, works particularly well for creating rows or columns of lines.*

To control format geometry, some individuals prefer to add dimensions and then delete the dimensions and use the fix constraint to keep everything from accidentally moving around. This is a good option if the format geometry is never going to change size or shape, but not a very good choice otherwise. Without dimensions, format geometry cannot be adjusted accurately.

Almost certainly it will be necessary to add text to the format as well. This can be accomplished by selecting Note from the Insert > Annotations menu. Adding notes is covered in greater detail later in this chapter, but briefly the process works as follows. Access the Note command, click on the sheet to position the note, type in your text, and then click on OK. Notes and other annotations can be dragged to reposition them.

Adding a Company Logo

When creating a customized format, it is often desirable to add a company logo to the format. The logo must be in electronic format (i.e., JPG or BMP files), and you must be able to open it for viewing in a photo editor or graphics art program. Once you are viewing the logo on your computer, the rest is easy. If you do not already have a logo and wish to create one from scratch in SolidWorks, that is an option as well.

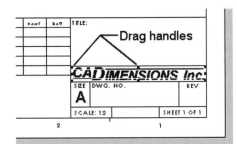

Fig. 12–7. Drag handles used for resizing.

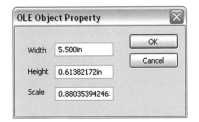

Fig. 12–8. Specifying precise values for an inserted image.

All graphics art programs are different, and therefore it is not practical to list the steps for adding a company logo with your specific arrangement. However, all you need to do is copy the logo to the Windows clipboard (typically by pressing Ctrl-C). Then, once you are editing your drawing format in Solid-Works click somewhere on the screen and paste the image into it by pressing Ctrl-V. Move the object by selecting and then dragging the object. Scale the object by dragging the handles that appear when the object is selected. There are eight drag handles available for scaling the object, all of which can be seen in Figure 12–7.

Basically, just about any type of object can be inserted into a SolidWorks document by selecting Object from the Insert menu. This includes files of all types, photos, charts, and so on. However, with simple images copying and pasting is the easiest method. Right-clicking on an inserted image and selecting Properties will open the window shown in Figure 12–8. Although all three values (height, width, and scale) are associated, it is possible to enter a precise value for one of the values. This allows the object, for example, to be made to fit nicely within a title block.

After making modifications to a sheet format and getting everything looking the way it should be, you will want to make sure to save the format. This will be required if it is your intention to use the format on other drawings, which will almost certainly be the case. There is a special process established for saving a format, and How-To 12–4 takes you through this process.

How-To 12-4: Saving a Sheet Format

To save a sheet format, perform the following steps.

1. Select Save Sheet Format from the File menu. The Save Sheet Format window will appear.

2. Navigate to the folder in which the format is to be saved. By default, this will be the *data* folder located in the SolidWorks installation folder.

3. Type in a name for the format.

4. Click on Save.

Although it is possible, you probably will not want to save your customized format on top of an existing SolidWorks standard format. This will overwrite the standard format with your own, and you will no longer have access to the standard format. Give your custom formats a unique name that is meaningful to you.

Drawing Templates

Templates were discussed in Chapter 3, and therefore will not be extensively dealt with here. Because templates have significant bearing on drawings, however, this topic is being discussed here in a somewhat different light.

Keep in mind that templates can exist for parts, assemblies, or design drawings. In the case of parts and assemblies, it would not be uncommon to have two or three templates for each. For example, you might have one part template for working in inches and one for working in millimeters. Some companies create part templates based on materials (such as aluminum or brass), in which case there may be dozens.

In the case of drawing templates, you might decide to create templates for various sheet sizes, various dimensioning standards (such as ISO or ANSI), metric or English dimensioning units, or any number of other reasons. It may also be necessary to create templates that contain various sheet formats.

Creating a template for a drawing is for the most part no different from creating a template for a part or assembly. A new SolidWorks drawing is begun, and then the Document Properties are modified to reflect the characteristics the template is to have. Once that is finished, save the drawing as a template file by clicking on Save As in the File menu. The complete steps are listed in Chapter 3 if you need to review this material further.

If a drawing was begun without a format (a blank sheet of paper, in other words), and it is now decided that a format should be used, the next section will show you how this can be done. The "Sheet Properties" section also examines how to change to a different sheet format.

It is very common that a company will already have drawing formats set up, and it would be to their benefit to import these formats into SolidWorks, where they can be reused instead of recreated from scratch. Importing drawing formats (in the form of DWG or DXF files) is discussed in Chapter 23.

Sheet Properties

Almost everything in SolidWorks has properties, whether it is a feature, line, dimension, or drawing sheet. The properties of an object are accessible via the right mouse button and can be used to modify the given object in various ways. In the case of a drawing sheet, properties can be used to modify sheet scale, the format used, paper size, and other options.

Accessing the properties of a drawing will open up the Sheet Properties window, the upper portion of which is shown in Figure 12–9. Because this window can be accessed via a simple right click on the drawing sheet (or by selecting Properties from the Edit menu), it is not necessary to list the steps for this process.

Fig. 12–9. Sheet Properties window.

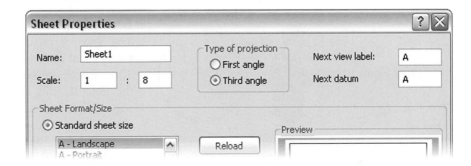

Most of the various properties within the Sheet Properties window are self-explanatory. If there are any items that may still be unclear, the following material provides some useful notes that should help in regard to the Sheet Properties window.

Name

Changing the sheet name will change the name displayed on the tab at the bottom of the screen. These tabs are visible in Figure 12–4.

Scale

Sheet scale is controlled from two settings. One setting is in the Sheet Properties window. It is simple enough to change and to understand. For example, a setting of 1:2 is half scale. All of the views on the sheet will be half their actual size. You need not worry about dimension values, as they will read correctly, regardless of the sheet scale.

The second setting that affects scale is an option titled *Automatically scale new drawing views*. This option, shown in Figure 12–10, will automatically set the scale of new views when they are inserted into the drawing. To access this option, Select Options

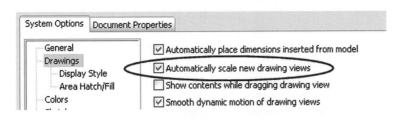

Fig. 12–10. Automatically scale new drawing views option.

from the Tools menu, and then select the Drawings category (see Figure 12–10). If you prefer to establish the sheet scale yourself, leave this option off.

If the scale of a drawing sheet is set to a particular value, and that drawing is saved as a drawing template, the Scale setting will carry forward to the next new drawing that uses that template. This assumes that *Automatically scale new drawing views* is turned off. To keep things simple, try leaving *Automatically scale new drawing views* on and let SolidWorks worry about the scale, at least until you become more familiar with the software.

Fig. 12–11. Third-angle projection.

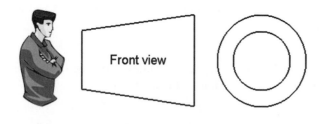

Fig. 12–12. First-angle projection.

Type of Projection

The *Type of projection* setting allows you to change between *First angle* and *Third angle* projection. *Third angle* is the default, and is what most industries in the United States use. The right view is projected off the front view as if looking at the front view while standing off to the right side of that view. This is depicted in Figure 12–11.

First-angle projection is used in many European countries. The right-side view is projected as if standing to the left of the front view, as depicted in Figure 12–12. With first-angle projection, the projected view is directly opposite the model with respect to the viewer's line of sight. Although front and right-side views are used to describe the types of projection, changing the type of projection will affect all projected views, not just right-side views.

Sheet Format/Size

This portion of the Sheet Setup window is nearly identical to the Sheet Format/Size window shown in Figure 12–6. It is possible to change the paper size of a drawing in midstream. A new format will automatically be selected to match the paper size, but this can be overridden by clicking on the Browse button and selecting a custom format.

To define a custom paper size of your own, select *Custom sheet size* and then type in values for the paper width and height. Formats are not available when specifying custom paper sizes, which might appear to mean that formats cannot be used with a custom paper size. This is not the case. Once the paper size has been specified and a new format has been created for that paper size, make it a point to save the format. Your custom sheet size will then be included in the *Standard sheet size* listing, and the custom format will be linked to it automatically.

Sheet formats are treated as OLE objects. OLE is an acronym for Object Linking and Embedding, a Microsoft name that describes the ability to insert data from one application into another. More importantly, what this means to you is that sheet formats are embedded objects, not linked to individual files. Once a sheet format has been specified, a copy of the original sheet format file is placed within the drawing. This is being explained because it is directly related to the Reload button, just visible in Figure 12–9.

The Reload button can be important for two reasons. If changes are made to a drawing format, and then you decide not to keep those changes, reloading the sheet format will remove any changes by reloading a fresh copy of whatever format was originally specified. This requires that the original format exists on the computer or network where the drawing is being edited.

A second reason the Reload button may be important is due to formats being embedded, and not linked. If a change is made to the original drawing format file, it will not automatically propagate to any drawing containing that format. This is the nature of embedded documents. However, if the Reload button is pressed the changes will transfer from the modified format to the drawing.

Additional Sheet Properties

The next time a section or detail view is created, the letter shown in *Next view label* will be used on that view. The next time a new datum is added, the letter shown in *Next datum* will be used. The option at the very bottom of the Sheet Properties window (not shown) relates to custom properties, discussed in detail in Chapter 17. Linking to custom properties is discussed later in this chapter, in the section titled "Notes" (see subtopic "Text Format Panel").

Adding Drawing Sheets

On a single drawing sheet you can mix and match parts and assemblies in any fashion. It is also possible to add as many sheets to a single drawing file as necessary. One scenario might be a model drawing that contains one sheet with the three standard views, a second sheet with section and detail views, and a third sheet with additional auxiliary or cut-away views. To relay a small piece of advice, it is probably not a good idea to place too many sheets in a single drawing, for the same reason it is not good to place all of your eggs in one basket.

✓ **TIP:** *It is perfectly acceptable to name drawings the same as the part or assembly they reference, because the file extensions will be different. This makes finding drawings that much easier later on.*

How you organize your drawings is a matter of personal or company policy. A drawing of a complex part might require half a dozen sheets, which is acceptable. A 20-part assembly drawing with multiple sheets for each component would be a bad idea. Use common sense when it comes to how many sheets you are adding to the drawing document. How-To 12–5 takes you through the process of adding drawing sheets.

How-To 12-5: Adding Sheets to a Drawing

To add sheets to a drawing, perform the following steps.

1. Select Sheet from the Insert menu, or right-click on the drawing sheet and select Add Sheet. The Sheet Properties window will appear.

2. Specify a name for the sheet, or leave the default name (e.g., *Sheet2*).

3. Specify paper size, template to use, scale, and other relevant information.

4. Click on OK to add the sheet.

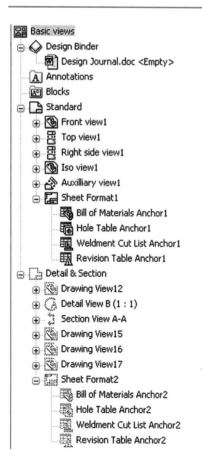

A new sheet tab will appear at the bottom of the drawing, and the new sheet will become the current (top) sheet. To switch between sheets, click on the applicable tab.

The drawing FeatureManager will reflect any sheets added to the drawing or any views on each sheet. Views can be renamed, though it usually is not necessary to do so. Figure 12–13 shows an example of FeatureManager. This particular drawing contains two sheets: *Standard* and *Detail & Section*. Inactive sheets appear with ghosted text, such as with the *Detail & Section* sheet shown in Figure 12–13.

Below the sheet format listing in FeatureManager is another listing of the available anchor points in the drawing. A sheet format must be used in a drawing, as the anchor points will reference geometry in the format. These anchor points can be seen in Figure 12–13.

Anchor points are used for anchoring various tables in a drawing. These include revision tables and hole tables, to name two. Tables are discussed elsewhere in this book. Anchor points, which can be saved with the drawing template, are discussed in this chapter. How-To 12–6 takes you through the process of establishing anchor points.

Fig. 12–13. Drawing FeatureManager.

How-To 12-6: Setting Anchor Points

To set any of the available anchor points in a drawing, perform the following steps. Note that a sheet format must be present in the drawing. (In actuality, the sheet format does not have to be displayed, just present in the drawing, and thereby listed in FeatureManager.)

1. Right-click on the anchor point you wish to set in FeatureManager and select Set Anchor. You will be placed in edit mode for the sheet format. Any existing views on the sheet will temporarily disappear.

2. Select near the endpoint of any line in the format to establish the anchor point. Any existing views will reappear and you will once again be editing the sheet.

3. Repeat the process to set additional anchor points for other tables.

At this point, you have learned quite a bit regarding the setup of formats and templates. Just don't forget to save your formats and templates. Although creating formats and templates usually takes a fair amount of work up front, the job only has to be done once. With that said, you are now ready to start adding more views to a drawing.

Inserting Views

Nearly any view type you could possibly want can be inserted into a Solid-Works drawing. This includes standard three-view layouts, section and detail views, and projected and auxiliary views, to name a few. This section explores creating these views, which is an automated process to a large degree. Section views, for instance, take a matter of seconds to create because SolidWorks generates the view from the solid model.

Bringing the standard three views (top, front, and right side) into a design drawing is often the first step of any basic design drawing. You learned one way to perform this action in How-To 12–1. There are a few other shortcuts for creating these views that can serve you well. Review the following material and use whatever method you find most useful at the time.

Inserting the Standard Three Views: Shortcut 1

If the model file you wish to generate the standard three views from is already open, you can perform a simple drag-and-drop operation. Drag the name of the model from the top of FeatureManager and drop it onto the sheet of the drawing, as shown in Figure 12–14. A top, front, and right-side view will automatically be generated.

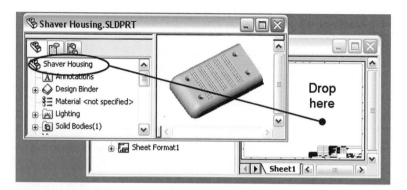

Fig. 12–14. Dragging and dropping from FeatureManager.

✓ **TIP:** *Use the Windows menu to tile any open document windows horizontally or vertically.*

Inserting the Standard Three Views: Shortcut 2

If the model is not currently open, drag the name of the model from Solid-Works Explorer onto the drawing sheet. This process works from the Windows Explorer as well, but the SolidWorks Explorer is more convenient because it is so easily accessible.

Knowing what you have learned from How-To 12–1, and being aware of the two shortcuts previously described, does it really matter if there is a more traditional way of generating the standard three views? No, it probably does not matter. But if you are a traditionalist, there is a Standard 3 View icon on the Drawing toolbar. You can also select Standard 3 View from the Insert > Drawing View menu. You can then browse for the model and insert the standard three views that way.

Moving Views

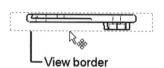

Fig. 12–15. Drag a view's border to reposition the view.

Once there are views on the drawing sheet, you will probably need to move them around to position them properly. Move the cursor over a view until the view border appears, shown in Figure 12–15. The cursor will then change into a four-pointed arrow, and the view can be repositioned as needed.

Drawing views exhibit behavior akin to parent/child relationships. For example, if a right-side view is projected from a front view, the right-side view appears aligned to the front view. Moving the front view moves the right view, and vice versa.

✓ **TIP:** *It is also possible to move a view by dragging an edge of the model in the view.*

If you wish to see the content of a view as the view is moved, turn on the *Show contents while dragging drawing view* option found in the Drawings section of the System Options (Tools > Options). If working with complex parts or large assemblies, or if your computer is showing its age, turn the *Show contents while dragging drawing view* option off.

View Alignment

It is sometimes desirable to place a view at some other location on the drawing sheet, rather than leave it in its default aligned condition. This operation requires breaking the alignment of a view. How-To 12–7 takes you through the process of breaking the alignment of a drawing view.

How-To 12-7: Breaking View Alignment

To break the alignment of a view, perform the following steps.

1. Right-click inside the border of the view whose alignment you want to break.

2. Select Break Alignment from the Alignment menu.

3. Move the view to the desired location.

Once a view's alignment has been broken, it is possible to change your mind and return its alignment to the default state. The procedure is the same, but you must select the Default Alignment option instead of Break Alignment. The view will then snap back into position to maintain its default alignment.

Another option you have is to align a particular view horizontally or vertically with another view. Most of the time, the software takes care of view alignments just fine. Normally you will not need to manually align to other views. However, it is nice to have the options available for those times you need them.

You probably already noticed the additional alignment options for aligning a view horizontally or vertically to another view, also found in the Alignment menu. The mechanics behind the procedure are the same as for breaking a view's alignment, with one extra step. How-To 12–8 takes you through the process of horizontally or vertically aligning one view to another.

How-To 12-8: Horizontally or Vertically Aligning a View

To horizontally or vertically align a view, perform the following steps.

1. Right-click inside the border of the view whose alignment you want to change.

2. Select Align Horizontal or Align Vertical from the Alignment menu. Alignment can be by center or by origin, whatever the task calls for.

3. Click on the view you want to align with.

Aligning by origin refers to the origin of the part. Aligning by center refers to the center of a bounding rectangle; that is, a rectangle just large enough to enclose the model. In most cases, either option provides the same result.

Right-clicking Precautions

It should be noted that some care should be taken when right-clicking while working on a drawing. When you are right-clicking, remember that you are accessing a *context-sensitive* menu. This means that the right mouse button is sensitive to what you click on.

Due to the context-sensitive nature of the menu that appears when right-clicking objects in a SolidWorks document, different commands are available depending on cursor position when right-clicking. Cursor position is always important in any SolidWorks document, but even more so in a drawing. This is due to the amount of information typically present in a drawing. There are often densely packed drawing views and annotations occupying a single sheet of paper.

The point of this discussion is that different menu selections will appear depending on what you right-click on. For this reason, make it a point to pay attention to cursor position prior to accessing the context-sensitive menu or you may not see the menu selections you are looking for.

Rotating Views

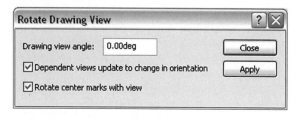

Fig. 12–16. Rotate Drawing View window.

The nice thing about rotating a drawing view is that the command used is the same Rotate command found on the View toolbar used for rotating a 3D model. If you click on the Rotate icon while in a drawing, the window shown in Figure 12–16 appears.

Once the command has been initiated, all that is necessary is to drag the desired view the desired number of degrees. You will find that the view being dragged has a tendency to snap to preset positions, making it easier to place the view in the proper orientation. If necessary, a specific rotational value can be typed in for the *Drawing view angle* setting if using the drag method does not offer enough control.

Another method of rotating a view is to make a particular edge horizontal or vertical. This is considered an alignment function in SolidWorks terms, though the view is not being aligned to another view. Rather, it is being aligned by an edge in the same view. How-To 12–9 takes you through this process.

HOW-TO 12-9: Aligning by Edges

To rotate a view by dictating that one of its edges should be either horizontal or vertical, perform the following steps.

1. Select the edge that should be either horizontal or vertical.

2. Select Horizontal Edge or Vertical Edge from the Tools > Align Drawing View menu.

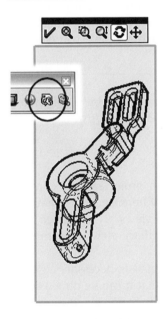

Fig. 12–17. Using the 3D Drawing View command.

You should find at this point that the selected edge becomes horizontal or vertical, depending on what was specified. To return the view to its original orientation, right-click on the view and select Default Rotation from the Alignment menu.

SW 2006: Models contained within a drawing view can be rotated using the 3D Drawing View command. The icon for this command, which can also be found under the View > Modify menu, is shown in Figure 12–17. The command allows for rotating, panning, and zooming the model contained within the view, but not for the reasons you might think.

Do not use the 3D Drawing View command to change the orientation of a model within a view permanently. To create a view of a model in a certain orientation, use the Model View command, discussed shortly. Where the 3D Drawing View command is useful is when trying to select certain model geometry. For example, trying to select certain edges for reasons of hiding those edges or dimensioning them may prove much easier after rotating the model a bit. The toolbar shown at the top of Figure 12–17 is used for manipulating the model in the view. As soon as the OK button is pressed on that same toolbar, the model snaps back to its original orientation.

Projected Views

Projected views are very similar in nature to the standard three-view arrangement. When you get right down to it, the three standard views are nothing more than two views projected from a front view: one top and one right-side projection.

A projected view can be projected from any view, not just orthographic views. This means that you can project from isometric or auxiliary views as well. In addition, isometric projections can be generated from standard ortho-graphic views. What type of projection is generated depends totally on cursor position. How-To 12–10 takes you through the process of projecting views.

How-To 12-10: Creating a Projected View

To create a projected view, perform the following steps.

1. Select Insert > Drawing View > Projected, or click on the Projected View icon found on the Drawing toolbar.

2. Select the view to project from.

3. Specify the side to project to by clicking the left mouse button. This can be north, south, east, or west (above, below, to the right, or to the left, respectively). Additionally, NE, NW, SE, and SW isometric projections can be generated. You will see a preview, which will help you position the new view.

NOTE: *If a view is selected prior to starting the Projected View command, SolidWorks will not prompt you to select a view first.*

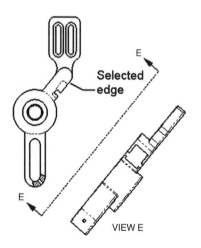

Fig. 12–18. Auxiliary view.

Auxiliary Views

Auxiliary views are similar to projected views in that they are also projected. The difference is in what they are projected from. Projected views are projected from an existing view, whereas auxiliary views are projected from an angled edge.

Create auxiliary views when it is necessary to see a particular face straight on so that it can be dimensioned parallel to the sheet. This might be some sort of oblique face that is not parallel in any of the typical ortho-graphic projections. Such a face can be seen in Figure 12–18. The edge the auxiliary view was projected from is also shown. How-To 12–11 takes you through the process of creating an auxiliary view.

How-To 12-11: Creating an Auxiliary View

To create an auxiliary view, perform the following steps.

1. Select Insert > Drawing View > Auxiliary, or click on the Auxiliary View icon found on the Drawing toolbar.

2. Select an edge to project the auxiliary view from.

3. Position the auxiliary view using the preview as a guide.

Fig. 12–19. Auxiliary view arrow options.

Auxiliary views use view arrows such as those shown in Figure 12–18. To turn these arrows off, uncheck the Arrow option shown in Figure 12–19. Likewise, the arrows can be flipped to point in the opposite direction. If the arrows are left on, the label employed by the arrows can be changed as well from the same panel.

Model Views

The Model View command is used to place a specific system view or user-defined view in a drawing. A model view can be any view listed in the Orientation window, including views added by the user. See How-To 3–9 for an explanation of how to add user-defined views to a model. How-To 12–12 takes you through the process of creating a model view using one of two methods.

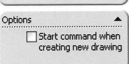

HOW-TO 12-12: Creating a Model View

To insert a specific model view of a part or assembly into a drawing, perform the following steps.

1. Select Insert > Drawing View > Model View, or click on the Model View icon found on the Drawing toolbar. The Model View PropertyManager, shown in Figure 12–20, will appear.

2. Specify the model from which the model view should be generated. This can be done by selecting an existing view of the model, selecting the model from the list shown in Figure 12–20, or clicking on the Browse button and selecting the model from a file listing. If selecting from the list shown in Figure 12–20, make sure to click on the Next button.

3. The second PropertyManager page will be displayed. From the Orientation list, choose the model view to be inserted into the drawing.

4. Click somewhere on the drawing sheet to position the view.

5. Click on OK or somewhere on the drawing to close the command.

Fig. 12–20. Model View PropertyManager (page 1).

As can also be seen from Figure 12–20, there is a preview of the model used to generate the model view. The option *Start command when creating new drawing* does just what the wording implies. Whenever a new drawing is begun, you will automatically be placed in the Model View command. Enable this option based solely on your personal preference.

Model views have the unique capability of having their orientation changed after they are created. To accomplish this, select the view, and then double click on the appropriate view in the Orientation panel (shown in Figure 12–21). The view orientation will then change to accommodate your selection.

SW 2006: The Orientation panel has a graphical interface as opposed to a text-based listing. The new orientation panel is shown in Figure 12–3.

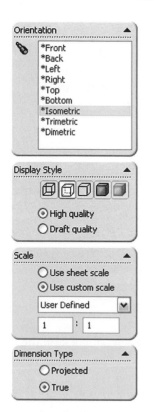

The properties displayed in Figure 12–21 are fairly generic, with the exception of the Orientation panel, which will only be visible when a model view is selected. The other properties are available for most views, though they may vary in subtle ways. For example, if a view contains cosmetic threads (discussed later in this chapter) there will be a setting for high-quality versus draft-quality cosmetic thread display.

Each view in the drawing can have its own display changed to any of the standard display styles via the view PropertyManager. These display styles include a shaded display mode, along with the typical settings for hidden lines visible or removed. Each view also has a high-quality or draft-quality setting. The general rule with "quality" settings is dependent on many variables, but boils down to one question: Are you happy with the performance of the drawing?

Drawings with many views of complex parts or large assemblies have a tendency to rebuild more slowly. It will take longer to perform what may sometimes seem like simple tasks. Of course, performance is highly dependent on machine speed as well. If you notice degradation in system performance, change the view settings (or other pertinent quality settings) to draft versus high quality.

Fig. 12–21. Model View PropertyManager (page 2).

Scale settings can be changed on a view-by-view basis. By default, views will use the sheet scale, which you learned how to set in the section "Sheet Properties." If a view should be something other than the sheet scale, set the scale in the Scale panel of the view's PropertyManager (this setting is visible in Figure 12–21). A preset custom scale can be selected from the list, or a User Defined value can be entered, which happens to be the selection made in Figure 12–21. Some view types (such as projected views) will have a setting for *parent view.* In this case, the view uses whatever scale is specified in the parent view's PropertyManager.

SW 2006: Drawing views can no longer be changed between draft versus high quality. All views are high quality by default. It is possible to force new views to use draft quality by selecting Draft Quality from the Display Style section of the System Options (Tools > Options). However, once a view is set to high quality (via PropertyManager) it cannot be changed back to draft quality. The one exception to this behavior relates to large assemblies, discussed in Chapter 19. When SolidWorks detects a large assembly, it will ask if you would like to use a draft-quality view to help increase performance.

Relative to Model View Method

Relative to Model is a view insertion option that lets you control how the part is oriented when you insert it into the drawing. The process involves selecting faces on the model and telling the software how you want each face oriented, such as toward the front or the top. How-To 12–13 takes you through the process of creating a view using Relative to Model.

How-To 12-13: Creating a View Using Relative to Model

To create a view using Relative to Model, perform the following steps. The model document should be open when implementing this function, so that faces on the model can be selected. It is suggested that these steps be read through completely before attempting this procedure, so that you have a better understanding of what this command accomplishes.

Fig. 12–22. Relative
View PropertyManager.

1. With the drawing file visible, select Insert > Drawing View > Relative to Model, or click on the Relative View icon found on the Drawing toolbar.

2. Switch to the model file. This can be done by pressing Ctrl-Tab, or by selecting the model file from the Window menu. The Relative View PropertyManager, shown in Figure 12–22, should be visible.

3. Select a face on the model and specify the direction that face should point. Choices are Front, Top, Right, and so on.

4. Select a second face on the model and again specify the direction that face should point.

5. Click on OK.

6. Select a position on the drawing sheet where the view is to be positioned.

7. Click on OK or somewhere on the drawing to close the command.

If you are new to the Ctrl-Tab hotkey combination specified in step 2, you should be aware that this hotkey works in any Windows program and is not particular to SolidWorks. As a matter of fact, Ctrl-Tab can be used to alternate between open documents in a Windows program, whereas Alt-Tab can be used to alternate between open programs.

Section Views

SolidWorks handles section views very nicely. You will find that the Section View command will save you countless hours of manually drafting such views. Section views require that a line be added to the view to be sectioned. This line, or set of lines, is what becomes the section line. Geometry added to a view is automatically associated with that view. This is important in that if the view is repositioned the geometry should move with the view.

You can begin by drawing a line that will become the section line. Section lines can consist of lines and arcs in a wide variety of combinations. They can also be construction entities or regular lines; it makes no difference. How-To 12–14 takes you through the process of creating a simple section view.

How-To 12-14: Creating a Section View

To create a section view, perform the following steps.

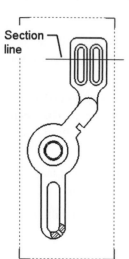

1. Select Insert > Drawing View > Section, or click on the Section View icon found on the Drawing toolbar. You will be placed in the Line command.

2. Create geometry that will represent the section line, such as that shown in Figure 12–23.

3. If prompted to create a Partial Section, click on No (this is explained shortly).

4. Click on the drawing to position the section view using the preview as a guide.

Fig. 12–23. Creating the section line geometry.

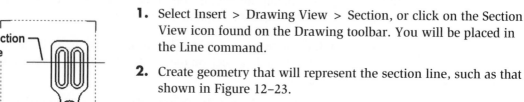

The first thing you will notice is that the geometry you added to the view to be sectioned now looks like an actual section line, as shown in Figure 12–24. The section view itself is also shown.

Selecting either the section line or the section view will cause the PropertyManager to display the properties for the section view, the upper portion of which is shown in Figure 12–25. Note that the label of the section line can be changed, as well as the direction the arrows are pointing. Double clicking on the section line also serves to flip its direction.

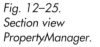

SECTION A-A

Fig. 12–24. Resultant section view and transformed section line geometry.

Fig. 12–25. Section view PropertyManager.

The section line itself can usually be dragged to another position, and the endpoints of the section line can also be dragged. If you have trouble with this, it could be because of existing geometric relations on the original section line geometry. Right-click on the section line and select *Edit sketch*. You can then use your existing knowledge to see what relations are on the geometry, and remove or add relations as needed. Rebuild the drawing when you are finished editing the section line sketch.

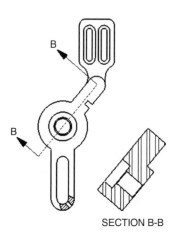

SECTION B-B

Fig. 12–26. Partial section.

In general, whenever a section line is altered in any way the section view will need to be updated. This can be accomplished by either rebuilding the entire drawing or rebuilding just the section view, which is much more efficient. Views in need of rebuilding are difficult to miss, as they will have crosshatch lines covering the entire view. Right-clicking on the view and selecting *Update view* will allow the view to be rebuilt.

Additional section view options include *Partial section*, which should be checked if a section line does not completely cut across the model geometry. Such is the case in Figure 12–26. During the creation of section line B-B in Figure 12–26, you would have been asked if a partial section should be created. In this case, it would have been wise to click on the Yes button when the prompt appeared. Otherwise, an error would have resulted, and the section view would not appear correctly.

It should also be noted that the arrows in Figure 12–26 have been repositioned to extend farther from the section line. This can be done by selecting the section line and then dragging the small green handles that appear at the tip of each arrow. The labels themselves can also be repositioned via dragging. Labels will have a tendency to snap to their ideal placement near the end of an arrow if the label is dropped on the arrow's tip.

Sometimes the partial section prompt (not shown) will appear, asking if you want to create a partial section even though the section line cuts across the geometry but not across the entire view. In a case such as this, it is perfectly safe to inform SolidWorks that you do not want to create a partial section. When the partial section prompt appears, just click on the No button. If a mistake is made, you can always change the *Partial section* option or modify the section line.

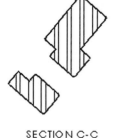

SECTION C-C

Fig. 12–27. Display only surface option enabled.

The *Display only surface* option hides any edges that do not touch the section-line cutting surface. An example of enabling the *Display only surface* option is shown in Figure 12–27. Compare section B-B in Figure 12–26 to section C-C in Figure 12–27. The effect of showing only the section view surface should be apparent.

The *Auto-hatching* option only pertains to assembly section views or section views of parts containing multiple bodies. Multiple bodies are explored in Chapter 20. To put it simply, automatic ("auto") hatching takes place when there are multiple chunks of geometry that require hatching. To differentiate between the individual chunks, hatch patterns are automatically rotated.

Complex Section Lines

The procedure outlined in How-To 12–14 shows how to create a section view in the most user-friendly manner possible. However, this process does not leave much room for creating more elaborate section lines utilizing multiple segments or arcs.

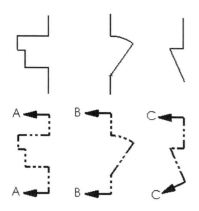

Fig. 12–28. Examples of multisegment section lines.

The solution to creating multisegment section lines is simple: create the section line geometry prior to starting the Section View command. Prior to entering the command, make it a point to select one of the line segments in the section line geometry. The segment you choose plays an important role in that the section view will be projected perpendicular from the selected segment.

Figure 12–28 shows three examples of section lines with multiple segments. Above each section line is the original geometry from which the section lines were defined. Figure 12–28 does not show the model geometry on which the section line geometry would have been sketched. Nonetheless, the image should give you a better understanding as to what can be accomplished when it becomes necessary to create a section view.

✓ **TIP:** *When creating section line geometry, sketch geometry used to create the model can be shown in order to help precisely position or constrain the section line. Right-click on the applicable sketch in the drawing FeatureManager in order to show it.*

Aligned Sections

Another type of section view is an aligned section. Aligned sections are "unfolded," and the resultant view is displayed as if it were projected perpendicular from each segment in the section line. In contrast, the standard section views examined in the previous material are projected from a single segment of the section line.

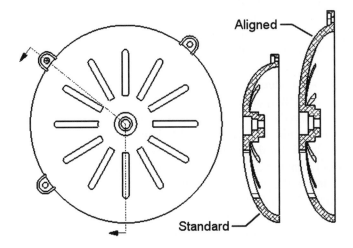

Fig. 12–29. Standard section versus aligned section views.

Figure 12–29—which shows the original model view on the left, with two possible outcomes on the right—will help elucidate how an aligned section works. Note that the standard section view appears malformed near the top. This happens because the view is projected perpendicular from the vertical portion of the section line geometry. The view is true to what a person would actually see if the geometry were indeed cut, but it is not very clear on a drawing.

The aligned section view displays much more clearly, as can be seen from the figure. Aligned section views are not projected from a single section line segment. Rather, the view is generated by looking perpendicular to each segment. How-To 12–15 takes you through the process of creating an aligned section view.

HOW-TO 12-15: Creating an Aligned Section View

To create and aligned section view, perform the following steps.

1. Select Insert > Drawing View > Aligned Section, or click on the Aligned Section View icon found on the Drawing toolbar. You will be placed in the Line command.

2. Create geometry that will represent the section line.

3. Click to position the aligned section view on the sheet.

It is easier to create the section line geometry up front. That is, create the section line geometry prior to performing step 1 of How-To 12–15. This will permit more flexibility when creating the section line, and you will not be limited to creating two segments. Just make sure to select the segment you wish to project the view from prior to initiating the Aligned Section View command or you will be placed in the Line command unnecessarily. The selected segment dictates how the resultant view will be aligned with the original view, but will otherwise not change the aligned section view's appearance.

Modifying Crosshatch

There are two places it is possible to modify the hatch style being used. You should first ask yourself if you want to modify the hatch pattern associated with the part file, or if you want to modify the hatch strictly for a particular view. The following material explains why this question is important.

If the hatch pattern is modified in the drawing, the section view will look correct according to your specifications. However, will the part being sectioned be used in assemblies? If so, it would be more prudent to change the hatch associated with the part itself. This will make the crosshatching automatically appear correctly in a drawing of the part. More importantly, if the part is used in an assembly and the assembly later appears in a drawing section view the crosshatch for the part will already be set.

What you should be taking away from this discussion is to correctly set the crosshatch properties in the part once and be done with it. You were taught how to do this in Chapter 8 (see How-To 8–6). Changes made to a part's crosshatch will immediately propagate to affect any drawing that part is in.

If crosshatch should be changed in a part file, why would it ever be necessary to modify crosshatch properties from the drawing view? One common reason is associated with the case of assembly section views. It is often necessary to rotate the hatch patterns to make it easier to discern individual components. Another reason is simply flexibility.

If you are interested in changing the crosshatch properties of a specific view, or even a single region within a view, the procedure is different than that for a part file. How-To 12–16 takes you through the process of changing the crosshatch properties of a view.

HOW-TO 12-16: Changing View Crosshatch

To change the crosshatch properties of a view, perform the following steps. Note that this process will override the crosshatch properties associated with the part file.

1. Click on the crosshatch you wish to modify. The Area Hatch/Fill PropertyManager, shown in Figure 12–30, will appear.

Fig. 12–30. Area Hatch/Fill PropertyManager.

2. If necessary, uncheck the Material Properties option to override the material properties associated with the part.

3. Make any necessary changes to the crosshatch pattern, scale, and angle.

4. In the Apply To drop-down list, specify View, Region, or Body. If you are in an assembly section view, applying changes to the part will also be an option.

5. Click on OK to accept the changes.

With regard to the Apply To list, selecting View will apply changes to hatching in the entire view. If Region is selected, only the hatch initially selected will have its pattern characteristics altered. Selecting Body applies changes to the affected body only, rather than to the entire part. This last selection only applies if working with multiple bodies (see Chapter 20).

The Part option is only available in the Apply To listing if you are working with a section view of an assembly. Specifying Part will alter any crosshatch associated with the entire part whose crosshatch you are changing in the specified view. All regions (hatch boundary areas) of the part will be affected.

Crosshatch can be solid, if desired. It can also be placed on a specific layer. In this way, crosshatch can be a specific color, taking on whatever color is assigned to the layer. For those not familiar with layers and their uses, there is a section devoted to layers later in this chapter.

Area Hatch

When section views are created, crosshatch is automatically applied where necessary. Area hatch is simply crosshatch applied manually. Area hatch can be applied to any surface or boundary area you would like hatch to appear. In some cases, area hatch can be applied to nonplanar faces, with the limitation that the face does not wrap around itself, displaying both inside and outside faces (e.g., a cylinder). Area hatch can be applied in drawings only. How-To 12–17 takes you through the process of adding area hatch.

HOW-TO 12-17: Adding Area Hatch

To manually apply area hatch to a drawing, perform the following steps.

1. Select Area Hatch/Fill from the Insert > Annotations menu.

2. Specify the desired hatch pattern, scale, and angle.

3. Select a face or faces on which to apply hatching. Sketch geometry that forms a closed boundary is also a valid choice.

4. Click on OK to accept the hatch pattern.

Hatch applied here ⟶

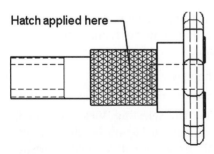

Fig. 12–31. Drawing view with area hatch.

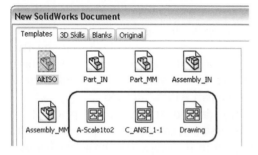

Fig. 12–32. Hatch will leave space around text.

Figure 12–31 shows a sample of an area on a drawing view where area hatch has been applied. The area is interesting in that it represents a cylindrical face on the model. Usually it is not possible to add hatch to such an area, so how was this accomplished? The solution is to use the Convert Entities command to convert existing edges into sketch geometry, or to simply sketch on top of the view geometry. The hatch can then be applied to the sketch geometry rather than the model geometry.

Hatch can be applied to sketch geometry of any shape. The geometry shown in Figure 12–32 shows another example of area hatch. The hatch pattern is smart enough to know it should not hatch within the circle, and it also knows enough to avoid text. If additional text is added, or existing text is moved, the existing hatch pattern will update accordingly.

Broken-out Section Views

Similar to section views, broken-out section views display a cross section of a model. In the case of a broken-out section, the cross section is taken by removing a portion (or "breaking out") a bit of material on the model, to see what is underneath. A depth must be specified in some way to indicate the depth of the broken-out section. Closed profiles are used rather than the open-profile sketch geometry that would be used with typical section views. To create a broken-out section view, perform the steps outlined in How-To 12–18.

HOW-TO 12-18: Creating a Broken-out Section View

To create a broken-out section view, perform the following steps.

1. Add sketch geometry in the shape of a closed profile (such as that shown in Figure 12–33), which will dictate the shape of the broken-out section.

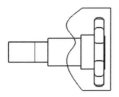

Fig. 12–33. Closed-profile sketch geometry added to a view.

Fig. 12–34. Broken-out Section PropertyManager.

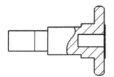

Fig. 12–35. Broken-out section view.

2. Select the profile geometry.

3. Select Insert > Drawing View > Broken-out Section, or click on the Broken-out Section icon found on the Drawing toolbar.

4. In the Broken-out Section PropertyManager, shown in Figure 12–34, either enter a value for the section depth or select an object that will dictate the depth. Examples of objects that can be selected are model edges and axes.

5. Click on OK to create the broken-out section view.

An example of a broken-out section view is shown in Figure 12–35. Crosshatch is added automatically by SolidWorks. Right-clicking over the broken-out section area gains access to a Broken-Out Section submenu that contains menu picks for deleting the broken-out section, editing the underlying sketch geometry, or editing the definition of the broken-out section view.

✓ **TIP:** *Broken-out section views can be used in place of standard section views if you do not wish to show the "parent" view the section is taken from. By creating a rectangle around the entire view from which the broken-out section is defined, no parent view is needed.*

Detail Views

Like section views, detail views require that you add geometry to the view you want to create a detail of. The geometry required can take any shape. The detail view itself can be a circle or some other geometric shape. The geometry created, which will represent the detail area, should be a closed profile.

Like any view, the scale of a detail view can be changed independently of the sheet. However, you can establish the default scale for detail views in the Drawings section of System Options (Tools > Options). This is known as the *Detail view scaling* option (not shown). There is also an option in the same Drawings section titled *Display new detail circles as circles*, shown in Figure 12–36.

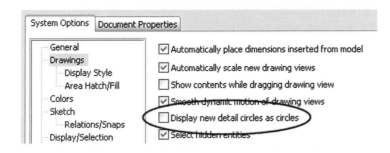

Fig. 12–36. Display new detail circles as circles option.

When cleared, the *Display new detail circles as circles* option is supposed to allow detail views to take on the shape of the original profile geometry. In other words, detail area sketch profiles do not have to be circles. They can be any shape you choose. With the option turned on (checked), all profiles used for detail views are converted to circles. This would be true even if you drew a rectangle, for instance. As it turns out, this setting is either broken or obsolete, as profile geometry display is actually dependent on another setting we will discuss in a moment.

The value set for the *Detail view scaling* is what the sheet scale will be multiplied by. For example, if the sheet scale is set to 1:2 and the *Detail view scaling* is 4X, the detail view will have a scale of 2:1. How-To 12–19 takes you through the process of creating a detail view.

How-To 12-19: Creating a Detail View

To create a detail view, perform the following steps.

1. Select Insert > Drawing View > Detail, or click on the Detail View icon found on the Drawing toolbar. You will automatically be placed in the Circle command.

2. Draw a circle (or some other closed profile) around the area to be detailed, as indicated in Figure 12–37.

3. Click on the drawing sheet to position the detail view. An example of a detail view is shown in Figure 12–38.

Fig. 12–37. Sketched circle that will represent the detailed area.

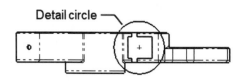

Fig. 12–38. Resultant detail view.

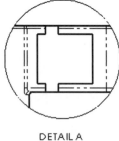

DETAIL A
SCALE 1 : 1

Similar to section lines and section views, selecting a detail view or the detail circle displays the detail view's PropertyManager. A portion of the Detail View PropertyManager is shown in Figure 12–39.

The Detail View PropertyManager provides access to a number of options, such as the detail view style, label, and font. With regard to the style setting, the Per Standard option will display the detail view circle per whatever drafting standard the drawing is set to (e.g., ANSI). Other style options include Broken Circle, With Leader, No Leader, and Connected. Figure 12–40 shows the first three styles respectively, from top to bottom.

Fig. 12–39. Portion of the Detail View PropertyManager.

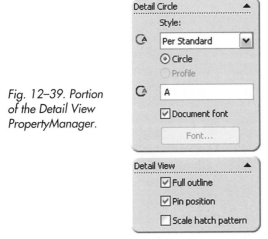

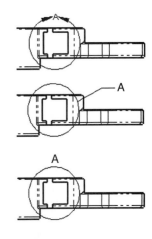

Fig. 12–40. Broken Circle, With Leader and No Leader styles.

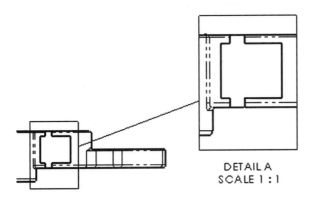

DETAIL A
SCALE 1 : 1

Fig. 12–41. Connected detail view using a rectangular detail area.

If a shape other than a circle is used to describe the detail area (e.g., a rectangle), that shape is automatically converted to a circle. This does not have to be the case. It is possible to use the original profile shape if Profile is selected rather than Circle. Although Profile is grayed out, these options can be seen in Figure 12–39. Only the With Leader, No Leader, and Connected styles allow using the Profile option. Figure 12–41 shows an example of using the Connected detail style with the Profile option selected.

✓ **TIP:** *View labels are bidirectionally associative with detail circle or section line labels. Changing one (which can be done by double clicking) automatically updates the other.*

Dragging a detail circle will serve to resize it, and dragging the centerpoint will alter its location. When moving or resizing detail circle (or profile) geometry, the detail view is updated automatically and does not require a rebuild. If the profile geometry is constrained to existing model geometry in the view, it may not lend itself to dragging. In this case, right-click on the profile geometry and select Edit Sketch to perform any editing on that geometry.

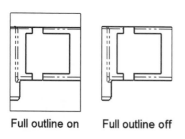

Full outline on   Full outline off

Fig. 12–42. Effects of the Full outline *option.*

Another option available in the Detail View PropertyManager is *Full outline*, which will draw a border around the detail view. This will work with any profile geometry. An example is shown in Figure 12–42.

The *Pin position* option forces the detail view to stay locked in its current position, even if the parent view changes size. The only time this might be a problem is when the parent view changes to a large extent and runs into the detail view. When creating a detail view of an area on a section view, checking *Scale hatch pattern* will increase the scale of the hatch using the same scale value as the detail view.

Broken Views

When a part is too long to fit on a drawing sheet, you would typically create what is known as a broken view. Broken views can be created in SolidWorks with minimal effort. The break lines can be adjusted or modified in appearance, and the break gap can be altered. You will learn how to perform all of these functions.

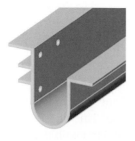

Fig. 12–43. A long, extruded part.

As an example, the extruded part shown in Figure 12–43 will be used. This part is 32 inches long, and could fit quite nicely on an E-size sheet of paper. However, an E-size sheet would be overkill for a part of this nature. There are no detailed features on this part, except for holes drilled in the center of the part and three holes on each end.

Figure 12–44 shows the same part on a piece of paper it is obviously too big for. Once the part is broken, it will fit just fine on the A-size sheet of paper, even at 1:2 scale. How-To 12–20 takes you through the process of breaking a view.

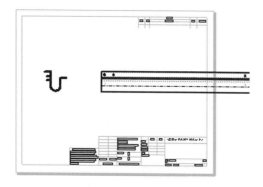

Fig. 12–44. Prior to breaking the view.

HOW-TO 12-20: Creating a Broken View

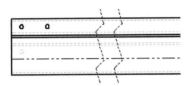

Fig. 12–45. After inserting break lines.

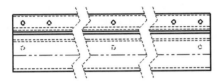

Fig. 12–46. Completed broken view.

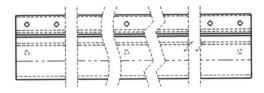

Fig. 12–47. Break line choices.

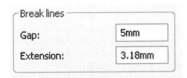

Fig. 12–48. Setting the break line gap and extension.

To create a broken view, perform the following steps.

1. Select the view to be broken.

2. Select Horizontal Break or Vertical Break from the Insert > Drawing View menu, depending on the situation. You can also click on either the Horizontal Break or Vertical Break icon found on the Drawing toolbar. Break lines will appear on the view, as shown in Figure 12–45.

Be aware that break lines can be added to a view to create more than one break. Simply repeat the first two steps as necessary. In our example, two sets of break lines will be added.

3. Drag the break lines where the break (or breaks) is to occur. Whatever is between each set of break lines will be removed when the view is broken.

4. Right-click on the view to be broken and select Break View.

5. Reposition the view or break lines as needed. A completed broken view is shown in Figure 12–46.

As mentioned in step 5, break lines can be repositioned by dragging them to a new location. The appearance of the break lines can be changed as well. Right-click on one of the individual break lines and you will see four options at your disposal. These options include straight, curved, zigzag, and small zigzag break lines, which are shown in this order in Figure 12–47.

You also have control over the break line gap distance and extension. The extension is how much the break lines extend out from the model geometry. To alter either of these values, select the Detailing section of the Document Properties (Tools > Options). Change the gap and extension settings as required, which are shown in Figure 12–48.

Incidentally, do not be concerned with dimensions, as they will display the proper value, regardless of whether they were added before or after breaking the view. Adding dimensions is covered in material to follow.

SW 2006: Bend lines in broken views of sheet metal parts are supported. Sheet metal parts that have been broken will display their bend lines appropriately, and the bend lines will be trimmed to the break lines.

Cropped Views

When a view is cropped, portions of the view not critical to the drawing are hidden. From the user's standpoint, creating a cropped view requires sketching a profile around an area of the view that is to remain. When the view is cropped, everything outside the profile is removed. How-To 12–21 takes you through the process of cropping a view.

How-To 12-21: Cropping a View

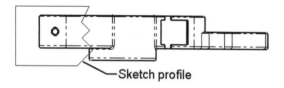

Fig. 12–49. Preparing to crop a view.

To crop a view, perform the following steps.

1. Sketch a closed profile around the area of the view that is to remain visible. An example of this process is shown in Figure 12–49.

2. Select any single segment in the crop profile sketch.

3. Select Crop from the Insert > Drawing View menu, or click on the Crop View icon found on the Drawing toolbar.

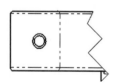

Fig. 12–50. Cropped view (enlarged for clarity).

A completed cropped view is shown in Figure 12–50. If it becomes necessary to edit a cropped view, possibly to alter the underlying sketch geometry and thereby change the cropped area, use your right mouse button menu. That is, right-click over the view and select Edit Crop from the Crop View submenu. The same menu also has an option for removing a cropped view. Actually, the underlying view will remain, but it will no longer be cropped. If you edit a cropped view, make sure to rebuild the drawing afterward.

Empty Views

At first you may wonder why you would ever need to create an empty view. The view would not so much be an empty view as it would a placeholder for geometry you create. An empty view allows for adding geometry to the view, thereby "grouping" the geometry. If the view is moved, the geometry all moves with it.

You may find it necessary to add geometry for a number of reasons. Possibly you would like to create a simple detail of a part, and it may actually be easier to just sketch it in by hand rather than create a new view. Creating an empty view gives you that option. It is also possible to sketch directly on the drawing sheet without adding an empty view first, but then the geometry is more difficult to reposition.

To create an empty view, select Empty from the Insert > Drawing View menu, and then use the preview to help position the view. Sketch the desired geometry on the view using any of the sketch tools you have learned how to use. To move the view, drag the view by its border as you have been taught, and you will find the geometry moves as a group, even without being fully defined. Blocks, which you will learn about later in this chapter, are another way of grouping geometry.

View Appearance

There are a number of options that can be used to change the appearance of views. These are options such as hiding tangent edges, controlling what lines are being shown, setting default display characteristics, and so on. Some views may need to have their hidden lines removed, others may require that hidden lines remain visible, and yet others need to be shaded.

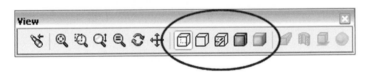

Fig. 12–51. Controlling view appearance from the View toolbar.

How is the appearance of individual views controlled? The answer is the View toolbar, which you should already be familiar with. Select a view, and then click on the applicable icon to show the view in the desired state. The display options accessible through the View toolbar were discussed in Chapter 3. These are shown in Figure 12–51 for your reference. Icons that control view appearance are circled.

Edge Display

One display option you may commonly need to change is the appearance of the view's tangent edges. Tangent edges may need to be hidden, or perhaps the tangent edges should be displayed with a lighter phantom-line type, for instance. This really depends on the part, though.

You have three choices when it comes to displaying tangent edges: display them, turn them off, or display the tangent edges with a specific style of line (also known as a line font). These options are found in the right mouse button menu. To access them, right-click on a view and access the Tangent Edge submenu. Figure 12–52 shows a part with Tangent Edges With Font activated (top), and Tangent Edges Removed activated (bottom). The one tangent edge display option not shown (Tangent Edges Visible) displays the tangent edges in a standard continuous line style rather than the phantom line style used by the Tangent Edges With Font display setting.

You can control the default behavior of how tangent edges are displayed for new views added to the drawing. This setting is in the Display Style section of System Options. There is also a setting for what display mode is used for new views. How-To 12–22 takes you through the process of changing the default display characteristics of new views (not existing views).

Fig. 12–52. Tangent edge display options.

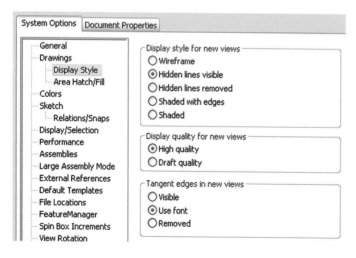

How-To 12-22: Setting Default Display Characteristics

To change the default behavior of tangent edges in new views, and to set how new views are displayed, perform the following steps.

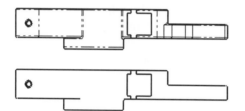

Fig. 12–53. Setting default view display characteristics.

1. Select Options from the Tools menu.

2. Select the Display Style section.

3. In the area labeled *Display style for new views*, shown in Figure 12–53, specify how new views should initially appear.

4. In the section labeled *Tangent edges in new views*, specify how tangent edges are displayed in new views.

5. Click on OK when finished.

Default settings should be settings that are most frequently used. View appearance can always be changed later, as you have recently discovered.

The *Display quality for new views* setting, which can be seen in Figure 12–53, controls whether new views are draft or high quality. Draft quality does not look as nice, but results in better performance.

Hidden edges in drawings are, by default, displayed with a thin dashed line type. Tangent edges are displayed with a thin phantom line. These settings work well, and it is not advised you make modifications to the line fonts used by SolidWorks unless you feel confident you understand what you are doing. How-To 12–23 takes you through the process of changing what line fonts are used for various objects, such as tangent edges and hidden lines.

HOW-TO 12-23: Modifying Line Type Associations

Line types are known as line fonts in SolidWorks, and are based on line style (such as dashed or continuous) and line weight (such as thin or thick). Any changes made to line font associations are saved with the document, and do not affect any previously created files. It is possible to incorporate line font changes into a drawing template if desired. To modify the line font associated with specific SolidWorks objects, perform the following steps.

✗ WARNING: *Changing line font settings indiscriminately can be detrimental to your current drawing. Use caution when performing these steps! It is recommended that you create a copy of a drawing and use that as a test case.*

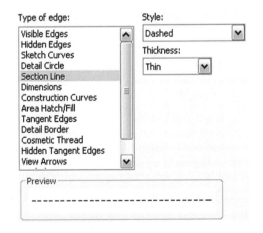

Fig. 12–54. Changing an object's line font.

1. Select Options from the Tools menu.

2. Select the Line Font section in the Document Properties tab. The Line Font section is shown in Figure 12–54.

3. In the *Type of edge* list, select the type of edge you wish to change the line font for. A preview of the line type for that object will be displayed.

4. Change the Style and Thickness (line weight) if desired.

5. Click on OK to accept the changes.

If you have followed through these steps, it is easy to see that much more than hidden and tangent edge display can be altered. It would not be a bad idea to print some test plots both before and after making any line font changes to see what the results of your modifications will look like. What you see on the screen may not be exactly what gets printed. This can be due to many variables, including the printer or plotter itself.

Line Style Creation

Line styles can be created if the choices SolidWorks gives you are not enough. This is accomplished through the Line Style section of the Document Properties tab, which is directly below the Line Font section you selected in step 2 of How-To 12–23. Instructions for creating new styles are included right on the Line Style page, so it is not necessary to repeat the steps here. However, a couple of tips might be beneficial.

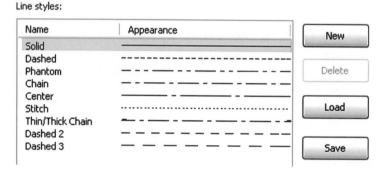

Fig. 12–55. Line Style section.

A portion of the Line Style section is shown in Figure 12–55, which is the area that lists the line styles available. Clicking on the New button copies whatever style is highlighted and adds it to the list. Use this fact to your advantage and prior to clicking on New select a style similar to the one you wish to create.

The "coding" used to generate line styles is very simple. Positive values indicate lines and negative values indicate spaces. The letter A indicates a normal line, and B indicates bold ends at either side of the line. In other words, entering the code

```
A, .5, -.5
```

will result in a standard dashed line type where the lines are the same length as the spaces between the lines. SolidWorks' online help will try to tell you that the values utilize the units specified in the document, but that is not true. Do not take the values as literal lengths. Rather, create new line styles using the values of existing styles as a starting point.

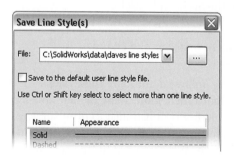

Make sure you save your newly created line styles so that they will be available for other drawings. Click on the Save button, and then make it a point to select all of the line styles displayed in the window that appears (the upper portion of which is shown in Figure 12–56). Only the selected styles will be saved, and it does not hurt to save them all.

Fig. 12–56. Saving line styles.

The Browse button (labeled with three dots in Figure 12–56) allows for navigating to a location to save the line style file and for naming the file. There is an option for saving your line style file to the default user line-style file, which can be seen in Figure 12–56. Feel free to check this option, which keeps you from having to come up with a name for the file on your own. The file will be saved in your SolidWorks *lang\english* folder with the name *userlines.sldlin*.

✓ **TIP:** *Use the Load button in the Line Style section of the Document Properties (Tools > Options) to load previously saved line style files into existing drawing templates. The templates can then be resaved and will contain all of your user-defined line styles.*

Hiding Individual Edges

Individual edges can be hidden when necessary. Simply right-click over the edge to be hidden. In the menu that appears, you will see the option Hide Edge. If you change your mind, position the cursor over where the edge used to be. The hidden edge will highlight, and you will be able to right-click over it. The menu that appears will contain a Show Edge option. The Hide Edge and Show Edge icons can also be used for this functionality, and are typically found on the Line Format toolbar.

To be able to show edges that are hidden, you must turn on the *Select hidden entities* option, found in the Drawings section of the System Options window (Tools > Options). The reason this option even exists has to do with the ability to hide multiple edges that may be on top of each other. Turn *Select hidden entities* off if it becomes necessary to hide multiple stacked edges.

SW 2006: Multiple edges can be hidden or shown at the same time. Hiding or showing stacked edges is easier due to the 3D Drawing View command discussed earlier in this chapter (see "Rotating Views").

Line Formatting

Fig. 12–57. Line Format toolbar.

The Line Format toolbar has options for changing line color, thickness, and style. Line formatting is not reserved strictly for lines, as the name would imply. Sketch geometry of any type can be formatted, including edges of feature geometry and even crosshatch. You can even change the color of annotations with the Line Format toolbar, shown in Figure 12–57.

From left to right, the icons on the Line Format toolbar pictured are Layer Properties, Line Color, Line Thickness, Line Style, Hide Edge, Show Edge, and Color Display Mode. These functions are fairly straightforward, with a few exceptions.

To change an object's color, thickness, or line style, select the objects that are to have their formatting changed, click on the appropriate Line Format icon, and then specify the desired trait (color, thickness, or style). The Hide Edge and Show Edge commands perform the same function as hiding or showing edges via the right mouse button menu, as described in the previous section. Layers are discussed later in this chapter.

✓ **TIP:** *An entire sketch in a part or assembly can have its color, thickness, or line style changed with the Line Format toolbar. These changes should only be applied when not editing the sketch, so that the formatting can be seen. Formatting can only be applied to the entire sketch, and will only appear when the sketch is not currently being edited.*

SW 2006: Using the Line Format toolbar, individual entities in a sketch can be formatted to have their own color, thickness, or style. Formatting is visible even while editing the sketch.

The Color Display Mode icon is a toggle switch. When the icon is depressed, sketch geometry in the drawing will display with the sketch color codes you are familiar with when creating sketch geometry in a part. This refers to the sketch color codes, such as blue or black, covered in Chapter 2. When the Color Display Mode icon is not toggled on, geometry will appear with whatever color formatting has been applied to it. This setting only affects drawings, and does not apply to part or assembly sketch geometry.

View Scale

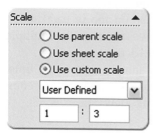

Fig. 12–58. Scale panel in PropertyManager.

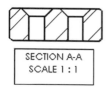

Fig. 12–59. Editing a view label.

You learned how to change the scale of a drawing sheet earlier in this chapter (see "Sheet Properties"). However, any view's scale can be altered independently of the sheet's scale. Simply clicking on a view will display that view's PropertyManager. Look for the Scale panel, shown in Figure 12–58. The Scale panel may look slightly different, depending on the type of view selected. For example, the selected view may not have a parent view, in which case the *Use parent scale* option would not be available.

The sheet scale would obviously be the scale associated with the drawing sheet. Selecting *Use custom scale* allows for picking a preset common scale value from the drop-down list. One of the selections in the drop-down list is *User Defined*, visible in Figure 12–58. This setting allows for typing in any scale desired for the view.

If the view whose scale you want to change is either a section or detail view, it is possible to change the scale of the view by modifying the associated note (otherwise known as the view label). This technique demonstrates the bidirectional associativity between a view and its label. Double clicking on the view's label gains access to the text, which can then be altered. A box will appear around the text to indicate it is now editable, as shown in

Figure 12–59. Clicking outside the label edit box will exit text edit mode. Adding and modifying notes in general are covered later in this chapter.

If a section or detail view is using the same scale as the sheet, no note will be displayed calling out the scale in the view's label. This is because the sheet scale is typically called out in the title block area. It is possible, though, to add a scale value (e.g., 2:1) directly to the label, which is different than the sheet's scale. SolidWorks will correctly interpret the scale entered and use it for the view. For this trick to work, it is not necessary that the value be prefixed by the word *scale*, but you should use a colon (:) to separate the values.

Dimensioning Drawings

When dimensioning a drawing, the dimensions that were added in the model can be imported and displayed on the drawing. This includes all of the dimensions added to sketch geometry, as well as any dimensions added by SolidWorks (such as a depth dimension generated automatically when performing an extrusion). The user's biggest task is to position those dimensions.

When inserting dimensions, ask yourself which view should be the main view in which most of the dimensions will be placed. If dimensions are inserted on all views, SolidWorks will attempt to place as many dimensions as it can on any detail or section views first. It will then place the rest of the dimensions on any other available views. Duplicate dimensions will typically not be added automatically. In addition, dimensions are only added to a view if they are parallel with the view. These facts should be considered when inserting dimensions.

More than just dimensions can be imported from the model into a drawing. This includes other annotations, such as weld or surface finish symbols, as well as the display of reference geometry. Therefore, SolidWorks uses the term *model items* to describe these elements. How-To 12–24 takes you through the process of inserting model items into a drawing.

HOW-TO 12-24: Inserting Dimensions

To insert into a drawing any dimensions or other annotations that were created in a part or assembly document, perform the following steps. Following this How-To, the numerous possibilities available when inserting model items are explored.

1. Select Model Items from the Insert menu, or click on the Model Items icon located on the Annotations toolbar. The Model Items PropertyManager will be displayed, a portion of which is shown in Figure 12–60.

Fig. 12–60. Model Items PropertyManager.

2. From the drop-down list in the *Import from* panel, specify the items you wish to import the dimensions for. This can be *Selected features* or *Entire model*. Options for *Selected component* or *Assembly only* will be available if inserting model items for an assembly.

3. From the Dimensions panel, specify what type of items you wish to insert dimensions from.

4. Select the desired views you wish to import dimensions on, or check the *Import items into all views* option, visible in Figure 12–60.

5. Click on OK to insert the model items.

This all probably seems pretty confusing, but once you get past the terminology it isn't too bad. Therefore, let's break down the various panels in the Model Items PropertyManager to give a better understanding of what can be accomplished. Table 12–1 outlines the various items in the Dimensions panel. See also Figure 12–60.

Table 12-1: Model Items Dimensions Panel

Panel Object	Object Type	Description
Marked for drawing	Button	All dimensions in a part that have been marked for import into a drawing (see Table 3-2).
Not marked for drawing	Button	Dimensions that have previously had the Mark For Drawing option turned off in the model file.
Instance/Revolution counts	Button	Dimensions indicating the number of instances in a pattern or the number of revolutions on a helix or spiral.
Hole Wizard profiles	Button	Dimensions associated with a Hole wizard profile.
Hole Wizard locations	Button	Dimensions associated with the locating points of a hole added with the Hole wizard.
Hole callouts	Button	Hole callouts can be used in place of, but not in addition to, Hole wizard profile dimensions.
Select All	Option	This option automatically selects all of the Dimensions panel buttons with the exception of Hole wizard profiles.
Eliminate Duplicates	Option	Takes slightly longer to calculate, but will attempt to prevent duplicate dimensions from appearing on the drawing.

Fig. 12–61. Annotations and Reference Geometry panels.

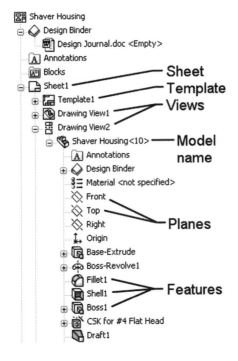

Fig. 12–62. Drawing FeatureManager.

Two other panels in the Model Items Property-Manager are the Annotations and Reference Geometry panels, shown in Figure 12–61. These panels are fairly self-explanatory. For example, the Annotations panel allows for selecting notes or datums (to name two) for importation into the drawing from the model. However, these types of annotations do not typically exist on a model. Usually, they are added to the drawing on an as-needed basis.

Reference geometry is frequently required in a model, but rarely should be displayed in a drawing. Drawings should be kept neat and clean, without superfluous geometry. In other words, for the most part you can safely ignore the Annotations and Reference Geometry panels.

As a special side note to showing reference geometry in a drawing, there is a more practical way of displaying reference geometry in particular views, which involves the drawing FeatureManager. A plane might need to be shown so that a datum can be established, for example. Figure 12–62 shows an example of the drawing FeatureManager.

Objects relative to a drawing are not the only objects shown in the FeatureManager. The entire hierarchy of the part contained in each view is displayed as well. This can be seen in Figure 12–62. This being the case, it becomes very easy to show a plane (for example) in one particular view. This is accomplished by right-clicking on the plane, as you have already learned.

✓ **TIP:** *Use the drawing FeatureManager to show a sketch in a particular view. This makes positioning detail circles or section lines much easier because those circles or lines can be constrained to the existing sketch geometry.*

If a view contains an assembly, all parts and subassemblies contained in the assembly will be accessible from FeatureManager. This is important when inserting model items (such as dimensions) because it is possible to select specific assembly components you want to insert the dimensions for. It should be noted that features or assembly components can also be selected directly from a drawing view when inserting model items. Sometimes this is much simpler, and may be the preferred method, depending on how cluttered a view is.

The only options in the Model Items PropertyManager we have not discussed are shown in Figure 12–63. Unchecking the *Include items from hidden features* option will prevent dimensions from being inserted if they are associated to features hidden by other geometry.

Fig. 12–63. Additional Model Items options.

Checking the *Use dimension placement from sketch* option essentially tells the software to not worry about trying to space dimensions around the view. Instead, it simply plops the dimensions down in the same place they were located in the model file. If you are meticulous when positioning dimensions while creating models, try turning this option on. The Layer setting allows for selecting a layer the imported dimensions will be placed on. There is a section devoted to layers later in the chapter.

SW 2006: Support for dimensioning to foreshortened dimensions has been added to drawings. Figure 12–64 shows an example of a foreshortened radius dimension (functionality available in SolidWorks 2005). The 8-inch foreshortened radius dimension was created by selecting the dimension value and then checking *Foreshortened radius* in PropertyManager. The 6.25-inch dimension correctly displays a zigzag, which is new to SolidWorks 2006.

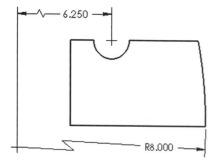

Fig. 12–64. Dimensioning to a foreshortened radius.

A related enhancement is the addition of foreshortened diameter dimensions. Figure 12–65 shows an example of such a dimension (left). The inset (right) shows a setting that affects the appearance of foreshortened diameter dimensions. Examples of how the dimensions can appear are shown in the setting itself, visible in the image. The setting is found in the Arrows section of the Document Properties (Tools > Options).

Fig. 12–65. Foreshortened diameter and related setting.

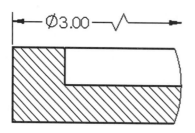

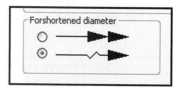

Moving and Deleting Dimensions

SolidWorks' method of adding dimensions to drawing views works well, but it is not perfect. This is the nature of solid modeling programs such as SolidWorks. When dimensions are added to the model, extension lines terminate at places that may not be suitable when those same dimensions are shown in a drawing.

To delete a dimension, simply select the dimension value and press the Delete key. This is no different than deleting anything else in SolidWorks. You can delete more than one dimension at a time by control-selecting multiple dimensions. Window-selecting dimensions is also an option.

To move a dimension, simply drag the dimension to a new location. You may notice what is known as inferencing lines trying to snap the dimension to a particular location. Inferencing lines are light dashed lines that appear when moving dimensions or text. These help to align dimensions or text horizontally or vertically with each other. This inferencing action is controlled by settings in the System Options, shown in Figure 12–66.

Fig. 12–66. Detail item snapping.

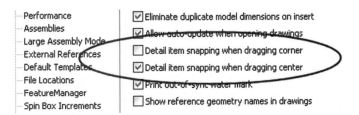

The inferencing action is better known as *detail item snapping*. The options are found in the Drawings section of the System Options window (Tools > Options). The two options are *Detail item snapping when dragging corner* and *Detail item snapping when dragging center*. Which you use, if either, is totally up to you. Experiment with the settings to see what is best for you. By the way, "corner" and "center" refer to the corner or center of a dimension value or a note. When a dimension value or note is dragged (for repositioning), either the corner or center of the object being dragged will align with other stationary dimensions or notes.

✓ **TIP:** *You can temporarily override detail item snapping by holding down the Alt key. This gives you the best of both worlds.*

10-second Topic: Moving/Copying Dimensions Between Views

SolidWorks gives you an easy way of moving or copying dimensions between views. Normally, a move would be the better choice, because you would not want to duplicate dimensions. However, both methods are examined.

The difference between moving and copying dimensions is actually quite minor. Both processes involve dragging the dimension with the left mouse button. To move a dimension, hold down the Shift key while dragging the dimension. To copy a dimension, use the Ctrl key. When dropping the dimension, make it a point to drop it directly on top of the view you are moving it to. Otherwise, the drag-and-drop operation will not work.

The order in which the mouse button and keyboard are depressed is not important. In addition, you cannot move any dimension to any view arbitrarily. If the dimension cannot be accurately displayed in its new view, SolidWorks will not allow it to be moved there.

Annotation Views (SW 2006 Only)

Rather than creating views on a drawing and then inserting the model dimensions, dimensions can be attached to certain views within the model. This functionality in essence "predefines" the drawing views and associates specific dimensions with those views. When it comes time to create the drawing, views known as *annotation views* are inserted into the drawing. These views contain the appropriate dimensions and there is much less work that needs to be done on the drawing.

New models created in SolidWorks 2006 will generate annotation views automatically, as long as the Automatically Place Into Annotation Views option is turned on. This option can be toggled on by right-clicking on the *Annotations* folder in FeatureManager and making sure there is a check in front of said option.

Assuming Automatically Place Into Annotation Views has been enabled (checked), new annotation views will be created as the model is developed. The views will be listed below the *Annotations* folder, an example of which is shown in Figure 12–67. By right-clicking on the *Annotations* folder and checking Show Feature Dimensions, the various annotation views can be double clicked, thereby showing the dimensions associated with that annotation view. In the figure, the Top annotation view is being shown (not to be confused with a Top system view, which would reposition the model).

✓ **TIP:** *Double clicking on the none annotation view (see Figure 12–67) hides all annotations currently associated with views, and shows only those dimensions not associated with any view, assuming Show Feature Dimensions in the* Annotations *folder is checked.*

Sometimes additional user-defined annotation views are required. Such might be the case if certain dimensions should be placed on a detail view or auxiliary view. It would also be the case if annotation views for parts developed in previous releases of the software were desired. How-To 12–25 takes you through the process of manually creating annotation views.

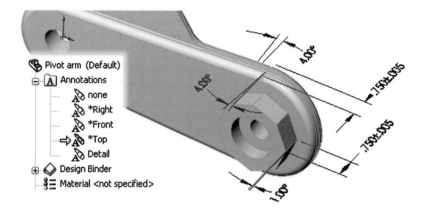

Fig. 12–67. Annotation views.

How-To 12-25: Creating Annotation Views

To manually create annotation views, perform the following steps. Annotation views can only be created in part and assembly files.

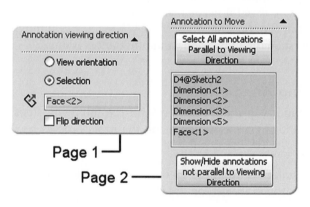

Fig. 12–68. Annotation View PropertyManager.

1. Right-click on Annotations in FeatureManager and select Insert Annotation View.

2. In the Annotation View PropertyManager, shown in Figure 12–68 and labeled Page 1, either select an existing view from which to define the new annotation view or click on Selection (option used in figure) and select a plane or face to define the annotation view.

3. Click on Next.

4. Dimensions parallel to the face, plane, or view selected in step 2 must now be associated with the new annotation view being developed. To do this, select features from the fly-out FeatureManager, or select dimensions and/or features from the work area, or use the buttons in the Annotation View PropertyManager (see Page 2 in Figure 12–68).

5. Click on OK to create the annotation view.

*Fig. 12–69.
Inserting
annotation views
into a drawing.*

Once the annotation views have been established in the model, it will eventually become desirable to utilize those views in a drawing. While creating a new drawing from an existing model or inserting model views, the opportunity for employing the annotation views will present itself. The Orientation panel of the Model View PropertyManager will contain an option titled *Annotation view,* visible in Figure 12–69. Aside from selecting the Annotation view option shown in the figure, there is nothing else different from inserting annotation views from what you have already learned.

✓ **TIP:** *Annotation views support 3D annotations based on ASME Y14.41-2003. This includes annotations such as datums and geometric tolerances, should you decide to include these in the model's annotation views.*

Hiding Annotations

After importing dimensions on various views, you will probably find that some of the dimensions are not needed. Deleting a dimension is one option, but that removes it completely. Hiding dimensions may be a better choice, in that they can be accessed if needed, such as when a dimensional change must be made from the drawing. How-To 12–26 takes you through the process of hiding and showing dimensions and other annotations.

How-To 12-26: Hiding and Showing Annotations

To hide and show annotations in a drawing, perform the following steps. Previously hidden annotations can be shown using the same process used to hide them.

1. Select Hide/Show Annotations from the View menu, or click on the Hide/Show Annotations icon located on the Drawing toolbar.

2. Select the annotations to be hidden. They will turn gray.

3. Select any previously hidden annotations to show them.

4. Select Hide/Show Annotations a second time to exit the command, or simply press the Escape key.

If it becomes necessary to hide numerous annotations for a particular reason, the best choice would be to place the annotations on a separate layer. That way, the layer can easily be turned off. Layers are discussed in material to follow.

✓ **TIP:** *Even cosmetic threads can be hidden using the Hide/Show Annotations command.*

Reference Dimensions

Reference dimensions are nothing more than driven dimensions. They will still update if geometry is altered (just like dimensions inserted via Insert > Model Items), but you cannot use them to modify geometry. Reference dimensions do not drive model geometry. Rather, they are driven by the model geometry.

You can add reference dimensions to the drawing the same way you add dimensions to a sketch. The mechanics involved are exactly the same. Reference dimensions appear differently than regular dimensions. By default, they appear gray and are enclosed in parentheses, which differentiates them from dimensions inserted with the Model Items command.

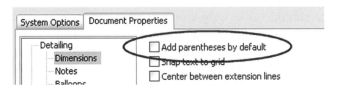

Fig. 12–70. Add parentheses by default *option.*

Reference dimensions can have their color changed through the Line Format toolbar, or by placing the dimensions on a particular layer (discussed shortly). To globally turn off parentheses when adding reference dimensions, uncheck the *Add parentheses by default* option found in the Dimensions section of the Document Properties (Tools > Options). This option is shown in Figure 12–70. To turn off parentheses on dimensions already added to the drawing, right-click on a dimension and uncheck the *Display Parentheses* option found in the Display Options menu.

Globally turning off the appearance of parentheses on reference dimensions in the document's properties will affect only the current document. You may want to incorporate this change into your drawing templates.

Extension Lines

Another important aspect of cleaning up design drawings is modifying where the dimensions' extension lines are terminating. Specifically, dimensions inserted as model items often require having their extension lines repositioned. Dimensions added to a sketch in a part may look fine on the sketch, but those same dimensions have to contend with the rest of the geometry on the part when a view of that part is shown on a drawing. Thankfully, modifying the extension lines is an easy task, albeit tedious.

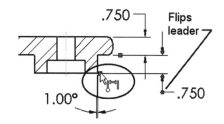

Fig. 12–71. Modifying where an extension line terminates.

To alter the terminating location of a dimension's extension lines, you must first select the dimension. You will then see small green "handles" attached to the extension lines. Place the cursor over a green handle and drag it to the desired location. This is illustrated in Figure 12–71. The circled area shows the cursor as it is dragging the extension line handle to a new location.

The green handles on the tips of the arrows are for flipping the arrows to either side of the extension lines. There is also a handle on dimensions that contain a bent leader, such as the two .750-inch dimensions shown in Figure 12–71. The lower of these two dimensions has been selected, and therefore the small handle at the leader's "elbow" can be seen. Clicking on this handle will flip the dimension value and bent leader line from one side to the other.

Breaking Extension Lines

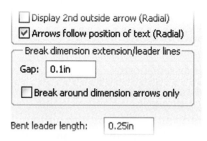

Fig. 12–72. Uncheck Break around dimension arrows only.

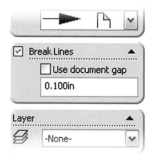

Fig. 12–73. Break Lines panel.

Current ANSI standards specify that dimension extension lines should break only around arrows. Default settings reflect this specification, so if you choose to break extension lines around dimension line leaders (for example) the default settings must be altered first. To accomplish this task, access Document Properties (Tools > Options) and select the Dimensions section. Locate the option *Break around dimension arrows only*, shown in Figure 12–72, and uncheck it.

Also shown in Figure 12–72 is the gap setting used for the break. This value defaults to .060 inch, but can be increased to create a more noticeable gap. To complete the process, it is necessary to specify exactly which dimensions should have their extension lines broken. Incidentally, dimension leaders can be broken as well. Selecting a dimension will display its properties in PropertyManager. Note the Break Lines panel near the bottom of PropertyManager, shown in Figure 12–73.

By turning on the Break Lines option for a particular dimension, that dimension will break around whatever other dimension it intersects. Note that this option will *not* work by itself, and that the *Break around dimension arrows only* option mentioned previously must be turned off first. An example of a dimension with a broken extension line is shown in Figure 12–74.

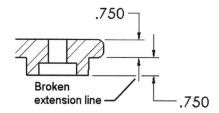

Fig. 12–74. Broken extension line.

Layers

Nearly anyone who has ever used a 2D CAD program understands what layers are. For those not in the know, layers are a way of keeping track of objects in a drawing. The biggest benefit of using layers is seen in the 2D drawing world, in which objects can be separated into various layers. For example, dimensions, notes, balloons, revision blocks, and so on can all be placed on their own individual layers.

Layers are not as important in SolidWorks as they are in nonsolid 2D programs, but they can still be useful. Layers are only available in drawings, not parts or assemblies. It may be convenient to place certain objects on a particular layer, because this makes turning the display of those objects on and off very easy.

Each layer has its own individual settings for color, line type, line weight, and whether or not the layer is visible. Objects can be moved from one layer to another. Layers can have different names and descriptions assigned by the user. Layer properties can be changed at any time, and layers can be deleted if they are no longer necessary. They can also be useful when exporting drawing files to other CAD systems.

Using layers is certainly not a requirement. If you find no advantage to using layers, do not feel obligated to make use of this functionality. If you decide to use layers, you should first know how to create some new layers and set up their properties. How-To 12–27 takes you through the process of creating a new layer.

How-To 12-27: Creating Layers

To create a new layer, perform the following steps.

1. Click on the Layer Properties icon, found on both the Layer and Line Format toolbars.

2. In the Layers window that appears (shown in Figure 12–75), click on the New button.

3. Type in a name for the new layer and press the Enter key.

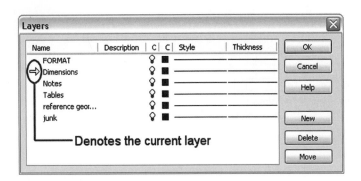

Fig. 12–75. Layers window.

4. Perform a slow double click in the area under the Description heading and type in a description if desired.

5. Click on the color square under the Color heading and select a color if desired. Black is the default.

6. Click on the line under the Style heading and select the desired line style.

7. Click on the line under the Thickness heading and select the desired line thickness.

8. Repeat step 3 for as many layers as you would like to create. Steps 4 through 7 are optional, as you can always use the default values for color, line style, and thickness.

9. Click on OK when finished.

The Layers window (shown in figure 12–75) in the example has a few layers that have been created. The *Dimensions* layer happens to be the current layer. When dimensions or any other annotation is added, or when geometry is added to the drawing, it will automatically be placed on the current layer. Therefore, it is important to know which layer is current at any point in time.

There are two methods in which a layer can be made to be the current layer. With the Layers window open, simply click to the left of the name of any layer. When a small yellow arrow appears to the left of the layer name, it is current.

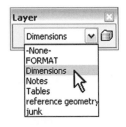

Fig. 12–76. Making a layer current.

The Layer toolbar contains a drop-down list (shown in Figure 12–76) that can be used to set the current layer. This is a quick and easy way to make a layer current without opening the Layers window, and is the preferred method. Note in this example the layer named *-None-*. This is the selection that would be used when adding geometry or annotations that should not be on any layer.

Obviously, the Delete button in the Layers window deletes a layer. Select a layer name from the Name column and press the Delete button. You will be asked if you are sure you want to delete the layer, in which case you can then make the applicable reply, Yes or No.

When a layer is deleted, the objects on that layer are not deleted. Any annotations or geometry on the layer will be transferred to layer *-None-*, which really is not a layer at all. Rather, it is a designation for any objects that do not belong to a particular layer.

Clicking directly on a light bulb in the Layers window will toggle the associated layer on or off. If the light bulb is yellow, the layer is on. If the light bulb is dimmed, the layer is off. Turning a layer off takes effect immediately. It is not necessary to click on the OK button to have these settings take effect.

✗ **WARNING:** *SolidWorks will allow you to turn off the current layer and will not issue a warning message. Make sure you do not turn off the current layer by accident, or objects added to your drawing will not be visible.*

Placing objects on the wrong layer is something that has happened to nearly every CAD operator on the planet; that is, assuming the CAD software uses layers. Sometimes objects need to be moved from one layer to another, for various reasons. Whatever the reason, you will invariably find the need to move objects from one layer to another. There are a few ways to accomplish this. The method used in How-To 12–28 makes use of the Move button in the Layers window.

HOW-TO 12-28: Moving Objects Between Layers

There is a button in the Layer window named Move. This button allows for moving objects onto a specific layer (if those objects are not currently on a layer), or from one layer to another. The implementation of this function is very simple, and is outlined in the following steps.

1. Click on the Layer Properties icon.

2. Make sure the current layer is the layer objects are to be moved to.

3. Select items from the drawing, such as annotations or sketch geometry, to be moved to the current layer. You may have to reposition the Layers window to accomplish this.

4. Click on the Move button.

5. Click on OK when finished.

✓ **TIP:** *For a quick and easy method of moving objects to a particular layer, select the desired objects and then select the layer to move those objects to from the Layers toolbar drop-down list..*

Annotations

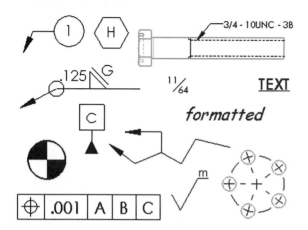

Fig. 12–77. Examples of annotations.

There are many items you might need to add to a design drawing besides dimensions. Notes, geometric tolerances, center marks, surface finish symbols, and other annotations all go into the making of a complete detailed drawing. A smattering of the annotations available can be seen in Figure 12–77. Most of these annotation types are discussed in this section. Any remaining annotations are covered in the next chapter, the main topic of which is assemblies. This is because certain annotations are better suited to assemblies, such as balloons.

It is not the intention in this book to teach what goes into making a good surface finish symbol or how to create a meaningful geometric tolerance (and other) symbols. That is up to you to learn if you or your employer deems it necessary. This section deals with employing the various annotation commands in SolidWorks. The general guidelines for creating annotations are presented, but it will be up to you to type in the correct parameters. More detailed descriptions are provided for certain annotations or annotation parameters if necessary. Some samples are provided as guidance.

You should be familiar with how to use the individual types of annotation symbols. Otherwise, they are meaningless. For example, if you have never seen a surface finish symbol, entering the proper parameters to create a surface finish symbol that makes sense will be difficult. The technical procedure for actually creating one on a SolidWorks drawing, however, is simple.

General Annotating Guidelines

How are annotations created? For the most part, the process consists of clicking on the applicable annotation icon, plugging in some parameters, and clicking on the drawing where the annotation should be placed. That is basically it, in a nutshell, and most annotations are added in precisely that manner.

There are a few details common to multiple annotations, which are consistent throughout the software. For example, most annotations can be repositioned by dragging them. If a leader is attached to an annotation, and needs to point to a different object, its arrow can be dragged. Leader arrows are intelligent and will display arrows if attached to an edge, or small dots if attached to a face of a model in a view. Right-clicking over the end of a leader displays the menu shown in Figure 12–78, and allows for changing what is displayed at the leader's end.

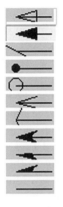

Fig. 12–78. Choosing what is displayed at the end of a leader.

When adding any annotation (e.g., notes, weld symbols, and so on) capable of displaying a leader, clicking on an object in a view (such as an edge) will cause a leader to automatically be added. A second click on the drawing will position the annotation itself. Annotations move with their associated view, remaining attached to it. Leader arrows stay attached to view geometry. This keeps leader arrows from becoming disassociated with the model if, for example, the model undergoes size alterations.

Editing an annotation is a matter of a single click in many cases, or a double click for certain annotation types. This is not anything that requires memorization. If you were to perform a single click on most annotations, the PropertyManager for that annotation will appear. You will be presented with the various parameters for that annotation, and the necessary alterations can be made.

If a single click does not give you the editing options you are looking for in PropertyManager, double click on the annotation instead. A perfect case in point would be a geometric tolerance. A single click on a geometric tolerance brings up PropertyManager, with basic formatting options, such as font style and layer choice. Double clicking on the same annotation opens the Geometric Tolerance window, which allows for editing the geometric tolerance itself.

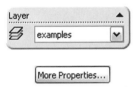

Fig. 12–79. Layer panel.

The Layer panel (shown in Figure 12–79) exists for all annotations, and will be visible in PropertyManager. Use the Layer panel to move an annotation to a different layer by selecting the desired layer from the drop-down list. If the More Properties button is available (see Figure 12–79), click on it to access additional panels titled Text Format, Leader, and Border, described in material to follow.

Certain annotation types have their little quirks, which is what the following section is meant to explore. Because notes are by far the most common annotations, they are the first annotation type to be examined.

Notes

As soon as you click on a drawing to begin entering text for a note (assuming you have already entered the Note command), the Formatting toolbar (shown in Figure 12–80) will appear. Any note can have a variety of formatting applied to it, such as color, font style, and so on. To format only a portion of text within a note, select the desired text first, and then apply the desired formatting to that text via the Formatting toolbar.

Fig. 12–80. Line Formatting toolbar.

SW 2006: Notes are defined by a bounding box that can be resized. Lines of text will wrap to match the width of the bounding box.

The Stack icon, circled in Figure 12–80, opens the Stack Note window (shown in Figure 12–81). Use this functionality to create fractional text, an example of which is shown on the right-hand side of Figure 12-81. The Alignment options (see Figure 12–81) refer to alignment with the line of text the stacked text is part of. Type in the text you wish to stack in the areas labeled Upper and Lower in the Stack Note window. To edit existing stacked text, select the stacked text (you must be editing the note to do this), and then click on the Stack icon. This will once again present you with the Stack Note window.

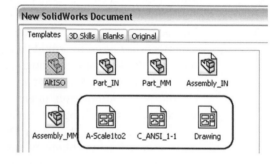

Fig. 12–81. Stack Note window.

Text Format Panel

Fig. 12–82. Text Format panel.

Aside from the Formatting toolbar, there are various panels available in the PropertyManager when adding notes. The first such panel is the Favorites panel. There is a section dedicated to Favorites later in this chapter, so we will not discuss it here. The Text Format panel, shown in Figure 12–82, can be used to justify or angle text. These options are self-explanatory.

SW 2006: An "anchor" icon has been added to the Text Format panel. It is used to lock a note in position so that it cannot be accidentally moved.

To change the font of the note being added to the drawing to something other than what has been established via the document template, uncheck the *Use document font* option (shown in Figure 12–82) prior to clicking on the drawing sheet. Otherwise, the font used for the note will be that specified in the Annotations Font section of the Document Properties (Tools > Options). If you have already begun typing in text for your note, using the Formatting toolbar it is possible to change the font used by specific portions of the note.

The row of icons in the middle of the Text Format panel (circled in Figure 12–82) require explanation. They are, in order from left to right: Insert Hyperlink, Link to Property, and Add Symbol. If you have ever surfed the Web, you should be familiar with what a hyperlink is. Clicking on the Insert Hyperlink button allows for adding a hyperlink to a note. The link can be to a file on your computer or company network, or it can be a web address, also known as a uniform resource locator (URL).

Drag from a corner ─<u>This is a hyperlink</u>

Fig. 12–83. A hypertext link and two of its drag points.

Hypertext links are commonly shown as underlined blue text, such as that shown in Figure 12–83. Solid-Works will make the text blue, but you will have to underline it yourself. Use the Formatting toolbar while editing the text to display the text as underlined. In this way, the hyperlinked text will appear as it typically does in most web browsers.

Moving a hyperlink is tricky, because if you click on it you will open the link. Try placing the cursor at one corner of the hyperlink, just out of range so that you cannot see the hand. You can then move the link by dragging it. You can also right-click on the link and access its properties for editing purposes.

Clicking on the Add Symbol button will open the Symbols window, shown in Figure 12–84. Many different symbols are available, so you should never find yourself lacking for a particular symbol.

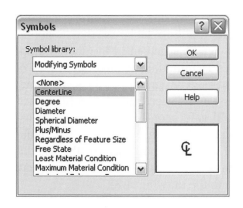

Fig. 12–84. Symbols window.

The Link To Property button (middle icon circled in Figure 12–82) opens the door to another topic. All SolidWorks documents contain file properties, as do most other files on your computer. SolidWorks files can contain information a user can specify. This information can be literally anything, such as a model's weight or size characteristics, cost, vendor, or any other information deemed important.

A file's properties can be linked to a note. If the properties change, the note updates accordingly. Notes on a sheet format, such as sheet size or scale, can be linked to system-generated file properties SolidWorks provides. These notes are intelligent and will update automatically to reflect accurate data representing the drawing. Creating user-defined custom properties within a file is explored in detail in Chapter 17. For now, this section will show you how to link to the file properties generated by SolidWorks.

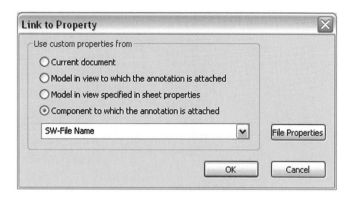

Clicking on the Link To Property button opens the Link To Property window, shown in Figure 12–85. When linking to a property, there are four options that control what document the properties are linked to. Table 12–2 outlines what the four choices in the Link to Property window represent.

Fig. 12–85. Link To Property window.

Table 12-2: Custom Property Link Choices

Link Choice	What It Links To
Current document	Properties in the drawing itself, rather than the model shown in any drawing view. Also available when creating linked notes in parts or assemblies.
Model in view to which the annotation is attached	Properties of the model file shown in the view the note is associated to (see "10-second Topic: Locking Focus" following).
Model in view specified in sheet properties	Properties of the model in the view selected from the list of views presented in a sheet's properties. This option is important if there are views of more than one model file on the same sheet.
Component to which the annotation is attached	Properties of the component in the drawing view to which the note leader is attached. This option only pertains to views of assemblies and notes that utilize a leader pointing to a component in said assembly.

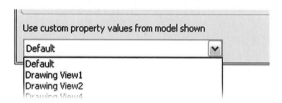

Fig. 12–86. Specifying a view from which properties will be linked.

If *Model in view specified in sheet properties* is selected, the link looks to the model in whatever view is specified in the sheet's properties. This is important if there are views of different parts or assemblies in the same drawing. Accessing the sheet's properties (described earlier) will gain access to the drop-down list from which a view can be selected (see Figure 12–86). If a view is not selected, the *Default* setting links to the model in the first view inserted into the drawing sheet.

10-second Topic: Locking Focus

When adding sketch geometry to a view for whatever reason, that geometry is automatically associated with the view. This is important for the reason that if the view is moved the geometry should move with it. This is significant with respect to annotations as well. For example, a note may be associated with a particular view. When the view is moved, you would want the note to move with it.

Sometimes notes or other annotations are not associated with a particular view as you would like them to be. On occasions such as this, focus can be locked on a particular view by right-clicking on a view and selecting *Lock view focus.* When focus is locked on a particular view, any geometry or annotations created will be associated with that view. Focus can be unlocked in the same way. Focus can also be locked on the sheet, which would be beneficial if geometry or annotations should remain stationary whether or not any of the views on the sheet are moved.

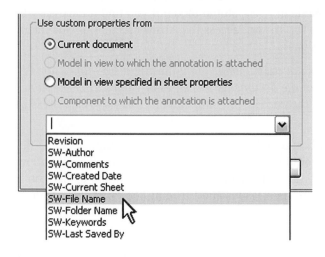

Fig. 12–87. Linking to a property.

SolidWorks includes a number of custom properties with every document. The property highlighted in Figure 12–87 happens to be *SW-File Name.* The *SW-* prefix signifies that this is a custom property created by SolidWorks for the convenience of the user. *File Name* obviously refers to the name of the file. In other words, if this link is used the name of the file will appear in the note. Considering that the *Current document* option is selected, the file name of the document (in this case, a drawing) is what will appear in the note. If the file name is changed, the note automatically updates, which is the advantage of linking to file properties.

Leader Panel

Fig. 12–88. Leader panel.

The Leader panel, shown in Figure 12–88, contains a lot of icons, but they make sense once you understand their layout. The first row of icons determines if there will be a leader attached to the annotation. Using one of the first two icons overrides the intelligent behavior normally inherent in annotations with leaders. A leader can be "turned on," for example, if one does not currently exist for the annotation.

✓ **TIP:** *Whenever a small yellow star is shown on an icon, it means a "smart" (or "automatic") mode is available. Using an automatic mode is preferable unless a particular setting must be overridden for some reason.*

The second row of icons determines if a straight or bent leader will be used for the note. Obviously, if there is no leader this setting will not matter. The third row of icons determines what side of the leader the text is on. Leave this set to Leader Nearest (automatic mode) unless it becomes necessary to specify one side or the other.

The type of arrow present on the leader can be specified. Here, as is typically the case, you should leave this setting in automatic mode. In other words, select from the drop-down list the arrow with the star attached in order to let SolidWorks automatically set the arrow according to what the leader is pointing to.

Arrow "smart" mode is only as smart as what the document defaults tell it to use. In other words, the arrows used in automatic mode are determined in the document's properties. If these default settings are changed, it is possible to change smart mode to, well, stupid mode! If you find it necessary to alter the default settings for arrows, access the Arrows section of Document Properties (Tools > Options).

✗   **WARNING:** *It is not recommended that the arrow size dimensions in the Arrows section of the documents properties be changed. These values are determined by the current drawing standard in use (i.e., ANSI or ISO).*

The *Apply to all* option pertains only to notes with multiple leaders. Because different leaders could have different arrows, *Apply to all* would allow for making them all the same. How does one obtain multiple leaders on a note? Select a note containing a leader and Ctrl-drag the tip of the leader to an additional attachment point.

Border Panel

The Border panel, shown in Figure 12–89, allows for placing a border around a note. The border style can take many shapes, including circles, diamonds, rectangles, five-sided flags, and others. The border size setting should typically be left at Tight Fit, so that it will automatically adjust to whatever note is present inside the border. Otherwise, the border size will be relegated to accepting a certain number of characters, and will stay at that size even if the number of characters in the note changes.

Adding lines of text containing spaces allows for increasing the size of the box to give the text more room. The example on the far right in Figure 12–89 uses this technique. You can experiment with font size

Fig. 12–89. Border panel and bordered text examples.

for the spaces to get the desired space above or below the text. Adding spaces to the beginning or end of the note will add space to either side.

Aligning Notes

If a drawing contains a series of notes that should be aligned in some way, SolidWorks provides you with plenty of options. It is suggested that the Align toolbar be used for this, as it is more convenient than accessing the Align menu via the right mouse button or Tools menu. After selecting the text to be aligned, click on the desired alignment condition, such as Align Left or Align Top.

Additional alignment tools, also present on the Align toolbar, are for spacing or grouping text. Text can be spaced tightly across or up and down. Of course, if there is text that belongs to the same note it can be justified as desired and will be evenly spaced automatically, so alignment and spacing commands only apply to separate lines ("strings") of text. All of the commands on the Align toolbar should be self-explanatory.

To show an example, the text in Figure 12–90 was left aligned and was also grouped. The fact that it has been grouped makes positioning all of the text equidistant from the left vertical line in the title block.

SW 2006: A spell-check utility has been added to drawings. SolidWorks spell-check functionality is extremely similar to that in Microsoft Office. It is also very user friendly, so examining the minutiae of the spelling checker is not necessary. To use the spelling checker, select Spelling from the Tools menu, click on the Spelling icon on the Tools toolbar, or simply press the F7 hotkey.

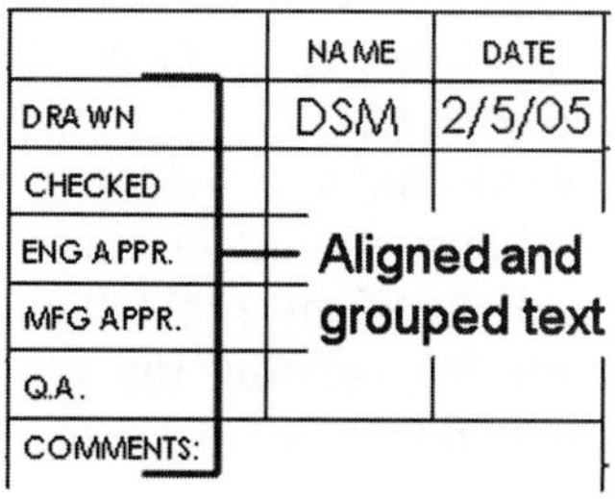

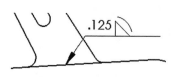

Fig. 12–90. Aligned and grouped text.

NOTE: *SolidWorks 2006 Spelling command only checks notes or dimensions on the drawing sheet. To check notes on the sheet format, it is necessary to be in Edit Sheet Format mode.*

Weld Symbols, Caterpillars, and End Treatments

A caterpillar is an annotation used to show on a drawing where a weld bead has been applied. Often, it is enough to call out the weld with a weld symbol, an example of which is shown in Figure 12–91. Clicking on the Weld Symbol icon opens a window that allows for entering the various parameters for the weld symbol. Although the weld symbol properties window is not shown here, the various parameters should be self-explanatory for individuals familiar with welding.

Fig. 12–91. Weld symbol.

An option that controls the size of weld symbols is shown in Figure 12–92. When *Fixed size weld symbols* is left unchecked, which is the default setting, weld symbols change size according to the size of the weld symbol font. Otherwise, the weld symbol stays the same size even if the weld symbol font size is altered.

The *Fixed size weld symbols* option is found in the Detailing section of Document Properties (Tools > Options menu). To change the weld symbol font, access the Annotations Font section of the Document Properties and click on the Weld Symbol annotation type presented in the list.

Creating a caterpillar is done through the Caterpillar PropertyManager, a portion of which is shown in Figure 12–93. Both full-length and intermittent weld bead caterpillars can be created. The various options associated with caterpillar creation are self-explanatory.

Once an edge is selected on which the caterpillar should be applied, an additional list box is provided that allows for selecting edges that can be used to trim the caterpillar. An example of a caterpillar is shown in Figure 12–94. Note the two edges selected for trimming purposes. This terminates the caterpillar quite nicely at either end of the edge where the caterpillar was applied.

An end treatment shows the end of a weld bead in a weldment. The End Treatment PropertyManager is shown in part in Figure 12–95, along with an example of an end treatment annotation. Once again, the various options available for defining the end treatment are self-explanatory, so we will not need to go into detailed descriptions.

It should be noted a final time that many people will choose to keep a drawing simple and call out welds with a simple weld symbol, rather than creating caterpillars or end treatments. However, for those who have an eye for detail, or would like the manufacturing department to have a very detailed drawing, the functionality is there for you to take advantage of. Weld beads can even be modeled in SolidWorks, which is discussed in Chapter 15.

Fig. 12–92. Fixed size weld symbols *option.*

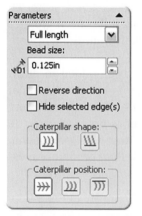

Fig. 12–93. Caterpillar PropertyManager.

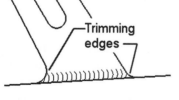

Fig. 12–94. A caterpillar.

Trimming edges

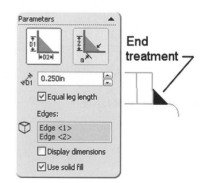

Fig. 12–95. End Treatment PropertyManager and example.

End treatment

Center Marks and Centerlines

Fig. 12–96. Center Mark PropertyManager.

Adding center marks is one of the easiest annotations to add to a drawing. There is very little involved, at least in its simplest form. However, center marks actually have a fair number of options that allow for creating center marks in widely varying appearances.

A portion of the Center Mark PropertyManager is shown in Figure 12–96. Adding a single center mark is accomplished by clicking on an arc or circular edge on a drawing view. However, note the three icons at the top of the Options panel visible in Figure 12–96. These are, from left to right, the Single, Linear, and Circular Center mark options.

Center marks and patterns get along very nicely together in SolidWorks. For example, if the Linear Center Mark option is used a linear pattern of holes can have center marks applied simultaneously. Figure 12–97 shows a pattern of countersunk holes with an option labeled *Connecting lines* turned on, which becomes available when the Linear Center Mark option is used.

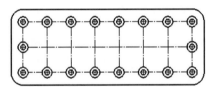

Fig. 12–97. Linear center marks.

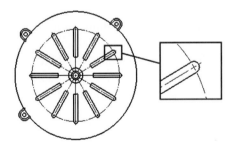

Fig. 12–98. Circular center marks.

When using either the Linear or Circular Center Mark options, and an instance in a pattern has been selected, a propagate symbol will appear next to the selected pattern instance. Clicking on the propagate symbol is what causes the entire pattern to have center marks applied to it. Figure 12–98 shows an example of employing the Circular Center Mark option. In this example, options titled *Circular lines* and *Radial lines* have been enabled. The enabled options cause the circular centerline and radial lines to be added to the view in addition to the center marks at the centers of the arcs. A portion of the model has been enlarged in Figure 12–98 to better show the arc edges at the end of each slot.

Centerlines can be added in a sketch, as you learned in Chapter 2, and they can also be added to drawing views of parts that have cylindrical or tubular features. Do not confuse the Centerline sketch tool with the Centerline annotation, as they are two different commands for accomplishing different tasks. The Centerline sketch tool is found on the Sketch toolbar, whereas the Centerline annotation command is found on the Annotations toolbar.

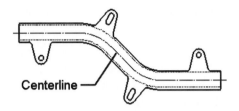

Fig. 12–99. A centerline annotation.

There are no options associated with adding a centerline annotation. It is a command of convenience more than anything else. When applying a centerline, it is only necessary to select a single cylindrical section on a part with a tubular construction, and a centerline will be applied to the entire part. This is how the centerline shown in Figure 12–99 was applied. It is also possible to select two edges, in which case a centerline will be applied between the edges.

Automatic Center Marks and Centerlines

Center marks and centerlines can automatically be added during the view creation process in drawings. The automated process often adds more center marks or centerlines than are desirable, as every arc or circle in a view will have center marks applied. This can actually be something of an annoyance.

Fig. 12–100. Automatic insertion options for center marks and centerlines.

To enable (or disable, for that matter) the automatic addition of center marks and centerlines, access the Detailing section of Document Properties (Tools > Options). There, you will see options for turning on the automatic creation of center marks and centerlines, as shown in Figure 12–100. Because these options are part of Document Properties, they can be incorporated into a template if desired.

You will notice there are also options for the automatic insertion of dimensions and balloons. These options are not for everybody, but can be used to automate the creation of drawings to a large degree. Not all models can benefit from these options, as they are better suited to parts of a more simple nature.

Cosmetic Threads

Cosmetic threads are treated differently than any other annotation. When added, cosmetic threads attach themselves to the associated feature in FeatureManager. Figure 12–101 illustrates this situation, and shows a bolt with a cosmetic thread. Cosmetic threads are usually added to part documents, and then imported into drawings via the Model Items command discussed earlier in this chapter.

Fig. 12–101. Cosmetic threads.

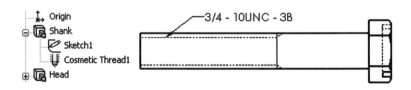

Many designers opt to place the cosmetic threads in the part file, as this relieves the drafter of the burden of having to call out the thread specifications. This also gives the designer control over what the thread specifications should be and is an ideal way of communicating the thread data to the drafter. In fact, the drafter does not even have to be concerned with the thread specifications.

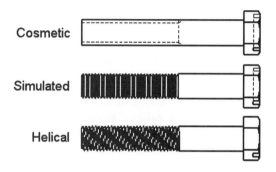

Cosmetic

Simulated

Helical

Fig. 12–102. Thread display in a drawing.

Cosmetic threads work a little differently than most other annotations. They are more substantial, in a manner of speaking, than most other annotations that would typically exist in a strictly 2D drawing. If a cosmetic thread is added to a drawing, it will traverse to the part and will actually appear in the part's FeatureManager.

When it is necessary to show threads in a drawing, it can be very cumbersome to show detailed threads (true helical threads). Even showing the illusion of threads with a linearly patterned V groove makes for too much ink on a printed drawing. It is likely the threads will simply appear as black areas (see Figure 12–102). For this reason alone, cosmetic threads are invaluable. How-To 12–29 takes you through the process of adding cosmetic threads.

How-To 12-29: Inserting Cosmetic Threads

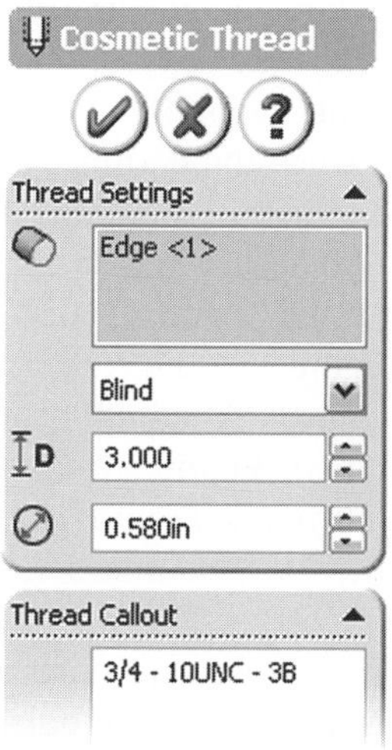

To add cosmetic threads to a SolidWorks part, assembly, or drawing, perform the following steps.

1. Select Insert > Annotations > Cosmetic Thread, or click on the Cosmetic Thread icon found on the Annotations toolbar.

2. Select a circular edge that defines where the cosmetic threads should start. In the case of a drawing, a circular edge may appear as a line, such as when viewing a cylinder from the side.

3. In the Thread Settings panel of the Cosmetic Thread PropertyManager, shown in Figure 12–103, specify the length of the thread or end condition (which will thereby determine the thread length).

4. Specify the Minor Diameter or Major Diameter, depending on whether the feature is a cylinder or hole, respectively.

5. Specify a callout, if desired.

6. Click on OK to create the cosmetic threads.

Fig. 12–103. Cosmetic Thread PropertyManager.

If a callout has been added per step 5, the callout will not appear in a part, but will appear in the drawing of the part. Cosmetic threads in a part are displayed with a bitmap texture that gives the illusion of threads. This is shown in Figure 12–104. There is currently no way to edit the texture associated with cosmetic threads, but the cosmetic thread as a whole can be hidden. Right-clicking on the cosmetic thread in FeatureManager will allow for hiding it.

Fig. 12–104. Cosmetic thread texture.

> ✓ **TIP:** *Hiding all cosmetic threads, or many other annotation types, can be accomplished by right-clicking on the* Annotations *folder at the top of FeatureManager and selecting Details. Use the display filter to check only what should be shown.*

Editing the text of a cosmetic thread callout in a drawing can be accomplished by double clicking on the callout. Editing the definition of a cosmetic thread is only available from within the part, not the drawing. This is true even if the cosmetic thread was added to the drawing initially.

Deleting a cosmetic thread must be done from within the part, although a cosmetic thread callout can be deleted from the drawing. Deleting the callout is not recommended because it is not possible to get the callout back if you change your mind. It would be much better to use Hide/Show Annotations, discussed earlier, which is much more forgiving.

Datum Feature Symbols and Targets

Fig. 12–105. Datum Feature PropertyManager and symbol.

Datum feature symbols are easily added, consisting of nothing more than clicking on the Datum Feature Symbol icon, entering the appropriate parameters, and clicking on the geometry where the symbol should be placed. Once the datum feature symbol has been added, it can be moved around, just like any other annotation. The Datum Feature PropertyManager and an example of a datum feature are shown in Figure 12–105.

Fig. 12–106. Two types of datum feature symbols.

Some companies still use the outdated style of datum feature symbols, shown in Figure 12–106. The symbol on the left is the current standard. The symbol on the right shows the 1982 standard for datum feature symbols. Which version of the datum feature symbol you use will depend on the standards dictated by your company. To use the old-style datum symbols, check *Display datums per 1982* in the Detailing section of Document Properties (Tools > Options).

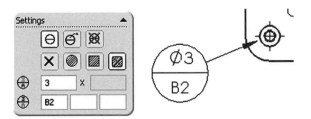

Fig. 12–107. Datum Target PropertyManager and example.

Fig. 12–108. Dowel pin symbols.

Create datum targets in much the same way as datum symbols. That is, enter the appropriate parameters and choose where to place the datum target. Figure 12–107 shows a portion of the Datum Target PropertyManager and an example of a datum target annotation.

Dowel Pin Symbols

Dowel pin symbols should win the award for "fewest number of options." After adding this annotation, the dowel pin symbol can be flipped, and that is about it. Of course, the usual panel for choosing what layer to put the symbol on will also be available. Figure 12–108 shows the option for flipping the symbol, along with examples. One of the symbols in the example has been flipped.

Geometric Tolerancing

The art of using geometric tolerancing could probably be a book in itself. As a matter of fact, it is, and is better known as ASME Y14.5M-1994, a standard established by the American Society of Mechanical Engineers. It is not the purpose of *Inside SolidWorks* to teach you how to use geometric tolerancing, but to show you how to implement geometric tolerancing through the SolidWorks interface.

Adding geometric tolerances works in much the same way as adding notes. Picking on an object first, such as a model edge, will automatically add a leader, whereupon a second pick is required to position the geometric tolerance. Picking in a blank area simply places the geometric tolerance on the drawing, with no leader.

Upon entering the geometric tolerance command, the PropertyManager will display the usual formatting tools for leaders, font style, and layer choices. In addition, the geometry tolerance Properties window, shown in Figure 12–109, will appear. In this window is where the various parameters for the geometric tolerance are set.

As long as you are familiar with the correct usage of creating a geometric tolerance, it is quite difficult to mess up this annotation type because there is a preview in the properties window. The preview will display exactly what will be displayed on the drawing. Just make it a point to click on the drawing to position the geometric tolerance after taking the time to enter all of the appropriate parameters.

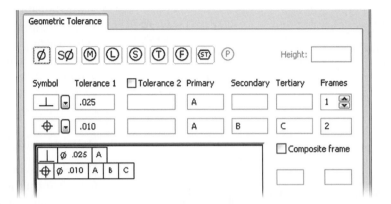

Fig. 12–109. Geometric tolerance Properties window.

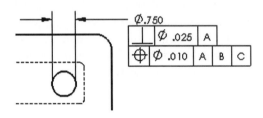

Fig. 12–110. A geometric tolerance attached to a dimension.

A geometric tolerance can be attached to a dimension. This is done by dragging the geometric tolerance to a dimension and dropping it on the dimension value. If the geometric tolerance has a leader, the leader will automatically be turned off. When the dimension now containing the geometric tolerance is moved, the geometric tolerance symbol will move with it. Figure 12–110 shows the result of performing this action.

✓ **TIP:** *If a geometric tolerance is accidentally attached to a dimension, copy (or cut) and paste it to a new location to effectively "detach" the tolerance from the dimension.*

Hole Callouts

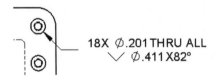

Fig. 12–111. Hole callout.

When a hole is created with the Hole wizard, information such as depth, diameter, counterbore dimensions, and any other data associated with the hole is stored in the part file. This information can be quite extensive. Selecting the outermost edge of the hole will serve to add a hole callout describing that hole (assuming you have already entered the Hole Callout command). Figure 12–111 shows an example of a hole callout.

Hole callouts are associative, but they do not drive the geometry. In other words, they will update if the original hole feature is modified. It is interesting to note that the callout shown in Figure 12–111 contains the text *18X*, which indicates the number of holes on the model. This occurs because the holes are part of a pattern.

Fig. 12–112. Hole callout
PropertyManager
Tolerance/Precision panel.

Fig. 12–113. Dimension Text
panel.

The hole callout PropertyManager contains a wide variety of optional parameters. This allows for creating a hole callout with all of the relevant tolerances, proper justification, and so on. The Tolerance/Precision panel is shown in Figure 12–112. By selecting a callout value from the drop-down list of the same name (visible in the figure), a tolerance can be set for that value. Precision can be set for any of the primary or tolerance values using this same panel.

Use the Dimension Text panel, shown in figure 12–113, to position the callout relative to the leader. Text can be added manually through this panel, as well as numerous symbols. However, you will probably find that adding text or symbols is not necessary, considering how extensive hole callouts are all by themselves.

If you do happen to find the need, the Variables button opens a window (not pictured) that allows for selecting specific dimensions associated with the hole you wish to appear in the callout. In this fashion, the callout can be customized quite extensively. It is suggested that any text in the Dimension Text panel that appears between the less than (<) and greater than (>) symbols be left alone. This text is code SolidWorks recognizes as a link to the true dimension values of the model. Modifying or deleting these codes will break the link and dimension values entered manually as text may not be accurate.

Multi-jog Leaders

Multi-jog leaders are leaders with multiple segments, typically with a note at one end. Multi-jog leaders can be attached to notes, surface finish symbols, geometric tolerances, and balloons (covered in Chapter 13). The key is to have the note or other annotation created prior to creating the multi-jog leader. How-To 12–30 takes you through the process of creating a multi-jog leader and attaching it to a note (or one of the other annotations mentioned previously).

How-To 12-30: Adding a Multi-Jog Leader to an Annotation

To add a multi-jog leader to an annotation, perform the following steps. Be aware that the annotation must already exist in the drawing.

1. Select Insert > Annotations > Multi-jog Leader, or click on the Multi-Jog Leader icon found on the Annotations toolbar.

2. Pick where the leader will be pointing to. This will be the arrow side of the leader.

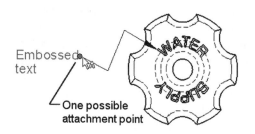

Fig. 12–114. Attaching a leader to a note.

3. Pick points to establish the various segments in the leader.

4. To end the leader at the appropriate annotation, place the cursor over the attachment point of the annotation and left-click. This attachment point will appear as a small blue dot, discernable in Figure 12–114.

Notes with multiple lines of text will have multiple attachment points, so take care when attaching the leader. Once a multi-jog leader has been attached to a note, the note can be moved to a new position, and the leader will follow. Right-clicking on the leader segment that touches the note will gain access to a menu option titled Add Horizontal Bent. Selecting this option adds a small horizontal leader line prior to connecting to the note.

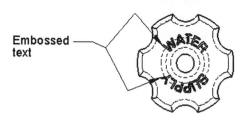

Fig. 12–115. Multi-jog leader.

To add a branch, right-click on a jog point and select Insert New Branch. Jog points can be added and deleted as needed. To delete a jog point, right-click on the point in the leader and select Delete Jog Point. To add a jog point, right-click somewhere on one of the leader segments and select Add Jog Point. To change what arrow style is displayed at the end of a leader, right-click on the arrow and choose a new arrow style. The arrow can also be turned off using this same technique. An example of a multi-jog leader with multiple branches and a horizontal bent is shown in Figure 12–115.

It should be noted that branches can be added to any annotation that already contains leaders. The leaders do not have to be multi-jog leaders. Selecting an annotation that contains a leader will result in small green handles appearing on the annotation and on the point that attaches to the model. Hold down the Ctrl key, and then drag the green handle at the "arrow" end of the leader. This results in a new leader that can be attached to some other object. Hole callouts are one exception. These cannot have multiple leaders.

Revision Symbols

Revision symbols are a special type of annotation link to a custom property automatically generated by SolidWorks. The symbol is also directly related to a revision table. In fact, revision symbols can only be added if a revision table has been inserted into a drawing and a revision has already been made to the drawing. Revision tables are discussed in the section "Tables" later in this chapter.

Surface Finish Symbols

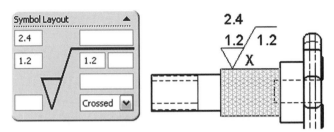

Fig. 12–116. Symbol Layout panel and surface finish symbol example.

As has been par for the course, it is not necessary to list the steps involved in creating a surface finish symbol. Like most other annotations, type in the appropriate parameters, and then pick some location in the drawing where the symbol should be placed. As is standard operating procedure, multiple symbols can be added without ever leaving the Surface Finish PropertyManager. The Symbol Layout panel from the Surface Finish PropertyManager, along with an example, is shown in Figure 12–116.

If you are familiar with surface finish symbols, other panels (not shown) in the Surface Finish PropertyManager should be meaningful to you. These are the by-now-familiar formatting panels, and a panel that allows for rotating a surface finish symbol. These functions are self-explanatory.

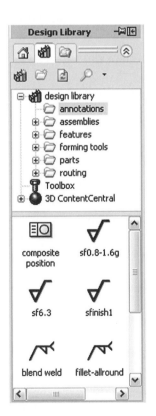

Fig. 12–117. Favorite annotations in the Design library.

Favorites

Favorites are certain annotation types that have been saved with a particular set of predefined attributes. For example, a dimension favorite may have a particular tolerance and precision, or a weld symbol favorite might have various parameters established that you use on a regular basis. Favorites are a way of transferring characteristics to other like annotations in the same drawing or even other drawings.

Favorites can exist for dimensions and weld symbols, as well as notes, surface finish symbols, and geometric tolerances. All favorites are created in the same manner, so learning how to create any favorite type means you will know how to create them all.

All favorite annotation types can exist as separate files. In fact, this is what allows them to be moved between drawings (or any SolidWorks document). Perhaps the most convenient aspect of favorites is that they can easily be accessed from the Design library, discussed in Chapter 9. Figure 12–117 shows the Design library opened to the *Annotations* folder. Some sample favorites are included with the SolidWorks software, a few of which are visible in the image.

Creating favorites and saving a favorite out as a separate file involve slightly different processes. How-To 12–31 takes you through both of these processes. Although the steps listed can be used to add a favorite of any type, a dimension will be used in this example.

HOW-TO 12-31: Adding Favorites

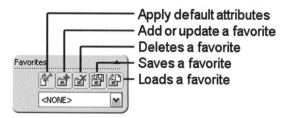

Fig. 12–118. Favorites panel in PropertyManager.

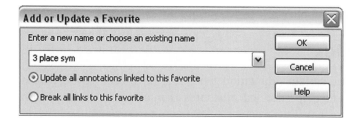

Fig. 12–119. Add or Update a Favorite window.

To create a favorite, perform the following steps. Although a dimension favorite will be used in this example, the same steps will apply for any favorite you wish to create.

1. Select the annotation that contains the attributes you wish to retain as a favorite.

2. In the annotations PropertyManager, click on the Add or Update a Favorite icon, shown in Figure 12–118.

3. In the Add or Update a Favorite window, shown in Figure 12–119, enter a name for the favorite. The name should be short but descriptive.

4. Click on OK to create the favorite.

At this point, the favorite will be saved, but will only be available in the document in which the favorite was created. This is not entirely true, because dimension favorites created within a drawing from model dimensions (driving dimensions rather than reference dimensions) will also be available in the part file. If you wish the favorite to be available for insertion into other unrelated SolidWorks documents continue with the following steps.

5. Click on the Save a Favorite icon.

6. Enter a name for the favorite.

7. Click on the Save button.

8. Click on OK to close PropertyManager.

Step 3 suggests that the name should be short but descriptive. Use whatever naming convention makes sense to you, but try to be succinct, as there is not a lot of room in the favorite panel drop-down list. Incidentally, favorites must be saved in the current document before they can be saved out as a separate file. In other words, steps 2 through 4 must be completed prior to carrying out steps 5 through 7.

The five types of favorites that can be saved as separate files all have different file extensions. Table 12–2 outlines each favorite and its associated file extension.

Table 12-2: Favorite File Types

Favorite Type	File Extension
Dimensions	.sldfvt
Notes	.sldnotefvt
Geometric Tolerance Symbols	.sldgtolfvt
Surface Finish Symbols	.sldsffvt
Weld Symbols	.sldweldfvt

Once you have some favorites established, applying those favorites to other annotations is a snap. Select the annotation you wish to apply the favorite to, and then select the applicable favorite from the drop-down list in the Favorites panel. To import a favorite previously saved out as a separate file, use the Load Favorites icon found on the same panel. The *Apply default attributes* icon (see Figure 12–118) removes any formatting previously applied with a favorite.

Most favorites can be dragged from the Design library and dropped on a drawing (or part or assembly). The exception to this would be dimension favorites, which cannot be added to the Design library. The reasoning behind this is that dimensions are specific to geometry and would not benefit from drag-and-drop behavior.

Blocks

Blocks are essentially collections of entities grouped as one object. They can be inserted into drawings, rotated, scaled, and positioned as needed as many times as necessary. Similar to favorites, blocks can be created and reused within a single document, or they can be saved as a separate file. When saved as a separate file, blocks use the file extension *.sldsym*. Think of a block as a custom-made symbol file.

If there are certain symbols used in your drawings on a regular basis, blocks will probably work in your favor. Blocks can be linked to external files and can be made to update if the original block is redefined. Blocks can also be dragged and dropped into a drawing from the Design library.

To take advantage of blocks, you should be familiar with how to create, insert, and edit a block. That being the case, let's start at the beginning and discuss how to create a block. How-To 12–32 takes you through the process of defining a block.

How-To 12-32: Defining a Block

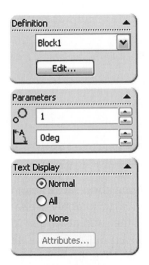

Fig. 12–120. Block Instance PropertyManager.

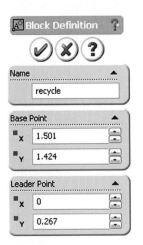

Fig. 12–121. Block Definition PropertyManager.

To define a block, perform the following steps. You should be working in a drawing file at this time.

1. Create the geometry that will make up the block symbol. Add text as required.

2. Select the geometry, including any text you may have added.

3. Select Make from the Tools > Block menu. The Block Instance PropertyManager, the majority of which is shown in Figure 12–120, will appear.

4. Click on OK to close the Block Instance PropertyManager.

At this point the block will have been defined and can be inserted into the drawing whenever needed (the process for which we will get to in a moment). Another way to create a new block is to select New from the Tools > Block menu. It would be necessary to select New rather than Make (such as in step 3 of the previous How-To) if the geometry for the block had not been created yet. Simply stated, use New to create a block completely from scratch, and Make to create a block from existing geometry. There is no difference in the final outcome of the block.

Clicking on the Edit button (visible in Figure 12–120) brings up the Block Definition PropertyManager (shown in Figure 12–121). This is the same interface seen when creating a new block from scratch. Note that when editing a block definition any drawing previously displayed on screen will temporarily disappear. Do not panic, as your drawing will reappear once you are finished editing the block.

Blocks can be given a name, as can be seen in the figure. This is really only necessary if there are numerous blocks in a drawing and it is necessary to distinguish between them during the insertion process. Blocks saved as files should also be given a meaningful name, so that they can be easily identified when brought into another drawing.

While editing a block definition, there are options for controlling the base point and leader point of a block. The base point is

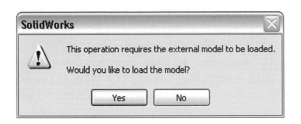

Fig. 12–137. Do you want to load the model?

There are a few things that cannot be accomplished when working with a detached drawing. For example, model items cannot be inserted into the drawing. If you happen to try to perform some action that requires loading the model, Solid-Works will inform you of this fact with the warning message shown in Figure 12–137. Click on Yes or No, depending on whether or not you wish to load the model.

If at any time it is desired that the detached drawing no longer be detached, the process can be reversed. Simply save the drawing using the Save As command and select *Drawing* as opposed to *Detached drawing*.

Summary

In this chapter you have learned that there is a lot you can do with design drawings. Many views can be created, with a good portion of the view creation process being automated by SolidWorks. A perfect example is the drag-and-drop functionality when creating the three standard views. You can drag a part or assembly file from Windows Explorer into a drawing, or drag the file from the top of FeatureManager if the file is already open.

When beginning a new drawing, specify the template you want to use to start your drawing. Templates contain various settings, such as dimensioning standards and work units (i.e., English or metric). Templates may also contain a drawing sheet format. If you do not want to use a format, you must at least specify a sheet size for the paper.

If creating a new sheet format, it is best to save the format as a separate file. This can be done by using the Save Sheet Format command found under the File menu. Once a format has been saved as a separate file, it can be used for other drawing sheets that may be added to the same drawing.

You can alternate between editing the sheet and editing the format by right-clicking in an empty space on the drawing or by using the Edit menu. Edit the properties of the sheet (again, by right-clicking) to change the sheet size, scale, or format on the fly.

To move a view, drag its border. A parent view will move with its dependent views. However, if the dependent view is moved it will move by itself. Views will remain aligned with their parent views unless you break the alignment. If you change your mind, you can return to the default alignment condition. Once again, this can all be done with the right mouse button.

Creating certain views requires that geometry be added to a view. For instance, detail or section views require circles or lines to be sketched on a view before the detail or section views can be created. Locking focus on a particular view often aids in this process. Geometry added when a view focus is locked will be associated with that particular view. This holds true for notes as well.

Auxiliary views require that you select an edge to project from when creating the view. Projected views require that you select a view to project from. Model views can exist in an almost limitless variety. Defining a view using the Orientation window while in a model helps make this possible.

Dimensions used to create a part can be brought into a drawing and reused. This is known as inserting model items. Any type of annotation can be inserted from a part or assembly into a drawing in this manner. Most annotations, however (other than dimensions), are typically created in the design drawing and not in the part.

Moving dimensions is a simple drag operation. You can also drag the ends of extension lines by selecting the dimension and then dragging the small green handles that appear at the ends of the extension lines. The green handles attached to the dimension arrows can be used to flip the arrows inside or outside the extension lines. To move a dimension to another view, hold down the Shift key.

Edit a dimension's properties (right mouse click) to change its appearance. Many characteristics (such as arrow display, diameter, radial options, symbols, and text) can be altered through a dimension's properties. Most common options, such as tolerance settings or precision, can also be modified in PropertyManager.

Other annotation types (such as hole callouts, center marks, and notes) can be added via the Annotations toolbar. Most annotations, such as notes, allow for placing multiple annotations without issuing the command repeatedly. You can add a leader to the note, or reposition the note or leader once it has been created. Attaching a leader to an edge or surface of the model associates the leader with the model. If the model changes size or shape, the leader will remain pointing to the desired location. Add multiple leaders by holding down the Ctrl key when positioning the leader arrow.

Use blocks when there are symbols or groups of sketched objects that need to be reused. Blocks can be defined in a drawing, and then added multiple times at various locations. If one block is edited, all related blocks will update as well. Blocks can also be externally linked to a separate file. Editing the block file results in drawings that update to reflect the change. All blocks can be scaled and rotated as required.

A detached drawing does not require the referenced model geometry in order to open and view the drawing. The two greatest benefits of detached drawings is that they will open extremely fast, even if the drawing is of very large assemblies, and that they can be sent to other individuals without the associated model files. However, some functions—such as inserting model dimensions—will require that the referenced model geometry be loaded. This can be done by right-clicking on a drawing view and selecting Load Model.

CHAPTER 13

Assemblies

PART FILES, AS YOU HAVE LEARNED, typically contain one contiguous solid model. Assembly files can contain more than one part. Assembly files give you the capability of assembling the parts you have created, putting the parts together as if you were actually building the assembly in real life. Figure 13–1 shows an example of an assembly. It is the example you will use to learn about assemblies in this chapter.

Fig. 13–1. A SolidWorks assembly.

Assemblies vary in complexity over a very wide range. They can be as simple as a two-part assembly or as complex as an assembly containing thousands of components. A limiting factor on the size of an assembly is the type and amount of computer hardware you have. The faster the processor the better off you will be, and a lot of computer memory is a very good thing. If you will be creating large assemblies with thousands of parts, you would be wise to purchase extra memory. On the order of a gigabyte of RAM is common, and 2 or more GB would not be considered overkill.

Assemblies fall into two categories. The first and most common assembly is known as a bottom-up assembly. A bottom-up assembly is what you will be reading about in this chapter. Think of a bottom-up assembly as a table with parts lying on it. You pick the parts up and place them together, building the assembly. The parts themselves have already been created.

The other type of assembly is known as a top-down assembly. Imagine that you have a partial assembly. A part that is needed for the assembly must be built by referencing other parts in the partially completed assembly. The new part is created from within the assembly, otherwise known as "in the context" of the assembly. (This topic is discussed in detail in Chapter 19.)

This chapter first deals with how to begin an assembly and then insert the components into the assembly. There are various methods you can use for component insertion. You will learn all of them, and then can decide for yourself which method is most convenient for you.

Starting a New Assembly

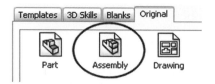

Fig. 13-2. Specifying an assembly template.

Start a new assembly just as you would start a new part. Select New from the File menu, or click on the New icon found on the Standard toolbar. The only difference is that you should specify an assembly template instead of a part or drawing template, as shown in Figure 13-2. You can then click on the OK button and be on your way to building an assembly.

You will not notice many differences between the part and assembly interfaces. There are only a few. First, you may notice that there is something new in FeatureManager, shown in Figure 13-3. This is known as the *Mates* folder, used as a storage area for the geometric relations (known as mating relationships, or simply mates) between components in an assembly. You should also see the Assembly toolbar, shown in Figure 12-3. Make it a point to turn this toolbar on if it is not visible.

Fig. 13-3. Mates *folder and Assembly toolbar.*

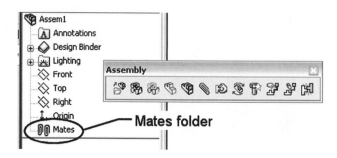

Rather than starting a new assembly devoid of components, it is also possible to begin a new assembly from within a part file. How-To 13-1 takes you through the process of creating a new assembly from a part.

How-To 13-1: New Assemblies from Parts

To begin a new assembly from a part that is currently open, perform the following steps. Note that this function is not limited to parts, and that assemblies can be created from within an existing assembly.

Fig. 13–4. Insert Component PropertyManager.

1. Select Make Assembly From Part from the File menu, or click on the Make Assembly From Part/Assembly icon found on the Standard toolbar.

2. Select the assembly template on which to base the new assembly.

3. Click on OK. The Insert Component PropertyManager (shown in Figure 13–4) will appear.

4. In the new assembly that appears, click in the work area to position the new component.

When positioning the component in the assembly, it is good practice to place the first component at the center of the assembly's coordinate system. This just makes good sense, rather than having the first component positioned at some arbitrary point off in space. To centrally locate the first component, position the cursor over the assembly's origin point and look for the system feedback (shown in Figure 13–5) while performing step 4 of How-To 13–1. This will essentially lock the origin point of the part to the origin point of the assembly.

Fig. 13–5. Positioning the first component centrally.

Fixed or Floating?

By default, the first component inserted into an assembly is fixed (locked) in position. This is not necessarily a permanent condition. Any component can be made to be fixed in space or floating. This is accomplished by right-clicking on a component in the assembly and selecting either Fix or Float. Both options will not be present at once, as this setting is a toggle switch and a component is always in one state or the other.

You will always want to have at least one component fixed in space. The fixed component serves as an anchor for the rest of the assembly. You can also think of the fixed component as a point of reference other components are positioned by or mated to.

Considering that the first component brought into an assembly is automatically fixed in position, it is good practice to bring the main assembly component into the assembly first. To offer an example, assume an assembly of an automobile is being created. It would make more sense to bring in the frame of the car first, rather than (for example) a lug nut. It is not always a cut-and-dried decision what should be the "main" component in an assembly, in which case you should just make an educated guess. As was just mentioned, the state of the part can always be changed later from fixed to floating, or vice versa.

FeatureManager Symbology

Fig. 13–6. FeatureManager symbols.

You can tell a few things about an assembly just by looking at its FeatureManager. For instance, if a component is fixed it will have the letter *f* prefixing the component's name. This can be seen in Figure 13–6. It is the letter *f* in parentheses (f) that indicates that the Cover component is fixed.

The Axle component is a subassembly, which can be discerned by the symbol to the left of the component's name. Subassemblies are assembly files in their own right, but are considered subassemblies when inserted into a higher-level assembly. If a component has not been fully mated to other components in the assembly, and is not fixed in position, you will see a minus sign (–) before its name. Such is the case with the Axle.

If a component is fully mated to other components, you will not see any symbol before its name. Such is the case with the Insert and Washer components shown in Figure 13–6. A plus sign (+) preceding a component is bad news, and means you have overdefined a component. This happens if you add too many mating relationships between components, or if you try to fix components already mated in position. If you encounter this situation, you should find a solution to the problem before going any further.

All of a part's features are accessible from FeatureManager (see Figure 13–6). You can see that the Slotted Bolt component has been expanded so that its features are viewable. You could go one step further and expand the features to gain access to the underlying sketch geometry. In short, the entire hierarchy of every component is accessible within the assembly's FeatureManager.

The same part can be inserted into the same assembly as many times as required. In this case, the components would all be unique from an assembly standpoint, but are actually all based on the same part file. To identify each component uniquely, SolidWorks uses a numbering scheme known as *instance identification numbers.*

Instance ID numbers are the numbers placed within angle brackets (< >) following the name of each part (as part of the full name of each component). In other words, the name of the Washer component in Figure 13–6 is really *Washer < 1 >*. If a second washer were inserted into the assembly, it would be named *Washer < 2 >*. SolidWorks users have no control over this, nor does it matter. For reference, Figure 13–7 shows the various aspects of a component's name.

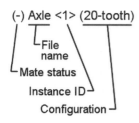

Fig. 13–7. Anatomy of a component's name.

It should be noted that there is no association between the number of components in the assembly and the instance ID number. Continuing with the previous example, let's imagine *Washer < 1 >* were deleted from the assembly. *Washer < 2 >* would retain its original instance ID. The point to be made is that the instance ID is not a true indicator of how many of a certain component are in an assembly.

The final detail we will look at regarding a component's name is the data found within parentheses following the instance ID number. This data indicates the name of the configuration used for that component in the assembly. You learned how to create configurations in Chapter 11, which showed that multiple versions of a part can exist in the same part file. The "version" (configuration) used in the assembly can then be controlled from within the assembly, the process for which is discussed later in this chapter. If there is no data in parentheses following the instance ID, only one configuration of that part exists.

Inserting Components

Now that you have a new assembly begun, it is time to bring additional components into the assembly. There are numerous ways of accomplishing this. One method is by way of Windows Explorer. This is literally a drag-and-drop procedure. For instance, by dragging and dropping a SolidWorks part from Windows Explorer into the work area of an open assembly the part will be inserted into the assembly. Part or assembly files can also be drag-and-dropped from SolidWorks File Explorer and from the Design library.

Another alternative requires that the part file already be open. Again, this second method is a drag-and-drop procedure. Place the cursor over the name of the part in the part's FeatureManager, and then drag the part to the assembly work area. You will need to tile the open document windows in order to use this procedure, which can be done from the Window pull-down menu.

There is a more traditional method of inserting components into an assembly. It is advisable to gain familiarity with this method of inserting a component before trying out some of the "shortcut" methods, as there are some additional points you should be aware of that go along with drag-and-drop functionality. These points you will learn, but all in good time. How-To 13–2 takes you through the standard method of inserting a component into an assembly.

How-To 13-2: Inserting Components Using the Insert Menu

To insert components into an assembly, perform the following steps. Note that this process allows for inserting part or assembly files into an assembly.

1. Select Insert > Component > Existing Part/Assembly, or click on the Insert Components icon found on the Assembly toolbar.

2. If the document to be inserted into the assembly is currently open, select the component from the Open documents list in the Insert Component PropertyManager (see Figure 13–4). If the document is not open, click on the Browse button, select the part or assembly, and click on Open.

3. Click somewhere in the work area to indicate where you want to position the component.

Fig. 13–8. Do not click on OK unless inserting the very first component!

You will find that it is not necessary to click on OK in the Insert Component PropertyManager. Simply clicking somewhere in the work area is enough to position the component and close out of the Insert Component command. If you were to click on OK, the component would be automatically dropped on the origin. This in itself is not necessarily a problem. The problem derives from the *fix* relation SolidWorks applies to any component inserted on the assembly's origin.

The very first component inserted in an assembly should be fixed in position, and placing the first component on the origin is also good practice. However, if additional components are inserted on the origin, components will overlap each other. Those components also cannot be moved, because they have been fixed, and will need to be floated first. Do not click on the OK button unless it is the first component you are inserting into an assembly (see Figure 13-8).

Moving and Rotating Components

The Assembly toolbar contains two icons specifically designed for moving and rotating components. They are titled, appropriately enough, Move Component and Rotate Component. The interesting point, however, is that you will not need to use them for basic move or rotate operations.

As long as a component is not fixed in position or fully mated in position, it can be moved with the left mouse button. Place the cursor over the component, and then drag the component to a new location (see Figure 13–9). Try this same operation with the right mouse button and you will find that the component can be rotated. It doesn't get much easier than this.

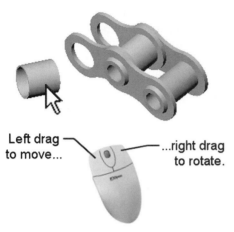

Left drag to move...

...right drag to rotate.

Fig. 13–9. Moving or rotating components.

The ability to move or rotate components via dragging is controlled by a solitary option titled *Move components by dragging.* This option is found in the Assemblies section of the System Options (Tools > Options).

Moving or rotating components proves invaluable when trying to build an assembly. In the case of adding mates, which you will learn about shortly, it is best to position the components close together and in an orientation that is similar to how they should wind up after adding the mates. This is so that you can zoom in to the faces being mated and more easily select them, but also to help Solid-Works pick the correct alignment condition (discussed in material to follow).

✓ **TIP:** *If a component is hidden behind other components and it is necessary to move it out into plain view, select the component from FeatureManager, click on the Move Component icon, and drag the component into view. You will not have to position the cursor over the component when dragging, because it has already been preselected via FeatureManager.*

Moving components is extremely helpful when trying to determine which mates have or have not already been added to the assembly. This is similar to dragging sketch geometry to see how it will behave. By moving components, you can discern how the assembly will behave and whether or not more mating relations should be added.

✓ **TIP:** *Multiple components can be moved or rotated with the Move Component or Rotate Component command as long as they are selected first and not fully mated in position.*

Moving with the Triad

If you want a little more control over how a component is moved or rotated, use the Move with Triad command found in the right mouse button menu. Right-click on a component and select Move with Triad and the triad shown in Figure 13–10 will "bloom" out of the center of the component.

Once the triad is visible, place the cursor over one of its axes. Note the small mouse displayed next to the cursor. This is informing you that by dragging a particular arrow with the left mouse button the component can be moved in that direction. By right-dragging an arrow, the component can be rotated about that arrow. It only takes a small amount of practice before feeling comfortable with this process.

If that were not enough, the center of the triad can be dragged to a new position. This alters the location of the triad axes, thereby changing the point around which the component can be rotated. While dragging the triad, it is also possible to "snap" it to certain edges or faces on the component, thereby realigning the triad axes with whatever it snaps to.

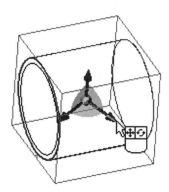

Fig. 13–10. Triad used for moving and rotating along or about certain axes.

Mating Relationships

Once you have begun bringing components into the assembly, you must then mate them to each other. It is not enough to just position the components. You must establish a set of conditions that describes how the components relate to one another. This should sound familiar to you. After all, it is the same thing as adding geometric relations when creating sketch geometry. However, in this case it is applied to 3D part geometry instead of 2D sketch geometry.

How do you know what types of mating relationships to add between components? Ask yourself how you would put the components together in a real-life assembly and you will have your answer. Generally, create an assembly with motion in mind. Unless, of course, the assembly has no moving components, in which case the assembly process is simplified even more.

There are six degrees of freedom of movement with regard to the individual parts in an assembly. Any component, assuming it is not fixed, can translate in the *x*, *y*, or *z* direction. Each component can also rotate along the *x*, *y*, or *z* axis. By restricting the direction a part can translate or rotate, you can control how the assembly behaves.

Different mates affect assembly components in different ways. A concentric mate, for instance, will allow a component to translate in one direction and rotate about the axis of translation. That is the nature of the concentric mate.

Adding mates between components in an assembly is similar to adding geometric relations in a sketch, but with two main differences. When creating a sketch, it is good technique to fully define the sketch. When dealing with an assembly, you may not want to fully define the components. On the contrary, you may want to leave components intentionally underdefined, thereby enabling the component to move or rotate in the desired direction, whatever that may be. On the other hand, fully mating components that should not move is perfectly acceptable.

A second difference is that you are now dealing with 3D geometry. When working with a sketch, you needed to select points or sketched entities to define relations. Because you are now dealing with 3D parts, it is usually best to select faces. This is not required, but is often good technique. Mate relationships can be added between many things, including the following.

- Faces

- Planes

- Axes

- Model edges

- Sketch geometry

- Origin or vertex points

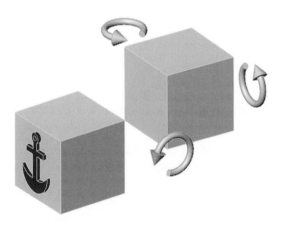

Fig. 13–11. Coincident mate between vertices.

Let's take a simple example to help illustrate what might happen when adding a coincident mate. Assume there are two cube-shaped building blocks in an assembly. What happens if a coincident mate is added between two corners (vertex points) of the blocks? Assuming also that one of the blocks is fixed (anchored in position), the remaining block will be free to rotate in any direction, but will pivot at the mating point. This is illustrated in Figure 13–11.

You could think of this mate as a ball-and-socket type of relationship. The upper block is free to rotate about each axis, but cannot be translated in any axis. Next, consider what would happen if a coincident mate were added between two edges. This is shown in Figure 13–12, and creates a hinge-type mate. This leaves only two degrees of freedom of movement: the ability to translate along the mating edges and to rotate about the mating edges.

Finally, what happens when a coincident mate is applied between two faces? Figure 13–13 shows the outcome of this mate. The upper block is free to rotate about one axis but can slide along the other two axes.

Hopefully, what you gather from these illustrations is that it is okay to mate between various objects, but care must be taken. A good general rule is to add mates between faces or planes when possible. Next, you will take a look at the mechanics behind adding mating relationships between components. How-To 13–3 takes you through the process of adding mate relationships.

Fig. 13–12. Coincident mate between edges.

Fig. 13–13. Coincident mate between faces.

HOW-TO 13-3: Adding Mate Relationships

To add mate relationships, perform the following steps.

1. Select Insert > Mate, or click on the Mate icon found on the Assembly toolbar.

2. Select the objects between which to add a mate relationship. For instance, select a face on each of the two components being mated.

3. From the Mate PropertyManager, shown in Figure 13–14, select the mate type to be added.

4. If adding a distance or angle mate, type in a distance or angle value for the mate.

5. Click on OK to add the mate.

6. Repeat steps 2 through 5 as needed, or click on OK a second time to close the Mate PropertyManager.

Fig. 13-14. Mate PropertyManager.

Fig. 13-15. Pop-up toolbar appears when adding mates.

It is very likely that when adding a mate you will see the pop-up toolbar shown in Figure 13–15. This toolbar is enabled by turning on the Show pop-up dialog in the Mate PropertyManager Options panel, visible in the same figure. The pop-up toolbar contains the mates that are capable of being added, which is dependent on the geometry selected. That is, if two planar faces are selected the tangent mate will not appear, as it would not be relevant. Likewise, the tangent mate would be unavailable in the PropertyManager as well.

Sure, it is possible to click on the mate to be added in the PropertyManager, but clicking on the same icon in the pop-up toolbar is much more convenient. This is due to the toolbar popping up wherever the cursor happens to be positioned at the time. Perhaps it is the epitome of laziness to carry out a command in a manner that means moving the mouse less far. Nevertheless, the pop-up toolbar achieves that objective, and makes adding mates very user friendly.

Components being mated will only move as far as they have to in order to satisfy the mating condition. A common question is, "Which component will move, and is the order in which the faces are selected important?" Selection order is not important. In addition, the component that will move depends on existing mates. If one component is fixed or already fully mated in position, it will not be able to move.

Mate Types

There are many mate conditions that can be applied between components, and most of them are self-explanatory, such as perpendicularity and parallelism. Not everyone may be familiar with all of the terms used to describe mates. If you fall into this category, Table 13–1 will help to clarify the various mate types. Advanced mates will be expanded upon later in this section.

In Table 13–1, the descriptions of the mates describe some of their more common applications. Common objects that are mated between (such as planar or cylindrical faces) are also listed. All possible objects that can be used for any particular mate are not listed, as this is not really necessary. As a general rule, a particular mate condition can be applied between two objects if it makes geometrical sense and seems logical.

Table 13-1: Mate Definitions

Basic Mates	Description
Concentric	Cylindrical faces become aligned along their axes. Conical and spherical faces, as well as axes, can be mated concentric.
Tangent	All types of faces can be made tangent to each other, with the exception of two planar faces, which would be considered a coincident mate.
Coincident	Planar faces become coplanar (flush against each other). Edges or points can also be used in coincident mates.
Parallel	Planar faces will be parallel, but the distance between them is not defined.
Perpendicular	Planar faces will be perpendicular, but will otherwise be free to move.
Distance	Similar to parallel, but the distance between faces can be specified. Use this mate for specifying clearances.
Angle	Places two planar faces at a specified angle.
Advanced Mates	
Symmetric	Makes two similar objects symmetric about a plane or planar face. Vertex points work very well with this mate type.
Cam	Similar to a tangent mate, a single cylindrical or planar face will become tangent to one or more tangent faces of a cam.
Width (SW 2006 Only)	Centers a tab or object between two other faces, such as the walls of a groove.
Gear	Two components will rotate with respect to each other using a specified gear ratio. Cylindrical or conical faces or axes can be used to dictate the rotation axis.
Distance Limit	Same as a distance mate, but with the addition of minimum and maximum distance values.
Angle Limit	Same as an angle mate, but with the addition of minimum and maximum angular values.

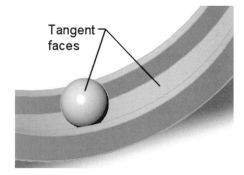

Tangent faces

Fig. 13–16. Adding a tangent mate between two faces.

Most of the basic mates in Table 13–1 should be fairly self-explanatory. Let's examine a tangent mate, for instance. Using the steps outlined in How-To 13–3, it is nothing more than a matter of selecting the two faces to be tangent (assuming you have started the Mate command) and clicking on the Tangent mate icon, preferably from the pop-up toolbar discussed earlier. The sphere shown in Figure 13–16 is being mated tangent to a cylindrical face. The components are on their way to becoming a bearing assembly.

An interesting note on concentric mates is that a spherical surface can be used in the mate, as mentioned in the table. For example, a spherical face can be mated concentric to a cylindrical face. This results in the sphere being able to travel along the axis of the cylinder, but the sphere can still rotate in every direction.

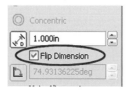 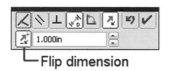

—Flip dimension

Fig. 13–17. Flip dimension option.

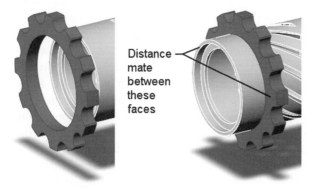

Distance mate between these faces

Fig. 13–18. Flipping a distance dimension.

70.0°

Fig. 13–19. Making use of an angle mate.

Distance and angle mates have an option for "flipping" the dimension associated with the mate. This option is present in PropertyManager and on the pop-up toolbar, as indicated in Figure 13–17.

What does flipping a dimension do? It moves the components so that the distance or angle mate is applied to the opposite side. Figure 13–18 illustrates this. There is a 1-inch distance mate between the two faces specified in the figure. By flipping the dimension (which is not shown), the components are reoriented and the dimension is applied to the opposite sides of the faces.

✓ **TIP:** *To access dimensions associated with any mate, double click on the mate in the* **Mates** *folder in FeatureManager.*

Angle mates can be used to create hinged assemblies. In Figure 13–19, the two faces of the hinge assembly had an angle mate added between them. Each hinge is also mated concentric about the hinge pin. Coincident mates are added to position the hinges correctly so that they mesh, and to seat the pin completely into the female hinge. The pin is fixed, and the female hinge has been completely mated in position.

Prior to adding distance or angle mates, SolidWorks often assumes that a coincident mate should be added, and animates the two components into position. As soon as the distance or angle mate is chosen, the components will revert back to their original positions. The original distance or angle value will also be displayed. This value can then be used as a gage to decide on what distance or angular value should be used. Pressing the Enter key after entering a value will update the preview, and a decision can then be made on whether or not to use that value or try something different.

Alignment Conditions

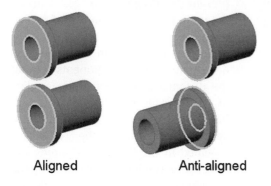

Aligned   Anti-aligned

Fig. 13–20. Aligned versus antialigned components.

When two components are mated, such as with a coincident mating relationship, the components could either be aligned or anti-aligned. This alignment refers to the side of the selected faces the part geometry is on. If the geometry is on the same side, the components are aligned. If the geometry is not on the same side, the components are antialigned. This is depicted in Figure 13–20.

With regard to the coincident mate, alignment conditions are very easy to understand. Regarding other mates, such as tangent, alignment conditions are not as straightforward. The best advice regarding alignment conditions is to not worry about them. If the alignment is wrong, reverse the alignment. This is done using the alignment options available in PropertyManager and on the pop-up toolbar (shown in Figure 13–21).

Fig. 13–21. Changing alignment.

The software is essentially lazy with regard to adding mates. It will only move a component as far as it has to in order to satisfy the mate condition. If you make it a point to position a component in the general position of where it should wind up anyway, the alignment condition will almost always be correct.

Finding and Editing Mates

Mates can be edited in the same way any other feature is edited: via Edit Feature in the right mouse button menu. However, finding a particular mate can be tricky without knowing the best way to track a particular mate down. Consider that it usually takes about three mates to fully define a component. If an assembly has 100 components, that means there will be somewhere around 300 mates in the *Mates* folder.

It should be noted that mates can be renamed. Performing a slow double click on a mate allows for renaming the mate. A distance mate might be renamed *Cover clearance* or some other meaningful name. Renaming makes it easier to locate the mate and thereby change the distance dimension associated with it.

✓ **TIP:** *Selecting a feature (or mate) from FeatureManager and pressing the F2 function key allows for renaming that feature.*

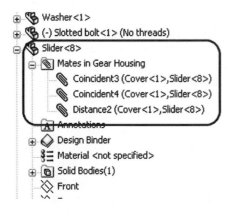

Fig. 13–22. Locating a component's mates.

The easiest way to locate a specific mate is to first ask yourself what component the mate is associated with. That component can then be expanded in FeatureManager, and the mates associated with that component can be found. Look for the *Mates* folder shown in Figure 13–22. In the figure, the folder is named *Mates in Gear Housing*, with Gear Housing being the name of the assembly the component is in.

Following the name of the mate are the components the mate is between, as shown in Figure 13–22. This helps in tracking down a particular mate. Component mate folders are only visible in an assembly, even though they appear to be part of each component's FeatureManager. The simple reason for this is that individual parts, when opened in their own window, have no need for mates. Mates, obviously, are only relevant in assemblies.

✓ **TIP:** *Selecting any two components, then accessing PropertyManager, will show all mates associated with those components. Mates in bold are the mates that exist between the two selected components.*

Advanced Mates

There are a total of five advanced mate types listed in Table 13–1. These advanced mates are not quite as cut-and-dried as the basic mates. A case in point would be the Symmetric mate, which can act very strangely under certain situations, depending on what entities are used for the mate.

The symmetric mate works best when two points (sketch points or vertex points) are selected for adding the mate. Other objects can be used, such as cylindrical faces, but the results are sometimes unpredictable. Points work best because a single point's x-y-z location can easily be interpreted by the software and its counterpart point on the other side of the plane of symmetry can be determined.

Cam mates require selecting appropriate faces on a cam and follower. The faces on the cam should all be tangent. If they are not, the follower will probably not track the cam properly. The follower itself must have a planar or cylindrical face at its end that can be selected to track the cam. Followers that terminate at a sharp point are also allowed, in which case the point should be selected.

The follower shown in Figure 13–23 has a small cylindrical face selected as the cam mate. Additional basic mates were required to keep all components in their proper orientation. When the cam spins, the follower moves up and down accordingly.

SW 2006: Width mates allow for centering an object (which SolidWorks refers to as a "tab") between two faces on another component (such as a groove). Valid selections for the groove are two planar faces, such as the faces on either side of the groove. The faces do not have to be parallel, and can in fact form a V shape. Valid selections for the tab are two planar faces, a single cylindrical face, or an axis.

Gear mates are easy to add and easy for the software to solve. There is no actual interaction between gears. Rather, a ratio is used to determine how two cylindrical faces or circular edges rotate with respect to each other. Axes can also be selected when applying a gear mate. The ratio is entered in PropertyManager in the area shown in Figure 13–24.

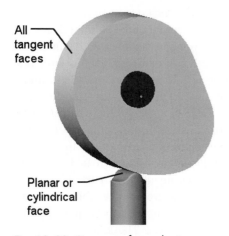

All tangent faces

Planar or cylindrical face

Fig. 13–23. Cam mate face selection.

If there are actual gears in the assembly, the effect can be quite dramatic. Gear teeth appear to be meshing and interacting with each other. In reality, all that is happening is that the components are rotating about the selected axis or axes of the selected cylindrical faces or edges. If one component is rotated, the ratio is used to calculate how much the other component is rotated.

Fig. 13–24. Adding a gear mate.

The reduction box shown in Figure 13–25 contains a gear mate. The ratio has been set to 4:3. When the vertical axle rotates 360 degrees, the horizontal axle rotates 270 degrees. This is true even though there are no actual gears in the assembly, as can be seen in the cutaway view shown in Figure 13–25.

Not showing meshing gears may reduce the dramatic visual effect, but there is a good reason to leave the gears out. If the reduction box were a purchased part, it would not be necessary to include the gears in a BOM. Therefore, there is no reason they need to be included in the assembly. The benefit gained from this is increased performance. Gears have a tendency to be fairly complex from a geometric standpoint, especially in the case of worm gears. Not placing them in the assembly will help make for a more efficient assembly from a system performance standpoint.

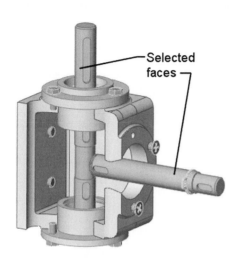

Fig. 13–25. Reduction box with a gear mate.

Limit mates come in two varieties: distance and angular limit. We will only examine a distance limit mate as the angular limit mate works in a very similar manner. Figure 13–26 shows a plunger assembly with a distance limit mate added. Double clicking on this type of mate gains access to its dimension, which is displayed as a limit value. Note that the upper and lower limit values conform to what is shown in the inset, which displays PropertyManager as it would appear when adding the mate.

The 3-inch value shown in Figure 13–26 is the initial positioning value that would be identical to a standard distance mate. The 5-inch and 1-inch values are the maximum and minimum distance values, respectively. These values are what control the plunger's "throw." The plunger can move, but only within the range determined by the maximum and minimum values.

Fig. 13–26. Adding a distance limit mate.

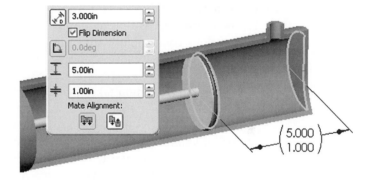

SmartMates

It is best that mates in general be understood before addressing this next topic. If you feel fairly comfortable with the various basic mate relationships, then you are ready to learn some shortcuts for automating the mating process.

Smart mates provide a means of automating the addition of mates between components. All of the basic mate types can be added using the SmartMate process. There are a number of combinations of entities that can be mated between when using SmartMates. Prior to understanding how to implement the SmartMate process, you should be aware of what Smart-Mates can accomplish. Table 13–2 provides a partial listing of objects that can benefit from SmartMates and the type of mate that would result between those objects.

Table 13-2: SmartMate Combinations

Entities Selected	Mates Added
Two points	Coincident mate
Two planar faces	Coincident mate
Two cylindrical faces	Concentric mate
Two conical faces	Coincident mate
Spherical and planar face	Tangent mate
Cylindrical and conical face	Concentric mate
Two edges	Coincident mate
Two circular edges	Coincident and concentric mates

SmartMates are implemented by holding down the Alt key, which acts as the hotkey for SmartMates. The component to be mated should be dragged by the entity used for the mate (such as a face on that component). As the component is dragged around the screen and the cursor is paused over various compatible mating entities, you will see the component animate into position. When this occurs, the cursor will also indicate through its system feedback the type of mate being added.

When the animation takes place and the system feedback is informing you of the mate about to be added, it is safe to let go of the left mouse button and Alt key. Initially pressing the Alt key or holding down the left mouse button, or letting up on either of these buttons, is not order sensitive. In other words, it doesn't matter which button you press first or let go of first.

After executing a SmartMate via the Alt key, the pop-up toolbar will appear. With this toolbar, the alignment can be changed, as well as the mate type. In this way, any of the basic mate types can be added with the SmartMate process. Incidentally, mates added using the SmartMate technique or using the Mate command result in the same mates. Only the process is different.

There is one SmartMate for which the pop-up toolbar does not appear when the mate is added. This is the SmartMate that involves dragging a circular edge to another circular edge, such as that shown in Figure 13–27. Each circular edge has an adjacent planar and circular face. This results in both a coincident and concentric mate being added between each similar set of faces.

Because there are no other possible mating combinations for the faces adjacent to the circular edges, the pop-up toolbar does not appear. Therefore, use a special process for changing the alignment condition, if necessary. If the alignment condition is incorrect, release the Alt key and press the Tab key once. In this way, you can make sure the alignment is correct prior to releasing the mouse button.

Fig. 13–27. SmartMating circular edges.

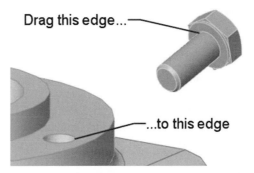

SmartMating While Inserting Components

If a part or assembly file has already been opened, and it needs to be inserted into the assembly, the SmartMate process can be implemented. The process is so ridiculously straightforward it is almost too easy. No geometry needs to be selected first and no commands need to be initiated. The process of inserting a component and simultaneously mating it in position involves nothing more than a simple drag-and-drop procedure.

The key to correctly inserting a component and taking advantage of the SmartMates functionality involves how you use the mouse. When the part is dragged into the assembly, it is important to drag it from the edge or face that will be referenced for the mate. When the part is dropped into the assembly, it is important to drop it on the edge or face the part will be mated to. See Table 13–2 for some examples of what can be used for SmartMating.

Mate References

When a file is dragged into the assembly window from Windows Explorer, SolidWorks' File Explorer, or the Design library, the ability to reference specific geometry is not available because the file is not open. A way to overcome this apparent shortcoming is to add what is known as a *mate reference*. A mate reference tells SolidWorks what object should be referenced for mating purposes when the part is inserted into the assembly.

To add a mate reference to a part, select Mate Reference from the Insert menu or Reference Geometry toolbar. The Mate Reference PropertyManager, shown in Figure 13–28, will appear. At this point, it is just a matter of selecting the object that should be referenced for mating purposes, and then clicking on OK. Both parts and assemblies can have mate references added to them.

Fig. 13–28. Mate Reference PropertyManager.

Once a mate reference has been added, it will appear near the top of FeatureManager. All of the usual rules that apply to Smart-Mate technology will apply when a model with a mate reference is inserted into an assembly. If the model is dragged into the assembly from Windows Explorer, it will appear in a translucent color. The model will "snap" into position when positioned over another object the mate reference can associate itself to.

The mate type can be specified for a mate reference depending on the entity selected. Likewise, the alignment type can be established. Unless you have a particular reason to do otherwise, leaving these settings at their default values is usually fine. It is only necessary to give the mate reference a name if you are adding multiple mate references in the same model file.

It is not typically necessary to add secondary or tertiary references, but the option is available if you wish to use it. If SolidWorks finds that it cannot make use of the primary reference, it will use the secondary. If that does not work either, it will attempt to use the tertiary reference.

Mate references are typically added to parts used numerous times in multiple assemblies. For example, it would not be worthwhile to add a mate reference to a component that is only going to be used once in one assembly. However, for parts used frequently mate references are a huge benefit.

Location-specific References

One aspect of mate references that can be derived from the previous section is, for example, adding a reference to hardware so that it can easily be drag-and-dropped into positions throughout an assembly. To take mate references to the next level, imagine a component that should always go to a particular location. This might be, for example, a particular hole among twenty holes in an assembly. Mate references can accomplish this. Let's look at a pictorial example so that you can better understand how to make this process work.

The male hinge shown in Figure 13–29 has had a mate reference added to it. The reference has been given the name *hinge*. Faces were selected for both the primary and secondary references. Specific mate types were chosen for each reference, as well as the alignment condition for the primary mate reference.

The female hinge has had an identically named mate reference added to it, as shown in Figure 13–30. Once again, specific faces and mate types were selected. This is crucial for the location-specific mate reference to work. A second set of mate references has been added and given the name *pin* (also shown in Figure 13–30).

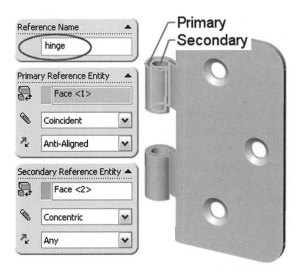

Fig. 13–29. Adding mate references to the male hinge.

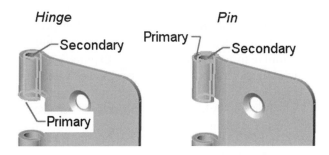

Fig. 13–30. Mate references on the female hinge.

Fig. 13–31.
Mate references
on the pin.

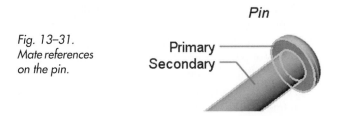

Finally, the pin model has had a set of mate references added to it. As you may have anticipated at this point, the name of the reference is *pin*. The faces used for this reference are shown in Figure 13–31.

How does all of this come together? The advantage of location-specific mate references becomes apparent when bringing the components into an assembly. Let's assume an assembly is started with the male hinge. Next, the female hinge is dropped into the assembly using Windows Explorer, SolidWorks File Explorer, or the Design library. The significant point to be made is that the female hinge can be dropped anywhere in the assembly and it will go to the correct location. Likewise, the pin can be dropped into the assembly and it will automatically position itself at the proper location with the appropriate orientation.

Location-specific references rely on the mate reference names. If names match up between two components, the mates will take effect regardless of where the component is dropped. If there is not enough information for the inserted component to snap into position, the mate reference will not be solved. For this reason it is important to select specific mate types when establishing the reference, and to specify the alignment condition for the primary reference.

Design Library Parts and Assemblies

The *Parts* folder in the Design library contains part files that can be conveniently inserted into an assembly. The *Assemblies* folder contains assemblies that could be inserted into a top-level assembly. They are generally items that will be used over and over again in various assemblies. Utilizing these parts and assemblies is a matter of dragging the item from one of the Design library folders into the assembly. (The Design library was discussed in Chapter 9.)

✗ **WARNING:** *Before Design library assemblies can be used, a folder must be designated an* **Assemblies** *Folder. To do this, right-click on the desired folder in the Design library and make sure Assemblies Folder is checked (as shown in Figure 13–32).*

Fig. 13–32. Designating an Assemblies *folder.*

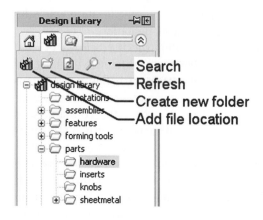

Fig. 13–33. Design library command icons.

It is very common to add mate references to Design library parts and assemblies. When an item is dragged into an assembly, drop the component on what it should be mated to. As is standard operating procedure, a translucent image of the part or assembly will be shown, along with system feedback, which will help you in positioning the component prior to releasing the mouse button.

The icons at the top of the Design library, shown in Figure 13–33, allow for performing functions such as refreshing the Design library (similar to refreshing Windows Explorer) or performing a search. Searching can be performed on file names or custom properties (discussed in Chapter 17). Adding a file location has the same effect as adding a path using the steps outlined in How-To 3–6.

New folders can be created and used to store additional parts or assemblies for later use. Parts or assemblies can then be added to those folders. Adhering to the common theme of dragging and dropping, parts or assemblies can be dropped into folders in the library, or dropped into the preview pane in order to add the model to the library. When adding models to the library, a Save As window appears, which allows for typing in a name for the copied file.

✗ **WARNING:** *Dragging and dropping assemblies into the Design library is not recommended due to assembly components not being copied in the drag-and-drop process. It is better to add a separate folder to the library and copy an assembly along with all components in that assembly to the new folder using Windows Explorer.*

Dragging files *into* the library for the reason of adding to the library has one major difference over dragging files *from* the library with the intention of adding components to an assembly. Files are always copied when added to the library via drag and drop. When dragging part or assembly files out of the library, you are really only copying a reference to that file. That is, the assembly in which the library part or assembly was dropped references the original design library part or assembly. The object dropped into the assembly is in fact the same object that is in the library.

If a library model is used in an assembly, and the original model is altered in some way, it will affect every assembly in which that model is used. This is sometimes a desirable situation, such as in the case where a revision is made to a part and the revised part should then be used in every assembly where it is located. The down side to this is that there may be some assemblies for which the revised part should not be used. In addition, if something were to happen to the original library part every assembly using it would be affected. The following section shows you how to keep this from happening.

Determining File References

Knowing where all of the components are in an assembly is important, and there are many reasons why this is so. If a part is moved, renamed, or deleted—and that part is in an assembly—it will affect the assembly. If a copy of an assembly is needed, it may be necessary to copy the components as well. These are just two reasons.

To determine where components in an assembly are located, select Find References from the File menu. This will open the window shown in Figure 13–34. Files may be labeled "read-only" if they are opened by some other user. Files labeled "not open" are suppressed in the assembly. Sometimes it is necessary to determine which files are being referenced by a drawing. Even part files can reference other files under certain conditions.

The main point of concern in this discussion are the files with a path leading back to the Design library. These files are circled in Figure 13–34. It is preferable to make a copy of the library model and have the assembly reference the copy. This can make for easier organization with regard to the assembly's components. If all parts being used in an assembly are located in one area, housekeeping becomes a much easier task.

If you wish to save a copy of a part that was just inserted from one of the library's part or assembly folders, open the model in its own window and perform a Save As. (Note that opening the file can be done most easily by right-clicking on the component and selecting Open Part or Open Assembly.) Assuming the original assembly is open, SolidWorks will issue a cryptic warning as soon as Save As is selected from the File menu.

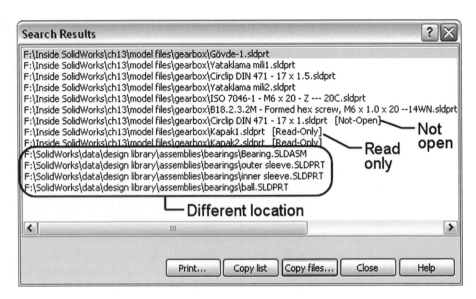

Fig. 13–34. Finding references.

It is very important that you understand exactly what this message is actually trying to say. First, go ahead and click on OK when you see the message. When the Save As window appears, decide where you want to save the file and give the file a name. You can change the name or leave it as is, whatever you choose. Now here is where you need to pay close attention.

If you do not check the Save As Copy option, shown in Figure 13–35, the assembly will reference the copy, not the original file. In other words, if you were copying a file named *Bolt*, and the copy is being named *Bolt Revision2*, the assembly that was using the file named *Bolt* will use the copy, *Bolt Revision2*, instead. If you do check the Save As Copy option, the assembly will continue to use the original file.

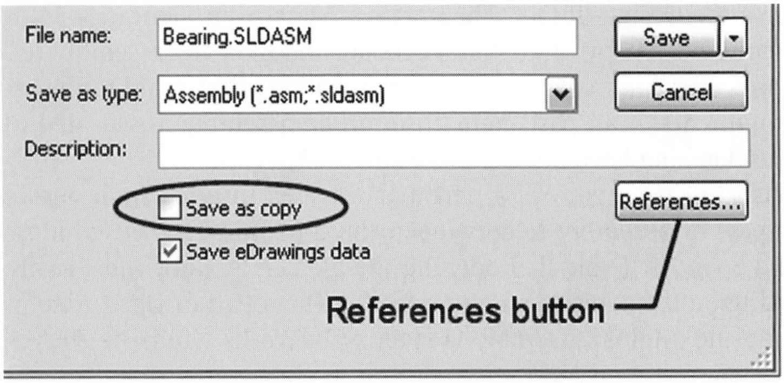

You will only see the warning message if the assembly referencing the file you are copying is open. If the assembly is not open, you will not see the message when performing a Save As (because the message does not apply). The assembly would not use the copy even if the Save As Copy option were not checked.

Fig. 13–35. Saving a copy.

In our particular case, we want the assembly to begin referencing the copy, so make it a point not to check the Save As Copy option. Once the file has been saved to the hard drive, you can feel free to close the part window and return to the assembly. You will find that FeatureManager now references the copy, and not the original part that was dragged and dropped out of the Design library.

If it becomes necessary to save a copy of an assembly brought in from the Design library, the assembly's components should be copied as well. To accomplish this task, click on the References button (shown in Figure 13–35) and check all of the components listed. For a more in-depth discussion on topics related to file references see Chapter 19.

Component Patterns

Component patterns are extremely similar to feature patterns. The only differences are what is being patterned and the command used to start the patterning process. Component patterns make for an efficient assembly, so they should be used whenever possible.

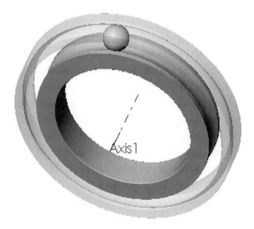

There are two flavors of component patterns: locally defined patterns (which are most like feature patterns) and feature-based patterns, which can be thought of as derived patterns. Examples of both component pattern types are shown in material following.

Figure 13–36 shows a bearing assembly. The outer component is shown in a transparent state for clarity. The ball has already been mated in position and requires patterning. A circular pattern will be used, and similar to circular feature patterns an axis will determine the center of rotation the ball will be patterned about. How-To 13–4 takes you through the process of creating a component pattern.

Fig. 13–36. Prior to creating a component pattern.

HOW-TO 13-4: Locally Defined Component Pattern

To create a component pattern, perform the following steps. In this particular example, you will create a circular pattern, but linear component patterns can also be created.

1. Select Circular Pattern (or Linear Pattern) from the Insert > Component Pattern menu.

Fig. 13–37. Circular (component) Pattern
PropertyManager

Fig. 13–38. Completed component pattern.

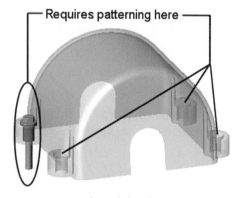

Fig. 13–39. Good candidate for a feature-based
component pattern.

2. In the Circular Pattern PropertyManager (shown in Figure 13–37), click in the Pattern Axis list box and select an axis to establish the pattern direction. If creating a linear pattern, select a linear edge to establish the pattern direction (linear dimensions can also be used).

3. Click in the Components to Pattern list box and select the components to be patterned.

4. Specify the Spacing and Number of Instances for the pattern. The Number of Instances is the total number of items for the pattern, including the original.

5. Click on OK to complete the pattern.

Figure 13–38 shows the completed component pattern. In our particular case, a circular pattern was created. It would have been just as easy to create a linear pattern. Basically, if you understand how feature patterns work component patterns should pose no difficulty.

Locally defined component patterns are just that—local patterns that are not based on external geometry. It would be best if a component pattern could be based on an existing feature pattern. This would be an ideal situation, because if the underlying feature pattern changes in any way the component pattern will automatically update. This includes changing the number of instances in the original feature pattern.

The type of component pattern we are discussing is known as a *feature-based pattern*. Figure 13–39 shows an example of where such a pattern would be very well suited. It is the type of pattern you should always use if circumstance permits it. How-To 13–5 takes you through the process of creating a feature-based pattern. The process is less involved than defining a local pattern.

How-To 13-5: Feature-based Componet Pattern

Fig. 13–40. Feature Driven PropertyManager.

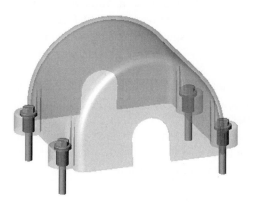

Fig. 13–41. Feature-driven pattern.

Because featured-based component patterns are based on existing feature patterns, an existing feature pattern belonging to a component in the assembly is a requirement. The steps that follow outline the process of creating a feature-based component pattern.

1. Select Feature Driven from the Insert > Component Pattern menu.

2. In the Feature Driven PropertyManager, shown in Figure 13–40, select the components to be patterned.

3. Click in the Driving Feature panel and select a feature pattern to base the component pattern on. Perform this task using FeatureManager.

4. Click on OK to create the pattern (see Figure 13-41).

An important benefit of using a feature-based component pattern is associativity. The component pattern is based on an existing feature pattern. Therefore, if the number of instances in the feature pattern changes the number of components in the component pattern will update accordingly. Any feature pattern can be used to drive a component pattern, with the exception of curve-driven patterns. As of this writing, curve-driven patterns cannot be used to drive a component pattern.

Deleting Patterned Components

Deleting a component in an assembly component pattern is accomplished in the same manner features are deleted from feature patterns. If a face is selected on a patterned component and the Delete key is pressed, SolidWorks will ask you to confirm whether the component should really be deleted.

Once a component has been deleted from a component pattern, it will appear in the pattern's definition. Specifically, the deleted components will be listed in the *Positions to skip* list box. By removing a listing from this box, the deleted pattern component can easily be recovered. To edit a pattern (or any feature), right-click on the pattern in FeatureManager and select Edit Feature.

Keep in mind that with a feature-based component it may be more prudent to delete an underlying feature pattern instance rather than a component pattern instance. In this way, the number of component instances will continue to match the number of instances in the feature pattern.

Dissolving Component Patterns

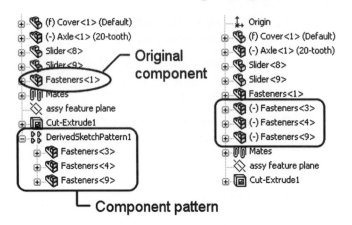

Fig. 13–42. Dissolving a component pattern.

If a number of components have been patterned, it is not possible to delete just one of the components from one instance in the component pattern. Applying this to our earlier example (see Figure 13–39), it would not be possible to delete just the washer from one of the patterned instances. The bolt and insert would wind up being deleted as well. You could, however, hide a component (such as the washer) from a single instance.

Figure 13–42 shows an example of what happens in FeatureManager when a component pattern is dissolved, which can be accomplished by right-clicking on the pattern in FeatureManager and selecting Dissolve Pattern. Each pattern instance becomes a separate component. Be forewarned that each component should be mated in position, which will take time. It is good practice to leave component patterns intact. Do not dissolve them unless you have a specific reason to do so.

Hiding Components

Sometimes it is difficult to add mating relationships because other parts get in the way. On other occasions it might be desirable to simply hide a component for whatever reason. This is easily accomplished by selecting a component and clicking on the Hide/Show icon found on the Assembly toolbar. Right-clicking on a component either in FeatureManager or in the work area will also gain access to the *Hide* command. Considering it will not be visible, you will have to right-click on the same component in FeatureManager in order to access the *Show* command.

When a component is hidden, its icon will appear ghosted, as shown in Figure 13–43. This behavior is consistent throughout the software for any object that is hidden. For example, the icon of the hidden plane in Figure 13–43 is ghosted as well.

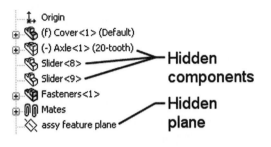

Fig. 13–43. Hidden component icons appear ghosted.

If you want to hide or show more than one component at a time, make it a point to hold down the Ctrl key in order to select more than one item. Under the Edit menu, you will find the options Hide and Show (selected components). There is also a selection titled Show with Dependents, which will show a hidden component, along with any dependent components hidden with it.

When using the Edit menu to hide or show components, there will be a submenu with the options *This configuration*, *All configurations*, and *Specified configurations*. These are made available because a component's visibility is a configurable trait. In other words, a component can be hidden in one configuration but not in another. (See Chapter 11 for additional information on configurations.)

✓ **TIP:** *To make a component transparent rather than hidden, use the Change Transparency icon found on the Assembly toolbar.*

Component States

There are three states an assembly component can exist in: suppressed, lightweight, or resolved. Components in assemblies can be suppressed using the same techniques learned in Chapter 11 to suppress features. These options included using the Suppress/Unsuppress icons, using the Edit menu, and accessing a component's properties. There is also a multifunction icon on the Assembly toolbar that can be used to suppress components (discussed in material to follow).

When a component is hidden, its visibility is turned off. It can, however, still be "seen" by SolidWorks. When a component is suppressed, it is invisible to both user and SolidWorks. Suppressing versus hiding can have its advantages and disadvantages. For example, suppressing components removes them from memory, which will reduce rebuild times and increase overall system performance. It will also suppress any dependent mates. This may cause the assembly to move apart unexpectedly if too many key components are suppressed. On the other hand, hiding a component does not affect its associated mates.

When an assembly is opened in SolidWorks, its components can be loaded in a state known as *lightweight*. A component loaded lightweight is not fully loaded into memory. Only the visual information needed to display the component is loaded. This situation results in two very important facts. The first is that assemblies opened lightweight can open very quickly. This has huge benefits to those who work with assemblies containing many components. When a component is loaded lightweight, it is displayed with a blue feather in FeatureManager, as indicated in Figure 13–44.

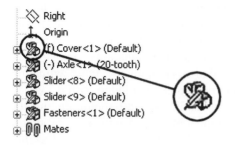

Fig. 13–44. Lightweight components appear with blue feathers.

→ **NOTE:** *An assembly must be saved in shaded display mode for it to be loaded lightweight.*

Another important aspect of lightweight components is that they must be loaded first before they can be edited. Simply clicking on a component in the work area is enough to load it (i.e., *resolve* the component). When a component is resolved, it has been fully loaded into memory and is neither suppressed or lightweight.

To load an assembly lightweight, select the Lightweight option present in the Open window (File > Open) whenever an assembly is selected for opening. Other options that affect the loading of lightweight components are shown in Figure 13–45. These are found in the Performance section of the System Options (Tools > Options).

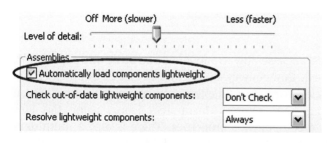

Fig. 13–45. Lightweight component options.

The first option, *Automatically load components lightweight*, speaks for itself. The *Check out-of-date lightweight components* option can be set to Don't Check, Indicate, or Always Resolve. A component is out of date if it has been changed since the last time the assembly was opened. Typically, you would want to know if something has been changed. If the Indicate option is used, the components will remain lightweight, but will be displayed with a red feather instead of blue. In such a case, it might be a good idea to resolve the component so that the changes can be seen.

The *Resolve lightweight components* option can be set to Always or Prompt. The Prompt option gets annoying because a small window will pop up asking if you really want to resolve a component every time SolidWorks needs to load it due to some action on your part. It is best to leave this option set to Always, which is what causes a component to be resolved simply by clicking on it.

→ **NOTE:** *When suppressed, components in FeatureManager appear gray, just as features do, and cannot be expanded to show the features or components they contain.*

Fig. 13–46. Change Suppression State icon and submenu.

The Change Suppression State icon on the Assembly toolbar, shown in Figure 13–46, functions slightly differently than other icons. Clicking on this icon opens a submenu with the three options shown in the figure. In this way, any component can have its state changed to suppressed, resolved, or lightweight.

Any component that requires editing must first be resolved. You have already discovered a few ways to accomplish this task. Right-clicking on components also gains access to the Set to Resolved option. If it becomes necessary to resolve the entire assembly, right-click on the assembly's name at the top of FeatureManager and select Set Lightweight to Resolved. This will resolve every component in the entire assembly.

Working with Subassemblies

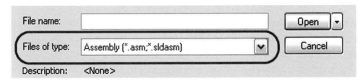

Fig. 13–47. Specifying the proper file type.

It is not always parts that are added to an assembly. Sometimes other assemblies must be added. Inserting an assembly works just like inserting a part, except that you must browse for the correct file type. Follow the same procedure you would for inserting any component. When the Insert Component window opens, specify assemblies in the *Files of type* drop-down list box (shown in Figure 13–47).

After an assembly has been inserted as a component into another assembly, it is then considered a subassembly. Subassemblies are no different than actual assembly files we have been discussing throughout this chapter. It is just a matter of hierarchy and terminology.

An important concept you should try to keep in mind is that a subassembly is considered a single component. Subassemblies will (by default) move and react as a single component. It typically takes just three mates to fully mate a subassembly into position, even if it has hundreds of components.

If you expand a subassembly, you can see the components that belong to it. This is shown in Figure 13–48. Note that the *Bearing* subassemblies are components that in part make up the *Axle* subassembly, which in turn is a component in the Gear Housing. FeatureManager can get quite complicated in an assembly. Parts can exist within subassemblies, which in turn exist within subassemblies, and so on. The level to which components exist within other components within the hierarchical tree is known as an assembly's "depth."

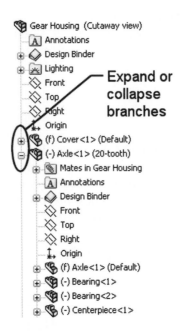

Fig. 13–48. Subassemblies and expanding or collapsing branches in FeatureManager.

Component Properties

There are a few functional parameters, as well as general information, that can be obtained from a component's properties. Accessing a component's properties is done via the right mouse button menu. By this time, you should be getting used to the right mouse button menu process. Right-click on a component and select Component Properties.

Whether you right-click on an individual part component or a subassembly component will make a slight difference as to what is shown in the Component Properties window. In Figure 13–49, only the upper portion of the Component Properties window is shown. This represents the properties for the *Axle* subassembly.

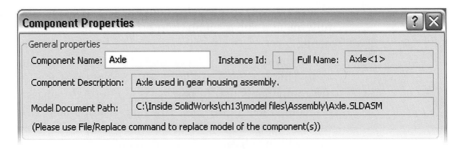

Fig. 13–49. Component Properties window, upper portion.

Table 13–3 outlines the options found in the General Properties section of the Component Properties window. Note that the term *component* refers to individual part components and subassemblies as well.

Table 13-3: General Component Property Definitions

Property Name	Definition
Component Name	By default, the file name of the component (see material following)
Instance ID	Instance ID of the component
Full Name	Component name with the instance ID suffix
Component Description	Text taken from the Description custom property (see Chapter 17)
Model Document Path	Path to where the model document is physically located

The component name is typically the same as the actual name of the part or assembly file being referenced. However, it does not have to be. In the External References section of System Options (Tools > Options), there is an option titled *Update component names when documents are replaced* (shown in Figure 13–50). When this option is unchecked, the component name can be anything typed in by the user.

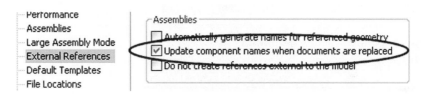

Fig. 13–50. Update component names when documents are replaced *option.*

You need to be very careful with this option! It is recommended that *Update component names when documents are replaced* be left on (checked). Otherwise, names of components in FeatureManager do not have to reflect the names of the actual files being referenced. This can cause a lot of confusion and make file tracking particularly nasty. Use this option only if you have a proper document tracking system in place, such as PDM software.

✓ **TIP:** *Descriptions that appear in the File > Open and Component Properties windows can display something other than the Description custom property. To change this setting, choose a different property in the drop-down list available in the General section of the System Options (Tools > Options).*

Other properties in the Component Properties window control visibility and other configuration-specific properties. Table 13–4 outlines these other properties and what they can accomplish. Material to follow expands on some of these properties. Figure 13–51 shows the lower portion of the Component Properties window.

Fig. 13–51.
Component
Properties window,
lower portion.

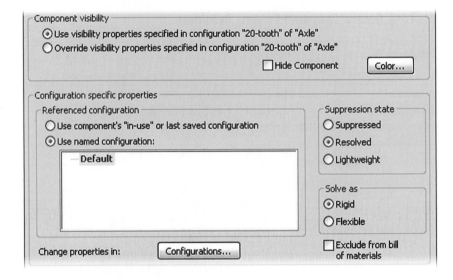

Table 13-4: Configuration-specific Component Properties

Property Name	Definition
"Use visibility…"	Only available in the properties of a subassembly component, this option forces the component to be hidden or shown based on the parent assembly.
"Override visibility…"	Only available in the properties of a subassembly component, this allows the component to be hidden or shown independently of the parent assembly.
Hide Component	Identical to the Hide/Show functionality discussed earlier in this chapter.
Color button	Allows for changing a component's color. Does not affect original part or assembly file.
Use components "in-use" or last saved configuration	Configuration shown in the assembly is whatever configuration was active when the component was last saved.
Use named configuration	Allows for selecting a particular configuration to be displayed in the assembly. This is the preferred setting.
Suppression state section	Allows for changing the state of the component. Identical to changing the state of a component, as discussed previously in this chapter.
Solve as section	Only available for subassemblies and contains the following two options:
	— Rigid: Solves the subassembly as a single rigid component (default setting).
	— Flexible: Allows for movement of components within a subassembly. Components can only move dependent on mate conditions placed on them within the subassembly.

Property Name	Definition
Configurations button	Any options changed in the configuration-specific properties section of the Component Properties window can be made to apply to specific configurations in the top-level assembly.
Exclude from bill of materials	Component will not appear in a BOM.

Many of the options listed in Table 13–4 you are already familiar with, such as changing a component's visibility or state. Other options are fairly new, and are actually very significant in the grand scheme of things. Of prime concern would be the *Use named configuration* option. It is extremely important to be able to switch what version of a particular component is being used in an assembly, for any number of reasons.

Likewise, being able to solve a subassembly as flexible as opposed to rigid is also very significant. Subassemblies, by default, will act as a single component. That is, movement between individual components in a subassembly is typically not possible. This behavior would be known as a *rigid* subassembly. By setting the Solve As option to Flexible, individual components within a subassembly can move. This option does increase the complexity level of an assembly, so use this only when necessary.

Component Replacement

It is typical to replace assembly components from time to time with a revised component, usually a component with a better design. Maybe the new component is lighter or stronger, or cheaper to build, or maybe it fulfills the stringent requirements set out by the customer. Whatever the reason, the component replacement function allows you to do this.

You might think that replacing a component with another in an assembly would wreak havoc with your mating relationships. If the new component originates from a completely different part file, mate relationships to the original component do require editing. However, if the replacement component is a modified version of the original, SolidWorks will reuse the existing mates if at all possible.

The Replace command can be used to replace just one occurrence of a component in an assembly, or all occurrences. It can also be used to replace, for example, an assembly with a part, or vice versa.

In the following example, one of the 20-tooth gears used in the axle assembly will be replaced with a 34-tooth gear with a wider profile for a different type of chain. The Replace command will be used to accomplish this. How-To 13–6 takes you through the process of performing a component replacement. Figure 13–52 shows a "before" screen shot of the assembly used in the example.

How-To 13-6: Replacing Components

To replace a component (or every instance of a component) in an assembly, perform the following steps. The steps listed are the basic steps you will need to perform. Other options are discussed in material to follow.

1. Right-click on the component to be replaced and select Replace Components, or click on the Replace Components icon typically found on the Assembly toolbar.

2. In the Replace Component PropertyManager, shown in Figure 13–53, click on the Browse button and select a replacement component.

3. Place a check in the *All instances* option if you wish to replace every instance of the component in the assembly.

4. Click on OK to replace the component or components.

At this point, assuming the *Re-attach mates* option (shown in Figure 13–53) was not checked, the new component should appear in place of the old one, and you may have some mate errors to clean up. If, on the other hand, you chose to check the *Re-attach mates* option, the Mated Entities PropertyManager (shown in Figure 13–54) will appear.

The Mated Entities panel can help in selecting replacement faces to repair the mates with errors. For example, if the concentric mate is selected from the list shown in Figure 13–54, both faces associated with the mate will highlight in the work area. However, it is likely one of the faces originally on the replaced component will be missing. It is your job to pick another face on the new component that can be used instead. Try not to let these types of errors intimidate you, because they really are not that difficult to fix.

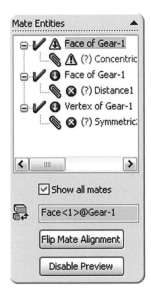

Fig. 13–54. Mated Entities PropertyManager.

Do what you can from the Mate Entities PropertyManager. Following that, if there are still mates with errors edit them independently. That is, right-click on any problematic mate in FeatureManager and select Edit Feature. It may be necessary to select replacement faces, flip mate alignment conditions, or perhaps even select a different mate. Figure 13–55 shows the same assembly with the new replacement gear.

There is one last option we need to examine. Select the *Match name* option if SolidWorks should try to use a configuration of the same name in the replacement component as the component being replaced. This is the default setting, and should be left alone if neither the original or replacement components have any additional configurations. If *Manually select* is chosen, you will be presented with a list from which a configuration can be selected for the replacement component prior to its insertion.

✓ **TIP:** *When replacing components, the Replace these components listing can be used to select multiple instances of a component for replacement. For example, if there are 20 hex-head bolts in an assembly, eight of those could be replaced with socket head cap screws simultaneously.*

Fig. 13–55. After component replacement.

Mate Troubleshooting and Repair

When replacing components, it is common to have mating problems arise. In a worst-case scenario, the old mates to the original component are no longer valid with the new replacement component. This has to do with the way SolidWorks recognizes the faces or other objects being mated to. For example, faces have internal identifiers, and a new replacement component will have faces with different internal identifiers than the original component. This is why mates often fail with replacement components and errors appear in FeatureManager.

Usually, the types of problems that arise from component replacement are easy to fix, once you know the drill. This is because it is rare to replace a component with one that is drastically different than the original. For instance, a hex-head bolt might get replaced with a square-head bolt, but a hex-head bolt would not get replaced with an external retaining ring. Well, at least not usually!

Chapter 7 first introduced you to error symbology. Mates use this same symbology. In addition, overdefined components will appear with a plus symbol (+), similar to how overdefined sketches appear in FeatureManager. You have also learned in this chapter that mates can be edited via the right mouse button, just as any feature can. In short, tracking down and repairing erroneous mates is a matter of right-clicking on the mates that display an error symbol and selecting Edit Feature.

There is more than one way to mess up mates in SolidWorks. Some of these problems are easier to fix than others. The following section categorizes and explores some of the problems that can occur with mates.

Missing References

Fig. 13–56. Fixing a mate with a missing reference.

When replacing a component with another, a missing reference is exactly the type of problem that generally crops up. A few mates in the assembly are looking for faces on a component that is no longer present. Look for error symbols in the *Mates* folder and edit the definition of the mates with problems. When a reference is missing, the Mates panel will look something like that shown in Figure 13–56.

There are three things that will help you in this situation. First, look for the invalid object in the Mates panel. You cannot miss it, because it will be labeled *Invalid*. Delete that object from the list because it is not in the model anymore. Second, note the type of mate you are editing. This is written in big letters at the top of the Mates panel. Note in Figure 13–56 the word *Coincident*. This will help in picking something to replace the invalid object you just deleted.

The third helpful aid is to look in the work area for the currently highlighted object. This will be one of the objects the mate is between, or in this case coincident with. Now all that remains is to pick something, such as a face, on the replacement component. Continue this process until you have repaired all of the mates with missing references.

Overdefined Mates

When too many mates are present, they overdefine components. This will be immediately noticeable due to the rebuild error symbols in the *Mates* folder. Look for the components in FeatureManager with plus signs preceding the component names and you will have found the overdefined components. Figure 13–57 shows an example of FeatureManager when there are too many mates present.

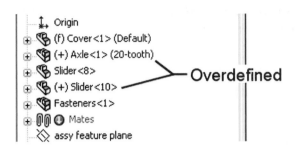

Fig. 13-57. Overdefined components.

As long as overdefining mates are caught early on and you do not let them go, they are not difficult to fix. If they are ignored and more accumulate, it gets more difficult to track down the problems. The moral: Do not put off fixing errors or they will just get worse.

Typically, an overdefining mate is found between a couple of components that have a few too many mates. Select the mates in the *Mates* folder and this will highlight in the work area the objects used for the mate. For example, selecting a concentric mate may highlight two cylindrical faces. Repeat this process for any mates containing error symbols. Once you get a handle on which mates are conflicting, delete the mate or mates you have decided you do not need.

Improper Alignment

When one mate has its alignment condition improperly set, the mate may show up as an overdefined mate. However, this often results in only one mate showing up in FeatureManager as being overdefined. How can one mate be overdefined? Doesn't it take two mates to cause a conflict and show up as overdefining mates? The answer is not necessarily, when it is a mate's alignment condition that is backward.

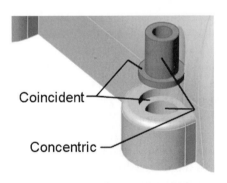

Fig. 13-58. Preparing to add a coincident mate. The concentric mate is already present.

If we use the insert and cover components shown in Figure 13-58 as an example, it is easy to understand how alignment conditions can cause mate errors. Assume that there is already a concentric mate between the cylindrical surface of the hole in the cover and the outer cylindrical surface of the insert. This reduces the movement of the insert to only two degrees of freedom.

Next, a coincident mate will be added between the surface of the insert and the top planar face surrounding the hole. This is also illustrated in Figure 13-58. The insert is currently flipped in the wrong direction, so the alignment condition is set to antialigned. Because the insert cannot flip around to be in the correct direction to satisfy the coincident mate, an error occurs. This is because the concentric mate has already reduced its degrees of freedom of movement. What to do?

To reiterate: the coincident mate needs to be antialigned. This causes the insert to want to flip, but it cannot because of the concentric mate. The solution: because the concentric mate is not technically causing errors, it contains

---

no error symbol as the coincident mate does. Nonetheless, changing the alignment condition of the concentric mate causes the insert to flip around so that the coincident mate can be satisfied. If the concentric mate had the proper alignment in the first place, the error would never have occurred.

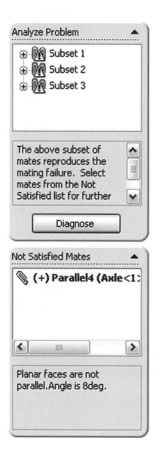

Fig. 13–59. Diagnostics PropertyManager.

Mate Diagnostics

With all of the problems that can occur with mate relations, it is good to know there is a little extra help available when needed. This extra help is known as mate diagnostics, which aids in troubleshooting assemblies with mate problems. A simple right-click on the problem mate will gain access to the Mate Diagnostics command.

Figure 13–59 shows the Diagnostics PropertyManager. The upper panel contains a Diagnose button. To be honest, most of the time clicking on this button does little to help find out what the problem actually is. It is the lower panel that is most beneficial.

The lower panel is titled Not Satisfied Mates. The primary mate causing the conflict is displayed in bold. If selected, this mate will display an additional message. In our case, the message is "Planar faces are not parallel. Angle is 8 degrees." This helps quite a bit. Now it is just a matter of deleting one of the mates causing the conflict, or editing the definition of one of the conflicting mates. The choice is yours.

Component Editing

If it is necessary to make a few dimensional changes to an assembly component, it is quite an easy process to access the dimensions. The process involves nothing you have not already learned. To access a dimension, double click on the part's feature whose dimensions you want to access, and then double click on the dimension to edit it.

↝ **NOTE:** *Altering a dimension on a component in an assembly will affect the original part.*

Components can be expanded in FeatureManager as needed to gain access to their respective features. Clicking on a component in the work area will highlight the component's name in FeatureManager. Turning on an option named *Scroll selected item into view* will cause FeatureManager to scroll up or down as necessary to show the component when it is selected from the work area. This makes hunting down a component in the feature tree much easier. Look in the FeatureManager section of System Options (Tools > Options) for this setting, shown in Figure 13–60.

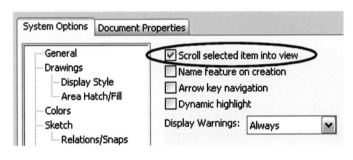

Fig. 13–60. Scroll selected item into view *option.*

Often it is necessary to go one step further than simply modifying dimensions on assembly components. There are occasions when feature geometry must be added to components. This requires editing the component and adding features to it, either as a separate part in its own window or in the context of an assembly. In either case, you would use the right mouse button to access these functions.

✓ **TIP:** *Right-clicking on a component in the work area or in FeatureManager allows for opening that file. Look for the Open Part command, or in the case of a subassembly the Open Assembly command.*

There are times when it is necessary to edit a part while still in the assembly, so that geometry can be extracted from other assembly components in order to help build new features on the part being edited. This can be accomplished by right-clicking on a component and selecting Edit Part or Edit Sub-assembly. Be aware that this process will very likely establish external references to other model geometry external to the part being edited. The process of editing a component in the context of an assembly carries with it many implications (examined in detail in Chapter 19).

Interference Checking

At certain times during the assembly process, or when the assembly is completed, you may want to see if there are components that interfere with each other. There is a tool that will allow you to perform an interference check, and it is aptly named Interference Detection. It is found in the Tools menu. Suspected components or the entire assembly can be tested for interference. How-To 13–7 takes you through the process of testing an assembly for interference.

How-To 13-7: Performing Interference Detection

To test an assembly for interference, perform the following steps.

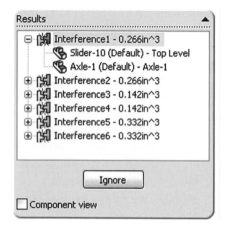

1. Select Interference Detection from the Tools menu, or click on the Interference Detection icon found on the Assembly toolbar.

2. Select the components (parts or subassemblies) to be checked for interference. By default, the entire assembly will be selected, but you can pick and choose.

3. Click on the Calculate button, which will be present in the Interference Detection PropertyManager (not shown).

4. To see individual interferences, select the interference in the Results panel, an example of which is shown in Figure 13–61.

5. Click on OK when finished.

Fig. 13–61. Interference detection results panel.

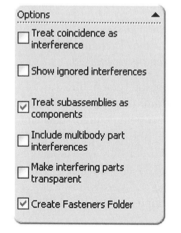

Fig. 13–62. Interference detection options.

Note in Figure 13–61 that each set of interfering components is listed in the Results panel. The interfering areas (or volumes) are indicated graphically as each incidence of interference is selected from the list. How the assembly is displayed is controlled from the various settings in the Interference Detection Property-Manager. The settings (Wireframe, Hidden, Transparent, and *Use current*) refer to the components that are *not* interfering. Experiment with these settings to see which you prefer.

The Options panel, shown in Figure 13–62, has a fair number of settings that will aid you in determining the interference. The Treat Coincidence as Interference option, if checked, will treat any faces on assembly components as interfering if they so much as touch each other. Typically, you will want to leave this setting off.

Certain interferences are intentional. Such would be the case with a press-fit fastener, for instance. Certain interferences are necessary, but it is not necessary to view these interferences. This is what the Ignore button is for, which can be seen in Figure 13–61. If certain interferences have been marked "ignored," you can choose to see them by checking the *Show ignored interferences* option.

all instances option been checked, and both bolts would have been excluded from being cut in the resulting section view. Be aware that the *Don't cut all instances* option can be set individually for each component in the Section Scope window.

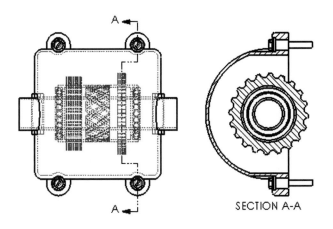

Fig. 13–76. Assembly section view.

If the *Auto hatch* option is selected, sectioned components will have alternating hatch patterns, making it easier to distinguish components from one another. *Exclude fasteners* is an option that forces fasteners to be ignored in the section view, similar to adding those fasteners to the section scope. Use the *IsFastener* custom property described previously to accomplish this task.

The final option, *Flip direction*, flips the direction of the section line arrows and the section view itself. To modify the section scope after the section view has been created, right-click on the section view and access its properties.

Hiding Components in Drawing Views

Sometimes it becomes necessary to hide a particular component for the sake of clarity, or for other reasons. The process is quite simple and can be accomplished via the right mouse button. Right-click on the view containing the components you wish to hide and select Properties. All drawing views have properties, just like most objects in SolidWorks.

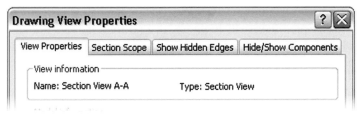

Fig. 13–77. Top of the Drawing View Properties window.

Once the view's Drawing View Properties window has been opened, note the tabs present. If the view is a section view, there is a tab named Section Scope, and the top of the Drawing View Properties window would look as it does in Figure 13–77. This is where the section scope for the assembly section view, discussed previously, can be modified.

In the same Drawing View Properties window is a tab named Hide/Show Components. Selecting this tab and then selecting components, either from FeatureManager or from the drawing view, will result in those components being hidden. To make a component visible once again, simply delete it from the Hide/Show Components list.

An option that goes hand in hand with hiding components in an assembly drawing view is an option titled *Automatic hiding of components on view creation*. This option, shown in Figure 13–78, is accessed in the Drawings section of System Options (Tools > Options). When enabled, all components behind other components will automatically be hidden. If you intend on showing the drawing views with hidden lines removed anyway, this option is an outstanding choice.

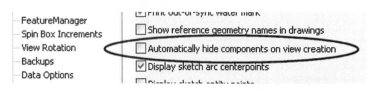

Fig. 13–78. Automatic hiding of components on view creation *option.*

Using *Automatic hiding of components on view creation* can decrease the amount of time it takes to generate assembly views containing many components. If you then choose to show the view with hidden lines dashed, it is possible to remove from the Hide/Show Components list only those components you wish to show the hidden lines for.

Show Hidden Edges

If you would rather not show any hidden edges, but still find it necessary to show the hidden edges for one or perhaps just a few components, use the Show Hidden Edges tab. Similar to using the Hide/Show Components functionality, showing the hidden edges of particular components in a view is just a matter of selecting the components. However, selecting the components may be a bit more difficult, in that they cannot be seen in the first place!

There is no secret process for selecting a hidden component in order to have its hidden lines shown. You will just have to select the component from FeatureManager instead of the drawing view. No big deal. If the view is selected first, its associated icon will highlight in FeatureManager. You can then drill down through FeatureManager to find and select the desired component or components.

Bill of Materials

A bill of materials (BOM) is a list of the components being used in the assembly. One of the basic BOM templates included with the SolidWorks software has four columns, one each for Item Number, Quantity, Part Number, and Description. Once a BOM is inserted into an assembly layout, it will keep track of components automatically if you add or remove components from the assembly.

There are two types of BOMs that can be created. The first type is a BOM generated using an Excel spreadsheet. You must have Microsoft Excel loaded on your computer to take advantage of this functionality. The second type of BOM is SolidWorks based. SolidWorks-based BOMs do not require an outside program.

The deciding factor on what type of BOM you choose to create lies in what will be done with the BOM data. Excel-based BOMs can be exported as Excel spreadsheets. SolidWorks-based BOMs can be exported as text or CSV (comma-separated value, otherwise known as comma-delimited value) files.

Exploded views work nicely with BOMs as there is plenty of space for the balloons, which are almost always used in conjunction with a BOM. Figure 13–79 shows the exploded assembly drawing we will be adding a BOM to. Both exploded views and adding balloons are explored in material to follow. How-To 13–12 takes you through the process of inserting a SolidWorks-based BOM. The section that follows How-To 13–12 explores the various options present during the insertion process.

Fig. 13–79. Preparing to add a BOM.

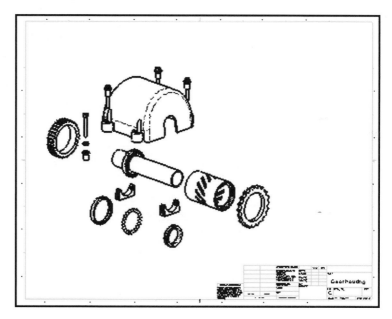

HOW-TO 13-12: Inserting a BOM

To insert a BOM, perform the following steps. Note that this How-To discusses the steps for inserting a SolidWorks-based BOM. If you wish to insert an Excel-based BOM, see How-To 13–13.

1. Select a view. (This is required because theoretically there may be different assemblies or parts within the same sheet and the software needs to know what it should generate the BOM from.)

2. Select Bill of Materials from the Insert > Tables menu, or click on the Bill of Materials icon found on the Tables toolbar.

3. In the BOM PropertyManager, a portion of which is shown in Figure 13–80, select the desired template to base the BOM on.

4. Select if an anchor point should be used; and if so, which corner of the table to be anchored.

5. Specify the desired parameters in the Bill of Materials PropertyManager. These parameters are explained in detail in material to follow.

6. Click on OK to create the BOM.

7. If an anchor point was not used, pick on the drawing to locate the BOM.

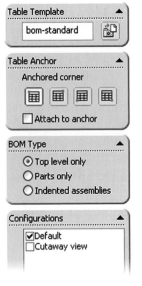

Fig. 13–80. Bill of Materials PropertyManager.

A portion of a sample BOM is shown in Figure 13–81. There are quite a few optional parameters that affect a BOM's appearance. If you have followed along with the steps in the previous How-To, your BOM will almost certainly look different. Let's explore these parameters so that you can gain a better understanding of how to customize the appearance of a BOM.

*Fig. 13–81.
Sample BOM.*

ITEM NO.	PART NUMBER	QTY.	DESCRIPTION
1	Cover	1	
2	Axle	1	
3	outer sleeve	2	
4	inner sleeve	2	
5	ball	44	
6	Centerpiece	1	
7	Gear	1	
8	Gear 20-tooth	1	

Step 3 in How-To 13–12 states that a template should be selected. SolidWorks-based BOM templates are stored, by default, in the SolidWorks *lang\english* folder, and will have the file extension *.sldbomtbt*. Review the section "Tables" in Chapter 12 if you need help customizing SolidWorks table templates.

Defining anchor points was discussed in Chapter 12 (see How-To 12–6), so step 4 in the previous How-To should be familiar to you. What requires further explanation is the BOM Type panel, which is visible in Figure 13–80. This panel contains three options, and they are very important because these options dictate exactly what is listed in the BOM. Table 13–5 outlines these options and indicates exactly what is shown in each case.

Table 13-5: BOM Options

BOM Type	Description
Top level only	All top-level components in the assembly are listed in the BOM. This includes part components and subassembly components. Components in subassemblies are not listed.
Parts only	All individual part components are listed in the BOM. Subassemblies are not listed, but any part components contained within those subassemblies are listed.
Indented assemblies	All top-level components in the assembly are listed and are assigned item numbers. Components contained within subassemblies are listed in an indented list and are not assigned item numbers.

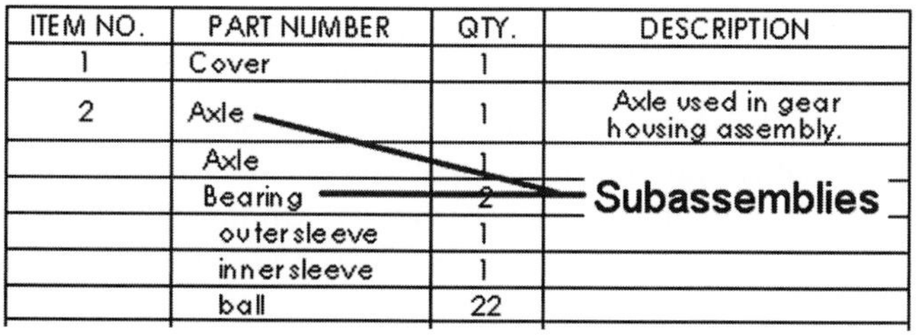

ITEM NO.	PART NUMBER	QTY.	DESCRIPTION
1	Cover	1	
2	Axle	1	Axle used in gear housing assembly.
	Axle	1	
	Bearing	2	Subassemblies
	outer sleeve	1	
	inner sleeve	1	
	ball	22	

Fig. 13–82. An Indented assemblies *BOM.*

The *Indented assemblies* BOM type is a bit easier to understand if it can be seen in action. Therefore, Figure 13–82 shows an example of one such BOM. The Axle is a subassembly that contains an axle part component as well as the bearing subassembly component. Being that the bearing is a subassembly in its own right, its components are also shown in an indented list. None of the components that are not top-level components have item numbers assigned to them.

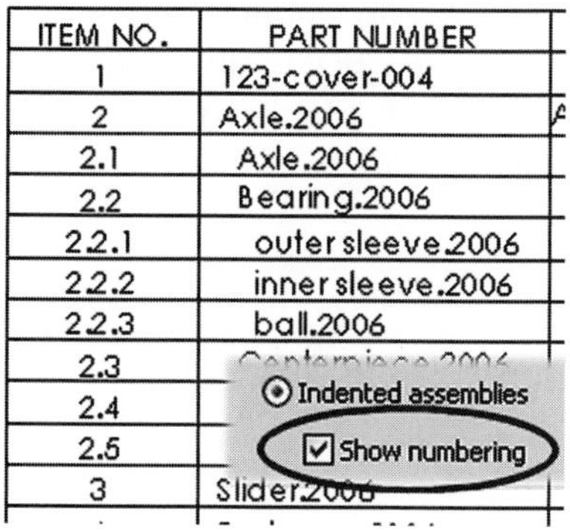

SW 2006: When BOMs are shown in an indented list, such as that shown in Figure 13–82, the indented items can be numbered. This is accomplished by checking the Show Numbering option in the BOM Type panel. The option is shown as an inset in Figure 13–83, which also shows what the numbering scheme would look like after enabling the Show Numbering option.

ITEM NO.	PART NUMBER
1	123-cover-004
2	Axle.2006
2.1	Axle.2006
2.2	Bearing.2006
2.2.1	outer sleeve.2006
2.2.2	inner sleeve.2006
2.2.3	ball.2006
2.3	Centerpiece.2006
2.4	⊙ Indented assemblies
2.5	☑ Show numbering
3	Slider.2006

Fig. 13–83. Results of using the Show Numbering *option.*

The Configurations panel (visible in Figure 13–80) is interesting in that multiple configurations can be selected to be displayed in a BOM, but only if *Top level only* is selected as the BOM type. If using one of the other two BOM types, select the desired configuration from a dropdown list. This will in turn generate the BOM based on the components present in the selected configuration. By the way, hidden components will appear in a BOM, but suppressed components will not.

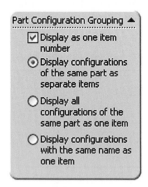

Fig. 13–84. Part Configuration Grouping panel

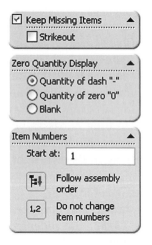

Fig. 13–85. Additional BOM PropertyManager settings.

The Part Configuration Grouping panel, shown in Figure 13–84, controls how a component is listed in a BOM when that component has multiple configurations that are used throughout the assembly. The options in this panel are very descriptive, so further elaboration isn't necessary. However, be aware that options in this panel are only relevant when the BOM type is set to *Top level only*.

When multiple configurations are shown in a BOM, some configurations may not contain all of the same components as other configurations. How these missing components appear in the BOM is controlled by the Keep Missing Items option and the Zero Quantity Display panel, shown in Figure 13–85. If Keep Missing Items is not checked, anything removed from the assembly will be deleted from the BOM. Additional possibilities are to leave the item in the BOM but show that item with a quantity of zero, to display the missing item with a strike-through, and so on.

Finally, there is the Item Numbers panel, also visible in Figure 13–85. As the name would suggest, this setting controls what item number the components in the BOM are numbered from. The default setting is one (1), which would typically be left as is. The *Follow assembly order* option assigns item numbers based on the order in which the components are listed in the assembly. Using the *Do not change item numbers* option forces items to retain their item numbers, even if alterations are made to the table.

When inserting an Excel-based BOM, the steps are very similar to inserting a SolidWorks-based BOM, though the interface is different. Excel-based BOMs utilize a conventional Windows-style interface, whereas SolidWorks-based BOMs utilize the by-now-familiar PropertyManager. How-To 13–13 takes you through the process of inserting an Excel-based BOM.

HOW-TO 13-13: Inserting an Excel-based BOM

To insert an Excel-based BOM, perform the following steps.

1. Select a view.

2. Select Excel Based Bill of Materials from the Insert > Tables menu, or click on the Excel Based Bill of Materials icon found on the Tables toolbar.

3. In the window that appears, select the desired template to base the BOM on and click on the Open button. The Bill of Materials Properties window will appear, the upper portion of which is shown in Figure 13–86.

4. Specify the desired parameters for the Excel-based BOM.

5. Click on OK to create the BOM.

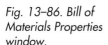

Fig. 13–86. Bill of Materials Properties window.

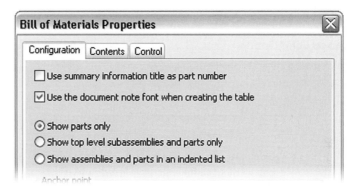

At this point there will be some commotion taking place on screen. Give SolidWorks a few seconds to generate the table and then display the table in the drawing. Once displayed, the BOM table can be repositioned by dragging it.

The summary information title has to do with the individual part file's properties. To view the Summary Information window, open any part and select Properties from the File menu. Data can be typed into the area labeled *Title* in the Summary tab of the Summary Information window. That data will be used in place of the file's name for the Part Number column in the BOM if *Use summary information title as part number* is checked.

If the Title area is not filled in, SolidWorks will use the part file name as the part number, rather than leave the part number column blank. Using the part file name as the part number is the default when *Use summary information title as part number* is not checked. (See Chapter 11, "Bill of Material Part Numbers," for the preferred method of controlling what is used as the part number in a BOM.)

By checking the *Use the document's note font when creating the table* option, through Document Properties (Tools > Options) you can specify what font will be used by the BOM. In the Document Properties tab, select the Note section, and then click on the Font button and specify a font, font size, and any other characteristics required for the note font. A benefit of this is that the BOM font can match the font in the rest of the document. If this option is not checked, the default font saved with the BOM template will be used.

There are three options when creating an Excel-based BOM that directly relate to the three BOM types outlined in Table 13–5. These options directly pertain to what is shown in the resultant BOM, and are therefore very important. Table 13–6 compares the BOM Type options in the Excel-based Bill of Materials Properties window with those available when inserting a Solid-Works-based BOM.

Table 13-6: Excel-based BOM Options

Excel-based BOM Option	*Equivalent SolidWorks-based BOM Type*
Show top-level subassemblies and parts only	Top level only
Show parts only	Parts only
Show assemblies and parts in an indented list	Indented assemblies

The anchor point settings are present when inserting an Excel-based BOM and function in exactly the same way as when inserting a SolidWorks-based BOM. The *Add new items by extending top border of table* option (not shown) is only available if an anchor point is not being used. It is one of the more commonly misunderstood BOM options. What it means is that if new components are added to the assembly the top of the BOM will be extended. What it does not mean is that newly inserted items will be placed at the top of the BOM.

This option works well if the BOM is positioned directly above the title block, or near the bottom of the drawing. Newly inserted items will always be inserted into the BOM automatically by SolidWorks at the bottom of the BOM list, but the BOM will "grow" upward, as if it were anchored at its base.

Editing a BOM

Editing a SolidWorks-based BOM can be accomplished through Property-Manager. Because the BOM table is a standard SolidWorks table, all of the formatting options and general table functionality you learned in Chapter 12 still apply.

To edit an Excel-based BOM, either double click on it or select Edit after right-clicking on the BOM. This action will open Excel within SolidWorks and allow for formatting or making manual text changes. There is a problem associated with adding text to the BOM in this manner. If the BOM is regenerated (such as when changing what is shown in the BOM), any text added manually to the BOM will be lost. There is a proper way of adding data to a BOM that will associate itself with a component and remain even through a regeneration.

✗ **WARNING:** *Do not manually change item number, part number, or quantity, as these items are directly controlled by SolidWorks. These items will update automatically if changes to the assembly are made.*

The proper process used to populate a BOM with information involves custom properties, which are discussed in detail in Chapter 17. To give you a basic understanding of the concept, we will touch on the topic here. Essentially, any type of custom data can be made to populate a BOM. This includes data such as a part description, cost, vendor, mass, size parameters, or anything else imaginable.

In Chapter 12, you learned that a table can have certain elements, and that those elements have certain modifiable parameters. For example, a table cell can have its text formatted a certain way, or a table column can be repositioned. Let's take this topic to the next level and look at the column properties of a BOM, as shown in Figure 13–86. It is the Custom Property setting we want to focus on.

If any of the components in an assembly have had custom properties added, the names of those properties will appear in the drop-down list shown in Figure 13–87. If a particular property is selected, the column will display that property. If there are any components in the assembly that have not had the specified custom property added, the cell in the column for that component will remain blank.

Fig. 13–87.
Column properties of a SolidWorks-based BOM.

✓ **TIP:** *To add generic items to a BOM, such as paint or glue, create a new part and save it with the name of the generic item. Insert that part into the assembly (no mating is required) and it will automatically appear in the BOM.*

It has been mentioned on more than a few occasions that adding custom properties is discussed in Chapter 17. The topic of custom properties is left to that chapter because of custom properties' close relationship with design tables, but that doesn't mean you can't skip ahead if your curiosity is getting the best of you!

Customizing Excel BOM Templates

The key to customizing an Excel-based BOM template is to make sure to name the cell at the top of a column in the BOM template exactly the same as the name of the custom property that will fill that column. This is not required with SolidWorks-based BOM templates.

It is very important to realize the difference between naming a cell and adding data to a cell. The cells must be named correctly, or the customized BOM template will not work. This book is not meant to teach Excel, but this topic deserves a bit of special attention. Figure 13–88 shows a small portion of an Excel window in which a BOM template is being edited.

Fig. 13–88.
Cell names
versus cell data.

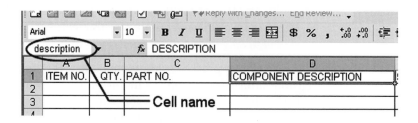

Focus on cell D1 in Figure 13–88. Note the data within the cell. The data in the cell is COMPONENT DESCRIPTION. This data is not the same as the name of the cell. By default, the names of cells follow cell positions. Therefore, the name of cell D1 was originally *D1*. The name of cell D1 in Figure 13–88 has been changed to *description*. It is the name of the cell that is most important. It is the name of a cell at the top of a column in the BOM template that must match the name of the custom property you wish to automatically populate that column.

Now you know the technique for creating your own Excel-based BOM templates. Be aware that formulas can also be added to increase a BOM's functionality even further. When designing your own template, it is best to begin with one of the default templates provided by SolidWorks and go from there. Once the template has been modified, save it with a different name so as not to overwrite the original.

✗   **WARNING:** *It is possible to create a custom property called* PartNo, *which is in fact the name SolidWorks uses to link to an Excel-based BOM. However, custom properties added by the user with the name PartNo will not transfer to a BOM.*

Excel-based BOM Properties

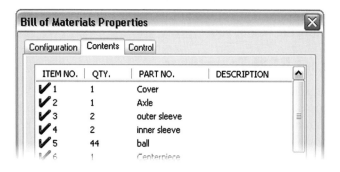

Fig. 13–89. BOM properties Contents tab.

Accessing the properties of an Excel-based BOM provides access to other, more powerful, options. Right-clicking over the BOM and selecting Properties will bring up the same Bill of Materials Properties window displayed when first creating the BOM. In that the options in the Configuration tab were already explained in previous material, we will focus on the other two tabs, Contents and Control. The Contents tab is shown in Figure 13–89.

Through the use of the Contents tab, the BOM can be sorted in various ways. For instance, clicking on the Part No. heading sorts the content alphabetically. A second click on the same heading reverses the order. Unchecking the *Display labels at top* option (not shown) places the BOM labels on the bottom. Unchecking one of the listed components hides that item from appearing in the BOM.

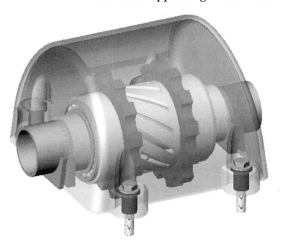

Fig. 13–90. BOM properties Control tab.

The Control tab, shown in Figure 13–90, provides functionality for customizing the appearance of the BOM even further. If *Row numbers follow assembly ordering* is checked, the order in which items are displayed in the BOM will be the same order the components are listed in the assembly's FeatureManager. This can be advantageous. By dragging components to different positions in the assembly FeatureManager, the order of items in the BOM can be altered.

When *Row numbers follow assembly ordering* is turned off, it becomes possible to reserve row numbers for missing components. The missing components can either have their rows removed entirely or be displayed with a zero quantity or strikethrough text.

If a BOM is long enough to flow off the bottom of the sheet, use the *Split tables into multiple sections* option. You could then specify how often the split should occur, such as every 15 inches. Of course, this setting will depend on the sheet size of the paper.

Exploded Views

An exploded view is a view of an assembly where the components have been separated, typically in the reverse order in which they were assembled. These view types are often created for illustrating how assemblies are built, or even just to show all components in a clear manner without using numerous cutaway views. Often assemblies are exploded for BOM drawings as well.

Before an exploded view of an assembly can be shown in a drawing, the assembly itself must be exploded. It is highly recommended that a user-defined view be saved in the assembly prior to exploding the assembly (see How-To 3–9). If the model happens to be rotated at some future point in time, this makes it easy to return the model to the same orientation it was in prior to exploding it. How-To 13–12 takes you through the process of exploding an assembly.

How-To 13-14: Exploding an Assembly

To explode an assembly, perform the following steps. This must be done in an assembly file, not a drawing.

Fig. 13–91. Explode PropertyManager.

1. Activate the configuration you wish to explode.

2. Select Exploded View from the Insert menu, or click on the Exploded View icon found on the Assembly toolbar. This will open the Explode PropertyManager, a portion of which is shown in Figure 13–91.

3. Select the component(s) to be exploded from either the flyout FeatureManager or the work area. A triad will appear near the last component selected, such as that shown in Figure 13–92.

4. Place the cursor over the arrow on the triad that represents the direction in which the component should be exploded, and then drag (explode) the component the desired distance.

5. Repeat steps 3 and 4 until all of the desired components have been exploded.

6. Click on OK to accept the exploded view and complete the explode process.

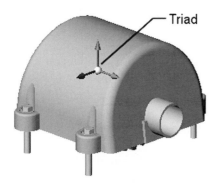

Fig. 13–92. Exploded view triad.

In its simplest form, creating exploded views is extremely user friendly, and is essentially nothing more than a series of drag operations. However, there is additional functionality you should be aware of. For example, each time components are exploded they are listed in the Explode Steps panel, shown in part in Figure 13–93. Components can be deleted from individual explode steps, thereby "unexploding" them, in a manner of speaking. Entire explode steps can also be deleted. It is even possible to reorder explode steps by dragging them to a new position within the Explode Steps panel.

When components are selected and the triad appears, it is not necessary to accept the default orientation of the triad. By clicking in the Explode Direction list box in the Settings panel (see Figure 13–91), it becomes possible to select an alternate edge or face with which to align the triad. Components can thereby be exploded in a particular direction. A precise explode distance can also be specified in the same Settings panel. Use the Apply button to preview any distance entered via the keyboard, and the Done button to accept those changes.

The Options panel (see Figure 13–91) contains beneficial functionality, but can be a bit confusing at first. First, if *Auto-space components after drag* is checked multiple components *must* be selected or nothing out of the ordinary will take place. Assuming multiple components have been selected, those components will evenly space themselves out as soon as the mouse button is released to complete the drag operation. Think of this as an auto-explode function, with the slider bar controlling the spacing between components.

Fig. 13–93. Explode Steps panel.

✓   **TIP:** *Every component in the assembly can be selected at once by dragging a window around the assembly. Dragging the triad a small amount and letting it go will then space every component in the assembly evenly along the drag direction.*

You may find yourself turning the *Select sub-assembly's parts* option on and off throughout the explode process. Turning this option off allows for picking an entire subassembly with one click, whereas leaving the option checked allows for picking and choosing which components within a subassembly should be exploded.

There is a button at the very bottom of the Explode PropertyManager (not shown) with the label *Re-use Sub-assembly Explode*. If a top-level assembly contains a subassembly that has already been exploded, and you wish to reuse that exploded view when creating an exploded view of the top-level assembly, use this button. Select the subassembly or one of its components and click on *Re-use Sub-assembly Explode* and the subassembly will explode in the same fashion it did in its own assembly exploded view.

10-second Topic: Expanding and Collapsing Exploded Views

Collapsing and exploding (expanding) an exploded view is accomplished by right-clicking on the *ExplView* object in ConfigurationManager. Collapsing and exploding is a toggle switch, so both commands will never appear simultaneously in the right mouse button menu. Along with the standard Collapse and Explode commands will be Animated Collapse and Animated Explode.

When animating an exploded view, the Animation Controller appears, which is the same controller used with simulations (discussed previously). Use it in exactly the same way as with a simulation to move through the explode or collapse process. This can be one of the more entertaining aspects of SolidWorks, so you should give it a try.

Incidentally, when collapsing an exploded view it is not actually necessary to right-click on the *ExplView* object. Right-clicking anywhere in the work area will provide access to the Collapse option. This is because when collapsing an exploded view you do not have to be specific. However, when you explode an exploded view it is necessary to specify exactly which view should be exploded, in that there can be more than one exploded view in an assembly.

Editing Exploded Views

Exploded views are stored in ConfigurationManager. Each configuration can have its own exploded view, but is limited to only one. Figure 13–94 shows the result of exploding the gear housing and what the exploded view looks like in ConfigurationManager. The name of the exploded view is *ExplView1*. The explode steps are listed below this name.

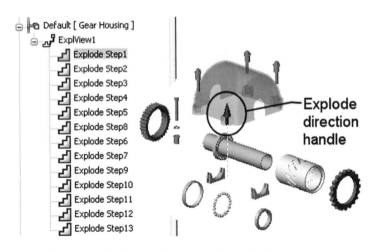

Fig. 13–94. Exploded view object in ConfigurationManager.

Any of the exploded view objects in ConfigurationManager can be edited via the right mouse button. Access the Edit Feature command, as is par for the course, and you will be returned to the original Explode PropertyManager used to create the exploded view.

If any of the explode steps listed in Configuration-Manager is selected, a small arrow will appear, which can be seen in Figure 13–94. The arrow acts as a handle that can be dragged. In this way, the part or parts exploded in that step can have their explode distance dynamically altered. The explode direction, however, can only be changed by editing the feature itself.

Exploded views usually contain explode lines, which are typically used to indicate the manner in which the components are assembled. Creating explode lines is a manual operation within SolidWorks that involves selecting objects between which the explode lines will be created. How-To 13–15 takes you through the process of creating an explode line sketch.

HOW-TO 13-15: Creating an Explode Line Sketch

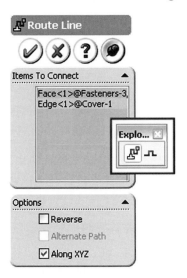

To create an explode line sketch, first make sure that the assembly is in its exploded state (not collapsed), and then perform the following steps.

1. Click on Explode Line Sketch in the Insert menu, or click on the Explode Line Sketch icon found on the Assembly toolbar.

   Two things should happen following step 1: the Route Line PropertyManager should appear (see Figure 13–95) and the Explode Sketch toolbar should appear (Figure 13–95 inset). Considering that the Explode Sketch toolbar contains just two icons, it is easy to locate. Check the corners of the screen where it may be hiding, and drag the toolbar out to a more convenient location. Then continue with the following steps.

Fig. 13–95. Route Line PropertyManager and Explode Sketch toolbar.

2. Select edges or faces on components that the explode line will connect between. Order of selection is important.

3. If an explode line is emanating from the wrong side of the component, click on the arrow to reverse its direction, as indicated in Figure 13–96.

4. Drag the route lines to reposition them, as indicated in Figure 13–97.

5. Click on OK to create the route line.

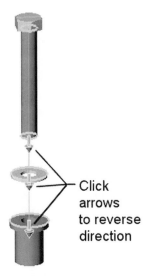

Click arrows to reverse direction

Fig. 13–96. Arrows indicate route line direction.

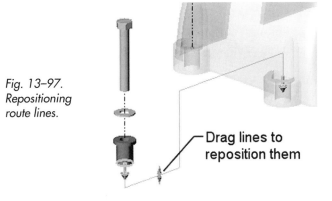

Fig. 13–97. Repositioning route lines.

Drag lines to reposition them

6. Repeat steps 2 through 5 to add as many route lines as necessary.

7. Click on OK a second time to close the Route Line PropertyManager, or click on the Explode Line Sketch icon.

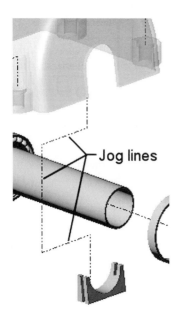

Fig. 13–98. Explode line sketch in ConfigurationManager.

Fig. 13–99. Jog lines.

Learning to explode assemblies and create an explode line sketch often takes a few attempts, and it is common to create the explode line sketch only to find out that the exploded view itself could use some revising. If necessary, exit the explode line sketch process and edit the exploded view. You can go back to the explode line sketch later to make alterations.

The explode line sketch is actually a 3D sketch, which you learned about in Chapter 10. The sketch will be located in ConfigurationManager directly below the *ExplView1* object, as shown in Figure 13–98. To edit the explode line sketch, right-click on the sketch and select Edit Sketch.

Once in edit mode for the explode line sketch, many of the segments in the sketch can be repositioned by dragging. It is also possible to delete segments, add geometric relations, or add new sketch geometry. It helps if you have a little knowledge in how the 3D Sketcher works, so it may be beneficial to review the section "3D Sketcher" in Chapter 10.

To add a new route to the explode line sketch, click on the Route Line icon on the Explode Sketch toolbar. Use the Options panel in the Route Line PropertyManager (see Figure 13–95) to try an alternate path, to reverse the arrow direction, or to try a route that does not follow the *x-y-z* axes of the assembly's coordinate system. Use the Jog Line icon on the Explode Sketch toolbar to create jogs in existing segments of a route line. This is accomplished by clicking on an existing route line segment where the jog should start, and then clicking a second time to control the width and length of the jog. An example of a jog (or rather, the lines that make up a jog) is shown in Figure 13–99.

Copying Exploded Views

There can be only one exploded view per configuration, but because it is possible to have a huge number of configurations the number of exploded views is essentially unlimited. (There actually is a limit to the number of configurations possible, but that limit is so high that it would never be reached.)

If another configuration exists and another exploded view is required, there is a way to copy one exploded view to another configuration. The duplicate exploded view can then be altered if necessary. The routine for copying an exploded view to another configuration is very particular; it must be carried out just right. How-To 13–14 takes you through the process of copying and exploded view.

HOW-TO 13-16: Copying an Exploded View

To copy an exploded view to another configuration, perform the following steps.

1. Activate the configuration that contains the exploded view to be copied.

2. Expand the configuration so that the *ExplView* object can be seen.

3. Hold down the Ctrl key, and with the left mouse button drag the *ExplView* object to the configuration the exploded view is to be copied to.

4. Activate the configuration containing the copied exploded view. Only then will the *ExplView* object be shown in the new configuration.

5. Right-click on the new *ExplView* object and select Edit Definition. Make any required changes.

If the configuration the exploded view was copied into is quite a bit different than the original exploded view configuration, it will be necessary to add or to delete some explode steps. This is due to the configuration possibly having either more or fewer components available for inclusion in the exploded view.

As a final word, an assembly cannot be edited while in an exploded state. SolidWorks will allow you to make modifications to the assembly, but as soon as you attempt to make any alterations to the assembly it will be forced to collapse. Because of this, it is best to collapse the assembly first manually, just to be on the safe side.

Exploded Views in Drawings

Exploded views are typically shown with the assembly in some sort of isometric viewing angle. This works best for showing the assembly components in their exploded state. In addition, a user-defined view could be created in the assembly that shows the exploded assembly in its best light. If you need to refresh your memory on how to create a user-defined view, see Chapter 3.

You do, of course, have to create the exploded view in the assembly before you can show it in the design drawing. How-To 13–17 assumes that you have done this, and assumes you have inserted a model view into the drawing. Ideally, the model view shown in the drawing should be of the same user-defined view saved in the assembly. See Chapter 12 for a discussion on inserting model views. How-To 13–17 takes you through the process of showing an exploded view in a drawing.

HOW-TO 13-17: Showing an Exploded View in a Drawing

To show an exploded view in a drawing, perform the following steps.

1. Right-click on the drawing view you wish to show in an exploded state and select Properties. The Drawing View Properties window will appear.

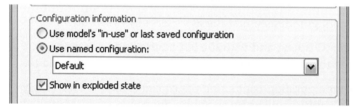

2. In the *Configuration information* section, shown in Figure 13–100, specify the configuration that contains the exploded view. This is only necessary if there is more than one configuration present.

Fig. 13–100. Configuration information section of the Drawing View Properties window.

3. Place a check in the option titled *Show in exploded state.*

4. Click on OK.

SolidWorks will show the view using the exploded state for that particular view. You may need to modify the scale so that the view will fit on the sheet, which you learned how to do in Chapter 12.

Balloons

A drawing that includes a BOM is almost always accompanied by balloons on one or more drawing views. The balloons usually incorporate leaders that point to the various components or subassemblies in the assembly layout. This correlates the items in the BOM with those items shown in the drawing.

Balloons can be customized to meet most requirements. This is done through the Document Properties tab of the Options window (Tools > Options). Figure 13–101 shows the Balloons section of Document Properties, which has many ways of customizing how balloons will appear.

Balloon style and size for both single and stacked balloons can be controlled independently. The section marked *Balloon text* allows for customizing what appears inside the balloons. For example, if a circular split line style is used the data contained within both the upper and lower areas of the balloons can be dictated. Figure 13–102 shows an example of circular split line balloons. In this case, the bolt, washer, and insert are items 12, 11, and 10, respectively, and there are four of each component in the assembly.

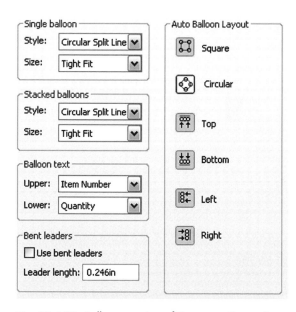

Fig. 13–101. Balloons section of Document Properties.

Fig. 13–102. Circular split line balloons.

When adding balloons, the order in which you place the balloons is not important. SolidWorks will number the balloons correctly, corresponding with the component's item number as designated in the BOM. For that matter, balloons can be added prior to inserting a BOM, and Solid-Works will still get it right.

The process of adding balloons is a simple matter that although we haven't specifically addressed you should already be familiar with. Review the section "General Annotating Guidelines" in Chapter 12 if necessary. In summary, once in the Balloon command click to position the leader, and then click to place the balloon. Like all other annotations, the Balloon command icon is found on the Annotations toolbar.

When adding balloons, the appearance of the balloons can be modified via the Balloon PropertyManager. However, it is better practice to set up default characteristics of how the balloons should appear in the Document Properties, as discussed previously. In this way, balloon appearance will remain consistent, and it will not be necessary to make modifications via the PropertyManager.

✓ **TIP:** *Reattaching a balloon's leader to another component will make the balloon's content automatically update.*

Stacked balloons, an example of which is shown in Figure 13–103, are added in a manner slightly different from that associated with single balloons. Therefore, How-To 13–18 takes you through the process of adding stacked balloons.

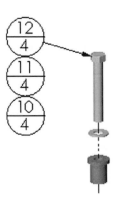

Fig. 13–103.
Stacked balloons.

How-To 13-18: Adding Stacked Balloons

To add stacked balloons, perform the following steps.

1. Select Insert > Annotations > Stacked Balloon, or click on the Stacked Balloon icon located on the Annotations toolbar.

2. Select a component in the drawing on which to attach the balloon's leader.

3. Click to position the balloon.

Fig. 13–104. Controlling
stack direction.

4. Select components to be added to the balloon's stack.

5. In the Stacked Balloon PropertyManager, specify the direction for the stack. Use the setting shown in Figure 13–104.

6. Optionally, specify how many balloons should be on a single stacked line before a second line of stacked balloons is created (see Figure 13–104). The default value is 10 balloons in a stack.

7. Click on OK when finished.

A single balloon can be deleted from a stack by selecting that particular balloon and pressing the Delete key. To add balloons to the stack, right-click on any of the stacked balloons and select *Add to stack*. A sample exploded assembly view, complete with explode lines and balloons, is shown in Figure 13–105.

✓ **TIP:** *To force a balloon to have multiple leaders, select the balloon and then Ctrl-drag the end of the leader to a new location or to another component.*

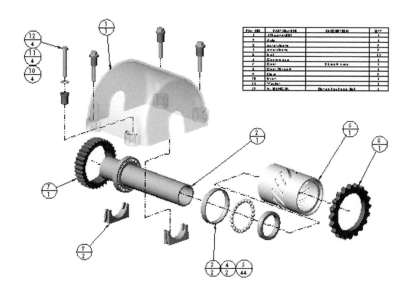

Fig. 13–105. Exploded view with balloons and explode lines

Auto-ballooning

Before adding every balloon manually, try auto-ballooning. The Auto-balloon command is found on the Annotations toolbar, and in the Insert > Annotations menu. What does auto-ballooning do? As the name implies, it automatically attaches a balloon to every component in any of the views you select.

When starting the Auto-balloon command, the PropertyManager will display the usual balloon formatting options for balloon style, size, and so on. In addition, the Balloon Layout panel (shown in Figure 13–106) will allow for setting the formation the balloons will take. Views can be selected before or after initiating the Auto-balloon command.

Balloons will not be duplicated on the same component, but if a component is used more than once multiple balloons with identical numbers may appear. Turn on (check) the *Ignore multiple instances* option found at the bottom of the Balloon Layout panel (see Figure 13–106) to keep this from occurring.

Auto-ballooning isn't perfect; balloons may not be positioned perfectly. However, it is a good place to start when there are plenty of balloons to add. Once the balloons have been added, move them as necessary using the by-now-familiar drag technique. Likewise, the leader arrows may need repositioning. This can be done by first selecting a balloon and then dragging the leader to a new position. Look for the system feedback shown in Figure 13–107 to know when the leader arrow can be repositioned.

✓   **TIP:** *Make it a point to note the item number in a balloon before reposi-*
tioning leader arrows. In this way, you can be certain that the balloon is
reattached to the same component.

Fig. 13–106. Auto-balloon
PropertyManager Balloon
Layout panel.

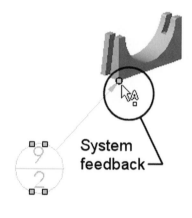

Fig. 13–107. Look for this system feedback
when dragging a leader arrow.

Checking Balloon Presence

With small assemblies, it is usually pretty easy to make sure all necessary
balloons are present and accounted for, but what to do when an assembly
has hundreds of components? There is a very easy way to check if every
balloon has been added to a drawing or not, but there is one requirement:
the drawing must contain a SolidWorks-based BOM.

Assuming the drawing does contain a SolidWorks-based BOM, click on
a cell in the BOM, and then click on the BOM Contents button found in
PropertyManager. A list of every component in the assembly will be displayed. If the component is missing a balloon, the balloon state column will
indicate this fact, as illustrated in Figure 13–108.

Fig. 13–108.
Checking whether
a balloon is present
and accounted for.

		7	Gear	34-tooth wide	1
		8	Gear 20-tooth	**Missing balloon!**	1
		9	Slider		2

From within this same window, it is also possible to hide a component from appearing in a BOM, hide an entire column by clicking on the column heading, move rows up or down in the list, or group components so that they will share the same item number. These last two tasks are performed using the command buttons below the Contents tab (shown in Figure 13–109). Select an entire row for moving or grouping (or ungrouping) by clicking on the button to the left of the row (see Figure 13–109). Hold the Ctrl key down to select multiple items.

SW 2006: A new column in the BOM Properties window (which is accessed by clicking on the BOM Contents button described in this section) allows for expanding or collapsing subassemblies in a BOM. This capability will only exist for BOMs that use an indented list for subassemblies. Click on the plus or minus symbols (shown in Figure 13–110) to expand or collapse (respectively) subassemblies shown in the BOM.

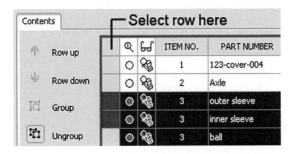

Fig. 13–109. Moving rows and grouping items.

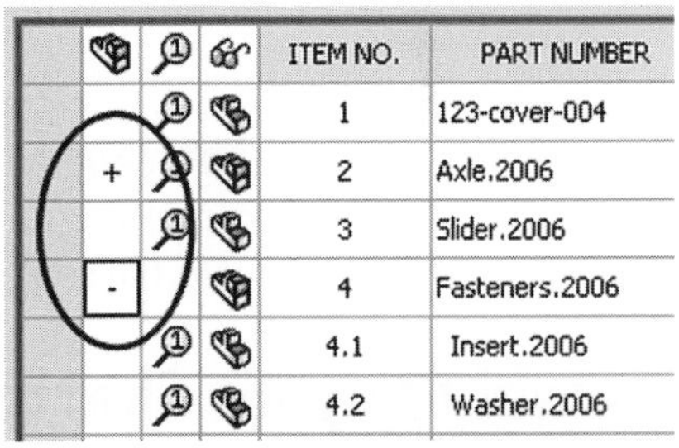

Fig. 13–110. Expanding or collapsing subassemblies in a BOM.

Balloons and Component Names

A common request among SolidWorks users is the ability to place the name of a component inside a balloon on a drawing. You have actually learned what is necessary to accomplish this task, which involves linking to the properties of a component (see "Notes," Chapter 12).

Assuming you are working in a drawing containing at least one view of an assembly, begin by adding a note. It is necessary to add a note and not an actual balloon mainly due to the ease in which the component name can be linked to a note. Make it a point to click on a component when adding the note so that the note will have a leader. Format the border of the note if you wish. Next, add the note as usual (you learned all of this in Chapter 12).

For the text of the note, use the Link to Property icon found in the Text Format panel of the Note PropertyManager. The Link to Property window, a portion of which is shown in Figure 13–111, will appear. Select the *Component to which the annotation is attached* option and then select the property named *SW-File Name* from the drop-down list.

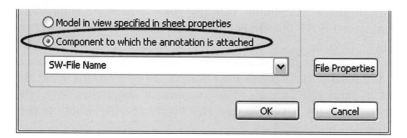

Fig. 13–111. Lower portion of the Link to Property window.

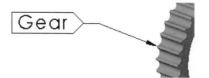

Fig. 13–112. Adding a balloon style note with a component name.

Once completed, you should have a balloon style note in the drawing that points to a particular component. The text will indicate the file name for whatever component the leader is attached to. For the example used in Figure 13–112, a five-sided flag was used as the border for the note. It would have been just as easy to use a circular border around the note to match the other circular balloons in the drawing.

✓ **TIP:** *Component balloon notes can be copied and pasted. If a leader from a copied note is reattached to another component, the text will update with the correct component name automatically.*

Assembly Layout Sketches

Using a sketch to position components within an assembly is nothing new to SolidWorks. The technique consists of creating a sketch in an assembly in the same way a sketch is created in a part. The difference lies in what the sketch is used for. Rather than use the sketch in the creation of a feature, the assembly sketch is used for mating purposes. If components are mated to the sketch, the components will move if the sketch is resized. All assemblies are not well suited to this technique, so use your best judgment as to whether this technique will work well for your particular application.

Sketch Block Layouts (SW 2006 Only)

SolidWorks 2006 expands on the idea of assembly layout sketches by allowing blocks to be created in part and assembly files. Previous to SolidWorks 2006, blocks could only be created in drawings. Blocks are collections of sketch geometry that react as a group. In the context of a part or assembly, blocks can be related to other blocks via geometric constraints and dimensions. Figure 13–113 shows an example of a sketch that contains blocks. The inset shows what FeatureManager would appear like, and the sketch itself has been annotated to show the name of each sketch block.

The block creation process was described in the previous chapter in the context of creating blocks in a drawing. The process for creating sketch blocks in a part or assembly is extremely similar, so it will not be reiterated here. However, there are a few details you should be made aware of.

Multiple sketches can contain multiple blocks, but if the blocks are to be related to each other they should exist within the same sketch. All that is needed is to begin a single sketch and then start creating geometry for each block. As the blocks are defined, they will be listed under the sketch that contains them.

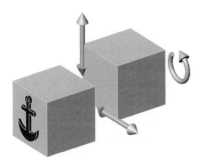

Fig. 13–113. Example of an assembly block layout.

To move a block, such as dragging the block to reposition it, the sketch that contains the block must be edited. It is for this reason that placing all blocks in a single sketch makes the most sense. Blocks can be geometrically related to each other, and can show motion, much like components can within an assembly. Think of the sketch containing the blocks as the assembly, and each block as a separate component.

When creating a block layout, be careful what you are editing, as it is easy to lose track. Stay within a single sketch, and do not edit the individual blocks by accident. As each block is created, use the Line Format toolbar to change the color of blocks, which makes it easier to distinguish individual blocks.

Assemblies from Block Layouts

Let's look at a hypothetical situation to better understand how block layouts can be beneficial. By creating blocks and using geometric relations to obtain the desired relationships and motion between the blocks, it is possible to obtain an idea of how the assembly components should relate to each other. The blocks are in essence a 2D prototype of an assembly. Once the layout is complete, the actual assembly components can be created.

A typical scenario would be to create the part files from scratch that will then be used as the components within the assembly. If a block layout has already been created, a second option would be to generate an assembly from the sketch containing the blocks. This can be done from either a part or assembly. How-To 13–19 takes you through the process of generating an assembly from a block layout sketch.

How-To 13-19: Creating an Assembly from a Block Layout

To create an assembly from a layout sketch containing blocks, perform the following steps. These steps can be performed in a part or assembly.

1. In FeatureManager, right-click on the sketch that contains the sketch blocks, and select Create Assembly From Layout Sketch.

2. In the window that appears (see Figure 13–114), check the blocks from which components should be generated.

3. Select *Create new top level assembly* to create a new assembly. This option is enabled automatically if the layout sketch was created in a part. If the layout sketch was initially created in an assembly, it is only necessary to check this option if you wish to create a new assembly and retain the original layout sketch for safe keeping.

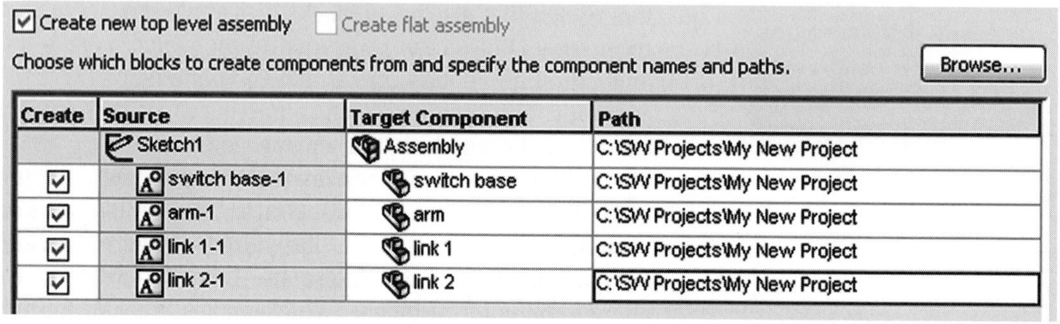

☑ Create new top level assembly   ☐ Create flat assembly

Choose which blocks to create components from and specify the component names and paths.   Browse...

Create	Source	Target Component	Path
	Sketch1	Assembly	C:\SW Projects\My New Project
☑	switch base-1	switch base	C:\SW Projects\My New Project
☑	arm-1	arm	C:\SW Projects\My New Project
☑	link 1-1	link 1	C:\SW Projects\My New Project
☑	link 2-1	link 2	C:\SW Projects\My New Project

Fig. 13–114. Create Assembly From Layout Sketch window.

4. Optionally, click in each cell in the Target Component column and enter a new name for each component.

5. Optionally, click in each cell in the Path column and specify a path for each component using the Browse button. A path can also be specified for the top-level assembly.

6. Click on OK to generate the new assembly.

Following step 6, the assembly should appear. The FeatureManager will contain a new component for each block in the original layout sketch. Figure 13–115 shows an example of this, along with the resultant assembly in a partial state of development.

It is very important to note that the original sketch containing the blocks will also be present in the assembly. It is acceptable to hide this sketch, but do not delete it. Each component in the assembly is linked to its original counterpart sketch block through a special mate called a *LockToSketch* mate. Two of these mates are visible in Figure 13–115. Deleting the original layout sketch will break the mates and the assembly will require repair.

To further develop each component, use the skills you have learned so far in this book. For instance, open one of the components and extrude the geometry, or add other features to it to create the desired model. To reposition the components, edit the layout sketch and move the blocks.

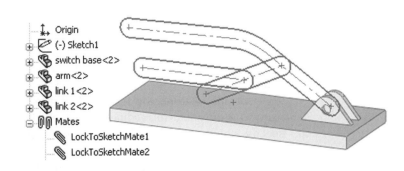

Fig. 13–115. Developing the newly generated assembly.

Summary

Looking back on this chapter, there is quite a bit of information that goes along with assemblies. The first aspect of dealing with assemblies is getting the components into the assembly. The most common method of accomplishing this is to use the Insert Components icon on the Assembly toolbar.

The first component brought into an assembly is fixed in location by default. Generally, this first component should be the main component other parts will be attached to. Any component in an assembly can be fixed in space or allowed to float freely in space. Use the right mouse button and click on the component to fix or float a component.

Components can be moved and rotated independently of other components in the assembly by dragging the component with the left mouse button or right mouse button, respectively. Any part can be moved or rotated, as long as the part is not fully defined. Right-click on a component and select Move with Triad to gain more control over the direction or axis a component is moved or rotated in.

Add mates to place components together and build the assembly. Always try to assemble with motion in mind if your assembly will have moving parts. Remember that every component has six degrees of freedom of movement prior to being mated, as long as it is not fixed. These degrees of freedom are in reference to translation and rotation about the x, y, and z axes. Never add a mate to a part that is fixed, and vice versa, as this will over-define the assembly.

Hide components by right-clicking on the component. You can also suppress components, but this will also suppress any mates associated with that component. When a component is suppressed, it is removed from memory, and any mates associated with the component will be suppressed also. Components can also be lightweight, which means only a portion of the component's database has been loaded into memory. Opening assemblies using lightweight components greatly reduces load times. Components that require editing can then be loaded on an as-needed basis.

Use interference checking to establish whether or not there is interference between selected components. If working with assembled components that can describe motion, use dynamic interference detection (also known as collision detection) to check for interference. Faces of components that collide and interfere with each other will highlight when collision takes place.

Assembly features are features that affect the assembly only. Assembly features are limited to revolved or extruded cuts. Use the feature scope to specify which components will actually be affected by an assembly feature. Patterns of assembly features can be created as well. Creating assembly feature patterns is very similar to creating part feature patterns, so the process should be familiar to you.

Components can be patterned in an assembly. These patterns can be locally defined patterns or patterns based on an existing feature pattern. If the feature pattern changes, so will the derived component pattern.

When creating assembly drawings, use the same techniques you learned for creating the various views of parts. Assembly section views are different only in the regard that you must specify what components you want to have excluded from being cut during the section, if any. This is known as the section scope.

To add a bill of materials, make sure you select a view first. This is because drawing sheets can contain views from different parts or assemblies. When adding balloons, the item number in the bill of materials will automatically match the item the balloon is attached to on the view geometry.

Exploded views can be added to an assembly through the use of the Insert menu. While in the Exploded View command, selecting a component causes a triad to appear, which can then be dragged by one of its three arrows. Existing exploded views are accessed in ConfigurationManager, where they can be expanded or collapsed, either instantaneously or as an animation. Show exploded assemblies in an assembly design drawing by accessing the drawing view's properties.

CHAPTER 14

Core and Cavity

SOLIDWORKS GIVES YOU THE ABILITY TO CREATE A MOLD using various methods. One mold-making method works by subtracting material (the design component) from another block of material (the mold). This process requires a minimum of two components and must be carried out in an assembly. The function being described is known as the Cavity command.

By its very name, it is easy to assume the Cavity command would be part of the process of creating a core and cavity. This is not necessarily true. There are much more appropriate tools that have automated the entire mold-making process, which we will be exploring in this chapter.

There are obviously a wide variety of parts for which molds must be developed. This chapter explores two parts with different characteristics. The first part is a cast part, which will show the basic routine of creating a mold. The second part is a more complex plastic part, which will require a side core, manually created interlocking surfaces, a core and cavity. The How-Tos are presented in an order that enables you to follow an entire mold-making (core-and-cavity) process.

Determining the Parting Line

Over the course of designing a part, the parting line is often "built in" while adding features to the model. For those not familiar with the term, a parting line is actually a series of connected edges on the model that describes where the top and bottom halves of a mold meet. In its simplest form, a parting line will be a planar edge around the perimeter of a part. In more complex parts, the parting line will consist of nonplanar geometry, and there may even be multiple parting lines. The molds themselves do not have to consist of strictly a top and bottom, but can be multiple pieces.

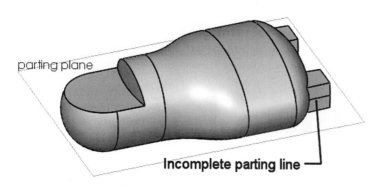

Fig. 14–1. No distinct parting line.

Two important commands that relate to parting lines have already been discussed. The Split Line command was discussed in Chapter 6. One aspect of the Split Line command, known as the Silhouette method, can be used to aid in the development of a parting line. Such would be the case in the part shown in Figure 14–1, where there is no distinct parting line. There are only a few edges where the parting line can be distinguished, one of which is shown in the figure.

A second command discussed in Chapter 6 was the Draft Analysis command, which allowed for analyzing a part to see if it would pop out of a mold. The Parting Line command, which we are preparing to explore, encompasses functions of both the Split Line and Draft Analysis commands.

If the Parting Line command were used on the part shown in Figure 14–1, the parting line could automatically be determined. The user is required to select a plane that dictates the direction the mold will be pulled away from the part. Figure 14–2 shows a part where the parting line is being developed. The plane in Figure 14–2 has been selected to indicate the direction of pull, and the parting line can then be automatically generated by the software.

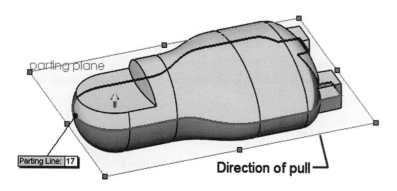

Fig. 14–2. Developing a parting line.

Whether you decide to use the Split Line command to help generate a parting line, or the Parting Line command itself, is really a matter of necessity. It would be common to use the Parting Line command in the model used as an example in the previous two illustrations. In a situation in which there must be additional draft added to the part so that it will come out of the mold, it would be typical to use the Split Line command, and then apply draft as required. The Draft command was discussed in Chapter 6 (see "Adding Draft").

A model may contain a clearly defined parting line, or it may not. If the parting line is not clearly defined, the Parting Line command will help you to define it. If the parting line is clearly defined, it is still necessary to dictate exactly where that parting line is as far as SolidWorks is concerned. The Parting Line command is required prior to using subsequent mold-making commands in SolidWorks. How-To 14–1 takes you through the process of using the Parting Line command.

How-To 14-1: Using the Parting Line Command

To define a parting line on a part, perform the following steps. This process is required prior to using other mold tools within SolidWorks, such as the Tooling Split command (discussed later in this chapter).

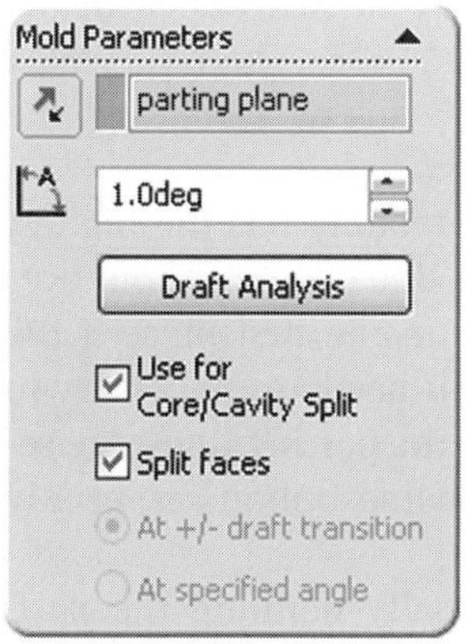

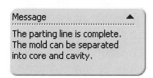

Fig. 14–3. Parting Line PropertyManager.

1. Select Parting Line from the Insert > Molds menu, or click on the Parting Lines icon found on the Mold Tools toolbar.

2. Select a plane that will dictate the direction of pull (direction of pull will be perpendicular to the selected plane).

3. Specify the minimum draft angle for the part in the Parting Line PropertyManager, shown in Figure 14–3.

4. If there are straddle faces (curved faces that will have both positive and negative draft) that will require splitting, select the *Split faces* option. An example of a part with straddle faces would be the model shown in Figures 14–1 and 14–2.

5. Select the *Use for Core/Cavity split* option if a mold will be created for the part. Do this even if the part will not require a core!

6. Click on the Draft Analysis button, visible in Figure 14–3.

Fig. 14–4. The parting line is complete.

At this point you will have to make a choice. If the part is a fairly simple part and SolidWorks was able to create the parting line, you are almost done. If the parting line is complete, PropertyManager will inform you of this fact with the message shown in Figure 14–4. Click on OK to create the parting line.

Even if the parting line is complete, it may not be satisfactory. Likewise, if the parting line is incomplete you will be advised of this in PropertyManager with a message. Sometimes there are small flat areas on a model that a parting line should flow across, but SolidWorks does not know this. In such a case, the Parting Line command needs a little assistance.

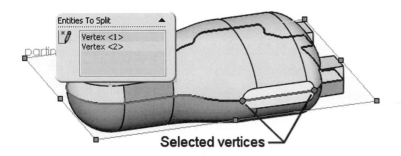

Fig. 14–5. Specifying vertex points for parting line development.

Figure 14–5 shows a model with small flat spots on either side. There is no draft on the flat spot shown in Figure 14–5, and therefore no parting line. Solid-Works will attempt to skirt this flat spot by developing the parting line around it, either around the top or bottom edges of the flat face.

If a parting line should not flow along certain edges, simply deselect those edges. This can be done by clicking on the edges from within the work area. Three of the edges at the top of the flat have been deselected in Figure 14–5. Next, click in the Entities to Split panel, shown as an inset in Figure 14–5. The two vertex points shown in Figure 14–5 were then selected, effectively "rerouting" the parting line.

One aspect of the Parting Line command that probably has the highest "cool" factor is the ability to manually select parting line edges without ever having to click on an edge. This is only necessary in the first place if the parting line cannot be automatically determined. If it becomes necessary to manually define where the parting line should flow, the parting line edge selection tools shown in Figure 14–6 will be a welcome option, especially when there are many edges to select.

Fig. 14–6. Parting line edge selection tools.

When manually defining the parting line path, keep a close eye on the preview arrow, shown in the inset in Figure 14–6. This arrow will point toward the next edge to be added to the parting line should you click on the button labeled *Select edge* in the figure. *Change arrow* points the arrow toward an alternative edge. The button labeled *Scrolling pan* in the figure acts as a toggle switch. When turned on, the model will continually be panned across the screen so that the preview arrow will remain in your field of vision.

⟶ **NOTE:** *Parting line edge selection tool buttons have tool tips that vary from the "labels" given them in this book. The reason for this is that the labels used in this book more accurately represent their functionality.*

✓ **TIP:** *For those who prefer using a keyboard, hot keys for the* **Select edge** *and* **Change arrow** *buttons are* Y *and* N, *respectively.*

Surface Bodies Folder

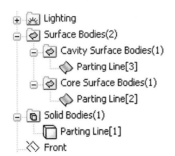

Fig. 14–7. Surface Bodies *folder.*

If the parting line has been successfully created, and there are no holes in the part that would require shut-off surfaces (see material to follow), the *Surface Bodies* folder should show the presence of new surfaces. This is indicated in Figure 14–7. The *Surface Bodies* folder appears whenever there are surfaces present in the model. Up to this chapter, we have only had to contend with a single solid body.

Bodies, both solids and surfaces, take on the name of the last feature that modifies them. Because this feature happens to be the parting line feature in the case of the image shown in Figure 14–7, the bodies take on the name *parting line*, followed by some unique identifying number. The naming convention is actually of little consequence to the person creating the model.

Of greater importance are the names of the subfolders that branch from the *Surface Bodies* folder. These subfolders are named *Cavity Surface Bodies* and *Core Surface Bodies*. The number in parentheses following each folder name indicates the number of surfaces present in the folder.

As time goes by, we will return to the *Surface Bodies* folder. It plays an important role in the creation of a mold. The creation of these folders and the placement of the various surfaces contained within them are automated for the most part. However, there will be occasions when it becomes necessary to manually relocate surfaces into the appropriate folders. When this need arises, it will be clearly stated in this book.

Shut-off Surfaces

When there are holes in the model, such as those shown in Figure 14–8, it is impossible to create the mold without first closing the holes. This is due to the fact that the mold itself needs to split into top and bottom halves. The split cannot occur if the mold halves are connected to each other by material that fills the holes. The shut-off surfaces are used to "close" the holes, thereby making a clear distinction between one side of the mold and the other.

Fig. 14–8.
*Shut-off surfaces
are needed.*

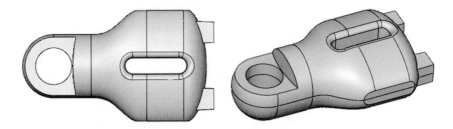

Creating shut-off surfaces used to be a manual process, but has now become automated like so many other aspects of the software. How-To 14–2 takes you through the process of creating shut-off surfaces.

HOW-TO 14-2: Creating Shut-off Surfaces

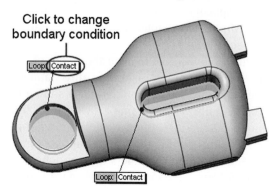

Click to change
boundary condition

Fig. 14–9. Creating shut-offs.

To automatically generate shut-off surfaces, perform the following steps.

1. Select Shut-off Surfaces from the Insert > Molds menu, or click on the Shut-off Surfaces icon found on the Mold Tools toolbar.

2. If necessary, click on a callout (see Figure 14–9) to change the boundary condition of the shut-off surfaces. This is explained in material to follow.

3. Click on OK to create the shut-offs.

Boundary conditions for shut-off surfaces can be Contact, Tangent, or No Fill. Boundary conditions are dependent on the surrounding geometry. When a tangent boundary condition is used, the tangency condition can be set to faces on one side of the parting line or the other. This is controlled by clicking on a preview arrow that will appear whenever a tangency boundary condition is used.

Fig. 14–10. Use for Core/ Cavity Split option.

The No Fill option is not really a boundary condition, but more like an option, as no surface is created when No Fill is specified. No Fill may seem like an odd option, but there are times when it would not be desirable to have SolidWorks automatically attempt to fill in an opening in the model. Such might be the case with a more complex model, in which the opening has a stepped parting line, areas exist where side cores are necessary, and so on.

In a situation involving complex holes in the model, use the Parting Line command discussed earlier. Just make it a point to uncheck the Use for Core/Cavity Split option found in the Parting Line PropertyManager (shown in Figure 14–10).

After an internal parting line is established on the model, a surface can be radiated outward (or rather, inward) from the parting line. Additional surfacing techniques can be used to completely close the opening. A command used to radiate a surface away from the parting line is called the Parting Surface command and will be explored shortly.

When creating shut-off surfaces, there are a few noteworthy options that should be discussed. A portion of the Shut-off Surfaces PropertyManager is shown in Figure 14–11. The *Show preview* and *Show callouts* options are self-explanatory and don't require further discussion. *Filter loops* will attempt to exclude model edges that form closed loops that are not valid holes through the model. Typically, just leave this option on, and then select or deselect any edges as necessary to create the appropriate shut-offs.

Fig. 14–11. Shut-off command options.

The Knit option should be left on unless there are additional shut-offs you will need to create manually. When the shut-offs are created, the shut-off surfaces are knitted to the existing surfaces in the *Surface Bodies* folders. If the Knit option is turned off, there will be multiple surfaces in the *Surface Bodies* folder. Additional surfaces could then be created as necessary, and the various surfaces could then be knitted manually. Don't worry about this process until you have a better basic understanding of how the entire mold-making process works.

Incidentally, if the term *knit* is new to you, think of it as combining surfaces, almost as if they were taped together edge to edge. Knitting can be an automated process performed through certain commands, but it is also a command in its own right. With the Knit Surfaces command, which you will explore later in this chapter, surfaces can be knitted manually.

The Reset All Patch Types panel allows for globally setting the boundary conditions of every shut-off surface. This can be useful if there is a large number of holes in the model and SolidWorks uses the incorrect boundary condition for all of them. Typically, though, you would click on an individual callout to change its boundary condition, as mentioned earlier.

➥ **NOTE:** *When creating a mold in SolidWorks using the processes in this chapter, surface bodies are overlaid on the solid body of the model. This causes graphics display to occasionally look odd, but does not otherwise affect the model. The odd display is not a result of a low-budget graphics card. Rather, it is the result of having two bodies inhabit the same position in space. To overcome this, you may wish to hide the original solid body while working on the surfaces needed for the tooling split, discussed in the next section. Use the Hide/Show Bodies command in the View menu to accomplish this.*

Tooling Split Development

So far you have learned how to define the parting line and how to create the shut-off surfaces. The final step prior to creating the mold is to define the surfaces that will in essence split the mold from top to bottom. This can sometimes be the most difficult step in the process, depending on the model geometry.

The ability to split a mold is dependent on surfaces that can physically bisect the mold. These surfaces can be created in a variety of ways, but there are certain commands that work better than others. One such command is the Parting Surface command. How-To 14–3 takes you through the process of creating a parting surface.

HOW-TO 14-3: Creating a Parting Surface

Fig. 14–12. Parting Surfaces PropertyManager.

To create a parting surface, perform the following steps.

1. Select Parting Surface from the Insert > Molds menu, or click on the Parting Surfaces icon found on the Mold Tools toolbar.

2. In the Parting Surfaces PropertyManager, shown in Figure 14–12, select *Normal to surface* if the parting surface should project away from the model perpendicular to each surface adjacent to the parting line segments. Alternatively, select *Perpendicular to pull* (default selection) if the parting surface should project perpendicular to the direction of pull specified when creating the parting line.

3. If there are multiple parting lines in the model, select the parting line you wish to use to generate the parting surface. If there is only one parting line, it will be selected automatically.

4. Specify the distance the parting surface should radiate.

5. Click on OK to create the parting surface.

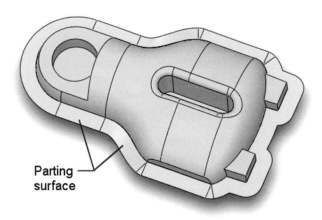

Parting surface

Fig. 14–13. Example of a parting surface.

Figure 14–13 shows an example of a parting surface. The surface in the example was radiated outward only a small distance. In other cases, which we will examine in another example later in this chapter, it is necessary to radiate the surface outward a greater distance. Keep in mind that the objective is to create a surface (or set of surfaces) that will physically split the entire mold.

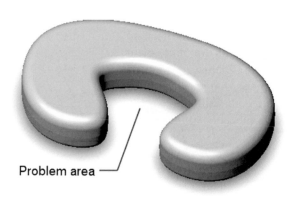

Problem area

Fig. 14–14. Parting surface command has problems in this area.

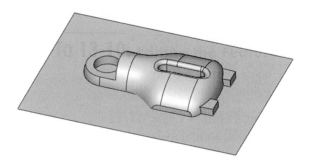

Fig. 14–15. What is needed to split the mold?

Some parts do not lend themselves very nicely to the Parting Surface command. A perfect example is the model shown in Figure 14–14. The model geometry contains an area that might be described as an inlet. It isn't a hole, so it can't be shut off, but the parting surface command doesn't like it either because the surface has a tendency to self-intersect. In situations like this, it often becomes necessary to use advanced surfacing techniques (which you will learn in Chapter 21).

Let's once again consider our objective, which is to create a surface that will physically split the mold. It would be logical to think that a surface similar to that pictured in Figure 14–15 would be needed, and you would not be incorrect to think this. So what are we to do with the parting surface that only radiates outward a small distance, as was seen in Figure 14–13?

There are two ways this story could play out. The first scenario is more user-friendly, with SolidWorks creating the remaining required surfaces and automatically generating the interlock for the mold. The second scenario involves creating all of the required surfaces manually. How-To 14-4 takes you through the process of creating a tooling split that mimics the first scenario. Use this method whenever possible, as it will save time.

HOW-TO 14-4: Creating a Tooling Split with Interlocking Surfaces

To perform a tooling split, complete the following steps. The model shown in Figure 14–16 is used for this example.

1. Create a plane at some distance above or below the original parting plane. The distance this new plane is offset will dictate the height of the interlock.

2. On the new plane, create a sketch that represents the size of the mold, such as that shown in Figure 14–16.

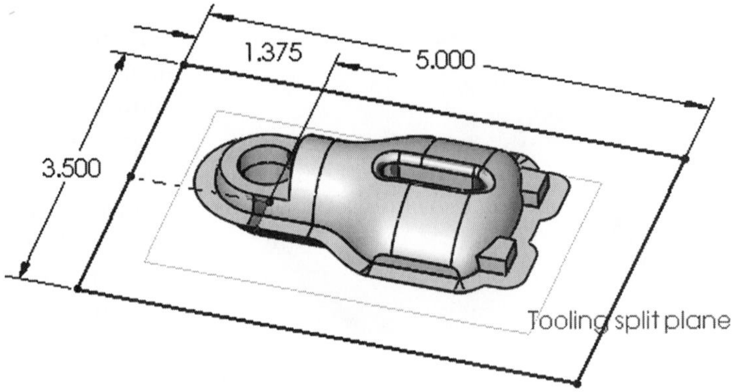

1.375   5.000

3.500

Tooling split plane

Fig. 14–16. Preparing to create a tooling split.

3. Exit the sketch.

4. Select Tooling Split from the Insert > Molds menu, or click on the Tooling Split icon found on the Mold Tools toolbar.

5. Select the sketch created in step 2.

6. In the Tooling Split PropertyManager, a portion of which is shown in Figure 14–17, specify the extrusion distances. The distances will determine the height (or thickness) of the top and bottom halves of the mold, and will be visible in the preview.

7. Check on the Interlock surface option.

8. Specify a draft angle for the interlock surfaces. This is typically greater than the part's draft angle so as to prevent gouging when the mold is closing.

9. Click on OK to complete the tooling split.

Fig. 14–17. Tooling Split PropertyManager.

Once the tooling split has been complete, the model won't look very impressive. It will be the model surrounded by the mold, which is not much to look at from the outside. The next step in the mold-making process would be to save the individual bodies out either as separate part files or as a complete assembly. You will learn how to do this following the second example. Until then, Figure 14–18 shows the top and bottom halves of the mold created during the tooling split operation. Of particular interest is the interlock, which was automatically generated.

If SolidWorks can create the interlocking surfaces automatically during the tooling split operation, that reduces the time spent manually creating the required surfaces. But what happens when SolidWorks cannot perform this feat, and the task is left in your hands? That is precisely what we will examine next, using a plastic molded part (shown in Figure 14–19).

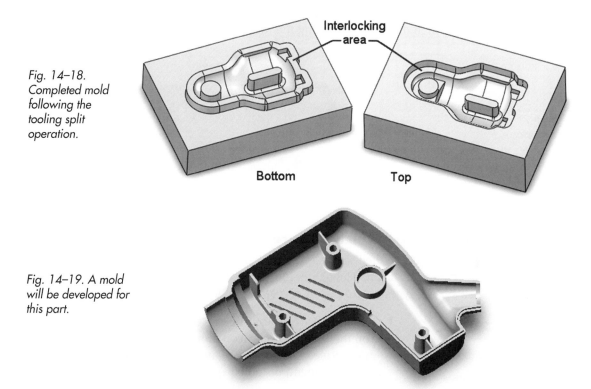

Fig. 14–18. Completed mold following the tooling split operation.

Fig. 14–19. A mold will be developed for this part.

Because you now understand how to develop a parting line and shut-off surfaces, we will not have to concern ourselves with those tasks. You have also discovered how to create the parting surfaces, at least up to a point. With regard to the first example used in this book, it was possible to radiate a parting surface outward a small distance and basically let SolidWorks do the rest of the work.

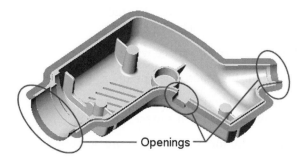

Fig. 14–20. Incomplete parting surfaces.

Note the circled openings in the part shown in Figure 14–20. It is largely due to these openings that the software has trouble automatically generating the interlocking surfaces and creating the tooling split. This part has had its parting line defined, shut-offs added, and the initial parting surface generated. The remaining surfaces will now be developed manually.

One surfacing command that plays a huge role in developing surfaces prior to creating the tooling split is the Ruled Surface command. It is what allows for creating surfaces such as that shown in Figure 14–21. How-To 14–5 takes you through the process of using the Ruled Surface command.

Fig. 14–21.
Ruled surface.

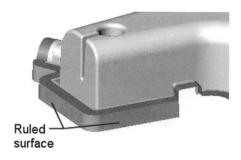

Ruled
surface

HOW-TO 14-5: Creating a Ruled Surface

To create a ruled surface, perform the following steps. The ruled surface type used in this How-To is the Tapered to Vector surface type. Additional surface types are discussed in material to follow.

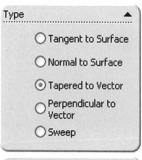

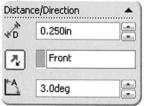

Fig. 14–22. Ruled Surface
PropertyManager.

1. Select Ruled Surface from the Insert > Molds menu, or click on the Ruled Surface icon found on the Mold Tools toolbar. (The Ruled Surface command can also be accessed from the Surfaces menu or toolbar.)

2. From the Ruled Surface PropertyManager, a portion of which is shown in Figure 14–22, select the type of ruled surface you wish to create.

3. Depending on the type of surface being created, it may be necessary to select a plane or planar surface the ruled surface will reference. For example, if Tapered to Vector is the type of surface chosen, a plane can be selected. The ruled surface will be tapered at some angle as measured from a vector perpendicular to the selected plane.

4. Specify a distance the ruled surface will extend.

5. Specify an angle for the taper, if required.

6. Click in the Edge Selection list box (not shown) and select the edges from which the ruled surface will emanate.

7. Click on the Reverse Direction button if the surfaces are pointing in the wrong direction.

8. Click on OK to finish creating the ruled surface.

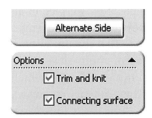

Fig. 14-23. Alternate Side button and additional options.

✓ **TIP:** *Use the partial loop selection technique (see Chapter 6) to make selecting a large number of edges on the model much easier.*

When creating a ruled surface, you may have noticed an Alternate Side button (shown in Figure 14–23). Sometimes this button is labeled Alternate Face or Alternate Direction, depending on the type of ruled surface being created. In the example shown, *alternate side* refers to which side the taper is added to, inside or outside. Edges can be selected from the Edge Selection list and the taper can be switched to the opposite side for those selected edges only.

✓ **TIP:** *Select multiple edges in a list box by selecting the first edge and then Shift-selecting the last edge. Every edge between the first and last selected edges will also be selected.*

It is not very common to want to switch the taper direction to the alternate side for different sections of the same ruled surface. However, if you choose to do this the *Connecting surface* option will play a role. Alternating taper direction produces splits in adjacent surfaces, such as that shown in Figure 14–24. The *Connecting surface* option closes these gaps. Also, leave the *Trim and knit* option checked, as this will force SolidWorks to knit the new ruled surfaces to the existing parting surface.

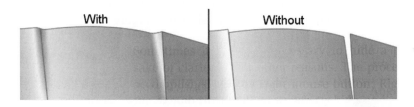

Fig. 14–24. With (left) and without (right) the Connecting surface option turned on.

What do all the other ruled surface types accomplish? They are just different ways of defining a ruled surface. For example, the ruled surface can be tangent or normal (perpendicular) to an existing surface. The ruled surface can be perpendicular to a user-specified vector (direction) or tapered (at an angle) to a vector. Finally, the ruled surface can be defined similar to sweeping a line along selected edges. The type of surface selected depends on what you are trying to accomplish.

So where does this leave our plastic part? The ruled surfaces have been created, but there are still openings that must be closed. One such opening is shown in Figure 14–25. It will be necessary to explore additional surfacing commands to complete this project. Some of the surface edges in Figure 14–25 will need to be extended outward, and other edges will need to have a lofted surface created between them. Finally, these surfaces will need to be trimmed to each other so that their edges do not overlap.

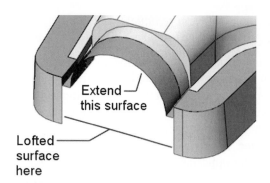

Fig. 14–25. Opening that must be closed.

To begin, let's use the Lofted Surface command to close the gap between the two edges shown in Figure 14–25. A lofted feature is typically a blend between two or more closed profiles. In our case, a lofted surface is used, which will blend between two edges of existing surfaces. You will learn a great deal about lofted features in Chapter 20, so all of the options surrounding lofting in general will not be discussed now. How-To 14–6 takes you through the process of creating a basic lofted surface.

How-To 14-6: Creating a Lofted Surface

To create a lofted surface, perform the following steps. Note that selection technique with all loft types (lofted bends, lofted surfaces, and lofted features) is critical, so pay special attention to step 2.

1. Select Insert > Surface > Loft, or click on the Lofted Surface icon located on the Surfaces toolbar.

2. Select sketches or surface edges to loft between. Make it a point to select near points that should match up during the lofting process, such as near the lower endpoints of each edge shown in Figure 14–25.

3. Click on OK to complete the lofted surface.

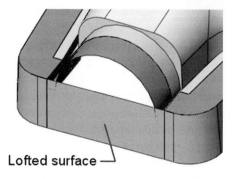

Fig. 14–26. Lofted surface.

When creating lofted surfaces, creating a sketch is not a requirement, though sketches can be used if necessary. Two surfaces can be lofted together to create a "bridge" between those two surfaces. This is exactly what was accomplished in the preceding How-To, with the final outcome shown in Figure 14–26.

There is a great deal of loft functionality and options that have not been touched on. The reason for this is twofold. First, most of the lofting options simply do not pertain to what we are trying to accomplish at this point in time. Second, surface lofting and lofting closed sketch profiles for the sake of creating solid geometry are two extremely similar functions. If you feel the need to learn more about creating lofted features, see Chapter 20.

Continuing with the example, it is now necessary to extend some of the surface edges outward so that they will fully contact each other. This is a fairly easy process, which How-To 14-7 takes you through.

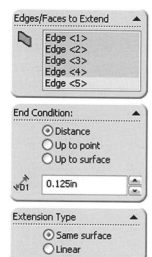

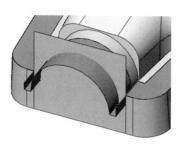

Fig. 14–27. Extend Surface PropertyManager.

Fig. 14–28. Results of extending surfaces.

HOW-TO 14-7: Extending Surfaces

To extend surfaces, perform the following steps.

1. Select Extend Surface from the Insert > Surface menu, or click on the Extend Surface icon found on the Surfaces toolbar.

2. Select edges on the surface to be extended.

3. In the Extend Surface PropertyManager, shown in Figure 14–27, specify a distance the surface should extend. Alternatively, specify an end condition for the extended surface.

4. Click on OK to create the extended surface.

The Extension Type panel, shown in Figure 14–27, controls whether the surface extends in a linear fashion or in a more "natural" fashion (*Same surface* option). For instance, if a curved surface is extended with the *Same surface* option enabled the resultant extended surface will continue to curve.

Figure 14–28 shows the result of extending a portion of the original parting surface. Incidentally, the lofted surface created earlier has also been extended. This creates intersecting surfaces that can then be trimmed to each other, with the outcome being a perfect edge, as will be seen shortly. Let's continue by examining how surfaces can be trimmed to other surfaces, or to each other. How-To 14–8 takes you through this process.

HOW-TO 14-8: Trimming Surfaces

To trim one surface to another, or to mutually trim multiple surfaces to one another, perform the following steps.

1. Select Trim Surface from the Insert > Surface menu, or click on the Trim Surface icon found on the Surfaces toolbar.

2. In the Trim Surface PropertyManager, shown in Figure 14–29, choose whether a Standard or Mutual trim should be performed.

3. Select the surface to use as the trim tool (select multiple surfaces if performing a Mutual trim).

Fig. 14–29. Trim Surface PropertyManager.

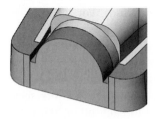

Fig. 14–30. Results of trimming the surfaces.

4. Choose whether the surfaces to be selected next will be surfaces that are retained (*Keep selections* option) or discarded (*Remove selections* option).

5. Click in the lower list box of the Selections panel (see Figure 14–29) and select the surfaces to keep or discard, depending on the choice made in the previous step.

6. Click on OK to complete the trim operation.

Trimming surfaces causes surfaces to be knitted. In other words, when two surfaces are trimmed to each other the result is a single surface. This works to your advantage, as the surfaces don't require knitting later on. Figure 14–30 shows the results of trimming the surfaces extended earlier.

Considering that there were multiple openings on this part originally, multiple loft operations were required. This holds true for the additional operations performed on this part, including multiple operations in which surfaces were extended and trimmed.

Even with all of the operations completed to this point, there is still more to do. Consider once again the original objective of creating a surface that will completely separate the mold. This would suggest that a final surface is necessary. This final surface will be planar, and will connect with all of the existing ruled and lofted surfaces. Figure 14–31 shows the model at its current stage of completion, with some additional preparatory work done in anticipation of creating a planar surface.

Fig. 14–31. Preparing to create a planar surface.

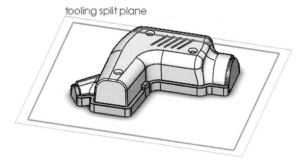

tooling split plane

The plane shown in Figure 14–31 was created a small distance above the bottom edge of the ruled surfaces created earlier (creating planes was discussed in Chapter 4). The reason for this is associated with the sketch also visible in the figure. The sketch will be used to define a planar surface, which will then be trimmed to the ruled and lofted surfaces. The trimming operation will go more smoothly if there is a little overlap to begin with. How-To 14–9 takes you through the simple process of creating a planar surface.

How-To 14-9: Creating a Planar Surface

To create a planar surface, perform the following steps.

1. Select Planar from the Insert > Surface menu, or click on the Planar Surface icon found on the Surfaces toolbar.

2. Select a 2D sketch that forms a single closed profile from which to generate the planar surface. Note that a series of surface edges can also be used, as long as the surfaces are all coplanar and form a closed boundary.

3. Click on OK in PropertyManager to create the surface.

Of all the surfacing commands, creating a planar surface is the easiest. Although either surface edges or sketch geometry can be used when creating planar surfaces, it is not possible to use a combination of the two types of geometry at the same time. Bear in mind that the Convert Entities command can be used to extract sketch geometry from model edges, which may aid in creating a planar surface.

Figure 14–32 shows the bottom of the model after the planar surface was created. The inset shows how the ruled surfaces protrude a small distance through the planar surface. As mentioned earlier, this will aid in the trim operation to follow.

Fig. 14–32. After creating the planar surface.

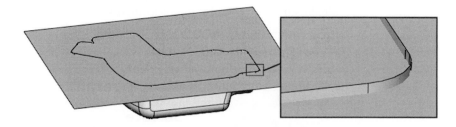

Performing the final trim operation will be easier if all of the ruled surfaces are a single surface that can be selected as a single object. Let's revisit the *Surface Bodies* folder mentioned earlier in this chapter, so that we can see what has happened while we have been creating new surfaces. Figure 14–33 shows the *Surface Bodies* folder in its current state. Note the new *Parting Surface Bodies* folder, which was added automatically following the implementation of the Parting Surface command.

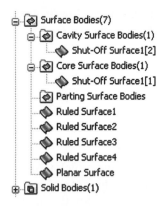

Fig. 14–33. Parting Surface Bodies *folder has been added.*

The *Cavity Surface Bodies* and *Core Surface Bodies* folders are as they should be. Each folder contains one surface that corresponds to all model faces on the inside of the part (used in the core creation) and all model faces on the outside of the part (used in the cavity creation).

What happens behind the scenes when the Tooling Split command is carried out is very interesting. SolidWorks will take the surface in the *Cavity Surface Bodies* folder and add it to the surfaces in the *Parting Surface Bodies* folder, and then use the resulting surface in the development of the mold cavity. This process is repeated to create the core. The surfaces in the *Parting Surface Bodies* folder are actually used twice: once for the cavity and once for the core.

If you have a SolidWorks model on which you have attempted to carry out the commands performed in this chapter, try hiding various surfaces and solid bodies. Right-click on the body listed in the *Solid Bodies* folder and hide the original design component. Hide various surface bodies so that you can examine the surface in the *Core Surface Bodies* folder, and so on. This should give you a better understanding of why we have carried out the various operations in the How-Tos to this point.

Note in Figure 14–33 that the surfaces listed below the *Parting Surface Bodies* folder are not actually in that folder. This is a problem, but we will take care of it in a moment. The fact that there are five surfaces listed should be taken care of first.

There are surfaces named *Ruled Surface1* through *Ruled Surface4,* followed by a single *Planar Surface.* Although it isn't required, it is good technique to knit these surfaces. Only the ruled surfaces will be knitted into a single surface. The subsequent knitted surface and planar surface will knit together automatically when they are mutually trimmed. How-To 14–10 takes you through the process of manually knitting surfaces.

Fig. 14–34.
Knit Surface
PropertyManager.

HOW-TO 14-10: Creating a Knitted Surface

To create a knitted surface, perform the following steps. The Knit Surface PropertyManager is shown in Figure 14–34.

1. Select Knit from the Insert > Surface menu, or click on the Knit Surface icon found on the Surfaces toolbar.

2. Select the surfaces to be knitted. This can most easily be accomplished by selecting the surfaces from the *Surface Bodies* folder in the flyout FeatureManager.

3. Click on OK to create the knitted surface.

It is worth noting that although the Knit Surface command is deceptively easy it can fail if improper surfaces are selected. For example, if the planar surface were selected in addition to the four ruled surfaces in our ongoing example, the Knit Surface command would fail. Surfaces to be knitted should contain edges that meet smoothly without overlaps or gaps.

Following the knit operation, a mutual trim is performed between the newly knitted surface and the planar surface. Because trimming has already been discussed, we won't repeat the process here. Figure 14–35 shows the example part from the top and bottom. The surface development for the tooling split has been completed at this time.

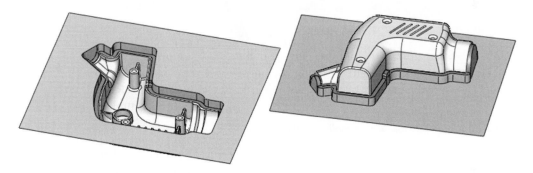

Fig. 14–35. Completed surfaces for the tooling split operation.

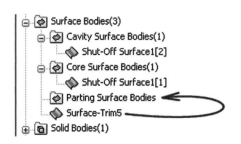

Fig. 14–36. Drag and drop the surface into the proper folder.

So, what's next on the agenda? It is necessary to place the trimmed surface into the *Parting Surface Bodies* folder. This is nothing more than a simple drag-and-drop operation, as indicated in Figure 14–36.

It is important that this final drag-and-drop operation be accomplished. With all surfaces in their proper folders, SolidWorks will be able to auto-select all of the appropriate surfaces when performing the tooling split.

When the Tooling Split command is implemented, it will be necessary to select a sketch that will dictate the size of the mold. For this reason, a sketch should be created beforehand. Such a sketch is shown in Figure 14–37.

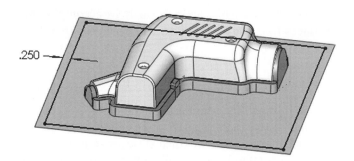

.250 —

Fig. 14–37. Sketch for the tooling split.

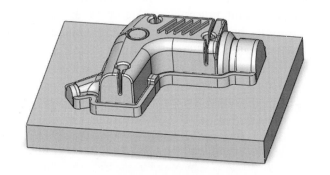

Fig. 14–38. Core body following a tooling split operation.

The sketch for the tooling split should be smaller than the outer perimeter of the surface shown in Figure 14–37. The sketch plane used is the planar surface itself, and the Offset Entities command was used to offset the perimeter of the surface inward a quarter inch. These are all operations you should be familiar with at this time. Once the sketch is complete, exit the sketch, and you will be ready to perform the Tooling Split command. Figure 14–38 shows only the completed core portion of the mold after completing the tooling split operation.

Because the Tooling Split command was covered in How-To 14–4, it isn't necessary to cover it again. The main point to keep in mind is that the interlocking surfaces have been created manually in the second example. This means that the *Interlock surface* option in the Tooling Split PropertyManager should *not* be checked.

How does a person know if interlocking surfaces can be created automatically? The one-word answer to that question is *experience*. Because that is not something most readers of this book will have a lot of regarding the SolidWorks software, the following list contains tips that will aid in the mold creation process.

- If you are unsure as to whether or not SolidWorks will be able to automatically create the interlocking surfaces, simply turn the *Interlock surface* option on and let it try.

- If attempting to use the *Interlock surface* option, the sketch for the tooling split must reside on a plane that is at some distance above the highest point (or below the lowest point) on the parting line.

- The Parting Surface command is usually the last step in the process of creating a mold prior to creating the tooling split if the *Interlock surface* option is checked.

- If the *Interlock surface* option is not used, you must ensure that the parting surface is larger than the sketch used for the tooling split.

- Depending on the geometry of the part, and if interlocking surfaces are not required, it is sometimes possible to specify a large value when using the Parting Surface command, thereby ensuring that the parting surface will be larger than the tooling split sketch.

- The sketch for the tooling split can be any shape, but should represent the footprint of the mold.

You should now have a much better understanding of how a mold is created in SolidWorks. However, there is often still more work to be done. Molds often require side cores, pins, lifters, and so on. These are covered in material to follow.

Core Pins

In the plastic part used for example 2, there were holes where screws would be placed in order to fasten the part to the other half of the final assembly. These holes are circled in Figure 14–39. Areas of the cavity portion of the mold that represent these holes have a tendency to wear out faster than the rest of the mold. One such area, a slender cylindrical feature, is enlarged in Figure 14–39.

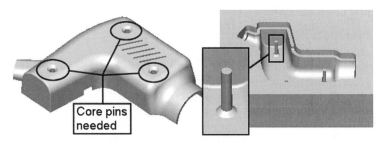

Rather than machine a new mold every time the slender features shown in Figure 14–39 wear out, core pins can be used. Instead of creating an entirely new mold (which could be a very expensive affair), only the pins are replaced.

Fig. 14–39. Where core pins would be beneficial.

Core pins, lifters, and side cores all involve the same basic creation steps. The processes also share the same command, which is the Core command. If we examine the process for creating the core pins, we find that the process begins like any other feature. That is, a sketch plane must be selected, and a sketch is created. Figure 14–40 shows the three circles that will be used to generate the core pins.

The top face of the mold is used as the sketch plane. Convert Entities (see Chapter 5) was used to convert the top edge of each countersunk hole on the original part into sketch circles. This is illustrated in Figure 14–40. A cutaway section view is shown for clarity. In practice, it would be necessary to change to a Hidden Lines Visible display mode so that the three edges of the countersunk holes could be selected prior to using the Convert Entities command.

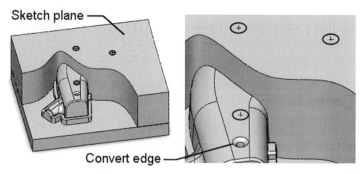

Sketch plane

Convert edge

Fig. 14–40. Sketch for core pins.

How-To 14–11 takes you through the process of using the Core command to create core pins. As mentioned previously, the Core command can be used in the creation of a variety of mold components. Following the How-To, we will examine a scenario in which a side core is created.

How-To 14-11: Using the Core Command

Fig. 14–41. Core PropertyManager.

To use the Core command (in this case to create core pins), perform the following steps. It is assumed that a sketch has already been created. Note that you must also exit the sketch prior to performing the following steps. Before attempting to use this command, tt is suggested that you read through the complete process to gain a better understanding of how SolidWorks creates core features.

1. Select Core from the Insert > Molds menu, or select Core from the Mold Tools toolbar.

2. Select the sketch that will be used to generate the core feature.

3. In the Core PropertyManager, show in Figure 14–41, specify an end condition or depth for each extrusion direction. (In the case of the core pins, only the first direction is used. See material to follow for a description of where it would be necessary to specify a second direction end condition or depth.)

4. Specify a draft value, if desired.

5. Select the *Cap ends* option if the core feature ends inside the tooling body. (As a general rule, this option can be left on, though unchecking it when creating the pins has no ill effect.)

6. Click on OK to complete the core feature (which in this case consists of the core pins).

What is happening when the Core command is used to create the pins? A number of events are taking place, and the top of the Core PropertyManager will give a clue as to what these events are. The Selections panel (visible in Figure 14–41) shows three areas that display items used in the core operation. The first item is the sketch selected in step 2. The second item dictates the extraction direction. This is perpendicular to the sketch plane, by default, but does not have to be.

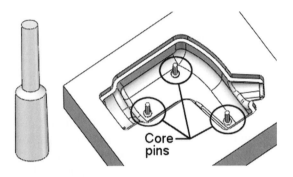

Fig. 14–42. Completed core pins.

The third item in the Selections panel is the body the core is extracted from. SolidWorks often picks the body on its own, but it isn't always correct, in which case you should make sure the proper body is selected. The important point to be made is that the software is acting on a single body. In the case of the pins, it is acting on the top half of the mold. The original component and the bottom half of the mold are left untouched. For this reason, the end condition for step 3 can be Through All (when creating the pins) and only the top portion of the mold is affected. Figure 14–42 shows what the top half of the mold would look like, along with one of the core pins shown separately.

Side Core

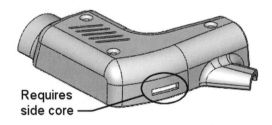

Requires side core

Fig. 14–43. Hole requires a side core.

A side core is a separate portion of the mold that would slide out in a direction differing from the direction the mold would pull away from the part. How is creating a side core different from creating core pins? The only differences are where the sketch plane resides and the shape of the sketch geometry. Imagine a hole in the side of the original part that would require a side core in the mold. Such an example is shown in Figure 14–43.

If a chunk of geometry in the side of the mold were sticking into a hole in the side of the molded part, there would be no way to pull the part out of the mold. Therefore, a portion of the mold is made to pull away in a different direction, thereby allowing the part to be pulled out of the mold.

The sketch used for the side core example is shown in Figure 14–44. The top half of the mold has been made transparent for clarity. Note the rectangular opening in the side of the original design component. When the sketch is used to create the side core, it will project perpendicular to the sketch plane, directly over the opening. The body described by the extrusion of the sketch will be subtracted from the top half of the mold. This is what allows the resultant core feature (in this example, a side core) to fit perfectly into the body (the top half of the mold) from which it was subtracted.

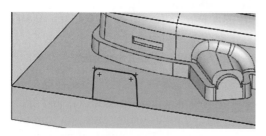

Fig. 14–44. Side core sketch.

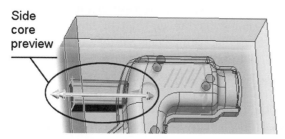

Fig. 14–45. Side core preview.

When stepping through the Core command, step 3 of How-To 14–11 states that end conditions must be specified for the extrusion directions. Often, such as when creating core pins or side cores, only a single direction is required. Such is the case in Figure 14–45, which shows the preview seen when creating the side core in our example. Once again, the top of the mold has been made transparent for clarity.

The second extrusion direction is pointing toward the left. This can be discerned by the double arrow visible in the figure. A blind end condition of zero was used because there is no material in that direction. The first extrusion direction points inward, toward the part. The primary and secondary extrusion directions can be reversed with the Reverse Direction button on the Selections panel of the Core PropertyManager.

One main point of confusion would be the primary extrusion depth as seen in Figure 14–45. A common concern is that the side core will extend too far beyond the original design component, or into the mold core (bottom half of the mold). This is not an issue. Once again, consider that the body acted upon is specified in the Core PropertyManager. If the extrusion depth of the side core sketch extends a reasonable distance beyond the wall of the original design component, no harm will come to the other pieces of the mold.

✓ **TIP:** *It is not necessary to sketch on outside faces of the mold. Create planes as needed and create sketch geometry that is embedded in the mold geometry. In this way, mold components such as lifters can be created. Make sure to check the* **Cap ends** *option in such cases, as the core feature (i.e., a lifter) is defined by end conditions that terminate within the mold geometry.*

10-second Topic: Multiple Bodies

Multiple bodies have been mentioned on more than a few occasions. Chapter 4 taught you what bodies were, such as solid bodies and surface bodies (see "10-second Topic: Solid Modeling Terms" in Chapter 4). In this chapter, we find that multiple surface bodies are being developed. There are even subfolders for surface bodies with specific functions automatically added to the main *Surface Bodies* folder.

When referring to multiple bodies, all we are really doing is stating that there is more than a single body in the part. For the most part, we have been dealing with parts that contain a single solid body. From this point forward, part files may contain more than multiple surface bodies. They might also contain multiple solid bodies. Chapter 20 explores design practices using multiple solid bodies that aid in modeling flexibility.

Mold Assembly Creation

Once a mold is completed, it is still a single part file, albeit composed of multiple bodies. Usually, it is necessary to machine the pieces of the mold. Some computer-aided manufacturing (CAM) programs run inside Solid-Works, whereas other programs require exporting files from SolidWorks that can then be read in by the CAM software. Exporting files is discussed in Chapter 23.

Before any machining can occur, the bodies of the mold part file should be saved out as an assembly file. This allows each body to be represented as a single part file. Tool paths could then be developed for each part, or they can be exported as mentioned previously. How-To 14–12 takes you through the process of saving multiple solid bodies in a part file as an assembly. These steps will work for any file with multiple solid bodies, and the process of which is not limited to bodies created as a result of developing a mold.

How-To 14-12: Saving Solid Bodies as an Assembly

To export solid bodies in a part file as an assembly, perform the following steps. Each body in the part will become a single part file. The subsequent parts can then be used as components in an assembly, though this is optional.

1. Right-click on the *Solid Bodies* folder in FeatureManager and select Save Bodies.

2. Select the bodies that are to be saved out as parts. This can be done by selecting the bodies in the work area, selecting the callouts, or placing checks in front of the names of the parts in the Save Bodies Property-Manager, shown in Figure 14–46. To save all bodies, click on the floppy disk icon circled in Figure 14–46.

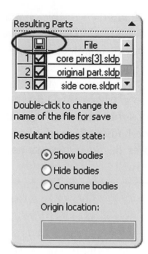

Fig. 14–46.
Save Bodies PropertyManager.

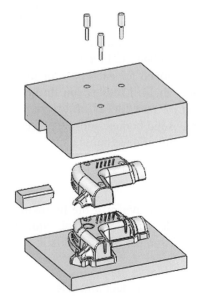

Fig. 14–47. Exploded view of the mold assembly.

3. Choose what you would like to have happen to the original bodies in the part file (Show, Hide, or Consume bodies, explained in material to follow).

4. Click on the Browse button (optional) if you wish to have the resultant part files placed into an assembly. You will be required to give the new assembly a name.

5. Click on OK to finish saving the bodies.

Step 3 allows for controlling what happens to the bodies in the original part file. There is usually no reason why the bodies should be hidden, as it serves no purpose. The Consume setting is synonymous with geometry that is "consumed," such as when creating a cut in a model. Using the Consume option consumes the original solid bodies. The only benefit is perhaps saving a little hard drive space due to a smaller file size. Unless you have a specific reason to do otherwise, set this value to Show.

Use the *Origin location* option to pick a vertex point on the model only if you want to establish an origin point on the new assembly that is different from the original origin point in the part file. You might want to do this if saving a body as a part for reasons of stereolithographic export, for which the geometry needs to be located in positive space with regard to a Cartesian coordinate system. If this information sounds meaningless to you, it is safe to ignore this option.

Note that step 4 is optional. Multiple solid bodies can be saved out as part files simultaneously, but they do not necessarily have to be saved out as an assembly. In the case of the mold used in the example, an assembly was saved, and then an exploded view was created. The final mold assembly is shown in Figure 14–47 in an exploded view.

✓ **TIP:** *Right-clicking on a single body and selecting Insert into New Part will allow for saving a single body of your choice as a separate part file.*

When saving bodies out as part files (regardless of whether an assembly is created), the resultant parts have an external reference back to the original part file containing the bodies. If changes are made to the original part, those changes will propagate to the new part files. Associativity is unidirectional only, meaning that changes made to the new part files will not affect the original part file the bodies were derived from. Figure 14–48 shows FeatureManager for an assembly generated from the original part's *Solid Bodies* folder.

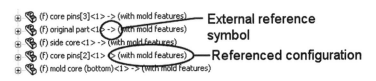

External reference symbol

Referenced configuration

Fig. 14–48. FeatureManager of the new assembly.

The text in parentheses following the external reference symbols in Figure 14–48 refers to the configuration referenced on the original part. See Chapter 7 for more information on external references and referenced configurations.

The original part from which the bodies were saved will have a Save Bodies feature in FeatureManager. It is important to remember that FeatureManager is a chronological history of events. Changes made to features created prior to saving the bodies will propagate to the new part files. However, any feature created after the bodies were saved will not propagate to the new part files.

10-second Topic: Rolling Back FeatureManager

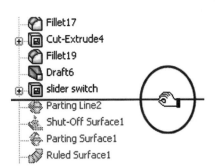

Fig. 14–49. Rollback bar.

The rollback bar provides a means of stepping back through time in FeatureManager. You will see the rollback bar at the bottom of FeatureManager. When the cursor is positioned directly over the rollback bar, a small hand appears, such as that shown in Figure 14–49. The rollback bar can then be repositioned simply by dragging it with the left mouse button. All features below the rollback bar will appear grayed out. These features are essentially ignored by SolidWorks and are not rebuilt.

It is also possible to use the rollback bar to insert features at a point in time other than that represented by the bottom of the FeatureManager list. Wherever the rollback bar is located, none of the features below it are rebuilt by SolidWorks. Essentially, the program ignores them. If a feature is created, it is added to FeatureManager at a position immediately above the rollback bar.

As a side note, it is worth mentioning that when a new sketch is started while in a rolled-back state the sketch will temporarily be positioned at the bottom of FeatureManager. However, if the sketch is exited, or if it is used to build a feature (which is more likely), the sketch will reposition itself to its proper place just over the rollback bar.

Scaling

A command that is very common to molded parts is the Scale command. As an example, it may be necessary to scale a model up in size prior to developing the mold. This would account for part shrinkage when the plastic cools, for instance.

Scaling is typically done early in the mold-making process, prior to creating the parting line. If you have already begun the mold-making process and decide the part should be scaled, it is possible to roll back prior to when the parting line was created. Scaling does not have to be uniform, but can be differing values in each of the three axes. How-To 14–13 takes you through the process of scaling a part.

How-To 14-13: Scaling a Part

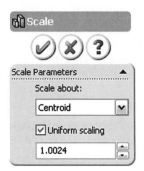

Fig. 14–50. Scale PropertyManager.

To scale a part, perform the following steps.

1. Select Insert > Features > Scale, or click on the Scale icon typically found on the Mold Tools toolbar or Features toolbar. The Scale PropertyManager, shown in Figure 14–50, will appear.

2. Specify what to scale about, such as the part's centroid or origin.

3. Specify a scale factor. If a nonuniform scale, uncheck the *Uniform scaling* option and specify the scale factor values for the *x*, *y*, and *z* axes.

4. Click on OK to scale the part.

If there are multiple bodies (solids or surfaces) in the part file, there will be an additional area in the Scale PropertyManager that will allow for selecting the bodies to be scaled. This list area is absent from Figure 14–50 because there is only a single body in the part file being scaled. In addition, the scale value is a factor, not a percent value. Therefore, if the final part will have a shrink factor of 2% the value to type in for the scale factor would be *1.02*.

Splitting Parts

Another tool SolidWorks users have at their disposal is a command aptly named Split. It allows for taking a part and breaking it into separate pieces. The Split command can be used in a number of situations, but it applies well to molded part creation due to the flexibility it gives the user. A model can be developed with both top and bottom halves intact, and then split at a convenient time. The top and bottom halves can then be developed separately, but both pieces update if a change to the original shape is made.

Splitting an existing part into separate chunks of geometry is not limited to two pieces. A single part can be split into any number of pieces. Planes, surfaces, or existing faces on the model can be used to define where splits occur. The subsequent bodies can then be further developed in the existing part file, or exported as separate parts or as an assembly (as discussed in How-To 14–12).

The example used in How-To 14–14 shows how a trackball housing in its early stages of development can be split into two pieces. Figure 14–51 shows the plane that will be used to split the part. The plane is named the Splitting plane in the figure, but could also be thought of as the parting plane.

Fig. 14–51. Preparing to split the trackball housing.

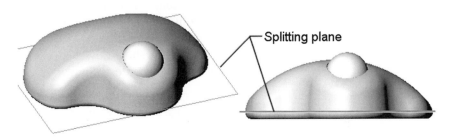

Splitting plane

Fig. 14–52. Split PropertyManager.

How-To 14-14: Splitting Parts

To split a part into multiple pieces (bodies), perform the following steps.

1. Select Insert > Features > Split, or click on the Split icon typically found on either the Features toolbar or the Mold Tools toolbar.

2. Select faces, planes, or surfaces to be used as the trim tool. In other words, the faces, planes, or surfaces that will determine where the part is split.

3. Click on the Cut Part button in the Split PropertyManager, shown in Figure 14–52.

4. Select the bodies that are to be split by selecting the bodies from the work area or placing a check in front of the body listed in the Resulting Bodies panel, shown in Figure 14–52.

5. Specify what should happen with the resultant bodies (shown, hidden, or consumed).

6. If the bodies should be saved out as parts, click on the callouts for selected bodies and enter a name for the part files.

7. Click on OK to create the bodies (and parts, if applicable).

When finished, a new feature (named *Split1*) will appear in FeatureManager. If you chose the Consume option in step 5, bodies split away from the original model will disappear from view. This is what is meant to happen. Whatever pieces are split from the model become separate part files, leaving behind whatever is left. If there is nothing left, there is nothing to display. Also note that selecting Consume will cause no bodies to be created. Only the separate part files will remain.

If the Hide or Show options are used, there will be multiple bodies present in the *Solid Bodies* folder. Each portion split from the original will be a separate body. You should have a separate body listed in the *Solid Bodies* folder in FeatureManager for each chunk split from the original part. Individual bodies can be hidden or shown by right-clicking on the body in the *Solid Bodies* folder and selecting Hide or Show.

✓ **TIP:** *To save parts generated with the Split command as an assembly, right-click on the Split feature in FeatureManager and select Create Assembly.*

The Split command can be used to save bodies as parts without specifying a trimming face, plane, or surface. How can this be possible? It is possible if the part being split already consists of multiple bodies. For example, the trackball marble in the trackball housing shown in Figure 14–53 was created as a separate body. The bodies have been separated in the image to stress the fact that they are indeed separate bodies.

If the SolidWorks model shown in Figure 14–53 is a part, and not an assembly, this raises the question as to how the part was created. A more technically accurate question would be: How were multiple bodies created in the part file without the Split command?

The easiest way to create multiple bodies in a part file is to extrude multiple closed profiles. Such an example is shown in Figure 14–54. The three closed profiles shown will create three chunks of geometry when extruded. This is an example of multiple bodies (solid bodies, in this case) in their most simple form.

Fig. 14–53. Separate solid bodies in a part file.

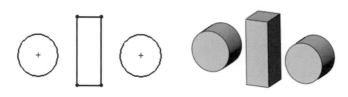

Fig. 14–54. A simple example of multiple solid bodies.

Fig. 14–55. Merge results *option.*

Another way to create multiple bodies is to uncheck the *Merge results* option found when creating various feature types, such as an extrusion or revolved feature. This option is shown in Figure 14–55. The feature being created in Figure 14–55 is a revolved feature, and in fact happens to be the trackball. Unchecking the *Merge results* option causes the trackball to not become part of the existing geometry. Instead, it becomes a separate body that (in this case) intersects the existing geometry.

Without going into the details, the Indent command was used to remove material where the sphere intersected existing geometry, and to add a small amount of clearance. You will learn about the Indent command in Chapter 20. For now, understand that by employing the Split command the solid bodies in a single part file can be saved as individual part files. It is not necessary to specify some object to split the solid geometry, because the bodies are already individual pieces of geometry.

Summary

The creation of a mold begins with a parting line. In many cases, a parting line is designed into a part. However, parting lines can be manually created even after the part is otherwise completed. Parting lines can be made to jump across flat faces where no edge previously existed. The Parting Line command can also split faces where no edge previously existed, thereby creating an edge on that face that is then made to be part of the parting line itself.

Shut-off surfaces can be automatically generated. The resultant surfaces will close any openings in the original part, thereby allowing the mold to separate into multiple pieces. Boundary conditions when creating shut-off surfaces can be of the type *tangent* or *contact*. A third boundary condition known as *no fill* will cause no shut-off to be created for a particular opening. This would be advantageous if there were complex internal openings in a part that might require a shut-off of a complex nature. In such a case, the Parting Line command can be used to generate an internal parting line.

Parting surfaces can automatically be generated wherever there is a parting line. Depending on how convoluted the parting line is, it may be possible to specify a large value to dictate how far the parting surface is extended. This can make generating the tooling split an easier process if interlocking surfaces are not required. As long as the mold footprint is smaller than the parting surface, a tooling split can be generated.

If interlocking surfaces are required, specifying a small value for the parting surface will suffice. When generating the tooling split, the *Interlock surface* option can then be employed, saving the designer a great deal of manual labor. If letting SolidWorks generate the interlocking surfaces, it is crucial to create a plane that is at some distance either above or below the highest or lowest points of the parting line, respectively, as measured perpendicularly from the tooling split plane and the parting line.

When developing interlocking surfaces manually, it will be necessary to use a variety of surface commands. One surface command that will prove useful, if not invaluable, will be the Ruled Surface command. Other surface commands and surface types that aid in mold design are the Extend Surface and Trim Surface commands, ruled and lofted surfaces, and planar surfaces.

Making sure all surfaces are in their proper place in the *Surface Bodies* folder is important. Surfaces can be rearranged within this folder by dragging and dropping the surfaces where desired. This makes it easy for Solid-Works to automatically select the proper surfaces when carrying out certain commands, such as when performing the tooling split. Although it is not required, it is good practice to knit surfaces so that there are single surfaces in each folder, as opposed to half a dozen surfaces (as an example).

Mold components can be generated using the Core command. Side cores and core pins are the easiest to create, as the sketch geometry for such items can often be created directly on an outside planar face of the mold. Other objects, such as lifters, require creating a sketch at some location internal to the mold. When creating such an object, the *Cap ends* option in the Core PropertyManager should be selected. This aids SolidWorks in separating the newly created body from the surrounding material.

You learned a number of topics in this chapter that pertain to mold making, but can be applied to other areas of design. For example, the ability to save multiple solid bodies out as part files, or even as an assembly, was explored. A part, or selected bodies, can be scaled. Scaling can be uniform about a centroid or origin point (among other things), and can be nonuniform if required. That is, scale values can differ in the x, y, and z axes. Parts can be split into multiple chunks of geometry and then saved out as separate part files or as an assembly.

An interesting aspect of FeatureManager was explored. Rolling back to a different point in time in FeatureManager can be accomplished with the rollback bar. This involves dragging the rollback bar and dropping it at the point you wish to roll back to. Features created at points that come after the rollback bar's position are essentially ignored by SolidWorks and are not considered during rebuild operations. The rollback bar also allows for determining how a part has been built, and can be beneficial in many ways, regardless of what type of model is being created.

Finally, you learned about creating multiple bodies through the use of the *Merge results* option. Unchecking this option is a way in which features can be kept from merging with existing intersecting geometry. The individual bodies can then be used in the development of the model, or even saved out as individual part files through the use of the Split command.

CHAPTER 15

Weldments

THERE ARE TWO METHODS FOR CREATING WELDMENTS in SolidWorks. One method consists of putting together an assembly and then adding weld beads. This is a very realistic real-world approach to creating weldments. The second method consists of creating a sketch or collection of sketches that define where structural elements of the weldment will be positioned. This second method is a very user-friendly approach and has downstream benefits, such as weldment cut lists.

Let's first explore how weld beads can be added in an assembly. Prior to adding weld beads, you should fully define the components in your assembly. There is no need to assemble with motion in mind when adding welds. It would be a very unique assembly, or a very poor job of welding, if you could move components after welding them. For this reason, components to be welded should be fully mated in position.

The mechanics behind adding weld beads is a very streamlined process. There are a series of four windows that step you through the entire process. Each window requires that you plug in some bit of information for one reason or another. Different weld bead types require different input from the user. There are quite a few types of weld beads that can be created. This chapter covers a few of them so that you can get a good feel for the weld bead creation process. For demonstration purposes, an assembly has already been built and the components fully mated in position. The assembly (*Weldment1*) is shown in Figure 15–1.

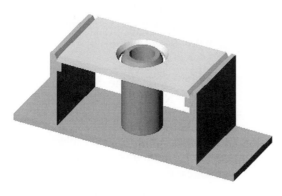

Fig. 15–1. An assembly in need of weld beads.

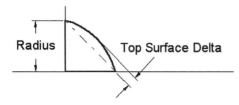

Fig. 15–2. Fillet weld bead dimensions.

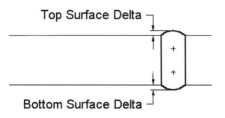

Fig. 15–3. Backing-run weld bead.

Before diving into the steps used to create a weld bead, you should first understand the way in which SolidWorks dimensions weld beads. Three different measurements are used. Depending on the type of weld bead you are creating, different combinations of these measurements will be required. The measurements are top surface delta, bottom surface delta, and radius.

A radius would be used for weld beads such as fillet welds. You would also need to specify the top surface delta for a fillet weld bead. Figure 15–2 shows a fillet weld bead as seen from the side.

Figure 15–2 shows how the radius and top surface delta dimensions are used to describe the size of the fillet weld bead. These measurements are used on other weld bead types in the same fashion. On occasion, as with a backing-run weld bead, a bottom surface delta must be specified. An example of a backing-run weld bead is shown in Figure 15–3.

These three measurements (radius, top surface delta, and bottom surface delta) are the only three needed to describe all of the weld bead types. With this knowledge, you are on your way to creating weld beads. In the following material, a fillet weld will be added. How-To 15–1 takes you through this process.

HOW-TO 15-1: Adding Weld Beads in an Assembly

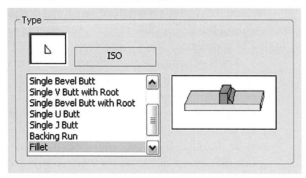

Fig. 15–4. Weld Bead Type window.

To add a weld bead between two parts in an assembly, perform the following steps.

1. Select Insert > Assembly Feature > Weld Bead. The Weld Bead Type window will appear, a portion of which is shown in Figure 15–4.

2. From the list, select the weld bead you wish to create, and then click on the Next button.

3. In the Weld Bead Surface window, shown in Figure 15–5, select the Surface Shape (such as convex or concave).

4. Specify the Top Surface Delta, Bottom Surface Delta, and Radius, as required and then click on the Next button.

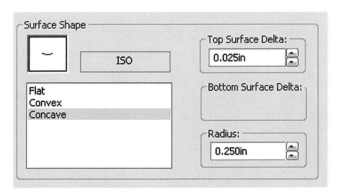

Fig. 15–5. Weld Bead Surface window.

5. In the Weld Bead Mate Surfaces window (not shown), specify the faces that will describe the location of the weld bead. This will include Contact Faces, and possibly Top Faces and/or Stop Faces, depending on the weld bead type. Figure 15–6 shows which faces were selected for the fillet weld bead being created in the current example.

6. Click on Next. The Weld Bead Part window will appear, a portion of which is shown in Figure 15–7.

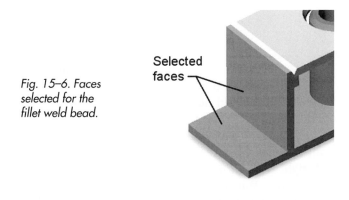

Fig. 15–6. Faces selected for the fillet weld bead.

Selected faces

7. The Weld Bead Part window will display the default file name and location given to the weld bead. This can be changed by clicking on the Browse button, but there is usually no reason to do so. Click on the Finish button to accept the SolidWorks default file name and location.

8. After adding the weld bead, you will find yourself in *edit part* mode for the new weld bead. It will be necessary to return to editing the assembly, rather than to the weld bead, so click on the Edit Part icon found on the Assembly toolbar.

Fig. 15–7. Weld Bead Part window.

A weld bead is an individual part file. As a matter of fact, you will be able to see it added to FeatureManager, just like any other assembly component. Figure 15–8 shows what the new fillet weld bead looks like, along with an insert that shows the new weld bead component in FeatureManager.

➥ **NOTE:** *When objects such as faces are selected in an assembly and you are working in one of the wireframe display modes, you will see the part outlined by a box to show that a face on that component is selected. If working in shaded mode, you will not see a box surrounding the part because the face turns green instead.*

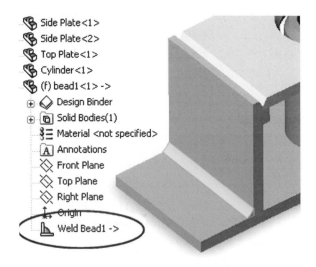

Side Plate<1>
Side Plate<2>
Top Plate<1>
Cylinder<1>
(f) bead1<1> ->
⊕ ◇ Design Binder
⊕ 🗀 Solid Bodies(1)
   ⅜≡ Material <not specified>
   Ⓐ Annotations
   ◇ Front Plane
   ◇ Top Plane
   ◇ Right Plane
   Origin
   Weld Bead1 ->

Fig. 15–8. New weld bead and its appearance in FeatureManager.

The name of the component in FeatureManager is *bead1*. As more weld beads are added, you will see the file names for those weld beads automatically increment to *bead2, bead3,* and so on. You can name the weld beads anything you want, and place the beads at any location on the hard drive. To do this, click on the Browse button found in the Weld Bead Part window. However, there are few reasons you would need to change the default name or file location for the bead. It is suggested that you accept the SolidWorks default settings for the weld bead file.

When a weld bead is created, it is fixed in position automatically. Common sense dictates that you would never want to move a weld bead. You certainly would not be able to move a weld bead if this were a real part. Furthermore, if you try to float a weld bead by way of the right mouse button you will find that SolidWorks will not let you. Weld beads are fixed, and are made to stay that way.

To edit a weld bead, right-click on the weld bead feature circled in Figure 15–8 and select Edit Feature. You can then cycle through the same four windows displayed originally, changing any parameters necessary. When the Weld Bead Part window appears (the final of the four windows), use the default name specified by SolidWorks. This file name will be the same as the original file name and will overwrite the existing file. There is no reason you would want to give the file a different name when you are editing an existing weld bead.

If you do decide you need to edit a weld bead's definition, remember that you are in an assembly. When you are finished editing the definition of the weld bead, you will still be in edit mode for that particular part. Make it a point to return to editing the assembly by clicking on the Edit Part icon on the Assembly toolbar.

Weld Bead Mate Surfaces

The trickiest part of mastering weld beads is determining what faces to select for the mate surfaces. Sometimes you must select what SolidWorks refers to as "stop faces," and sometimes "top faces" are required. These faces, known as mating surfaces, are used to determine where certain types of weld beads will be positioned. SolidWorks knows that selecting these faces can be confusing, and therefore provides descriptions of the specific faces required. If another weld bead type is added, you can see for yourself how SolidWorks handles the act of selecting the required faces.

In the example that follows, a single V-butt weld will be added. The following data is supplied for this particular weld bead.

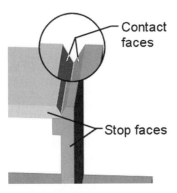

```
Convex

Top Surface Delta = .025 inches
```

Eventually, you will find yourself at the Weld Bead Mate Surfaces window. For the single V-butt weld type you will be required to select both contact faces and stop faces. Figure 15–9 shows the faces that were selected to satisfy these requirements.

Fig. 15–9. Faces selected for a single V-butt weld.

The stop faces are the faces where the weld bead will terminate. Because the weld bead has two ends, you must select two sets of stop faces. The first set of stop faces is shown in Figure 15–9. The second set of stop faces will be the same set of corresponding faces on the opposite side of the part. Make sure you select the second set of stop faces as well, or the weld bead function will not work. The completed single V-butt weld is shown in Figure 15–10.

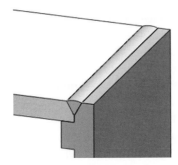

Because it is sometimes difficult to understand what faces should be selected for the various weld beads, SolidWorks gives you a description of what those faces are. Take your time and read through the descriptions, because they can be helpful.

Fig. 15–10. Completed single V-butt weld.

Weld Bead Special Situations

Because weld beads are basically defined by sets of faces, you will find certain locations between components that lend themselves poorly to defining weld beads. Take the case of where the cylindrical part protrudes up into the top plate, as shown in Figure 15–11. This would appear to be an area in which you would have little trouble adding a weld bead. It is, however, impossible to add a weld bead between the cylinder and top plate without making a small modification first.

The geometry between the cylinder and top plate appear to be likely candidates for a "single V-butt with root" weld bead. This type of weld bead requires two sets of contact faces: two faces for the V portion of the weld and two for the root. The two contact faces for the root are what pose the problem. The cylindrical face below the chamfer would have to be selected as a contact face, but this face does not create the desired results. The weld bead winds up extending all the way to the bottom of the cylinder.

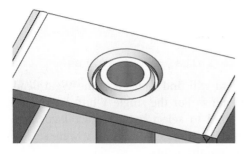

Fig. 15–11. The cylinder and top plate would prove troublesome for adding a weld bead.

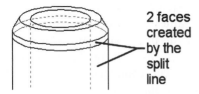

Fig. 15–12. After using the Split Line command on the cylinder.

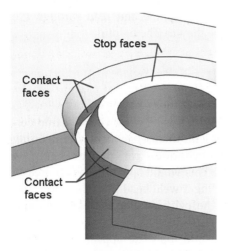

Fig. 15–13. Selecting the applicable faces.

For the weld bead to see the contact face as ending .125 inch below the chamfer, a split line is added to the cylinder (see "Split Lines" in Chapter 6). As a quick reminder, a split line will split a face into two faces by projecting a sketch onto a face. Figure 15–12 shows the cylindrical face separated into two faces via the Split Line command.

The odds of successfully creating a weld bead between the cylinder and the top plate have just drastically increased. Let's examine the process of creating one last type of weld bead. This time the "single V-butt with root" weld bead will be created. You already know how to begin the command. For the weld bead type, select Single V Butt with Root. For the weld bead surface, select Flat. You will not have to specify a surface delta if the surface is flat.

Next, supply the mate surfaces. The cutaway view shown in Figure 15–13 illustrates the faces that should be selected. Keep in mind that you should click in the applicable list box prior to selecting each set of faces. In other words, click in the Top Faces list box in the Weld Bead Mate Surfaces window prior to selecting the top faces, and so on.

For a "single V-butt with root" weld bead, it is usually necessary to also select stop faces. Stop faces would normally be the faces that tell the weld bead where to stop. In this case, there are no stop faces because the weld bead is circular. Instead of issuing a warning message telling you to select the stop faces, SolidWorks decides it can create the weld bead without any stop faces and finishes the task without complaining. Figure 15–14 shows the completed assembly with all weld beads added.

Like almost everything else in SolidWorks, weld beads are associative. This means that if the underlying geometry on which the weld bead was applied is modified the weld bead will update as well. Take for example the "single V-butt with root" weld bead just recently added as an example. It is partially between two chamfers: one on the top plate and one on the cylinder. What happens to the weld bead if the chamfers are modified? The weld bead updates. What happens if a component is repositioned? Again, the weld bead will update.

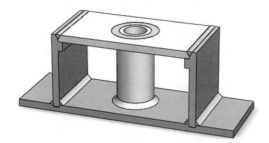

If drastic changes are made, a weld bead will not update. For example, if a component is mated to a different face on another component any weld beads associated with that component will not update. For that matter, the weld bead would fail because the faces being used as mating surfaces for the weld bead definition are no longer valid. It would be necessary to redefine the weld bead at that time, or to simply delete it from the assembly and create a new weld bead.

Fig. 15–14. Assembly with weld beads.

10-second Topic: Update Holders

There are certain functions you can perform in an assembly that will create update holders in FeatureManager. Adding weld beads is one of those functions. An example of update holders is shown in Figure 15–15. It is Solid-Works' way of keeping track of things, and a way of letting the user know that external references are being developed in the model.

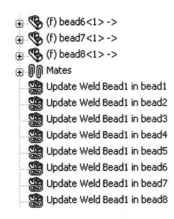

Fig. 15–15. Update holders.

External references can be a good thing at times. In our case, the external references are what will allow the weld beads to update if model geometry is altered. These same references can hinder you at other times. For example, creating a component in the context of an assembly may be desirable because the component should update automatically if the assembly changes. However, these same references can get in the way when performing other actions (such as assembly restructuring, discussed in Chapter 19).

Aside from understanding that the update symbols are warning you of the external references being developed in an assembly, there is nothing of value the SolidWorks user can accomplish with update holders. Therefore, they are best left alone. SolidWorks will not let you delete them, although they can be hidden. If you right-click on the name of the assembly at the top of FeatureManager, you will see the option Hide Update Holders. Likewise, if they are hidden, you will see the option Show Update Holders.

Weld Bead Files

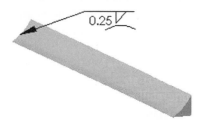

Fig. 15–16. Weld bead part file.

If you were so inclined, you could open one of the weld bead files in its own window. After all, weld beads are individual part files in their own right. There are not many good reasons you would want to open a weld bead part file. You really would not want to modify it because it is dependent on the rest of the assembly in which it was created. Figure 15–16 shows the first weld bead added to the welding assembly.

Note that the weld bead has a weld symbol attached to it. SolidWorks adds these symbols automatically. You can leave these symbols as is, or you can edit them. To edit the weld symbol, simply double click on it.

Typically you would add a weld symbol in a detailed design drawing, but weld symbols, like almost any annotation, can be added in a part or assembly file as well. The process for adding weld symbols is the same whether in a part or drawing. The mechanics of adding weld symbols were covered in Chapter 12. It is not the purpose of this book to explain welding terminology or weld symbols, so further elaboration is not justified.

The Join Command

To summarize the Join command, you might say that joining components in an assembly adds those components together. Components that had been touching in the assembly can be made into one contiguous solid part. This new part is literally a part file representing the sum of all original assembly components, or perhaps specific components only. There are a number of reasons a SolidWorks user might want to employ the Join command, a few of which are examined in this section.

One reason you may want to join components has a lot to do with welded components. If you were to weld components, you would normally want to show that welded assembly in a drawing layout as a single component. Many of the edges and hidden lines that would have been present in the assembly would not be shown if that same assembly had been welded. Joining the welded assembly components removes the edges not normally seen on a welded assembly and allows you to create a drawing layout that is more representative of the welded assembly. You will see an example of a joined component's hidden lines later in the chapter.

Another important reason to join components has to do with communicating and performance issues. For example, assume an assembly has been joined and the newly created part is named *Phantom Assembly*. Now also assume that the external references back to the original assembly are broken. This leaves us with a standalone part (with no external references) that is an exact duplicate of the original assembly.

One use of the *Phantom Assembly* would be to insert it into a higher-level assembly. If the *Phantom Assembly* were a purchased item, or if it were a completed unit that would not change, inserting it into a higher-level assembly would increase performance. The *Phantom Assembly* would appear as a component. It could be mated to like any other component, but the top-level assembly would have a much reduced overhead on computer system resources.

A second use of the *Phantom Assembly* scenario is that it could be sent to a customer, client, or perhaps an international sister company overseas (whatever the case may be) without the need to send every individual component. Keep in mind that if it became necessary to send an assembly via e-mail, all components contained in the assembly would also have to be sent. With a *Phantom Assembly*, this is not the case.

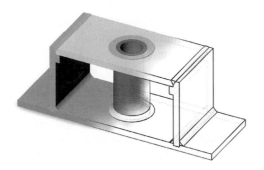

Fig. 15–17. This assembly will be joined.

The Join command is only available while in an assembly. In addition, a new component must be added to the assembly before attempting the Join command. This new component will become the joined component. You must also be editing this newly added component. This all may sound somewhat confusing, but in practice the procedure is very straightforward.

How-To 15–2 takes you through the process of inserting a new component into an assembly. Following this, How-To 15–3 takes your through the process of joining components. As an example, the assembly containing the weld beads shown in Figure 15–17 will be joined. Every component in the entire assembly will be joined in this procedure. The figure is shown in part with hidden edges because the hidden edges will change once the assembly is joined.

HOW-TO 15-2: Inserting New Components

Inserting existing components into an assembly was covered in Chapter 13. Not to be confused with that process, this How-To shows you how to insert a new part into an assembly.

1. Select Insert > Component > New Part.

2. Type in a name for the new part (*Weldment1-joined* will be used for the example used in this book).

3. Click on the Save button.

4. Select a plane or planar face to sketch on for the first feature of the new part.

Once a plane or face is selected on which to sketch, the new part will appear in FeatureManager. Note also that the new part is actively being edited. This is indicated by the title bar of the SolidWorks window, and from the colored text in FeatureManager. Typically, the part being edited will be the same color as the text, but there are no features present in the part yet. For all intents and purposes, the newly inserted part is an empty part file. That will change shortly.

In-context parts are typically created as single-use parts. They are usually not used outside the assembly they were created in. This stands to reason, considering the part will be based on geometry (i.e., other components) in the assembly. The typical scenario is that you would want the part to update if changes are made to the assembly. This is the most common reason for creating new parts in the context of an assembly.

In the example used in this book, the reason for inserting the new component is to create a joined component. This situation is somewhat atypical, in that it is likely that the newly joined component may be used in some other assembly.

There are a number of settings that affect the display when editing a component in an assembly. One such setting is the *Use specified color when editing parts in assemblies* option. This option can be found in the Colors section of the System Options (Tools > Options). When checked, other components will turn gray when a component is being edited.

Another such option is titled *Assembly transparency for in context edit* (shown in Figure 15–18). This option controls whether or not assemblies should remain opaque, retain their original transparency settings, or become transparent when a component is being edited. The latter setting is what is used in the image. When *Force assembly transparency* is selected, the slider bar controls the amount of transparency. A value of about 90% usually works well. You should use whatever setting works best for you.

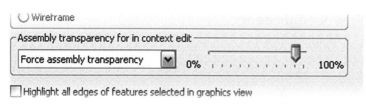

Fig. 15–18. Assembly transparency for in context edit option.

To move back and forth between editing a component and editing the assembly, use the Edit Part icon found on the Assembly toolbar. If a particular component is selected, and the Edit Part icon is clicked on, you will be placed in edit mode for that component. Likewise, clicking on the Edit Part icon while editing a component will return you to editing the assembly. Right-clicking on components or the assembly in FeatureManager will also gain access to the Edit Component or Edit Assembly command.

10-second Topic: InPlace Mates

When a new part is added to the assembly, a special mate relationship is added by SolidWorks. This is known as an InPlace mate, and will exist between the Front plane of the new part and the plane or planar face selected during the insertion of the new component. The reason for the InPlace mate has to do with external references. Considering that the new component will be developed from existing geometry within the assembly, the last thing you would want is to have the new part moving around. This would affect the new part in drastic ways. For this reason, the InPlace mate is added, which essentially locks the new part in position.

What happens if movement is a desired trait of the new part? If that is the case, it is best to design the part outside the context of the assembly, and then insert the part as an existing part file. The part can then be mated with the desired motion in mind.

If the part has already been created in the context of the assembly, and you change your mind about how the part should have been created, the InPlace mate can be deleted. In this case, the external references should also be deleted, which could primarily be accomplished via the Display/Delete Relations command. The component can then be mated to existing components in the assembly. Deleting an InPlace mate, breaking external references, and remating the component should only be done as a last resort. It is additional work that can be avoided by a little planning on the part of the designer. How-To 15–3 takes you through the process of joining components.

HOW-TO 15-3: Joining Components

To join components, perform the following steps. Note that if a new component was just inserted you will be in edit sketch mode for that component. Make sure sketch mode is not activated prior to performing these steps. In addition, you must be editing the component that will become the joined component. Typically, this will be the blank part file recently inserted into the assembly.

1. Select Insert > Features > Join. The Join PropertyManager, shown in Figure 15–19, will appear. (The image in Figure 15–19 has been edited to reduce its height.)

2. Select the components to be joined.

3. Click on OK to join the components.

Fig. 15–19. Join
PropertyManager.

4. Click on the Edit Part icon on the Assembly toolbar to return to editing the assembly.

✓ **TIP:** *When selecting components to be joined, all components can easily be selected by selecting the first component in FeatureManager and then Shift-selecting the last component in FeatureManager.*

After joining the components, assuming the default options were used you will be viewing the newly joined component. Figure 15–20 shows the *Weldment1-joined* component. It is all one color because it is all one component. The hidden lines shown are fewer than in the image shown of the original assembly (see Figure 15–17). Try comparing the illustrations and you will see the differences.

Fig. 15–20. The
Weldment1-joined
component.

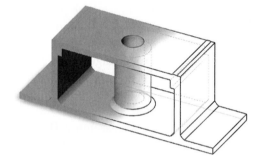

✗ WARNING: *When selecting components to be joined, do not select the component that will become the joined component. It is not possible to join a component to itself.*

The Join window is very basic, containing just two options. The first option is Hide Parts, which refers to the design components being joined. The general rule of thumb regarding this option is that if you have more work to do on the assembly do not hide the input components. If you are finished with the assembly, as is usually the case when joining components, you will want to leave the Hide Parts option checked (which is the default setting).

If you decide to hide or show the input components later, you can always change your mind. You can hide or show assembly components with the Hide/Show Components command (which you learned about in Chapter 13) or by right-clicking on components.

The other option, Force Surface Contact, will add a small bit of material to components that just touch each other. For example, if a cylinder is tangent to a planar face a miniscule amount of material will be added so that the components can physically be joined. Otherwise, a zero thickness error would occur. Without delving too deeply into technicalities, zero thickness is what a surface would have. Solids must have some thickness, even if very small.

Multiple Bodies in Joined Parts

Because multiple bodies can coexist in the same SolidWorks part file, joining components in an assembly that do not physically touch is not a problem. This particular fact makes it much easier to create phantom assemblies from assemblies that contain large numbers of parts.

When joining components and implementing the Force Surface Contact option discussed earlier, components that touch will be formed into a solid body. If there happens to be other components that do not touch any other components, they will become separate solid bodies. This is not an issue, but you should be aware of how SolidWorks is handling the situation.

Disassociating Joined Components

Joined components are dependent on the assembly in which they were created. On occasion, you may want to break this associativity. You can do this very easily by breaking the external references (see "Breaking External References" in Chapter 7) or by exporting the joined component as a file other than a SolidWorks part file, such as a Parasolids or STEP file. These are translation options discussed in more detail in Chapter 23. For now, bear in mind that exporting a file removes all intelligence from the file (such as parametric capabilities) but leaves the solid geometry intact.

Exporting a joined component can be a better solution than breaking external references. Once broken, external references cannot be reestablished. Leaving the references intact means that if the original assembly is modified the joined component will automatically update. Likewise, a new file encompassing these changes could be exported. This provides the best of both worlds yet still allows for the creation of a disassociated phantom assembly.

Saving Assemblies as Part Files

An alternative to joining assembly components is to save the assembly as a part file. There are benefits to this method if your sole objective is to create a phantom assembly. For example, only exterior components can be saved as a part, thereby limiting what might be numerous components housed within the assembly.

A variation of the gear housing assembly used as an example in Chapter 13 is used here. The gear housing has been placed on a stand, and drive chains have been added to the pulley gears, along with some other hardware. The assembly is shown in Figure 15–21, with the cover semitransparent so that interior components are visible.

There is a total of 237 components in the gear housing assembly: 206 parts and 31 subassemblies. Imagine an individual designer creating a factory layout with pieces of machinery at various locations on the factory floor. It would be very unwieldy to be forced to work with assembly files of this nature when all that is really needed are phantom assemblies. Working with single part files that represent the completed assemblies would be much more efficient.

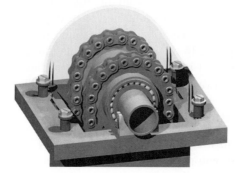

Fig. 15–21. Gear housing assembly.

An assembly file does not have to be joined into a single part to obtain a single part file. Rather, the assembly can be saved directly as a part file. The Save As command would be used to accomplish this task. How-To 15–4 takes you through the process of saving an assembly as a part file.

How-To 15-4: Saving an Assembly as a Part

To save any assembly file as a single part, perform the following steps. The possible results of performing this operation are discussed in material to follow.

1. With the assembly open, select Save As from the File menu.

2. In the *Save as type* drop-down list, select Part.

3. Enter a name for the new part file. By default, the part file name will be the same as the original assembly, but this can be changed.

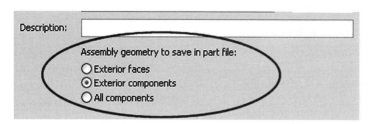

Fig. 15–22. Selecting the type of geometry to be saved in the part file.

4. Specify the type of geometry to be saved in the part file. Use one of the three selections shown in Figure 15–22.

5. Click on Save to complete the process.

Depending on the type of geometry saved and the size of the original assembly file, it may take a moment to save the part file. When SolidWorks is finished saving the file, you will find yourself back in the original assembly.

The important factor when saving an assembly as a part is what type of geometry is to be saved in the part. There are three options for saving: *Exteri-*

or faces, Exterior components, and *All components.* If saving just the exterior faces, the resultant part contains surfaces that basically represent all exterior components in the assembly. The resultant part file will not contain any solid geometry, so this may not be the most suitable choice in most cases.

Exterior components will save a part that displays only those components that could originally be seen in the assembly. The resultant part file contains solid bodies, rather than surfaces. This is the option that is most common. Figure 15–23 shows the same assembly seen in Figure 15–21 after it has been saved as a part with the *Exterior components* option.

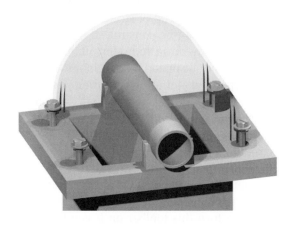

Fig. 15–23. After saving the assembly as a part.

The original assembly contained 237 parts and occupied approximately 9 megabytes of hard drive space. The single part file that resulted from saving the exterior components contained 21 solid bodies and was only 1.2 megabytes in size. The 21 bodies relates to the 21 components that could be seen from the outside of the assembly.

The last option not yet discussed is to save the assembly as a part while employing the *All components* option. This results in a part file containing the same number of solid bodies as there were parts in the assembly. The part file can turn out to be quite large, and there are few reasons for making use of this option.

If an assembly is saved as a part for reasons of creating a phantom assembly, and the mass properties of the top-level assembly are a concern, make sure the mass properties of the phantom assembly are correct. By default, SolidWorks automatically assigns the mass properties of the original assembly to the newly saved part, so the mass properties shouldn't be an issue. However, it doesn't hurt to double check. (For more information, see "Mass Properties in Assemblies" in Chapter 8).

Weldment Part Files

Weldments can be created in a part file from 2D or 3D sketch geometry. The sketch geometry is not used in the traditional sense to create features. Rather, the sketch geometry dictates where various structural members will be positioned. Any combination of 2D and 3D sketches can be used in the development of a weldment part.

Don't let the fact that weldment part files are parts and not assemblies scare you away from this functionality. Weldment part files contain a great deal of functionality. They are easy to create, they can be placed on a drawing with a cut list that is automatically generated, and they can be ballooned in the same fashion as an assembly.

Weldment part files need not be reserved strictly for weldments. The functionality crosses into other areas. For example, wooden two-by-fours can be used instead of steel structural members and the frame of a building can be constructed. Another example might be to use PVC to lay out a route for plumbing. We will talk about how to define specific materials later in this section.

Regardless of what type of material or structural members will be used, it is always necessary to start out with a sketch. For this next example, some exercise equipment will be created. The sketch geometry for this project is shown in Figure 15–24. The sketch geometry is actually a collection of five 2D sketches.

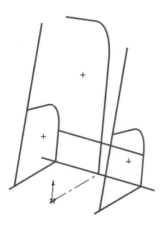

Fig. 15–24. Sketch geometry for the exercise equipment project.

Once the sketch geometry has been created, it is time to start adding structural members. Although there is a Weldments submenu under the Insert menu, it is easier to use the Weldments toolbar. Make sure to turn this toolbar on if you have not already done so. How-To 15–5 takes you through the process of defining the structural members in the weldment.

How-To 15-5: Creating a Weldment

To create a weldment in a part file, perform the following steps. It is assumed that the sketch geometry has already been created. If you attempt to step through this How-To and run into problems, read the material that follows.

1. Click on the Weldment icon on the Weldments toolbar. This defines the part as a weldment and enables additional functionality.

2. Click on the Structural Member icon on the Weldments toolbar. The Structural Member PropertyManager, shown in Figure 15–25, will appear.

3. Select the Standard to use, such as ANSI or ISO.

4. Select the Type of structure, such as pipe or tubing.

5. Select the Size for the structure.

6. Select path segments in the sketch where the specified structural members should be placed.

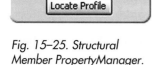

Fig. 15–25. Structural Member PropertyManager.

7. If the structural members should be rotated, modify the rotation angle setting as needed.

8. If the structural members need to be repositioned relative to the path segment, click on the Locate Profile button and select a new reference point on the structural member profile sketch. The sketch profile will be moved so that the selected point will fall on the path segment.

9. Click on OK to finish creating the structural members.

There are a few places where trouble can be encountered when creating weldments. Not having the sketch geometry set up properly will cause the most problems. The main point to remember is that the sketch geometry describes the location of structural members. Therefore, sketch lines (for example) should end where the structural member should end. Lines that cross can cause problems because the structural members will intersect each other. There is a way to trim structural members, but intersecting lines cannot be trimmed.

Another issue is that the right size is not available for a particular structural member. The fix for this is to create a custom size. This is discussed shortly.

The order in which sketch path segments are selected is important. When the first set of structural members were added to exercise equipment weldment, the segment that was the main segment was selected first. This can be explained better with reference to Figure 15–26. Note that there are two structural members that "T" into the main structural member. If either one of the secondary members had been selected first, the command would have failed.

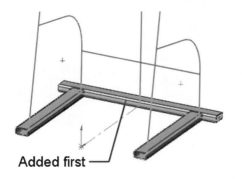

Fig. 15–26. Order of selection is important.

Added first

Structural members cannot usually be added to every single path segment in the sketch geometry, unless the geometry is very simple. This is true even if all structural members are of the same type. For example, the support braces on either side of the exercise equipment example must be added separately. Figure 15–27 shows these support braces.

An interesting aspect of the support braces is that they contain arcs. When arcs are selected as part of the path segments, the *Merge arc segment bodies* option will become available. Checking this option, visible in Figure 15–27, results in the structure being all one piece. The alternative would be to leave the option unchecked, thereby creating structural members that are individual pieces. Whether you choose to use this option or not depends on your application.

When selecting path segments for perfect corners, such as those shown in Figure 15–28, how the resultant structural members meet must be determined. This is done by what is referred to as *corner treatment*. Corner treatment can be set to a default value in PropertyManager when creating the structural members, and can also be defined at each particular corner if a specific treatment is needed. Clicking on the dot visible in Figure 15–28 brings up a toolbar (also shown in the image) for specifying the corner treatment (also known as the end condition).

Fig. 15–27.
Support braces.

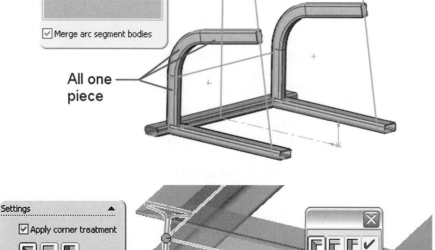

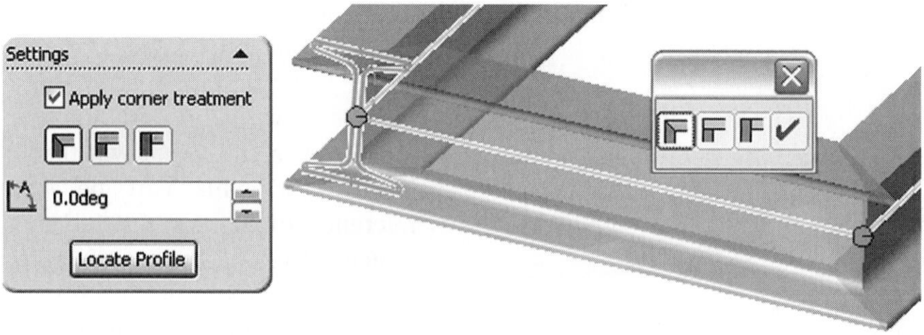

Fig. 15–28.
Applying
corner
treatment.

> ✓ **TIP:** *After adding structural members, hide the underlying sketch geometry by right-clicking on the sketches in FeatureManager.*

Custom Weldment Profiles

If the proper type of structural member does not exist, or another size is needed, it is time to create your own. By default, weldment profiles are stored in the main SolidWorks installation folder under *data\weldment profiles*. A portion of the file directory structure is shown in Figure 15–29. This is shown side by side with the listing of the structural member types (on the left). The types in the drop-down list match up perfectly with the directory structure.

If new folders are added to the directory structure shown on the right in Figure 15–29, they will automatically appear in the Type drop-down list, assuming the folder actually contains at least one weldment profile. To create a custom weldment profile, it is best to start with an existing profile. It is suggested that you use a profile that most closely resembles the profile you wish to create. How-To 15–6 takes you through the process of creating a custom weldment profile.

Fig. 15–29.
Structure types (left)
and weldment
profiles directory
structure (right).

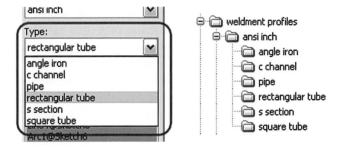

How-To 15-6: Creating a Custom Weldment Profile

To create a custom weldment profile, perform the following steps. Note that weldment profiles are actually library feature part files, which were discussed in Chapter 9.

1. Select Open from the File menu, or click on the Open icon on the Standard toolbar.

2. Select Library Feature Part file as the *Files of type* you wish to open.

3. Navigate to the *weldment profiles* folder and locate the profile that most closely resembles the profile you wish to create. Open that file.

4. Make any dimensional modifications as required.

5. Select Properties from the File menu.

6. Make alterations to the custom properties description field as desired, and then click on the OK button.

7. Rebuild the file by clicking on the Rebuild icon found on the Standard toolbar.

8. Select Save As from the File menu.

9. Enter a name for the new library feature part file.

10. Click on Save.

11. Close the file.

Many of the weldment profiles included with SolidWorks are tubular in nature. In other words, they have some wall thickness. However, this does not have to be the case. Single closed profiles can be used. Such might be the case if it were necessary to create various sizes of lumber, for instance.

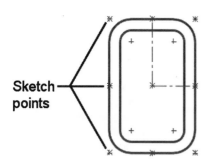

Sketch points

Fig. 15–30. Sketch points in a weldment profile.

Saving a copy of a weldment profile and making simple dimensional changes is easy, but creating a brand new profile from scratch will require further elaboration. For instance, it is necessary to establish sketch points at various locations that can be used to position the profile relative to the weldment path segments. Figure 15–30 shows such sketch points in a weldment profile.

Step 6 mentions custom properties, a subject covered in detail in Chapter 17. The Description custom property is important in the case of weldment profiles because the description will be used in a cut list. Any text can be entered and used for the description, but weldment profiles take advantage of a special technique to create the description.

When adding a custom property to a SolidWorks document (i.e., a part file), selecting a dimension while entering the custom property value results in a dimension value code being inserted into the value string. The code takes the following form.

```
"dimension_name@sketch_or_feature_name@part_name.SLDPRT"
```

The code may look a little intimidating, but it is nothing more than the name of the dimension, followed by the name of the sketch or feature the dimension is associated with, followed by the file name of the part, assembly, or drawing. The "at" (@) symbols act as separators between names.

The important thing is that the code will translate into the dimension value itself when the custom property is displayed in a cut list. If a dimension value is altered, the description automatically updates because it is directly linked to specific dimension values.

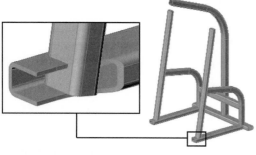

Fig. 15–31. An area that requires trimming.

Now that the basics of creating a weldment have been covered, it is time to perform some clean-up operations. In particular, some of the structural members intersect one another. This problem can be remedied by trimming members so that they do not intersect, similar to cutting them to the proper size. Figure 15–31 shows an area of the exercise equipment that requires trimming. A cutaway view is used for clarity in the image. How-To 15–7 takes you through this process.

How-To 15-7: Trimming Structural Members

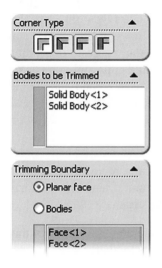

Most of the time, a structural member will require trimming to another structural member, if anything. However, the following steps will also allow for extending a structural member that may be too short.

1. Click on the Trim/Extend icon on the Weldments toolbar.

2. In the Trim/Extend PropertyManager, shown in Figure 15–32, select the Corner Type. This will most commonly be End Trim if the bodies to be trimmed do not meet at a perfect corner.

3. In the Trimming Boundary panel, select if objects should be trimmed to bodies or planar faces.

4. Click in the Trimming Boundary list box and select the bodies or planar faces (dependent on the choice made in step 3) to use as the trimming tools.

5. Click in the Bodies to be Trimmed list box and select items to be trimmed.

Fig. 15–32. Trim/Extend PropertyManager.

6. Click on OK when finished.

Fig. 15–33. Following the trim operation.

Figure 15–33 shows the same structural members as seen in the previous figure after the trim operation. Trimming, similar to when adding structural members in the first place, often needs to be performed a number of times. For the best results, do not try to trim every single member in the same operation. Generally, it is usually possible to trim a number of objects to the same structural member feature, or the same structural member feature to multiple objects.

When trimming to planar faces, structural members do not have to fully intersect the planar face. The face will extend outward and trim the member accordingly. When trimming to bodies, members must completely intersect the bodies. For this reason, it is often easier to use planar faces when trimming. The benefit of trimming to bodies is that the end of the body being trimmed will conform precisely to the trimming body.

Figure 15–34 shows an example of a weldment in which the legs of the table have been trimmed to the tabletop. The legs were trimmed to bodies, and the results are shown in the close-up of one of the table legs. The top has been hidden in the close-up so that the remaining geometry can be seen. Incidentally, the tabletop was created as a single structural member, so it was possible to trim all four legs simultaneously.

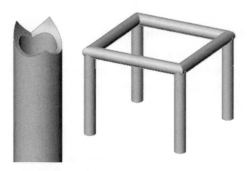

Fig. 15–34.
An example
of trimming
to bodies.

Continuing with the exercise equipment project, we will next explore how to create caps on the ends of the open members. How-To 15–8 takes you through this process. How-To 15–9 takes you through the process of adding gussets. End caps and gussets are both very easy to add and should be a very straightforward process.

HOW-TO 15-8: Adding End Caps

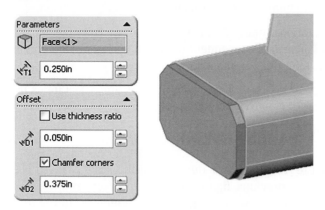

Fig. 15–35. End Cap PropertyManager.

To add end caps to openings at the end of structural members, perform the following steps.

1. Click on the End Caps icon on the Weldments toolbar.

2. In the End Cap PropertyManager, shown in Figure 15–35, specify a thickness for the end cap. An example of an end cap is also shown in the figure.

3. Select an end face of a structural member on which the cap should be placed.

4. In the Offset panel, specify how far the edge of the cap should be from the outer edge of the structural member. If the *Use thickness ratio* option is checked, the offset value will be some percentage of the thickness of the member being capped.

5. Optionally specify a value for chamfers on the corners of the end cap.

6. Click on OK to create the end cap.

Only one end cap can be created at a time, but fortunately they are easy to create. A filtering process is used to make it easier to select the end face where a cap should be placed. When creating caps, it is possible to specify an offset distance of 0. This will make the caps edges flush with the outside of the structural member. If the *Use thickness ratio* option is checked, the ratio value can be anywhere from 0 (identical to specifying an offset of 0) to 1.

How-To 15-9: Adding Gussets

To add gussets between structural members, perform the following steps.

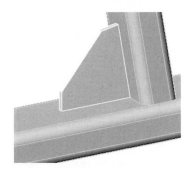

Fig. 15–36. Lower portion of the Gusset PropertyManager.

1. Click on the Gusset icon on the Weldments toolbar.

2. Select adjoining faces that the gusset should be positioned between.

3. Specify the dimensional parameters for the gusset via the Gusset PropertyManager. Use the preview as an aid, as it will update as dimension values are changed.

4. In the lower portion of the Gusset PropertyManager, shown in Figure 15–36, specify the thickness for the gusset, as well as what side the thickness should be applied to.

5. Specify a location for the gusset using the Location icons visible in Figure 15–36.

6. Optionally specify an offset value to help locate the gusset.

7. Click on OK to complete the gusset.

An example of a gusset is shown in Figure 15–37. Gusset position and size are updated in the preview whenever changes are made in the PropertyManager. As long as you are paying attention to the preview, it is not difficult to create the desired gusset.

A gusset (such as that shown in Figure 15–37) isn't going to stay in place very well unless welded into position. At the beginning of this chapter, the topic of adding weld beads in an assembly was examined. However, a weldment part file is not an assembly. Can weld beads still be added? Yes, they can, but a different command is used.

To add weld beads to a weldment part file, the Fillet Bead command should be used. As with all other commands explored so far in this section, the Fillet Bead command can be found either in the Weldments menu or on the Weldments toolbar. Unlike adding weld beads in an assembly, using the Fillet Bead command does not create a separate part file, although it does create a separate body.

Fig. 15–37. Example of a gusset.

Fillet beads can be full length, intermittent, or staggered. The ability to create a staggered fillet weld implies that the weld bead can be applied to both sides of the object being welded. The staggered weld can indeed be added to both sides of an object. In addition, the weld beads can be added simultaneously as part of the same feature definition. How-To 15–10 takes you through the process of creating all of these fillet weld bead types.

HOW-TO 15-10: Adding Fillet Weld Beads

To add full length, intermittent, or staggered fillet weld beads to a weldment part file, perform the following steps.

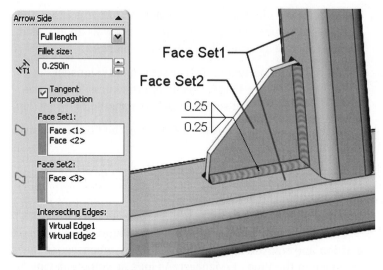

Fig. 15–38. Fillet Bead PropertyManager.

1. Click on the Fillet Bead icon on the Weldments toolbar.

2. Specify from the drop-down list in the Fillet Bead PropertyManager (shown in Figure 15–38) the type of weld bead that should be created.

3. Specify the fillet size.

4. Select the faces for *Face Set1*. Use Figure 15–38 to help determine what theses faces should be.

5. Select the faces for *Face Set2*. Again, refer to Figure 15–38.

6. Optionally check the Other Side option (not shown), and repeat steps 2 through 5 for that fillet bead as well.

7. Click on OK to complete the fillet bead.

As can be seen in Figure 15–38, fillet beads look more realistic than weld beads created in an assembly. The ability to create staggered or intermittent weld beads is also very attractive. The applicable weld symbol, visible in Figure 15–38, will also be added.

If it is necessary to add a weld bead on the opposite side of the object being welded (such as in step 6 of How-To 15–10), the same faces selected for *Face Set1* can be selected a second time to define the Other Side weld bead. The face for *Face Set2* would be the face on the opposite side of the gusset, facing away from the viewer as seen in Figure 15–38.

✓ **TIP:** *If it is not necessary to show weld symbols after adding fillet weld beads, right-click on the* Annotations *folder in FeatureManager and uncheck Display Annotations.*

Adding Traditional Features

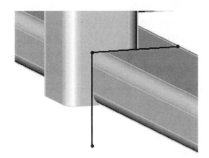

Fig. 15–39. A cut is needed.

As nice as the weldment commands are that have been examined so far, they don't always get the job done. There are often cuts that need to be added, or additional pieces are required to complete the weldment project.

Figure 15–39 shows an area on the back of the exercise equipment that requires a cut. The Trim command could trim the large vertical member, but it would result in a horizontal cut, eliminating more than is required. The design requirements call for a cut that will conform to the existing member that rests on the floor.

Cutting with an Open Profile

Using the sketch that is also visible in Figure 15–39, a simple cut can be performed. Note that the sketch contains nothing more than two simple lines. It is possible to use an open profile to cut geometry. However, the only end condition available in such a situation is Through All. Think of sawing a piece of wood in half with a hand saw. If the saw only cut halfway through, there would still be a single piece of wood (albeit with a cut in it). Likewise, stopping halfway through solid geometry would confuse SolidWorks. The cut must slice through all in order to get two pieces.

When using an open profile to create a cut, one side of the part must be discarded. Check or uncheck the *Flip side to cut* option in the Direction 1 panel to alter what side of the cut is discarded. The Split command would be required if we wanted to keep both pieces of the part after the cut was completed. The Split command was explored in Chapter 14.

In the example shown in Figure 15–39, the intent is to throw away the small portion of the vertical member that intersects the horizontal member resting on the floor. What complicates matters is that the piece on the floor should not be affected by the cut. The problem is resolved by modifying what is referred to as the *feature scope*, described next.

✓ **TIP:** *It is also possible to cut with planes. Select a plane and use the With Surface command found in the Insert > Cut menu.*

Multiple Bodies and the Feature Scope

When there are multiple solid bodies, and a new feature is created, the new feature can be merged with specific bodies. Likewise, when a cut is created certain bodies can be selected that the cut will affect. Which bodies are merged or cut is completely up to the user and is controlled in the Feature Scope panel when creating a cut or an extrusion (and various other feature types).

When creating a boss extrusion, and the *Merge results* option is checked (discussed in the previous chapter), the Feature Scope panel in Property-Manager will become available. Note that the Feature Scope panel will only appear if there are multiple bodies in the part file in the first place. In the case of creating a weldment, multiple bodies will exist, but you wouldn't really want to merge the results into a single solid (at least not in most cases). Therefore, if creating another piece of geometry for the weldment project, such as a plate or something similar, leave the *Merge results* option unchecked.

The Feature Scope panel, shown in Figure 15–40, allows for specifying which solid bodies are cut (or merged) when a feature is created. In the case of the exercise equipment, only the vertical member has been selected, which appears as *Solid Body < 1 >* in the list box.

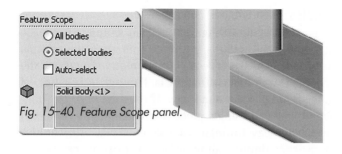

Fig. 15–40. Feature Scope panel.

The *All bodies* option will cause every body in the part to be considered when the feature is created. The default setting is the *Selected bodies* option with *Auto-select* checked (turned on). *Auto-select* causes only those bodies in the path of the feature to be selected, which makes for greater efficiency.

To select specific bodies, uncheck *Auto-select* and select the bodies to be affected by the feature you are about to create. This is what was done when creating the cut feature at the base of the vertical member shown in Figure 15–40. Only the vertical member was affected by the cut, even though there were other bodies in the path of the cut.

Weldment Cut List

Standard mechanical part files do not typically contain multiple bodies. In addition to multiple bodies, weldment part files typically contain a cut list and balloons. Balloons are usually found on assembly drawings, not part drawings. This raises the question whether or not ballooning is even possible on a weldment drawing. Ballooning is possible, but there is a little preparation that must be accomplished first.

Weldment part files contain a folder in FeatureManager similar to a *Solid Bodies* folder. It is called the *Cut List* folder, shown in Figure 15–41. The folder lists all bodies in the part file, such as structural members, gussets, end caps, and so on. The total number of bodies is listed to the right of the folder name, identical to a *Solid Bodies* folder.

Before a cut list can be created, cut list items must be generated. These "items" are folders that contain structural members of the same length, as well as other similar bodies. Each cut list item is then given properties that will populate an actual cut list on a drawing.

The manual process for creating cut list items is to right-click on an item in the cut list and select *Create cut list item*. It is not recommended that you use this process. Instead, right-click on the *Cut List* folder itself and make sure Automatic is checked. If it is, select Update from the menu and let SolidWorks create all of the cut list items for you. Figure 15–42 shows cut list items after updating and after renaming the various items.

Fig. 15–41. Cut List folder.

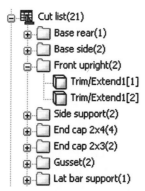

Fig. 15–42. Automatically generated cut list items.

Perform a slow double click on cut list items to rename them, which is the same technique used to rename anything else in SolidWorks. List item names are what will appear in the cut list once the list is placed on a drawing (which we will get to shortly). Other details that will appear in a cut list are controlled through the properties of each list item.

Right-clicking on a cut list item (one of the folders, and not an item in a folder) and selecting Properties brings up the Custom Properties window. A portion of the Custom Properties window for the Side support structural members is shown in Figure 15–43. Custom properties can be added to cut list items, but we will reserve that topic for Chapter 17.

	Property Name	Type	Value / Text Expression	Evaluated Value
1	LENGTH	Text	"LENGTH@@@Side support@exercise	45.325
2	ANGLE1	Text	"ANGLE1@@@Side support@exercise	0.0
3	ANGLE2	Text	"ANGLE2@@@Side support@exercise	80.0
4	DESCRIPTION	Text	TUBE, RECTANGULAR "V_leg@Structur	TUBE, RECTANGULA

Fig. 15–43. Custom Properties window for the Side supports.

If new structural members or other items are added to a weldment, it will be necessary to repeat the process of right-clicking on *Cut List* in FeatureManager and selecting Update. It is suggested that you wait until the weldment is completed and then update the cut list items all at once. Once the *Cut List* folder has been updated, a cut list can be added to a drawing. How-To 15–11 takes you through the process of adding a cut list table to a drawing.

Fig. 15–44. Weldment Cut List PropertyManager.

HOW-TO 15-11: Adding a Cut List Table

To add a cut list table to a drawing, perform the following steps. It is assumed that the cut list items have already been generated in a weldment part file. If they have not been, you will want to read the previous section and create the cut list items.

1. Select Weldment Cut List from the Insert > Tables menu, or click on the Weldment Cut List icon found on the Tables toolbar.

2. From the Weldment Cut List PropertyManager, shown in Figure 15–44, select the template to use for the cut list, or use the default template.

3. Specify an anchor point if desired (see How-To 12–6).

4. Click on OK to generate the cut list table.

5. If no anchor was used, click on the drawing to position the table.

Additional options in the Weldment Cut List PropertyManager allow for selecting a configuration other than the default configuration, and what to do with missing items, such as display with a strikethrough. Item numbers can start at some number other than 1, and item numbers can be made not to change even when the table is resorted. Figure 15–45 shows a portion of a weldment cut list table.

ITEM NO.	QTY.	NAME	DESCRIPTION	LENGTH
1	1	Base rear	TUBE, RECTANGULAR 4.00 X 2.00 X .25	48
2	2	Base side	TUBE, RECTANGULAR 4.00 X 2.00 X .25	34
3	2	Front upright	TUBE, RECTANGULAR 3.00 X 2.00 X .25	59.247
4	2	Side support	TUBE, RECTANGULAR 3.00 X 2.00 X .25	45.325
5	4	End cap 2x4	STEEL END CAP, 2 X 4 X .25	
¿	⌐	End cap 2x2	STEEL END CAP, 2 X 2 X .25	

Fig. 15–45. Weldment cut list table.

Certain items, such as end caps and gussets, will not contain a description or length in the table. To see a description in the table, it will be necessary to add a custom property named Description to that cut list item in the part file. Only then will it appear in the cut list table in the drawing. As mentioned earlier, right-click on a cut list item and select Properties to bring up the Custom Properties window.

If the default template was used when adding the table, a NAME column will probably not exist. Columns can be added to the table, and other formatting techniques can be used to customize the table's appearance. See "Table Dynamics" in Chapter 12 if you need assistance with tables.

All that remains is to add balloons. Adding balloons in a weldment drawing is exactly the same as adding balloons in an assembly, which you should already know how to do. If not, see "Balloons" in Chapter 13. Be aware that adding balloons in an assembly drawing can be done prior to adding a BOM, but adding balloons in a weldment drawing requires that the cut list table (which is very similar to a BOM) be added first.

✓ **TIP:** *To add views of individual bodies within a weldment or any part file, use the Relative To View command. Select the body by clicking on the body in an existing drawing view.*

Summary

Weld beads can be added to assemblies, and come in a variety of types. When creating a weld bead, you would typically define the weld bead by a specific set of dimensions and faces that determines where the weld bead will be positioned. Each weld bead type requires different sets of faces that define its boundaries.

The Weld Bead command, although found in the Assembly Features menu, is actually a separate part file SolidWorks creates. It is actually much more efficient to add weld bead symbols to a drawing rather than model weld beads in an assembly. Model weld beads in an assembly only if you need a very true-to-life representation of the assembly model geometry.

The Join command allows for creating a new part that is a conglomerate of components in an assembly. The new part may consist of specific components, or even every component in the assembly. Use joined components as phantom assemblies, for showing welded components, or for various other reasons.

Assemblies can be saved as part files, resulting in a file that is much more simplistic in nature. This is a better solution than the Join command if the objective is to create a phantom assembly. Of the three options available when saving an assembly as a part, the *External components* option is usually the optimal choice. Only the visible external components of the assembly will be saved as solid bodies in a new part file.

Weldment part files are an excellent choice for modeling weldments. It is much easier to model weldments in a part than it is to do so in an assembly due to the automation built into weldment part files. A combination of 2D and 3D sketch geometry can be used to dictate the paths structural members will take.

If a particular structural member profile does not exist, feel free to create custom profiles. Weldment profiles do not necessarily have to describe weldment structural members. The weldment functionality can be used for other disciplines beyond weldments, though all of the functionality typically associated with weldments may no longer apply.

Weldment part files can include gussets, end caps, and fillet beads. Other traditional features, which you are already familiar with, can be added. Because weldment part files contain multiple bodies, be careful to specify what bodies will be affected by any traditional feature types created. Examples of such features would be extruded bosses or cuts, to name two. Use the Feature Scope panel in PropertyManager when creating such features to dictate what bodies will be affected by the feature.

Add cut list tables by first creating cut list items in a weldment part file. Use the Automatic method of creating cut list items by right-clicking on the *Cut List* folder and selecting Update. A cut list table can then be added to a weldment drawing, and balloons can be added in the same fashion they are added to an assembly drawing.

CHAPTER 16

Assembly Configurations

IN CHAPTER 11 YOU LEARNED HOW TO SUPPRESS FEATURES within a part file. SolidWorks also gives you the capability to apply this same functionality to assemblies. Instead of features, though, it is components that are usually suppressed. It is possible to add configurations to an assembly and control which components are suppressed for each configuration, allowing for different "versions" of the assembly within the same file.

If a part file has multiple configurations, which version of the part is being used in the assembly can be specified in assembly configurations. You will also explore how certain dimensions that belong to the assembly can be controlled via configurations, and how components can have their visibility state controlled.

The process of adding a configuration in an assembly is exactly the same as adding a configuration in a part. As a quick refresher, the following is a synopsis of how it works.

1. Click on the ConfigurationManager tab at the top of FeatureManager.

2. At the top of ConfigurationManager, right-click on the name of the model.

3. Select Add Configuration, type in a name, and click on OK.

Assume for a moment you had an assembly on your hands such as the sliding brace shown in Figure 16–1. The inner component is a post, which can be positioned at various heights with the aid of a pin. Now imagine that you wanted a separate "version" of the assembly with certain components removed so that the post could be shown more clearly.

Fig. 16–1. Sliding brace assembly.

Two options come to mind immediately. You could hide certain components, or you could suppress them. Hiding components would remove them from view. Suppressing components would remove them from memory. Suppressing components also has the by-product of removing the components from view as well. Because of this, suppressing and hiding seem very similar on the surface, but actually they are not (as you learned in Chapter 13).

The main point to keep in mind is that suppressing a component will also suppress any dependent items, such as mates. If you do not care that the mates are suppressed, and the assembly will not be adversely affected, suppression is a better alternative. This is due to the suppressed components being removed from memory and increasing overall system performance.

Fig. 16–2. Assembly with a few components suppressed.

Every assembly is different, and assemblies should be considered on a case-by-case basis. Often, if the assembly contains movable components you may want to hide certain components instead of suppressing them. This is so that the assembly will not fly apart unexpectedly if other components are put through their motions. You may find that a certain combination of hidden and suppressed components works best for a particular assembly. Figure 16–2 shows the upper portion of the sliding brace assembly with a few components suppressed.

In the case of our example, it would be a simple task to suppress the components and then reverse the process (resolve them) when you wanted to see them. However, it would be much more convenient if configurations were used. Because configurations remember which components have been hidden or suppressed, it would be very easy to revert to the original assembly (where nothing is hidden or suppressed) simply by switching which configuration were being shown.

Assembly Configuration Properties

If you were to add a new configuration to an assembly, you would find that the PropertyManager for an assembly configuration's properties is more complex than that for a part. There are more options in this PropertyManager (as can be seen in Figure 16–3, which shows a portion of an assembly's Configuration Properties PropertyManager). To access the Configuration Properties PropertyManager, right-click on the configuration's name and select Properties.

The various options available in the Configuration Properties PropertyManager for an assembly are outlined in Table 16–1. Some properties common to both part and assembly configurations will not be expanded upon here, as they have already been explained in Chapter 11. Options that have the same function in an assembly configuration as they do in a part configuration are noted.

Fig. 16–3. Configuration Properties PropertyManager.

Table 16-1: Assembly Configuration Properties

Option or Setting	Description
Part number displayed when used in a bill of material	Shows what will appear in a BOM. This function is identical to part configurations.
Drop-down menu containing the following selections:	This function is identical to part configurations. The following data will be displayed in a BOM:
Document Name	Name of the file (default setting)
Configuration Name	Name of the configuration
User Specified Name	User can enter text to be displayed
Don't show child components in BOM when used as subassembly	When checked, none of the components in the subassembly will appear in the BOM.
Suppress new features and mates	When checked, new assembly features and mates added to other configurations are suppressed in this configuration (the configuration whose properties you are editing).
Hide new components	When checked, new components added to other configurations are hidden in this configuration (the configuration whose properties you are editing).
Suppress new components	When checked, new components added to other configurations are suppressed in this configuration (the configuration whose properties you are editing).
Use configuration specific color	Allows for changing the color of a single configuration, as opposed to every configuration (see material to follow).

Using Configuration-specific Colors in Assemblies

This has the same meaning as it does when working with a part file's configuration properties. If checked, a button for modifying the color of this specific configuration becomes active. Clicking on the Color button allows for specifying a color for the configuration. This makes it easier to differentiate between configurations that have minimal differences.

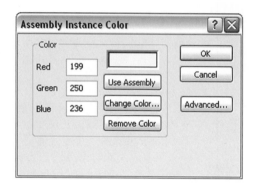

Fig. 16–4. Assembly Instance Color window.

Applying a specific color to an assembly configuration requires some extra effort from the user and is not as easy to implement as with solitary part configurations. If *Use configuration specific color* is implemented for an assembly and a new color is chosen, no change will be seen unless another action is taken. Specifically, access Component Properties for a particular component and click on the Color button (see "Component Properties" in Chapter 13). The Assembly Instance Color window, shown in Figure 16–4, will appear.

By default, the color of an assembly component will be whatever the color of the actual part is. The Color button can override this. If the Use Assembly option is selected, the component will use the colors set forth for the assembly via the Colors section of the assembly's Document Properties (Tools > Options). The Use Assembly option also allows the *Use configuration specific color* option in the Configuration Properties to work.

You may very well find that the *Use configuration specific color* option is not worth the bother when working with assembly configurations. It is just too much work to edit the properties of every component whose color needs to be configuration dependent, and is not necessary in most cases. Configuration-dependent colors may be occasionally useful if certain key components are being modified in various assembly configurations. Having the key component change colors in the various configurations is sufficient to draw attention to it. This helps to remind the user which configuration is being viewed.

Controlling Component Configurations

A very important aspect of assembly configurations is the ability to control which configuration of a part is being used in an assembly. This capability allows you to use one part file containing different configurations in different assemblies. It even allows for using different configurations of the same part in the same assembly. These conditions would also apply to subassemblies. You learned in Chapter 13 (see "Component Properties") how to change which configuration of a component is used in a top-level assembly. Here we will see this functionality in practice.

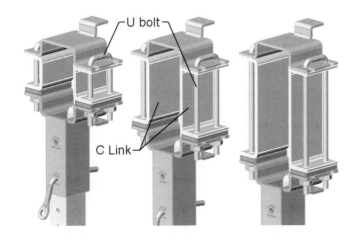

Fig. 16–5. Three different configurations of the sliding brace assembly.

As a simple example, imagine a scenario in which different configurations of U-bolts and C-links are used in the sliding brace assembly. The C-link is a component consisting of a single part, whereas the U-bolts are subassemblies consisting of the U-bolt itself, two washers, and two nuts. Figure 16–5 shows three versions of the sliding brace. They are all the same assembly, but are different configurations.

It is important to understand that both part configurations and assembly configurations are being used. For example, the C-link part contains multiple configurations that control the height of the C-link. The configurations used by the C-link can be called out from the assembly in which the C-link component resides. This is particularly significant because individual component feature dimensions cannot be directly controlled within top-level assembly configurations. Component feature dimensions can be changed, but they are not configurable items.

The entire concept of configurations, when related to assemblies, takes on a new complexity. However, once it is understood how component configurations can be controlled from within an assembly, the situation becomes easier to manage. How-To 16–1 takes you through the process of specifying a configuration for a particular component within an assembly.

HOW-TO 16-1: Controlling a Component's Configuration

To control a component's configuration within an assembly, perform the following steps. The term *component* refers to both individual part components and subassemblies.

1. In FeatureManager, right-click on the component whose configuration is to be specified and select Properties.

2. In the area titled *Referenced configuration,* shown in Figure 16–6, select *Use named configuration.*

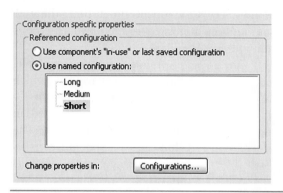

3. From the configuration listing, select the configuration to be used.

4. Click on OK.

Fig. 16–6. Using a named configuration.

If there are numerous configurations present for a particular component, a scrollable list of those configuration names will be presented. Having a large number of component configurations is one matter, but let's examine a different scenario. Imagine that there is a large number of assembly configurations. How could a particular component configuration be specified, but only in (for example) 5 of the 20 assembly configurations? This is where the Configurations button, visible in Figure 16–6, becomes critical.

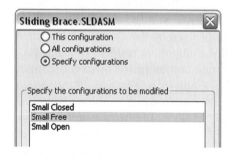

Fig. 16–7. Specifying which assembly configurations should be affected.

The window shown in Figure 16–7 appears after clicking on the Configurations button. The underlying functionality is identical to that described in Chapter 11 regarding changing feature suppression states. The difference is in what is being configured. Rather than controlling suppression states, it is the configuration of a particular component that is being controlled.

Selecting the *Specify configurations* option allows for selecting which configurations at the top-level assembly are changed. Any selected configurations will begin using the version (configuration) of the component as selected in the *Referenced configuration* listing (shown in Figure 16–6).

↝ **NOTE:** *All settings in the* Configuration specific properties *section of the* Component Properties *window, such as suppression state, can be applied using the* Configurations *button.*

Showing Alternate Components—A Procedural Outline

Showing certain components in one configuration and different components in another configuration is accomplished using techniques you have already learned in this chapter and in Chapter 11. The general process could be performed using the following steps as a guide. Each step is not could be performed using the following steps as a guide. Each step is not fully described, as you should already be familiar with performing the operations.

1. Add a new configuration to the assembly.

2. Suppress the component or components you do not wish to see in the new configuration.

3. Insert a new component into the assembly.

4. Mate the new component into position.

Remember that if the newly inserted component should not be resolved in the original configuration both the *Suppress new components* and *Suppress new features and mates* options should be checked in that configuration's properties. Otherwise, it will be necessary to activate the original configuration and manually suppress the new component added to the new configuration.

Configurations in Drawings

It is possible to control which configuration is being shown for a part or assembly within a design drawing. The configuration can be specified for any particular view, so design drawings containing multiple views can show different configurations for each view.

There are various reasons it might be necessary to show a specific configuration of a part or assembly in a particular drawing view. The most basic reason would be that a design drawing needs to be created for a particular configuration. Rather than spend time examining the reasons for showing a specific configuration, let's jump right to the point, which is that there is a technique involving configurations that allows for showing a cutaway view in a drawing.

Using configurations to create a cutaway view can be performed with parts or assemblies, but an assembly will be used in this discussion. Imagine an assembly for which it is necessary to show the internal components in one of the drawing views. By creating a new configuration in the assembly, and then creating an extruded cut assembly feature (discussed in Chapter 13), the cut need be shown in the new configuration only.

The benefit of using configurations in this manner is that there is no limit to the type of sketch geometry that can be used. This would not be the case if creating a standard section view. Broken-out section views also fall short, in that they cut down into a view from the top. Assembly features can affect specific components via a cut of any shape in any view orientation.

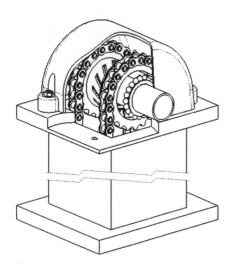

Figure 16–8 shows an example of an assembly cut-away view. Although the figure shows the view with hidden lines removed, the results can be fairly dramatic when shaded. This book has already discussed the processes involved in creating assembly features, configurations, and drawing views. How-To 16–2 takes you through the process of specifying which configuration to show for any particular drawing view.

Fig. 16–8. Assembly cutaway view using configurations.

How-To 16-2: Drawing View Configuration Selection

To specify which configuration is to be shown in a particular drawing view, perform the following steps. This assumes, of course, that the part or assembly already contains more than one configuration. The process is exactly the same whether working with a part or assembly drawing view.

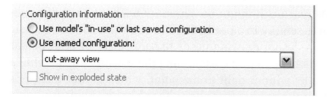

Fig. 16–9. Using a named configuration for a drawing view.

1. Right-click on the view you want to specify the configuration for and select Properties.

2. In the *Configuration information* section of the Drawing View Properties window, shown in Figure 16–9, select *Use named configuration.*

3. From the drop-down list, select the configuration to be shown in that view.

4. Click on OK.

You may have noticed the *Show in exploded state* option in Figure 16–9. This option, discussed in Chapter 13, is grayed out in Figure 16–9 because the selected configuration does not contain an exploded view. If the configuration did contain an exploded view, the option would be available.

Parent/Child Configuration Options

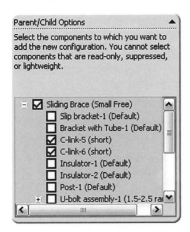

Parent/Child Options

Select the components to which you want to add the new configuration. You cannot select components that are read-only, suppressed, or lightweight.

☐ ☑ Sliding Brace (Small Free)
　☐ Slip bracket-1 (Default)
　☐ Bracket with Tube-1 (Default)
　☑ C-link-5 (short)
　☑ C-link-6 (short)
　☐ Insulator-1 (Default)
　☐ Insulator-2 (Default)
　☐ Post-1 (Default)
＋☐ U-bolt assembly-1 (1.5-2.5 ra

Fig. 16–10. Parent/Child Options panel.

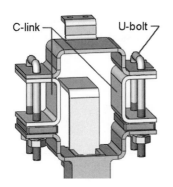

Fig. 16–11. Larger version required.

When adding a new configuration to an assembly, there will be a panel named Parent/Child Options at the bottom of the Add Configuration PropertyManager. This panel is shown in Figure 16–10. The panel will not appear in the properties of an existing configuration. It only appears for new configurations.

A common reason for creating a configuration in an assembly in the first place is often because different versions of components need to be shown in different versions of the assembly. This chapter has already taught you how to do this. However, the components used in this chapter have already had configurations added to them. What if they did not?

The Parent/Child Options panel allows for creating configurations in components at the same time a configuration is added to the top-level assembly. This is extremely convenient. Let's return to the sliding brace assembly and pretend that a larger version of the assembly is needed. The C-link components, pictured in Figure 16–11, will need a configuration added. The U-bolts, also pictured, already have a longer configuration available, so these will be left alone. Examine Figure 16–10 and you will see that the two C-link components have been checked, along with the top-level assembly.

After adding the configuration to the assembly, the new configuration appears in ConfigurationManager as expected. This is shown on the left in Figure 16–12. The interesting aspect of all of this is that the C-link components have new configurations as well. This can be discerned by looking at the assembly's FeatureManager, shown on the right in Figure 16–12.

Fig. 16–12. New configurations have been added.

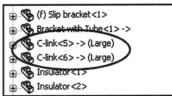

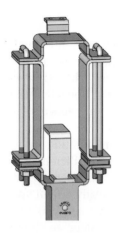

It is now a very simple matter of opening the C-link part file and making a dimensional change to the new large configuration of that part. Substitute the existing U-bolt component with the longer version and the new assembly is completed. The completed assembly with the new configuration is shown in Figure 16–13.

✓ **TIP:** *Any component can have a configuration added on the fly from the top-level assembly. Right-click on a component in FeatureManager and select Add Configuration. The new configuration will be used by that component in the assembly. The component can then be opened and modifications made.*

Fig. 16–13.
Completed assembly.

Additional Configurable Items

We have not discussed every object that can be configured in an assembly, though we have certainly discussed the most common objects. This includes controlling the suppression state of components, component visibility, and what configuration is being used by a component. Make a particular configuration current, and then change suppression states, visibility, or component configuration settings. The changes will be remembered and used the next time you return to that configuration.

Table 16–2 outlines all items configurable in an assembly. All of the topics associated with the listed items have been discussed in previous chapters, or in this chapter.

Table 16-2: Configurable Items in Assemblies

Configurable Item	Value
Component suppression state	Suppressed or resolved
Component visibility	Shown or hidden
Component configuration	Configuration name
Mates	Suppressed or unsuppressed
Distance and angle mates	Dimension values
Assembly features	Suppressed or unsuppressed
Assembly feature parameters	Dimension values and end conditions
Assembly feature sketch relations	Suppressed or unsuppressed
Mass properties	Assigned mass property values
Simulation elements	Suppressed or unsuppressed

Display States (SW 2006 Only)

A display state is similar to a configuration, but deals strictly with appearances rather than model geometry. The way in which a component appears can be saved in that particular state, and multiple states can exist for each configuration. Although display states can exist for part files, display states are much more useful in assemblies because each component can appear in a specific way, such as wireframe or shaded display modes.

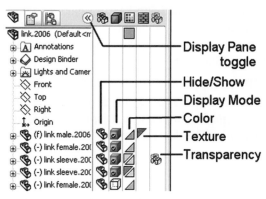

Fig. 16–14. Display pane.

Display Pane toggle
Hide/Show
Display Mode
Color
Texture
Transparency

The first point in understanding display states is to know how the display pane functions. The display pane can be shown by clicking on the display pane toggle switch shown in Figure 16–14. Also shown are the various identifying symbols within the display pane. These tell the SolidWorks user which display characteristics are associated with a particular component.

The various symbols found in the display pane can be clicked on in order to change a display characteristic for a particular component. For instance, clicking on the Hide/Show symbol to the right of a particular component allows for toggling between that component being either hidden or shown.

Clicking on the Display Mode symbol associated with a particular component allows for changing the display mode for that component. All display modes are available for selection from a list that appears when you click on the symbol. There is also a Default Display option, which forces the component to use whatever display mode is currently in use for the assembly.

The Color and Texture columns each can have two swatches associated with any one component. The top left-hand swatch shows the color or texture applied to the component. The bottom right-hand swatch shows the color or texture associated with the part file. Component color overrides part color, so the top left-hand swatch will always take precedence.

Finally, there is the Transparency column, which shows if a component has been made transparent. In all, the display pane makes it very easy to quickly discern which display characteristics are associated with a component. But how does the display pane relate to display states?

If we examine ConfigurationManager, the display states can be seen. Figure 16–15 shows ConfigurationManager with a number of display states already created. Only one configuration can be active at a time, and similarly only one display state can be displayed at a time. To create a new display state, right-click on the *Display State* folder and select Add Display State. Display states can be renamed like most everything else in SolidWorks, so give each display state a meaningful or descriptive name.

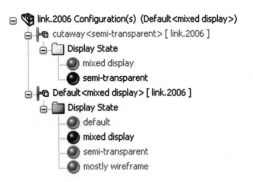

Fig. 16–15. Display states in ConfigurationManager.

To switch between display states, double click on the display state you wish to use. Active display states are displayed with a blue sphere, and inactive display states are displayed with gray spheres. To edit a display state, switch to that display state (activate it) and make changes to component display characteristics via the display pane or by using any of the methods learned in Chapter 8.

✓ **TIP:** *To copy a display state from one configuration to another, drag and drop an active display state from an existing* **Display State** *folder to another* **Display State** *folder.*

Summary

The configuration a part is using while in an assembly can be controlled from within the assembly. This is accomplished by accessing the properties of the part while in the assembly. You can then specify a named configuration to use for the component within the assembly. A similar situation exists in which you are in a drawing and need to see a particular configuration for a part or assembly in a view. By accessing the view's properties, you can specify a named configuration to be shown in that view.

Adding configurations in assemblies works the same as it does for part files. You reviewed this process early in this chapter, and explored what the various options mean when accessing the properties of an assembly configuration. Two very important options are *Suppress new components* and *Suppress new features and mates*. By manipulating these options, it becomes possible to easily add new components and mates while developing multiple configurations.

Configurations can be added simultaneously to multiple components within an assembly. This functionality has a very high convenience factor. Considering that each component will begin using the new configuration, all that remains is to open each part and make the desired modifications.

CHAPTER 17

Design Tables

THE ABILITY TO CREATE VARIOUS CONFIGURATIONS for parts or assemblies is very convenient. The configurations you have learned so far involve controlling feature or component suppression and dimension values on a relatively small scale.

Of course, there is nothing stopping you from creating 100 configurations, but it would be exceedingly time consuming using only the tools you have been given to this point. For greater flexibility and to control dimensions and suppression states on a larger scale, a family-of-parts table is necessary. This functionality requires the use of a design table, which you can create via an Excel spreadsheet.

Be aware that you must have Microsoft Excel installed on your computer to take advantage of SolidWorks' design table functionality. Check current requirements on the SolidWorks web site to see which versions of Excel are supported in regard to your version of SolidWorks (*www.solidworks.com*), or contact your SolidWorks reseller.

Adding a design table to SolidWorks is very sensitive to typographical errors. You will learn all of the dos and don'ts regarding design tables in this section. SolidWorks can generate a design table with a minimal amount of human intervention, but there are preparations that should be carried out prior to inserting a design table.

Design Table Preparation

When creating a design table, the only preparation you really need to be concerned with is renaming your dimensions or features. You already know how to rename features. As far as renaming dimensions is concerned, you need to rename only the dimensions you want to drive through the spreadsheet. Renaming dimensions is not a requirement, it is just good practice. The process of renaming dimensions was covered in Chapter 9, but will be reviewed here. The wavy washer pictured in Figure 17–1 is used as an example in the first portion of this chapter.

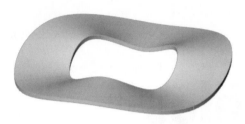

Fig. 17–1. Wavy washer.

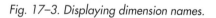

Fig. 17–2. Changing a dimension's name.

Because dimension names must be supplied to the spreadsheet, it helps to have descriptive names for the dimensions instead of the nondescriptive names SolidWorks uses. For instance, names such as *Length* or *Depth* have a lot more meaning than do SolidWorks' names of *D1* or *D2*.

To rename a dimension, first access the dimension in the usual way (double clicking on a feature), so that it is displayed on screen. Second, right-click on the dimension value and access its properties. The area in the Properties window containing a dimension's name is shown in Figure 17–2. It is the area labeled *Name* that can be altered. The area labeled *Full name* is for reference purposes only, and includes the feature or sketch associated with the dimension.

The name of the dimension in Figure 17–2 has been changed to *Hole diameter* as opposed to one of SolidWorks' default names. Although the full name of the dimension cannot be changed from this window, you already know how to rename a feature or sketch using a slow double click.

✓ **TIP:** *Selecting an object in FeatureManager and then pressing the F2 function key allows for renaming that object.*

The dimensions that will be driven by the design table have all been renamed in our example so that they will make more sense when placed in the design table. Figure 17–3 shows a screen shot of the wavy washer with the dimension names displayed. See the next section to discover how dimension names can be displayed.

Fig. 17–3. Displaying dimension names.

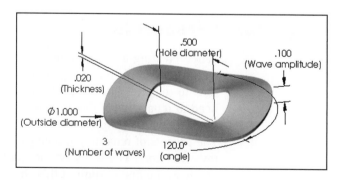

Displaying Dimensions with Names

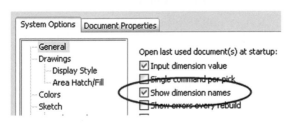

Fig. 17–4. Show dimension names *option*.

There is an option that exists specifically for displaying dimension names. The option is titled *Show dimension names*, located in the General section of System Options (Tools > Options). This option is shown in Figure 17–4. Simply check it to turn on dimension names, and then access dimensions as you normally would.

Turning on dimension names is not a requirement when creating design tables. It is just a convenience that some users may prefer. Sometimes it helps to be able to see the dimension names when creating the design table.

Getting the dimension names into the design table is of prime importance. Considering this fact, it is often beneficial to be able to see all dimensions at the same time. This makes the dimensions easily accessible. In contrast, however, if the model is a complex one and there are many dimensions it is not be advisable to show all dimensions at once. Too many dimensions on the screen at the same time makes distinguishing between dimensions difficult, if not impossible.

Fig. 17–5. Enabling *Show Feature Dimensions*.

The Annotations object in FeatureManager is a shortcut for performing a few simple functions. One of these functions is the ability to show all dimensions in the model. By right-clicking on Annotations and selecting the option Show Feature Dimensions, shown in Figure 17–5, you can turn on all dimensions in the model. To turn the dimensions back off, simply uncheck the option.

The other options shown in Figure 17--5 relate to the display of various types of annotations. Show Reference Dimensions refers to all driven dimensions in the document. If there are no driven (reference) dimensions in the model, this option will have no effect. Display Annotations must be checked in order to show any type of annotations, including feature dimensions. To pick and choose what type of annotations to show, click on the Details menu item. It is then possible to show (or hide) specific annotations, such as cosmetic threads, weld symbols, and so on.

Design Table Anatomy

As a final step prior to the design table creation process, let's examine exactly what type of data is going to be added to the table so that we have a better idea of what to expect. Figure 17–6 shows a design table that has been started, with the four main areas of the table labeled. The first area, known as the Title, can safely be ignored. It represents the name of the file in which the table resides. It is created automatically by SolidWorks and should be left alone.

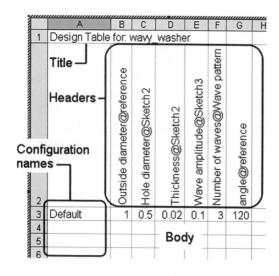

Fig. 17–6. Design table format.

The most intimidating portion of a design table is the Header area. This area is where the items being driven by the table will be added. The most common headers added to a design table are dimension names (which can be seen in Figure 17–6) and a header for controlling suppression state, which we will get to shortly. There are other headers as well, and how to add them is discussed in this chapter.

The first column of the design table contains the names of the configurations that will automatically be generated by the table. These are the configurations that will appear in ConfigurationManager, which you should already be familiar with. Adding configuration names to the design table is a simple matter of typing them in.

Finally, there is the body of the table, which contains the dimension values for the various dimensions in the headers. Data in the body is also typed in by the person creating the table. As you would expect, dimension values can be different for each configuration listed in the first column. Other data besides dimension values will be present in the body, depending on the headers used.

For your reference, headers that can be used in a design table are outlined in Table 17–1. The table includes descriptions of header functionality. Headers are not case sensitive. However, for the sake of this book portions of headers that are UPPERCASE are steadfast and should not be changed by the user. Portions of the headers in lowercase will change, dependent on the name of the item called out in the header.

The code-like appearance of headers can be intimidating. The dollar sign ($) and at symbol (@) are special characters SolidWorks understands and uses internally. After we examine some examples of design tables, you will probably find that they are much easier to implement than what you might think at first appearance. Design table functionality is very powerful, and you should give it a chance.

Table 17-1: Design Table Headers for Part Files

Header	Function
dimension_name@object_name	Controls dimension value.
$STATE@object_name	Controls the suppression of features and light sources via the controlling characters S or U.
$CONFIGURATION@part_name	Controls which configuration of a part is referenced when that part has been inserted into another part.
$PARENT	Allows for creating derived ("nested") configurations.
$PARTNUMBER	Allows for adding text in a column. The text will be used as the part number in a bill of materials.
$COMMENT	Allows for adding text in a column. Text will appear in the configuration's properties in the Comment area.
$USER_NOTES	Allows for adding text in a column or row.
$COLOR	Allows for controlling the color of the model in each configuration.
$PRP@property_name	Defines a custom property (see "Custom Properties" in this chapter).
$STATE@equation_number@EQUATIONS	Controls the suppression of equations via the controlling characters S or U.
$STATE@relation_name@sketch_name	Controls the suppression of sketch relations via the controlling characters S or U.
$SW-MASS	Assigns a mass property to the part.
$SW-COG	Assigns a center of gravity to the part using the syntax x-y-z.
$TOLERANCE@dimension_name	Assigns a tolerance to a dimension.

Creating a Design Table

There are a couple of ways in which a design table can be created. Much of the process involves getting dimension names into the table. Much of this tedium has been automated. You should find that creating design tables in general is a fairly streamlined process.

Once you have renamed relevant dimensions, you are ready to begin the design table. The aspects of working with Excel are outside the scope of this book, but the process of creating a design table is simple and you should be able to get by if you have even the most basic understanding of Microsoft Excel software.

✗ **WARNING:** *Once a design table is displayed on screen, clicking in the work area outside the table will serve to close the table. If this happens by accident, right-click on Design Table in FeatureManager and select Edit Table.*

How-To 17–1 takes you through the easiest method of creating a design table, which tells SolidWorks to automatically generate the table using dimensions selected by the user. There are a number of options that go along with creating a design table, explained in material to follow. The wavy washer is used in How-To 17–1.

How-To 17-1: Auto-creating a Design Table

Fig. 17–7. Design Table PropertyManager.

To add a design table to a SolidWorks part or assembly document, perform the following steps.

1. Select Design Table from the Insert menu.

2. Select Auto-create from the Source panel of the Design Table PropertyManager, shown in Figure 17–7.

3. Click on OK.

4. In the Dimensions window that appears, shown in Figure 17–8, select the dimensions to be added to the design table. (It may be necessary to hold the Ctrl key down.)

Fig. 17–8. Dimensions window.

5. Click on OK.

6. Add parameters or configurations to the design table as necessary.

7. Click once outside the design table to complete the process of adding the design table to the model.

8. A confirmation window should appear, stating that configurations were added to the model. Click on OK to close the window.

In step 4, you noticed that the Dimensions window appeared. If you renamed the dimensions you wished to drive from the design table, you discovered that selecting those dimensions from what can be a very extensive list becomes an easy task. Clicking on OK places any selected dimensions into the design table for you.

Step 6 is where the majority of the work gets done. Filling out the design table can be tedious, so use the standard copy-and-paste functionality available in any Windows program. Excel can also be made to automatically fill data ranges with minimal effort. This will greatly reduce the time it takes to fill in the body of the table. Excel functionality is outside the scope of this book.

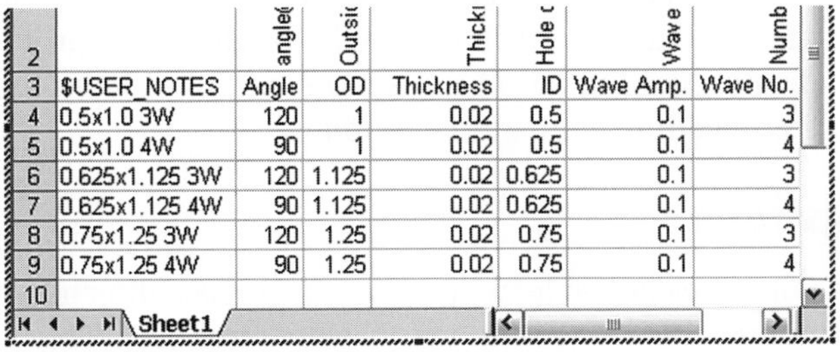

Figure 17–9 shows an example design table for the wavy washer. Only the configuration name area and body are shown, to conserve space. One point worth noting is the use of the *$USER_NOTES* header, which can be used as a column or row header (see Table 17–1). Use this header whenever it becomes desirable to add text, for whatever purpose. SolidWorks will ignore the text and will not attempt to process it.

	angle	Outsi	Thickr	Hole c	Wave	Numb
2						
3	$USER_NOTES	Angle	OD	Thickness	ID	Wave Amp.	Wave No.
4	0.5x1.0 3W	120	1	0.02	0.5	0.1	3
5	0.5x1.0 4W	90	1	0.02	0.5	0.1	4
6	0.625x1.125 3W	120	1.125	0.02	0.625	0.1	3
7	0.625x1.125 4W	90	1.125	0.02	0.625	0.1	4
8	0.75x1.25 3W	120	1.25	0.02	0.75	0.1	3
9	0.75x1.25 4W	90	1.25	0.02	0.75	0.1	4
10						
	Sheet1						

Fig. 17–9. Design table for the wavy washer.

Place (top to bottom) in the first column the names of the configurations you want to create. The default configuration will be added automatically because it is already present in the model (unless it was deleted). You do not have to retain the default configuration if you do not want to, though it would not hurt anything to leave it there. In our example, the default configuration was renamed *0.5x1.0 3W* (which refers to hole diameter, outside diameter, and number of waves).

✓ **TIP:** *Formulas can be used within the design table to control dimension values. For example, one cell can always equal half another cell by typing in the Excel formula for that operation (i.e.,* =SUM(360/G4)*).*

Any formatting can be used when creating a design table. This includes formatting cells for a particular decimal place precision, adding color to the table, rotating text, and so on. Feel free to use a little artistic license to add a little flair to your design table.

Once the design table has been generated, a new Design Table object will appear in FeatureManager. This is shown in Figure 17–10. For your reference, commands related to design tables found when right-clicking on the Design Table object are summarized in Table 17–2.

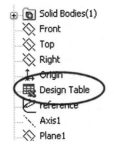

Fig. 17–10. Design Table object in FeatureManager.

Table 17-2: Design Table Menu Commands

Menu Command	Description
Edit Feature	Opens the Design Table PropertyManager
Edit Table	Opens table for editing within SolidWorks
Edit Table in New Window	Opens table for editing in Microsoft Excel
Save Table	Allows for saving the table as an Excel Spreadsheet

It should be noted that the Save Table command saves the design table as a separate disassociated file that is not linked to the original. Design tables are embedded by default, not linked. There is more on this topic in the section "Linking Versus Embedding" in material to follow.

Design Table Options

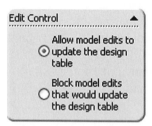

Fig. 17–11. Edit Control panel.

Whether or not a design table will update to reflect dimension changes in the model is something you have control over. This is known as edit control. There are only two options in this particular category, and they can be accessed in the Design Table PropertyManager. The Edit Control panel in the Design Table PropertyManager is shown in Figure 17–11.

Edit control is very straightforward. Allowing model edits to update the design table means that any dimensional changes made to the model will propagate to the table. Obviously, this assumes that the dimensions being edited are in the table. If they are not, whether or not they update in the table is not an issue.

The alternative to seeing dimensional changes propagate to the table is to lock dimensions so that they cannot be changed if they are already being driven via the design table. After all, you would not want to develop a conflict of interest, which is exactly what would happen if dimensions could be changed from two different locations (the table or the model). If you choose to block model edits that would update the design table, you must edit the design table to change the dimension values.

Fig. 17–12. Changing the color of dimensions controlled by design tables.

Dimensions driven by a design table have a different color than other dimensions. The color defaults to bright purple (at the time of this writing), so you may very well decide to change it. To do so, access the Colors section of the System Options (Tools > Options). In the system colors listing, shown in Figure 17–12, select *Dimensions, Controlled by Design Table*. It is then possible to click on the Edit button, also shown in the figure, and select a color of your choice.

It is recommended that a color other than any of the default system colors for dimensions be used. For example, try to avoid red (overdefined dimensions), blue (dimensions added by SolidWorks), gray (reference dimensions), or black (dimensions added by the user). Although the color cannot be discerned, a dark orange color was selected in Figure 17–12.

Adding Rows and Columns Automatically

Aside from dimensional changes updating in a design table, you can also have new rows or columns added to the table automatically if the software deems it necessary. For example, if a new configuration is manually added in ConfigurationManager, a new row will be added to the design table listing the new configuration. Furthermore, all of the dimension values in the body of the table, along with any other parameters, will also be filled in automatically.

If a dimension is altered that is not currently being driven in the design table, or if some other parameter is modified that was not previously being driven by the table, new columns can be added to reflect these additional parameters. The parameters being referred to include those listed in Table 17–1, some of which are explored in material to follow.

Fig. 17–13. Design table Options panel.

Whether or not new configurations or parameters are automatically added to the design table is controlled through the Options panel in the Design Table PropertyManager. The Options panel is shown in Figure 17–13. To have newly added configurations automatically added to the design table, check the *New configurations* option. To have modified dimension or other parameters added to the table automatically, check the *New parameters* option.

↝ **NOTE:** *If the* Warn when updating design table *option (Figure 17–13) is checked, a warning will be displayed if dimensions driven by the table are modified. This assumes that model edits have been allowed, as discussed in the previous section.*

Looking back at the wavy washer, imagine that a new configuration is added. The new configuration is called *0.5x1.0 3W perf*, which refers to the size of the washer, number of waves, and that the washer is perforated. A circular cut is then created on the washer and patterned in order to create the perforations. The features added to the washer are shown in Figure 17–14.

Fig. 17–14. Adding perforation features.

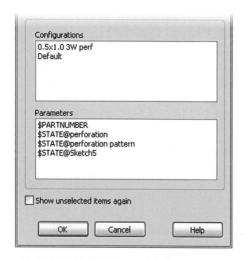

Fig. 17–15. Add Rows and Columns window.

Consider what has just happened: a new configuration was added to the model, and a new parameter has also been added. The parameter, in this case, is not a dimensional change. Rather, it is the suppression state of the features creating the perforation. The features are suppressed in every configuration except the new one. Because of these additions to the model, the Add Rows and Columns window (shown in Figure 17-15) appears.

Note that the name of the new configuration appears in the Configurations listing (upper list box). The name Default also appears. The Default configuration was renamed in the design table during the table's creation, but the Default configuration still existed in the part file. For this reason, Default is seen as a new configuration and appears in the list. For this example, only the newly added *0.5x1.0 3W perf* configuration will be selected from the list, thereby adding it to the design table. The Default configuration will be deleted from ConfigurationManager at a later time.

The Parameters listing (lower list box) lists four parameters. The *$PARTNUMBER* parameter is explored later. The three *$STATE* parameters refer to the suppression state of three items. These items are *perforation pattern*, the *perforation* feature (which is the original circular cut), and *Sketch5* (the sketch for the perforation feature). Suppressing the perforation feature will also suppress the underlying sketch, so the sketch need not be considered. Therefore, for this example, *$STATE@perforation* and *$STATE@perforation pattern* will be selected.

•◦ **NOTE:** *If certain items are not selected (such as* $PARTNUMBER*), those items will not be presented the next time the Add Rows and Columns window appears unless the Show unselected items again option is checked.*

Clicking on OK at this point allows the selected items to be placed in the design table automatically. Figure 17–16 shows the *0.5x1.0 3W perf* configuration added to the bottom of the table. Excel tools could be used to sort the configurations alphabetically, if this is desired.

Not only has the new configuration been added, but the new parameters have been added as well. The two parameters are shown in Figure 17–17, and appear upright, rotated 90 degrees. SolidWorks rotates headers automatically to conserve space.

2		Angle
3	$USER_NOTES	Angle
4	0.5x1.0 3W	120
5	0.5x1.0 4W	90
6	0.625x1.125 3W	120
7	0.625x1.125 4W	90
8	0.75x1.25 3W	120
9	0.75x1.25 4W	90
10	0.5x1.0 3W perf	120
11		

Fig. 17–16. Configuration added to the table.

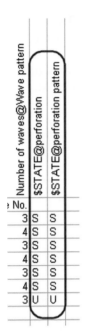

Fig. 17–17. Two new parameters added to the table.

Aside from dimension names appearing in the header area of a design table, we have not discussed additional headers to any great extent. The following section will familiarize you with the header used for controlling suppression state, along with a few other header types. This should give you a better feel for using design tables in general.

Feature Suppression in Design Tables

You have already learned how to suppress or unsuppress features in a part. You can also control the suppression of features from a spreadsheet, and the process for doing so is quite simple. The process involves the *$STATE* heading, followed by the name of the feature, with the at (@) symbol used as a separator. The entire string would look as follows.

`$STATE@perforation`

The controlling values when using the $STATE heading to control feature suppression are the words *UNSUPPRESS* or *SUPPRESS*, or more simply the letters U or S. If you examine Figure 17–17, you will notice these letters present in the table. Only two cells contain the letter U. This is saying, quite simply, that the perforation and perforation pattern features are unsuppressed in a particular configuration (namely, the *0.5x1.0 3W perf* configuration). In all other configurations, those features are suppressed.

✓ **TIP:** *If cells below a $STATE heading are left blank, they will be considered unsuppressed.*

Some of the headers listed in Table 17–1 should begin to make more sense now. For example, light sources (see Chapter 8) can have their suppression state controlled via a design table. Certain headers are very uncommon. For instance, you will probably not find a great need to control the suppression state of individual sketch relations from a design table (*$STATE@relation_name@sketch_name header*), as sketch relations are not commonly configured anyway. More common headers are described in material to follow.

BOM Part Numbers in Design Tables

The ability to control the part number displayed in a BOM is a very common request. Usually, when design tables are not involved the BOM part number is controlled through a configuration's properties (see "Bill of Material Part Numbers" in Chapter 11). If a design table is involved, the value displayed as the part number in a BOM is controlled through the *$PART-NUMBER* header.

Typing *$PARTNUMBER* at the top of a column will force whatever text is in the cell below it to appear in a BOM in the part number column. This assumes that the part configuration is used in the assembly the BOM is being generated for. By default, an assembly component's file name is used as the part number. All values that can be used in the cells below the *$PARTNUM-BER* heading are listed in Table 17–3. Values preceded by a dollar sign ($) are special values that return specific strings of text for use in the BOM.

Table 17-3: Part Number Heading Parameters

Cell Value	What the Bill of Material Will Display
Any text	Text is displayed in the BOM as a custom part number
Cell is left blank	Configuration name
$DOCUMENT	Document file name
$CONFIG	Configuration name
$PARENT	Parent configuration name when configuration is a nested configuration

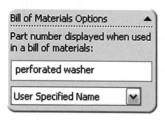

Fig. 17–18. Bill Of Material Options panel.

The *$PARTNUMBER* header will actually drive the Bill Of Material Options panel in a configuration's properties. Although discussed in Chapter 11, this panel is shown again in Figure 17–18. In the image shown, "perforated washer" was the text string entered into the cell in the column containing the *$PARTNUM-BER* header.

✓ **TIP:** *As a shortcut, $D and $C can be used in place of $DOCUMENT and $CONFIG, respectively.*

Unless making use of nested (derived) configurations, do not be concerned with the *$PARENT* value. Nested configurations are comparatively rare (as opposed to standard configurations), so there is not much need for this special *$PARTNUMBER* cell value. Nested configurations were discussed in Chapter 11.

Other Design Table Headings

If the *$COMMENT* heading is used, any text added in the cells below the heading will automatically appear in the appropriate configuration's properties. The *$USER_NOTES* heading is a more generic header and is used strictly for adding text. It is a way of warning SolidWorks not to try to process the data in a row or column, as the data is only text for the user's reference.

If the *$COLOR* heading is used, a 32-bit integer must be specified to dictate the red, green, and blue (RGB) values for the color. To convert an RGB value to a 32-bit integer, the following formula must be used.

```
Red + (Green * 256) + (Blue * 256²)
```

In other words, red plus green times 256 plus blue times 256 squared. In short, it is much easier to use configuration-specific colors from within SolidWorks (see Chapter 11) rather than trying to determine the 32-bit integer to drive the color from a design table.

The *$TOLERANCE@dimension_name* header, as the name would imply, controls what tolerance type is displayed on a particular dimension (see "Dimension Tolerance and Precision" in Chapter 5). Tolerance values are typically added to a drawing, rather than a part. However, if the tolerance display for a dimension is changed at the part level, it will automatically update in the drawing (and vice versa).

If you wish to control tolerance values from a design table, the range of available values that can be entered in the cells below the *$TOLERANCE@ dimension_name* header is fairly extensive. Rather than list all of the values here, it is suggested that you perform a search for and print the list of available values in the SolidWorks help. To accomplish this task, select Solid-Works Help Topics from the Help menu, and then perform a search for the phrase "tolerance keywords and syntax" (including the quote marks).

Creating a Blank Design Table

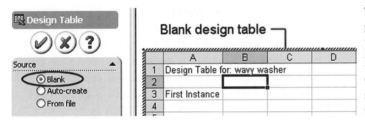

Fig. 17–19. Source panel and blank design table.

The Blank option found in the Source panel of the Design Table PropertyManager is shown in Figure 17–19 (left). It allows for creating a design table in a somewhat more manual manner than the auto-create functionality discussed previously. An example of a blank table is also shown in Figure 17–19.

If adding a blank table, the Add Rows and Columns window will appear after clicking on OK. As described previously, new configurations or parameters can be added to the table by selecting the configurations or parameters from those listed in the window. SolidWorks populates the Configurations listing with all existing configurations in the model, if any. Similarly, all dimensions or other parameters will be listed in the Parameters listing in order to facilitate the creation of the design table. In Figure 17–19, no configurations or parameters were selected, which is why the table is still essentially blank.

The name First Instance is a dummy name for a new configuration. Think of it as a placeholder. Whether this configuration is renamed or kept the way it is depends strictly on the user's preference. Almost certainly, more configurations will need to be added, which can be done by typing them in the first column. Make sure not to skip rows, as SolidWorks stops processing data as soon as it hits a blank row. For that matter, skipping columns is also prohibited.

Manually Adding Parameters

Obviously, a blank design table will not do you much good unless you complete it by adding some dimensions (or other parameters) to the table. This can be accomplished by double clicking on a dimension name. Double clicking on a dimension places the name of the dimension into the first available column, along with the current dimension value directly below it. Make absolutely certain the second cell down in the first available column is highlighted, or this process will not work!

If the dimensions have been shown, double clicking on the dimension value to be placed in the design table is a simple matter. If the dimensions have not been shown, double clicking on a feature will show the dimensions related to that feature. However, when a design table is being edited, double clicking on a feature places that feature into the design table, along with the *$STATE* heading. This is fine if it is your intention to control the suppression state of a feature. On the other hand, you may be trying to access the dimensions associated with the feature.

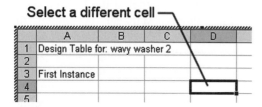

Fig. 17–20. Select a different cell prior to double on clicking a feature.

If you do not wish to use the *$STATE* heading, temporarily access an alternate cell prior to double clicking on a feature to access its dimensions. For example, the second cell down in the first available column will always be selected automatically by SolidWorks. This is necessary, but if cell D4 is selected (such as in Figure 17–20) nothing happens within the design table when double clicking on the desired feature. In this way it becomes possible to trick SolidWorks into not adding anything to the table while you access the appropriate dimensions.

After accessing the desired dimensions on the model, select the appropriate cell. Once again, this would be the first available cell in the header area of the table. This would correspond to cell B2 in the example shown in Figure 17–21. It is then possible to double click on the desired dimension to have it automatically entered into the appropriate cell. SolidWorks then automatically selects the next cell (B3 in the figure) in anticipation of the next dimension or parameter to be entered.

Select the appropriate cell

	A	B	C	D
1	Design Table for: wavy washer 2			
2		Thickness@Sketch2		
3	First Instance	0.02		
4				
5				

Fig. 17–21. Select the appropriate cell and then double click on the desired dimension.

Double clicking on items to add parameters to a design table works regardless of how the table was originally created (Blank or Auto-create option). This is essentially how a design table is edited. Right-click on Design Table in FeatureManager, select Edit Table, and make the necessary adjustments. Feel free to rename or add configurations, modify dimension values, or add new dimension names for other dimensions you want to control within the design table. You can also add new column headings, change the color formatting of the cells, or basically anything else required.

During the process of editing the design table, make sure you do not click outside the table in a blank portion of the work area. If you do, you will be kicked out of edit mode and placed back into the SolidWorks environment. Also avoid creating blank rows or columns. SolidWorks will stop processing data as soon as it hits a blank row or column.

Linking Versus Embedding

Design tables are embedded objects. After inserting a design table into a part file, for instance, it will not be possible to find a separate Excel spreadsheet on the hard drive. It simply does not exist. The spreadsheet exists in binary format inside the part file, and is saved whenever the part file is saved. With this said, it is possible to save a copy of the spreadsheet (the design table, in other words) as a separate file.

To save the design table as a separate file, right-click on Design Table in FeatureManager and select Save Table. By default, the name of the table (Excel spreadsheet) will be the same as the name of the SolidWorks document. The spreadsheet will also be placed in the same folder as the model. It is a good idea to accept these defaults, as it makes finding the table much easier.

Once the design table has been saved as a separate file, you may want to link the model to that file so that the table will retain its associativity. Otherwise, the newly saved spreadsheet and the SolidWorks model will in no way be associated with each other. How-To 17–2 takes you through the process of creating a linked design table.

HOW-TO 17-2: Creating a Linked Design Table

To create a linked design table, perform the following steps. The steps for saving a design table out as a separate file are also included, as these steps are a prerequisite to the linking operation.

1. Right-click on Design Table in FeatureManager and select Save Table.

2. Give the table a name or accept the default, and then click on the Save button.

Fig. 17–22. Select From file.

3. Right-click on Design Table in FeatureManager and select Edit Feature.

4. Select *From file* in the Source panel, shown in Figure 17–22.

5. Click on the Browse button.

6. Select the file previously saved out as a separate Excel spreadsheet in step 2, and then click on the Open button.

7. Check the *Link to file* option, shown in Figure 17–22.

8. Click on OK.

9. If the design table opens in SolidWorks, click outside the table to close it.

Editing Linked Tables

If a design table has been linked to a model using the *Link to file* option, editing the design table will not necessarily update the model. There are some intricacies to the relationship between the model and table you should be aware of.

Assume that a design table has already been created and linked to the model using the skills you have learned in this chapter. As you are already aware, making a change to the model will update the design table. This assumes that the edit control has been set to permit this action. But what happens if the process is reversed? What happens if a change is made to the Excel spreadsheet outside the SolidWorks model file?

Fig. 17–23. Should the changes made in the design table propagate to the model?

It is easy enough to edit a design table from within a model file, and if the design table is linked it is expected that the changes will propagate to the separate Excel spreadsheet. Opening the Excel spreadsheet in Excel and making changes to dimension values will also propagate to the associated model file, if you choose to let them. Make it a point to save the Excel spreadsheet and close the document before opening the SolidWorks model file containing the spreadsheet.

When the SolidWorks model is opened, the window shown in Figure 17–23 will be displayed. You can choose to accept the changes made in the design table and have the model updated or you can choose to have the model update the table. If the latter choice is made, changes made manually in the design table will be lost.

Where you should be careful is in trying to open the design table in Excel if the model containing the design table is already open. In theory, if the design table is being edited in SolidWorks Excel should not let the same table be opened for purposes of editing in a second application. Unfortunately, Excel allows the design table to be opened, which means that the file can be opened in two places at once. This is an accident just waiting to happen, so try not to let such a situation take place.

When saving a SolidWorks document with a linked design table, the table will be saved as well. You will see a message stating this fact. Give the software a moment to save both files prior to moving on.

Created manually

Design table driven

Fig. 17–24. Symbols in ConfigurationManager.

Incidentally, configurations generated via a design table are displayed with a different symbol in ConfigurationManager than those added manually. Figure 17–24 shows the symbols associated with configurations, enlarged for clarity. As you can see, configurations generated and driven from a design table display with the Microsoft Excel symbol.

Deleting Design Table Configurations

If you remove a configuration from a design table, such as by deleting an entire row, SolidWorks will realize this and ask if you want it deleted from the ConfigurationManager list as well. If a design table was not used to create configurations, deleting a configuration can be accomplished by selecting it from ConfigurationManager and pressing the Delete key.

If a design table is used, deleting the configuration from ConfigurationManager will not permanently remove the configuration from the model file. The act of editing the design table forces SolidWorks to reevaluate the data, thereby reinstating the previously deleted configuration. This is true even if nothing was changed in the design table. To permanently delete a configuration, delete the row containing that configuration in the design table.

If a row is deleted from a design table, SolidWorks asks whether or not you want to retain the associated configuration. All that is necessary at this point is to click on the Yes button, and the configuration will be permanently removed from the model file. If No is selected, the configuration will be retained, but it will not be driven via the design table.

✓ **TIP:** *If you want to temporarily delete a configuration from the design table, replace the configuration name with the heading $USER_NOTES. This will disable the configuration in the design table (and ConfigurationManager if desired) yet retain the data in the table in case it is needed* *in the future. The configuration can then easily be reinstated by reverting to the original, or another, configuration name.*

Assembly Design Tables

When used in an assembly file, design tables have a different functionality than they do when used in a part file. A very good place to start would be with a quick review of what design tables can accomplish within a part. In contrast, what can be accomplished within an assembly design table will also be listed. Following this comparison is an in-depth look at the various controls used in an assembly design table. This includes examples of an assembly design table that will help to reinforce the topics learned.

Table 17–4 will give you a better understanding of just what can be accomplished in part design tables in comparison to assembly design tables. The differences are significant, so you will want to understand the capabilities of both types of design tables to have a better understanding of what your options are. When a particular parameter does not apply, that parameter is noted as being not applicable (N/A) for the part or assembly design table.

Table 17-4: Part and Assembly Design Tables Compared

Part Design Tables	*Can control feature dimensions*
Can control suppression of features	Can contain user-defined notes
Can add comments to configurations	Can control part number used in a BOM
N/A	Can define custom properties
Can control part color	N/A
N/A	N/A
N/A	N/A
N/A	Can control part configuration usage of base parts (parts inserted into other parts)
Assembly Design Tables	Cannot control feature dimensions associated with parts
Cannot control suppression of features associated with parts	Can contain user-defined notes
Can add comments to configurations	Can control part number used in a BOM
Can control if assembly components are listed in a BOM	Can define custom properties
N/A	Can control mate dimension values (i.e., distance and angle)
Can control assembly feature dimension values	Can control component suppression state
Can control suppression state of assembly features	Can control suppression state of mates
Can control component visibility	Can control component configuration usage

Assembly design tables cannot be used to control the individual feature dimensions within a part, nor can they control the suppression state of individual features. However, an assembly design table can control what configuration of a component is being used in an assembly. In combination, part and assembly design tables can offer some extremely powerful tools for creating a family of parts or assemblies. Assembly design tables can also be used to control many aspects of an assembly above and beyond what can be accomplished in a part.

Design tables are created in an assembly in the same fashion they are created within parts. Assembly design tables are also edited in the exact same fashion as part design tables. The primary difference between part and assembly design tables centers on the parameters that can be used. Table 17–4 outlines these parameters. The section that follows describes the headings and control characters that can be used within an assembly design table.

Controlling Assembly Feature and Mate Dimensions

The only types of dimensions that can be controlled from an assembly design table are mate dimensions and any dimensions associated with the assembly itself. Examples of assembly dimensions (other than mate dimensions) would be dimensions associated with assembly features or perhaps a layout sketch created in the assembly. The syntax used when adding a dimension to an assembly design table is the full dimension name. Therefore, the dimension name might appear as follows.

```
D1@Distance1
```

Because dimension names can be renamed, and because mate relationships can also be renamed, you need not suffer with the nondescriptive default dimension name. As an example, a distance mate could be renamed to something such as *clearance*. In addition, the name of the dimension associated with this same mate could be altered.

There are a few dimensions that can have the value of zero, and this may aid you when defining the design table. For instance, when creating an offset plane the offset can be 0. When creating a distance mate, the distance can be 0. This can be very useful, because 0 can be used in a distance mate to create what amounts to a coincident relationship. If necessary, a value can be specified in place of 0 to establish an offset distance.

Controlling Component Suppression

The controlling characters used to indicate whether a component is being suppressed or not in an assembly are different from those when specifying feature suppression in a part design table. The heading is still the same. The proper terminology to use when indicating suppression of features in a part is *suppressed* or *unsuppressed*. The terminology with regard to assembly components is *suppressed* or *resolved*.

When the suppression state of a component needs to be controlled within an assembly design table, the *$STATE* heading must be followed by the full name of the component being controlled, which includes the component's ID number. An example would be as follows.

```
$STATE@Bolt<1>
```

The < *1* > suffix is the ID number that tells SolidWorks which component is being controlled in the table. If there are multiple components in an assembly, and only certain components are to be controlled, the numbers between the < > symbols can reflect those components using dashes and commas. For instance, < *1-12,14,18* > would mean that components 1 through 12 and components 14 and 18 will be controlled in the table. Controlled how? That depends on your intentions. This formatting works with any of the headings that require that a component name be included in the heading. The following is an example regarding this topic.

```
$STATE@Bolt<1-12,14,18>
```

This example would control the suppression state of all components named *Bolt* in the assembly whose ID numbers are 1 through 12, 14, and 18. An asterisk can also be used to control every occurrence of a component in the assembly, such as < * >. An example of the *$STATE* heading in use is shown in Figure 17–25.

	$STATE@Pin<1>	$STATE@Open1
Small Open	R	U
Small Free	S	S
Small Closed	R	S
Medium Opened	R	U
Medium Free	S	S
Medium Closed	R	S
Large Open	R	U
Large Free	S	S
Large Closed	R	S

Fig. 17–25. Controlling the suppression of the pin.

The controlling character when using the *$STATE* heading for components is either the letter S for *suppress* or R for *resolve*. Do not use the letter U, for *unsuppress*, as this will not work. The letter U is reserved for unsuppressing assembly features or mates in assembly design tables. This is discussed in the next section.

Controlling Assembly Feature and Mate Suppression

The heading used to control the suppression state of assembly features and mates is exactly the same as that used to control the state of components. In that the *$STATE* heading is the same heading used to control component suppression, care should be taken to use the proper controlling character, which is S for *suppressed* and U for *unsuppressed*. Examples of the proper syntax would look like the following.

```
[$STATE@assembly_feature_name
```

```
$STATE@mate_name
```

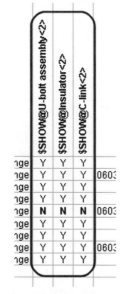

$STATE@Pin<1>	$STATE@Open1	$STATE@Open2	$STATE@Closed1	$STATE@Closed2	
R	U	U	S	S	sh
S	S	S	S	S	sh
R	S	S	U	U	sh
R	U	U	S	S	mec
S	S	S	S	S	mec
R	S	S	U	U	mec
R	U	U	S	S	lo
S	S	S	S	S	lo
R	S	S	U	U	lo

Fig. 17–26. Controlling mate suppression.

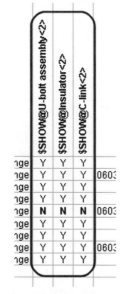

	$SHOW@U-bolt assembly<2>	$SHOW@Insulator<2>	$SHOW@C-link<2>	
ge	Y	Y	Y	
ge	Y	Y	Y	0603
ge	Y	Y	Y	
ge	Y	Y	Y	
ge	N	N	N	0603
ge	Y	Y	Y	
ge	Y	Y	Y	
ge	Y	Y	Y	0603
ge	Y	Y	Y	

Fig. 17–27. Hiding some components in a configuration.

Remember, the letter R is reserved for components only. Figure 17–26 shows an example of mates being controlled in various configurations of an assembly. What is occurring in this example is that different combinations of mates are either suppressed or unsuppressed, resulting in certain components being mated in different positions. The mates are actually four concentric mates, but have been renamed *open1*, *open2*, *closed1*, and *closed2*.

Controlling Component Visibility

What is the difference between visibility and suppression? When a part is suppressed, it is removed from memory. When it is hidden, it is only removed from view. If hidden, a component's mates are still present and accounted for. This is not the case when suppressing a component. This can be significant, because you would not want your assembly to fall apart, which could be the case if too many components are suppressed. (Of course, the assembly would not literally "fall apart," but components could move unexpectedly.)

The heading used to control a component's visibility is *$SHOW*. The controlling character is either Y for *yes* (show the component) or N for *no* (do not show the component). The words *yes* and *no* can also be spelled out, if desired. An example of the *$SHOW* heading in use is shown in Figure 17–27.

There are three components in Figure 17–27 having their visibility controlled. The three components are visible in all configurations (whose names are not being shown) except one. The *U-bolt assembly<2>*, *Insulator<2>*, and *C-link<2>* components will all be hidden in one configuration. Their mate relationships, however, will still be active and holding things in place.

SW 2006: The *$DISPLAYSTATE* heading controls which display state is shown for a particular configuration. This makes the *$SHOW* heading obsolete, in that display states now contain information regarding whether or not components are hidden or shown.

Controlling Component Configurations

This is an extremely important aspect of assembly design tables. It is very important to be able to specify what configuration of a part an assembly is using. This is even more important considering the fact that component feature dimensions cannot be controlled within an assembly design table, nor can component feature suppression states.

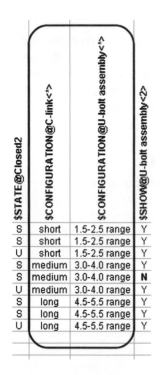

$STATE@Closed2	$CONFIGURATION@C-link<*>	$CONFIGURATION@U-bolt assembly<*>	$SHOW@U-bolt assembly<2>
S	short	1.5-2.5 range	Y
S	short	1.5-2.5 range	Y
U	short	1.5-2.5 range	Y
S	medium	3.0-4.0 range	Y
S	medium	3.0-4.0 range	N
U	medium	3.0-4.0 range	Y
S	long	4.5-5.5 range	Y
S	long	4.5-5.5 range	Y
U	long	4.5-5.5 range	Y

Fig. 17–28. Controlling the C-link and U-bolt assembly configurations.

The heading used to control which configuration of a component is being used is *$CONFIGURATION*. This should be followed by the full name of the component. An example of the complete heading might look like the following.

```
$CONFIGURATION@C-link<*>
```

This heading will control which configurations are being used for all C-link components in the assembly. In the cells below the heading, the actual names of the configurations to be used in the assembly must be spelled out. Spelling is critical here. If a configuration name is misspelled, SolidWorks will not recognize it.

It should go without saying that the part configurations must be created before this function will work. In Figure 17–28, the configurations of the C-link components are being called out in the design table. There are three different configurations being specified. These are *short*, *medium*, and *long*. Configurations for the U-bolt assembly are also being controlled.

Where do the names *short*, *medium*, and *long* come from? They are the names of the configurations in the *C-link* part. If the *C-link* part is opened and the ConfigurationManager examined, these are the names of the configurations that would be found. This is shown in Figure 17–29.

Fig. 17–29. The C-link's configurations.

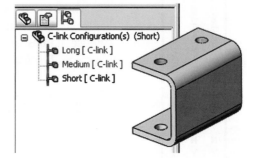

Additional Parameters

For the most part, nearly all of the headers listed in Table 17–1 can be used in an assembly design table, with a few exceptions (which we have already discussed). In particular, part feature dimensions and part suppression states cannot be controlled from an assembly design table. Certain headings, such as the *$CONFIGURATION* and *$STATE* headings, cross over between both file types (parts and assemblies). However, these headings control different objects in each file type. For example, the *$STATE* heading would be used to control suppression of a feature in a part, but might control a component in an assembly.

There are only four headings that are particular to assembly design tables only, and three of those we have discussed. They are the aforementioned *$STATE* heading for controlling component suppression, the *$CONFIGURATION* heading for controlling component configuration, and the *$SHOW* heading for controlling component visibility.

The fourth assembly heading has to do with a BOM and is the heading *$NEVER_ EXPAND_IN_BOM*. It should be used in a situation in which a subassembly should not be expanded to show its components in a BOM. You would not want to use this heading when it is desirable to show the components contained within the subassembly.

To implement this function, place the *$NEVER_EXPAND_IN_BOM* heading at the top of a column and use the controlling characters Y for *yes* or N for *no*. The full words *yes* and *no* can be used as well. Specifying *yes* will turn on the option and will never expand the subassembly's components, regardless of any options used when inserting the BOM.

Design Tables in Drawings

Design tables can easily be shown in a drawing. As a matter of fact, the process is so simple that the steps need not be spelled out as a How-To. The process is a matter of selecting a view in the drawing and selecting Design Table from the Insert > Tables menu. SolidWorks places the table in the drawing, whereby it can be scaled or moved as required. To move the table, drag it to a new location. To scale it, drag the small black handles on the perimeter of the table that appear after it has been selected.

There is one problem in particular that is often faced when displaying a design table in a drawing. There is a size limitation to embedded objects, and this limitation can sometimes present itself with design tables because they can often be quite large. There is not much you can do in this case, but there are a few things that might help.

Double clicking on a design table in a drawing will open the table in the part or assembly it is associated with. It could then be edited, but that is not the issue. You will see small black handles on each side of the table's border and at each corner. Use them to drag the border so that only the relevant information is being shown, and then close the table and save the model. Back in the drawing, click on the Rebuild icon and see if that helps.

If you still cannot see the entire table, you have a few other options. Try compacting the columns as much as possible. Rotate text 90 degrees in the row containing the headings so that the columns can be made smaller. Use letters, such as S instead of spelling out *Suppressed* or Y instead of *Yes*. Finally, if that does not work reduce the font size of the spreadsheet and re-compact (auto fit) the column widths. If you are creating a lot of configurations, you may need to re-compact the row heights as well. Use the on-line Excel help to aid you in these tasks if necessary.

Tabulated Drawings

A tabulated drawing is a drawing in which dimension values have been replaced with labels (often simply letters of the alphabet). A table is then created, in which the dimension labels are called out at the top of each column and dimension values can be referenced via the table. Drawings of this type are often created when it becomes necessary to list dimensions for a family of parts.

	A	B	C	D
1	Design Table for: tabulated			
2		$comment	A@Sketch5	B@Sketch2
3	.750-fill-051	3/4 fillister 50	2.75	2.25
4	.750-hh-051	3/4 hex head 50	2.90	2.10
5	.750-sock-051	3/4 socket 50	3.10	1.95
6	.750-fill-031	3/4 fillister 30	2.75	2.25

Fig. 17–30. Design table will be used in a tabulated drawing.

If a design table has already been created, it can be used as the table for a tabulated drawing. The design table shown in Figure 17–30 will be used as an example in this section. Only the top left-hand corner of the table is shown, as that is the only important area we need be concerned with at this point in time.

The first problem faced is where the labels for the tabulations should go. We do not want to keep the design table from working properly, so what choices do we have? The key is something known as the Family cell, which is a special cell SolidWorks uses to distinguish where the "meat" of the design table is.

In Figure 17–30, the Family cell would be cell A2. This is always the case when the auto-create method of creating a design table is used. The same holds true when inserting a blank design table. In fact, the name of cell A2 will be *Family*, even though no data will reside in the cell. (Understanding cell names and other Excel functions are outside the scope of this book.)

How does knowing where the Family cell is located help us? It helps us because rows or columns can be added above or to the left of the Family cell and the design table will still work! Figure 17–31 shows the same design table with a row and column added to make room for the labels and other text.

While comparing Figures 17–30 and 17–31, let's analyze what has taken place. The column used for the comments (column B) has been moved off to the end of the design table, out of view (we will get back to this column later). Row 1 has been inserted and now contains the labels for what will appear in the table. Column A has been inserted and contains user comments or other notes, such as company name and address.

Columns C through G have been narrowed down, which makes the full dimension names in row 3 illegible, but so what? At this point, we are only concerned with the appearance of the table in the drawing. The design table creation process is done, as far as the part is concerned. What happened to the Family cell? It is now in cell B3, and will function just fine at that location.

	A	B	C	D	E	F
1	Labels >>>		A	B	C	D
2		Design Table for: tabulated				
3			A@Sk	B@Sk	C@Sk	D@Sk
4		.750-fill-051	2.75	2.25	50	3.00
5	Company name	.750-hh-051	2.90	2.10	50	3.10
6	Company address	.750-sock-051	3.10	1.95	50	3.25
7		.750-fill-031	2.75	2.25	30	3.00
8		.750-hh-031	2.90	2.10	30	3.10
9		750-sock-031	3.10	1.95	30	3.25

Fig. 17–31. Adding a row and column.

	B	C	D	E	F	G
1		A	B	C	D	E
4	.750-fill-051	2.75	2.25	50	3.00	1.50
5	.750-hh-051	2.90	2.10	50	3.10	1.53
6	.750-sock-051	3.10	1.95	50	3.25	1.67
7	.750-fill-031	2.75	2.25	30	3.00	1.50
8	.750-hh-031	2.90	2.10	30	3.10	1.53
9	.750-sock-031	3.10	1.95	30	3.25	1.67

⏮ ◀ ▶ ⏭ \Sheet1 / ◀ | ▶

Drag handles

Fig. 17–32. Using drag handles.

All that remains is to hide what does not need to be seen. The Hide function in Excel can be used to hide rows 2 and 3. It can also be used to hide column A. Figure 17–32 shows the design table after hiding the rows and column. Use the drag handles, also shown in the figure, to hide rows or columns below or to the right of the main body of the table (respectively). This is how the final column containing the comments was hidden, though the Excel Hide function could also have been used.

At this point, make sure to save the model. You are then ready to create the tabulated drawing. Insert the design table into a drawing as discussed in the previous section. The table should appear nearly identical to how it appeared in the part file.

To complete the drawing, it will be necessary to place at least one view of the model in the drawing and add dimensions to it. Insert dimensions as model items or as reference dimensions, and then change the text for the dimensions. To change the text, select the dimension and delete the < *DIM* > code from the Dimension Text panel in PropertyManager used to link to the actual dimension value. You will probably receive a warning regarding tolerance display, which can be safely ignored. Type in a letter that relates to the labels in the table and you are done. Figure 17–33 shows a close-up of a view and a table that might be seen in a tabulated drawing.

Fig. 17–33. Model view and table in a tabulated drawing.

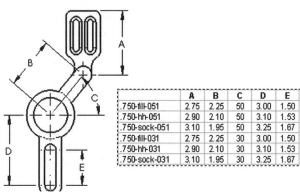

	A	B	C	D	E
.750-fill-051	2.75	2.25	50	3.00	1.50
.750-hh-051	2.90	2.10	50	3.10	1.53
.750-sock-051	3.10	1.95	50	3.25	1.67
.750-fill-031	2.75	2.25	30	3.00	1.50
.750-hh-031	2.90	2.10	30	3.10	1.53
.750-sock-031	3.10	1.95	30	3.25	1.67

Design Table Summary and Recommendations

Design tables are arguably a very efficient way of generating configurations, which in themselves are a powerful function of SolidWorks. In light of this, what really is the best way to put a design table together? The following summarize the content to this point in this chapter, with recommendations as to how a design table can most easily be created.

- *Rename dimensions.* By renaming dimensions, you make it much easier to pick those dimensions from a list when generating the design table later. You should also rename any mates (if an assembly), features, or other objects that will be driven from the table. This also helps to document what is happening within the table for future reference or for others working with the model.

- *Use Auto-create.* Let SolidWorks do as much of the work as possible when creating the table. Inserting a blank table means that you will have to manually add dimensions to the table. Selecting dimensions from a list is much easier with the auto-create method.

- *Know your Excel shortcuts.* Taking advantage of a few simple shortcuts in Excel can make short work of a design table. An example would be selecting a series of cells and dragging that selection downward to completely populate the rest of the table. You could easily populate the body of a table representing 100 configurations in a matter of seconds.

Custom Properties

Although custom properties can be generated via a design table, the functionality of custom properties extends far beyond design tables. To begin, let's first come to understand exactly what custom properties are.

A SolidWorks document can contain certain information above and beyond model data. For example, a part file can contain relevant information such as who created the file, or even that person's phone number and work extension. This information is known as the file's properties. You can create your own custom properties, which can contain literally any information you want them to.

The really beautiful thing about custom properties is that they can be transferred to other documents. For example, linking a note to a part file's properties gives you an easy way of updating title block information in a drawing. When the properties of the original part document are altered, these changes are reflected in the notes linked to these properties.

There are two types of custom properties: those created by the user and those automatically generated by SolidWorks. Both types are examined in this section. First, however, you should understand how to create your own custom properties within a SolidWorks document. How-To 17–3 takes you through the process of creating custom properties.

HOW-TO 17-3: Creating Custom Properties

To create custom properties, perform the following steps. Note that custom properties can be created for all SolidWorks document types (parts, assemblies, and drawings). It is also possible to create configuration-specific properties, meaning that a property for one configuration can have a value different than the same property for another configuration.

1. With a SolidWorks document open, select Properties from the File menu. This opens the Summary Information window, a portion of which is shown in Figure 17–34.

Fig. 17–34. Summary Information window.

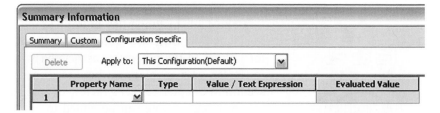

2. Select the Custom tab to create a custom property, or the Configuration Specific tab to create a custom property for a particular configuration.

➖ **NOTE:** *If adding a custom property to a drawing, there will be no Configuration Specific tab, as drawings cannot have multiple configurations.*

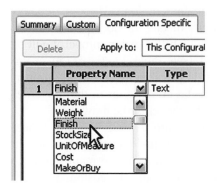

Fig. 17–35. Selecting a property name.

3. Click in the first cell in the Property Name column and type in a name for the property you wish to create. It is also possible to select a name from the dropdown shown in Figure 17–35.

4. Specify the property Type, such as a text string or number. This is akin to formatting, and the property will be a text string by default.

5. Click in the first cell in the Value/Text Expression column and type in a value for the new property you are creating. In Figure 17–36, the word *Polished* was entered.

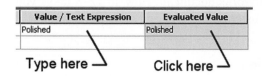

Type here ⟍   Click here ⟍

Fig. 17–36. Entering a value for the property.

6. Click in the first cell in the Evaluated Value column to see the end result of the custom property just added.

7. Repeat steps 3 through 6 for each new property you wish to add.

8. Click on OK to create the properties and close the Summary Information window.

The example shown in the images when performing the previous steps creates a custom property named *Finish* with a value of *Polished*. This is the most basic type of custom property, in that the user simply types in a value for the property. Custom properties can be much more interesting. Before exploring additional types of custom properties, let's examine how a custom property can be put to good use.

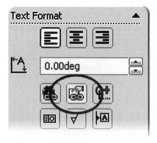

Fig. 17–37. Link to Properties button.

The Link to Properties button, found in the Note Property-Manager, is shown in Figure 17–37. This button is what allows for linking to the properties associated with the file. If you need to refresh your memory on how to create notes, see the "Notes" section in Chapter 12. Let's review the process.

Although a note can be added to any SolidWorks document type, imagine a note is being added to a drawing. Clicking on the Note icon opens the Note PropertyManager. Click somewhere on the drawing to position the note, and then enter text. Simple enough, so far. However, rather than type in text click on the Link to Properties button.

Fig. 17–38. Link to Property window.

The Link to Property window that appears (shown in Figure 17–38) will contain four options. These options essentially dictate the document or model from which the custom property will be linked (see Table 12–2). As an example, it is possible to link to the properties in the drawing, or to the properties from the model in the drawing.

In Figure 17–38, it is the model specified in the sheet's properties to which the link will be made. Because this is the case, any custom property created in the model file will appear in the drop-down list (also shown in Figure 17–38). As can be seen in the image, the Finish property was selected. Once the OK button is clicked on and the Link to Property window closes, the following text will appear in the note.

```
$PRPSHEET:"Finish"
```

The text in quotes is the name of the custom property. What the note actually displays, however, is the value of the property. In keeping with the example, the text would read *Polished*. What allows this to happen is the *$PRP* code, which may appear differently, depending on what the note is linking to. For your reference, Table 17–5 outlines the various code names and what selections in the Link to Property window they relate to.

Table 17-5: Link To Property Cross Reference

Property Code	Link Choice
$PRP	Current document
$PRPVIEW	Model in view to which the annotation is attached
$PRPSHEET	Model in view specified in sheet properties
$PRPMODEL	Component to which the annotation is attached

∞ **NOTE:** *When selecting a property to link to from the Link to Property window drop-down list, any property preceded with* SW- *is a property automatically generated by SolidWorks for the convenience of the user.*

Chapter 15 touched on how dimension names can be used in a custom property. This is a very important part of weldment profile descriptions, but can be used throughout SolidWorks. Custom properties can contain both text and links to dimension values. The wavy washer will be used in an example so that we might better understand how the linking is accomplished.

There are a number of configurations of the wavy washer. Depending on which configuration is being shown in a drawing, a note will need to show a description of the washer. The description will contain the hole diameter (which we will call ID), the outside diameter of the washer (OD), and the thickness of the washer (THK). The entire text string might therefore look something like the following.

```
.500 ID X 1.000 OD X .020 THK
```

Of course, if the text is typed in it would not update if a different version of the washer were shown in the drawing, or if the model dimensions were changed. To create a true link to the dimension values, a custom property containing those dimensions will be created.

You would begin creating the property in the usual fashion (see How-To 17–3). However, rather than typing in the text *.500* click on the dimension value representing the .500-inch dimension. It may be necessary to move the Summary Information window out of the way to accomplish this task. The string of text placed in the Value/Text Expression column might appear as follows.

```
"Hole diameter@Sketch2@@0.5x1.0 3W@wavy washer.SLDPRT"
```

The previous string of text could be thought of as saying "Show the dimension value for the dimension named 'Hole diameter' in 'Sketch2' of configuration '0.5x1.0 3W' that resides in part 'wavy washer.SLDPRT'." The configuration name is included because in this case the custom property is a configuration-specific custom property.

✗ **WARNING:** *File extensions (such as .SLDPRT) must be uppercase when appearing in custom properties.*

What happens next? Add to the cell in the Value/Text Expression column as you see fit. Be careful not to accidentally delete any portion of the code within the quotations, or the evaluated value will not read correctly. The End key is particularly useful in this situation. For instance, after clicking on the dimension click in the Value/Text Expression column cell, press End, type a space, type in the text *ID X*, and type another space. Next, click on the dimension that represents the 1.000-inch value, and continue in that manner.

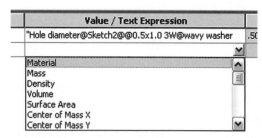

Fig. 17–39. Value/Text Expression column selections.

In addition to dimension values, it is possible to link to other physical aspects of a model. This is all done through the drop-down list (shown in Figure 17–39) that appears in the Value/Text Expression column. Selections include Material, Mass, Volume, and every possible value typically associated with the mass properties of a model. Of course, the beauty of all of this is that the custom properties will update when changes are made to the model. In turn, any note linked to the property will update.

Custom Properties in Design Tables

Now that you have a better understanding of custom properties, let's get back to the main topic of this chapter. How do custom properties relate to design tables? The answer is that they can be defined within the design table. The heading used is *$PRP*. This would then be followed by what you want the name of the property to be. In all, the heading might look as follows.

```
$PRP@cost
```

Cost, in this case, is the name of the custom property. Below this heading, in the same column of the design table, would be the values of what the property should have. If cells are left blank, the property will contain a blank (empty) value for that configuration. An example of a design table with a custom property defined within it is shown in Figure 17–40. Format the cells however you see fit.

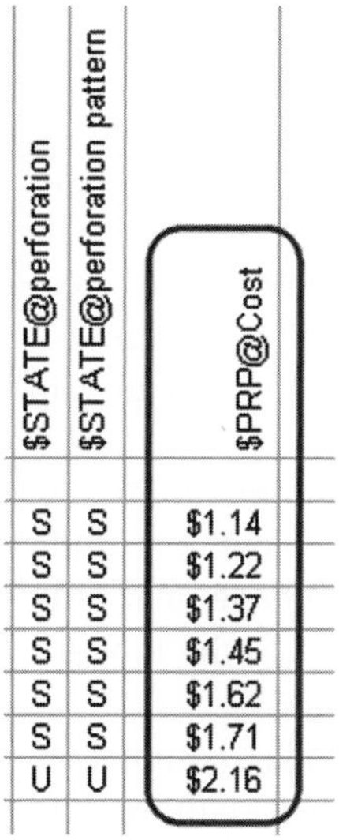

If you were to activate one of the configurations after closing the design table, and then access the files properties, you would see the property appear in the Configuration Specific tab. An example of this is shown in Figure 17–41.

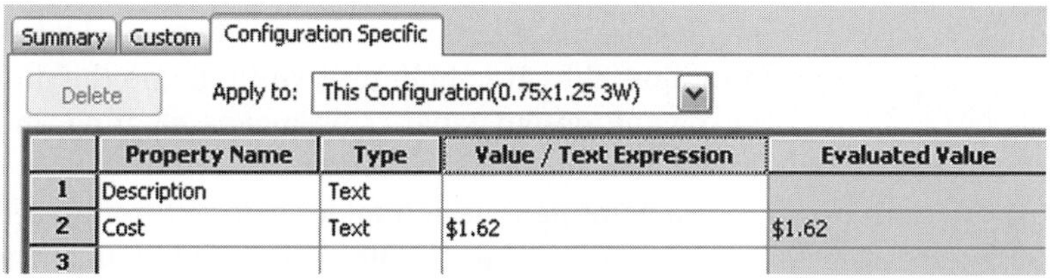

	Property Name	Type	Value / Text Expression	Evaluated Value
1	Description	Text		
2	Cost	Text	$1.62	$1.62
3				

Summary | Custom | Configuration Specific

Delete   Apply to: This Configuration(0.75x1.25 3W)

Fig. 17–41. The custom property appears in the Configuration Specific tab.

It is okay to drive simple custom properties from a design table, but don't try to drive custom properties containing dimension values. For example, if *$PRP@Description* were added as a heading to the wavy washer design table, the table would read the evaluated value of the property existing in the model's custom properties. Upon closing the table, this value would be fed back to the model's custom properties. The string of code containing the dimension names in the Value/Text Expression column would be overwritten, and the custom property would become a property with a static text value.

$STATE@perforation	$STATE@perforation pattern	$PRP@Cost
S	S	$1.14
S	S	$1.22
S	S	$1.37
S	S	$1.45
S	S	$1.62
S	S	$1.71
U	U	$2.16

Fig. 17–40. A custom property in a design table.

10-second Topic: Custom Properties and BOMs

In that BOMs can be customized (which you learned about in Chapter 12), we can now take custom properties to the next level and use them to populate a BOM. All you need to know regarding this process has actually already been covered. All that is left is to tie all of the information together.

Imagine a scenario in which a BOM template contains all of the column headings you find useful. There may be columns for cost, weight, material, and so on. Let's also imagine parts (or even assemblies) that contain design tables. It is easy to create custom properties from the design table, as you have recently discovered.

Believe it or not, that is all you need to do. SolidWorks does the rest. The custom property will be read from the design table and fed into the model's custom properties. The BOM in a drawing automatically reads the custom properties in the various parts and populates itself with that data. Considering that all properties can be defined in a design table, it isn't even necessary to actually define any custom properties manually from within each part file. The entire process is almost completely automated.

✓ **TIP:** *Use formulas in a BOM to multiply a cost by quantity to obtain a total cost value.*

Summary

Design tables are used to generate configurations in a part or assembly file. Feature dimensions and suppression states can be controlled in part design tables. Families of parts can be created and modified through the use of a single spreadsheet. You must use Microsoft Excel with SolidWorks to create a design table. It is helpful if you rename the dimensions that will be driven by the design table. This makes creating and editing the design table easier later on.

Generating a design table is done through a largely automated process. The Auto-create option allows for picking dimensions from a list to be added to a design table automatically in the proper format. Once generated, a design table can be saved as a file that can retain a link back to the model in which it was created.

Design tables are associative, meaning that changes made to dimensions in the model file will update in the design table. Conversely, changes made in the design table will update the model file. Through settings in the PropertyManager when inserting a design table, it is possible to lock dimensions being driven by a design table so that they can only be changed in the table and not in the model file. Rows or columns can be added automatically to a table if additional configurations or parameters are altered in the model.

Assembly design tables are an excellent means of controlling various parameters of an assembly and of adding configurations. Assembly design tables differ from part design tables in that they cannot control component feature dimensions or part feature suppression. Assembly tables can control suppression of components and the dimensions associated with mating relationships.

An important aspect of assembly design tables is the ability to specify which configuration of a component is being used in any particular assembly configuration. Component visibility can also be controlled. When a part is hidden, it is only hidden from view, but still exists in memory. When a part is suppressed, it is removed from memory, thereby disabling its respective mating relationships.

Use custom properties to create intelligent text in parts, assemblies, and especially drawings. Notes can be made to automatically update if certain model or file parameters change. Information in title blocks, such as sheet scale and drawing name, will update automatically if the notes containing that information are linked to custom properties. Users can also create their own custom properties that will propagate to BOMs, and they can define custom properties from within design tables.

CHAPTER 18

Equations

OFTEN IT IS DESIRABLE TO PLACE MATHEMATICAL RULES ON DIMENSION VALUES. Equations allow this to take place. A simple example would be in the case of a bracket whose width should always be half its length. By establishing this rule through the process of equations, it becomes possible for the model's width to automatically update if the length is changed. Equations can be used within parts or assemblies.

As powerful as equations are, there is not a lot to them. In that there is quite a bit that can be accomplished with equations, a couple of examples of their usefulness are included to help you better understand how this functionality works.

This chapter builds on topics discussed previously, and therefore assumes that you understand fundamental concepts and know how to perform certain basic functions within SolidWorks. These functions and knowledge include such things as changing dimensions, renaming dimensions and features, accessing a dimension's properties, and what is meant by suppressing or unsuppressing objects.

Link Values

Equations are a good tool to have for certain situations, but they can be overkill for simple applications. Because equations by their very nature create dimensions that are driven, they can sometimes reduce the amount of flexibility present in a model. For simple equalities, there are other options.

Figure 18–1 shows a sketch that will be used to create a bracket. There are two dimensions that according to the design intent of the engineer should always be the same length. This being the case, it would be best to eliminate one of the dimensions and add an equal relation between the two lines instead.

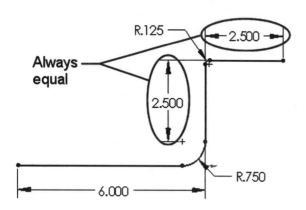

Fig. 18–1. Two dimensions should always remain equal.

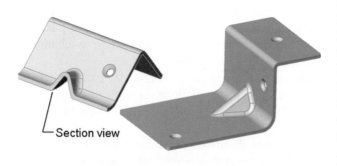

Fig. 18–2. Bracket should be the same thickness throughout.

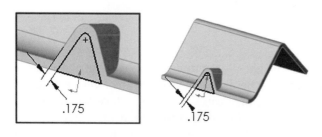

Fig. 18–3. Sketch used for cutting the gusset.

It could be argued that the deleted dimension is required to manufacture the part. This is not an issue, as the dimension could be added as a reference dimension in the drawing. The dimension could also be left on the part and made to be a driven dimension as opposed to a driving dimension. This can be accomplished by right-clicking on a dimension and selecting Driven from the menu. In this way, the dimension becomes a reference dimension in the model.

Adding an Equal geometric relation between sketch objects is very efficient. However, sometimes a slightly more powerful approach is necessary. In keeping with the previous example, Figure 18–2 shows the bracket at a future point of development. A gusset has been created for vertical stiffness. Design requirements call for the gusset to be the same thickness as the rest of the bracket. Figure 18–2 shows a section view of the gusset for clarity.

An equal relation cannot be applied here for the main reason that geometric relations exist between geometry, not dimensions. The thickness of the bracket is controlled through a dimension associated with the thin feature originally used to create it. The gusset thickness is controlled through a dimension as well. In the case of the gusset, a boss extrusion was performed first, and then a cut was made. The sketch for the cut was partially created by offsetting edges of the extrusion. The sketch is shown in Figure 18–3, once again using a section view for clarity.

The objective behind this example is to link the thickness of the original thin feature used to create the bracket with the offset dimension used in the development of the gusset. This can be accomplished through the use of the Link Values command. Linking values allows for linking two or more similar dimensions so that their values are always the same. When dimensions are linked, any of the linked dimensions can be modified, and all of the linked dimensions update.

Two or more dimensions can be linked, but the dimensions must be similar. In other words, multiple angular dimensions could be linked, but an angular dimension could not be linked to a linear dimension. How-To 18–1 takes you through the process of linking two or more similar dimensions.

HOW-TO 18-1: Linking Dimension Values

To link two or more dimension values so that they are always equal, perform the following steps. Note that part feature dimensions cannot be linked in an assembly. Dimensions native to an assembly can be linked if they are similar.

1. Right-click on a dimension and select Link Values. The Shared Values window, shown in Figure 18–4, will open.

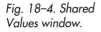

Fig. 18–4. Shared Values window.

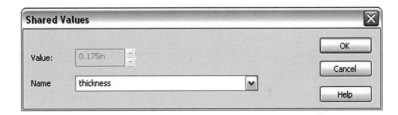

2. Type in a name that will be used to describe the linked dimensions. In the example used here (see Figure 18–4), the name *thickness* has been typed in.

3. Click on OK.

4. Right-click on another dimension, to be linked to the first, and select Link Values.

5. Either type in the same name used in step 2 or select the name from the drop-down list.

6. Click on OK.

7. Repeat steps 4 through 6 for every dimension to be linked to the original.

It is not really necessary to right-click on each dimension to be linked one at a time. If all dimensions to be linked are selected ahead of time, right-clicking on any one of them and supplying a name in the Shared Values window will link all values in one operation. The result is the same, which is to say that changing any one of the dimensions will change any of the dimensions linked to it.

Another place where dimensions can be linked is in the Modify window, which appears when double clicking on a dimension value. Note the drop-down arrow in the Modify window shown in Figure 18–5. Clicking on the arrow gains access to the Add Equation or Link Value command. Clicking on Link Value opens the Shared Values window, shown previously. Equations are discussed in material to follow.

How can one tell if dimensions have been linked? Numerous indicators will inform you of linked dimensions. Any dimension that is linked is displayed with a small red chain-link symbol. This symbol, shown in Figure 18–6, appears in front of the dimension value. If you happen to not notice the chain-link symbol, another symbol (also shown in Figure 18–6) is present in the Modify window.

Fig. 18–5. Linking through the Modify window.

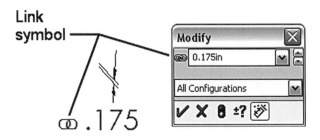

Fig. 18–6. Linked dimension indicators.

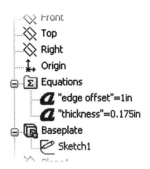

Fig. 18–7. Equations *folder in FeatureManager.*

Another indicator of linked dimensions is the *Equations* folder, which appears in FeatureManager. This folder (shown in Figure 18–7) only appears when employing linked values or equations. The names used for linking dimensions are listed in the *Equations* folder.

Unlinking dimensions can be done via the right mouse button. Right-clicking on a previously linked dimension gains access to the Unlink Value command. If selected, the link to that particular dimension will be broken, but only for that single dimension. All other dimensions previously linked will have to be unlinked in the same manner. It is possible to have a single dimension that is essentially linked to nothing. This will not cause any harm, but is not good practice.

Linking dimensions changes the name of those dimensions. Changing the name of a dimension, as you have learned, can also be accomplished by accessing the dimension's properties. However, it should be mentioned that simply renaming dimensions so that they have the same name is not enough to link them. Conversely, if dimensions are linked, changing one of the dimensions' names via its properties will rename all dimensions linked to it.

Writing Equations

The act of creating an equation is very similar to writing a sentence. Start on the left and work your way to the right. Getting dimensions into the equation is just a matter of selecting them. To select a dimension, it must be shown on screen. Therefore, you have two choices: right-click on the *Annotations* folder in FeatureManager and select Show Feature Dimensions or access dimensions a feature at a time by double clicking on the applicable features.

Accessing dimensions a feature at a time has the advantage of keeping the screen from becoming too cluttered. If it is possible to show all dimensions at once and still find the dimensions you are looking for, go ahead and do so, as this will make the task of getting the dimensions into the equation easier.

Equations are similar to design tables in the sense that there is a bit of preparation that could take place prior to creating them. The preparatory work is essentially the same in both cases. Specifically, it is best to rename dimensions and features prior to creating equations. The reason for this is identical to that for design tables. Because dimensions will be used in equations, it makes sense to give dimensions (and related features) meaningful names. This makes it easier to understand what is taking place within the equation when it is viewed at a later time.

SolidWorks is smart enough to know when a dimension has been renamed when equations are being used. That is, if a dimension is renamed SolidWorks will change the name in the equation accordingly. Nonetheless, good practice is to perform any renaming prior to creating equations.

It helps tremendously to physically jot down a rough draft of the equation. This is not necessarily true for very simple equations, but it definitely helps when trying to work out something a bit more complex. Give yourself an idea of how the equation should be formed, and the chances of the equation working correctly will improve.

Equation Syntax

An extremely important aspect of equations relates to which dimensions will be solved for, and which dimensions will be changeable by the user.

Any dimensions you want direct control over should be on the right-hand side of the equation's equal sign. The dimension value being solved for (controlled by the equation) should be placed on the left-hand side of the equation's equal sign. In other words, driving dimensions appear on the right, and the driven dimension should appear on the left. An example follows.

```
"D1@BaseExtrusion" = "D2@Sketch3" - .25
```

In this equation, the *D2@Sketch3* dimension could be altered. Double clicking on the dimension value would result in the typical Modify window, which would allow you to enter a different value for the dimension. If you were to try this on the *D1@BaseExtrusion* dimension value, a window would appear informing you that the dimension is being driven in an equation and that its value cannot be changed.

•◦ **NOTE:** *Only one dimension value is allowed on the left-hand side of an equation.*

At this stage, you have been taught everything you need to know in order to write your own equation. To test your knowledge, try creating a simple block whose height is half its width. This will give you a feel for the equation creation process. How-To 18–2 takes you through the process of creating an equation.

How-To 18-2: Creating an Equation

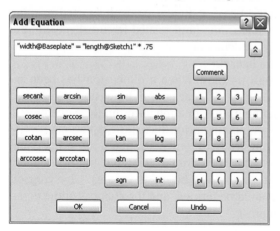

Fig. 18–8. Add Equation window.

To create an equation, perform the following steps.

1. Select Equations from the Tools menu, or click on the Equations icon found on the Tools toolbar.

2. In the Equations window that appears, click on the Add button. This will open the Add Equation window, shown in Figure 18–8.

3. Click on (select) the dimension you wish to drive in the equation. You may need to double click on the feature the dimension is associated with to gain access to the dimension.

•◦ **NOTE:** *It is possible to drag the Equation and Add Equation windows to different locations on the screen, which you may very well need to do in order to select dimensions.*

4. Using either the keyboard or the Add Equation window, add an equal sign to the equation.

5. Continue to click on dimensions, or add mathematical operators or numeric values, in order to finish writing the equation, depending on your requirements.

6. Click on OK to close the Add Equation window and create the equation.

7. Click on Add to create additional equations, or click on OK to close the Equation window.

The Equations window, shown in Figure 18–9, will display any equations that have been created, along with any names used in linking dimension values. Any equations will also be displayed with what the equation evaluates to. In the case of the example shown in Figure 18–9, this value happens to be 4.5 inches.

Fig. 18–9.
Equations
window.

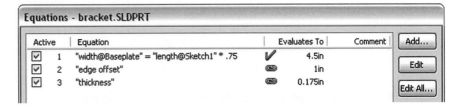

✓ **TIP:** *When adding equations, feel free to add comments. Anything in an equation line that falls after a single quotation mark is considered a comment. Clicking on the Comment button in the Add Equation window does nothing more than add a single quotation mark.*

A green check in the Equations window means that the equation is functioning normally. A red exclamation point indicates that the equation cannot be solved for one reason or another and you may have to perform some troubleshooting. The checkbox at the far left of the equation allows for turning the equation on or off. Equations that are turned off (unchecked) are considered deactivated (or suppressed).

Let's take a short moment to review the equation shown in Figure 18–9. The right-hand side of the equation reads *"length@Sketch1" * .75*. This equation serves to set the dimension on the left-hand side of the equation, *width@Baseplate*, to 75% of whatever the *length@Sketch1* dimension happens to be. For your reference, the equation reads as follows.

```
"width@Baseplate" = "length@Sketch1" * .75
```

The quotation marks that enclose the dimension names are critical, and are added automatically. If dimension names are manually typed in, make certain to add the quotation marks. Manually entering dimension names is not recommended, as it leaves too much room for human error. A single typographical error will result in the equation not working. Selecting dimensions and allowing SolidWorks to input the dimension names into the Add Equation window is much easier.

Once an equation has been created, the *Equations* folder will be listed in FeatureManager, as was shown earlier. This folder is nothing more than a shortcut to certain operations involving equations. These operations, accessed via the right mouse button, consist of deleting, editing, and adding equations.

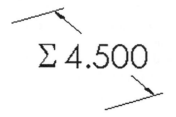

Dimensions placed on the left-hand side of an equation will display in the work area and will be preceded by the Sigma symbol, which happens to be the eighteenth letter of the Greek alphabet. This symbol will also be present in the Modify window when attempting to

Fig. 18–10. Sigma (equation) symbol.

change the value of such a dimension. As mentioned earlier, changing dimensions driven by equations is not possible.

Editing and Deleting Equations

Most of the buttons on the right-hand side of the Equations window (see Figure 18–11) are self-explanatory. OK and Cancel (not shown) perform the usual functions, and Add opens the Add Equation window, as you have discovered.

To delete an equation, select the equation in the Equations window and click on the Delete button. To edit an equation, select the equation and click on the Edit button. The Edit Equation window will appear, which is identical in every way to the Add Equation window (shown in Figure 18–8).

Clicking on the Edit All button opens what amounts to a very rudimentary text editor. Every equation in the model will be listed, and the window can be resized for convenience. Selecting a dimension with this window open will serve to add that dimension name to wherever the flashing cursor happens to be, identical to what happens when the Add Equation or Edit Equation windows are open.

Fig. 18–11. Buttons in the Equations window.

The Configs button allows for specifying which configurations will be affected if you decide to change the suppression state of any of the equations. As mentioned earlier, unchecking the box in the Active column (shown in Figure 18–12) serves to suppress an equation. The Configs button utilizes the standard configuration options, which should be familiar to you (see Chapter 11). You have the options This Configuration, All Configurations, and Specify Configurations.

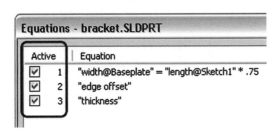

Fig. 18–12. Suppressing equations.

The Help button will provide you with instructions and tips for creating equations. The help also contains an excellent reference on supported functions. Which functions are available can be determined by looking at the buttons on the Add Equation window. This includes trigonometric functions and others, such as calling out pi or absolute values in equations.

Table 18–1 is supplied as a quick reference guide to equation functions. Use the syntax supplied in the Function column when writing your equation. Not all trigonometric functions are listed, as they will follow the sine or arcsine (inverse sine) formats. For example, for the Tangent function simply substitute the word *sine* with the word *tangent* in the sine function shown in the table.

Table 18-1: Equation Functions

Function	Name	Description
sin (α)	sine	Returns the sine ratio of angle α
arcsin (α)	arcsine	Returns the angle of sine ratio α
abs (α)	absolute value	Returns the absolute value of α
exp (n)	exponential	Returns e raised to the power of n
log (α)	logarithmic	Returns the natural log of α to the base e
sqr (α)	square root	Returns the square root of α
int (α)	integer	Returns α as an integer
sgn (α)	sign	Returns the sign of α as 1 or -1
pi	pi	3.14159...

✗ **WARNING:** *Angular units default to radians in equations. If you wish to use degrees, make certain to change the Angular Equation Units setting to degrees in the Equations window.*

Variables

If you find the need to use variables in equations, you will find them very easy to create. An example of a simple variable might be to assign the numeric value of 4 to the letter *A*. Another example would be to set the letter *B* to equal some dimension value. In both cases, the variables would be the letter *A* or *B*.

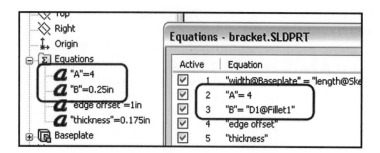

Fig. 18–13. Creating variables.

Variables always should be placed on the left-hand side of the equation. Creating a variable is done by clicking on the Add button in the Equations window. You would then simply type in the letter to be used as the variable, add an equal (=) sign, and then either type in a value or click on a dimension. SolidWorks will add the quotation marks around the variable automatically. Figure 18–13 shows how variables might appear in the *Equations* folder and the Equations window.

If Statements

One of the more powerful capabilities of equations is their ability to use *If* statements. Readers familiar with programming will be familiar with *If* statements. For the rest of us, a simple explanation is in order. An *If* statement defines actions that should take place in an equation based on certain conditions. An equation containing an *If* statement might read "If a dimension is greater than a particular value, do this. Otherwise, do that."

Creating an *If* statement actually uses the designation *IIF*, not *IF*. Case is not important, though the term is often capitalized. To create an *If* statement, use the following syntax.

```
IIF (logical condition, action if true, action if false)
```

The logical condition can use the logical operators of equal to (=), less than (<), or greater than (>). Dimension names can be used in defining the logical condition, as well as numeric values. The actions carried out, dependent on the outcome of the logical condition, can be simple numeric values or mathematical operations. Let's look at an example to get a better idea of how *If* statements work.

Figure 18–14 shows a wheel with spokes. Each of the two spokes on the left is a feature. These are patterned as a pair. There are two instances in the pattern, which is why there are only two sets of spokes currently shown in the wheel. To control the number of spokes, an equation will be used.

The current inside diameter of the wheel is 4 inches. The hub will remain a constant size, so we will not worry about it. The inside diameter of the wheel itself may change, anywhere from 3.5 inches to 7.5 inches. If the inside diameter is greater than 5 inches, we would like the wheel to have 17 spokes, but any diameter of 5 inches or less should have only 13 spokes. Therefore, the equation should read as follows.

```
"No of Spokes@Pattern" = IIF ("ID@Sketch1" > 5, 17, 13)
```

Fig. 18–14. An equation will be created for the spokes.

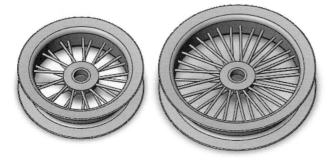

Fig. 18–15. The same wheel at different sizes after adding an If statement.

No of Spokes is the name given to the value associated with the number of instances in the circular pattern. The circular pattern itself was renamed *Pattern*. Therefore, the full name of the value associated with the number of instances in the pattern is *"No of Spokes@Pattern"*. The full name of the dimension associated with the inside diameter of the wheel is *"ID@Sketch1"*. Examining the equation, we find that if the inside diameter (*ID*) of the wheel is greater than 5 the number of spokes will be 17. If the *ID* value is less than 5 there will only be 13 spokes. What would happen if *ID* is exactly 5? There will be 13 spokes, because the *ID* value is not greater than 5. Make sense? Figure 18–15 shows the original wheel with an ID of 4 inches after adding the equation. A slightly larger wheel (with an ID of 5.25 inches) is shown beside the original.

✓ **TIP:** *If statements can be nested, meaning that another If statement can be used as the action to perform based on the logical condition set forth in a parent If statement.*

Changing Units in Models with Equations

If you have spent a lot of time on a model and have added equations to it, and then decide that the model's working units need to be changed, you are going to have a problem on your hands. However, if you think there is a chance the units will need to be changed sometime down the road, perhaps due to a customer or client overseas that may be working with the model, there is a precautionary measure that can be taken.

An equation used earlier in this chapter will be used again as an example. The equation reads as follows, and assumes that the working units are in inches.

```
"width@Baseplate" = "length@Sketch1" * .75
```

If it is assumed that *"length@Sketch"1* equals 3 inches, the solution to the equation becomes 2.25. If the working units are switched to millimeters, this ruins the equation. Keep in mind that *"length@Sketch1"* is a dimension value. Switching to millimeters means that *"length@Sketch1"* becomes three times the number of millimeters in an inch (approximately 25.4), or 76.2. This means that the equation's solution becomes 57.15, which is a far cry from 2.25!

How can we keep this from happening? The solution is to create a sketch with a line in it. Dimension the line to 1 inch. Rename the sketch to something such as *constant* if you want. For the sake of argument, assume that the dimension's full name is *"D1@constant"*. Hide the sketch, because it will not be used for anything else, and then plug the new dimension name into your equation. The equation might look as follows.

```
"width@Baseplate" = "length@Sketch1" * (.75 * "D1@constant")
```

As long as the units are in inches, *"D1@constant"* will equal 1, but as soon as you switch to a different set of units (such as millimeters) *"D1@constant"* equals 25.4. The .75 value is multiplied by *"D1@constant"* and everything works out. Depending on the equation, you may have to apply the "constant" in a different manner, but that should not be too difficult to figure out. Incidentally, this works when starting out in any set of units, not just inches.

Common Equation Errors

The order in which equations are created is significant. Equations are solved in the order in which they appear in the Equations window. Equation order can be changed by copying and pasting via the Edit Equation window (accessed by clicking on the Edit All button, discussed earlier).

A circular reference can be created with equations, and SolidWorks will not stop you from doing so. This is a nasty type of equation problem because it does not manifest itself as an error. There is no indication that the circular reference even exists, except that the model may behave strangely. Consider the following three equations.

```
"Length@Block" = "Width@Sketch1" * 2
"Height@Sketch1" = "Length@Block" / 2
"Width@Sketch1" = "Height@Sketch1" + 1
```

These three equations drive the size of a block. For the sake of discussion, assume that the width starts out at 2 inches. This means that the length will equal 4 and the height will equal 2. But then the third equation is solved, which resets the width to 3 (not 2).

This odd scenario causes a serious problem. The equations have the odd effect of causing the part to grow every time it is rebuilt. Without performing any other action other than to click on the Rebuild icon, the model grows larger and larger until it reaches the maximum size limitation of the software (which happens to be 1 kilometer). This is a circular reference, and is obviously not a desirable situation.

Another problem arises when equations are created in the wrong order. Equation ordering is probably one of the more common problems found in equations. Let's take another set of three equations and break down what is occurring in each of them.

```
"D1@Base-Extrude" = "D2@Sketch1" * 2

"D3@Sketch1" = "D4@Sketch1" + .5

"D4@Sketch1" = "D2@Sketch1" + 1
```

If we understand what SolidWorks does when it solves these equations, we will gain an appreciation for the problem that arises. First, the software takes dimension *D2* (for brevity, only the first names of the dimensions will be used) and multiplies it by 2 to solve for *D1*. It then attempts to take dimension *D4* and add .5 to it to solve for *D3*, but it cannot because *D4* is also being driven by the third equation. This forces SolidWorks to place the second equation on "hold," in a manner of speaking. It then adds 1 to *D2* and solves for *D4*.

The result of this scenario is that the second equation is not solved for after the first rebuild. It takes a second rebuild to show the model geometry in its proper state, with all equations solved. The true problem is that there is no indication that a second rebuild is required. No rebuild symbol appears in FeatureManager to warn you that a rebuild is needed. No warning of any type is given.

This problem can be remedied by reordering the equations. If the second equation were solved last, the issue would go away and the model would need only a single rebuild to solve all three equations. A rule of thumb for avoiding this situation is: *Do not create equations that drive dimensions that also appear as driving dimensions in earlier equations.*

It was possible to create a circular reference with only three equations (or even just two, for that matter), and it was also possible to create a rebuild issue with only three equations. Just imagine what could happen with a dozen or more equations. If you do not pay attention to equation ordering, it might take multiple rebuilds to solve all of your equations and show the model correctly.

You should not be afraid of using equations, as long as you are careful and as long as you follow one simple rule: simplicity breeds efficiency. In other words, use the easiest tool available to accomplish your design intent.

If the Equal relation will suffice, use it. If that does not quite accomplish what you require, perhaps use Link Values instead. If Link Values falls short, use equations (but be aware of and knowledgeable about the issue of equation order, such as the "rule of thumb" previously cited). Use the simplest tool that will get the job done, and your models will be less prone to errors and will rebuild much more quickly.

Summary

In this chapter, you have discovered the ability to link dimensions and create equations. Linking dimension values can be accomplished by right-clicking on a dimension value. Only similar dimension types can be linked, and linked dimensions must all be given the same name. Link Values only works for equalities (when two or more dimensions must be equal).

To set more complex relationships between (or among) dimension values, equations are required. The left-hand side of the equation can only contain one dimension, which will be the driven dimension. The right-hand side of the equation will contain the driving dimensions. Access to linking dimensions or adding equations can also be gained by double clicking on a dimension and using the drop-down list found in the Modify window.

Even with the Add Equation window open, it is still possible to double click on features to access their dimensions. One click on a dimension value is all it takes to get that dimension placed into the equation. This is the preferred method of inserting dimension names into an equation. Manually typing dimension names is allowed, but leaves plenty of opportunity for typographical errors.

Equations solve in the order in which they are listed in the Equations window. Use caution when creating equations so that circular references are not created. Also, try not to create equations that drive dimensions that also appear as driving dimensions in earlier equations. This will cause your model to require multiple rebuilds in order to solve all equations.

Use *If* statements and variables if you find these functions useful in writing an equation. Try not to use equations when they are not really necessary, though. More equations mean a more complex part file (or assembly), with a greater likelihood of something going wrong. A simplistic approach often works best. Therefore, use Equal relations (or relations in general) or Link Values if they will get the job done without the need for equations

CHAPTER 19

Advanced Assembly Modeling

THERE ARE SOME VERY IMPORTANT ASPECTS OF ASSEMBLY FILES that have not been covered to this point. This includes working with large assemblies and how to best optimize computer performance in such cases. This is an extremely important topic for anyone working with assemblies with more than a few hundred components.

In this chapter you will also explore ways in which an assembly can be restructured. Subassemblies can be created on-the-fly, and components can be organized into groups if the structure of the assembly needs modification. Advanced selection techniques when working with large assemblies are also explored, such as using volumetric envelopes for controlling component selection.

File references are looked at in depth. The topic of copying files and maintaining references is a topic faced by every SolidWorks user. External references to other files within an assembly can be very cumbersome when files need to be copied, moved, or renamed. The proper methods for performing all of these tasks are explored in this chapter.

Editing Components in Context

The act of editing a component within an assembly is known as editing in context. *In context* means that a feature or part is created *in association with* another component or assembly. Parts can be opened *out of context*, which means the assembly the part was created in is not currently open. Features and parts can be *out of context* or *in context* with respect to the original assembly in which they were created.

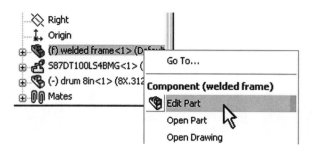

Fig. 19–1. Editing a part in an assembly.

When you wish to edit a part in the context of an assembly, it is usually because some other component must be referenced. For example, a hole in one component must be aligned with the axle of another component in the assembly. In such a case, it would be necessary to use the Edit Part command, located in the right mouse button menu. Right-click on the component in the work area or in FeatureManager, and then select Edit Part (shown in Figure 19–1).

It is also possible to select a component and click on the Edit Part icon found on the Assembly toolbar in order to edit that part. Clicking on the same icon a second time will return you to editing the assembly.

When editing a part in the context of an assembly, there are indicators that inform you of this fact. Components in the assembly turn either gray or become semitransparent, except for the part being edited. The part being edited will typically turn a different color (salmon or blue is common). In addition, the title bar of the SolidWorks window states the name of the part being edited, followed by the assembly name, whereas before it stated the assembly name only. The following text shows an example of what might be seen in the SolidWorks window title bar.

`Sketch5 of Welded Frame -in- Winch.SLDASM`

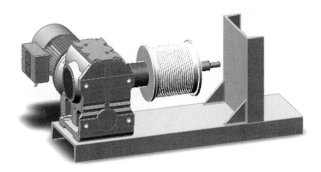

Fig. 19–2. Winch assembly.

This text not only informs the user that the welded frame is being edited but that it is *Sketch5* in the welded frame that is being edited. All of this editing is taking place in the winch assembly, also visible in the text. The title bar is an excellent way of determining what is being edited at any point in time. It may not seem very important, but understanding what is being edited at any particular time is critical. Let's look at an example to understand why this is so. Figure 19–2 shows the winch assembly used in our example.

The winch consists of a welded frame, a motor, and a drum. As the motor turns the drum, the drum winds a rope around a groove cut into the drum. This assembly is a generic setup used by a manufacturer. Different drums may be used, and motors of varying power (and size) may also be used. Considering that this is the case, the welded frame will require its vertical member to be at various positions. It will also be necessary to cut a hole through the vertical member dependent on the position of the drum axle.

If you began a sketch with the intention of creating a feature on a part, but realized too late that you were actually editing the assembly, the sketch would belong to the assembly, not the part. The sketch could not be used to create the feature you had intended. Similarly, when editing a part in the context of the assembly the Insert menu (and other menus, to a lesser degree) will change to reflect the same menu you would see when working on a part in its own window. In other words, depending on what is being edited (part or assembly), certain commands will be available or inaccessible.

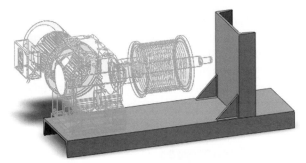

Fig. 19–3. Editing the welded frame.

As you have learned, right-clicking on the welded frame and selecting Edit Part allows that component to be edited. Figure 19–3 shows what you might see while editing a component in the context of an assembly.

A few settings that affect an assembly's appearance while editing components were discussed in Chapter 15 (such as assembly transparency). Other display settings, previously discussed in this book, can also affect model appearance. Sometimes these display settings can adversely affect a person's ability to see, and therefore edit, components in context. For reference, Table 19–1 outlines the main settings that will affect model appearance while performing in-context edits. When the Location column refers to System Options settings, access those options through the Tools > Options menu.

Table 19-1: Assembly Appearance While Editing Components

Setting	Location	Description
Shaded With Edges and Shaded display modes	View toolbar	Using Shaded With Edges can "clog" the display when using transparency, in which case use Shaded instead.
Assembly transparency for in-context edit	Display/Selection section of System Options	Controls whether components other than the one being edited are displayed transparent or opaque.
Use specified color for Shaded With Edges mode	Colors section of System Options	When checked, uses the color (black by default) specified in the System colors listing (also in the Colors section) for the edges of components.

Setting	Location	Description
Use specified colors for editing parts in assemblies	Colors section of System Options	When checked, uses the color specified in the System colors listing (also in the Colors section) for the component being edited.
Assembly, Edit Part	System colors listing in Colors section of System Options	Controls the color of the component being edited in context.
Assembly, Non-edit Parts	System colors listing in Colors section of System Options	Controls the color of all components not being edited in context.

Of the settings listed in Table 19–1, the most important is the *Use specified colors for editing parts in assemblies* option. It is highly recommended that this option be turned on (checked). If it is not checked, all components in the assembly will retain their original colors even when a component is being edited. This makes it more difficult to tell if a component is being edited in context. When checked, components not being edited become gray, and the component being edited becomes blue. Gray and blue are default color settings that can be changed per your preferences (as indicated in the table).

Transparency Selection Techniques

When working with transparency enabled, sooner or later you will find it necessary to either select something that is transparent or something behind a transparent object. There are special techniques that can be used to accomplish this.

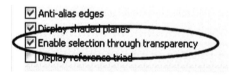

Fig. 19–4. Controlling selection through transparency.

The option that will have the greatest impact on how objects are selected when transparency is in use is shown in Figure 19–4. This option, *Enable selection through transparency*, is found in the Display/Selection section of System Options (Tools > Options).

When *Enable selection through transparency* is checked, faces on the component being edited can be selected through transparent objects. If turned off, the transparent faces are selected. Holding down the Shift key will temporarily reverse the *Enable selection through transparency* option, whatever it happens to be currently set at.

It is recommended that the *Enable selection through transparency* option be left on. This will force SolidWorks to exhibit the behavior that might be described as "if you can see it, you can select it." This type of behavior is logical for most users, and works out well. Remember, you can always hold down the Shift key if it becomes necessary to select something on a transparent component.

Editing In Context in Practice

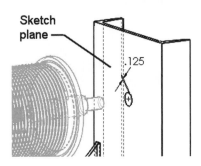

Fig. 19–5. Creating an in-context feature.

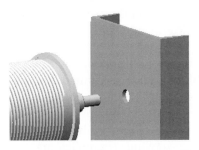

Fig. 19–6. Hole cut through the welded frame.

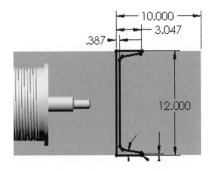

Fig. 19–7. Editing the sketch for the vertical member.

We could talk about in-context editing all day, but the effects of in-context editing don't hit home until it is seen firsthand. Figure 19–5 shows a feature being added to the welded frame. The welded frame is currently being edited, and a sketch was started on the face indicated in Figure 19–5. The circle was created by offsetting the outside circular edge of the drum axle. By using Offset Entities, a clearance value can be entered, such as the .125-inch value used in the example.

When editing a part and creating a feature in the context of an assembly, you are not doing anything you have not already been taught in this book. In the case of the example, the cut can now be created for the hole. This is done using the same cut feature discussed in Chapter 4. It just happens to be that the cut is being created on the part while in an assembly, rather than in the part's own window. Figure 19–6 shows the completed cut on the welded frame.

There is still another problem to be dealt with. The hole in the vertical member of the welded frame is in the correct position, but the vertical member is too far away from the drum axle. To correct this problem, it will be necessary to edit the sketch for the vertical member. This is shown in Figure 19–7. The 10-inch dimension is used to position the sketch along the length of the horizontal member.

One solution would be to increase the value of the 10-inch dimension to the point where the sketch would fall over the axle of the drum. However, it makes more sense to dimension the sketch to the drum axle in order to maintain a proper position to the axle. This latter solution protects the assembly if alterations are made to other components, such as the motor or some other part of the drum. Figure 19–8 shows the same sketch with the 10-inch dimension removed. A new .500-inch dimension has been added between the sketch and the axle.

Looking at the assembly from a slightly different vantage point, Figure 19–9 shows how the welded frame's vertical member is now at the proper position to accommodate the drum axle. Some holes have been added for a bearing that will added later

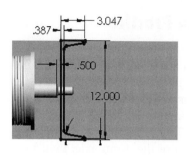

*Fig. 19–8.
Repositioning
the sketch.*

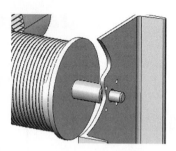

*Fig. 19–9.
Assembly
after in-
context edits.*

(hypothetically speaking). The welded frame is shown with a portion cut away for clarity.

The biggest advantage of in-context features can now be seen. If the drum axle is altered in diameter, or some other dimension on the drum is altered such that it forces the drum axle to change position, the welded frame will adapt to those changes. The holes will reposition themselves, the large hole diameter will change, and even the vertical member will move forward or back if necessary.

With benefits come risks. External references do not come without a price. External references limit how assemblies can be restructured. Copying assemblies or saving assemblies to other locations is more perilous when the assemblies contain external references. As long as you are aware of the proper way to handle assemblies with external references, no harm is done. Therefore, a number of scenarios involving copying assemblies are discussed later in the chapter.

Hole Series

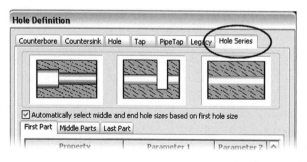

Fig. 19–10. Hole Series tab.

A series of holes can be created simultaneously through two or two hundred components within an assembly. This (known as a hole series) is accomplished via the Hole wizard (discussed in Chapter 6). The Hole Series tab, shown in Figure 19–10, is only available when accessing the Hole wizard while in an assembly.

SW 2006: The Hole Series command has been separated from the Hole wizard. To access the Hole Series command, select Hole Series from the Insert > Assembly Feature > Hole menu. In addition, there is a Hole Series icon, typically found on the Features toolbar.

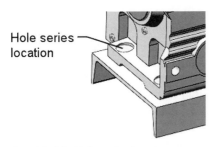

Hole series location

Fig. 19–11. Hole series location.

Although the Hole Series function is very similar to an assembly feature, it is a special-case assembly feature. Unlike other types of assembly features (which were discussed in Chapter 13), a hole series propagates to and affects individual part files. This is in stark contrast to a standard assembly feature, which only affects the components within the assembly and not the individual part files.

Figure 19–11 shows one of four locations where a hole must be drilled through the motor and welded frame. For this operation, a hole series will be used. How-To 19–1 takes you through the process of creating a hole series.

HOW-TO 19-1: Creating a Hole Series

To create a series of holes through multiple components within an assembly, perform the following steps.

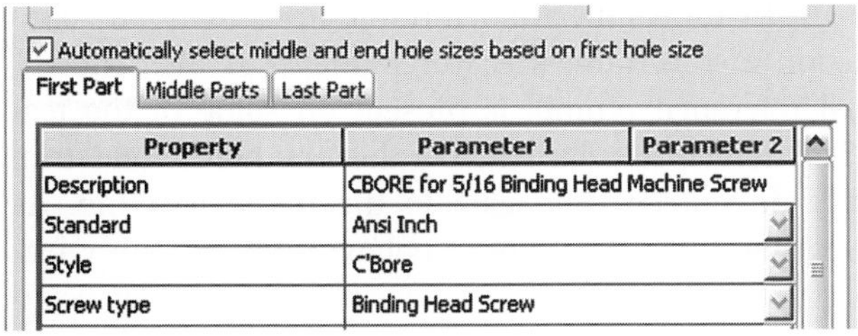

Property	Parameter 1	Parameter 2
Description	CBORE for 5/16 Binding Head Machine Screw	
Standard	Ansi Inch	
Style	C'Bore	
Screw type	Binding Head Screw	

☑ Automatically select middle and end hole sizes based on first hole size

First Part | Middle Parts | Last Part

Fig. 19–12. Specifying parameters for the hole series.

1. Select a face on the first component where the hole series should start.

2. Select Wizard from the Insert > Assembly Feature > Hole menu, or click on the Hole Wizard icon found on the Features toolbar.

3. Select the Hole Series tab, shown in Figure 19–10.

4. Select the First Part tab, visible in Figure 19–12, and specify the parameters for the hole that will cut through the first part.

5. Select the Middle Parts and End Part tabs, repeating the process of specifying the parameters for these holes.

6. Click on Next.

7. Add points as needed to create additional holes.

8. Add relations or dimensions to correctly position the points added in step 7.

9. Click on Finish to create the hole series.

To have SolidWorks automatically select the parameters for the last component and any components sandwiched between the first and last components, check the option titled *Automatically select middle and end hole sizes based on the first hole size.* This will keep you from having to plug in all applicable parameters in the Middle Parts and End Part tabs per step 5.

Fig. 19–13. Threaded hole with a clearance cut in the upper component.

Aside from a slightly different set of parameters to plug in, creating a hole series is implemented in the same fashion as other holes created with the Hole wizard. It is an excellent way in which to create a large number of holes in a very short time. Figure 19–13 shows a hole series in which the upper hole is a clearance hole and the lower hole is threaded for a 3/4-inch hex bolt. This is a very common usage of the Hole Series function.

There is one oddity that occurs when creating a hole series. Actually, it is an oddity that occurs after a hole series has been created. For some reason, SolidWorks does not know enough to hide the sketch containing the locating points. This is simply one of those quirks one takes in stride after a while. Manually hide the sketch for the hole series by right-clicking on the sketch in FeatureManager.

Figure 19–14 shows the welded frame in its own part window. A portion of the frame's FeatureManager is also shown. The external reference symbol is present on some of the features, and is enlarged for clarity. Note that the holes created in the assembly do indeed propagate to the frame part file. As mentioned earlier, this is not typical behavior for assembly features, and is the reason a hole series is considered a "special-case" assembly feature.

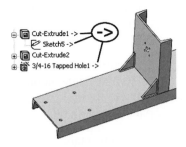

Fig. 19–14. Holes propagate to the welded frame part.

SW 2006: The Hole Series command interface utilizes PropertyManager, rather than sharing an interface with the Hole wizard. Multiple PropertyManager pages allow for specifying hole parameters for the start, middle, and end components.

In-context Parts

Creating features in the context of an assembly is one thing, but creating an entire part is another matter. The process is essentially the same while creating the various features that make up the part, but to begin the process a new component must first be inserted into the assembly. Inserting a new component was covered in Chapter 15 (see "How-To 15–2), so we won't repeat the steps here.

Once the new component has been inserted, it is possible to base geometry on existing components in the assembly. Some SolidWorks users prefer to create their components in the context of an assembly, but do not necessarily want to create all of the external references that go along with that design technique. An option that aids in this endeavor is shown in Figure 19–

15. This option, *Do not create references external to the model*, keeps any external references from being created.

Fig. 19–15. Do not create references external to the model *option.*

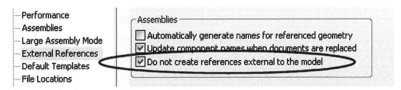

When the *Do not create references external to the model* option is checked, the option works as advertised: absolutely no external relations are added. If this option is employed, it carries with it its own dangers. It is more difficult to fully define geometry, because all of the geometry visible on screen will be off limits unless it belongs to the part being edited. This can be deceptive, so take extra care to fully define sketch geometry created in context.

When the *Do not create references external to the model* option is not used, external references will abound. It is not in your best interest to create external references unnecessarily. Aside from the risks mentioned earlier, external references can have a tendency to reduce system performance if used extensively.

When parts must be created in the context of an assembly, it is not usually necessary to develop every feature on the part in context. When the design process permits it, open the part in its own window to further develop the part. This reduces the overhead experienced when working in the assembly. It also helps to keep external references to a minimum. To review locking or breaking external references and external reference symbology, see "External References" in Chapter 7.

Derived Parts

So far, you have learned that features or entire parts can be created in the context of an assembly. A part created in context can be opened and modified in its own window. Holes created as a hole series will also propagate to the individual part files affected by the holes. However, assembly features (discussed in Chapter 13) do not propagate to the individual part files. Keeping this in mind, how would it be possible to create an accurate design drawing of a part showing the features created in the assembly?

Figure 19–16 shows the winch with a new component created in context. This new component is called the *hanger plate*. Because there are so many variables that can change in the winch assembly (different motors, drum sizes, and so on), it is more con-

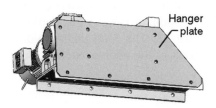

Fig. 19–16. Hanger plate component.

venient to create the hanger plate in context, after the rest of the assembly has been completed.

The four holes used to bolt the hanger plate to the motor were created in context. Therefore, these holes show up in the hanger plate part file. The additional holes were created as assembly features for the simple fact that it is not known where the holes should be drilled until the assembly is completed.

Think of a derived part as a copy of the original part used in an assembly, but with the addition of any assembly features that affect it. The process of creating a derived part takes place from within an assembly. How-To 19–2 takes you through the process of creating a derived part.

How-To 19-2: Derived Component Parts

To create a new part derived from an existing component within an assembly, perform the following steps. Be aware that you must be editing an assembly in order to employ this command.

1. Select the component from which to create a derived part from Feature-Manager.

2. Select Derive Component Part from the File menu.

3. You may be prompted to select a template. If so, select the part template you wish to use and click on OK.

4. The derived component part will be visible on screen. It will still be necessary to save the new file, so you should do so at this time.

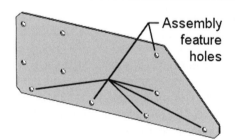

Assembly feature holes

Fig. 19–17. Derived component.

The hanger plate (shown in Figure 19–17) has been opened in its own part window. If the original hanger plate were opened in its own window, it would be missing the holes (indicated in Figure 19–17) that were created as part of the assembly feature.

Derived parts have external references to the component from which they are derived, and the base feature will inherit the name of the part it is referencing. The name of the base feature in a derived part takes on the name of the component from which is was derived, followed by the name of the assembly the component is in. An example of this naming convention is shown in Figure 19–18. The *Solid Bodies* folder, also shown, is SolidWorks' way of relating back to the referenced geometry. This folder can be safely ignored.

Fig. 19–18. Derived part base feature name.

Any changes made to the original part file will propagate to the derived part. Likewise, any future assembly features added to the assembly that affect the component will propagate to the derived part. Features added to the derived part will not affect the component in the assembly. The file association is strictly unidirectional.

When using the Derive Component Part command, the key word to keep in mind is *component*. The new part will reference the component in the assembly. Therefore, the new part will inherit any features created at the assembly level. In other words, the new part is derived from (and externally references) the component.

The Cavity Command

With the Cavity command, it is possible to subtract the solid geometry of one or more components from another. In earlier releases of the software, the Cavity command played an important role in mold making. With the capabilities of the mold tools discussed in Chapter 14, this is no longer the case.

There are some interesting uses that can be found for the Cavity command. For instance, odd-shaped interferences can be eliminated by subtracting one part from another. Cradles or nests can be made for fixturing purposes. You may come up with other inventive uses for the command. How-To 19–3 takes you through the process of creating a cavity, regardless of what you decide to use it for.

HOW-TO 19-3: Using the Cavity Command

Fig. 19–19. Cavity PropertyManager.

To create a cavity, perform the following steps. Be aware that the Cavity command can be used to subtract the volume of one or more components from another. It does not necessarily have to be used strictly for the sake of creating a cavity. You must be in an assembly for this function to be available.

1. Right-click on the component that will contain the cavity, and select Edit Part. This is the component from which material will be removed.

2. Select Cavity from the Insert > Features menu. The Cavity PropertyManager will appear, as shown in Figure 19–19.

3. Specify the *Design component* with which to create the cavity. More than one component can be selected.

4. Specify a value for scaling the design component, if desired. The scale value should be a percentage (%). For no scaling, specify a value of zero (0).

5. If scaling the design component, specify what to use as the scaling origin by selecting from the drop-down list labeled *Scale about*.

6. Click on OK to create the cavity.

7. To return to editing the assembly, click on the Edit Part icon.

The *Scaling about* option allows for scaling about Component Centroids, Component Origins, Mold Base Origin, or Coordinate System. Different parts and materials cool differently, so use your best judgment. If the Coordinate System option is used, a user-defined coordinate system must be selected. Coordinate system creation was discussed in Chapter 7.

A nonuniform scale can be performed by unchecking the *Uniform scale* option. In this manner, the scale factor for all three axes can be determined. Regardless of whether a uniform scale is performed, scale percentages have a maximum value of plus or minus 50%. To specify a negative scale value, type in a minus sign.

Assembly Restructuring

In a perfect world, all subassemblies in an assembly would contain all components required, and no components would ever have to be moved from one subassembly to another. No top-level components would ever have to be moved into a subassembly. However, when reality intervenes, we sometimes need to jog components around after the assembly has been built.

Restructuring components in an assembly could not be easier. The process is extremely easy to perform, and subassemblies can be created or dissolved on-the-fly. There are limits, however. For example, restructuring components may cause explode steps or assembly features to be deleted. In such a situation, a warning is issued, and you will always have the option to back out.

Incidentally, the terms *reorganize* and *restructure* are both used in this chapter. They both apply, and both terms have similar meaning. *Restructuring* pertains more to dissolving or creating assemblies, whereas *reorganizing* pertains more to moving components between assemblies. At certain times, one term may be used over another to more accurately reflect the action being taken.

When restructuring components in an assembly, breaking up a subassembly into top-level components offers the least resistance and is the least prone to problems. Therefore, let's tackle that capability first. The process is known as dissolving a subassembly, and is performed via the right mouse button. How-To 19–4 takes you through this process.

HOW-TO 19-4: Dissolving Subassemblies

To dissolve a subassembly into top-level components, perform the following steps.

1. In the assembly FeatureManager, right-click on the subassembly you wish to dissolve.

2. Select Dissolve Sub-assembly.

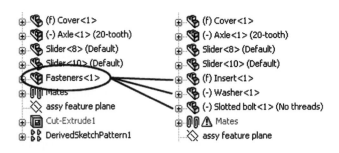

Fig. 19–20. Before and after; dissolving a subassembly.

The original assembly file (considered a subassembly in this case) will still exist on the hard drive after being dissolved. However, the subassembly's components will become top-level components. Figure 19–20 shows a before (left) and after (right) image of FeatureManager while dissolving a subassembly, which helps to explain this process. It should also be noted that any mates in the subassembly's *Mates* folder will transfer to the top-level assembly's *Mates* folder.

One common problem you should be aware of is how the fixed attribute of a component will move with the component during restructuring. It is common practice in assemblies to have one component that is fixed (locked) in position. During restructuring, this attribute is maintained. However, if one component in the top-level assembly is already fixed you would not want an additional component moving up from a subassembly to also be fixed. This is precisely what has transpired in Figure 19–20.

The problem is easily solved. Make it a point to mentally note which component is fixed in the subassembly prior to dissolving it. Once dissolved, right-click on the component and select Float, as opposed to Fixed. This will typically clear up any overdefining situations that arise from dissolving subassemblies.

Creating a subassembly on-the-fly is similar to dissolving a subassembly, only in reverse. When a subassembly is created, any mates belonging to the components being added to the subassembly will be transferred to the new subassembly's *Mates* folder. The new subassembly must be given a name. This is because SolidWorks is actually creating a new assembly file during the process. There are a few methods by which a subassembly can be created within a top-level assembly. One method is presented in How-To 19–5.

HOW-TO 19-5: Creating New Subassemblies

To create a subassembly from existing top-level components in an assembly, perform the following steps.

1. Select New Assembly from the Insert > Component menu.

2. If prompted, select a template to use for the new assembly, and then click on OK.

3. Type in a name for the new assembly.

4. Click on Save.

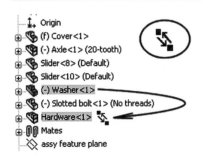

Fig. 19–21. Adding components to a newly inserted subassembly.

Following these steps, a new subassembly will be inserted at the bottom of FeatureManager. Then it is just a matter of inhabiting the new subassembly with components. This is done by dragging and dropping the desired components into the newly created subassembly. Look for the symbol shown in Figure 19–21 (enlarged for clarity) while dragging components over the new subassembly prior to letting go of the left mouse button.

The benefit of creating a new subassembly in the previously described manner is that the subassembly is created as an empty subassembly. It contains no components to start with, and you can decide at your convenience which components are to be added to the subassembly.

Another method of creating a subassembly is to use the right mouse button. By right-clicking over a component in FeatureManager, the option Form New Sub-assembly Here will appear (it may be necessary to expand the menu by clicking on the arrows at the bottom of the menu). If this option is selected, SolidWorks will ask you to give the new assembly a name, just as with the previous How-To. The part that was right-clicked on will occupy the new assembly. You can then add other components to the new assembly as you see fit.

✓ **TIP:** *By Ctrl-selecting multiple components, a new subassembly can be created that contains all of those components.*

The Fixed attribute is something to watch out for when creating new subassemblies, just as it is when dissolving them. It may be necessary to either fix a component in the top-level assembly or to fix a component in the new subassembly, depending on your situation.

Reorganize Components Command

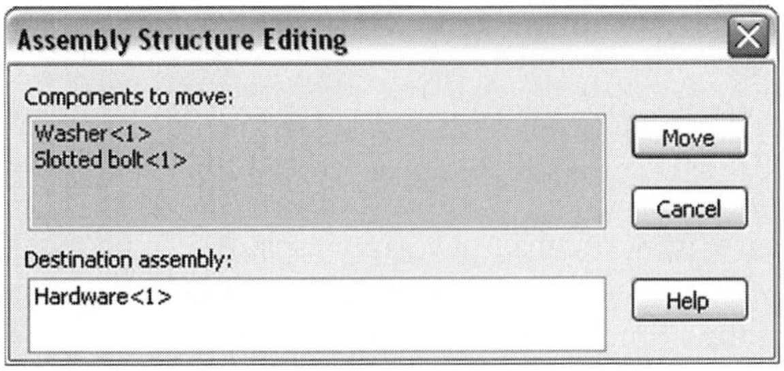

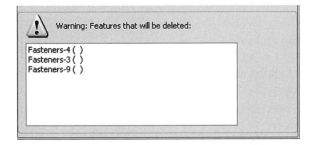

Fig. 19–22. Assembly Structure Editing window.

The Reorganize Components command, located in the Tools menu, opens the Assembly Structure Editing window (shown in Figure 19–22). The command itself is very straightforward. Select the component or components to be moved, select a destination assembly, and then click on the Move button. Remember that the salmon-colored list box area is the one that is active.

Although the Reorganize Components command is simple enough, certain problems can arise when reorganizing. This is true whether the Reorganize Components command is used or whether you simply drag and drop components into or out of subassemblies.

If a component is associated with an assembly feature, and you attempt to reorganize it, a warning message will appear. This is to warn you that if you proceed with reorganizing the component certain

Fig. 19–23. A reorganization warning.

assembly features will need to be deleted for the restructuring to take place. An example of this warning is shown in Figure 19–23.

If you decide to continue with the reorganizing process, any of the items listed in the window will be deleted. You will have to decide for yourself if the price is too high just to move a component to another assembly.

There are other factors that will make restructuring more difficult and prone to problems. External references often get in the way when trying to restructure assemblies. Equations can also have a tendency to fail, because component instance identification numbers often change during restructuring. Your safest bet is to perform any restructuring as early as possible to limit any unforeseen problems from cropping up.

Copying Files

When files contain external references, or when a multi-user environment is in place, the SolidWorks file structure can become an intimidating foe. Issues such as how to keep from overwriting other people's work arise. How can one file present in multiple assemblies be opened at the same time? Even the simple act of copying files becomes more complicated. This section addresses the complications involved with copying files. Sections to follow discuss how to successfully work in a multi-user environment.

Before continuing, let's first make sure some basic terms are fully understood. *References* (or *referenced files*) refers to files an assembly or drawing requires in order to be opened. For example, a drawing references a particular part file. *External references* refers to geometry in another model file that needs to be accessed in order to perform a proper rebuild of a part or assembly file. For example, a feature on a part may have an external reference to a feature on another part in an assembly. For the most part, the process of copying files can be broken down into the following three categories.

• Copying standalone files, such as parts

• Copying files with references, such as drawings or assemblies

• Copying files with external references, such as parts or assemblies

Because copying standalone files is not an issue, it will not be discussed here. It is a simple matter of using Windows Explorer to copy a file or files from one location to another, typically for purposes of backup.

Assemblies that reference individual part files, and drawings that reference part or assembly files, are considered files with references. The referenced files are required in order to open the assembly or drawing. For example, if an assembly containing 10 components were sent to a client, the client would not be able to open the assembly unless also sent the 10 part files corresponding to the components in the assembly.

As long as the assembly contains no components with external references, and sometimes even if it does, the Copy Files function located in the Find References command works just fine. How-To 19–6 takes you through the process of copying a file that includes references.

HOW-TO 19-6: Copying a File with Referenced Files

To copy a drawing or assembly file and include the files that drawing or assembly references, perform the following steps.

1. Select Find References from the File menu.

2. Click on the Copy Files button in the Search Results window, the lower portion of which is shown in Figure 19–24.

3. Specify whether or not you want to preserve the original directory structure. Clicking on No means that files scattered about can be saved in one location.

4. Browse for and specify a folder in which to save the copies of the referenced files.

Fig. 19–24. Search Results window.

5. Click on OK.

6. Click on Close.

This process will save a copy of the assembly or drawing and all files referenced by the assembly or drawing. Note that to open the copy you should close out of the current document first. Never try to open at the same time two files that have the same name.

If you click on Yes when asked to preserve the directory structure (step 3), the original directory structure of the assembly or drawing and its referenced files will be recreated below the directory where the assembly or drawing is being copied to. Be forewarned that SolidWorks has an easier time of finding referenced files if they are in the same directory as the assembly or drawing.

✓ **TIP:** *Placing all files associated with a particular project in the same directory is an ideal way of organizing files because moving or copying the project at a later time requires no special procedure. This is true even if external references are present.*

Copying Files with External References

This is where most people get into problems. If we use an assembly as an example, the problem will be easy to see. If an assembly containing a part with even so much as one seemingly innocuous external reference is opened, and a Save As is attempted on the assembly, the warning shown in Figure 19–25 appears. The warning appears as soon you click on the Save button in the Save As window.

Fig. 19–25. Warning regarding external references.

This message occurs for a reason most easily explained through example. Assume for a moment that *Part1* is externally referencing another component in *Assembly1*. Now imagine that a copy of *Assembly1* is going to be created. Using Save As, the name *Assembly2* is given to the new assembly. When the Save button is clicked on, the message shown in Figure 19–25 is displayed. Clicking on OK would result in creating an assembly named *Assembly2* containing the same components as *Assembly1*. However, *Part1* in *Assembly2* now contains an out-of-context external reference. Why?

The answer is that the feature in *Part1* contains an external reference that was created in the context of *Assembly1*. This same feature no longer recognizes the name of the new assembly. Because the feature was not created in the context of *Assembly2*, it fails. This results not in error symbols appearing in FeatureManager but as geometry that no longer updates correctly. Any external reference symbols (->) are now displayed with question marks that never go away (->?).

The solution to this problem is given in the same message that warns you about the problem. Specifically, use the References button found in the Save As window. How-To 19–7 takes you through the process of copying assemblies with external references.

HOW-TO 19-7: Copying Assemblies with External References

To copy an assembly file that contains components that include external references, perform the following steps.

1. In an assembly, select Save As from the File menu.

2. Using your existing Windows skills, navigate to where you wish to save the copy and type in a name. You do not have to check the Save As Copy option (explained later).

3. Click on the References button. This opens the Edit Referenced File Locations window, shown in Figure 19–26.

4. At a bare minimum, select any component that contains an external reference. You can also use the Select All button to select every file in the list.

5. Optionally, rename any selected component under the *New pathname* heading. This is accomplished with a slow double click.

6. Click on the Browse button and decide where to place the copied components. By default, this will be the same location as the main assembly being copied, which is the typical choice.

7. Click on OK. Note that the *New pathname* column has updated for the selected components.

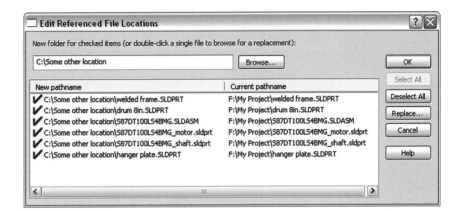

Fig. 19–26. Edit Referenced File Locations window.

8. Click on OK to close the Edit Referenced File Locations window.

9. Click on Save.

At this point, as long as the Save As Copy option was not checked (see step 2) you will be editing the copy. If Save As Copy was checked, the files will be on your hard drive someplace and you will be editing the original. This option is explained in more detail in material to follow.

To confirm that the external references are correct in the copied assembly, right-click on any of the components that contained an external reference and select List External References. You should find that the component references the new assembly, not the original. In other words, the external reference now refers to the copy. This is as it should be.

Save As Copy Option

Try this experiment. Open up any SolidWorks assembly, and then open up one of the assembly's components in its own window. With the component (part) window active, click on Save As in the File menu. The warning message shown in Figure 19–27 will appear.

The message is cryptic and has been known to cause a lot of confusion for new users. Let's use an example so that you may better under-

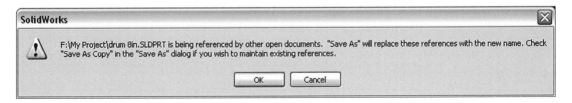

Fig. 19–27. The "Save As" warning message.

stand this message. Assume that the assembly you opened is named *Assembly1*. Assume that the component you opened in its own window is called *Part1*. Plain and simple, if you check the Save As Copy option (shown in Figure 19–28), *Assembly1* will continue to reference *Part1*. In other words, you will still see *Part1* listed in the assembly's FeatureManager.

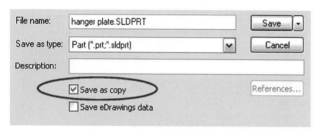

Fig. 19–28. Save As Copy option.

If the Save As Copy option is left unchecked, you could save a copy of the part file (let's use the name *Part2*) and the assembly would begin using the copy. In other words, *Assembly1* will now list *Part2* in FeatureManager where previously it was listing *Part1*. *Part1* would still reside on the hard drive, but would no longer be a component in *Assembly1*.

This scenario indicates the more complex behavior of the Save As Copy option. Be aware that this behavior occurs only if the referencing assembly (or drawing) is open. If the referencing document is not open, there is no warning message, and the Save As Copy option does not exhibit the same behavior. Rather, the option reverts to its more primitive and mundane behavior of deciding whether or not you will be editing the copy or the original after saving the copy.

Most Windows programs will contain a Save As Copy option in one form or another, and they all function similarly. First and foremost, however, it should be stated that whether the Save As Copy option is used or not a copy is still being saved. If the Save As Copy option is checked, a copy is placed on the hard drive and the original remains on screen. If not checked, the original is saved to the hard drive and the copy appears on screen.

Copying Drawings with Referenced Models

A common desire when creating design drawings is the capability to create a copy of a drawing along with a part (or assembly) that is in no way connected to the original drawing and part. This is often the case when minor changes have to be made to, for example, a part being sold to another customer and a second drawing is needed. The SolidWorks user does not want to spend the time to detail a completely new drawing from scratch, so a copy is created and the changes are applied to the copy.

Creative use of the Save As Copy option can produce a copy of a drawing and the model it references, neither of which are linked back to the original model or drawing. However, there is an easier method. How-To 19–8 takes you through the process of saving copies of drawings with models.

How-To 19-8: Save Copies of Drawings with Models

To save a copy of a detail drawing along with the model it references, perform the following steps. The copy of the drawing will reference the copy of the model, but neither the copied model nor the drawing will be related in any way to the original drawing or model.

1. Open the drawing to be copied.

2. Click on Save As in the File menu.

3. Type in a name for the new drawing.

4. Click on the References button.

5. Click on Select All to place checks in front of all of the model files. If the drawing is of a part, there will probably be just one file listed.

6. Click on the Browse button and choose a destination for the copies of the model files.

7. Click on OK to close the Browse window.

8. Click on OK again to close the Edit Referenced File Locations window.

9. Click on the Save button.

If you were to click on the Find References command in the File menu, you should find that the drawing is now referencing the copied model file (or files, if the model is an assembly). The newly created drawing or model can now be edited with no fear of modifying the original.

Occasionally it is necessary to change which model a drawing is referencing. For instance, perhaps a model has gone through a series of design changes and multiple copies of the model were saved. A design is finally settled upon, and the original drawing should reflect the final design.

How-To 19–9 takes you through the process of making a drawing reference another model. The other model should be some version of the original, obviously. If a completely different model is referenced, none of the annotations will be valid and the drawing will essentially be ruined.

How-To 19-9: Altering Drawing References

To change which model a drawing is referencing, perform the following steps.

1. Select Open from the File menu, or click on the Open icon.

2. Select the drawing, but do not click on the Open button yet.

3. Click on the References button.

4. Double click on the name of the model.

5. Select the new model the drawing should reference and click on Open.

6. Click on OK to close the Edit Referenced File Location window.

7. Click on Open to open the drawing.

8. Select Find References from the File menu to ensure that the drawing is now referencing the copied model rather than the original.

9. Save the drawing.

Knowing what you now know about copying drawings, it should be easy to save yourself quite a bit of time and effort. Don't make the mistake of creating a copy of a drawing without copying the referenced files. All that will give you are two copies of a drawing that are both linked to the same model.

Multiple User Environments

Opening files over a network is not recommended unless your network can handle the load. This is not to say that you should not move files over the network. It just is not advisable to open files that do not reside on your local computer. This holds true for many applications and is not isolated to SolidWorks. However, because SolidWorks files often contain references to other files, the resultant network traffic can be very high.

Even most product data management (PDM) software will copy files to a local machine prior to opening them in SolidWorks. The PDM software is written this way in order to limit the performance bottlenecks inherent in networks. Internal computer data transfers are much quicker than any data transfers that occur over a network, especially considering how SolidWorks files are continually communicating with each other.

Real-world circumstances dictate that under present situations your company may be forced to work on files located on the server. Any network running SolidWorks should be either a 100-Mbps (Megabits per second) or 1-Gbps network. A 10-Mbps network simply will not suffice for running SolidWorks in a multi-user environment. Let's discover what happens when multiple users are working in a networked environment.

Probably one of the most common problems associated with multiple users working on the same project simultaneously is the issue of read-only file access. Using two hypothetical people (Ray and Frank), the situation plays out as follows. Frank opens a file named *Part1*. Ray opens an assembly that contains *Part1*. When the assembly gets to the point where it loads

Part1, the file gets loaded without Ray ever knowing it is in use by Frank. As a matter of fact, Ray may never know *Part1* is being edited by another person unless he attempts to edit that particular component. When that happens, the message shown in Figure 19–29 will appear.

Fig. 19–29. Write access warning.

The name of the file in the message in Figure 19–29 is something other than *Part1*, but that's not what is important. The message states that you must obtain write access to the document to complete the operation, with the operation being that you want to edit the document. So, what do you do?

Do not attempt to edit a part someone else is already working on. If it is your responsibility to be completing this particular part, it would be a good idea to find out why someone else has your file open and is making changes to it. There are some legitimate reasons two different people may need to have access to the same part. The point of the matter is that you cannot have more than one person editing a part, no matter what the program. Communication is the key in this case.

Opening a File Already in Use

Another common warning message occurs when trying to directly open a file already in use (open) by another. The same warning can occur if opening a part or assembly that is a component in another assembly already open by another person. The message that displays on screen is shown in Figure 19–30. The situation is very straightforward and does not require much elaboration. Two people simply cannot be working on the same file at the same time.

Fig. 19–30. The file is already in use.

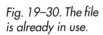

Usually the wrong thing to do in this situation is to click on Yes. Do you really want to make a copy? Bear in mind that a copy will not be linked to the original, nor will it be referenced by any assembly the original file was in. However, it will be referenced if you save the copy with the same name as the original and in the same location, in which case you will overwrite any changes made to the original file. This would not be a good situation!

Nonetheless, let's say you do click on Yes and open a copy. Some changes are made, and now it is time to save the file. Figure 19–31 shows what you will see if you simply click on the Save icon.

Fig. 19–31. A warning about overwriting a file.

This message is pretty clear because it tells it like it is. If you try to save the file with the same name and the original file is no longer being edited, the save operation will work but the original will be overwritten with your copy. In this case, you run the risk of incurring the wrath of whoever modified the original, because you just destroyed all of their modifications.

If you attempt to save the copy with the same name as the original and the original file is still open (in use), the save operation will fail and Solid-Works will inform you of that fact. Whether the original file is still in use or not, you will always be given one more warning whenever an attempt is made to overwrite another file. In either case, this is almost always a lose-lose situation.

When opening a copy, you would typically have no future intention of overwriting another person's work. In fact, the copy should be saved in some other project folder with a different name. After all, it is a copy and is probably going to get used somewhere else. When saving the copy, use the Save As command instead of Save and the warning message displayed in Figure 19–31 will not be shown.

Opening Referenced Files Read-Only

You are in a multi-user environment, everyone is working on related projects and no one knows from day to day what files they will be working on next. Sound familiar? To describe a situation a little more precisely (once again using a hypothetical SolidWorks user named Frank): Frank will be working on an assembly, including adding components to it and mating them in place. Other users will be making modifications to components already added to Frank's assembly. How can everyone be working on these interrelated files without worrying about a warning message cropping up every time a file is opened or edited?

There is an option that should help. Frank can turn on an option found in the External References section of System Options (Tools > Options) titled *Open referenced documents with read-only access*. This option is shown in Figure 19–32. Even though the title of the System Options section is External References, the *Open referenced documents with read-only access* option does indeed refer to referenced files *and* externally referenced files.

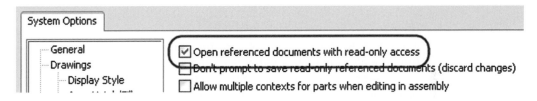

Fig. 19–32. Open referenced documents with read-only access *option.*

The *Open referenced documents with read-only access* option can be useful in a number of ways. Because all referenced files in an assembly are loaded as read-only, no problems arise if another user attempts to open one of the components present in the open assembly. The user opening the component part file will see no warning messages and may not even be aware that another user has an assembly open that contains that file.

If you were working on the assembly and attempted to edit one of the components in the assembly, you would be faced with the message shown in Figure 19–33. Clicking on Yes and accessing the file with write access means that you can then modify the component as you see fit. If another user happened to be working on that same component unbeknownst to you, the "You must obtain write access..." message (shown in Figure 19–29) would appear instead of the message shown in Figure 19–33.

Fig. 19–33. Asking if you want to enable write access.

Multi-user Environment Settings

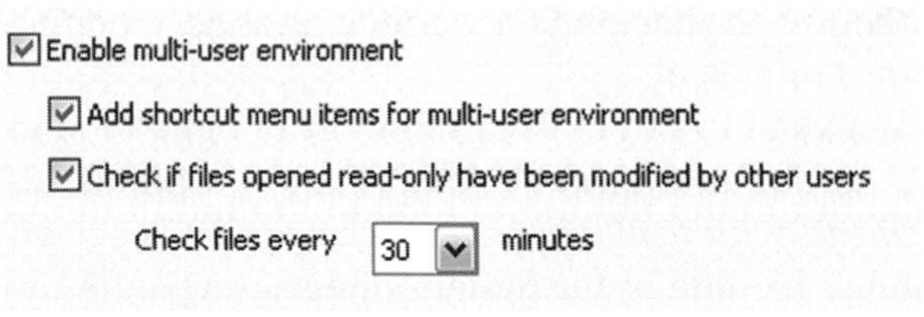

☑ Enable multi-user environment

   ☑ Add shortcut menu items for multi-user environment

   ☑ Check if files opened read-only have been modified by other users

   Check files every [30 ▾] minutes

Fig. 19–34. Collaboration options.

In the system options (Tools > Options), there is a section titled Collaboration. The settings contained in the Collaboration section are shown in Figure 19–34. Checking *Enable multi-user environment* makes available the other options below it.

Checking the *Add shortcut menu items for multi-user environment* option serves to add a few extra commands to the File menu and the right mouse button menu. One of the commands added is Make Read-only. For example, selecting a component and selecting Make Read-only from the File menu will remove write access from the file and allow another user to open the file for editing.

Another command that is added is Get Write Access. Right-clicking on a component and selecting this command will give the user write access to the file, assuming some other person does not already have the file open. If the *Open referenced documents with read-only access* option is utilized, everybody else can easily gain access to the files open in other people's assemblies. If a person with an assembly open decides they need to edit a particular component, it is a simple matter to gain write access to that component for editing purposes.

The *Check if files opened read-only have been modified by other users* option checks for two occurrences. It will check to see if a file you have opened as read-only is saved by another user. This would make your file out of date, and you would probably want to reload the file (described in material to follow), thereby allowing you to see what changes were made.

Another check that is made is whether another user has relinquished write access to a file you currently have loaded read-only. This would give you the ability to gain write access and make changes to the file should you choose to do so.

Fig. 19–35. Tool tip warning of a change.

Checks are made every so many minutes, depending on the length of time selected from the drop-down list (visible in Figure 19–34). If a change is detected, a tool tip will appear in the bottom right-hand corner of the screen, informing you of the change. This tool tip is shown in Figure 19–35. Clicking on the small icon shown in the image will call up the Reload window, which is described in material to follow.

Editing Read-only Documents

It is typically best practice to gain write access prior to editing a file. However, it is possible to make certain changes without ever gaining write access. Early on in this book you discovered that double clicking on a feature gains access to that feature's dimensions. In this way, dimensional changes can be made to components while you still have read-only access to the components. This is not recommended.

Double clicking on a dimension to modify it usually brings up the warning message shown in Figure 19–36. However, the warning can be disabled, so it might not be seen. There is no way to "reenable" the message, but SolidWorks will enable the message the next time the software is run.

*Fig. 19–36.
Read-only
document
warning.*

If a change is made to a read-only document, and the assembly is saved, you will be presented with the warning shown in Figure 19–31. At that point, clicking on Cancel will abort the operation, and clicking on OK will allow for saving the read-only document with a different name. Considering that clicking on OK results in the Save As command being executed, another warning message is displayed. The "Save As" warning message was discussed previously (see "Save As Copy Option" section).

Save yourself from having to click through the warning messages by getting write access to files first. The section on reloading documents in material to follow shows you how to accomplish this.

Special Note on Read-only Attributes

When files are opened read-only in SolidWorks, the Windows operating system does not see the document as read-only. Internal to the SolidWorks software, *read-only* is a term that dictates whether or not that file can be overwritten. A person with read-only access to a file cannot theoretically make changes to that file. Yet, under many circumstances a person with read-only access to a SolidWorks document can easily gain access to that document and change it any way they see fit.

When a file is marked read-only by the operating system (not SolidWorks), there is an actual attribute that gets set on that file that allows only those with proper permission to edit the file. We will not go into depth regarding file permissions and operating system security, as these topics are outside the scope of this book. However, it is important to understand that operating system read-only file attributes and opening SolidWorks files as read-only are two different things.

The information presented in this section is really for those "in the know." System administrators need to understand how SolidWorks handles situations in which associated documents are accessed by multiple users. Be aware that opening read-only files in SolidWorks is more of a precautionary function and not a function of security.

Reloading Documents

Put yourself back in the shoes of the user with an open assembly where the referenced files have been loaded read-only. If other individuals are editing some of the files present in the assembly, you may very well wish to see what changes have been made to those files. For that matter, it may be necessary to change the read or write status on certain documents, thereby giving another worker (or yourself) the ability to make changes to a file.

One method of changing the read/write status of a file is to take advantage of the multi-user environment settings discussed previously. Use the *Make read-only* command in the File menu, or access the *Get write access* option by right-clicking on a component. In addition, multiple components can be reloaded simultaneously using the Reload command. How-To 19–10 takes you through the process of reloading components, regardless of their current read/write status.

HOW-TO 19-10: Reloading Components

To reload components in an assembly, perform the following steps. Use the Reload command if the read/write status of files needs to be changed, or if it is necessary to see changes made to a document in your assembly because another worker is editing and has made changes to that document.

1. Select Reload from the File menu. The Reload window, shown in Figure 19–37, will appear.

Fig. 19–37.
Reload window.

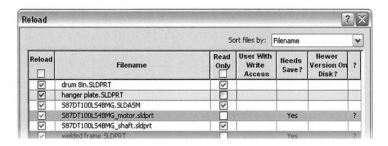

Reload	Filename	Read Only	User With Write Access	Needs Save?	Newer Version On Disk?	?
☐		☐				
☑	drum 8in.SLDPRT	☑				
☑	hanger plate.SLDPRT	☐				
☑	S87DT100LS4BMG.SLDASM	☑				
☑	S87DT100LS4BMG_motor.sldprt	☐		Yes		?
☑	S87DT100LS4BMG_shaft.sldprt	☑				
☑	welded frame.SLDPRT	☐		Yes		?

2. Click on the Show References button to see all of the files in the assembly and to choose which of them should be reloaded.

3. Select the documents that should be reloaded from the Reload column. You must reload a document to change its read/write status.

4. Change the Read-only status of files being reloaded, as required. Clearing a Read-only checkbox gains write access to that document.

5. Click on OK to finish reloading documents.

There is excellent information that can be obtained from the Reload window. The *User With Write Access* column will list the person who currently has access to a document. You could then communicate with that person and coordinate on what needs to occur next, such as having that person close a file or give up write access. Make a document read-only in order to give another person the ability to gain write access on that document.

The *Needs Save?* column informs you if the file requires a save. Think twice about reloading any documents that should be saved, as reloading them will destroy any changes made to that file since it was last saved. Conversely, the *Newer Version On Disk?* column is stating that a document probably should be reloaded if the column contains the word *Yes*.

After clicking on the Show References button, the documents can be sorted using the drop-down list that appears (see Figure 19–37). Color codes in the window help to show which rows have had some setting changed (green), such as changing the state of one of the checkboxes. If a row turns red, there is a potential for losing data, and you should not reload that file unless you are absolutely certain the changes can be discarded.

Right-clicking on components in an assembly and selecting Reload will open the Reload window shown in Figure 19–37, but only for that component. In this manner, a single component can easily be reloaded.

✓ **TIP:** *If a component is edited in its own window, and that component is in an open assembly, changes made to the component appear in the assembly. This is true even if the component is closed and the changes are discarded. In this situation, reload the component to show the component in its original state in the assembly.*

Directory Structure and Search Order

Search criteria play a very important role in how a company (or individual) implements directory structure. There are good ways in which to set up a directory structure, and then there are methods that can result in extremely inflexible file management options. Let's look at the recommended options first.

Every company is going to have its own file-naming conventions and preferences with respect to the directory structure of where files are saved. When possible, one of the best ways to categorize a directory structure would be to use the project-naming convention. In other words, create folders named after projects and save all documents relating to that project in that folder.

Project names might reflect the name of a customer, or perhaps a model number of a project, or some other criterion. What is important is that all assemblies, parts, and drawings reside in the same directory. With large assemblies, it might be prudent to break up the main project folder to include subdirectories for subassemblies and their related components and drawings.

The reason the project-naming convention works so well has to do with how SolidWorks searches for referenced files. One of the very first places it searches is in the same directory as the file just opened. This means that if an assembly is opened SolidWorks will look in the same folder for the components as well. The following list is the search order SolidWorks uses when it tries to find a referenced file.

- Paths specified in the File Locations section of System Options, if the *Search file locations for external references* checkbox is selected (see material to follow)

- The last path you specified to open a document

- The last path the system used to open a document (in the case of the system opening a referenced document last)

- The path where the referenced document was located when the parent document was last saved

- The path where the referenced document was located when the parent document was last saved *with the original disk drive designation*

Finally, if a referenced file still is not found you are given the option to browse for it. Once the referenced file is found, all updated reference paths in the parent document are saved when the parent document is saved.

The previous listing is a simplified version of the actual search criteria used by SolidWorks. The true search criteria consist of over a dozen steps and include recursive path searching. It just is not necessary to go into that level of detail in this book. The listing presented here is enough to get the point across, which is that SolidWorks will probably not find files if they are stored in different folders and are then moved.

The following is a hypothetical situation that might help explain the importance of directory structure. An assembly or drawing looks in a particular place for the files it references. The referenced files are moved. The next time the assembly or drawing is opened it cannot find the referenced files and asks if you would like to browse for them. If the assembly or drawing were kept with its referenced files in the first place, those files could be moved anywhere and SolidWorks would never have a problem finding the referenced files.

The scenario just mentioned refers to step 2 in the search path listing. For example, if an assembly were opened from the path *C:\My Project*, Solid-Works would look in *C:\My Project* for all referenced files. This makes life much easier for the SolidWorks user.

Most SolidWorks users do not use the option mentioned in the first search path. If you do wish to set up a specific search path that SolidWorks will always use to look for referenced files, use the method outlined in Chapter 3 (see How-To 3–6). Consider the implications, though. Unless all of your referenced files are in particular locations and all of the referencing files are someplace else, there is no advantage to setting up a particular search path.

Probably the worst directory structure that can be employed is to save parts, assemblies, and drawings all in their own separate locations. This makes moving the files at a later time extremely difficult. Even relocating all of the files to a different server for archival reasons (or perhaps due to a server upgrade) becomes very problematic. The simple answer to this is *don't do it*!

System Performance

When working with large assemblies, or even complex part files, performance is always a factor. Not all of us (or all companies) can afford the most expensive, state-of-the-art computer systems with huge amounts of memory. Computer hardware has come down in price drastically, and will continue to do so. For the time being, though, some of us have to make due with what we have.

There are a number of ways in which part or assembly performance can be optimized, above and beyond putting in a requisition for faster hardware. In this section, you will take a look at some methods that can be used to speed things up a bit. Some of these methods are directly related to graphics card and video performance. Other options relate to settings that can be changed in the SolidWorks software.

Large Assembly Mode

The primary way in which performance can be increased when working with assemblies is known as large assembly mode. This mode of operation is highly customizable, and can be triggered when a certain user-definable threshold is crossed. The threshold is defined as the number of components you feel makes an assembly a "large" assembly. This value will differ, depending on the complexity of the individual components in an assembly or the computer's hardware.

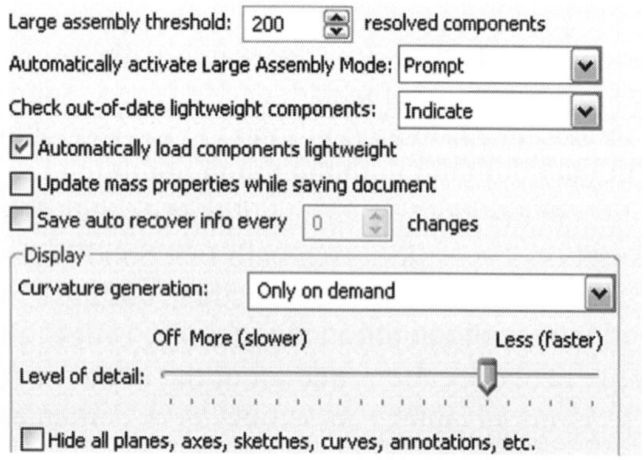

Fig. 19–38. A portion of the Large Assembly Mode section.

Large assembly mode is essentially a list of options that appear scattered throughout the system options. Certain options will change when large assembly mode becomes activated. The section titled Large Assembly Mode, found in System Options (Tools > Options), is dedicated to this mode of operation. A portion of these settings is shown in Figure 19–38. The threshold setting is at the top of the figure, and happens to be set to 200.

A portion of the Large Assembly Mode section is devoted to drawings, as shown in Figure 19–39. The settings contained therein pertain to drawings of assemblies whose component count is above the large assembly mode threshold.

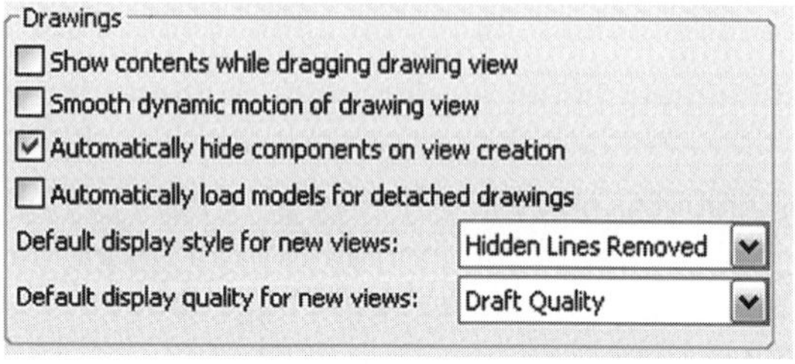

Fig. 19–39. Drawing settings for large assembly mode.

Nearly all of the settings found in the Large Assembly Mode section are duplicated in other areas of System Options. When large assembly mode is activated, these options cannot be accessed within System Options unless it is via the Large Assembly Mode section. This can be a point of confusion for new users, and even for some seasoned veterans. If certain options you feel should be accessible are grayed out, it is probably because large assembly mode has been triggered and is currently activated.

Large assembly mode can be triggered automatically, or turned on and off manually. How-To 19-11 takes you through the process of performing both of these tasks. It will also show you how to set the threshold level for large assembly mode.

How-To 19-11: Implementing Large Assembly Mode

To set the threshold for large assembly mode or to manually turn this mode of operation on or off, perform the following steps. Note that it is best to implement large assembly mode automatically rather than manually. This is due to lightweight components playing a large role in this mode of operation. If the threshold is triggered automatically, components can be loaded lightweight. If large assembly mode is activated after an assembly is already loaded, the components will not revert to lightweight.

1. Select Options from the Tools menu.

2. Select the Large Assembly Mode section in the System Options tab.

3. Set the *Large assembly threshold* option to what you feel would be an appropriate value.

4. Set Automatically activate Large Assembly Mode to either Prompt or Always, depending on your preference.

5. Check the option for *Automatically load parts lightweight*. This will ensure that lightweight mode is used anytime an assembly is opened that contains greater than the number of components specified for the threshold.

6. Click on OK to close the System Options window.

7. Optional: to manually activate or deactivate large assembly mode, check or uncheck Large Assembly Mode in the Tools menu.

If you were to open an assembly that contains more components than specified in the threshold setting, the assembly would load using lightweight components and large assembly mode would be activated. If an assembly was already open prior to enabling large assembly mode and you wish to activate it manually, use step 7. Alternatively, Large Assembly Mode can be unchecked within the Tools menu as well, if you wish to turn it off manually.

It is usually necessary to fine tune the threshold after a bit of experimentation. For instance, you may find that the assembly performance is better than expected with 200 components, so the threshold is set to 300 instead. There are a lot of factors that affect system performance, so use a threshold that works best for you.

➠ **NOTE:** *If an assembly is saved with large assembly mode activated, this mode will be active when the assembly is next opened, whether or not the trigger threshold is reached.*

SW 2006: A number of large assembly mode options were not commonly used, and have been removed. The Large Assembly Mode section of System Options has been removed, and the few remaining options associated with large assembly mode have been moved to the Assemblies section of the System Options. Options previously available for tweaking by the user are now set automatically when the large assembly mode threshold is reached.

Performance Options

It is not the intention of this book to serve as a reference guide for every option in the System Options window. Therefore, the options in the Large Assembly Mode section will not be listed. The SolidWorks on-line help serves very well as a reference guide to the individual options found in the Large Assembly Mode section and System Options in general.

The Large Assembly Mode section itself can be used as a listing for system options that will affect system performance. As a matter of fact, most of the options found in the Performance section of System Options are duplicated in the Large Assembly Mode section. There are some additional options or actions that can be taken to increase system performance that are not in the Large Assembly Mode section, and that is what this section is devoted to.

It should be noted that the options or actions discussed in the sections that follow will affect both part and assembly performance unless otherwise noted. Depending on your system hardware and your work habits, the following material may have a varying degree of impact on overall system performance.

General Transparency Settings

There are a few options that affect transparency in one form or another that can reduce graphics performance. Most 3D graphics cards contain enhancements that permit transparency to be accelerated, thereby diminishing any negative results of enabling options associated with transparency. You will have to experiment with these settings to see if they adversely affect system performance; in particular, performance regarding model rotation.

The *Display shaded planes* option allows planes to be displayed as translucent objects. This option is located in the Display/Selection section of System Options. When shaded planes are used, the fronts of planes appear green, and the backs of planes appear red. When shaded planes are not used, planes appear with gray borders only. Shaded planes can slow down model rotation on the screen, depending on the computer's graphics card.

Another transparency-related option is titled *Assembly transparency for in context edit*. This option is also located in the Display/Selection section of System Options. It only affects assemblies and only takes place when editing components in the context of an assembly. Change this setting to *Opaque assembly* if you find transparency slows down graphics performance on your system to unacceptable levels.

Verification on Rebuild

The *Verification on rebuild* option is found in the Performance section of System Options. It is an often-misunderstood option. Turning it on does not mean SolidWorks will ask you for verification prior to performing a rebuild. Rather, it means that SolidWorks will perform extra error checking during a rebuild.

It is rarely necessary to turn *Verification on rebuild* on. It will increase the time it takes to perform rebuilds. It is suggested that you only enable this option while working with problem geometry and suspect there may be errors in the model.

Image Quality

The image quality settings can make a big difference in display performance and even load times for large assemblies. Image quality settings are found in the Image Quality section of the Document Properties tab (Tools > Options). To review the various settings available in this section of the Document Properties, see "Image Quality" in Chapter 3.

When saving a part in shaded mode, the tessellation quality of the part will make a difference in the size of the file. Tessellation has to do with the number of polygons used to create the shaded image. It stands to reason that if a file contains more polygons the file will be larger. This is not always significant when working with one part file. However, parts with many curved faces are affected to a greater degree when changing the display quality.

Bear in mind that if the *Apply to all referenced part documents* option is used, changing the shaded display quality setting will affect every component in the assembly at once. This shaded display quality setting will be saved with each part file used in the assembly.

Add-ins

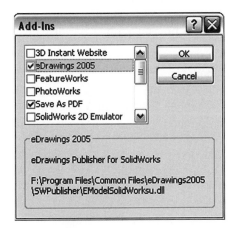

Fig. 19–40. Add-Ins window.

There are many add-in (also called add-on) programs that can be run with SolidWorks. Examples of add-in programs are the PhotoWorks rendering package and SolidWorks Toolbox, just to name two. These add-in programs often add a great deal of functionality to the SolidWorks program, but they do not necessarily need to be turned on if they are not being used.

The Add-Ins window, shown in Figure 19–40, is accessed by clicking on Add-Ins in the Tools menu. Add-in software should be disabled if it is not in use. This reduces the system resources required by the software. It just stands to reason that if less software is loaded or is running at the same time more processor power will be left for SolidWorks to take advantage of. To make a long story short: if you are not using it, turn it off.

Suppression and Configurations

Suppressing features or components and part or assembly configurations are topics that go hand in hand. These previously examined topics will not be reiterated here, but with regard to assembly performance you should be aware that component suppression and configurations can play a significant role.

When working with very large assemblies, it may not always be necessary to show every component in the assembly while working with the assembly file. If this is the case, create a configuration in which noncritical components can be suppressed. If you would rather have the advantage of being able to see certain components but do not want the memory overhead, make it a point to work with lightweight components as much as possible (see "Component States" in Chapter 13).

With regard to configurations, they can be quite advantageous when used with large assemblies. Take the case of a part with many intricate or complex features. Such is the case with the *Knurled Handle* component, shown in Figure 19–41. This part contains true helical threads, knurling, raised text, and a smattering of cosmetic fillets.

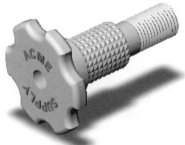

Adding a configuration and then suppressing features that do not need to be shown in the assembly result in the part shown in Figure 19–42. The part rebuilds much more quickly because the complex features have been suppressed.

Creating simplified configurations of parts that will be inserted into large assemblies can have a cumulative effect. Assembly rebuild times will be significantly shortened, as will assembly load times. Graphics performance will also improve.

Fig. 19–41. Knurled Handle.

Loading Simplified Components

If simplified versions of components have been created, but have not necessarily been utilized in an assembly, there is an easy way to force an assembly to load simplified versions of all components. To accomplish this task, the Advanced option must be selected when opening an assembly. This option is shown in Figure 19–43.

Fig. 19–42. The knob after suppressing noncritical features.

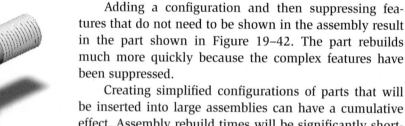

Fig. 19–43. Advanced option in the Open window.

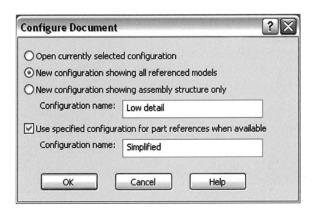

Fig. 19–44. Configure Document window.

When the Open button is clicked on, the Configure Document window (shown in Figure 19–44) appears. In the image shown, SolidWorks is being told that a new configuration should be created at the top-level assembly named *Low detail*. If there is a configuration named *Simplified* in any of the components, this configuration should be used. Incidentally, the option *New configuration showing assembly structure only* will create a new configuration in which every component in the assembly is suppressed.

The Advanced option has obvious advantages, but to really take advantage of it each component should have a configuration of the same name. *Simplified* was used in the example, but any name will do. The key is to be consistent. Whatever configuration name is used, make it a point to create a configuration using that name for every part you create. Suppress any unnecessary features in that configuration that need not be shown in the top-level assembly.

System Maintenance

There is a lot to be said for general system maintenance when it comes to increasing assembly performance in SolidWorks. Do not underestimate the power of defragmentation. Most Windows operating systems now contain a file defragmenter, which takes files on your hard drive and keeps them from being scattered over different physical locations. Instead, each file is stored in contiguous sections on the hard drive. This greatly reduces the time it takes for an assembly to locate and load individual part files.

Envelopes

An envelope, to put it simply, is a way of selecting components in an assembly. The envelope is a volumetric area that in actuality is really nothing more than a part. The area the envelope defines can be used to select components. For example, all components falling completely within the envelope can be selected. What happens with the selected components is up to you. Perhaps the components need to be hidden or suppressed, or perhaps some other action needs to be carried out.

Where envelopes are most useful is when working with large assemblies. Envelopes simplify the selection process by making it so that each component does not need to be separately selected, whether from the work area or FeatureManager. Hundreds of components can be selected with ease.

To use an envelope, one must first be created. An existing part can be inserted as an envelope, or a new envelope can be created in much the same fashion as a new part can be added to an assembly. To insert an existing envelope into an assembly, select From File in the Insert > Envelope menu. The envelope can then be mated into position using the skills you learned in Chapter 13.

Envelopes are not something you typically have lying around. Usually, it will be necessary to create an envelope on-the-fly, within the context of the assembly. This is because the size of the volume can more easily be decided on when the rest of the assembly is being viewed at the same time. How-To 19–12 takes you through this second method.

HOW-TO 19-12: Inserting a New Envelope

To insert a new envelope into an assembly, perform the following steps. The process is extremely similar to inserting a new part file, and in fact involves the same steps. Only the menu picks are different. If you attempt to insert a new envelope and are experiencing difficulties, see How-To 15–2.

1. Select Insert > Envelope > New.

2. Type in a name for the new part file.

3. Click on Save.

4. Select a plane or planar face to begin sketching on.

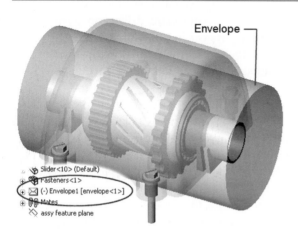

Envelope

Slider<10> (Default)
Fasteners<1>
(-) Envelope1 [envelope<1>]
Mates
assy feature plane

Fig. 19–45. Example of an envelope.

Once these steps have been completed, it will be necessary to create a sketch and subsequently create the feature that will define the volume of the envelope. Envelopes often contain only one simple feature, such as an extrusion, but this is by no means a requirement. An example of an envelope is shown in Figure 19–45. It is the cylindrical part that encompasses the axle component and surrounding area.

The envelope is automatically given the name *Envelope1*, as can be seen in the inset in Figure 19–45. The name of the part file is shown in brackets, which also happens to be *envelope*. The part file is nothing more than an extruded circle, and thus the act of creating a part file to use as an envelope can be a no-brainer. It simply has to be the right size for whatever it is that needs to be selected within the assembly. Size and position are most important when working with envelopes.

Envelope Selection Techniques

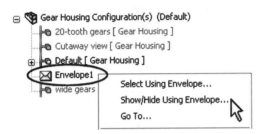

Fig. 19–46. An envelope displayed in ConfigurationManager.

There are a few simple methods that can be employed in order to select components with an envelope. The entire process is quite simple. Once an envelope has been created within your assembly, take a look at ConfigurationManager. You will notice the new envelope listed there. An example of this is shown in Figure 19–46, along with the menu selections available when right-clicking on the envelope.

One command found when right-clicking on the envelope in ConfigurationManager is titled, logically enough, *Select using envelope*. This opens the Apply Envelope window, shown in Figure 19–47. Most of the options speak for themselves. The *Select components in top assembly only* option will select an entire subassembly if any of the subassembly's components falls within the selection criteria.

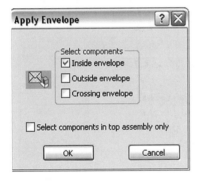

Fig. 19–47. Apply Envelope window.

The other command found in the right mouse button menu directly beneath *Select using envelope* is titled *Show/hide using envelope*. This opens another window with the same title of Apply Envelope, but with a slightly different purpose in mind. The window is shown in Figure 19–48.

Whereas the *Select using envelope* command is used for selection purposes, *Show/hide using envelope* is used strictly for showing or hiding components. The interface needs no further elaboration, as it is self-explanatory. Simply check the desired criterion, specify the desired visibility setting, and click on OK.

Select using envelope can be used for functions other than hiding or showing components. Use it to select components for any reason, such as obtaining mass properties, suppressing components, or whatever the case. The section that follows explores ad-

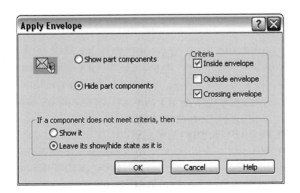

Fig. 19–48. Another variation of the Apply Envelope window.

vanced search functionality, which goes beyond envelopes and includes custom properties.

Advanced Selection Techniques

Have you ever thought that it would be nice to search for every component in an assembly that is made of a specific material? How about searching for any component that costs more than $100? Better yet, let's perform a search for any component made of copper that costs more than $100 and weighs more than 8 pounds. All of these hypothetical searches can be performed with the Advanced Select command.

Clicking on Advanced Select in the Tools menu opens the Advanced Component Selection window (shown in Figure 19–49). There is a variety of selection criteria that can be added to the Advanced Component Selection window. For example, a property can be selected from the Property drop-down list, or a user-defined property can be entered. The property of an object directly relates to the custom properties of a file, accessible through the File menu (see "Custom Properties" in Chapter 17).

Fig. 19–49. Advanced Component Selection window.

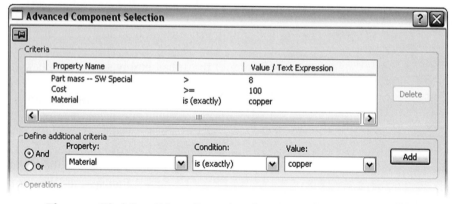

The area titled Condition allows for plugging in logical operators. For instance, if you are searching for a file that contains a property for *Date completed*, and the file being searched for must match a specific date, the condition should be set to *is (exactly)*. The Value would be that of the date being searched for. The value is something that would get typed in by the person performing the search.

SolidWorks has some special search properties it adds to the Property list. These special properties are *Part mass, Part volume, Configuration name,* and *Document name*. The *Document name* would be the actual name of the file. Special properties can be used as a search criterion without adding them to a file's custom properties. When a search is performed using any property other than one of the SolidWorks special properties, the file being searched for must contain relevant data in its properties. If not, the search will turn up nothing. How-To 19–13 takes you through the process of adding selection criteria.

HOW-TO 19-13: Adding Selection Criteria

To add a selection criterion to the Criteria listing, perform the following steps. At least one criterion must be added.

1. Select a property from the Property list, or type in a name of a property to search for.

2. Select a condition from the Condition list.

3. Select or type in a value in the section named Value, if required.

4. Click on the Add button.

5. If additional criteria are required, select either the And or the Or option and then repeat steps 1 through 4.

6. Click on Apply.

7. Click on Close to accept the selection.

Selection criteria can be saved as a file once created. This same file can then be called up at a later time, as when performing another advanced selection in another assembly. When a selection criterion is saved as a file, a SolidWorks property criterion is created. This type of file has an *.sqy* file extension.

After applying the selection criteria to the assembly and clicking on the Close button, you will want to carry out whatever command you had in mind, such as Hide, Show, Suppress, Delete, *Check the mass properties of*, or anything else you have learned how to accomplish to this point.

Summary

This chapter taught you how to use assemblies more efficiently. A number of settings were explored that should help you maximize computer power when working with large assemblies or complex part files. When using lightweight components, load times for assemblies can be significantly reduced. Components that have been loaded lightweight only load the data necessary to view the component.

Large assembly mode is a method of turning off certain resource-draining options when large assemblies are loaded. What is considered to be a large assembly is up to the user. By setting the large assembly threshold, you can determine the number of components allowable in an assembly before large assembly mode kicks in. Large assembly mode can be manually enabled or disabled through the Tools menu.

Subassemblies can be created on-the-fly while in an assembly. They can also be dissolved. Components can be dragged from one assembly to another within FeatureManager. There is a great deal of flexibility that exists, which allows for the reorganization or restructuring of components and subassemblies. Components with external references, however, often cannot be restructured without deleting the references. SolidWorks will always warn you if problems are encountered when performing any type of restructuring.

An important topic discussed in this chapter had to do with copying assemblies and components that exist within assemblies. If saving a copy of a referenced model, whether the model is referenced by an assembly or a drawing, make sure the Save As Copy option is checked. Otherwise, the assembly or drawing will begin referencing the copy, rather than the original.

When copying assemblies and components that exist in the assembly, the Save As command can be used. This is especially important when external references exist in the assembly. Use the References button on the Save As window to copy any of the files containing external references. If this procedure is not followed, the files containing references may not update correctly.

Envelopes are nothing more than part files whose volume is used to select components. For example, if components fall within the envelope's volume they will be selected. Advanced selection criteria can be performed by searching for selected file properties using the Advanced Select command. Logical operators can be used to help define the criteria. Once the components have been selected, commands can be exercised on the selection set. Envelopes are not required for using the Advanced Select command.

CHAPTER 20

Advanced Features

THERE ARE A NUMBER OF FEATURE TYPES worth mentioning that have not been discussed in previous material. These are features such as those created by the Indent or Flex commands, among others. There is also extended functionality related to topics previously discussed, such as the Fillet feature. Advanced filleting functions include variable radius fillets and full round fillets, to name two.

This chapter also explores the lofted feature, which is a feature type that allows the blending of two or more profiles into a single feature. Another main topic in this chapter is working with multiple bodies. Any part file that contains separate chunks of solid geometry is considered to have multiple bodies. This you learned in Chapter 14 (see "10-second Topic: Multiple Bodies" and "Splitting Parts").

Multi-body parts are utilized internally by SolidWorks when certain types of parts are created. For example, weldments often contain multiple bodies, as do parts where a core and cavity have been generated. There are also modeling techniques that can be employed by a SolidWorks user that are well suited to multiple bodies. One such technique is bridging, which often involves lofting between two bodies.

A common technique that lends itself particularly well to multiple bodies is patterning and symmetry. For example, Figure 20–1 shows a wire basket that makes very good use of body patterns. Let's examine how this part was made to gain a better understanding of how multiple body design techniques might prove useful. Examining the wire basket will also help to draw together a number of topics previously discussed that have not been used to create a single part to this point.

Fig. 20–1. Wire
basket uses
patterned bodies.

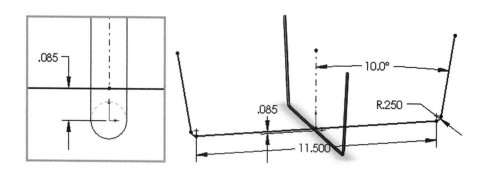

Fig. 20–2.
Wire basket at
an early stage.

The wire basket consists almost entirely of swept features and patterned bodies. The first two features created are the two perpendicular wires in the basket's center. Figure 20–2 shows the basket at an early stage of development. The first wire has been created as a swept feature, and the sketch for the second wire is under way.

The inset in Figure 20–2 shows the .085-inch dimension used to control how much the wires will intersect. The path for the first wire went directly through the origin. As can be seen from the image, the path for the second wire is offset .085 inch from the origin. Because the wires used to manufacture this basket will have a diameter of .100 inch, the overlap will be .015 inch.

Obviously, parts can be created in SolidWorks that would be very difficult, if not impossible, to manufacture. In the case of the wire basket, the wires could be modeled so that they pass directly through each other. However, that would be a horrible rendition of what happens in the physical world. Realistically speaking, a process is used that in essence welds the wires together, melting a small portion of each wire where the wires touch. Therefore, .015 inch works out very nicely.

When the second wire is modeled, the *Merge result* option is unchecked in the Sweep PropertyManager. This forces the second feature to become a second body. Patterning bodies is easy, and is often less prone to problems than trying to pattern complex features. Just keep in mind that bodies are being patterned, not features or faces. When patterning or mirroring, be sure to click in the proper list box, such as the Bodies to Pattern list box (shown in Figure 20–3).

All of the patterns used in the creation of the basket were linear patterns that patterned the seed only. (See Chapter 7 if you need to review pattern features.) Each of the original wires was patterned twice to create the additional wires near either end of the basket. The second pattern for the first wire is shown in a partial stage of completion in Figure 20–4.

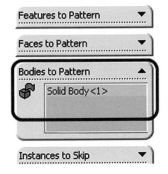

Fig. 20–3. Bodies to Pattern list box.

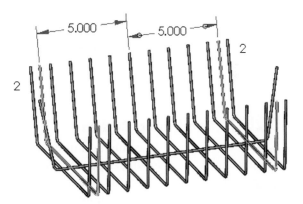

Fig. 20–4. Creating the second pattern for the first wire.

The circumferential wire poses the biggest challenge because it requires a 3D sketch. In that the 3D Sketcher was discussed in Chapter 10, we won't go into a detailed account of the process here. Figure 20–5 shows the circumferential wire after being swept along the 3D sketch, with the other wires in the model in a transparent state for clarity. Portions of some of the wires will need to be cut away, as indicated in the image.

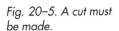
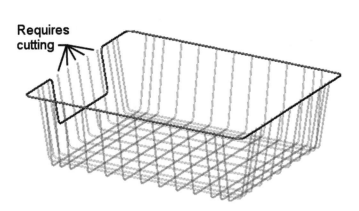

Fig. 20–5. A cut must be made.

Fig. 20–6. Feature Scope panel.

Because the wire basket is still made up of numerous bodies (25 in all), the cut used to trim away the wires indicated in Figure 20-5 should only affect a small number of those bodies. The Feature Scope panel, available in PropertyManager when creating the cut, allows us to select only the bodies that should be affected by the cut. The Feature Scope panel, shown in Figure 20-6, was discussed in Chapter 15 (see "Multiple Bodies and the Feature Scope").

Although construction of the wire basket consists of topics covered in previous material, this example brings together all of these topics in a single design scenario. Can we now say that the wire basket part is completed at this point? No, not if it still consists of multiple bodies. Sometimes it is necessary to leave a part as separate bodies (such as with a core and cavity). In most cases, it is best to combine the bodies into a single contiguous piece of geometry. How-To 20-1 takes you through the process of combining bodies into a single solid.

HOW-TO 20-1: Combining Bodies

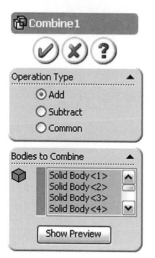

To combine multiple bodies into a single solid, perform the following steps. Bodies must be touching for this command to work. Bodies do not have to overlap, but they must not be tangent or touching at an isolated point. Otherwise, the feature will fail due to geometric conditions.

1. Select Combine from the Insert > Features menu, or click on the Combine icon found on the Features toolbar.

2. Select the Operation Type to perform from the Combine PropertyManager, shown in Figure 20-7.

3. Select the bodies to be combined.

4. Click on OK to complete the operation.

Fig. 20–7. Combine PropertyManager.

Bodies can be "combined" (added) and subtracted from each other, or the common intersection of bodies can be found. When subtracting bodies, a single body must be selected as the main body from which to subtract. More than one body can be selected, which would then be subtracted from the main body.

When finding the common area between bodies, all selected bodies must share a common intersection or the feature will fail. When finding the common area between two bodies, it is easy to tell if those bodies intersect. Finding the common area between three or more bodies is rare, but it makes for some interesting modeling experiments.

Lofted Features

Lofted features are one of the four main feature types possible in Solid-Works. So far you have discovered extruded, revolved, and swept features. Lofted features can be thought of as blends between two (or more) closed-profile sketches. Lofting can blend the shape of one profile into another, creating material between the profiles to form a boss or removing material to create a cut.

Although closed-profile sketch geometry is most commonly used for lofting, it is not always required. Open-profile sketch geometry is not allowed, though open profiles can be used in the construction of surface lofts, discussed in Chapter 21. Points can be used when lofting. For example, lofting from a point to a circle, and then to another point, results in a football shape.

Probably one of the most interesting aspects of lofting is that sketch geometry is not required at all in some cases. Lofts can occur between faces on solid geometry, between surfaces, or between edges of surfaces. An example of this capability is shown later in this section.

Let's revisit the wavy washer from Chapter 17 to see how a lofted feature is created. Figure 20–8 shows the geometry used to create the loft, which is the base feature for the washer. The great majority of the work in creating the washer is in the setup process and creating the sketch geometry. A layout sketch helped to position angled planes for the profile sketches. Each profile consists of a simple rectangle. The intermediate profile is slightly higher than the other profiles (with respect to the top, or x-z plane), which is what causes the "wave" to occur.

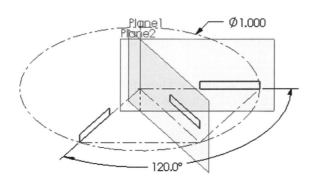

Fig. 20–8. Sketch geometry for the loft.

A total of four planes were used in the creation of the sketch geometry. Only two of these planes are shown in Figure 20–8. The three rectangular profiles all have the same number of segments, which is four, in that they each have four sides. This is significant. It is usually best to make sure all profiles have the same number of segments. This is not a requirement, but it is good technique. Otherwise, undesirable blending might occur.

When the sketch profiles used in the loft have a different number of segments, they will also invariably have a different number of vertex points. In such a case, SolidWorks must make a guess as to how it should blend the extra vertices. The feature will be created, but it will probably look odd. There are ways of controlling the blending, which we will explore shortly.

An important aspect of creating a lofted feature is how the profiles are selected, and in what order. It is best to select the profiles from the work area, not from FeatureManager. Select the profiles near a common point in each profile. In addition, select the profiles in the order in which the loft should occur. How-To 20–2 takes you through the process of creating a lofted boss feature. The process for creating a lofted cut is the same, except that the Loft command is selected from the Cut menu rather than the Boss/Base menu.

How-To 20-2: Creating a Lofted Feature

To create a lofted feature, perform the following steps. It is assumed that the sketch geometry has already been created. Note that similar to creating a swept feature it is necessary to exit sketch mode prior to creating a lofted feature.

1. Select Loft from the Insert > Boss/Base menu, or click on the Lofted Boss/Base icon, found on the Features toolbar. The Loft PropertyManager will appear, a portion of which is shown in Figure 20–9.

2. Select the sketch profiles from the work area in the order you want the loft to occur.

3. If the preview looks good, and the correct vertex points are in alignment, click on the OK button to create the lofted feature.

Fig. 20–9. Loft PropertyManager.

During the creation process of a lofted feature, connectors will appear on the loft preview. The connectors are directly related to the point nearest the selection of any given profile. The connectors, visible in Figure 20–10, serve to display which vertices will match up during the loft. Connectors can also be dragged in order to change how the profiles blend.

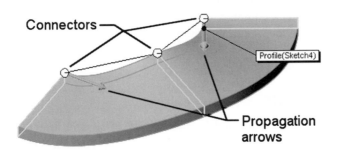

Connectors

Propagation arrows

Fig. 20–10. Connectors and propagation arrows.

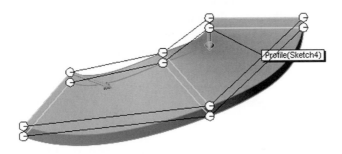

Fig. 20–11. Displaying all connectors.

Fig. 20–12. Sketch Tools panel.

Selecting sketch profiles from the fly-out FeatureManager as opposed to the work area often results in undesirable twisting during the loft. Dragging the connector points can eliminate twisting, but it is best to start off on the right foot to begin with, which is why it is important to select the sketch profiles from the work area.

Additional connectors can be shown if more control is needed over the loft. To show all connectors, right-click in the work area and select Show All Connectors. Right-clicking on a segment in one of the profiles also allows for creating an additional connector. Figure 20–11 shows four connectors, with each connector containing three connector points.

The order of the profiles in the Loft Profiles list box can be altered. It is relatively easy to select the profiles in the correct order, and thus reordering profiles should not usually be necessary. If this does become necessary, select the profile that is out of sequence and click on the Up or Down button as required. These buttons are found in the Profiles panel, shown in Figure 20–9.

SW 2006: A new panel in the Loft PropertyManager adds greater flexibility to the process of selecting loft profiles. The new panel is named the Sketch Tools panel, shown in Figure 20–12. The greatest benefit of the Sketch Tools panel is realized when employing new functionality involving 3D sketch geometry. Specifically, a single 3D sketch can be used to create a lofted feature.

When creating a lofted feature from a 3D sketch, use the Sketch Tools panel to aid in the selection process. For example, chained contour selection allows selecting open or closed profile geometry with a single click. Single contour selection allows for selecting individual segments in what might otherwise be considered a closed profile.

Specific segments in a profile can be selected, and then joined to form a single profile. This can be accomplished by selecting multiple profiles from the Profiles panel (shown in Figure 20–9) and clicking on the Join button.

The Drag Sketch button applies to 3D sketch geometry only, and allows for dragging underdefined sketch geometry while in the Loft command. This, in turn, reshapes the lofted feature. The loft preview will update accordingly, providing dynamic feedback and showing the results of dragging and re-shaping the underlying sketch geometry.

Start/End Constraints

The start and end constraints for a lofted feature essentially control how the loft propagates from the first and last profiles in the loft. The Start/End Constraints panel, shown in part in Figure 20–9, is also used to control how a lofted feature relates to other model faces adjacent to the loft.

A loft in its simplest form will blend from one profile to another with absolutely no regard for the orientation of the profile planes. This can be altered by selecting Normal To Profile in the start or end constraint drop-down list. When Normal To Profile is selected, the profile is lofted perpendicular (normal) to the profile sketch plane before it begins to conform to the next profile in the loft.

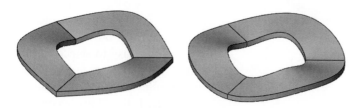

Fig. 20–13. Normal to Profile setting; with (right) and without (left).

Regarding the loft for the wavy washer, it was important to set both the start and end constraints normal to the profile sketches. Otherwise, there would be no continuity between each "wave." Figure 20–13 shows the washer after patterning the loft. The washer on the left uses no special settings for the start and end constraints, whereas the washer on the right uses the Normal To Profile setting. Edges have been highlighted for clarity.

Another available constraint is Direction Vector. When used, the loft can be forced to propagate in a particular direction dictated by an edge or sketch line. Planar faces can also be used, in which case the direction vector is perpendicular to the face. Propagation arrows help to illustrate the direction of propagation, regardless to the constraint being used. Propagation arrows also illustrate the propagation force being exerted either on the entire profile or at each individual connector point.

To examine the finer details of start and end constraints, a more complex part will be needed. We will use the ice-cream scoop, shown in a partial stage of completion in Figure 20–14. The part requires a lofted feature between the scoop and the handle. Incidentally, the modeling technique being used on this part is known as bridging.

Fig. 20–14. Ice-cream scoop.

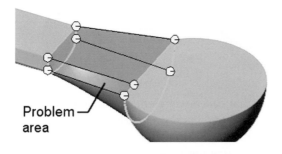

Fig. 20–15. Loft preview with connectors.

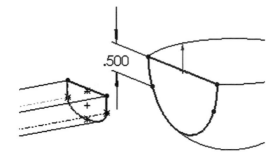

Fig. 20–16. Sketch profiles for the loft.

First, let's try to create a loft between the two bodies without developing any sketch geometry. When creating the loft, the two faces can be selected, and the loft will span between them. Similar locations on each face should be selected, similar to when selecting near-similar vertex points when selecting sketch profiles.

The first problem we see is that the faces do not seem to blend well. By showing all connectors, the problem becomes clear. Figure 20–15 shows the loft preview with all connectors visible. Note that the side of the handle appears to be blending into the top of the scoop. This is not a desirable situation.

Dragging the connectors eliminates the problem. However, the shape of the loft is still not right. The loft should propagate from the handle in a perpendicular fashion. When the Normal To Profile constraint is selected, the loft fails. This is not to say that a loft always fails when connectors are repositioned and various constraints are applied. Lofts are simply more problematic in such cases. What else can be done?

The most common reason lofts fail is due to the number of segments in the loft profiles. By selecting faces to loft between, rather than sketch profiles, the ability to more closely control segmentation is lost. By utilizing sketch profiles, controlling segmentation becomes easy. Figure 20–16 shows sketch geometry that will be used in place of the faces.

Each sketch was created using the Convert Entities command, which is extremely convenient in such cases. The sketch on the handle contains an elliptical arc and three lines. The sketch on the scoop originally contained one line and one arc. However, it now contains one line and three arcs. This was accomplished using the Split Entities sketch tool, discussed in Chapter 5. Because there are now four segments in each profile, the loft operation should go more smoothly.

Figure 20–17 shows the loft preview (left), which now looks much better. Note that all vertex points match up very nicely. In addition, the Normal To Profile constraint has been applied to the profile on the handle. This works well for the handle, but what of the scoop? After completing the loft, there is a hard edge that can clearly be seen in the image (right). Another end constraint will be needed to remedy this problem.

Fig. 20–17. Making improvements to the loft.

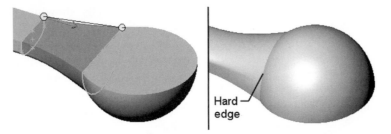

The two other constraints that can be applied are Tangent To Face and Curvature To Face. The constraints are similar in that each constraint causes the loft feature to be tangent to the faces surrounding the profile sketch. Curvature To Face additionally causes the curvature of the loft to gradually change to match that of the surrounding geometry, thereby resulting in continuous curvature. If you are not familiar with the term *continuous curvature*, there is a section that will help explain this concept later in this chapter.

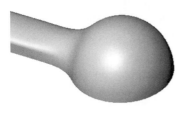

Fig. 20–18. The hard edge has been eliminated.

Figure 20–18 shows the ice-cream scoop after changing the constraint from None to Curvature To Face. As can be seen in the image, the transition is much smoother, with absolutely no hard edge.

To conclude the discussion on loft start and end constraints, a few final details will be examined. For instance, if either Normal To Profile or Direction Vector constraints are specified, there will be options for setting the draft angle and tangent force. These settings are shown in Figure 20–19. SolidWorks actually refers to the tangent force as *tangent length*, although the setting is not representative of an actual length value.

Fig. 20–19. Additional start/end constraint settings.

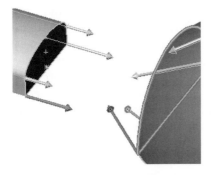

Fig. 20–20. Turning off Apply to all.

The button to the left of the draft angle setting reverses the draft angle. Likewise, the button to the left of the tangent force setting reverses the tangency direction. The propagation arrows will update to reflect changes made to the tangency direction or force. The Tangent To Face and Curvature To Face constraints only have tangent force settings, as a draft angle setting would not be relevant.

If the *Apply to all* checkbox is unchecked, regardless of which constraint is applicable, multiple propagation arrows will be displayed. Such is the case in Figure 20–20 (the preview has been turned of for clarity). With *Apply to all* turned off, a force can be applied independently to each point on the profile. This can be done either by selecting an arrow and entering a value in the tangent force setting or by simply dragging the arrow. Tangent force settings default to 1, but can be anywhere from .1 to 10.

Guide Curves and Centerlines

Similar to swept features, lofted features can take advantage of guide curves. In that guide curves were discussed at length in Chapter 10, they will not be discussed again here. However, you should be aware of how they differ from using the Centerline parameter.

Fig. 20–21. Centerline Parameters panel.

Centerlines are used to control loft direction, in much the same way a sweep path controls the direction of a swept feature. Centerlines do not require a pierce relation. Guide curves, on the other hand, usually do. Guide curves do as much to control the shape of loft profiles as they do to control the direction of the loft. Centerlines do nothing to change the shape of the profile geometry.

The Centerline Parameters panel is shown in Figure 20–21. To implement this functionality, click in the list box within the panel and select a sketch. That is about all there is to it. The remaining options in the Centerline Parameters panel are used to preview the cross sections of the lofted feature. This really isn't necessary, in that the translucent preview typically shown while performing a loft is much nicer anyway.

Why is the Centerline Parameters panel named as it is? That is something of a mystery, especially considering the sketch used for this function must not consist of centerlines! Rather, the sketch should consist of standard lines, arcs, or some other sketch entity or entities. The sketch can be 2D or 3D, but should for best results start on the same plane as the first loft profile and end on the sketch plane of the last loft profile.

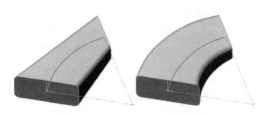

Fig. 20–22. Without (left) and with (right) a centerline.

Fig. 20–23. Flexible airflow guide.

Figure 20–22 shows the centerline function in use. The first loft (left) shows the arc, but does not make use of it. The loft blends from a larger rounded rectangle to a smaller rounded rectangle. The second loft (right) makes use of the arc as a centerline. Using the Normal To Profile constraint for both the start and end profile, without the centerline, would result in a similar outcome, albeit not quite as refined.

The feature shown in Figure 20–22 was used in the creation of a flexible airflow guide, shown in Figure 20–23. This part will make another appearance in Chapter 21, when surfaces are discussed. For now, there are still a few other options available when using the Loft command. These are explained in the following material.

Merge Tangent Faces

When checked, the Merge Tangent Faces option will merge faces created as a result of sketch segments that are tangent in the loft profiles. If you are lofting from a profile with tangent entities (such as a rounded slot) to a profile that does not contain tangent entities (such as a rectangle), the profile with the tangent entities will obviously not be able to remain tangent throughout the entire loft, and the faces would not be merged.

Not all faces are merged even if they are tangent. Complex curvature faces are merged into a single face, but planar, cylindrical, or conical faces are left alone. Do not expect to see all of the tangent edges in a lofted feature disappear when turning this option on, because it is not meant to function in that manner.

Close Loft

If you are creating a loft that uses three or more profiles, and you want the last profile to connect back up with the first, check the Close Loft option. This will create, in effect, a closed-loop loft. You must have a minimum of three closed profiles you are lofting between for this option to work. Realistically, you would probably use more than three profiles. A good example might be four profiles, 90 degrees apart, used to create the shape of a ring.

Show Preview

Checking the Show Preview option will show a nice shaded preview, but sometimes at the expense of time. Lofts can take a while if complex in nature. If that is the case, you may want to uncheck this option and nix the preview.

Merge Result

The Merge Result option is only available after the first base feature has been created. When this option is checked, the new feature will become part of the existing model. Otherwise, multiple bodies are created.

Inserting Parts

Entire part files can be inserted as bodies into other part files. They can then be used to aid in the development of a new part. A body created in this fashion is dependent on the original part, and in fact has an external reference back to the original part file. When inserting a part, you have the opportunity to position the part at a particular location. Rotating the inserted part is also possible.

What can be done with the inserted part once it is there? Considering that combining bodies is possible, one thought would be to reposition the inserted part and combine it to the geometry already present. Another modeling technique is to base a family of parts on the single basic shape constituting the inserted part file.

A very important concept related to inserting parts into other parts is that the process should not be confused with creating an assembly. You would not want to build an "assembly" in this way, as it would not contain the functionality inherent in SolidWorks assemblies (such as mating relationships). With that said, How-To 20–3 takes you through the process of inserting a part into another part file.

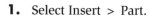

HOW-TO 20-3: Inserting Parts as Bodies

To insert a part into an existing part file, perform the following steps.

Fig. 20–24. Insert Part PropertyManager.

1. Select Insert > Part.

2. Select the part and click on Open.

3. [In the Insert Part PropertyManager, shown in Figure 20–24, select any additional objects you wish to have imported into the existing part file (it is acceptable to leave all items unchecked).

4. Check the Launch Move Dialog option if you wish to translate or rotate the part you are inserting (explained in material to follow).

5. Click on OK to insert the part.

If the Launch Move Dialog option was checked, the Locate Part panel will appear. It will almost certainly be necessary to locate the inserted part geometry, so this option should typically be left on. If Launch Move Dialog is unchecked, the part will be inserted per step 5. The inserted part will be positioned so that its origin point is at the same location as the origin point of the part it was brought into.

The ability to translate or rotate a body is a function unto itself. This capability plays a very important role when working with multiple bodies. The command is named Move/Copy, which is found in both the Features and Surfaces menus. Although the Move/Copy command can be carried out separately, How-To 20–4 will also benefit those who insert a part and check the Launch Move Dialog option.

HOW-TO 20-4: Moving, Rotating, and Copying Bodies

Fig. 20–25. Move/Copy Body PropertyManager.

Fig. 20–26. Rotate panel.

To move or rotate a body, and to optionally copy the body in the process of moving or rotating it, perform the following steps. The steps apply to both solid and surface bodies, regardless of how the Move/Copy command is accessed.

1. Select Move/Copy from either the Insert > Features or Insert > Surface menu.

2. In the Move/Copy Body PropertyManager, shown in Figure 20–25, select the body or bodies to be moved or copied.

3. Select the Copy option if you wish to copy bodies. If creating copies, specify the quantity.

4. If creating copies, specify the quantity.

5. Expand either the Translate or Rotate panel, depending on the action you wish to perform on the bodies (use the small arrows to the right of the panel names).

6. If translating, specify the desired translation values for the x, y, and z axes.

7. If rotating, specify a point for the rotation origin, and then enter a rotation angle. By default, the rotation origin is the model's origin point. Changing the distance values in the Rotate panel (see Figure 20–26) will alter this location.

8. Click on OK to finish translating or rotating the bodies.

Both the Translate and Rotate panels will not be open simultaneously. If a body must be both translated and rotated, it will be necessary to execute the command twice. When translating, geometry can be selected to help position the inserted part. For example, if a linear edge is selected, a single distance value can be specified to indicate the distance the body will move along that edge.

SW 2006: In addition to the functionality described in this section, mates can be used to position bodies. Use the Constraints and Translate/Rotate buttons in PropertyManager to move between PropertyManager panes.

An excellent way of specifying the translation distance is to pick two vertices. This action will position the inserted part along the vector established by the two vertices. To put it more simply, the first vertex selected will position itself directly on the second vertex selected.

When rotating bodies, the first three settings dictate the *x-y-z* point (with reference to the origin point) the body will rotate about. The rotation origin will be indicated on screen via the triad, shown in Figure 20–27. The second set of three settings (see Figure 20–26) determines the rotation, in degrees, about each axis.

Fig. 20–27. Move/ Copy Body triad.

It is much easier to use geometry to indicate the rotation origin. This can be in the form of a single point, which would dictate the rotation origin, or a linear edge the body would rotate about. If an edge is selected, a single rotation value can be supplied. The triad arrows can be dragged to translate or rotate the part, and the triad's centerpoint can be dragged to change the center of rotation. When dragging the triad arrows for rotational purposes, use the right mouse button.

Regardless of whether bodies are being moved, rotated, or copied, a preview will always be visible in the work area. Use this to your advantage. It is easy to figure out what the software is doing by experimenting with the settings and watching what the preview is doing. When finished with the Move/Copy Body command, a new *Body-Move/Copy* feature will appear in FeatureManager. Editing the definition of this feature allows for modifying the values specified for the translation or rotation, or for changing the number of copies created.

Indent Feature

Imagine a situation in which a part must be positioned in an assembly, but there are other components in the way. It might not be possible to start chopping off pieces of the part that must go into the assembly, but it might be possible to "indent" them. Indenting changes the shape of a part to accommodate some other object. Trying to get a part to fit into an assembly is only one possible scenario. The Indent feature can be used in other situations as well.

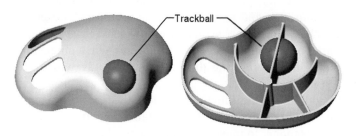

Fig. 20–28. Mouse and trackball bodies.

The Indent command will be useful in any situation in which one body must conform to another. To help explain how this command functions, the part model shown in Figure 20–28 will be used. The trackball is not a separate component. Rather, it is a separate body in the mouse cover part file. It has been created as a separate body so that the indentation where the trackball will reside can be further developed.

The Indent command can be performed either in a part or assembly file. If in an assembly, it is necessary to edit the component being indented, and another component can be selected by which to create the indentation. In our case, the Indent feature will be carried out within a part file. The body of the trackball will be used to indent the mouse cover. How-To 20–5 takes you through the process of creating an indent feature.

How-To 20-5: Creating an Indent Feature

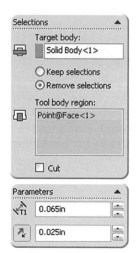

Fig. 20–29. Indent PropertyManager.

To create an indent feature, perform the following steps.

1. Select Indent from the Insert > Features menu, or click on the Indent icon found on the Features toolbar.

2. In the Indent PropertyManager, shown in Figure 20–29, select the Target Body. This will be the body receiving the indent feature.

3. Click in the Tool body region list box and select the body that will create the indentation.

4. Select whether to keep or remove selections. This will cause material to be created on one side or the other of the tool body. Pay attention to the preview.

5. Specify a value for the wall thickness developed around the tool body.

6. Optionally, specify an offset value. This will be the distance between the tool body and the wall thickness.

7. Click on OK to complete the indent feature.

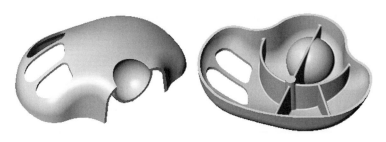

Figure 20–30 attempts to show the resultant indent feature. The trackball body has been hidden from view, and the mouse cover is shown in a section view (left). The section view helps to show the wall thickness added by the Indent command. The mouse cover can

Fig. 20–30. After completing the indent feature.

now be developed further, with the end result being that the trackball will fit perfectly in the final assembly.

Often, it is important to select bodies in a specific manner, or the indent feature will not turn out as expected. Even just selecting bodies from the opposite side of where the indent is being applied will make a difference. Be aware of the preview when making selections, and watch what the preview is doing.

There is an option titled Cut that is present in the Indent PropertyManager. Turning this option on results in material being subtracted from the target body with no wall thickness created. The result would be similar to using the Combine command with the Subtract option, but without the ability to supply an offset distance (such as for clearance).

Dome Feature

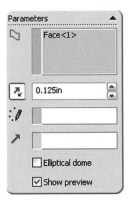

The mechanics behind the Dome command are very simple, and there is seemingly no limit to the geometric shapes that can be used to create a domed feature. The simplicity of the Dome PropertyManager, shown in Figure 20–31, belies its complexity. Some very interesting and complex shapes can be created with the Dome command. How-To 20–6 takes you through the process of creating a domed feature.

Fig. 20–31. Dome PropertyManager.

How-To 20-6: Creating a Domed Feature

To create a domed feature, perform the following steps.

1. Select Dome from the Insert > Features menu, or click on the Dome icon, found on the Features toolbar.

2. Select a face on which to create the domed feature.

3. Specify a height for the domed feature.

4. Click on the Reverse Direction option if necessary.

5. Click on OK to create the dome.

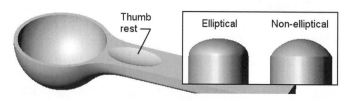

Fig. 20–32. Ice-cream scoop thumb rest.

Different options are available depending on the face selected in step 2. For instance, when an elliptical face is selected the *Elliptical dome* option is available. Such was the case when the thumb rest feature (see Figure 20–32) on the ice-cream scoop was created. The inset shows examples of both an elliptical and nonelliptical dome.

How was the thumb rest defined within a specific elliptical area? The Split Line command (Chapter 6) was used to first break out a portion of the top face of the ice-cream scoop. The Dome command was then employed on the newly created face. By default, the Dome command will add material to a part. However, clicking on the *Reverse direction* option will cause a domed cut to be created.

Direction vectors can be used to dictate the direction a dome will be applied. Edges must be selected for a direction vector, in which case the dome attempts to point in that direction. It is also possible to supply a sketch point with which to constrain the dome shape. In such a case, the dome surface will be constrained to the sketch point, thereby controlling the dome's shape.

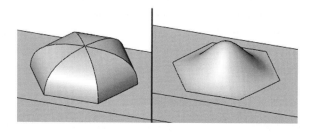

Fig. 20–33. Dome features on polygonal faces.

When polygonal faces are selected on which to create a dome, an option for creating a *Continuous dome* is available. Figure 20–33 shows a dome created from a hexagonal face. The left-hand side of the image shows a standard dome feature. The right-hand side of the image shows what happens when the *Continuous dome* option is checked.

Flex Feature

Fig. 20–34. Ladder model.

Fig. 20–35. Choose a flex operation.

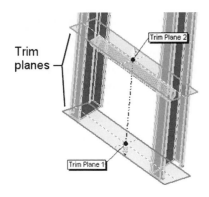

Fig. 20–36. Trim planes.

Sometimes adding various shapes to a model or creating cuts of one form or another just isn't enough. Sometime you want to take a piece of geometry and bend or twist it in some direction. This is exactly what the Flex command allows you to do.

There are four main operations that can be performed with the Flex command. These operations are bending, twisting, tapering, and stretching. This section looks at a few of the examples so that you can get a feel for how the command works. A model of the ladder shown in Figure 20–34 will be used for the examples shown throughout this section.

Due to the intricacies of the Flex command, it isn't feasible to use How-To steps. Therefore, we will tackle this command a little differently. If you would like to experiment with a flex feature on your own, create a simple bar of material. Create a rectangle, perhaps .25 x 1 inch, and extrude it 10 inches. Now you have something to work with, and can follow along.

Choose an Operation

The Flex command can be found in the Features menu (Insert > Features). Once in the command, the Flex Property-Manager will appear, which allows you to select what operation you wish to perform. This is done by selecting the operation from the top panel labeled Flex Input, shown in Figure 20–35. Note the four operations listed.

This is also a good time to select a body on which the operation will be performed. Nothing special here: simply click on the body in the work area. Don't worry about the rest of the options in the Flex Input panel. We will get to those shortly.

Position the Planes

Next on the agenda is to position the trim planes, shown in Figure 20–36. Whatever operation is performed on the part occurs between the trim planes. Trim plane position, therefore, is obviously important.

The arrows at the center of each trim plane can be dragged to reposition the planes. Take care not to drag the edges of one of the planes, as that will perform the flex operation on the model geometry. The time will come for that in a moment.

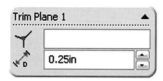

Fig. 20–37. Trim plane panel.

Panels in PropertyManager are labeled Trim Plane 1 and Trim Plane 2, one of which is shown in Figure 20–37. Precise distance values can be specified in each panel, thereby positioning the trim planes. The trim planes default to the outer boundaries of the model, which relates to a distance value of zero.

It often helps to be able to position a trim plane at a precise geometric point. This can be accomplished by clicking in the trim plane panel list box and then selecting a point on the model. The trim plane will position itself at that selected point. (See the next section for additional trim plane options.)

Position the Triad

Whatever operation is performed occurs within the trim planes, but the operation is centered about the triad. For best results, the triad should be located between the trim planes. There may be rare exceptions to this rule.

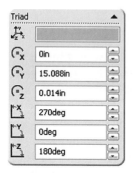

Fig. 20–38. Locating the triad.

The triad can be repositioned by dragging either its centerpoint or one of its axes arrows. Be very careful if dragging the triad's centerpoint, because it can easily be dropped on a model edge, thereby realigning the flex axis of operation, repositioning the trim planes in the process. If this happens, right-click somewhere in the work area and select Reset Flex. This will set the trim planes to their default position and you can try again without restarting the command.

If a coordinate system has been set up ahead of time, the triad can be attached to the coordinate system via the Triad panel list box (see Figure 20–38). This isn't usually necessary, so don't be concerned with creating a coordinate system. Dragging the arrows of the triad is safe enough, or the triad's position can be altered via the Triad panel.

Rotating the triad can be dangerous, because the various operations are affected in one way or another by the orientation of the triad's axes. Generally, only reposition the triad as is absolutely necessary. Figure 20–39 shows what would be the safest plan of attack for properly locating the triad. Drag the axis arrow that is already aligned with the trim plane axis and position the triad somewhere near the midpoint between the trim planes.

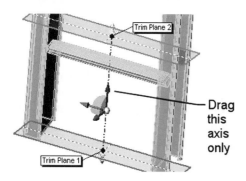

Fig. 20–39. Center the triad between the trim planes.

Even after moving the trim planes and the triad into position, the trim planes may still be skewed or at an odd angle. This can often be remedied through the right mouse button menu. By right-clicking on the center of the triad a number of options are available for orienting the trim plane axes and triad. Table 20–1 outlines these options and what they accomplish.

Table 20-1: Flex Triad Menu Options

Menu Option	Description
Align Trim Plane Axis to Selection	Aligns the blue Z axis with an edge or normal to a plane or face. Select an edge, plane, or planar face after choosing this option.
Align Bend Axis to Selection	Attempts to align the red X axis with an edge or normal to a plane or face by rotating the triad about the blue Z axis.
Center and Align to Component	Moves the triad to the center of gravity and aligns the trim planes to the part's coordinate system.
Center and Align to Principle	Moves the triad to the center of gravity and aligns the trim planes to the part's principal axis.
Move Triad to Plane 1	Repositions the triad to trim plane 1.
Move Triad to Plane 2	Repositions the triad to trim plane 2.

Perform the Operation

Fig. 20–40. After bending the ladder.

Depending on the operation being performed, it will be necessary to supply the applicable parameters for carrying out the operation. For instance, if performing a bend operation supply an angular value or radius. The trim planes can be dragged, but parameters cannot be accurately specified in this manner. Figure 20–40 shows the ladder bent at an angle of 180 degrees.

The slider bar in the Flex PropertyManager controls the accuracy of the flex feature. It is a good idea to leave this set to a low value unless the flex feature fails. The reason for this is that flex features require a huge amount of processing, and increasing the accuracy will substantially increase how much time you are waiting for the operation to finish.

If performing a twist operation, an angular value must be supplied. Stretching requires a distance value, and tapering requires a taper factor. Flex features can be added more than once, but multiple operations cannot be carried out simultaneously. For example, it is possible to perform first a taper and then a bend operation. The operations just have to be separate features. Figure 20–41 shows two other examples of flex features utilizing the twist (left) and taper (right) operations.

The *Hard Edges* option in the Flex Input panel causes what might be thought of as standard surfaces in the flex region whenever possible. These are actually called analytical surfaces, such as planar, cylindrical, or conical surfaces. Leaving Hard Edges checked might result in literal edges being created on the model, such as those shown in Figure 20–42.

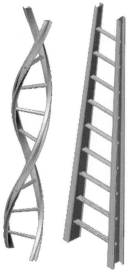

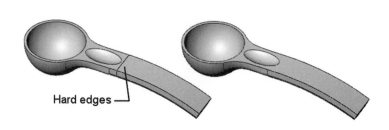

Hard edges ⎯

Fig. 20–41. Twisted and tapered flex features.

Fig. 20–42. Hard edges.

When the Hard Edges option is inactive, surfaces in the flex region are created from splines, which may result in smoother surfaces. There should be no hard edges visible after completing the flex operation. Occasionally, the flex feature may fail, depending on whether the Hard Edges option is set one way or another. If the feature does fail, try changing this option.

Deforming and Reshaping

There is more than one way to reshape geometry in odd fashions. You have just learned about one of the more straightforward methods, which is to use the Flex command. Other, less user-friendly, methods consist of employing the Deform and Shape commands. We will explore the Deform command first.

Similar to the Flex command, the Deform command has many permutations. The command can be used in so many ways that it would be nearly impossible to explore them all, at least not without adding another 20 pages to this already thick book! Therefore, a rough overview of some of the more commonly used applications of deforming are discussed.

Another similarity between flexing and deforming is that different operations can be performed. In the case of deforming, these operations are selected in the Deform Type panel, and consist of Point, Curve to Curve, and Surface Push. Depending on the operation selected, the Deform Property-

Manager will change drastically. That is, there is a completely different set of options depending on what operation is selected.

Of the three operations, the Point deform type is the most straightforward. Regardless of which operation is performed, it is never necessary to utilize all of the options present in PropertyManager. As a matter of fact, a good rule of thumb with the Deform command is to only use the options absolutely necessary to get the job done. Trying to plug in too many parameters can overly complicate the situation and result in shapes you probably are not looking for anyway. Let's look at an example to understand this behavior.

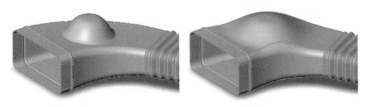

Fig. 20–43. Deform command in action.

Figure 20–43 shows the airflow guide used in an earlier example. The particular operation being performed on this model is the Point deform type. The operation requires no additional geometry and can act directly on a model with a minimal amount of user intervention. The top face of the part was selected, and the deformation occurs at the point where the face was selected.

Fig. 20–44. Deform PropertyManager when using the Point deform type.

The only difference between the two examples shown is that the model on the right has the *Deform region* option checked (shown in Figure 20–44). All other settings are identical. Values shown in the image include deform distance (which controls the height of the deformation) and deform radius, which controls the area affected. By checking the region deforming option, the entire selected surface is affected by the deformation. It is worth noting that none of the other parameters (list boxes) are utilized.

It is possible to specify a deform direction if the deformation should be something other than perpendicular to the selected surface. Select the body to be acted upon if there is more than one body in the model. Neither of these options were used in either of the examples shown in Figure 20–43. Figure 20–45 uses the *Deform axis* option, which will play a significant role in the outcome of the feature. It is one of the more useful options in the Deform command.

Fig. 20–45. Using the Deform axis option.

When available, the Stiffness settings (three buttons in the Shape Options panel, shown in Figure 20–45) help control the shape of the deformation. The *Shape accuracy* slider bar, present in all deformation types, will increase the resolution of the resultant shape, but at a serious cost in performance.

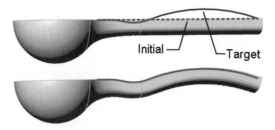

Fig. 20–46. Curve to Curve deformation.

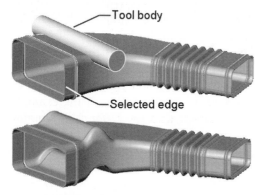

Fig. 20–47. Selecting deform regions.

Fig. 20–48. Performing a surface push.

The Curve to Curve deformation type offers a good amount of control when accuracy is important. Two sketches are typically used, which dictate the initial shape of the geometry and the targeted shape for the geometry. Figure 20–46 illustrates this concept perfectly.

Because the curve-to-curve deformation can act on various faces or an entire body, it is important to select either the body or faces before any action is taken on the part of Solid-Works. For instance, no preview will be shown until something is selected. In the case of the ice-cream scoop, this was done by clicking in the *Bodies to be deformed* list box (shown in Figure 20–47) and then selecting the ice-cream scoop body.

Rather than selecting the entire body, individual faces can be acted upon (middle list box of the Deform regions panel, shown in Figure 20–47). This greatly impacts the outcome of the deformed shape. Likewise, selecting edges or faces to be fixed (uppermost list box) will anchor those edges or faces in position during the deform operation.

One of the settings in the Shape Options panel (not shown) when performing a curve-to-curve deformation is titled *Curve direction*. When used, the deformation will be mapped to the normals (perpendicular vectors) of the target curve. What this relates to in simplified terms is that the area being deformed will have a tendency to more closely match the target curve.

Surface push is the final deform type, which allows for deforming based on various shapes called tool bodies. The tool body can be one of the predefined bodies available or another body already present in the model. A body is often created with the express reason of using it to deform model geometry.

Figure 20–48 shows what happens when a surface push is performed. A cylindrical body was created to deform the model. The vertical edge shown in the image was selected to determine the push direction. A preview arrow (not shown) will indicate the push direction, which can be reversed if necessary. Selecting faces versus bodies will have as much impact with a surface push as it does when performing a curve-to-curve deform type, so keep this in mind when selecting what should be deformed.

In the example shown in Figure 20–48, a body created by the designer was selected. The body was created in the position where it was needed. However, if a predefined shape is chosen from the drop-down list in the Deform Region panel (see Figure 20–49), it will be necessary to move the shape to the desired position. This can be accomplished by dragging the axes of the triad or by specifying values in the Tool Body Position panel (not shown).

Fig. 20–49. Choosing a shape for the surface push.

The *Deform deviation* setting (present in the Deform Region panel) controls how the deform region blends with the rest of the model. This can in turn have an affect on the remainder of the deformed region. Small values will make the transition between deformed and undeformed areas very sharp. Larger values will make the transition more gradual. Figure 20–50 shows two deformations where the only difference is in the setting for the deform deviation, which is indicated in the image.

Fig. 20–50. Changing the deform deviation.

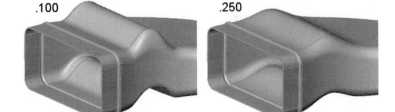

Using the Deform command is not an exact science. This holds true for the Shape command as well. These commands are used to change the shape of a model in ways that are not accurately represented with precise dimensional values. This is often reason enough to avoid these commands. In addition, shape-changing commands (Flex, Deform, and Shape) can take a serious toll on system performance. If there is an easier way to obtain the desired shape for a particular model, you are better off taking that route. Otherwise, you may be wishing you had a faster computer.

The Shape command, in its simplest form, will allow for creating an effect of a container being pressurized, with its sides bowing outward. It can also be used to "suck in" the sides of a plastic part to give it an indented effect. Think of the Shape command as being able to deform a face as though it were a rubber membrane. This can add some realistic effects to certain types of parts, but is not something you would use every day.

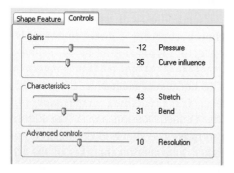

Fig. 20–51. Slider bar adjustments for the Shape command.

An advanced application of the Shape command is to use curves to control the shape a face takes on. The control curves can be used to "draw out" areas of the face being shaped. The amount of control the curve (or curves) exerts on the shaped face can be adjusted.

The slider bar adjustments (see Figure 20–51) used when shaping a face require some time to act on the model. Once an adjustment is made, SolidWorks recalculates the new shape of the face. How long this takes depends on your computer and some of the settings within the Shape command. The Resolution setting will increase the recalculation time, with higher resolutions requiring more time. How-To 20–7 takes you through the process of creating a shape feature.

HOW-TO 20-7: Creating a Shape Feature

To reshape a face on a model using the Shape command, perform the following steps.

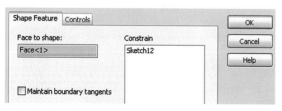

Fig. 20–52. Shape Feature window.

1. Select Shape from the Insert > Features menu, or click on the Shape icon, typically found on the Features toolbar.

2. Select a face to be reshaped. The face should appear in the *Face to shape* list box in the Shape Feature window, a portion of which is shown in Figure 20–52.

3. Select the *Maintain boundary tangents* option if the shape should try to maintain tangency to the boundary of the selected face.

4. Optionally, click in the *Constrain to* list box and select one or more curves that will help constrain the shape of the face to be deformed.

5. Optionally, click on the Preview button, or select the Controls tab to see a preview, as well as the current control settings.

6. Use the slider controls to change settings for Pressure, Stretch, Bend, and Resolution.

7. If any curves were selected to constrain to in the Shape Feature tab, additionally specify a setting for the Curve Influence slider bar.

8. Click on OK to accept the shape shown in the preview.

The Shape command is one you should experiment with. It can be a frustrating command due to the shape wanting to update whenever a minute adjustment is made. It can also be frustrating when a slider is moved to a position that really blows the shape out of the water.

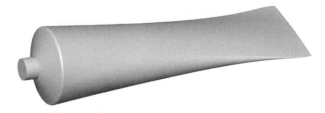

Fig. 20–53. Toothpaste tube.

Fig. 20–54. After applying the Shape command to the tube.

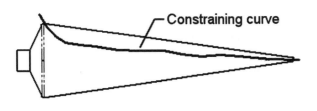

Fig. 20–55. A spline is created for shape control.

This can almost be taken literally, because the shape can take on a big "splash" appearance, seemingly gaining a mind of its own. Luckily, there is an Undo icon built into the Control tab portion of the Command window. However, you will have to wait for the last adjustment to settle down before clicking on Undo. The following deals with an example of what might be accomplished with the Shape command. Figure 20–53 shows a tube of toothpaste modeled with a lofted feature and variable radius fillets (discussed in material to follow).

Using the Shape command, the tube can be made to appear as though it is being squeezed, to some extent. This is shown in Figure 20–54. The effect is not overly dramatic because the deformation is applied fairly equally over the entire top face. For a more dramatic squeeze effect, we need some way of making the larger end of the tube appear as though it is being squeezed more than the bottom end, which is already fairly flat to begin with.

To achieve better control over the shape the deformation takes, a curve will be used. In the case of the example being used here, the curve is a free-form spline that was sketched on the Front plane of the model. The spline is shown in Figure 20–55, along with the toothpaste tube in wireframe display mode for reasons of clarity.

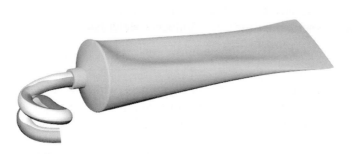

Fig. 20–56. Using the spline as a constraining curve and adding a swept feature.

When creating the shape feature, the spline can be selected per step 4 of How-To 20–7. The spline curve will exert a force on the face selected to be reshaped, thereby making the face conform to the curve. Additional curves could be selected to obtain a higher degree of control. In Figure 20–56, a swept feature was added (the toothpaste) to increase the squeeze effect even more.

Variable-radius Fillets

Fig. 20–57. Variable Radius Parameters panel.

Typically, when creating a fillet, edges are selected. This is not a requirement. Faces and features can also be selected, as was learned in Chapter 6. In the case of a variable-radius fillet, however, edges must be selected. This is a requirement, and for good reason. Selecting edges allows SolidWorks to populate the Vertex list box with vertex points (the plural being *vertices*) associated with the selected edges.

It is not possible to click inside the Vertex list area and select vertex points. Only by selecting edges will the software place the appropriate vertices within the list box. Once there, the individual vertex points can be selected and assigned a radial value. All of these actions are carried out within the Variable Radius Parameters panel of the Fillet PropertyManager, shown in Figure 20–57.

Variable-radius fillets come in two varieties: smooth transition and straight transition. Examples of each of these transition types are shown in Figure 20–58. As can be seen in the image, straight transitions are linear. Smooth transitions propagate tangent to the edge being filleted at both the start and end of the fillet.

When creating a variable-radius fillet, the prime concern is establishing radial values for each vertex point. There are two options that make this

Fig. 20–58. Difference between a smooth and straight transition.

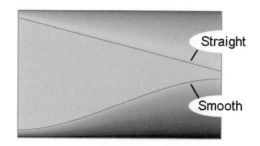

task easier, especially if there are numerous vertices listed. These options are in the form of buttons labeled Set Unassigned and Set All, which can be seen in Figure 20–57.

By specifying a value for the radius and clicking on the Set All button, a radial value can easily be assigned to every vertex point in the list. Likewise, the Set Unassigned button will apply the specified radius to only those vertices that do not yet have a radius established. How-To 20–8 takes you through the process of creating a variable-radius fillet.

How-To 20-8: Creating a Variable-radius Fillet

Fig. 20–59. Fillet Type panel.

To create a variable-radius fillet, perform the following steps.

1. Select Fillet from the Insert > Features menu, or click on the Fillet icon.

2. For Fillet type, select *Variable radius* from the Fillet Type panel (shown in Figure 20–59).

3. Select the edges to be filleted.

4. Select a vertex from the vertex list in the Variable Radius Parameters panel. The vertices will be labeled V1, V2, and so on.

5. Specify a radial value for the selected vertex.

6. Repeat steps 4 and 5 for the remaining vertices. Feel free to use the Set Unassigned or Set All buttons if it will make this process easier.

7. Select either Smooth or Straight transition.

8. Click on OK.

If you find it easier, use the callouts attached to the model in the work area to change the radial values for each vertex point. Click once inside a callout, type in a value, and then press Enter. When a variable-radius fillet needs to be edited, use techniques previously learned. When you double click on the fillet, all dimensions associated with the fillet will be displayed. That is, every dimension for each vertex point will appear on screen, and the dimensions can then be altered as required.

Movable Fillet Points

When an edge is selected that you wish to apply a variable radius fillet to, only two vertices appear in the vertex list. What happens if you want to dictate what the radius should be at certain positions along the edge? In a situation such as this, you would select one or more of the movable points on the edge, which you could then supply radial values for. An example of what you would see on screen during this process is shown in Figure 20–60. The inset shows a portion of the panel related to utilizing the movable points.

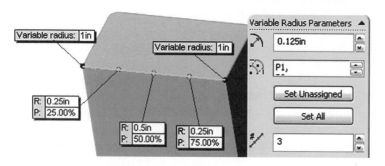

Fig. 20–60. Utilizing movable points.

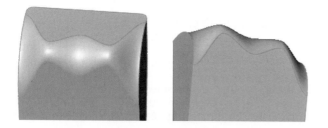

Fig. 20–61. Three movable points were added when defining this variable-radius fillet.

As can be seen in Figure 20–60, there is a setting that dictates the number of points that will be added to the edge. This is what you would set first. Following that, select the points themselves, directly from the work area. Clicking on the points places them in the vertex list, where they are listed as P1, P2, and so on. You can then specify radial values for the points in the same manner values were applied to the vertices.

To optionally specify a position for the movable points, enter a percentage value in the callout associated with the point. Unfortunately, length values cannot be specified. By default, the points will be evenly spaced along the edge. You will be able to change the position of the points after defining the fillet, but the number of points used is not something that can be altered once the fillet has been created. A variable-radius fillet defined with three movable points is shown in Figure 20–61. The fillet is shown from two angles so that you can better discern its shape.

✓ **TIP:** *To add vertex points at specific locations, rather than the percentile values used by movable points, use the Split Line command.*

Setback Fillets

Setback fillets are a special fillet type that can be applied to an area where filleted edges meet. The area where the fillets blend can be pulled back into the model, and a smoothly blended transition can still be maintained. This leaves the impression of a filleted corner that has been sanded down to give a nice rounded finish.

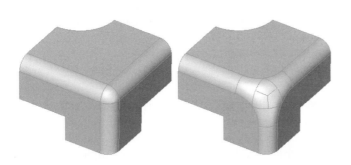

Figure 20–62 shows three edges, with a generic fillet on the left and a setback fillet on the right. The tangent edges have been highlighted so that the patchwork of the blend can be seen. How-To 20–9 takes you through the process of creating a setback fillet.

Fig. 20–62. A regular fillet versus a setback fillet.

HOW-TO 20-9: Creating a Setback Fillet

A setback fillet, by definition, should be created at a place where edges are coming together on the model. The setback area will be applied to a vertex point where the edges coincide. Keep this in mind when selecting edges where the setback fillet will be created. To create a setback fillet, perform the following steps.

Fig. 20–63. Setback Parameters panel.

1. Select Fillet from the Insert > Features menu, or click on the Fillet icon.

2. Leave the fillet type set to *Constant radius*.

3. Select the edges to be filleted. The edges should terminate at a common vertex.

4. Specify a radius for the fillets (creating a multiple-radius fillet is optional).

5. Click in the Setback Vertices list box found in the Setback Parameters panel, shown in Figure 20–63, and select the vertex point where the edges meet. The edges common to the vertex point will be listed in the Setback Distances list box as E1, E2, and so on.

6. Select an edge from the Setback Distances list box and specify a distance (setback value) for the selected edge.

7. Repeat step 6 for every edge in the Setback Distances list box.

8. Click on OK.

When double clicking on a setback fillet to access the feature's dimensions, all of the relative dimensions will appear, including the setback values. There should be a dimension displayed for each of the setback distances, along with the radial dimension for the edges being filleted. If the *Multiple radius fillet* option is used (Chapter 6), dimensions will appear for each edge filleted. This falls in line with standard SolidWorks operating procedure.

Face Fillets

A face fillet can often be used where no other fillet type will work. Face fillets can also be used in very difficult areas, where geometrically a fillet would be impossible to incorporate. Face fillets can completely remove faces of existing geometry in order to add a fillet. Consider a simple example.

In creating the fillet, the software must completely eliminate two faces. Generic fillets often will not allow for this type of behavior. That is why a face fillet can often prove a blessing when trying to complete a design in a situation in which a standard fillet just will not work. The same part with the fillet added is shown in Figure 20–65. How-To 20–10 takes you through the process of creating a face fillet.

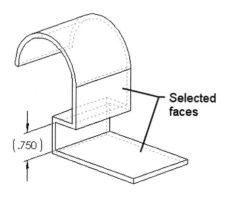

Fig. 20–64. Prior to creating a face fillet.

Fig. 20–65.
After adding
the face fillet.

HOW-TO 20-10: Creating a Face Fillet

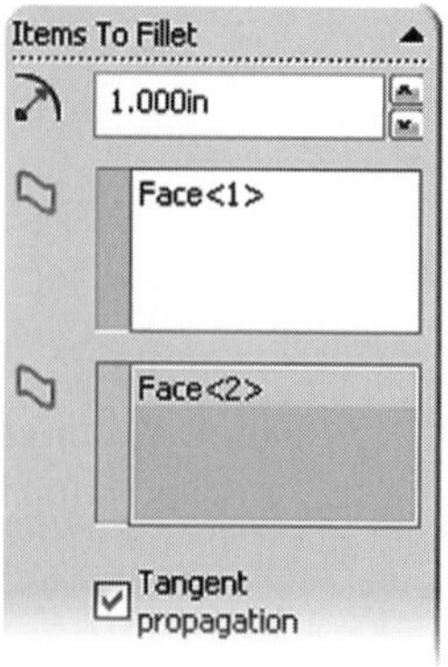

To create a face fillet, perform the following steps.

1. Select Fillet from the Insert > Features menu, or click on the Fillet icon.

2. For the Fillet type, select *Face fillet*.

3. Click in the Face Set 1 list box and select the first set of faces to be filleted (see Figure 20–66).

4. Click in the Face Set 2 list box and select the second set of faces to be filleted.

Fig. 20–66. Selecting face sets.

5. Specify a value for the radius.

6. Click on OK.

During steps 3 and 4, faces are selected that the fillet will be blending between. One set of faces should reside on one side of where the fillet will be defined, and the other set of faces should reside on the opposite side. Accidentally placing a face in the wrong face set list box will likely cause the face fillet to fail.

Using Hold Lines

An important aspect of face fillets is the ability to use what are known as hold lines. Hold lines control the fillet by establishing where the fillet will be tangent to the faces being filleted, thereby determining what the radius will be. For this reason, a radial value for the fillet does not need to be supplied. For that matter, the Radius setting is not even available.

Figure 20–64 shows a simple part that serves as a good example of what a face blend fillet can accomplish. The object is to blend the two faces (shown in Figure 20–64), thereby filling in the opening, which measures .750 inch.

When performing a face fillet, the hold lines can be supplied via the Fillet Options panel, shown in Figure 20–67. After running through a couple of examples, you will have a much better idea of exactly what can be accomplished with hold lines when filleting.

Figure 20–68 shows an example of a free-form shape part that will have a face fillet applied to it. Note the edges that will be selected for the hold lines (including an edge on the other side of the part not visible in the image). The hold lines will determine the radius of the fillet. The fillet will be tangent to the faces in Face Set 2, where the fillet meets the hold line. With *Tangent propagation* enabled, only a single face will need to be selected for Face Set 2.

Fig. 20–67. Where hold lines are specified.

Fig. 20–68. Geometry that will be selected for the face fillet with hold lines.

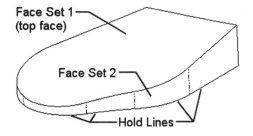

Figure 20–69 shows the same part with a face fillet added. Because hold lines were used, the radius of the fillet was made to vary around the top out-

er perimeter of the part. As stated previously, the fillet is tangent to where the perimeter faces of the part used to be. Because of this, a part such as the one pictured would make a nice cover for some sort of container with vertical walls. The transition between the cover and container would be perfectly smooth.

The *Help point* option is very rarely used. The only time a help point is needed is when an ambiguous situation arises while creating a face fillet. Sometimes it is not clear to the software where the face blend should take place. In other words, SolidWorks is not sure where to position the blend. When the *Help point* option is used, a point can be selected that is positioned near where the blend should occur. The faces closest to the help point are filleted. This is an option that most SolidWorks users need not concern themselves with.

Fig. 20–69. After adding the face fillet with hold lines.

Split Lines as Hold Lines

A split line can also be used as a hold line when creating a face fillet. The results are very interesting. The fillet will wind up being tangent to the face being filleted at the hold line. This situation is similar to that discussed in the previous section. Examine Figure 20–70, which shows a simple part with a split line (left). The split line was created using an ellipse, and is used as the hold line when creating the fillet.

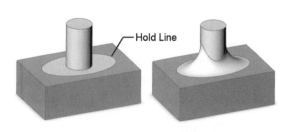

Fig. 20–70. Using a split line as a hold line.

When the face fillet is performed on the part, the cylindrical face is one face set, and the face within the boundaries of the elliptical split line is the other. The split line is selected as the hold line. Figure 20–70 (right) shows the outcome of the face fillet.

It should be noted that hold lines can be used for both face sets. This situation only works under certain conditions. Use common sense when employing multiple hold lines. If the fillet is physically impossible to define, SolidWorks will obviously fail to create it.

Constant Width Fillets

Constant width is a fillet option only available when creating face fillets. The option is found in the Fillet Options panel. Turning the option on results in a fillet as shown in Figure 20–71. As a comparison, a standard fillet and a fillet created with a hold line are shown. All fillets are the same radius, and have been applied to the intersection of where a small cylinder intersects a larger cylinder.

Fig. 20–71. Fillet comparison.

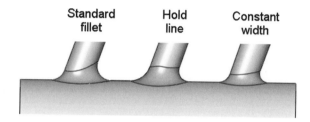

Continuous Curvature Fillets

When performing a face fillet, the radius of the fillet is consistent. Its curvature does not change. This is true for generic edge fillets as well. In certain cases, often for aesthetic reasons, it is desirable for the curvature to change gradually. This can be done through the use of the *Curvature continuous* option found in the Fillet Options panel. The option is only available when performing a face fillet.

In SolidWorks terms, when a fillet is said to be "continuous" the meaning is that the curvature is continuously changing in a gradual manner, rather than abruptly. This fact can be checked in a number of ways. We will learn how curvature can be examined shortly. First, there is one last fillet type to explore.

Full Round Fillets

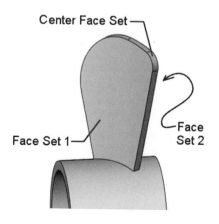

Fig. 20–72. Face selection for a full round fillet.

As the name implies, a full round fillet imparts a fully rounded face on an existing face of the model. This is similar to a face fillet, both in physical aspects and in how the model faces are selected. Three face sets must be selected when creating a full round fillet.

The model shown in Figure 20–72 is of a plastic fan, and the figure shows a single fan blade. Also shown are the faces that will be selected when applying the full round fillet. Face Set 2 is the face on the back of the fan blade. The Center Face Set represents all faces between Face Set 1 and Face Set 2. As long as the Tangent propagation option is turned on, it is only necessary to select the one face for the center face set.

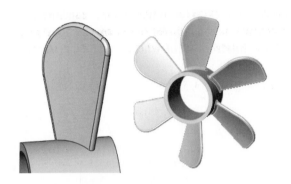

Full round fillets do not have to be placed on material that has a uniform thickness. A radius value is not required, as the software calculates what the radius must be as defined by the thickness of the center faces. Figure 20–73 shows the outcome of creating a full round fillet on the fan blade. The completed fan is also shown.

Fig. 20–73. Completed full round fillet.

Inspecting Curvature

What is curvature? It is a way of expressing the shape of nonplanar faces. Planar faces have a curvature of 0, as they are flat. Nonplanar faces will have a consistent curvature if they are, for example, cylindrical. More exotic surfaces, such as those often created by lofting, may contain a curvature that changes over its entire face. Curvature is expressed as the reciprocal of the radius. Therefore, if a fillet has a radius of .25 its curvature is 4.

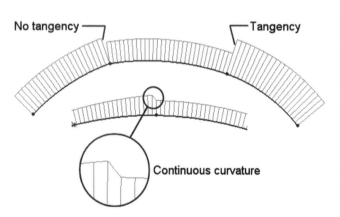

Fig. 20–74. Evaluating the curvature of a spline.

To inspect the curvature of a sketch entity, right-click on the entity and select Show Curvature Combs. Figure 20–74 shows the curvature combs for a number of objects. The curvature comb size can be adjusted via a slider bar in PropertyManager (not shown). Combs update dynamically as sketch entities are moved or repositioned. Try this for yourself by dragging some sketch geometry while the combs are visible.

What do the combs actually tell us? If tangency is present, the combs will touch side by side, though there may be a step between the combs. This can be seen in Figure 20–74. If there is no tangency, there will be an angular gap between the combs, or they may overlap. If combs have a gradual transition, rather than a step, the two sketch entities exhibit continuous curvature.

It should be noted that any sketch entity that has curvature can have its curvature information displayed. This includes arcs, circles, and parabolas. Entities such as lines and rectangles do not contain curvature, so obviously there will be no option for displaying their curvature information.

Face Curvature

When it becomes necessary to know what the curvature is of faces on a model, you have two choices. One is to use the Curvature option in the View > Display menu. This results in the model being displayed in a multicolor gradient, which indicates the degree of curvature present on the model. Moving the cursor over the model displays the curvature and radius of curvature at any given point. This is shown in Figure 20–75, though the gradient colors are lost to the grayscale image. Areas of zero curvature are displayed as black.

Fig. 20–75. Displaying curvature data.

Curvature: 1.55657 Rad. of Curvature: 0.642437

Be forewarned that turning on curvature display can result in lengthy calculations. The math involved is extensive, and it may take some time to calculate the model's curvature. Once calculated, however, the data is saved with the model. Displaying curvature a second time will take little time unless the model has been edited. To control the colors associated with the curvature values, use the Curvature button found in the Colors section of Document Properties (Tools > Options).

A second display option with regard to curvature is known as zebra stripes. Use the View > Display menu once again, and select Zebra Stripes. The Zebra Stripes PropertyManager (not shown) is self-explanatory and can be used to control the number of stripes, the width and accuracy of the stripes, and even the color of the stripes.

Zebra stripes can be cube mapped or spherically mapped on the model. Think of the model as having a reflective mirror coating, and there are black and white stripes painted on the ceiling. The stripes on the ceiling can either be straight or circular.

With zebra stripes, a number of factors regarding model faces can be determined. If the stripes do not continue unbroken across two faces, this indicates that the faces are not tangent. If the stripes continue unbroken, the faces are tangent. Furthermore, if the stripes bend gently across two faces the curvature between those two faces is changing gradually. Figure 20–76 illustrates this.

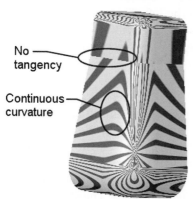

No tangency

Continuous curvature

Fig. 20–76. Utilizing zebra stripes.

It is difficult to properly show zebra stripes in a static image. Often, the model must be rotated in order to make a proper determination of the status of the curvature between two model faces.

Reordering

In the early chapters of this book, you became aware of the nature of FeatureManager, and the dependencies created therein. Think of FeatureManager as a chronological or historical list of events. Features created early in time are at the top of the list, and features created most recently will be at the bottom.

It is possible to change the place in time a feature was created. In other words, if you wanted a feature to exist earlier in the chronological history of events you could position the feature higher up in the FeatureManager list. This is known as reordering, which works in either direction. You can reorder a feature up or down in the FeatureManager list. Parent/child relationships, discussed in Chapter 11, play a very important role in the reordering process.

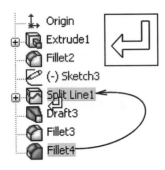

It is not a requirement to check the parent/child relationships of a feature prior to reordering the feature, but if you run into problems it is possible that checking the parent/child relationships will help. Be aware that a child feature cannot be reordered before its parent, and vice versa. After reordering, SolidWorks will rebuild the part automatically, with the features in the new order you specified.

Reordering is done via a simple drag-and-drop operation. This is illustrated in Figure 20–77, which has been enlarged for clarity. When preparing to drop the feature being reordered, there will be a small arrow pointing down and to the left. This is a reminder that the feature you are reordering will be placed immediately after whatever feature you drop it on top of.

Fig. 20–77. Reordering a feature.

Summary

This chapter began by teaching how multiple bodies can be utilized while creating a part. Bodies can be combined, thereby adding the geometry of one body to another. Bodies can also be subtracted from each other, or from the common intersection found between two bodies.

Lofted features can be used to create a wide variety of shapes. A lofted feature requires a minimum of two closed sketch profiles. The sketch planes do not need to be parallel. Quite the contrary, the sketch planes can be at any angle or position. The loft will blend the profiles to create a solid. The sketch geometry used to create a lofted feature should have the same number of segments per sketch. The Split Entities sketch tool is an excellent tool for accomplishing this.

When lofting, pay special attention to where you select the geometry to be lofted. One click per sketch in the work area is all it takes to select the geometry, but that click must be near the vertex points you are attempting to match during the loft operation.

Guide curves can be used while lofting to control the shape of a profile. The Pierce command should be used whenever a guide curve is used. Good technique is to create the guide curve first, and then create the profiles to be lofted. The profiles can then be pierced by the guide curve.

The Centerline option can be used while lofting to guide the direction the loft takes. No pierce relation is needed when using the Centerline option. In addition, the centerline can be any sketch entity, and should not literally be a centerline. Guide curves and the Centerline option can be used simultaneously.

Loft start and end constraints allow for controlling how the loft propagates from the start and end profile planes. The loft can be made tangent to the profile plane's normal (perpendicular vector), a specified vector (which can be a sketch line or existing edge), or tangent to existing faces. It is also possible to create lofted features that exhibit continuous curvature between the new lofted feature and existing faces.

Dome features allow for creating domed shapes on faces. An elliptical dome can be created, which means that the dome will be tangent to the selected plane's normal. The Flex command allows for performing various operations on a model, including bending, twisting, tapering, and stretching. The Deform command will deform geometry based on points, curves, or bodies. The Shape command is used for creating pressurized effects on parts. A pressure value can be applied to a face of a model, giving it a look as though some force were pushing outward on the face.

The Flex, Deform, and Shape commands all require a fair amount of processor power. If there are other, more simple, ways of creating an effect similar to what you are trying to achieve it is usually better to take the simplistic approach. The Flex, Deform, and Shape options will typically significantly increase rebuild times, and they have a tendency to create parts that are more troublesome to annotate.

A number of advanced fillet types were covered in this chapter. Variable-radius fillets were discussed, along with setback fillets, face fillets, and full round fillets. When creating a variable-radius fillet, edges must be selected. Selecting edges populates the vertex list box with vertex points that can then be given radial values. Optionally, movable points can be added to an edge to gain more control over radial values at particular locations along that edge.

Face fillets can often be used to create a fillet where no other fillet type will work. A face fillet has the power to completely absorb other faces in the model, which makes it a very powerful filleting command. Hold lines can be used that will determine the radius of the faces between which the fillet is applied. When hold lines are used, the fillet will be tangent to the faces adjacent to the hold line where the faces meet the hold line. Split lines can be used as hold lines, which can make it very easy to control the area where a fillet occurs.

Continuous-curvature fillets can be created when a face fillet is performed. Fillets of this nature have a gradually changing curvature. Use the Curvature or Zebra Stripes display options to check curvature. Zebra stripes can also be used to determine tangency conditions between faces.

CHAPTER 21

Surfaces

LET'S START THIS CHAPTER BY ASKING A QUESTION: What are surfaces, and why do we need them? The first portion of this question is easiest to explain. Surfaces are the faces that make up a solid model. A single face on a model can be any shape, from a planar face to a complex, wavy shape. Surfaces have no thickness, and can almost be thought of as pieces of paper.

A solid model is a set of surfaces (faces) that form a closed volume. This volumetric area is understood by the software. When the density of the model is considered, information such as mass properties can be obtained. The mass properties are just a single aspect of the downstream benefits of solid modeling you have been discovering throughout this book.

If solid modeling contains more information and has more downstream benefits, why even bother with surfaces at all? There are certain shapes that are very difficult to create without some sort of surface manipulation. The end result may be a single solid model, but sometimes it takes surfacing commands to get there. Models sometimes require manipulation at a more rudimentary geometric level, and that is where surfaces come into play.

Basic Surface Commands

There are four basic surface commands that are almost identical to the four basic feature commands. The four commands are found in the Surface menu (Insert > Surface) and are named Extrude, Revolve, Sweep, and Loft. These four basic surface commands are nearly identical to their solid feature counterparts of the same name found in the Features menu (Insert > Features). The only difference is in what is created; that is, surfaces versus solid geometry.

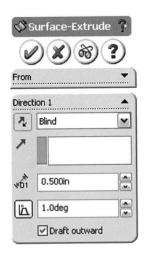

Fig. 21–1. Surface Extrude PropertyManager.

Figure 21–1 shows the Surface Extrude PropertyManager for creating an extruded surface. If the image looks familiar, that is good because it should. For all intents and purposes, we may as well be looking at the Boss Extrude PropertyManager. To reiterate, revolved, swept, and lofted surfaces exhibit characteristics identical to their counterpart solid feature commands.

Considering the similarities between the four basic surface and solid feature commands, it is not necessary to examine the processes of these surface commands. If necessary, review the solid feature commands as needed. Extrusions were discussed in Chapter 3, revolved features in Chapter 5, and swept features in Chapter 10. Lofted surfaces were actually explored in Chapter 14, with lofted features being discussed in Chapter 20. On a final note, planar surfaces can be thought of as a fifth basic surface type. See Chapter 14 for a discussion of the Planar Surface command.

Offset Surfaces

Offset surfaces are easy to create, and can be very beneficial when combined with the ability to cut with a surface. Cutting with surfaces is explored in a moment. First, let's look at the Offset Surface PropertyManager, shown in Figure 21–2. It has very few options and is straightforward and easy to use.

Fig. 21–2. Offset Surface PropertyManager.

Once the command is implemented, you would select a surface (or face) to be offset, specify the offset distance, and specify the direction for the offset through the use of the Flip Offset Direction button, which can be seen to the left of the offset distance parameter in Figure 21–2. It is possible to offset more than one surface or face at a time, and surface offset distances can be zero.

Cutting with Surfaces

Surfaces and planes can both be used to cut solid geometry, and the same command is utilized in either case. The command is appropriately named Cut With Surface, and was used to create the ridge pattern at the top of the hook shown in Figure 21–3. Let's explore how this was done.

Fig. 21–3. Hook with ridge pattern.

The ridges need to be at a uniform height of .060 inch, arranged radially around the top portion of the ring. They should also exist on both sides of the ring. There are a few problems faced in trying to create the ridge pattern. For instance, using the Offset From Surface extrusion end condition does not work, because the ring surface is a single surface. SolidWorks cannot determine where the offset should be taken from because the surface exists both in front of and behind the sketch being extruded.

There are a few other modeling dilemmas faced when trying to create the ridges. Rather than focus on various issues that could arise, let's focus on how to avoid the issues in the first place. The ridges can be created with greater flexibility if the problem of controlling the ridge height is solved first. Therefore, an offset surface will be created that will later be used to cut the ridges to the proper height.

Fig. 21–4. The ridge pattern prior to cutting to the proper height.

The offset surface must be created first, for the simple reason that the ridges change the surface geometry once they are created. It will be necessary to offset the ring's face the same amount as what the height of the ridges should be. The ridges can then be developed and patterned. Figure 21–4 shows the offset surface, along with the ridge pattern. The offset surface is shown semitransparent for clarity.

Next, the cut can be made, so this is a good time to step through the process. How-To 21–1 outlines the steps required for cutting with a surface. Keep in mind that any surface (or plane) can be used as a cutting tool. It just so happens that an offset surface is used in this case.

HOW-TO 21-1: Cut with Surface or Plane

To cut with a surface or plane, perform the following steps. It is assumed that a surface or plane has already been created.

1. Select Insert > Cut > With Surface, or click on the Cut With Surface icon, typically found on the Features toolbar.

Fig. 21–5. Surface Cut PropertyManager.

2. Select the surface or plane to cut with.

3. Click on the Flip Cut button if necessary, found on the Surface Cut PropertyManager (shown in Figure 21–5). The Flip Cut button determines what portion of the model will be discarded, which will be indicated by the preview arrow displayed in the work area.

4. Click on OK to complete the cut.

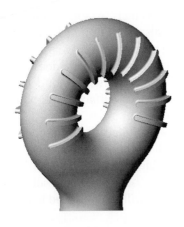

Figure 21–6 shows the hook ring after performing the cut. The image is identical to that shown in Figure 21–3, except that the cosmetic fillets have not been added. The original offset surface has been hidden for clarity, but also because there would be no reason to show it anymore anyway. The offset surface had one purpose, and that purpose has been fulfilled.

Fig. 21–6. Completed ridge pattern.

Radiated Surfaces

A surface command that once played a large role in the making of a mold is the Radiate Surface command. The mold tools discussed in Chapter 14 are better suited for that job now, but radiated surfaces can still be useful.

A radiated surface is named for the way an edge can be made to radiate outward in a particular direction, thereby defining a surface. The radiate direction is defined by selecting a plane or planar face, and is parallel to that selected object.

A good example of where a radiated surface could prove useful is in the creation of a nest. Think of a nest as a holding area where another part can rest while it is being machined, almost like a fixture. Such might be the case with the wedding vase shown in Figure 21–7. The odd shape of the vase and its relatively thin wall make it an unlikely candidate for clamping into a fixture.

Because the radiated surface requires edges, the Split Line command (Chapter 6) is often a prerequisite. Faces, such as the planar face at the bottom of the vase (not shown), were split so that an edge would run across the middle of the face. How-To 21–2 takes you through the process of using the Radiate Surface command.

Fig. 21–7. Wedding vase.

How-To 21-2: Creating a Radiated Surface

1. To create a radiated surface, perform the following steps.

2. Select Insert > Surface > Radiate, or click on the Radiate Surface icon, found on the Mold Tools toolbar and the Surfaces toolbar.

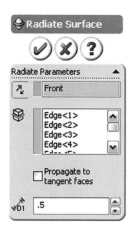

Fig. 21-8. Radiate Surface PropertyManager.

3. In the Radiate Surface PropertyManager, shown in Figure 21-8, select a plane or planar face that will determine the direction of radiation. The radiate direction will be parallel to the selected object.

4. Click in the *Edges to radiate* list box and select the edges from which the surface will radiate outward.

5. Click on the Flip Radiate Direction icon if necessary. This will reverse the radiate direction indicated by the small preview arrows.

6. Optionally, select the *Propagate to tangent faces* option. This reduces the number of edges the user has to pick. SolidWorks will select all tangent edges automatically.

7. Specify the radiate distance.

8. Click on OK to create the radiated surface.

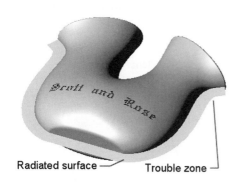

Fig. 21-9. Radiated surface.

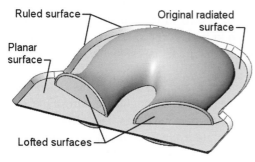

Fig. 21-10. Surfaces have been added.

The option Flip Radiate Direction is typically not needed because by default the selected edges will radiate outward from the part. The large preview arrow shows the normal of the face selected, which will be perpendicular to the radiate direction. An example of a radiated surface is shown in Figure 21-9.

Figure 21-9 shows an area where problems can occur if the radiate distance is too large. There are actually two such areas, one on either side of the model. If an inside curve is made to radiate beyond its zero radius point, the radiated surface will likely fail. That is, if an inside arc has a radius of 1 inch the radiated surface cannot exceed 1 inch.

In keeping with the original idea of creating a nest for this part, it is fairly obvious that more surfaces are in order. Figure 21-10 shows the reverse side of the vase after creating additional surfaces.

Although you have learned all of the skills necessary to create the surfaces shown in Figure 21-10, it might help to understand some of the finer details of how they were put together. For example, the planar surface was created through a mix of converted edges (using the Convert Entities command) and a few additional lines and arcs. The lofted surfaces are a bit more interesting.

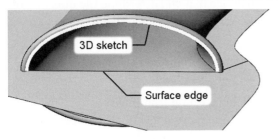

Fig. 21–11. Building the lofted surface.

To create each lofted surface, a 3D sketch was used, along with an existing surface edge. The 3D sketch was created by first clicking on the 3D Sketch icon and then converting the edge shown in Figure 21–11 (once again, using Convert Entities). The edge has been enhanced in the image so that it will stand out more clearly.

At that point, exit the sketch and create the lofted surface by selecting the 3D sketch and surface edge near the same common endpoint. This same process was repeated for the other opening in the vase. After creating all surfaces shown in Figure 21–10, a final planar surface was created. This final surface was created on a plane strategically positioned a small distance below the ruled surface. The reason for this becomes clear in Figure 21–12. Note that there is a small sliver of the ruled surface visible above the large planar surface (left).

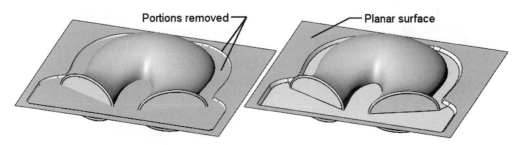

Fig. 21–12. Preparing for a trim operation.

When performing the trim operation, the ruled and planar surfaces are mutually trimmed to each other. This results in the outcome shown in Figure 21–12 (right). All that is left is to knit all of the surfaces, along with any faces on the reverse side of the model (the model faces visible on the right in Figure 21–12).

After knitting the appropriate surfaces and model faces, the outcome is very similar to a thermal-formed piece of plastic. Imagine drawing a piece of hot plastic over the vase and that is basically what has been accomplished in SolidWorks. The final task is to create a plane some distance from the "plastic," create a sketch, and extrude it up to the newly created knit surface. Remember to uncheck the *Merge results* option, as the nest should remain separate from the original vase. Figure 21–13 shows the completed nest, with the body of the vase hidden.

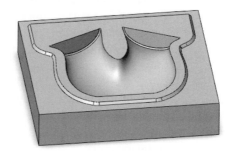

Fig. 21–13.
Completed nest.

Trimming and Untrimming

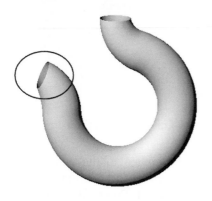

Fig. 21–14. Hook in its infancy.

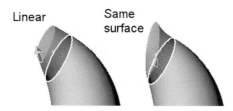

Fig. 21–15. The surface should be extended linearly.

Chapter 14 gave you the basics of the Trim Surface and Extend Surface commands. Here we will explore a few additional techniques for trimming and extending surfaces. The hook will be used again, but at a much earlier stage in its development. Figure 21–14 shows the hook after creating a swept surface. The area we will be focusing on is the end of the hook that should come to a soft, rounded point.

Because the end of the hook does not extend quite as far as we would like it to, the Extend command could be used to draw it out a short distance. Figure 21–15 shows what the results would look like using the Extend Surface command. In the case of the hook, the Linear option will be used because the results are more desirable. Sometimes the *Same surface* option can cause the surface to start wrapping into itself, in a manner of speaking. There is usually a maximum limit to how far a surface can extend, and this is totally dependent on the surfaces being extended and extension option used. Often, the option used is dependent on aesthetics.

✓ **TIP:** *Dragging the preview arrow with the left mouse button can determine the extension length.*

Before moving on, let's examine the newly extended surface a bit more closely. Note in Figure 21–16 that the juncture between the original surface and the extended surface can be seen. The imperfection is not huge, but it is enough that you would not want it to appear in the finished product. What can be done about this?

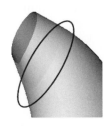

Fig. 21–16. An area of incongruity between two surfaces.

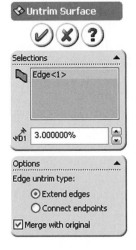

Fig. 21–17. Untrim Surface PropertyManager.

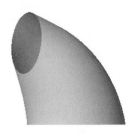

Fig. 21–19. After untrimming the end of the hook.

Rather than extending the surface, the surface can be "untrimmed." Extending a surface and untrimming a surface are very similar. Extending a surface actually creates a new surface, whereas untrimming a surface does not. The Untrim Surface PropertyManager, shown in Figure 21–17, is accessed via the Insert > Surface menu.

Untrimming a surface works much the same way extending one does. That is, an edge or edges to be untrimmed can be selected, and a distance specified. The distance is specified as a percentage value when untrimming, as opposed to a length value when extending. The options *Extend edges* and *Connect endpoints* will produce the results shown in Figure 21–18. The *Connect endpoints* option is used in the example on the right.

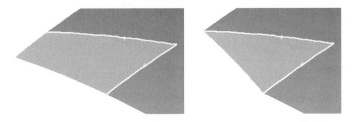

Fig. 21–18. Extend edges *versus* Connect endpoints.

If the Untrim Surface command is applied to the end of the hook, the results are much cleaner than when using the Extend Surface command. Figure 21–19 attests to this fact. Compare this to Figure 21–16 and the difference in the outcome is clear to see. There is no incongruity between surfaces because there is only one surface.

Let's move forward slightly in the development of the hook. Figure 21–20 shows the hook after adding a revolved surface that will become the ring. Two sketches have also been created, and are visible in the image. The two sketches will each be used to trim away portions of surfaces that are not needed.

Chapter 14 taught how surfaces could be trimmed to other surfaces, but it isn't necessary to create a surface for this task if the trimming surface is not otherwise needed. Trimming a surface with a sketch, for all intents and purposes, works similar to a cut-extrude command on a solid. Create a sketch that will describe the areas to be trimmed away, and then perform the trim. The sketch geometry cuts (trims) the surface, and it is just a matter of deciding on which piece of the surface you wish to keep.

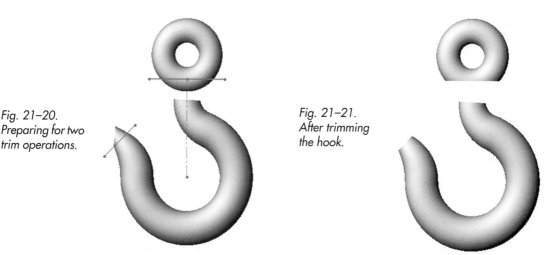

Fig. 21–20. Preparing for two trim operations.

Fig. 21–21. After trimming the hook.

When trimming surfaces with sketch geometry use the same Trim Surface command discussed in Chapter 14. Nothing changes in how the command functions. The only difference is in what is selected: a sketch rather than a surface or plane. Figure 21–21 shows the hook after the trim operation has been performed.

Surface Fill

A lofted surface can be used to fill the gap between the ring and hook surfaces, but what to do with the opening at the end of the hook? If you recall, the design intent was to create a smooth, rounded surface for the hook's point. For this feature, the Fill Surface command will be employed.

When employing the Fill command, all of the edges surrounding the opening should be selected. This can often be accomplished quite easily by using selection techniques discussed in Chapter 6, such as Select Loop. You are then given the ability to specify whether the new surface to be created should be tangent to these edges or simply make contact with them. Mixing things up is allowed, so some edges can be contact edges and some can be tangent. In our case, we would want all tangent conditions. Let's put the Fill command to the test in How-To 21–3.

HOW-TO 21-3: Using the Surface Fill Command

To fill in gaps in surface or solid geometry, use the Fill command as follows.

1. Select Insert > Surface > Fill, or select the Filled Surface icon, located on the Surfaces toolbar.

Fig. 21–22. Fill Surface PropertyManager.

2. From the drop-down menu in the Fill Surface PropertyManager, a portion of which is shown in Figure 21–22, select either Tangent or Contact as the default setting for the edges to be selected in the next step. Edge settings can be changed on an edge-by-edge basis later, if required.

3. Select the edges that surround the area to be filled in.

4. If a Tangent edge setting is used, the new surface will be tangent to a face on one side of a selected edge or the other. If the wrong face is highlighted for a particular edge, it needs to be changed. To most easily accomplish this task, as needed, click on the various edges in the Patch Boundary listing and then click on the Alternate Face button.

5. Using the same technique as in step 4, additionally change the edge settings to either Contact or Tangent for specific edges as required.

6. Click on OK to complete the process.

Fig. 21–23. After creating the filled surface.

Figure 21–23 shows the hook's point after creating the filled surface. The newly created surface is tangent to all of the surrounding faces. The surface is smooth, and blends perfectly with the surrounding geometry.

For sake of argument, let's say that the end of the hook is too pointy. Imagine that our design intent is to more tightly regulate the curvature of the surface used for the hook's point. To accomplish this next task, a small arc will be created at the end of the hook prior to creating a new filled surface. The arc is shown in Figure 21–24, along with the lower portion of the Fill Surface PropertyManager. The arc has been selected and is listed in the Constraint Curves panel.

*Fig. 21–24.
Selecting a
constraint curve.*

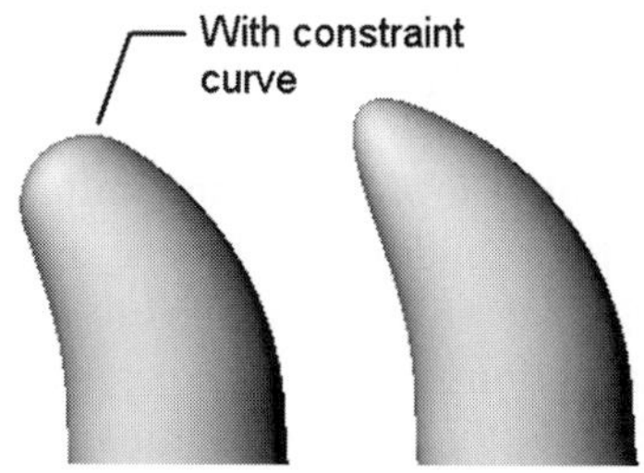

Fig. 21–25. Comparing filled surfaces.

Adding the constraint curve changes the outcome quite visibly. Figure 21–25 shows a side-by-side comparison. The difference is obvious. What is not obvious is whether the hook is solid geometry or surfaces. One option available when creating a filled surface is *Merge result*, which knits the resultant filled surface to the surrounding geometry. When *Merge result* is checked, *Try to form solid* is also available. This keeps us from having to turn the surfaces into solid geometry later on our own, a process which discussed shortly.

The *Optimize surface* option (shown in Figure 21–22) creates a simplified surface that should result in faster rebuild times. It would be beneficial to check this option whenever possible. If the resultant surface does not meet your demands, uncheck *Optimize surface* and use the Resolution Control slider bar instead (not shown).

NOTE: *At this point, the only surface command not discussed in this book is the Mid-Surface command. Mid surfaces have greater importance relative to finite element analysis, which is outside the scope of this book. Therefore, they will not be discussed.*

Solidifying Surfaces

If surfaces were used to create a model, you would find that certain functions typical of solid geometry are not available. The mass properties of the model, for example, could not be determined. Other problems may arise as well, such as when attempting to add fillets to surface geometry. It would be in your best interest to turn the surfaces into a solid when the first opportunity presents itself. The solidification process involves the Knit Surfaces command.

✓ **TIP:** *If you are trying to add a fillet between surfaces, and a warning message appears regarding laminar edges, knit the surfaces prior to filleting them.*

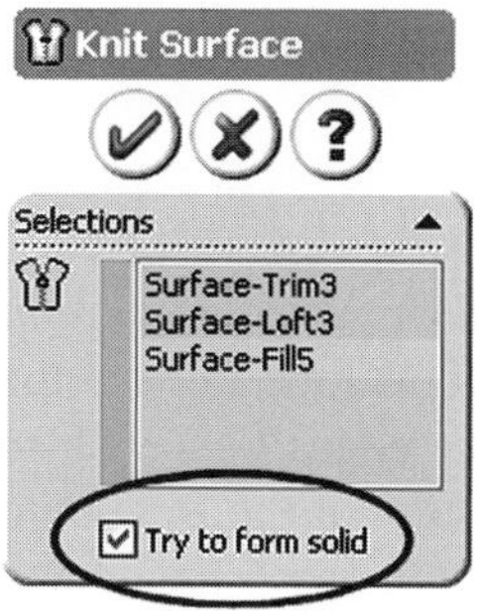

Knitting is a simple matter, but it is a requirement if you wish to solidify a surface model. The model should form a completely closed volume, or the model will refuse to solidify. The Knit Surface command was discussed in Chapter 14, so it is not necessary to revisit the command. If you wish to turn the resultant knit surface into a solid, simply check the *Try to form solid* option in the Knit Surface PropertyManager (shown in Figure 21–26).

Fig. 21–26. The Try to form solid *option.*

Perhaps you have a model made up of a single surface. Perhaps the model does not need to be knitted, it just requires being turned into a solid. If this is the case, use the Thicken command.

Thicken can be used in two ways. In its simplest form, the command allows for adding thickness to a surface. This creates a solid from a surface, adding a thickness as specified by the user. This thickness can be applied to one side of the surface or the other, and there is an option for mid-plane as well. How-To 21–4 takes you through the process of using the Thicken command.

HOW-TO 21-4: Thickening Surfaces

To thicken a surface and turn it into a solid, perform the following steps. Also use these steps to solidify a surface model. Surface models must form enclosed volumes with no gaps in order for the solidification process to work.

Fig. 21–27. Thicken PropertyManager.

1. Select Thicken from the Insert > Boss/Base menu.

2. Select the surfaces to be thickened.

3. Specify a thickness for the surfaces in the Thicken Property-Manager, shown in Figure 21–27.

4. Specify what direction the wall thickness should be applied using the Thickness buttons visible in Figure 21–27.

5. If the surface to be thickened is a closed volume surface, optionally select the *Create solid from enclosed volume* option. This will negate the need to perform steps 3 and 4.

6. Click on OK to complete the process.

The *Create solid from enclosed volume* option will not even be present unless the surface forms a closed volume. This can serve as a good check to see if there are gaps in a model. There are also tools available to troubleshoot surface geometry. Tools of this nature are related more closely to working with imported geometry, and therefore will be discussed in Chapter 24.

Deviation Analysis

To see what the deviation is between two faces, a tool known as Deviation Analysis can be used. This command is found in the Tools menu. Once the Deviation Analysis PropertyManager appears, select the edges you wish to check and click on the Calculate button. The upper portion of the Deviation Analysis PropertyManager is shown in Figure 21–28.

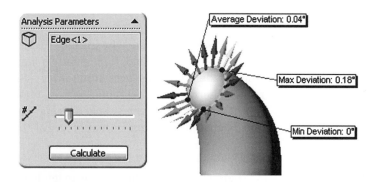

Fig. 21–28. Deviation Analysis PropertyManager.

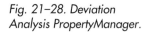

Once the Calculate button is pressed, multicolored arrows will appear on the selected edge or edges. The effect cannot be appreciated in a gray-scale image, though the arrows can be seen in Figure 21–28. To increase or decrease the number of sample points where the deviation is measured, move the slider bar. It will be necessary to click on the Calculate button again to redisplay the deviation arrows. The callouts (also visible in Figure 21–28) show what the average deviation is, along with the minimum and maximum deviation.

Hiding and Showing Bodies

Similar to planes, sketch geometry, or other objects, bodies can be hidden or shown. This refers to both solid and surface bodies. The method used is the same in either case. Simply right-clicking on a body in FeatureManager gains access to an option to hide or show the body. The available option obviously depends on its current state of visibility.

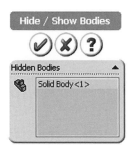

Fig. 21–29. Hide/Show Bodies PropertyManager

Right-clicking on a feature in FeatureManager and selecting Hide will hide the entire body associated with that feature. Individual features cannot be hidden, though they can be suppressed. (Review the section "Suppression States" in Chapter 11 if this is your wish.) Another way of hiding and showing bodies is to use the Hide/Show Bodies command. This command is accessible from the View menu. Once activated, the Hide/Show Bodies PropertyManager (shown in Figure 21–29) appears.

The best part of using the Hide/Show Bodies command is that anything currently hidden appears as a transparent object. You can see exactly what is hidden and what is not. Knowing what is currently hidden is impossible to tell simply by looking at FeatureManager. Use the Hide/Show Bodies command and you can tell everything from a single glance.

Summary

This chapter examined why surfaces can be important in the creation of certain models. Although it is usually advantageous to have a solid as a finished model, using surfaces along the way helps to obtain some shapes that might otherwise prove difficult or even impossible to achieve.

Basic surfaces can be created using surface extrude, revolve, sweep, or loft commands that are nearly identical to their solid counterparts. Lofted surfaces can be created between existing surfaces without the need to first create a sketch.

Surfaces can be trimmed, or extended to other surfaces. Untrimming a surface is a separate function, which closely resembles the Extend Surface command but whose end results differ in that an untrimmed surface does not create a new surface.

Extending a surface creates a new surface, which may deviate from the original. Surfaces can be made to extend naturally or linearly. Use the Deviation Analysis command found in the Tools menu to see what the deviation is between adjacent faces.

To turn a surface into a solid, the surface can be thickened. To create a solid from a series of surfaces that form an enclosed boundary, the surfaces must first be knitted. Knitted surfaces can be solidified during the knitting process.

Many commands that can be used on features can be used on surfaces. This includes patterning, scaling, and mirroring. Surfaces and solid bodies can also be moved, rotated, and copied. Surfaces can also be used as cutting tools to not just trim other surfaces but to cut solid geometry. If working with surfaces in conjunction with solid geometry, use the Hide/Show Bodies command found in the View menu to hide surfaces you are finished with.

CHAPTER *22*

Engraving and Embossing

THIS CHAPTER DISCUSSES METHODS of adding engraved text to a part. The term *engrave* could be taken literally here, but that does not mean to say you could not use other methods to create text on a part in your area of manufacture. For instance, if your part contains text that has been stamped into the part the methods learned in this chapter will still serve you well.

Other closely related scenarios, such as creating raised (embossed) text on a part, are covered as well. Creating raised or stamped text on a curved or irregular surface will also be looked into. You will be able to walk through the steps with the book to see exactly how to create these features. You will also learn how to make text follow an arc or even a free-form curve.

There is a distinction that should be made between adding text as a note and adding text as a feature. The methods (and reasons) used are very different. Notes are typically added to a drawing layout, but can be added to part and assembly files as well. Notes, which were discussed in Chapter 12, are used to convey information and annotate documents. Sketched text, on the other hand, is typically used to create a feature.

Regardless whether engraved or embossed text is your final objective, the place to begin is the same. Create the text using the Text sketch tool, and then create the feature as usual. Letters can be modeled with lines, arcs, and splines, but this is usually not necessary. How-To 22–1 takes you through the process of creating text in a sketch.

HOW-TO 22-1: Creating Text in a Sketch

This How-To represents the first phase in the process of creating a text feature. Throughout the rest of this chapter, you will examine the process for creating the actual feature geometry. Start out just as you would create any other sketch. To add text to a sketch, perform the following steps. Options available when adding text to a sketch will be explored in material to follow.

1. While in sketch mode, select Tools > Sketch Entity > Text, or click on the Text icon found on the Sketch toolbar.

2. Type the appropriate text into the Text panel of the Sketch Text PropertyManager, shown in Figure 22–1.

3. Pick a location for the text. This will be the lower left insertion point of the text string. Picking a location is only to roughly position the text. It can be more accurately positioned later.

4. Click on OK when finished.

5. Use dimensions or geometric relations to properly position the text.

Fig. 22–1. Text panel of the Sketch Text PropertyManager.

Fig. 22–2. Dimensions for the text location.

Normally you would want to position the text string at some specific location. It is possible to drag the text from its lower left insertion point to a new location, but this method is not accurate unless you are satisfied with eyeballing the text location. What works best is to place dimensions to the insertion point of the text from some other point or edge on the model. Figure 22–2 shows an example of this.

When text is used for creating a feature, SolidWorks takes the original font and "vectorizes" it. Microsoft Windows fonts are not geometric sketch entities by nature. The text has to be turned into something that SolidWorks can use, such as lines, arcs, or splines. Although any font can be used to create sketch text, not all fonts will successfully translate into a solid feature. It really depends on the style and complexity of the font used.

When sketch text is used to create a feature, the computations can be extreme. Your computer may balk at being pushed so hard, and the time it takes to create the feature can be lengthy. If there are a lot of text features to add, consider breaking the text up into separate features. The size of each text feature will vary, depending on your system hardware and the font used.

Sketch text does not usually follow the same color code rules as other sketch geometry. Don't worry if the text does not turn black, even after fully defining it. Also, text will align itself with the *x* axis of the origin point. It is possible to rotate text, or create backward text. You can accomplish this using the Modify Sketch command, discussed in Chapter 9. If you need to edit the text, either double click on it or right-click on the text and access its properties. You can do this only while editing the sketch that contains the sketch text.

Aligning Text on a Curve

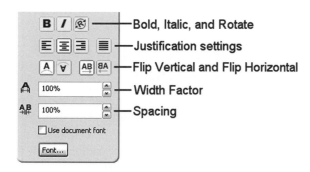

Fig. 22–3. Using curves for text.

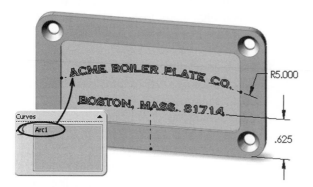

Fig. 22–4. Sketch Text options.

Sketch text often benefits from construction geometry. This helps to align or dimension the text, or to force the text to follow a path. If a curve is selected and placed in the Curves list box, any text entered in the Text panel will follow that curve. The Curve list box is shown as an inset in Figure 22–3. There is a construction arc with a 5-inch radius, and two construction lines, all of which are there for the sole reason of positioning the text. The Text command was used twice to create the two lines of text in the image. This was necessary because each text string needed to follow a different curve.

Text can follow any type of sketch entity, including splines. Even if you are only adding straight lines of text to a sketch, construction geometry will still be beneficial. The reason for this is that there are various alignment options that are only available when the text is made to follow a curve. All of the options present in the Sketch Text PropertyManager are explained in the following material. Figure 22–4 shows the options being discussed.

Use Document's Font

Uncheck the Use Document's Font option if you want to use a font other than the font you are currently using for your basic notes in the part file. This will activate the Font button. Clicking on the Font button will allow you to see a list of all fonts currently installed on your computer. This is the standard font window, containing the options Font Style (such as regular, bold, or italic), Font Height, and Effects (such as Strikeout or Underline).

Spacing and Width Factor

The Spacing and Width Factor settings control, respectively, the spacing between characters and the width of each character. Character spacing is also referred to as kerning. SolidWorks uses a percentage value to dictate the amount of kerning and character width. By altering these values, it is possible to space out a line of text so that it fills a particular area. The same effect can often be accomplished with full justification, but this can sometimes cause the distance between characters to be too great. It often comes down to tweaking settings until the text has the desired appearance.

Figure 22–5 shows a comparison of text where the character spacing and width has been altered (right). In Figure 22–5, the character width factor has been set to 196% and the spacing has been set to 120%. Note that you must uncheck the Use Document's Font option before the spacing and width factor settings become available. The text on the left is the original text with no spacing or width factor adjustments.

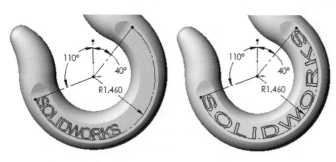

Fig. 22–5. Altering the spacing and width factor settings.

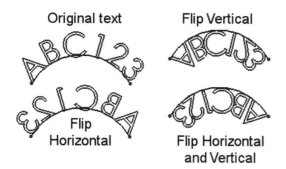

Fig. 22–6. Flipping text.

Flip Vertical and Flip Horizontal

Any text can be made to flip in a horizontal direction. This has the effect of creating text that reads backward (text that could be read if held up to a mirror). Only text made to follow a curve can be flipped vertically. This makes sense, in that it is not necessary to flip text vertically otherwise. Without a curve, flipping text horizontally and then rotating it 180 degrees has the same effect. The need to flip text becomes more apparent when using curves to guide text. Some sample text is shown along a simple arc in Figure 22–6, which shows the various effects that can be achieved by flipping text.

Alignment and Justification

In terms of a sheet of paper, alignment refers to the positioning of the text horizontally. Text can be left, right, or center aligned. In other programs, this is often referred to as left, right, or center justification. Full justification means that the text is spaced out to fill the entire width of the page.

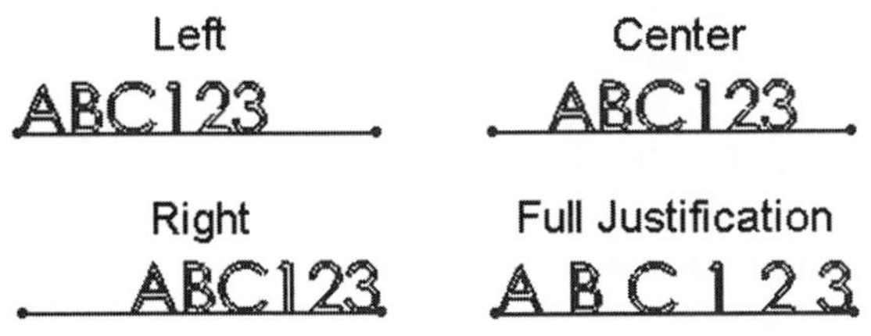

Left   Center

ABC123   ABC123

Right   Full Justification

ABC123   A B C 1 2 3

Fig. 22–7. Alignment conditions.

Because there is no "page" with regard to a SolidWorks model, there can be no alignment or justification unless there is something to gauge it by. That is, the alignment options are not available unless the sketch text is based on a curve. Figure 22–7 shows the three available alignment conditions and an example of full justification.

There is one other aspect of full justification you should be aware of. If you are used to working with publishing software, full justification will increase the spacing between words but leave the kerning (spacing) of the characters in each word alone. This is not the case in SolidWorks.

Figure 22–8 shows two examples of a line of sketch text with full justifi-

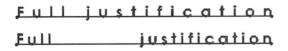

Fig. 22–8. Full justification of a line of text.

cation. The upper example has a single space between the two words in the text string. As can be seen in the upper line of text, the kerning between characters has changed, making it difficult to distinguish each word. Admittedly, this is annoying behavior, but has a simple enough workaround. To overcome this problem, spaces were added between the two words in the lower example, thereby making it easier to distinguish between the words.

Formatting and Rotating

Adding italics or bold formatting to text works much the same way as employing hypertext markup language (HTML) tags. However, if you have never created a web page this fact would mean nothing. Without going into an explanation of HTML tags, let's look at the way formatting is applied to sketch text.

If there is a string of text you wish to apply formatting to, highlight the portion of the string (or the entire string) that requires the formatting. Once that is done, click on the applicable formatting icon, such as Bold, Italic, or Rotate. In the Text box, you will see that "tags" have been placed around the selected text. This allows for adding formatting to specific words in a string of text, an example of which is shown in Figure 22–9.

Figure 22–9 shows what would be seen in the Text box, as well as the end result. The tags themselves can even be typed in manually if you wish. The Rotate tag operates in the same manner as the Bold and Italic tags, but has an additional numeric component. The number associated with the Rotate tag dictates the angle the characters (not the text string) will be rotated. Positive or negative values can be used. Examples of Rotate formatting are shown in Figure 22–10, along with the corresponding text that would appear in the Text box.

Italicized and Bold

Italicized and Bold

Fig. 22–9. Adding formatting to sketch text.

ROTATED   <r-15>ROTATED</r>

ROTATED   <r10>ROTATED</r>

Fig. 22–10. Rotating text.

It should be noted that rotating an entire string of text, rather than just individual characters, can be done with the Modify command, found in the Tools > Sketch Tools menu. The Modify command was discussed in Chapter 9.

Text Features

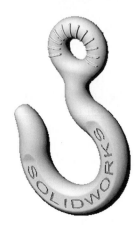

Fig. 22–11. Engraved text.

The simplest type of text feature to create is one in which the text is on a simple planar surface. Both engraved and embossed text features can be created in a similar fashion, with the only difference being that engraved features are created as a standard cut feature and embossed text is created as a boss.

Sketch text can be used to create revolved features, though actual applications requiring this would be limited. For that matter, sketch text can be used for swept or lofted features, but all of the basic rules you have already learned regarding these feature types must still be maintained. Maintaining these rules can be difficult with sketch text entities.

With regard to using sketch text for a boss or cut extrusion, all steps for creating the feature are identical to creating any other boss or cut extrusion. Therefore, it is not necessary to show the steps for this process. Figure 22–11 shows an example of an engraved text feature. The text was cut into the model to a depth of .030 inch.

Text Features on Curved Surfaces

Now that you have a better understanding of surface geometry (specifically, offset surfaces), you can take text features to the next level. Placing engraved text on a curved or irregularly shaped surface is fairly easy, but raised text is another matter. Because of this, engraved text will be tackled first.

There is really only one stipulation to placing engraved text on a non-planar surface, and that is to use the Offset From Surface end condition type. By employing this end condition, it is possible to extrude the text into the solid geometry by a specific amount while retaining a uniform depth. Although this procedure will work in many cases, there is an inherent problem associated with the procedure that cannot always be ignored.

The problem associated with performing an extruded cut into nonplanar faces is that the cut only flows in a single direction (which is typically perpendicular to the sketch plane). Figure 22–12 shows why this is an issue. The cylinder in the image has been sectioned through the top of the word *text* to more clearly show the cut. Note the sharp angles where the cut approaches the silhouette edges of the cylinder. In most cases, the desirable situation would be for the text to cut perpendicular to the surface at every point where it touched the surface.

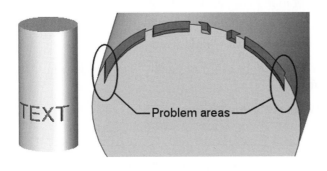

Fig. 22–12. Cut is unidirectional only.

Fig. 22–13. Sketch text about to be wrapped around a cylinder.

The geometry in Figure 22–12 is an exaggerated situation, but it helps to illustrate what occurs when adding cuts to curved surfaces. Where surface curvature is minimal, the cut-extrude method of creating engraved text is sufficient. In cases where text should wrap around a cylinder, the Wrap command works best.

The Wrap command accomplishes what cannot be accomplished with an extruded cut. The one main limitation of the Wrap command is that text can only be wrapped around cylindrical or conical faces. However, wrapping is not limited to text, and any closed-profile sketch can be used. The Wrap command will often work in situations that may surprise you. For example, the text shown in Figure 22–13 will be used for the next example. Even though the text flows beyond the sides of the cylinder, wrapping can still be accomplished. How-To 22–2 takes you through the process of using the Wrap command.

How-To 22-2: Using the Wrap Command

To create engraved text on a cylindrical or conical surface, perform the following steps. Note that wrapping is not limited to sketch text, and that any closed-profile sketch can be used. Also be advised that the sketch plane containing the sketch to be wrapped must be tangent to the cylindrical or conical surface. To review how tangent planes can be created, see How-To 4–10. The following steps assume that a sketch has already been created.

Fig. 22–14. Wrap PropertyManager.

1. Exit from sketch mode.

2. Select Wrap from the Insert > Features menu, or click on the Wrap icon typically found on the Features toolbar.

3. From the fly-out FeatureManager, select the sketch to be wrapped.

4. In the Wrap PropertyManager, shown in Figure 22–14, choose the wrap operation to be performed. Operations can be Emboss, Deboss, or Scribe, explained in material to follow.

5. Select the face the sketch will be wrapped around.

6. Specify a depth (if debossing) or height (if embossing) for the wrapped feature.

7. Click on OK to create the feature.

Fig. 22–15. Debossed (engraved) text.

Figure 22–15 shows an example of the completed wrap feature. The Deboss option was used in the example shown. The section view (upper image) clearly shows how the text points inward toward the axis of the cylinder.

There were a few options that we did not concern ourselves with when creating the wrap feature in the previous How-To. One such option is *Reverse direction*, which would put the text feature on the opposite side of the cylinder. Another option is the *Pull direction* option, which can be used to dictate the cut direction. Specifying a pull direction eliminates the most beneficial aspect of the Wrap command (cutting perpendicular to a surface), so chances are you will not want to employ this option.

The Scribe operation will work similar to the Split Line command, discussed in Chapter 6. That is, no cut or boss is created. Rather, faces are split off from the main face initially selected when performing the wrap. This operation results in

Fig. 22–24. Preparing to create embossed text on a thin-walled part.

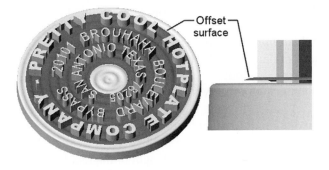

Fig. 22–25. Preparing to cut with the offset surface.

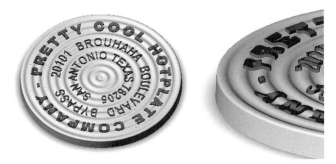

Fig. 22–26. Hotplate with completed text feature.

With a thin-walled model, extruding outward from the inside of the model isn't practical, because there would be extra material left over between the sketch plane and the inside surface of the model. It is much easier to extrude up to the target surface from outside the model, and then cut away any extra material. Figure 22–24 will help to clarify this process. The cutaway view shows the thin wall structure of the part, which happens to be a hotplate. Also note the wavy surface of the part, the sketch text, and the distance the sketch text is from the wavy surface.

The main concept behind creating the embossed text is to create an offset surface first. The surface is used to cut away the excess material after the text feature is created. The offset surface distance should be equal to whatever the height of the text feature will be. Figure 22–25 shows the offset surface (the dark region on the hotplate). The text has also been extruded. All that is left is to complete the final touches.

To complete the project, the Cut With Surface command can be performed, which you learned in Chapter 21. The offset surface can then be hidden, as it has performed its duty. Finally, the text feature can have its color edited so that it will stand out. The easiest way to color the text would be to change the color of both the extruded text and the surface cut features. Figure 22–26 shows the completed hotplate. Note in the close-up (right) that the text feature conforms to the curvature of the original geometry.

Summary

In this chapter you discovered that adding text to a sketch is a simple matter of selecting the Text command from the Sketch Entities menu. Sketch text and construction geometry can be used in unison. Sketch text can be made to follow a curve or set of curves, thereby allowing for curved text in a variety of forms.

Creating engraved text is quite straightforward with regard to flat, planar surfaces. Engraved text is nothing more than a cut extrusion, whereas raised text is a boss extrusion. It is possible to use any font on the computer when creating sketch text, but not all fonts are usable when it comes to creating features.

The Wrap command allows for creating both engraved (debossed) and embossed text features, though the Wrap command is not limited to sketch text. Any closed-profile geometry can be utilized. The Wrap command also has a Scribe function, which breaks out (or separates) portions of a surface, much like the Split Line command would. The separate surfaces can then be colored to make the surfaces stand out.

Creating wrapped features with the Wrap command only works on planar, cylindrical, or conical faces. Creating embossed text on a surface with more complex curvature often requires that first an offset surface be created, which can be used to cut the text feature down to the desired height. When this technique is used, it is possible to adjust the height of the text by adjusting the offset distance of the surface.

Use the Modify Sketch command to help manipulate sketch text. In this way, sketch text can be rotated and mirrored. If using a curve to guide the text, use the Flip Vertical and Flip Horizontal options to help position the text. When using a curve to guide sketch text, there will be more options available in the Sketch Text PropertyManager. Use construction geometry to your advantage, even if creating a simple straight line of text.

CHAPTER 23

Importing and Exporting Files

WHEN WORKING WITH COMPUTER-AIDED DESIGN or drafting software, there is almost always a need to import or export files. This is true whether you are in an area of manufacturing, surveying, mapping, architecture, or any other field. There are always customers, clients, or co-workers you will need to export files to or import files from.

SolidWorks has always been very strong in the area of translating files. So much so that companies have purchased SolidWorks for its translation capabilities, and have then decided to use the software for its modeling capabilities as well. There are numerous file translation options that allow for reading from or writing to almost any other CAD program. File translation is included with the SolidWorks program and is integrated in the command structure to make translating a very user-friendly process.

This chapter covers the processes involved in importing various file types and exporting SolidWorks documents. The various options available when importing or exporting files are explained in detail so that you will be able to make intelligent decisions regarding the translation options. The first half of this chapter discusses importing files, with exporting files discussed in the second half.

Importing Files

There is no special command for importing a file into SolidWorks. Simply use the File > Open command. The type of file to be imported is dictated through the *Files of type* drop-down list, shown in Figure 23–1.

Although different files will have different import options associated with them, the process of importing a file is always the same. Most of the

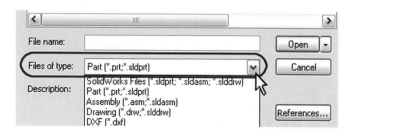

Fig. 23–1. Files of type *drop-down list.*

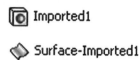

Fig. 23–2. Imported objects
in FeatureManager.

available import options associated with various file formats are discussed
in material to follow. The type of file being imported determines which document type SolidWorks will translate the file into. For example, files with a
.dwg or *.dxf* extension will typically be imported as drawings, whereas an
.igs file will be imported as a part or assembly.

FeatureManager will indicate what SolidWorks did with the imported
file. For example, if the imported geometry was successfully knitted into a
solid model the *Imported1* object will appear in FeatureManager. This object
is shown (enlarged) in Figure 23–2. If *Surface-Imported1* is shown in FeatureManager, this means that the imported geometry was not, or could not
be, knitted into a solid.

If multiple surfaces are listed in FeatureManager, this means that the individual surfaces could not be knitted into a single surface, not to mention a
solid. This is one of the worst possible scenarios, because it likely means
there will be quite a bit of repair work necessary. It could also mean that
your options are set incorrectly. Options that affect imported files are addressed in sections to follow.

It should be noted that feature data is typically not transferred during
any file translation process. This is the case in any CAD program. Occasionally, feature data can be transferred (such as with Pro/ENGINEER files), but
that functionality is not the norm. An add-on program named FeatureWorks
will help to recognize feature data from imported geometry. Ask your Solid-
Works reseller for more information on this product.

Table 23–1 outlines all of the file types SolidWorks can import. Although every effort has been made to keep Inside SolidWorks as up to date
as possible, there may be new file types or versions supported by Solid-
Works that are not listed in Table 23–1. Formats with supported versions
listed as N/A (not applicable) are industry standards and the version is essentially irrelevant.

SW 2006: SW 2006 supports AutoCAD DWG and DXF import and export through AutoCAD R2005.

Table 23-1: Supported Formats for Import

File Extension	Description	Imports Into	Versions Supported
DWG	AutoCAD drawing file	Drawing or part	R13 through 2002
DXF	AutoCAD Drawing Exchange Format	Drawing or part	R13 through 2002
DXF (3D solids)	AutoCAD DXF solid geometry	Part	R14 through 2002
DWG, DXF	Mechanical Desktop file (see following)	All	MDT 4.0 and above
X_T, XMT_TXT	Parasolid solid geometry text file	Part or assembly	All
X_B, XMT_BIN	Parasolid solid geometry binary file	Part or assembly	All
IGS, IGES	Initial Graphics Exchange Specification standard	Part or assembly	N/A
STP, STEP	International Standard for the Exchange of Product Data	Part or assembly	AP203 and AP214
SAT	ACIS solid geometry file	Part or assembly	14 and earlier
VDA	VDAFS (German auto industry standard)	Part	N/A
WRL	Virtual Reality Modeling Language	Part or assembly	Yes (not recommended)
STL	Stereolithography file	Part	N/A
PRT, PRT.*, *.XPR	Pro/ENGINEER part file	Part	V17 through 2001
PRT, PRT.*	Pro/ENGINEER Wildfire part file	Part	Versions 1 and 2
ASM, ASM.*, *.XAS	Pro/ENGINEER assembly file	Assembly	V17 through 2001
PRT	Unigraphics Parasolid file	Part or assembly	10 and higher
IPT	Autodesk Inventor part file	Part	7 and earlier
PAR	UGS SolidEdge part file	Part	All versions
PSM (SW 2006 only)	UGS SolidEdge sheet metal file	Part	All versions
ASM	UGS SolidEdge assembly file	Assembly	All versions
PRT	Cadkey part file	Part or assembly	V19
CKD	Cadkey multibody part file	Part	V21
EMN, BRD, BDF, IDB	IDF (Intermediate Data Format) file	Part	Versions 2 and 3
TIF	Tagged Image File Format image file	Part or assembly	N/A
CGR	Catia Graphics image file	Part or assembly	V5 R1 through R13

Special Note on Mechanical Desktop Files: When importing DWG or DXF files, there is an option for importing Mechanical Desktop (MDT) data into part, assembly, or drawing files. Mechanical Desktop (an AutoDesk product) must be installed on the system for this option to be present. Figure 23–2 shows this option, though it is grayed out in the image.

Most software programs will try to use a file extension that is unique to that particular application. For reasons of marketing and logic, however, this is not always the case. A perfect example happens to be the extension *.prt*. It is debatable which company first began using this extension. The *.prt* extension is used by both Unigraphics and Cadkey. It is also used by Pro/ENGI-NEER (Pro/E), but Pro/E files will typically have an additional numeric value following the *.prt* extension, so that it may look like *filename.prt.3*, or something similar.

SolidWorks used a *.prt* file extension in its earlier days (SolidWorks 97 and earlier versions). Assemblies and drawings used *.asm* and *.drw* extensions, respectively. Fortunately, SolidWorks Corporation realized that this was causing problems for some of their customers and switched to the six-character extensions you are now familiar with.

If you attempt to open a file type with the extension *.prt* without first specifying the type of file you are trying to open, the software will think you are trying to open an older SolidWorks file. The software program will not automatically detect the file type, and an error message will be issued. Of course, if the file actually is an old SolidWorks file you need not specify the file type.

Certain file types have special considerations regarding how they import, and there are things you should know about importing certain types of files. There may be other functionality regarding the importation of files that requires further explanation. The following section describes aspects of individual file types that will help you make informed decisions about importing files into SolidWorks.

10-second Topic: DWGEditor

Starting with SolidWorks 2005 and continuing through SolidWorks 2006, a native editor for AutoCAD DWG and DXF files is included with the Solid-Works software. This additional standalone program is known as DWGEditor. The history of this software is an interesting topic of discussion unto itself, but is outside the scope of this book, as is the operation of DWGEditor.

SolidWorks Corporation is obviously trying to help its customers break their ties with Autodesk products, primarily AutoCAD. DWGEditor is not a translator. Rather, it can edit DWG and DXF files natively, and will read or write nearly any version of AutoCAD files. Supported versions range from release 2.5 right up to the most recent version of AutoCAD.

DWGEditor is an AutoCAD clone. It looks and functions much the same as AutoCAD. Its purpose is to give existing AutoCAD users a way to break free from maintaining existing AutoCAD licenses or maintenance fees and still have a way to support their legacy data. With DWGEditor, SolidWorks Corporation has given users the freedom to move forward into a solid modeling world without being mired in the past.

AutoCAD Drawing and DXF Files

DWG and DXF files can be imported into either drawings or parts. They can also be opened in the DWGEditor. If you are already familiar with AutoCAD, and you have a few changes to make to an existing AutoCAD drawing, you may choose to use DWGEditor. If you are not an ex-AutoCAD user, you would be better off importing the files into SolidWorks and saving the files as native SolidWorks drawings. When specifying a DWG/DXF file for importing, the DWG/DXF Import window will present choices as to how the file should be opened. These choices are shown in Figure 23–3.

Fig. 23–3. DXF/DWG Import window choices.

Select the method to open this DXF/DWG file:

○ Edit / view in DWGeditor
◉ Create new SolidWorks drawing
  ◉ Convert to SolidWorks entities
  ○ Embed as a sheet in native DXF/DWG format
    ☑ Link to original file
○ Import to a new part
○ Import MDT data from file (imports as parts, assemblies and/or drawings)

It is recommended that you import files into a drawing rather than a part, even if the file contains geometry that will be used to create a part. If you import a DWG/DXF file into a drawing (rather than a part), layers can be turned on or off as needed. Therefore, only what you need to see is shown. In contrast, part files do not recognize layers.

If in a drawing file, sketch geometry can be copied and pasted into a part file if you wish to use that geometry in developing a solid model. Chapter 24 deals further with issues related to working with imported geometry. This chapter is meant to focus on various file formats and how they can benefit the SolidWorks user.

If embedding a DWG/DXF drawing into a SolidWorks drawing (see Figure 23–3), whether the imported drawing is linked or not there is basically nothing that can be done with the geometry. There will be a button for editing the geometry (not shown), which serves to open the file in DWGEditor. Plainly stated, there are very few reasons to embed a DWG/DXF file into a SolidWorks drawing.

Because DWG/DXF files can consist of many types of geometry, you should understand the limitations in SolidWorks DWG/DXF translator capabilities. Standard 2D geometry will translate well into a SolidWorks drawing. SolidWorks installs AutoCAD style fonts, which helps when importing AutoCAD drawings.

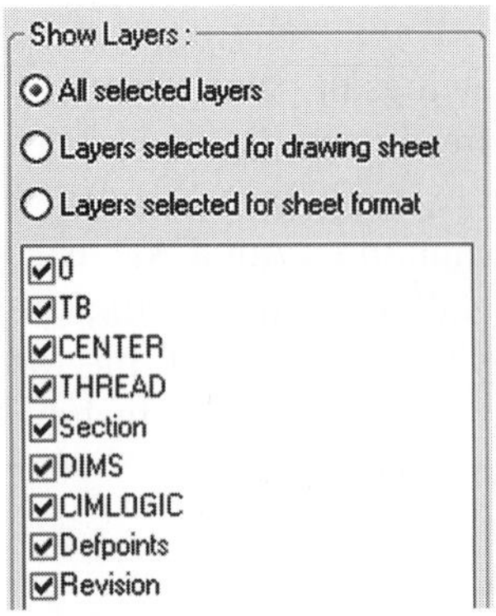

AutoCAD DWG/DXF files can contain solid geometry. Solid-Works can read solid geometry from DXF files only. If you must read in AutoCAD solids, it is preferable to use an IGES or SAT file type when exporting from AutoCAD. AutoCAD wireframe geometry will translate into a SolidWorks drawing, but it will be "flattened," losing its 3D aspect.

The most common option when importing DWG/DXF files is to convert the geometry to SolidWorks entities, which is the option selected in Figure 23–3. Clicking on the Next button opens the Drawing Layer Mapping window, a portion of which is shown in Figure 23–4. The main function of this window is to choose which layers are imported and whether each layer gets placed on the sheet or the sheet format, as indicated in Figure 23–4.

If the drawing being imported contains multiple layouts (typically described as paper space layouts in AutoCAD terms), those layouts appear as tabs in the layer mapping window. These same tabs, shown in Figure 23–5, will become the tabs in a SolidWorks drawing used to navigate between drawing sheets. Name these tabs as desired, or uncheck the checkbox shown in the figure if a particular layout is not needed.

Fig. 23–4. Specifying which layers to place on the sheet or sheet format.

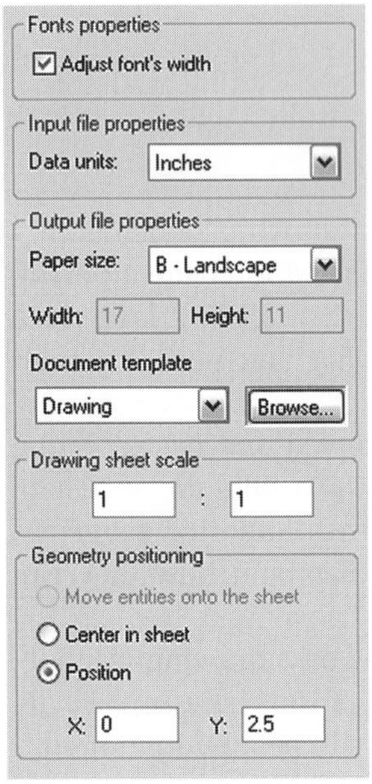

Fig. 23–6. Portion of the Document Settings window.

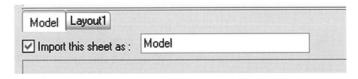

Fig. 23–5. Rename sheet tabs or choose which sheets to import.

Clicking on the Next button brings up the final import wizard window, titled Document Settings. The main purpose of this window is for establishing the drawing units, scale, and paper size. The portion of the Document Settings window that relates to these settings is shown in Figure 23–6. This is all very self-explanatory, and thus details regarding these options are unnecessary.

The Move entities onto the sheet option, grayed out in Figure 23–6, is available only if SolidWorks detects that the x or y coordinate of the lower left-hand corner of the drawing being imported has an x-y location other than 0,0. If you are familiar with AutoCAD, this could mean that there is geometry to the left or below the User Coordinate System (as an example).

If you are going to be importing existing AutoCAD formats, it would be beneficial to first create some SolidWorks templates that do not contain a sheet format. As can be seen in Figure 23–6, one of the options is to select what Document template you wish to use. Obviously, if you will be importing AutoCAD formats you would not want to use a SolidWorks template that already contains one (see "Using Templates" in Chapter 3).

If an AutoCAD file was imported with the intention of saving the data as a SolidWorks sheet format, you should save the newly imported geometry using the File > Save Sheet Format command. Saving a format was discussed in Chapter 12 (see How-To 12–4).

Parasolid Files

The Parasolid translator is probably one of the best methods for translating solid files into (or out of) SolidWorks. It is fast and efficient, and results in small files that can easily be moved from one location to the next. The Parasolid kernel is the solid modeling kernel used by SolidWorks. If you do need to import solid geometry, and the CAD program at the other end of the line can send a Parasolid file, you are better off using this method than any other.

Always use the highest version that both the sending and receiving machines can translate to ensure the highest possible quality and compatibility. This holds true no matter what file format is being used. In addition, binary files are more efficient than text. Thus, if you have a choice, use binary.

IGES Files

It is not difficult to find a technical support person who has received many questions regarding IGES translation. The reason for this is in large part the many IGES translation "flavors" available. There are always a number of settings or options that go along with IGES translation, and if those settings do not coincide with the requirements of the software on the receiving end, the translation does not work well or is incomplete.

When importing an IGES file, you do not have any choice but to hope the translation works. Hopefully the person sending the file knew what type of program you were running and tweaked their export settings correctly. If your goal is to import an IGES file as a solid model, the person sending you the file should make sure they send you IGES trimmed surfaces. You will not want IGES wireframe, as there is very little you will be able to do with the resulting geometry.

IGES curves will translate into SolidWorks, but become 2D or 3D reference curves. This is normally not a desirable situation. On occasion, it may be desirable to import a curve for use as a guide curve or sweep path, but this is rare.

A situation that can develop when importing IGES trimmed surfaces from other programs is that SolidWorks refuses to knit the surfaces. Programs that create strictly surface (as opposed to solid) geometry will probably not translate well into SolidWorks. This has to do with how solid modelers differ from surface modelers.

If this is a troublesome topic for you, try to look at it this way. A solid can be broken down into surfaces, but surfaces do not necessarily have to form a solid. Solid geometry has faces (surfaces) that all meet edge to edge. Surface modelers can create surfaces that do not meet edge to edge. This is the crux of the problem. If the surfaces do not form a closed volume without overlapping edges, the solid modeler cannot knit the surfaces and create the solid geometry.

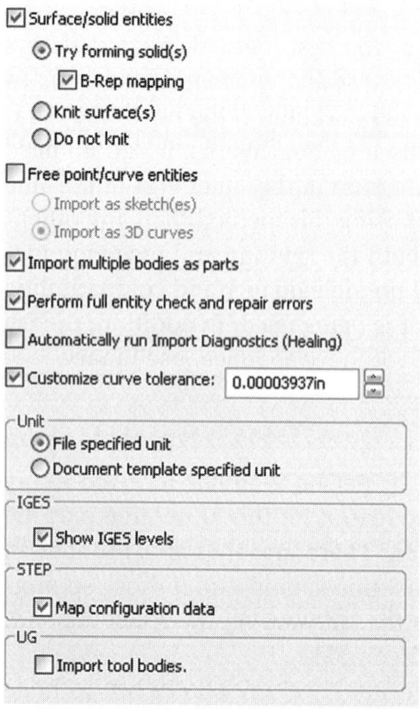

SolidWorks will typically make numerous attempts to knit surfaces to form the solid if at first it does not succeed. Each time another attempt is made, the tolerance used to knit the surfaces is loosened to a less accurate level. If eventually the solid cannot be formed, a series of reference surfaces is created. At this point there is little you can do except request from the person who gave you the file that they try to clean up things at their end. Unfortunately, this is not always a politically feasible situation.

You do have some choices on your end as to how an IGES file is handled during importation. Figure 23–7 shows these options, which can be accessed by clicking on the Options button after selecting IGES as the type of file you wish to open. The Surface/solid entities option should be checked, assuming that is what is being imported, which is almost certainly the case. The General options shown in Figure 23–7 will affect not only IGES files but the import of ACIS, STEP, VDAFS, UGII and Inventor files.

Importing surfaces without knitting them results in reference surfaces that could then be manipulated using surface commands. Knitting the surfaces results in one surface (usually), but not a solid. The most common option is to let the software attempt to form a solid.

Fig. 23–7. General import options.

The B-Rep mapping option is okay to leave checked, as it may be faster, especially for complex models. Boundary representation mapping is a bit different than knitting. Rather than knitting individual surfaces, the topology of the entire model is mapped out.

If *Import multiple bodies as parts* is checked, individual bodies are mapped out to individual part files and placed in an assembly. With the option cleared, individual bodies are imported into a single part file. This setting affects STEP and ACIS file formats only.

The *Perform full entity check and repair errors* option will force Solid-Works to perform extra checks, which will slow down the import process. It will also repair errors where possible, which can be a good thing when importing problem IGES files. However, the extra error checking and repairing may not be necessary.

Import diagnostics can be run automatically during import, but diagnostic tests can also be performed manually after completing the import process. Therefore, it really is not necessary to check the Automatically run Import Diagnostics (Healing) option. Chapter 24 discusses how to diagnose problem geometry once it has been imported. (This option is specific to SolidWorks 2006.)

In summary, when importing IGES files leave Surface/solid entities checked, and use the *Try forming solids* option. Turn on B-Rep mapping, because it is usually faster. Leave the *Perform full entity check and repair errors* option on unless your computer is older or imports are taking a long time. It is almost never necessary to check *Free point/curve entities*, and it usually is not necessary to customize the tolerance.

STEP Files

The STEP file format has been constructed as a multipart ISO standard (ISO 10303). It is a neutral file format constructed for the exchange of data that encompasses product development life cycles rather than strictly CAD data. A number of application protocols have been developed, two of which are AP203 and AP214. These two protocols are supported by SolidWorks.

Its basic parts have been completed and published, but more aspects of the STEP format are still in development. There is only one option associated with STEP importation, which is the *Map configuration data* option. This option can be seen at the bottom of Figure 23–7. Checking this option will result in any configuration data associated with the file being mapped out as custom properties. The Map configuration data option is in no way related to SolidWorks configurations.

ACIS Files

The ACIS solid modeling kernel is used by AutoDesk for its AutoCAD and Mechanical Desktop products. If being sent a solid model from either of these products, asking for an ACIS file is a valid request. Do not ask for solid geometry to be sent via a DWG file, as SolidWorks will not import the data. The DWG format is best suited for 2D geometry and should be used as such.

Virtual Reality Markup Language

VRML, as it is better known, was first thought to be a great new way to view the Internet, with 3D worlds taking the place of the flat 2D and multi-media web pages most of us see on the Web today. Huge 3D vistas and exciting worlds were imagined, in which "virtual reality" could be explored by daring individuals. The entire concept of virtual worlds was further glamorized by Hollywood and the media (Stephen King's *Lawnmower Man* being a perfect example).

Virtual reality on the Web was an interesting concept at the time, but the idea fell flat. Viewing VRML models is often akin to viewing a construction paper model in a dark closet with a flashlight. If your primary concern is to transmit 3D images of your solid files over the Internet, there are much better ways. For instance, SolidWorks offers a free viewer that can be obtained from their web site at *www.solidworks.com*. Specifically, look into SolidWorks' eDrawings and you will not be disappointed. There are other viewers for other file types available as well. Ask your local SolidWorks reseller for more information regarding viewers.

If your primary concern is to transfer solid files from one computer program to another, VRML files (having a *.wrl* extension) should be your last choice. VRML was not designed with file translation in mind. It is recommended that you do not use it for this purpose.

Stereolithography Files

Most individuals would want to export stereolithography (STL) files for reasons of rapid prototyping (discussed in the second half of this chapter). However, STL files can also be imported into SolidWorks. The ability to import an STL file into SolidWorks is a benefit to those operating prototyping machines, because it means they have the capability to make last-minute modifications to a model per a customer's request without the need for other file types. The other benefit is the ability to simply view the file.

Options available when importing STL files are shown in Figure 23–8. Options that apply to STL files also apply to VRML files, but as was already mentioned VRML is not the best choice for file translation. The STL format is also a very poor choice for importing solid geometry.

Opening an STL file as a graphics body allows for viewing the file, but not much else. This is the quickest import option, so if the file needs no editing and you just want to see what it looks like this is the option to use.

Fig. 23–8. STL import options.

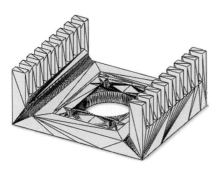

Fig. 23–9. Imported STL file.

Bringing in an STL file as a surface body will result in a single knitted surface, but it will consist of thousands of polygons. This is the nature of STL files, which is to say they are polygonal representations of a solid model. Importing an STL file as a solid body will still result in a polygonal model, but at least it will be a solid. The solid can be edited, but there will be a lot of separate faces to contend with. Figure 23–9 shows an imported heat sink STL file. The polygons, easily visible throughout the model, are why STL is such a poor choice for ordinary data translation.

Pro/E and Other Proprietary Formats

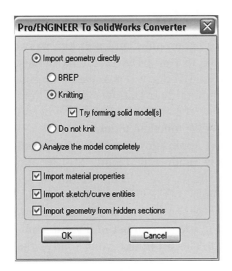

Fig. 23–10. Pro/E to SolidWorks converter.

SolidWorks can open up a whole range of file types, including proprietary file formats from other CAD programs. These proprietary formats include (but are not limited to) files from AutoDesk Inventor, Unigraphics, and Pro/ENGINEER. Recognized proprietary formats were listed in Table 23–1.

Of particular merit is SolidWorks' capability of importing encrypted Pro/E files. What is even better is that in many cases SolidWorks recognizes feature data from these file types. When importing Pro/E files, the Pro/E to SolidWorks Converter window shown in Figure 23–10 will appear.

It is advisable to use *Analyze the model completely,* unless feature data is simply not needed. If the model is analyzed, SolidWorks will attempt to reconstruct the feature data, making the part more amenable to editing. Otherwise, the data is brought in as any other solid model, with no features listed in FeatureManager.

After analyzing the imported geometry, another window will appear, shown in Figure 23–11. This second window is essentially a summarized report stating how many features and surfaces were recognized. Click on the Features button to have the features recreated.

When the summary states that 11 out of 14 features were recognized (as an example), or that all surfaces were not recognized, caution should be used. It would not be advisable to click on the Features button in such a case, as there will most certainly be missing

Fig. 23–11. Pro/E converter summary.

geometry. Click on the Body button instead. Furthermore, it would be advisable to reimport the same file without analyzing the geometry, and then compare both models. In this way, you can visually compare whether anything is being left out when analyzing the geometry.

There is an actual Compare Geometry command available for Solid-Works. The command is part of an add-on program called SolidWorks Utilities. The add-on comes standard with SolidWorks Office Professional. Similar to Microsoft Office, SolidWorks Office Pro is a suite of programs that most engineers and designers would find very useful. Discussing these add-ons in detail is beyond the scope of this book.

➤ **NOTE:** *A translation report window will appear when analyzing the model during import if clicking on the Features button. The report can be used to determine if geometry is missing.*

Dynamic Link Libraries and Add-Ins

Fig. 23-12. Add-Ins window.

Those who dabble in programming will probably realize the benefit of being able to run a Dynamic Link Library (DLL) file from within SolidWorks. It is possible to run custom applications by running their DLL files directly from the File > Open window. This might be a custom program for performing some specific function, for example.

Third-party add-on applications are typically listed in the Add-Ins window (shown in Figure 23-12), and should be started from there. The Add-Ins window is accessed through the Tools menu. To enable an add-in, place a check in front of the add-in and click on OK. Add-ins are available for a large number of diverse applications, and can be purchased through your SolidWorks reseller.

Exporting Files

The list of file types that can be exported is nearly identical to those that can be imported, but with a few extra formats included that are primarily viewing formats. The file types present in the *Save as type* list are dependent on whether you are exporting a part, assembly, or drawing. Options available for file types that can be exported are discussed in material to follow.

To export a SolidWorks document, select Save As from the File menu, and then select the appropriate file type from the *Save as type* drop-down list in the Save As window. The mechanics for saving the file are quite easy. However, when it comes to deciding on the file type or how that file type's export options should be set, the situation can become more complicated.

Most file types that can be exported have their own set of options that can be tweaked by the user. In many cases, the default settings will work fine. In other cases, you may find it necessary to modify the optional export settings. When exporting a file, an Options button will appear on the Save As window, which will gain access to the Export Options window. Screen shots of this window's various sections are included throughout this section for your reference.

The seemingly vast array of export options will not be as intimidating once you understand what many of the options accomplish. The following section is intended to make you feel more comfortable when deciding on an export option for a SolidWorks file. Whether you decide to read through the entire section or merely use it as reference, you should find it helpful.

Some export file types are available for 2D design drawings only, whereas others are available for part and assembly documents. This is noted throughout for the various export file types. In addition, the most common file types and optional settings are recommended. Table 23–2 outlines the available file formats that can be exported from SolidWorks.

Table 23-2: Supported Formats for Export

File Ext.	Description	Exports From	Versions Supported
DWG	AutoCAD drawing file	Drawing	R12 and above
DXF	AutoCAD Drawing Exchange Format	Drawing	R12 and above
X_T	Parasolid solid geometry text file	Part or assembly	All
X_B	Parasolid solid geometry binary file	Part or assembly	All
IGS	Initial Graphics Exchange Specification standard	Part or assembly	5.3
STEP	International Standard for the Exchange of Product Data	Part or assembly	AP203 and AP214
SAT	ACIS solid geometry file	Part or assembly	1.6 through 14
VDA	VDAFS (German auto industry standard)	Part	N/A
WRL	Virtual Reality Modeling Language	Part or assembly	VRML 1.0 and 2.0
STL	Stereolithography file	Part	N/A
PRT	Pro/ENGINEER part file	Part	V20
PRT	Pro/ENGINEER Wildfire part file	Part	V20 (Pro/E)
ASM	Pro/ENGINEER assembly file	Assembly	V20
PDF	Portable Document Format (must enable in Add-Ins)	All	1.4 (Acrobat 5.x)

File Ext.	Description	Exports From	Versions Supported
TIF	Tagged Image File Format image file	All	N/A
JPG	Joint Photographic Experts Group image file	All	N/A
CGR	Catia Graphics image file	Part or assembly	N/A
HCG	Catia CATWeb Highly Compressed Graphics	Part or assembly	V5 R9
HSF	HOOPS streaming Internet graphics	Part or assembly	Version 10
MTS	Viewpoint graphical data (must enable in Add-Ins)	Part or assembly	Version 3.0.11
ZGL	RealityWave streaming Internet graphics	Part or assembly	Version 2

DWG and DXF Export Options

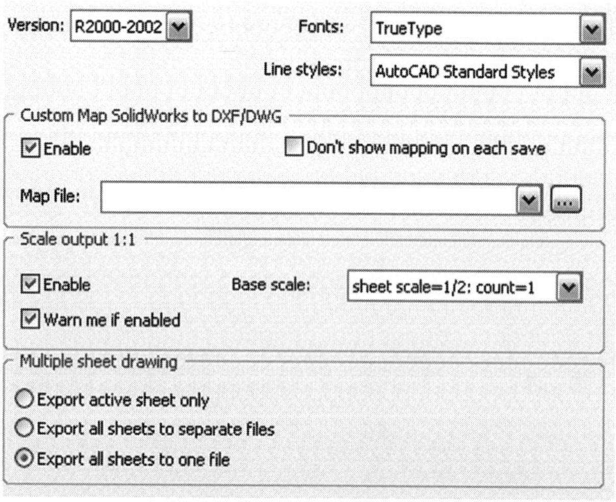

Fig. 23–13. DXF/DWG Export Options category.

The DWG and DXF file formats are available for 2D drawings only. When saving a drawing as one of these file types, click on the Options button in the Save As window to gain access to the optional settings, shown in Figure 23–13. The export options for these two file types are explained in the following material.

Version

When presented with a choice for a version, you should always try to export the file at the highest version possible. Lower versions may not support as many objects as more recent versions. Make sure to check which version of AutoCAD the recipient of the exported drawing is using. Otherwise, they may not be able to read the file.

Fonts and Line Styles

Your two choices for mapping fonts are AutoCAD Standard only and True-Type fonts, either of which are selectable from a drop-down list. If the AutoCAD Standard only option is used, fonts are mapped to standard AutoCAD fonts using the file *drawfontmap.txt*. This file is found in the *SolidWorks\data* folder, and can easily be modified by the user.

If TrueType is specified, the TrueType font used in SolidWorks will be used in AutoCAD if it is available on the computer importing the file. Otherwise, fonts will be mapped out according to AutoCAD's system variables. Consult your AutoCAD User's Guide for more information regarding font mapping and system variables.

AutoCAD version 12 used its own set of fonts, and did not use Windows fonts. In other words, TrueType fonts were not available in that version. It is suggested that you select AutoCAD Standard only when exporting to R12.

If SolidWorks Custom Styles is selected for the *Line styles* setting, Solid-Works line styles are retained. If AutoCAD Standard Styles is selected, Solid-Works line types are mapped out to the stock line styles used by AutoCAD. If saving as R2000 or later, line weights will be mapped out as well.

Custom Map SolidWorks to DXF/DWG

By checking the Custom Map SolidWorks to DXF option, you enable file mapping, and the Map File area is activated. Mapping allows you to specify layers for specific entity types. In other words, you can define layer names for specific SolidWorks entity types, such as text, dimensions, and sketch entities. You can also specify colors and line types for the layers you define. See the section "Creating a Map File," which follows, to see how this is done.

Don't Show Mapping on Each Save

The SolidWorks to DXF/DWG Mapping window can appear every time you use the Save As command and select DXF or DWG as the file type. If you do not want to have the SolidWorks to DXF/DWG Mapping window appear every time, check the *Don't show mapping on each save* option. Mapping will still take place, but the mapping window is not displayed.

Map File

To create a custom map file, you must first save a drawing as a DWG or DXF file. The SolidWorks to DXF/DWG Mapping window will appear at that time, and you will then be able to define the map file and save it. Because it is possible to have more than one map file, the Map File area allows for selecting the map file that will be used whenever you save as a DXF or DWG file.

Scale Settings

AutoCAD users typically design models at a scale of 1:1. This holds true for drawings of those models as well. Sheet formats, often inserted as blocks, are typically scaled up or down to fit the size of the model drawing views. Variables for controlling line type, text, and dimension scale are all changed to adjust for whatever the final print scale will be.

SolidWorks handles things differently. Parts and assemblies are always modeled 1:1. When creating a drawing, a sheet scale is set once and that is all the drafter need worry about. If a drawing with a sheet scale of 1:4 is exported as a DWG file, however, a decision must be made.

By not checking the Enable option, the sheet scale is left intact. This means the drawing will be opened in AutoCAD without the typical 1:1 scaling most AutoCAD users are accustomed to. On the other hand, checking Enable allows for then choosing either the sheet or drawing view to base the 1:1 scale setting on.

If the Enable option is checked, select from the *Base scale* list to determine what view or views the 1:1 scale will be based on. The drawing is saved with a model geometry scale of 1:1 based on the views chosen from the *Base scale* list, and everything else is scaled accordingly. The sheet scale is not used in such a case.

Multiple Sheet Drawings

This setting is self-explanatory, but make sure you know it is there. Otherwise, not all of the data in the original drawing will be included in the DWG or DXF file. Specifically, if you want all sheets to be exported, whether as separate files or all in the same file, make sure you say so.

Creating Map Files

Map files are not for everyone. They are used to specify, for example, what entity types get placed on which layers. It is not necessary to use a map file to maintain layers, line types, and colors of your drawing during exporting. Layer, line type, and color data will export just fine without mapping.

Mapping can override existing layer data, so that exported layers have a particular line type or color. This is an important aspect of mapping you should be aware of. Unless there is a specific need to use mapping, it is a good idea to not even bother with it.

If you decide to use mapping, it is probably because you would like certain layers to have certain properties in the exported file. This could be the case if sending a file to a customer accustomed to working with the typical black background of older versions of AutoCAD. It might be nice to have dark layer colors mapped out to lighter colors for the convenience of the person reading in the file.

Before you can define your first map file, first make sure you have the Custom Map SolidWorks to DXF/DWG option selected (described in the previous section). Once you have completed your SolidWorks drawing, save it as a DXF or DWG file. The SolidWorks to DXF/DWG Mapping window, the upper portion of which is shown in Figure 23–14, will appear automatically.

Typically, the first step in creating a custom map file is to create some layers using the Define Layers tab, shown in Figure 23–14. Following this, the Map Entities tab allows for assigning specific properties (such as color, line type, and layer) to various objects. These objects might be geometric tolerances, center marks, or any other object type in SolidWorks.

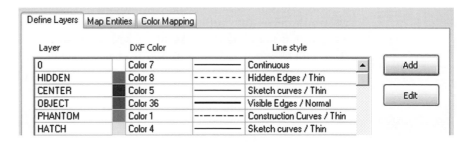

Fig. 23–14.
SolidWorks to
DXF/DWG
Mapping
window.

Finally, there is the Color Mapping tab, which allows for mapping Solid-Works colors to standard DXF colors. For example, you could choose to have the colors blue, red, and green map out to the color black in the exported file. Settings made in the Color Mapping tab will override settings made in the Map Entities tab, which in turn override settings in the Define Layers tab. Therefore, be careful.

There is an option at the bottom of the SolidWorks to DXF/DWG Mapping window (not shown) that is labeled Keep existing SolidWorks drawing layers for entities. Checking this option will ensure that existing layer data is left intact, with only those objects not currently assigned to layers custom mapped. Be forewarned that not checking this option will cause all mapping data to overwrite existing layer data and entity attributes in the exported file.

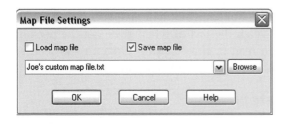

Fig. 23–15. Map File Settings window.

If you click on the Map File Settings button, you will bring up the window of the same name, shown in Figure 23–15. To save a map file, uncheck *Load map file* and check *Save map file*. Map files are saved as simple text files, so they can be edited in a basic text editor. To load an existing map file, check the *Load map file* option and uncheck *Save map file*.

For first-time map file creators, make it a point to save the map file. Specify a name for the file, and a location. Use the Browse button to determine a location if that is easier for you.

Give the file a descriptive name and a .txt extension. The map file will be a text file, but SolidWorks does not apply an extension automatically. This is an extremely rare exception, because file extensions are almost never added by the user. Typically that job is left to the operating system or software.

Once you have established a name and location for the map file in the Map File Settings window, click on the OK button. This will bring you back to the SolidWorks to DXF/DWG Mapping window.

✗ **WARNING:** *Be careful with the Reset All button, because it will wipe out your settings and you will have to start over.*

The entire process for mapping out objects is fairly straightforward, so we won't delve any farther into this topic. However, you should be forewarned that mapping can take place without your knowledge if the *Don't show mapping on each save* option is checked (Figure 23–13). By default, the last map file used is loaded automatically. Considering that map files can cause all of the original layer data to be overwritten, you will want to be very cautious with this setting.

Parasolid Export Options

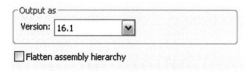

Fig. 23–16. Parasolid export options.

There are few options when it comes to saving a part or assembly as a Parasolid file, as can be seen from Figure 23–16. This is good, because fewer choices mean fewer places to make mistakes. Any solid modeler using the Parasolid kernel (such as SolidEdge and Unigraphics) should be able to read a Parasolid file. Because SolidWorks also uses the Parasolid kernel, this is the translator most highly recommended.

Always try to specify the highest version supported by the target system. Check Flatten assembly hierarchy if you wish all subassembly components to become top-level components in the exported file.

When exporting to a Parasolid file, always go with Parasolid Binary unless you have a good reason not to. Binary files will be smaller than text files, which makes for easier transfer of data, and will load faster during import. Of course, you must also weigh the fact of whether or not the person you are exporting the file to can read it. If a software program can read a Parasolid text file, it should be able to read a Parasolid binary file.

Export Coordinate System

Specifying a user-defined coordinate system is an option when exporting a model, but this is rarely necessary. The option is available for all solid model file formats, such as STEP, IGES, and so on. If a coordinate system has been defined ahead of time, it can be selected from the list shown in Figure 23–17. Unless you have a specific reason to dictate a new coordinate system for the exported model that is different than the default coordinate system, it is safe to ignore this option.

Fig. 23–17. Output coordinate system selection.

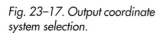

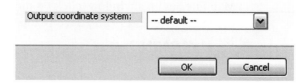

IGES Export Options

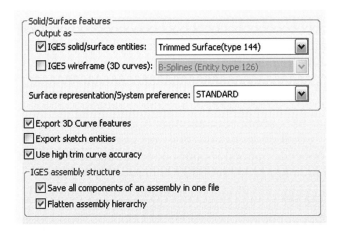

Fig. 23–18. IGES export preferences.

IGES is one of the most frequently used file export options and one of the most common neutral file formats used for file translation. Both parts and assemblies can be exported as IGES files. There are many types of IGES entities. Not all software programs can read all IGES entity types. There are numerous preset selections for specific programs in a drop-down listing, shown in Figure 23–18. Specifically, this listing is labeled *Surface representation/System preference.*

If the target system application is listed in this drop-down list, you should be all set. Programs or named settings include STANDARD (the default setting), Nurbs, Alias, and quite a few others. If the target application is not listed, use the default STANDARD setting. The following sections describe the various options present when exporting IGES files.

Trimmed Surfaces Versus Manifold Objects

Always use the Trimmed Surfaces option if you want to export surfaces that can be imported as solid geometry and are not sure if the target system can read manifold solid B-rep objects. Trimmed surfaces can be imported and knitted into a solid in just about any solid modeling program that accepts IGES files. Not all programs can read in manifold solid B-rep objects, so check the documentation for the target system software prior to using this setting. (Dassault Systemes' CATIA software and EDS's I-DEAS software can read manifold solid B-rep object data.)

IGES Wireframe (3D Curves)

Select the 3D Curves option only if you want to export 3D wireframe geometry. Wireframe geometry can be thought of as a bunch of wires glued end to end. They can be curved, and they can exist in 3D space, but they are not solid geometry. If you select this option, it is highly unlikely the target system will be able to convert the data to solid (or even surface) geometry automatically. It is not suggested you use this option unless the target system owner specifically requests wireframe geometry.

Export 3D Curve Features

Turning on this option allows for features such as composite curves or helical curves to be exported. Typically you would not want to include curves when exporting trimmed surfaces. It just is not necessary, and additional curves may make it difficult to distinguish between the various IGES surface entities once imported into the target system.

Export Sketch Entities

Checking the *Export sketch entities* option will enable SolidWorks to export all 2D and 3D sketch objects, along with all other IGES surfaces being exported. This option is turned off by default, and you will more than likely want to leave it that way under normal circumstances. Similar to the *Export 3D Curve features* option, it just is not necessary to export sketch data.

Use High Trim Curve Accuracy

Checking the *Use high trim curve accuracy* option will result in the exported IGES file being more accurate. If file size is not an issue, which it usually is not these days, leave this setting checked. A higher trim curve accuracy can reduce possible import problems on the target system. If, however, you are trying to keep the size of the file as small as possible uncheck this option.

IGES Assembly Structure Options

If the *Save all components of an assembly in one file* option is checked, an assembly will be saved as a single file as opposed to separate files. In other words, SolidWorks exports an assembly as separate IGES files for each component in the assembly when this option is disabled. It should be noted that checking *Save all components of the assembly in a single file* does not turn the assembly into a single contiguous solid file. It simply exports the assembly as a single file containing multiple components, rather than separate files for each component.

When the *Flatten assembly hierarchy* option is checked, any subassembly components in the original assembly will become top-level components in the exported assembly. The assembly hierarchy will have no depth, as all components will be top-level components. This is known as "flattening"; hence, the name *Flatten assembly hierarchy*.

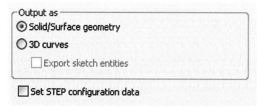

Fig. 23-19. STEP export options.

STEP Export Options

STEP export options include exporting solid and surface geometry, or 3D curves, as can be seen in Figure 23-19. Always export solid and surface geometry whenever possible, as this data is the most complete. Not much can be accomplished on the target system with curve data.

The only time the *Set STEP configuration data* option will be available is when you are exporting STEP Application Protocol 203 (AP203). AP214 does not support configuration data. By checking this option, you gain the ability to enter data relative to the part or assembly you are saving. Data regarding the design or about yourself or your company can be added as well. The Step Configuration Data for Export window will appear when saving the file, a portion of which is shown in Figure 23–20.

Fig. 23–20. Step Configuration Data for Export window.

It certainly is not mandatory to enter STEP configuration data. It is there if you want to use it, but can be ignored if the configuration data is not something that concerns you. STEP configuration data is in no way associated with SolidWorks configurations. Checking the *Set STEP configuration data* option will not affect the model being saved.

ACIS Export Options

For those exporting to Autodesk products such as AutoCAD and Mechanical Desktop, be aware that these products often used old versions of the ACIS kernel. Depending on the version of the Autodesk product being used, it may be necessary to export an ACIS file using a very early version of the kernel. The options associated with ACIS export are shown in Figure 23–21.

ACIS export versions can be anywhere from version 1.6 to the most recent versions in use by CAD systems. If you are exporting to AutoCAD or Mechanical Desktop, be aware of which ACIS file type is being exported. AutoCAD releases 13 or 14 can read ACIS versions 1.5 or 1.6 only. AutoCAD 2000 can only read up to ACIS version 4.0. Mechanical Desktop, on the other hand, can read higher versions of ACIS, but make sure to check what version of Mechanical Desktop the person on the target system is using and what versions of ACIS they can import.

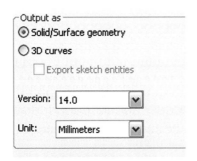

Fig. 23–21. ACIS export options.

If exporting ACIS files, specify the units you want to export into. This is not necessarily the units you are working in but the units you want to be used when the file is imported to another program. Options are millimeters, centimeters, meters, inches, and feet. Additional ACIS output options are the same as for STEP export. If you have any questions about these options, see the section "STEP Export Options."

VRML Export Options

When exporting as VRML, it is most likely that you will be using the exported geometry for reasons of web site development. It has already been stated that VRML should not be used for reasons of file data transfer between two systems. There are much better choices available for accomplishing that task, such as Parasolid or IGES.

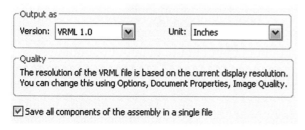

Fig. 23–22. VRML export options.

VRML export options, shown in Figure 23–22, are limited. Choices exist for version (VRML 1.0 being the most recent) and units. If the *Save all components of the assembly in a single file* option is not checked, SolidWorks will create a directory with the same name as your assembly and then create a separate file (with a *.wrl* extension) for every component in the assembly. Otherwise, all assembly data is saved in one file. The option is irrelevant if exporting a part file.

Be aware that image quality settings will affect the size of the VRML file by changing how many polygons are used to represent the VRML model. Adjust display quality via the Image Quality section of Document Properties (Tools > Options).

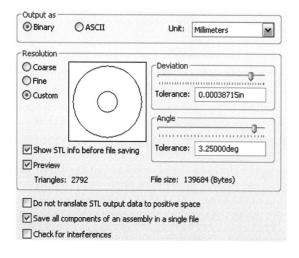

Stereolithography Export Options

Stereolithography (STL) files are polygonal representations of parts or assemblies, and are used for rapid prototyping. After specifying that you wish to save a model as an STL file, click on the Options button. You will then be able to make adjustments that control the size of the polygons that make up the STL file. Export options for STL files (shown in Figure 23–23) are a bit more involved than other file types.

Fig. 23–23. Exporting STL files presents many choices.

STL file export is an important topic, and the rapid prototyping business is a huge industry. Changes made in the STL export settings will directly affect how accurate the STL file is. Accuracy and other settings are discussed in material to follow. Refer to Figure 23–23 when reading the descriptions of the various STL options.

Output Format and Units

The output format can be either binary or ASCII. Most rapid prototyping machines can read binary files, and you should always choose binary files if you have a choice. ASCII files are text based and are much larger than binary. You should be safe creating a binary file unless your rapid prototyping service specifically requests ASCII. As far as the units are concerned, it would be most common to use the same units as the model being exported. However, if the model were modeled in mils or angstroms (as examples), select a more common unit of measurement to minimize problems with the rapid prototyping process.

Resolution

The Resolution section has settings that control the overall accuracy of the STL file. Parts with flat surfaces do not require as high a quality setting as parts with small, curved surfaces. If the quality is set too low for parts with curved surfaces, the curved surface geometry of the prototype may require a good deal of sanding or rework, and small curved areas (such as holes) may not form correctly. Both Coarse and Fine have their quality settings preset. Selecting Custom allows you to use the slider-bar settings Deviation and Angle Tolerance (see Figure 23–23).

Adjusting the Deviation or Angle Tolerance setting too high will result in a very large file. It might take some trial and error before finding the best settings for a particular model. It is better to err on the side of accuracy and wind up with a large STL file. Use the *Show STL info before file* saving option to preview the file size just prior to saving the file. The window shown in Figure 23–24 will appear immediately after clicking on the Save button, and you will still have a chance to back out of the save process and readjust the tolerance quality settings.

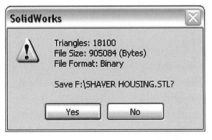

Fig. 23–24. Showing the STL information.

Perhaps a better choice would be to check the Preview option. This will give triangle and file size information in the STL Export Options window. This makes it easier to experiment with the settings and see what the results are right away. If the file is complex and the tolerance settings are set too high, however, you may be in for a wait.

If you do check the Preview option, it will probably be necessary to move the Save As window and the Export Options window prior to clicking on the Options button. Otherwise, those windows will probably block your view of the model, and therefore the preview. Changing the Deviation and Angle tolerance settings will update the preview, triangle, and file size values automatically.

Do not translate STL output data to positive space Option

Due to the nature of rapid prototyping machines, the polygon files must exist in the positive quadrant of an *x-y-z* Cartesian coordinate system. This is a technicality that you as a SolidWorks user should not have to worry about. Most STL machines can move the geometry if it does not already exist in positive space. However, you might as well let SolidWorks move the geometry for you if the geometry needs to be moved. Your rapid prototyping service will have one less detail to worry about, and it takes no effort on your part. Just make sure you leave this option unchecked.

Save all components of an assembly in a single file Option

Typically, STL files are exported one part at a time. You can choose to export an entire assembly as an STL file if you prefer. If you decide to do this, you can pack all of the assembly components into one STL file by checking this option. If you do not check this option, each component will be written to an individual file, using the name of the component as the file name.

Check for interferences Option

This option is only available if the previous option to save an assembly as a single file is checked. Some rapid prototyping systems do not take very kindly to interference. You should probably have already checked your assembly for interference anyway, in which case it would not be necessary to check it again.

Exporting Proprietary Formats

Not all of the proprietary file formats recognized by SolidWorks for importation are also available when exporting files. For example, Inventor, Solid-Edge, Mechanical Desktop, Unigraphics, and Cadkey formats are not available for exportation. Use one of the neutral file formats discussed previously if exporting files for use in these applications.

Pro/ENGINEER parts and assemblies can be exported directly from SolidWorks. This results in a native Pro/E file that can be read in directly by the Pro/E software. SolidWorks currently saves Pro/E files as nonencrypted Pro/E version 20 parts or assemblies.

Other export formats exist, but only pertain to graphics files or streaming web media. They would not be useful to those looking to transfer solid model data to other systems. An example would be saving an assembly as a ZGL file, which is compatible only with the RealityWave viewer.

➥ **NOTE:** *TIF and JGP image formats, discussed in the next section, are not considered proprietary.*

Saving Image Files

Images of SolidWorks documents can be captured from the screen in a couple of ways. One often-neglected method is to simply press the Print Screen button on your keyboard to capture a screen image. After performing this action and starting any basic photo-editing software, you will find that pasting in whatever is in the Windows clipboard results in an image of whatever was on the screen when the Print Screen button was pressed.

There are other more formal methods for saving images of SolidWorks documents. Two types of image formats are supported, which are JPEG and TIFF. The JPEG format (pronounced "jay-peg") derived its name from the committee that wrote the standard, the Joint Photographic Experts Group. JPEG files utilize a *.jpg* file extension.

JPEG files are often related to web page creation, as they are compressed files that transfer well via the Internet. JPEG images are recognized by web browsers and e-mail programs. It is a very common graphics format, which makes it a great choice for sharing graphics data.

TIFF files (rhymes with cliff) derive their acronym from Tagged Image File Format, and utilize a *.tif* file extension. One advantage of TIFF files is that they are platform independent, meaning that they can be viewed and edited on both Windows and Macintosh operating systems, as well as UNIX systems. TIFF files also provide advanced options for various compression schemes (including no compression) and color models, including RGB (red-green-blue) and CMYK (cyan-magenta-yellow-black). TIFF is the format many graphics professionals use. As a matter of fact, TIFF images were used in the development of this book.

Image files can be exported from any SolidWorks document: parts, drawings, or assemblies. Use the Save As command to export a JPEG or TIFF file using the same technique you would use to export any other file type. When saving a JPEG image, there are no export options. What you see in the work area of your SolidWorks screen is exactly what the image will look like. This holds true for any document type, including drawings. Therefore, if the text is not legible in the drawing when saving a JPEG image it will not be legible in the exported image. TIFF images, explored in the following section, are a different matter.

TIFF Export Options

The main reason for saving files as TIFFs is because this file type produces a high-resolution image, required for archival or publishing purposes. If it is your desire to create an image that can be sent to a customer for reasons of adding notes or redlining a drawing, image files work particularly well for this purpose. Anyone with photo-editing software can open and edit a TIFF file.

Reasons for exporting an image of a drawing are often different from those for exporting images of parts or assemblies. Images of drawings are often used to create electronic hard copy for the purpose of archiving. Images of parts or assemblies are saved as high-resolution images for reproduction in a brochure, perhaps for reasons of advertising or company literature. Unfortunately, saving high-resolution images of parts or assemblies is not possible unless additional software is purchased.

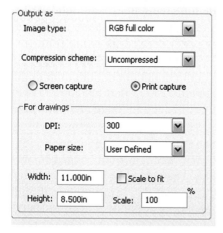

NOTE: *For photorealistic images, look into the PhotoWorks product available from your Solid-Works reseller. PhotoWorks, which offers high-resolution output, can make your SolidWorks models nearly indistinguishable from photographs of real-world objects.*

Regardless of your reason for wanting to save a high-quality image of a SolidWorks document, the following section will help familiarize you with the options available. TIFF export options are accessed by clicking on the Options button in the Save As window, once you have specified TIFF as the file type to save. The TIFF export options are shown in Figure 23–25.

Fig. 23–25. TIFF export options.

Image Type

Always select RGB full color for parts or assemblies. You may also want to use RGB full color with drawings that contain shaded views or colored lines or layers. Select *Black & white (bilevel)* for monochrome images. Do not try to use the *Black & white (bilevel)* setting for anything other than drawings.

Keep in mind that bilevel is strictly black or white. That is, there are no 256 shades of gray. Therefore, if you try to use the black-and-white setting for a shaded part or assembly, or on drawings with shaded views, the results will not be desirable.

Compression Scheme

There are actually more than a few types of compression available for TIFF files, and SolidWorks offers three. This is quite satisfactory for most purposes. Packbits is the compression you should use if saving a color image. Use Group 4 Fax if saving a black-and-white image. Use Uncompressed (no compression) to ensure the highest quality and compatibility.

The Uncompressed setting offers the highest possible quality, although Packbits is a lossless compression scheme. This means that if you wish to use Packbits compression the image should suffer no ill effects in the slightest degree. When the file is viewed by a TIFF image viewer, the file will be uncompressed into an exact duplicate of the original.

Screen Capture Versus Print Capture

If you are saving a TIFF image of a part or assembly, your only option will be Screen Capture. This means that you cannot export the file to a higher resolution even if you wanted to. For this reason, it is best to turn on anti-aliasing and zoom in to the part or assembly as far as possible while keeping all portions of the model on screen. This will make for the best possible image, which can then be downsampled if necessary. It is usually not an issue to reduce an image in size and retain quality, but it is impossible to increase the size of an image without losing quality.

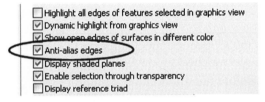

Fig. 23–26. Turning on anti-aliasing.

Anti-aliasing is a method of removing jagged lines from graphic images. Intermediate colors are interpolated between the background and the model color in order to give the appearance of very smooth model edges. To turn anti-aliasing on, select the *Anti-alias edges* option found in the Display/Selection section of System Options (Tools > Options). This setting is shown in Figure 23–26.

If you are saving a TIFF image of a drawing, make sure to use Print Capture. If set to Screen Capture, your drawing will be saved with the same resolution as that displayed on the computer screen, with no regard for scale or dots-per-inch (dpi) settings. This will make for a very poor print if you decide to try to print the image at a later time. About the only benefit of using Screen Capture with drawings is that it is quick.

✓ **TIP:** *Turn on shadows and perspective prior to saving an image of parts or assemblies for a more realistic-looking model.*

When using Print Capture, you can adjust many aspects of the saved drawing file. The settings described in the following material are only available if Print Capture is selected, and as noted previously Print Capture is only available for drawing files.

DPI

DPI stands for dots per inch. The higher the DPI setting, the greater the clarity of the image and its resolution. However, file size will increase as the DPI setting is increased. Use some experimentation to determine what is right for you. You may want to start with a DPI setting of 100. If the file size is not too large for you to work with, and the processing time for the image is reasonable, you can gradually increase the DPI setting to a higher level.

If creating an RGB full color image type with a large sheet size and high DPI setting, the processing time for the image may be quite long. Depending on your computer speed, this could take quite a few minutes. In addition, file sizes can be extreme. As an example, a full-color uncompressed TIFF image of a C-size drawing at a DPI setting of 300 will be right around 100 MB in size. Be reasonable when adjusting the settings for your TIFF file.

Paper Size and Scale

This speaks for itself. All of the typical sheet sizes are available. You would typically want to use a sheet size to match the sheet size of your drawing. If necessary, you can also specify a setting under User Defined, which will make the Width and Height settings available to you. The unit used for the Width and Height options is always millimeters, but you can type in inches as long as the "in" suffix is attached to the end of the numerical values.

If scale is important, such as if people will be taking measurements from the print, you will want to leave the *Scale to fit* option unchecked. If, on the other hand, you simply want to fit your drawing on a particular sheet size and do not care about scale, place a check in the *Scale to fit* option.

If the *Scale to fit* option is not checked, the Scale area will become available so that you can specify a scale. Type in an appropriate value to scale your drawing to the sheet size you want to save the image to. This setting is set to 100% by default. Fifty percent, of course, would be half-scale, and 200% would be 2:1 scale. Enter whatever percentage value you require.

PDF Files

PDF files are not really images in the usual sense, but they may as well be. Adobe is the company that invented the portable document format (PDF) file, and the format has become something of a standard. Many companies use PDF files for archival purposes. To enable the export of PDF files, make sure to check the Save As PDF option in the Add-Ins window. Access Add-Ins from the Tools menu.

To view PDF files, a special viewer is required. The viewer is free, and can be obtained from Adobe's web site at *www.adobe.com/*.

eDrawings

Electronic drawings, or e-drawings, are an excellent means of communicating with others that do not have SolidWorks. Drawings can be animated, displaying all of a drawing's views and dimensions, much to the benefit of the individual on the receiving end. The model can be rotated and examined by the user, who does not need any other software to accomplish this. Best of all, the eDrawings viewer is free.

The eDrawings publishing software is all that is necessary to create an e-drawing. The eDrawing viewer can be obtained by anyone from Solid-Works' web site at *www.solidworks.com/edrawings*. You must have a copy of SolidWorks or AutoCAD (see web site for currently supported version information) to publish eDrawings. Once created, however, the published eDrawing can be viewed by anyone with a computer running Windows.

SW 2006: eDrawings are supported on Apple Macintosh computers running the Panther operating system version 10.3 and higher.

eDrawings are not limited strictly to drawings, but can be generated from parts and assemblies as well. Feature data is not included in an eDrawing. The only data sent via an eDrawing is the data necessary to view the model. This makes eDrawings small and easy to send via e-mail. eDrawings, when saved, can embed the viewer directly into the eDrawing itself. In this way, the file can be self-contained.

There are a few options available when saving an eDrawing, and these options are shown in Figure 23–27. The options contain their own descriptions, but there are a few details you should be aware of. For instance, if you do not wish for the person receiving an eDrawing to be able to take measurements from the document, make sure to uncheck the first option, *Ok to measure this eDrawings file.* On a similar note, the second option makes it possible for a person to create an STL file from an eDrawing of a part or assembly. If sending somebody an eDrawing of proprietary data, it would probably not be in your best interest to give that person the ability to reverse engineer the model or create a prototype of it!

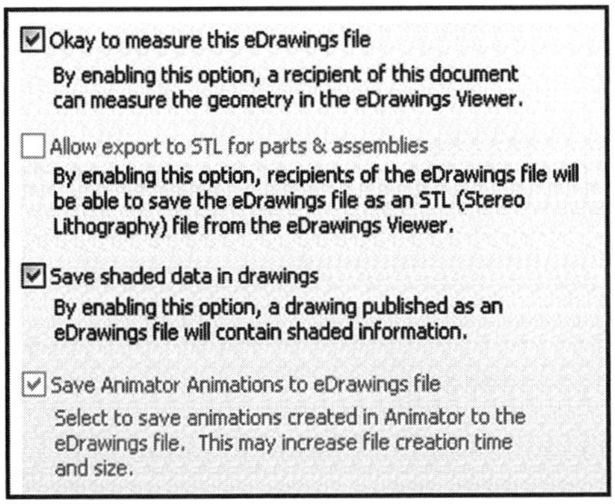

Fig. 23–27. eDrawing export options.

The third option, *Save shaded data in drawings,* can safely be left on. In that way, any views in a drawing that are shaded will appear as you intended them to. The final option, *Save Animator Animations to eDrawings file,* only pertains to those people who have purchased the SolidWorks Animator add-on software, and is self-explanatory.

An example of an eDrawings screen is shown in Figure 23–28. The image does not do the eDrawings viewer justice, as the program really needs to be seen to appreciated. Do yourself a favor and download a free copy. The eDrawings viewer is not limited to eDrawings. It can also view SolidWorks part, assembly, and drawing files. It will even open AutoCAD DWG and DXF files. This is amazing, considering the viewer is free.

If you find it necessary to communicate your drawings to others that do not have the good fortune to own a copy of SolidWorks, consider the eDraw-

Fig. 23–28.
Viewing an
eDrawing.

ings publishing software. It is an excellent collaboration and communication tool. The eDrawing software is included on the SolidWorks installation disk.

➥ **NOTE:** *eDrawings Professional is an enhanced version of eDrawings, and can be purchased as an add-on. The professional version includes redlining and mark-up capabilities, measure tools, and section views.*

Summary

In this chapter you learned that the mechanics behind importing or exporting files are quite easy. The settings behind file translation can be somewhat confusing. When importing files, you are dependent on the person sending you the file. You hope that the person on the sending end is making the appropriate adjustments so that you will be able to read the file. You have learned what types of files SolidWorks can read, so you can now relate this information to the sender and receive a readable file. Use the File > Open command to import a file. Make sure you specify the file type you are importing.

When exporting a file, you would use the Save As command and specify the file type. Exporting a file requires that you are knowledgeable of the applicable settings for specific file types. To make changes to export settings, click on the Options button, which will appear in the Save As window after you have specified what type of file you are exporting.

This chapter is intended to help serve as a guide to sending and receiving files. It should help you make the correct decisions regarding export settings when exporting a file to another party. There are always inherent difficulties when translating files, and no translator is perfect. Always communicate with the party you are translating files to or from, as an awareness of the software on the other end of the line will aid in the translation process.

CHAPTER 24

Working with Imported Geometry

AFTER IMPORTING A MODEL INTO SOLIDWORKS, the first order of business is to make sure the model geometry does not have defects. The model should be a solid model, but this is not always the case. Defective geometry may have gaps or holes in it. Sometimes the geometry can be automatically repaired, and other times the geometry requires repair work from the user.

In a worst-case scenario, geometry not only has gaps but surfaces that fail to translate at all or that are erroneously repositioned. Such is the case with the plate shown in Figure 24–1. Repairing errors this extreme is outside the capabilities of the software to repair automatically. User intervention is required.

Imported geometry with errors similar to the plate leave few choices. One approach is to delete surfaces and manually repair the holes using surface commands learned in previous chapters. Lofted surfaces and the Filled Surface command are extremely useful in such situations. It may also be feasible to move surfaces back into position using the Move/Copy Bodies command discussed in Chapter 20.

Smaller errors are often more difficult to catch, but a sure sign of problems is seeing surfaces in FeatureManager. If there are solid bodies listed in the *Solid Bodies* folder, and no surfaces in the *Surface Bodies* folder, that is a good sign. However, it still doesn't hurt to perform diagnostics. If there are no solid bodies, performing diagnostics is a good idea. How-To 24–1 takes you through the process of performing diagnostics on imported geometry.

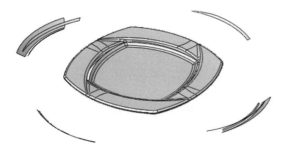

Fig. 24–1. Imported file with serious flaws.

How-To 24-1: Diagnosing Imported Geometry

To diagnose errors in imported geometry, perform the following steps.

Fig. 24–2. Import Diagnosis PropertyManager.

1. Right-click on one of the imported objects in FeatureManager and select Diagnose.

2. In the Import Diagnosis PropertyManager, shown in Figure 24–2, click on the Diagnose button in the Gaps panel to check for gaps.

3. If gaps are found, turn on the Healing Preview option. This will enable the Accept button.

4. Click on Accept to allow SolidWorks to attempt to fix the gap, or click on Close All to try and heal all gaps.

5. Click on Next and repeat step 4 to attempt repair on any remaining gaps.

6. Click on the Diagnose button in the Faces panel (not shown) to check for faulty faces.

7. Click on Fix to attempt the repair of a faulty face, or click on Remove to delete the faulty face.

8. Click on OK when finished diagnosing and attempted repair of the imported geometry.

The Diagnose command is only available on unmodified geometry. If a single feature is added, including the deletion of a single face, the Diagnose command will not be available. In such a case, the Check command found in the Tools menu can be used to diagnose errors on a model. However, unlike Diagnose, the Check command does not allow for the attempted automatic repair of geometry, and provides no tools for repair procedures.

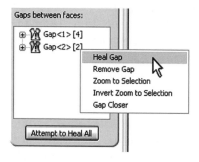

Fig. 24–3. Import Diagnostics in SW 2006.

Repairing problem geometry is an imperfect science. It often takes numerous attempts before an acceptable solution can be found. Manual repair using surface commands is the only solution if automated repair with the Diagnose command fails.

SW 2006: The Diagnose command was revamped in SolidWorks 2006 and renamed Import Diagnostics. The interface has been simplified and repairs are attempted via the right mouse button, an example of which is shown in Figure 24–3. The Gap Closer, available through the menu and visible in the figure, enables handles similar in functionality to connector points when creating a lofted feature (see "Lofted Features" in Chapter 20).

Fig. 24–4. Heal Edges PropertyManager.

In addition to faulty faces or gaps on a model, edges can also be made up of multiple segments. As a simple example, imagine a linear model edge made up of 20 segments rather than a single segment. When trying to repair geometry, it is usually easier to deal with a smaller number of segments on edges. Merging small edges to form a single edge is known as healing.

The process of healing edges is quite simple, and the Heal Edges command can be found in the Insert > Face menu. The Heal Edges PropertyManager is shown in Figure 24–4. Once a face has been selected, and an angular and length tolerance has been specified, click on the Heal Edges button. A report at the bottom of the Faces panel (visible in the image) will list how many edges there were before healing, and how many there are after healing.

Multiple faces can be healed simultaneously. The tolerance settings will cause any edges with values less than those set to be healed. For instance, using the settings in the image as an example, edges whose angle at the adjoining vertex is less than 5 degrees and whose length is less than 1 inch will be healed.

Face Manipulation

The ability to work with faces plays a very large role in editing imported geometry. The main reason this is so is due to the lack of features typical of imported geometry. Adding new features is not an issue, but the only way to manipulate existing geometry is to use face commands.

Patterning and mirroring faces is something you have already learned how to do, though it may not be obvious. Faces are patterned and mirrored using the same Mirror command and pattern commands explored throughout this book. The only difference is in what is selected, and where the selected objects are placed.

Figure 24–5 shows the Mirror PropertyManager, but all of the pattern commands have a similar arrangement. There are three list boxes where selected objects will reside. When patterning or mirroring, a decision must be made as to what type of geometry will be acted upon. Most readers will be accustomed to selecting features, and bodies can be selected as well. Figure 24–5 draws attention to the list box where faces would be selected.

When manipulating faces in one way or another, it is most important to select all relevant faces. If a face or two are mistakenly left unselected, mirror or pattern commands will likely fail. This fact is also true in relation to other face commands, which is explored in material to follow.

Fig. 24–5. Selecting faces.

Deleting and Patching

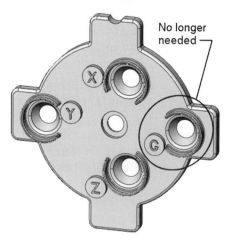

Fig. 24–6. Imported geometry to be modified.

Fig. 24–7. Deleting faces.

What can be done about areas of an imported model that just are not needed any more? Portions of a model can be cut away, and holes could be filled in by adding material to the model. However, there might be a better way.

Figure 24–6 shows an imported model in need of a few modifications. Specifically, the geometry circles in the image are no longer required. They must be removed from the model. The most efficient way to accomplish this task is to use the Delete Face command. To access this command, select Delete from the Insert > Face menu.

There are three modes of operation available when deleting faces, as can be seen from the inset in Figure 24–7. Also shown is the imported part after deleting and patching some of the existing geometry. The countersunk hole has been eliminated, as well as the boss with the raised letter G. All that is left is the thin wall, which originally surrounded the countersunk hole.

When the Delete option is used, faces are removed from the model and no other action is taken. If the model was a solid to start with, it will be reduced to surface geometry. This cannot be avoided, considering that the resultant geometry is not a completely enclosed volume.

The Delete and Patch option was used in the example shown in Figure 24–7. After removing the selected faces, the resultant holes are automatically healed, and the integrity of the solid model is maintained.

Delete and Patch and Delete and Fill are very similar options, and which option you use ultimately depends on the model geometry. Patching extends and trims surfaces to fill the holes resulting from the deleted faces, and works well in most cases. Filling holes creates a single surface and is sometimes more well suited to curvy geometry. A *Tangent fill* option can be enabled that appears only when using Delete and Fill, which can force the resultant surface to conform better with surrounding geometry. Experiment with these settings to see which one works best for particular situation.

10-second Topic: Filtering Techniques

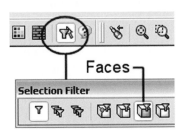

Fig. 24–8. Using the Filter toolbar.

Filters are built into many commands and are employed automatically in many cases. A manual filtering process can also be employed, making it much easier to select certain objects. To enable the Filter toolbar, click on the Toggle Selection Filter Toolbar icon found on the Standard toolbar (shown in Figure 24–8). A small portion of the Filter toolbar is also shown.

Clicking on icons on the Filter toolbar allows whatever object is associated with that icon to be filtered through. For example, the Faces icon is toggled on in Figure 24–8, which means that faces can be selected. Filters can be used in any combination.

Combining window selection with filters makes for a powerful selection tool. When the Hidden Lines Visible display mode is used, window selection of faces will select all faces visible to the user, as well as those usually hidden behind other objects. This makes it much easier to select a large number of what could very well be very small faces. What would otherwise be a tedious process has just become an easy task.

Moving Faces

Fig. 24–9. Move Face PropertyManager.

One last way of manipulating imported geometry we will examine is the ability to move faces. The name of the command, Move Face, is misleading, considering that the command allows for offsetting and rotating faces as well. The Move Face PropertyManager is shown in Figure 24–9, and can be accessed by selecting Move from the Insert > Face menu.

Depending on what action is carried out on the geometry (Offset, Translate, or Rotate), it will be necessary to supply the appropriate parameters. Rotating requires selecting an axis or model edge to rotate faces about. Translating requires selecting an edge to indicate the translation direction. Faces or planes can also be used, in which case the translation direction is normal to the selected face.

Offsetting faces is similar to using the Offset Surface command, but rather than creating a completely new surface the offset faces are merged back into the model and a single solid body is the result. Of course, all of this happens transparently to the designer. Figure 24–10 shows an example of where faces have been moved (or to be more specific, translated). All faces on the raised letter Y were selected, as well as the faces that make up the cylindrical boss on which the letter resides.

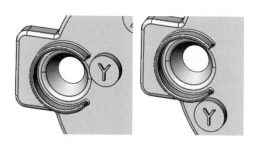

Face selection is critical. When moving the Y boss in Figure 24–10, accidentally neglecting to select one of the small faces on the letter would have caused the command to fail. When rotating or translating faces, it is often necessary to select adjacent faces; that is, faces belonging to surrounding geometry.

Fig. 24–10. Moving faces.

Growing Parts from 2D Drawings

Considering the extremely large number of companies with vast numbers of 2D drawings they would like to turn into solid 3D models, it is no wonder there are individuals looking for a push-button solution to this problem. The fact of the matter is, there is no push-button solution, and there will probably never be one. Creating a 3D model from 2D geometry is a semiautomated process at best. In most cases, it is often just better to create the 3D model from scratch.

Chapter 23 discussed importing drawings, the large majority of which are DWG files, into SolidWorks. Bringing drawings directly into a part file is possible, but not practical. The reason for this is that too much unnecessary data has a tendency to come over with what is needed, which is usually just the geometry required for creating the model. Importing (for example) a DWG file into a SolidWorks drawing, and then bringing over just what is needed, makes much more sense.

How-To 24-2 takes you through the process of moving sketch geometry from a drawing to a sketch in a part file. It should be noted that a valid shortcut for this process is to use the Windows clipboard. In other words, copying and pasting geometry from a drawing to a sketch in a part is possible.

How-To 24-2: Inserting Sketch Geometry from a Drawing

To insert sketch geometry from a drawing, perform the following steps. You must be editing a sketch in a part or assembly before performing this process. It is also recommended that unnecessary layers be turned off in the drawing, or that superfluous geometry and annotations be deleted prior to performing this process.

1. Select Sketch From Drawing from the Insert menu.

2. Using Ctrl-Tab or the Windows menu, activate the window that contains the drawing.

3. Drag a rectangle around the objects you wish to bring into the sketch.

4. Using Ctrl-Tab or the Windows menu, return to the original document. You will find that the geometry from the drawing is now present in the sketch.

It may be necessary to zoom out in order to see the newly inserted geometry. Common practice in AutoCAD is to create drawings with the user coordinate system at the bottom left of the screen. In SolidWorks, this translates to the sketch geometry being positioned above and to the right of the origin point. It is often best to move the inserted sketch geometry to the origin and thereby anchor it in position. However, if attempting to drag the geometry to the origin point the geometry will likely change shape. Read on to see how this can be prevented.

Constrain All

It isn't absolutely necessary to fully define sketch geometry, but it is a good idea, especially if there is any possibility that modifications will be made at some point in the future. With that said, there is a tool that will aid in fully defining imported sketch geometry. The command is called Constrain All, and is located in the Tools > Relations menu.

The Constrain All command has no interface. It simply adds as many geometric relations as it possibly can. This includes most of the basic relations, such as horizontal, vertical, parallel, perpendicular, and tangent. The only stipulation in using Constrain All is that absolutely no existing relations, including dimensions, are present. For this reason, Constrain All should be the very first command performed after inserting sketch geometry from a drawing.

Following utilization of Constrain All, attempting to drag geometry should offer clues as to what additional relations should be added. Use the Undo command (found on the Standard toolbar) to reverse the dragging process if geometry changes shape, and then add relations as needed. Of course, dimensions will be required also, and the perfect tool for this would be the automatic dimensioning capabilities of SolidWorks, discussed in the following section.

Automatic Dimensioning

Automatic dimensioning, otherwise known as autodimensioning in Solid-Works lingo, can save time in certain situations. It is not a command that would be very beneficial if creating a sketch from scratch, largely due to the fact that automatic dimensioning does not do a very good job of imparting your design intent. It also does not know what dimension values you want to use, so it would be necessary to double click on every dimension and alter its value after the fact.

Where autodimensioning is beneficial is sketch geometry imported from other programs. If we assume that the geometry is the right size, all you really want to do is get dimensions on the geometry. Let's look at an example, such as that shown in Figure 24–11. This is a simple stamped metal plate that was imported from a DWG file. Geometric relations have already been added using Constrain All, and some were added manually.

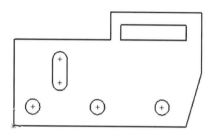

Fig. 24–11. Sketch for a stamped metal plate.

Dimensions need to be added at this point, and then the feature can be created. Autodimensioning makes adding the dimensions easy. Using dimensioning schemes—such as baseline, chain, or ordinate—we can tell SolidWorks what types of dimensions should be added. How-To 24–3 takes you through the process of using the Autodimension function.

How-To 24-3: Autodimensioning

To add dimensions automatically to a sketch, perform the following steps.

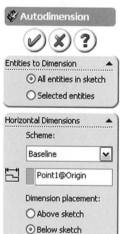

Fig. 24–12. Autodimension PropertyManager.

1. Select Autodimension from the Tools > Dimensions menu, or click on the Autodimension icon typically found on the Dimensions/Relations toolbar.

2. In the Autodimension PropertyManager, a portion of which is shown in Figure 24–12, specify the dimensioning scheme to use for both horizontal and vertical dimensions.

3. Specify a point or line where the dimensioning should start. By default, this is the leftmost point or line for horizontal dimensions and the lowermost point or line for vertical dimensions.

4. Specify where the dimensions should be placed, such as above or below the sketch.

5. Click on OK to create the dimensions.

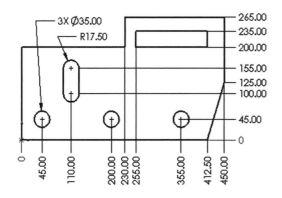

To see how well autodimensioning works, examine Figure 24–13. SolidWorks added 17 dimensions in the blink of an eye. Not only that, but it knew enough to add the proper ANSI standard text *3X* before the hole diameter. This is because the three holes were constrained to be equal prior to autodimensioning.

Fig. 24–13. After autodimensioning the sketch.

Contour Select Tool

Occasionally you may find it convenient to create a sketch containing more geometry than SolidWorks typically allows for feature creation. Reasons vary, but the point is that it isn't necessary to use an entire sketch for a feature. Rather, you can choose specific regions, or contours, with which to "grow" the desired feature. Contours can be open or closed, and can be made up from any geometry in an otherwise invalid sketch.

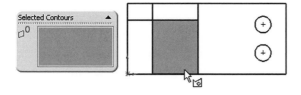

Fig. 24–14. Selecting a region or contour.

In essence, the Contour Select tool makes it possible to use what would otherwise be considered an illegal sketch. This makes working with geometry imported from, for example, an AutoCAD drawing much easier. Examine Figure 24–14, and note that the Contour Select tool highlights a region made up of the surrounding self-intersecting sketch geometry. The highlighted region could then be extruded, as an example.

Multiple contours can be used from the same sketch. If attempting to create a feature from a sketch that would otherwise be considered illegal geometry, the Selected Contours panel (visible as an inset in Figure 24–14) automatically becomes activated.

To access the Contour Select tool prior to creating a feature, right-click in the work area or on a sketch in FeatureManager and select Contour Select Tool. Moving the cursor over the sketch geometry will cause various contours to dynamically highlight. Select the desired contour and create the feature as usual.

⟶ **NOTE:** *Although the Contour Select tool makes it possible to work with otherwise invalid sketch geometry, this tool should not be used as a* *crutch. Contours are limited with what they can be used for, so it is not a good idea to become dependent on this functionality.*

2D-to-3D Conversions

For certain types of parts, there are a series of commands that may make it easier to create a 3D model from existing 2D drawings. There are those who would argue whether these commands are actually worth using, and that recreating the part from scratch using a paper drawing to go by would be easier. In this section, we explore the 2D-to-3D sketch tools, and then you can make a decision for yourself whether or not they are beneficial to you.

The tools that will be explored in this section rely on having a good drawing in DWG or DXF format to begin with. The drawing should have all views necessary to manufacture the part. Import the drawing into a Solid-Works drawing and turn off layers that contain annotations and geometry not directly related to the actual geometry of the part.

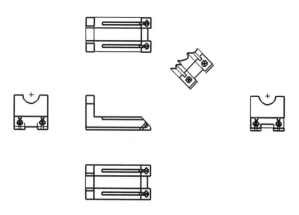

Fig. 24–15. Views inserted from a drawing.

Views that will aid in growing the 3D part out of the 2D geometry are standard top, front, and right side views. Other beneficial views may include additional projected views and auxiliary views. Bring all of these views into a sketch in a part file using the Insert From Drawing command discussed in How-To 24–2. Isometric, section, and detail views should typically be left on the drawing and are not required for building the 3D part. Figure 24–15 shows what might be seen in a part after inserting the appropriate views from a drawing.

Assuming that you basically have a large portion of a 2D drawing in a sketch right now, it may be necessary to do a little extra cleanup work. Get rid of center marks, centerlines, and extra little bits and pieces of leftover 2D geometry not directly related to model geometry. Following this, dictating what view is the front view is what puts the ball in motion for turning the 2D views into a solid. In the part file, turn on the 2D to 3D toolbar, shown in Figure 24–16. The prepara-

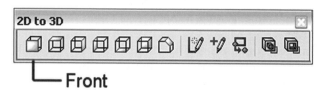

Fig. 24–16. 2D to 3D toolbar.

tory work is completed, and now you are ready to begin. How-To 24–4 takes you through the process of converting the 2D imported geometry into a 3D model.

How-To 24-4: Using 2D-to-3D Sketch Tools

To use the 2D-to-3D sketch tools to convert 2D drawing geometry into a solid model, perform the following steps. It is assumed that you have completed the preparatory work discussed in the material preceding this How-To.

1. Select all of the geometry that makes up the front view of the imported 2D geometry.

2. Click on the Front icon, shown in Figure 24–16.

3. Select geometry that relates to other views of the model, such as the top view, and click on the corresponding view icon on the 2D to 3D toolbar. Repeat this process for all orthographic views.

4. To define an auxiliary view, select the geometry that makes up the auxiliary view, as well as a line from another sketch that could have been used to project the auxiliary view from. Then click on the Auxiliary icon on the 2D to 3D toolbar.

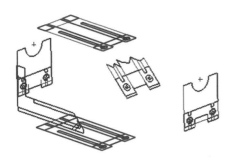

Fig. 24–17. After defining views for 2D-to-3D conversion.

While performing steps 3 and 4, you will undoubtedly have noticed the animation sequences taking place. Views rotate to take up position at their respective locations in 3D space. Figure 24–17 shows what might be seen after performing these first few steps.

What is happening is known in the industry as the "glass box" effect. The views are placed on what seems like a glass box, and are then used to help define the solid model. The next step is to get the views aligned properly.

5. Select a line in a sketch that will be used to align that particular view to another.

6. Select a line in another sketch that the previous view will be aligned to.

7. Click on the Align Sketch icon on the 2D to 3D toolbar.

8. Repeat steps 5 through 7 to align all of the views orthogonal to each other. View geometry does not have to be touching.

With all of the views aligned, the solid model can be created. There are Extrude and Cut icons on the 2D to 3D toolbar that provide additional functionality over the standard Extrude and Cut commands found on the Features toolbar. Use the commands on the 2D to 3D toolbar for the following steps.

9. Select geometry from one of the views from which to create an extrusion. The Select Chain command found by right-clicking on sketch geometry works well.

10. Ctrl-select a point from a second view from which to start the extrusion.

11. Click on Extrude on the 2D to 3D toolbar.

12. Select a point to dictate the extrusion depth. This second point should be from the same sketch the point was selected from to dictate where the extrusion would begin.

13. Supply additional parameters as necessary in the Extrude PropertyManager, and then click on OK to complete the extrusion.

The How-To will end here, as it should give you a good basic outline of what needs to take place to create a part using the 2D-to-3D sketch tools. Additional techniques can be discussed without necessarily listing steps. For example, creating cuts in the model can be accomplished in much the same way extrusions can be. Points on a second sketch can be selected to dictate where the cut terminates. (With regard to the 2D-to-3D sketch tools, the Extrude command refers strictly to boss extrusions, whereas the Cut command can be thought of as a standard extruded cut.)

There are two other tools found on the 2D to 3D toolbar that are convenient tools in their own right. These tools are the Repair Sketch command and the Create Sketch From Selections command. Both commands, discussed in the following sections, can be used outside the 2D to 3D conversion process.

Create Sketch from Selections

There will be a series of sketches created in FeatureManager when the 2D-to-3D sketch tools are used. A new sketch is generated every time a new view is defined from the imported 2D geometry. It is sometimes advantageous to break out some of the geometry from the individual view sketches and then create a new feature using that new sketch.

Selecting geometry from a sketch and then clicking on the Create Sketch From Selections command automatically generates a new sketch. Feel free to use this command anytime it is convenient, even if you never use the rest of the tools on the toolbar.

Repair Sketch

The Repair Sketch command is a command every CAD program should have. Clicking on this command will take a messy sketch and clean it up for you. If a sketch contains multiple collinear line segments, Repair Sketch will convert all of the segments to a single line. If a sketch contains overlapping lines, Repair Sketch will eliminate the overlapping geometry.

Use the Repair Sketch command anytime it becomes necessary to bring geometry from an imported drawing into a part. The command can also be used if troubleshooting a sketch with a problem that is proving elusive. Repair Sketch doesn't fix everything, but it can't hurt to try. If using Repair Sketch to try to fix a problem with a sketch, and the problem remains, use the Check Sketch For Feature command to troubleshoot the sketch (discussed in Chapter 3).

Converting to Sheet Metal

SolidWorks can take model geometry from other CAD programs and turn it into a sheet metal part. This includes unrolling or unbending areas of the model to show it in its flattened state. It also includes the ability to rip out corners in order to flatten a part. This added capability gives the user greater flexibility in designing sheet metal parts because the model does not have to start out as sheet metal.

Geometry, whether imported from another CAD system or created in SolidWorks using non sheet metal design techniques, can be "turned into" a sheet metal part by employing the Bends command. However, not every model will lend itself to the Bends command. Use common sense and be practical with what you attempt to convert to a sheet metal part. For example, shelled parts can often be converted to sheet metal parts once the corners are ripped out. Lofted or swept parts usually cannot.

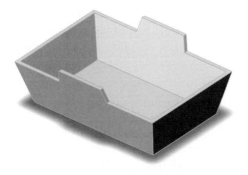

There are two important requirements that must be met by a model that is about to have bends added to it. First, it should be of a uniform thickness. Second, the model must physically be capable of flattening. Picture a box with four sides and open at the top. If the sides are connected, such as with the model shown in Figure 24–18, they cannot be unfolded. If there are openings between the sides, essentially making each side a "flap," the box can be unfolded. Creating these openings is addressed in the following section

Fig. 24–18. This model cannot be unfolded.

Rip Features

Occasionally it is easier to design a sheet metal part as a shelled part to begin with. This is really more of a convenience than anything born out of necessity. Figure 24-18 shows a shelled part. There would be no way to flatten this part, not with the walls connected the way they are. However, performing the Rip command on this model first would allow the part to be flattened.

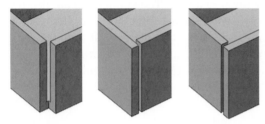

Rip can be found as a separate command, and it is also built into the Bends command. How the Rip command is accessed makes no difference as far as the end result is concerned. When a rip is performed, a small portion of material is automatically removed from where two walls meet. It is possible to specify which side of the corner (which wall) will be modified during the rip, or the entire corner can be removed. Figure 24-19 shows three possible outcomes of performing the Rip command on a model. How-To 24-5 takes you through the process of adding a rip feature.

Fig. 24-19. Three possible outcomes of the Rip command.

How-To 24-5: Adding a Rip Feature

To add a rip feature, perform the following steps.

Fig. 24-20. Rip PropertyManager.

1. Select Rip from the Insert > Sheet Metal menu, or click on the Rip icon found on the Sheet Metal toolbar.

2. Select the edge or edges to be ripped (internal edges only can be selected).

3. Click on the Change Direction button in the Rip PropertyManager (shown in Figure 24-20) as required. Preview arrows will be shown for each edge selected. The arrows will point in the direction of the rip. Two preview arrows indicate that the entire corner will be removed.

4. Uncheck *Use default gap* to specify a value larger than the default value.

5. Click on OK when finished.

SW 2006: The *Use default gap* setting has been removed. If you wish to use a value other than the default value, simply type in the desired gap distance. Also, the limitation on selecting only internal edges has been lifted.

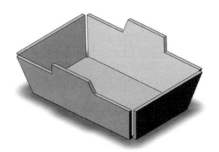

Fig. 24–21. After adding a rip feature.

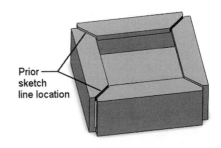

Fig. 24–22. Using sketch lines and edges to rip geometry (SW 2006 only).

All rips do not have to be the same with regard to which side of the edge is ripped out. In other words, rip direction can be mixed in the same model. However, if ripping a series of edges connected end to end the rip direction should be consistent. Figure 24–21 shows the part from Figure 24–18 after performing the Rip command on it. The part is still not considered a sheet metal part as far as SolidWorks is concerned. That detail will be remedied in the next section.

SW 2006: Sketch lines can be used to rip a model. Figure 24–22 shows an example of how four sketch lines, in combination with four model edges, were ripped to create the model shown.

Inserting Bends

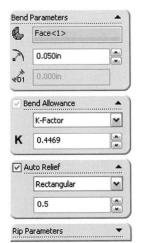

The act of converting a standard SolidWorks part file into a sheet metal part is known as inserting bends. When inserting bends, it is important to remember that a face must be selected that will be held stationary. It is this face that will remain stationary when the part is flattened or folded back up. This is just one of the steps in adding bends to a part. How-To 24–6 takes you through the process of inserting bends. Figure 24–23 shows the Bends PropertyManager displayed when inserting bends.

Fig. 24–23. Bends PropertyManager.

HOW-TO 24-6: Inserting Sheet Metal Bends

To insert sheet metal bends, perform the following steps.

1. Select Bends from the Insert > Sheet Metal menu, or click on the Insert Bends icon located on the Sheet Metal toolbar.

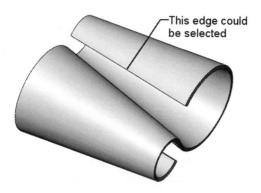

This edge could be selected

Fig. 24–24. Select an edge when there are no flat faces.

2. Select a face to remain stationary when the part is unfolded. If the model is a rolled part without a flat face, such as a cylindrical or conical part, an edge can be selected. See Figure 24–24 for an example of where edge selection would prove useful.

3. Specify a Bend Radius that will be applied to the inside of any sharp corners.

4. Specify a default bend allowance using the method of your choice.

5. Specify whether or not to incorporate Auto Relief, and if so, what type of relief cuts should be made.

6. Click on OK to accept the settings.

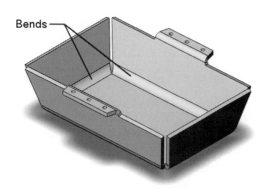

Bends

Fig. 24–25. Sample model after inserting bends.

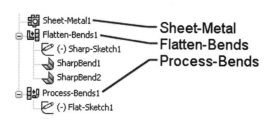

Sheet-Metal1 — **Sheet-Metal**
Flatten-Bends1 — **Flatten-Bends**
(-) Sharp-Sketch1
SharpBend1
SharpBend2
Process-Bends1 — **Process-Bends**
(-) Flat-Sketch1

Fig. 24–26. FeatureManager after inserting a bends feature.

Bends will be added to any sharp corner that existed in the model. The bend radius specified in the Bend Parameters panel (see Figure 24–23) will be applied to the inside of any sharp corner in the original model. Figure 24–25 shows an example of the sample part after adding bends with a fairly large radius. Additional holes and sketched bend features were added to the tabs.

Sheet metal functionality experienced a major overhaul in SolidWorks 2001. Inserting bends is how sheet metal design was implemented prior to SolidWorks 2001. Inserting bends still plays a role in converting imported geometry into a sheet metal part, and also gives users more flexibility in how they design sheet metal parts (such as from shelled models). However, the sheet metal commands discussed in Chapter 9 are more advanced, and should be used when possible for sheet metal design.

The features added to FeatureManager when inserting bends are different than when extruding a base flange (discussed in Chapter 9). Figure 24–26 shows an example of what FeatureManager might look like after adding bends to a model. The *Sheet-Metal1* feature shown in the figure has the same purpose when inserting bends as it does when extruding a base flange, so its description is not repeated here.

Think of the *Flatten-Bends* feature as the place where the bends and the appropriate bend allowances get defined. The part is still in a flattened state (hence the name *Flatten-Bends*). Its definition allows you to change the default bend radius and bend allowance. The *Process-Bends* feature is where the forming of the part takes place. By suppressing this feature (or by simply clicking on the Flattened icon), you can see the sheet metal part in its flattened state.

Examining the outdated processes of editing sheet metal models created by inserting a bends feature would not be a very productive use of time. To get directly to the point, if a sheet metal part is to be created by inserting bends and there are additional sheet metal features that need to be added, use the functionality described in Chapter 9. The *Flatten-Bends* and *Process-Bends* features can then essentially be ignored.

Measure Tool

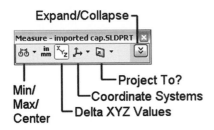

Fig. 24–28. Measure command buttons.

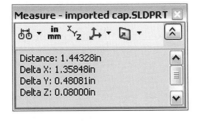

Fig. 24–27. Measure window.

The ability to measure distances on a model or between components in an assembly can be beneficial for obvious reasons. Measuring a model allows for double checking the design and making certain all is as it should be. The ability to measure geometry also plays a role in reverse engineering imported geometry, which is why we are discussing the Measure tool at this point in time.

It should be stated that measuring certain items, such as the length of an edge or the radius of an arc (to name two) does not require the Measure tool. In many cases, all that is required is to select an edge on the model and read what is displayed on the status bar, which is typically visible at the bottom of the SolidWorks window. If the status bar is not visible, turn it on by checking Status Bar in the View menu.

If it does become necessary to utilize the Measure command, select Measure from the Tools menu or Tools toolbar. A small window will appear that will display information depending on the object or objects selected. An example of the measure window is shown in Figure 24–27.

Quite a bit of information can be obtained from the Measure tool. For example, a face's area and perimeter, whether faces are parallel or perpendicular, minimum distance between two selected faces, and so on. Buttons on the Measure windows add to the functionality of the command. These buttons are shown in Figure 24–28. They are described in material following.

The Expand/Collapse button is self-explanatory, and is convenient if the Measure command is a tool you use on a regular basis. Leave the Measure window somewhere easily accessible, and then expand the Measure window when needed. If leaving the Measure command on screen, the cursor will continue to look like a small ruler, which is standard operating procedure when the Measure command is active. To return to the standard cursor and exit measure mode, right-click somewhere in the work area and click on the Select option. This will allow for performing other commands. To return to measure mode, click on the title bar of the Measure window.

The Min/Max/Center switch allows for controlling how distances are measured between circles or circular edges. Delta XYZ values shows changes in distance in the x, y, and z directions when measuring between (for example) two points. When selected, the Coordinate System button will show a list of available coordinate systems, assuming additional coordinate systems have been defined. This might be useful (to cite yet another example) if it becomes necessary to find the x-y-z coordinates of a particular point while referencing something other than the origin point.

The Project To button controls how distances are measured. For instance, an edge's length could be projected onto a particular plane, and the length of the edge as projected could be determined. This is a less common option for most individuals, as the true distance is the value most people will be looking for.

Sketch Pictures

Sketch pictures are images that can be inserted onto a sketch plane, and the functionality has some interesting uses. Images can be used to enhance a model, similar to placing a label on an actual physical part. The limitation is that only planar faces can have images applied to them. To apply an image to a nonplanar face, PhotoWorks is needed. (See your SolidWorks reseller for more information on PhotoWorks.)

Sketch pictures can also be inserted onto a sketch plane for reasons of tracing. This type of functionality would be about as close as one could get to putting a piece of onion skin paper over an object and tracing the outline with a pen or pencil. By inserting an image into a sketch, the shape of the image could be "traced" with sketch tools, such as lines, arcs, splines, and so on.

To insert a picture into a sketch, first start a sketch, and then select Sketch Picture from the Insert > Sketch Tools menu. After selecting the image, the Sketch Picture PropertyManager (shown in Figure 24–29) will appear. Also shown is an example of a sketch picture. The image does not do the Sketch Picture command justice, because being able to rotate a 2D picture in 3D space is a bit strange. You will have to try it to appreciate the effect.

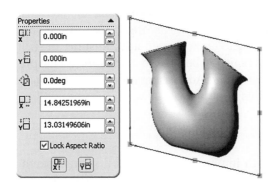

Fig. 24–29. Sketch Picture PropertyManager (left) and sketch picture (right).

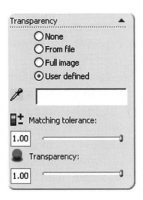

Fig. 24–30. Transparency panel.

Fig. 24–31. Areas that a loose tolerance can't make up for.

The Sketch Picture PropertyManager is very straightforward and does not require much explanation. Images can be resized or moved by dragging the green handles visible when the image is selected. Double clicking on the image will bring up PropertyManager, which also allows for moving and resizing the image, as well as rotating the image.

Sketch pictures appear in FeatureManager beneath the sketch they are attached to. It is not necessary to edit a sketch in order to edit the picture's parameters through PropertyManager, which is very convenient. If a sketch is hidden, its associated image will be hidden as well.

Getting rid of a sketch picture can be tricky, unless you know the "secret." Hopefully, SolidWorks will make this process easier in the future. Until then, deleting the sketch will delete the associated image. If you prefer to delete the image only, double click on the image to bring up the Sketch Picture PropertyManager, and then press the Delete key. The image will be removed from the sketch (and model) at that time.

↦ **NOTE:** *Image files supported by the Sketch Picture command are* .bmp, .gif, .jpg, tif., *and* wmf *files only.*

SW 2006: Image transparency can be controlled through the Transparency panel in the Sketch Picture PropertyManager. The Transparency panel is shown in Figure 24–30. When *From file* is selected, the transparency information present in the image file is used. Relative to most graphic art programs, transparency is defined for a particular color and the degree of transparency is defined as a percentage. For example, if the color yellow is used for the transparent color, any area of the image that is yellow will be invisible at 100% transparency.

The *Full image* option allows for controlling transparency for the entire image, rather than a specific color. *User defined*, which is the option selected in the image, allows for picking a color directly from the image in the work area, thereby dictating what color will be transparent. The Transparency slider bar controls the degree of transparency. The Matching Tolerance slider bar is used when the area that should be transparent does not consist of strictly one color.

Areas in an image may appear to be one color, but in reality are many different similar shades of a color. This is where it may be important to loosen the tolerance for the transparent color. Figure 24–31 shows such an example. A ring has been placed behind the image of the vase. Note the areas surrounding the vase image.

The original background color surrounding the vase has, for the most part, been made completely transparent. However, areas close to the vase had too much of a color variation the tolerance could not adjust for. In such a case, graphic art programs could typically be used to modify the image prior to the image being used in SolidWorks. Addressing functionality of those programs is outside the scope of this book.

Summary

An important part of working with imported geometry is first understanding what form the geometry takes. Imported geometry often has gaps and holes, missing faces, and other errors. This is almost always due to settings or modeling techniques from which the geometry was exported from, and which the SolidWorks user has no control over. Communicate with the party exporting the file and let them know they should send you Parasolid solid geometry when possible, which will yield the best results. Other neutral file formats (such as IGES) are satisfactory, but the exported file should consist of trimmed surfaces.

The usual goal is to have a solid body that can then be edited in Solid-Works. Trouble geometry that cannot be knitted into a solid can be diagnosed and repaired in SolidWorks using a variety of tools. Surfacing commands are often required, but face-manipulation commands may also be beneficial. Face commands also benefit the solid model itself, and allow for making changes to geometry without the luxury of features. In other words, even though FeatureManager contains only a single item (i.e., *Imported1*) individual portions of the model can be modified.

Although there are commands that allow for generating a 3D model from imported 2D drawings, it is often easier to create the solid geometry from scratch. The 2D to 3D conversion tools work best for simple prismatic type parts, but not for complex shaped parts containing much more than a dozen or so features. The Contour Select tool may help when trying to work with what would otherwise be illegal sketch geometry.

Imported sheet metal parts can be recognized by SolidWorks as sheet metal parts by inserting bends. After inserting bends, and there are other edits to be made, it is best to add a sheet metal feature first. This will update FeatureManager to reflect the sheet metal features discussed in Chapter 9. The model can then be edited like any other sheet metal part. If geometry contains corners that do not contain gaps, ripping those corners out will be a necessary course of action prior to flattening. Use the Rip command prior to inserting bends if this is the case.

CHAPTER 25

Customization

THERE ARE MANY WAYS YOU CAN CUSTOMIZE SOLIDWORKS. What appears in the menus, which command icons appear on the toolbars, or keyboard hotkey shortcuts are just the beginning. For instance, one could also modify what tabs appear in the New SolidWorks Document window when beginning a new part, assembly, or drawing.

The Customize command is one of the main ways in which customization can take place. Through this command, it is possible to make modifications to the toolbars, create keyboard shortcuts, and so on. The Customize window is separated into tabs that reflect areas of customizability. The functionality of each of these tabs is described in the sections that follow.

✗ **WARNING:** *Before reading further, a note of caution: If you are very new to the SolidWorks program, be careful when customizing the interface. It is possible to make changes that will impede your productivity when using the software. Always use a good deal of discretion when customizing SolidWorks.*

Prior to performing any customization, make it a point to save your current interface settings through the use of the Copy Settings wizard. In this way, if changes are made that adversely affect your SolidWorks workstation it will be possible to reinstate the previous settings as if nothing were changed. The Copy Settings wizard is also beneficial for those who wish to "clone" their settings and move them to a different machine. System administrators may find the wizard useful for setting up workstations with a preset interface customized to meet their company's demands.

How-To 25–1 takes you through the process of saving system settings. The settings that will be saved include settings established through the System Options (Tools > Options), keyboard shortcuts, menu customization, and toolbar modifications. Settings can be restored using this same process.

HOW-TO 25-1: Saving SolidWorks Settings

To save any combination of system options, keyboard shortcuts, menu or toolbar settings, perform the following steps. Be advised that any combination of these settings can be restored using this same process.

1. Make sure SolidWorks is not currently running.

2. Click on the Copy Settings Wizard command found in the SolidWorks Tools menu of the SolidWorks program group. This is accomplished through the Start button of the Windows operating system.

3. Click on the Save Settings button. This portion of the wizard is also where the system settings could be restored as well.

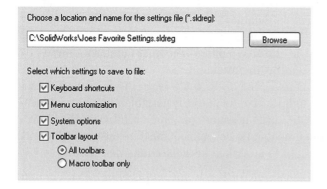

Fig. 25–1. Checking which settings to save.

4. Click on the Next button.

5. Click on the Browse button and type in a name for the file that will contain your settings. Select a location for the file as well, and then click on the Save button.

6. Place a check mark in front of the settings that are to be saved. It is recommended that you use the settings shown in Figure 25–1, though the file name and location will be different.

7. Click on the Finish button.

The file that is saved will have the file extension SLDREG, but actually the file is nothing more than a registry file. System administrators and IT professionals will recognize this fact if the file is opened in a text editor such as Notepad. Double clicking on a file with the SLDREG extension will serve to run the Copy Settings wizard and will allow for restoring the settings saved in the file.

Now that you have successfully saved your settings, it will be much more safe to explore some of the ways in which SolidWorks can be customized. Let's begin with the Customize command itself. To access the Customize window, you must have a SolidWorks document open (part, drawing, or assembly). Select Customize from the Tools menu to open the Customize window. Customize is also available from the View > Toolbars menu. The various tabs available are detailed in the following sections.

Customizing Toolbars

The Toolbars tab is what will appear when first initiating the Customize command. Figure 25-2 shows a portion of the Toolbars tab. You cannot get into much trouble here. Turn various toolbars on or off by clicking on the toolbar's name to toggle the check marks on or off. If a check mark is present, the toolbar is turned on.

Fig. 25-2. Lower portion of the Toolbars tab.

Fig. 25-3. Example of a large tooltip.

The lower portion of this tab has three options, which can be seen in Figure 25-2. The first option, *Large icons*, speaks for itself. The toolbars are not different; the icons are just larger. The second switch, *Show tooltips*, turns the toolbar icon tips on or off. Tooltips are those small yellow boxes that appear when you hold the mouse motionless over an icon. Finally, the *Use large tooltips* option includes a nice description of what the command icon does. A sample of a large tooltip is shown in Figure 25-3. It is highly recommended that you take advantage of large tooltips while learning the software. Once you become familiar with the software, you may decide to turn the tooltips off.

The Reset button will remove any changes made to the Toolbars tab. This will work only for the current session of modifications made. This means that if you make a change and click on OK to exit the window you will not be able to use the Reset button. The Reset button will remove only those modifications made since the Customization window was open.

Resize toolbars by dragging a toolbar's border. This can be done at any time and does not require that you be in the Customize window. Toolbars can be made into different shapes and sizes (as long as the shape is rectangular) by dragging their edges. In Figure 25-4, the Sketch Tools toolbar has been resized and is floating, rather than docked to one side of the screen.

Fig. 25–4. After resizing a toolbar.

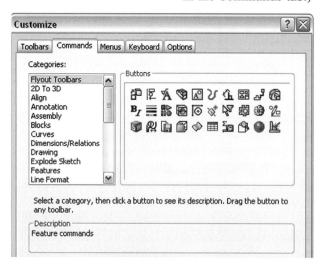

Fig. 25–5. Commands tab.

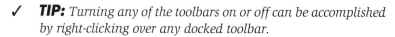

✓ **TIP:** *Turning any of the toolbars on or off can be accomplished by right-clicking over any docked toolbar.*

Toolbar positions are remembered by the software, meaning that after shutting SolidWorks down for the night toolbars should be in the same location after booting up the program the next day. Which toolbars are turned on or off is determined on a document type basis. Chapter 12 discussed a method for setting toolbar positions (see How-To 12–2).

The Commands tab of the Customize window, shown in Figure 25–5, is where toolbar icons can be added to toolbars. Be careful that you do not accidentally delete icons from toolbars when using this tab. (Actually, you can accidentally delete icons whenever the Customize window is open. You do not have to be in the Commands tab.)

It is possible to add, remove, or relocate icons to different positions on most of the toolbars. To add an icon to a toolbar, first select the toolbar category from the Categories list (see Figure 25–5). The icons for that particular toolbar are displayed in the Buttons area. Click on an icon to see a description of what the command does. To add an icon to a toolbar, drag the icon from the Buttons area to the toolbar located elsewhere on the screen. Any icon can be moved from one toolbar to another. To delete an icon from a toolbar, drag the icon off the toolbar and drop it in the work area.

The CommandManager

SolidWorks does not give you the capability of creating your own toolbar per se, but in essence you have that functionality anyway. For example, you could take a toolbar you do not often use and add all of the icons you would like to see present on that toolbar. Not all toolbars can be customized, however, and therefore all icons for all toolbars will not be accessible.

After examining the CommandManager, you may very well decide that having your own personally customized toolbar is no longer necessary. Think of the CommandManager as an all-encompassing intelligent toolbar.

It can be turned on or off, and docked or repositioned, like any other tool-bar. Figure 25–6 shows the CommandManager docked at the top of the screen.

*Fig. 25–6.
CommandManager.*

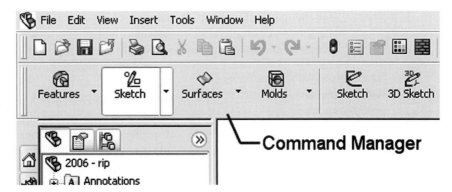

Right-clicking on the CommandManager allows for turning descriptions on or off, or for customizing the CommandManager. Figure 25–6 happens to show the CommandManager with descriptions turned on. When the descriptions are turned off, the CommandManager looks like any other toolbar.

What really differentiates the CommandManager from other toolbars is how it functions. For example, when the Sketch button is depressed (see Figure 25–6), all of the command icons typically associated with the Sketch toolbar will appear in the CommandManager. If the Features button is depressed, all of the command icons typically associated with the Features toolbar will appear, and so on.

Customizing the CommandManager is quite easy, and is done via the right mouse button. Right-click on the CommandManager, select Customize CommandManager, and select which fly-out command icons you would like to see available on the CommandManager. What are fly-out icons? They are basically entire toolbars in single icon format. Clicking on a fly-out icon gains access to all commands that would otherwise be found on that toolbar.

Utilizing the CommandManager is very efficient, and is in fact an ideal way to work. Fly-outs for all of the major toolbars you regularly use can be added to the CommandManager. Turning off the descriptions will make the CommandManager very compact so that it will take up very little screen real estate. You will also find that holding down the left mouse button over an arrow to expand a fly-out will gain access to the commands within. Position the cursor over the desired command and then release the mouse button to initiate that command. In other words, commands can be access with a single well-executed mouse click.

Customizing Menus

All menus can be customized by selecting the Customize Menu command at the bottom of any menu. This will gain access to a list of all items available for that particular menu. In the case of Figure 25–7, the Help menu is being customized. Simply check items you would like to see in the menu, or uncheck items you do not use and would like to see removed from the menu.

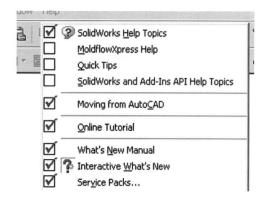

Another way to customize menus is to use the Menus tab of the Customize command. The Menus tab, shown in Figure 25–8, is the area in which you can wreak the most havoc. It is strongly suggested that new users steer clear of this area. There are few reasons most users would want to change menus using this procedure. Generally, leave menu customization in the Menus tab to those who are very experienced with SolidWorks.

Fig. 25–7. Customizing the Help menu.

Fig. 25–8. Menus tab.

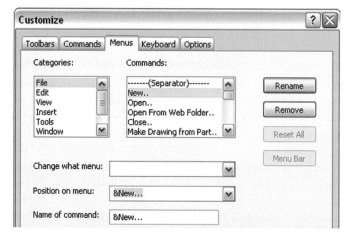

You have the capability of adding, deleting, repositioning, and renaming menu items. The buttons in the Menus tab will change depending on what menu item has been selected. For example, the Rename button will change to an Add or Add Below button in certain cases.

Not all menu items can be renamed, including those menu items added by the user. Top-level menus, such as File and Insert, cannot be renamed—not that you should want to anyway. Although deleting menu items is also possible, it is not advised. It is much easier and safer to uncheck an item using the Customize Menu command found at the bottom of every menu.

Considering the amount of trouble that can be caused by modifying the SolidWorks menus via the Customize command, the steps for using the Menus tab within this command will not be shown. It is also fair to assume that if your level of expertise is such that modifying the SolidWorks menu structure is necessary figuring out how to use the Menus tab should pose no challenge.

Creating Keyboard Shortcuts

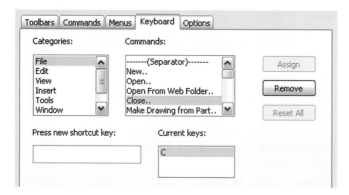

Fig. 25–9. Keyboard tab.

A fairly safe way of customizing SolidWorks is to set up your own keyboard shortcuts. If there is one command you use often, assign a keyboard shortcut to it so that you can easily call up that command. You use the Keyboard tab of the Customize command, shown in Figure 25–9, for this purpose.

The Commands section of the Keyboard tab will reflect whatever document type happens to be open. Therefore, if you wish to add a keyboard shortcut for use with an assembly make sure an assembly document is open. This also holds true for parts and drawings. How-To 25–2 takes you through the process of adding a keyboard shortcut.

HOW-TO 25-2: Adding a Keyboard Shortcut

To add a keyboard shortcut, perform the following steps.

1. Select a menu category from the Categories iist.

2. From the Commands list, select a menu item to assign a keyboard shortcut to.

3. Depress a key on the keyboard that will be used as the shortcut key. You can use key combinations, such as Ctrl + A, or some other suitable key combination. If using a single key, such as the V key, you may need to depress the key more than once.

4. Click on the Assign button.

5. Click on OK when finished.

Fig. 25–10. Keyboard shortcuts are displayed in menus.

After adding a keyboard shortcut, the shortcut key will be listed to the right of the menu item. An example of this is shown in Figure 25–10. Note for example the letter C, which has been assigned to the Close command.

A warning will be issued if an attempt is made to add a keyboard shortcut already in use. The software simply will not allow it. To remove a keyboard shortcut, complete the first two steps listed previously and click on the Remove button.

Macros

In the world of computer software, a macro can be described as a collection of commands that can be recorded and then played back at will to simplify a task. In many programs, a macro is a collection of key strokes that can be saved and then run back to repeat a process that would otherwise be tedious and time consuming. If a macro does not save you time and effort, it is not worth creating in the first place.

In the case of SolidWorks, macros are a collection of SolidWorks API (Application Program Interface) calls. In other words, they are lines of computer code that talk to the SolidWorks program and ask it to do certain things. SolidWorks macros do not remember key strokes. If they did, it would be easy for the majority of SolidWorks users to create their own macros. Because this is not the case, macros have been reserved for those who have some basic understanding of the SolidWorks API and how it works.

There are a number of resources an individual can use to better understand the SolidWorks API. The best method most easily within your reach is to use the Help menu in SolidWorks to access SolidWorks API Help Topics. This is an excellent place to start. For more information regarding the SolidWorks API, contact the reseller where you purchased the software.

Because learning and implementing the SolidWorks API is a specialized field within itself, that area of expertise is outside the scope of this book. You will need to have experience in Microsoft Visual Basic or C/C + + programming languages in order to fully understand SolidWorks macros. Assuming this is the case, *Inside SolidWorks* can at least show you how to create macros.

One way of picking up quite a bit of information regarding the API is to record a macro and then edit it to better understand the internal language

SolidWorks is using to carry out commands. By piecing this information together, you might be surprised how much you can pick up. The Macro toolbar is shown in Figure 25–11. It has been modified a bit, so yours may look slightly different.

The two icons on the right of the toolbar in Figure 25–11 were added to run custom macros that were previously recorded. Customizing the Macro toolbar is covered in upcoming material. How-To 25–3 takes you through the process of recording a macro.

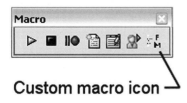

Custom macro icon →

Fig. 25–11. Macro toolbar.

How-To 25-3: Recording a Macro

To record a macro, perform the following steps.

1. Make sure you are at a point in time where you want the macro to start recording. For example, this might be prior to creating a new part, prior to starting a new sketch, and so on. This will make editing the macro easier when the time comes.

2. Click on the Record/Pause Macro button.

3. Carry out any tasks to be recorded. Perform the tasks in the correct sequence, because the order will be reflected in the macro.

4. Click on the Stop Macro icon when finished recording.

5. Type in a name for the macro.

6. Click on Save to save the macro.

At any point during the recording phase, click on the Record/Pause Macro icon to pause the recording process if necessary. Click on the same icon to start recording once again. These actions can be carried out in the previously listed step 3. Once recording has ended and the macro is saved, you will notice that the macro is saved with an *.swp* extension.

Older versions of SolidWorks created text files that had an *.swb* extension (SolidWorks Basic file). These were text files, unlike modern-day macro files. To edit a macro file, you must open the file in an editor capable of reading it. SolidWorks includes the Visual Basic for Applications (VBA) editor for this purpose. How-to 25–4 takes you through the process of editing a macro.

HOW-TO 25-4: Editing a Macro

To edit a macro, perform the following steps.

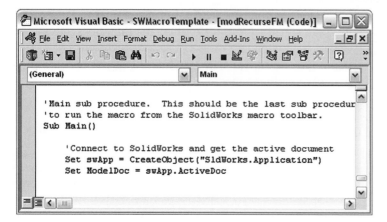

Fig. 25–12. VBA editor.

1. Click on the Edit Macro icon.

2. Select the macro to be edited.

3. Click on Open. This will open the VBA editor, shown in Figure 25–11.

4. Edit the macro as necessary.

5. Select Close and Return to SolidWorks from the File menu to save the macro when finished.

If an old macro with an *.swb* file extension is opened, it will automatically be saved with an *.swp* file extension. As mentioned earlier, *Inside SolidWorks* will not go into the finer aspects of the SolidWorks API, nor will we delve into how to use the VBA editor, as these topics are outside the scope of this book. The VBA editor is not a part of SolidWorks per se, but is licensed by SolidWorks from Microsoft Corporation and installed during a SolidWorks software installation.

For the sake of argument, let's assume you now have a fully functional macro designed to perform some sort of function that will save time and increase productivity. Alternatively, perhaps it does not increase productivity at all, but carries out some specific action, such as saving the FeatureManager as a text file. In any event, now you need to run it. That is the easy part. How-To 25–5 takes you through the process of running a macro.

HOW-TO 25-5: Running a Macro

To run a macro, perform the following steps.

1. Click on the Run Macro icon.

2. Select the macro to be run.

3. Select Open.

Assuming the macro being run is a functional macro without any glitches, the macro will run. You may have to interact with the macro in some way, depending on the nature of the macro. After all, a macro is really just a miniature program. It may include dialog boxes or request information from the user.

If the macro does not work as intended, it is back to the drawing board, or at least the VBA (macro) editor. Assuming the macro works, the Commands tab of the Customize window can be used to assign a macro to an icon. How-To 25–6 will show you how to accomplish this task.

HOW-TO 25-6: Assigning a Macro to an Icon

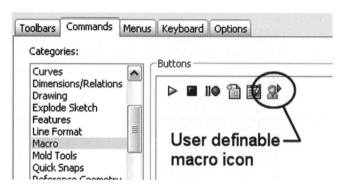

Fig. 25–13. Icon used for customizing the Macro toolbar.

Fig. 25–14. Custom Macro Button window.

Macros can only be assigned to certain icons reserved just for that purpose. Those icons, shown in Figure 25–13, are found in the Commands tab of the Customize window. Actually, there is only one icon, but it can be used over and over. Custom bitmaps (images) can also be assigned to macro icons. To create a macro icon and assign a macro to it, perform the following steps.

1. Select Customize from the Tools menu.

2. Click on the Commands tab.

3. Drag the icon shown in Figure 25–13 to the Macro toolbar (or any other toolbar of your choice). The Custom Macro Button window, shown in Figure 25–14, will appear.

4. Optionally, select a bitmap file to use as the image on the icon (macro button). Images must be 16 by 16 pixels in size, contain only 16 colors, and be saved in the Windows Bitmap file format (*.bmp* file extension).

5. Optionally add a tooltip. This is the yellow box that appears when holding the cursor over the icon.

6. Optionally add a prompt. This is the text that appears in the Status bar when the cursor is held over the icon.

7. Select a macro to run when the icon is clicked on. Use the button adjacent to the Macro listing to browse for a macro, if necessary.

8. Optionally select a method to run from your macro.

9. Optionally establish a hot key to run your macro.

10. Click on OK to finish creating your custom macro button.

Clicking on the corresponding icon on the macro toolbar (or whatever other toolbar you may have added the icon to) will now run the macro. If you added a hot key when defining the new macro button, you will also be able to use that hot key to run the macro. To edit an existing macro icon, right-click on it while the Customize command (Tools menu) is open.

It is not necessary to assign a custom bitmap to a macro icon, but macro icons will be difficult to tell apart if you do not. Some icons are included with the software. Look in the *SolidWorks\data\user macro icons* folder for predefined icons, or feel free to create your own.

Gradient or Image Backgrounds

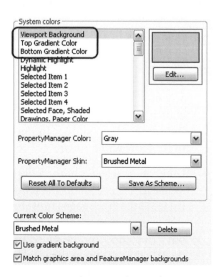

Fig. 25–15. Changing the work area background color.

If you are tired of the default background color, it is possible to change it. Light colors work best, such as an off-white or pastel color. The background can even be made to appear as a color gradient. The settings for changing these colors, located in the Colors section of System Options (Tools > Options), are shown in Figure 25–15.

Many CAD users migrating to SolidWorks from other programs (such as AutoCAD) often ask about changing the background color. The common request is to change the color to black, but it is highly recommended that black or dark backgrounds be avoided. Changing the background to a dark color makes it difficult to see the various color codes SolidWorks employs.

If the gradient function is used, try to avoid using dark colors for both top and bottom gradient colors. You may notice a loss in performance when rotating models on the screen when using gradient backgrounds. It really depends on the graphics card and computer you are running SolidWorks on. Experiment to see if a gradient background degrades display performance on your particular computer. How-To 25-7 takes you through the process of establishing a gradient background, or changing the background color in general.

HOW-TO 25-7: Altering Background Colors

To change the background color of a part or assembly, or to specify a gradient background, perform the following steps.

1. Select Options from the Tools menu.

2. Select the Colors section in the System Options tab.

3. In the listing titled *System colors*, select Viewport Background, Top Gradient Color, or Bottom Gradient Color, depending on what it is you wish to change.

4. Click on the Edit button and select the desired color.

5. Click on OK when finished selecting a color.

6. Repeat steps 3 through 5 for other color settings as necessary.

7. Check the option *Use gradient background* if you wish to use a gradient background. This option is visible near the bottom of Figure 25–15.

8. Click on OK.

Although there are plenty of additional objects that can have their colors modified through the *System colors* listing, it is not a good idea to do so unless you are familiar with the software and have a good idea what it is you are changing. However, most of the other options in the Colors section of the System Options are fairly safe to change and won't cause any harm.

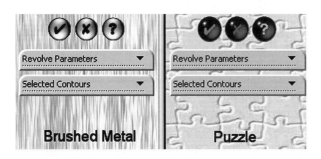

Fig. 25–16. PropertyManager skins.

The *Match graphics area and FeatureManager backgrounds* setting (also visible near the bottom of Figure 25–15) causes FeatureManager to use the same background as the graphics area (work area). This is a nice effect, but experiment to see what you like best. FeatureManager and PropertyManager can have their color changed independently if the aforementioned option is unchecked, and PropertyManager can also be made to use certain skins. Skins are bitmap images that are used similar to using wallpaper in Windows. Figure 25–16 shows a couple examples of skins. Note that the buttons in PropertyManager appear differently as well, which will happen automatically depending on what skin is chosen.

Changing the FeatureManager background color or PropertyManager skin is done through the same Color section of the System Options we have been discussing. The options are named PropertyManager Color and PropertyManager Skin (PropertyManager Color affects both PropertyManager and FeatureManager colors). Likewise, there are predefined schemes that can be chosen from the Current Color Scheme drop-down list. Schemes remember color and skin settings and can be saved, thereby adding to the existing predefined schemes included with the software. How-To 25–8 takes you through the process of saving color schemes, which is a good idea if you have taken the time to get all of your color settings just right.

HOW-TO 25-8: Saving Color Schemes

To save color settings, perform the following steps. Settings that will be saved include any adjustments made to items in the *System colors* list and color or skin settings made to the PropertyManager.

1. Select the Colors section of the System Options tab (Tools > Options).

2. Make changes to system colors as desired. Use the procedure outlined in How-To 25–7.

3. Make changes to the PropertyManager Color or PropertyManager Skin options, as desired.

4. Click on the Save As Scheme button.

5. Type in a name for the scheme to be saved, and then click on OK.

6. Click on OK again to close out of the System Options window.

If you would rather have PropertyManager use the same background as FeatureManager, select None from the PropertyManager Skin drop-down list. To force PropertyManager and FeatureManager to use the same color background as specified for the Windows operating system, select Windows from the PropertyManager Color drop-down list.

PropertyManager skins can be created by those with artistic inclinations. Image files can be created and used as backgrounds (skins) for PropertyManager, but they must be in Windows Bitmap format (*.bmp* file extension). Skins should be placed in the *data\skins* folder under the SolidWorks main installation folder. Color depth is typically 256 colors, but can be 16 million with no ill effects. Tiled images work best, but are not required.

PropertyManager buttons can also be customized. It is not terribly difficult to take a look inside the *SolidWorks\data\skins* folder to see what is already available, and then to adapt some of the existing button bitmap files to suit your needs. The button backgrounds should have an RGB value of 255-0-255 in order to have a transparent background.

If the preceding two paragraphs make absolutely no sense, you should not attempt to create your own PropertyManager skins or custom buttons without performing further research into the matter. This type of customization is not for everyone. If you feel up to the task, the SolidWorks Help file has additional information that will help. Otherwise, further elaboration on this topic is outside the scope of this book.

Using Background Images

Background images can be inserted into parts or assemblies, though there are few technical reasons why this would be necessary. Background images can be used as a backdrop if performing a screen capture, or if saving an image file. If rendering an image, there are much better backdrop alternatives that are present in the PhotoWorks add-on software.

To insert a background image, select Picture from the Insert menu and then select the picture you wish to use as your SolidWorks background. The picture will not be used for every part or assembly, but only the one in which the image was inserted.

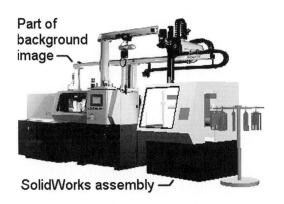

Part of background image

SolidWorks assembly

Fig. 25–17. Using a background image.

Figure 25–17 shows an example of using an image as a background to an assembly file. The image in this case is of a large robotic assembly. The SolidWorks assembly is on the right. Photorealistic rendering software is much more well suited to this sort of task, but for those who do not own such software a background image is a nice alternative.

SolidWorks will only accept TIFF files (*.tif* file extension) as the file format used for a background image. The image will scale with the size of the window, and you will have no control over this behavior. Background images are embedded objects that are saved with the file. In other words, the original image will no longer be needed in order to display the SolidWorks document with the image background.

Be advised that only Packbits compression (a lossless compression scheme) is recognized by SolidWorks. Uncompressed TIFF files are also recognized. If inserting a TIFF image as a background and an *Unexpected File Format* error is received, it is probably because the TIFF file uses a compression scheme unrecognized by SolidWorks.

If you have added a background image and then decide you want to turn the image off, uncheck Picture in the View > Display menu. Checking the same option brings the picture back. To replace the picture being used with another, select Replace from the View > Modify > Picture menu. To delete the image, select Delete from the View > Modify > Picture menu.

Summary

In this chapter, a number of methods of customizing the SolidWorks software were explored. One such method was to use the actual Customize window. This allows for customizing a number of aspects of the software. It can also be used for turning toolbars on or off.

Icons can be moved to or from toolbars using the Command tab of the Customize window. Icons can be added to toolbars you may find useful that are not otherwise present on a toolbar. SolidWorks does not stop a person from placing icons on toolbars where they do not really belong, so some care should be taken when customizing the toolbars.

Menus can be customized to a large extent, but this is an area in which you should tread very carefully. If you would like to customize the menus, it is advisable to avoid the Menus tab of the Customize command. Rather, it is much safer to use the Customize Menu command found at the bottom of every menu. This latter method is easy to implement and easy to reverse should it become necessary to bring a command back.

Hot keys can be established using the Keyboard tab of the Customize window. This allows for adding shortcut keys to make certain tasks more easily accessible. The software will not allow for adding a shortcut key that has already been assigned to another task. Macros can be created in SolidWorks, but they require some basic understanding of software programming and knowledge of the SolidWorks API.

Changing the background color of the work area can be accomplished quite easily, but try to avoid using dark colors, as they will make it difficult to distinguish some color codes used by SolidWorks. Background colors can be chosen for FeatureManager, and bitmap files can be used as backgrounds for PropertyManager. Various schemes that employ these different backgrounds can also be saved by the user.

Finally, TIFF files can be inserted as background images in part and assembly files. Backgrounds can serve to enhance the model aesthetically if performing a screen capture, or simply to show an added degree of realism in the model.

Index

Toolbar
Quick Reference

LEARNING SOLIDWORKS MAY BE FAIRLY EASY, but as with learning any software figuring out and remembering where everything is can often be a daunting task. To help make that task easier, this section depicts in alphabetical order every SolidWorks toolbar and labels the icons (commands) found on each toolbar. This provides you with a convenient quick reference for learning the various toolbars and for efficiently accessing the functionality the toolbars provide. If you cannot find a particular icon, this is the place to do so quickly.

Toolbar layout can change from workstation to workstation, and toolbars are customizable. It should also be noted that the toolbars pictured here have every conceivable icon added to each toolbar. The sole reason for this is to name each icon for reference purposes. Most users would not want to add every icon to every toolbar for a number of reasons. Too many icons result in screen clutter, and there would invariably be icons you would probably never use anyway.

The Toolbars

Throughout the book, such as during the step-by-step How-To sections, you will be asked to access a command via the menus or by clicking on an icon. If you choose to access the command by clicking on the applicable icon, the toolbar the icon is typically found on will be specified in the How-To. Simply look up the toolbar name here and the icon you are seeking will be pictured and labeled. Some icons are found on more than one toolbar.

Align Toolbar

Used to align dimensions and notes, this toolbar is only needed when working in a drawing.

Annotation Toolbar

This toolbar is used to add various annotations to a document. Annotations are most frequently added to drawings, though they can also be added to a part or assembly.

Align

- Group
- Ungroup
- Align Collinear/Radial
- Align Parallel/Concentric
- Align Left
- Align Right
- Align Top
- Align Bottom
- Align Horizontal
- Align Vertical
- Align between Lines
- Space Evenly Across
- Space Evenly Down
- Space Tightly Across
- Space Tightly Down

Annotation

- Note
- Balloon
- AutoBalloon
- Stacked Balloons
- Surface Finish
- Weld Symbol
- Caterpillar
- End Treatment
- Geometric Tolerance
- Datum Feature
- Datum Target
- Hole Callout
- Cosmetic Thread
- Revision Symbol
- Center Mark
- Centerline
- Multi-jog Leader
- Dowel Pin Symbol
- Area Hatch/Fill
- Block
- Model Items
- Hide/Show Annotations

Assembly Toolbar

Contains commands common to assembly modeling and can be turned off while working in part or drawing documents.

Block Toolbar (SW 2006 Only)

Allows for creating and manipulating sketch blocks within part and assembly files (not to be confused with drawing blocks or symbols).

Curves Toolbar

Contains commands associated with defining curves.

Assembly

- Insert Components
- New Part
- New Assembly
- Large Assembly Mode
- Hide/Show Components
- Change Transparency
- Change Suppression State
- Edit Component
- No External References
- Smart Fasteners
- Mate
- Move Component
- Rotate Component
- Replace Components
- Replace Mate Entities
- Exploded View
- Explode Line Sketch
- Interference Detection
- Assembly Transparency
- Simulation Toolbar

Block

- Make Block
- Edit Block
- Insert Block
- Add/Remove
- Rebuild
- Save Block
- Explode Block

Curves

- Split Line
- Project Curve
- Composite Curve
- Curve Through XYZ Points
- Curve Through Reference Points
- Helix and Spiral

Dimensions/Relations Toolbar

Contains commands associated with adding various dimensions and geometric relations to geometry. This toolbar will be useful when working in either part or drawing documents.

Drawing Toolbar

Contains commands common to drawing documents only and is useful in developing various drawing views.

Explode Sketch Toolbar

Explode Sketch is a very specialized toolbar used in creating explode lines in an exploded view and is only needed when working in an assembly document.

Dimensions/Relations

- Smart Dimension
- Horizontal Dimension
- Vertical Dimension
- Baseline Dimension
- Ordinate Dimension
- Horizontal Ordinate Dimension
- Vertical Ordinate Dimension
- Chamfer Dimension
- Autodimension
- Add Relation
- Automatic Relations
- Display/Delete Relations
- Scan Equal

Drawing

- Model View
- Projected View
- Auviliary View
- Section View
- Aligned Section View
- Detail View
- Relative View
- Standard 3 View
- Broken-out Section
- Horizontal Break
- Vertical Break
- Crop View
- Alternate Position View
- Empty View
- Predefined View
- Update View

Explode Sketch

- Route Line
- Jog Line

Features Toolbar

The Features toolbar is needed when creating parts and contains commands for adding features of all types.

Features			Features
Extruded Boss/Base			Revolved Boss/Base
Swept Boss/Base			Lofted Boss/Base
Thicken			Extruded Cut
Revolved Cut			Swept Cut
Lofted Cut			Thickened Cut
Cut With Surface			
Fillet			Chamfer
Rib			Scale
Shell			Draft
Move Face			Simple Hole
Hole Wizard			Dome
Shape			Deform
Indent			Flex
Wrap			Move/Size Features
Suppress			Unsuppress
Unsuppress With Dependents			Linear Pattern
Circular Pattern			Mirror
Curve Driven Pattern			Sketch Driven Pattern
Table Driven Pattern			Split
Combine			Join
Delete Solid/Surface			Heal Edges
Imported Geometry			Insert Part
Move/Copy Bodies			

Formatting Toolbar

The Formatting toolbar appears automatically when adding a note in any document. This toolbar cannot be customized.

Layer Toolbar

Used only when working with a drawing document, the Layer toolbar can dictate what layer is currently active. This toolbar cannot be customized.

Line Format Toolbar

The Line Format toolbar can be used to change the appearance of geometry in a drawing, but also of annotations and sketch geometry in a part or assembly.

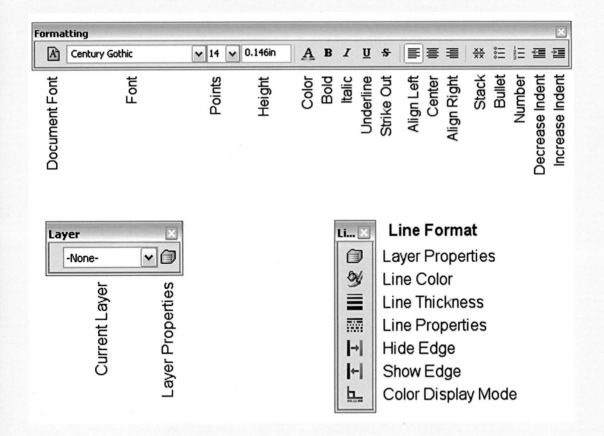

Macro Toolbar

Display this toolbar only if you wish to utilize macro functionality in SolidWorks. The Macro toolbar is the only toolbar with an icon (titled New Macro Button) whose picture can be customized.

Mold Tools Toolbar

The Mold Tools toolbar contains commands specific to mold making.

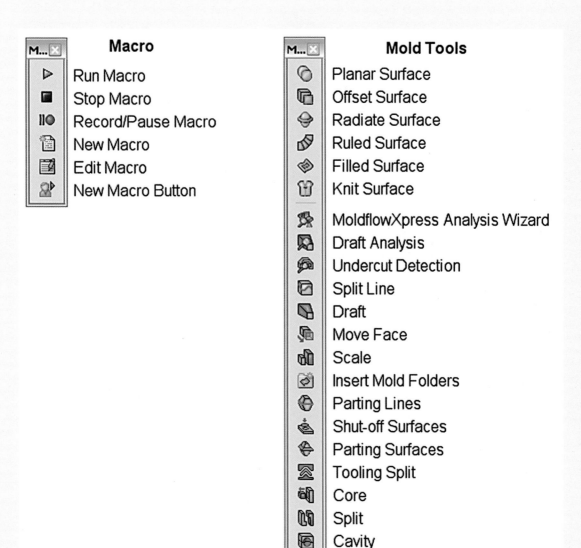

Macro

▷	Run Macro
■	Stop Macro
IIO	Record/Pause Macro
	New Macro
	Edit Macro
	New Macro Button

Mold Tools

	Planar Surface
	Offset Surface
	Radiate Surface
	Ruled Surface
	Filled Surface
	Knit Surface
	MoldflowXpress Analysis Wizard
	Draft Analysis
	Undercut Detection
	Split Line
	Draft
	Move Face
	Scale
	Insert Mold Folders
	Parting Lines
	Shut-off Surfaces
	Parting Surfaces
	Tooling Split
	Core
	Split
	Cavity

Quick Snaps Toolbar

Use the Quick Snaps toolbar to turn specific snap functions on or off. It is very common to leave most snaps on all the time, in which case it is not necessary to display this toolbar.

Reference Geometry Toolbar

Creating reference geometry in part or assembly documents is very common. Therefore, this toolbar should be turned on when working with those document types.

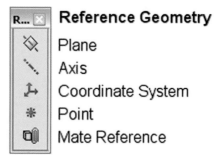

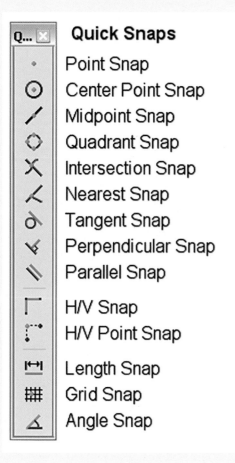

Selection Filters Toolbar

This toolbar toggles on or off the ability to select certain types of objects. Because the Selection Filter toolbar itself can easily be toggled on or off via an icon on the Standard toolbar (aptly named Toggle Selection Filter Toolbar), it is not necessary to leave this toolbar on all the time.

Selection Filter		Selection Filter
Toggle Selection Filters		Clear All Filters
Select All Filters		
Filter Vertices		Filter Edges
Filter Faces		Filter Surface Bodies
Filter Solid Bodies		Filter Axes
Filter Planes		Filter SKetch Points
Filter Sketch Segments		Filter Midpoints
Filter Center Marks		Filter Centerlines
Filter Dimensions/Hole Call-outs		Filter Surface Finish Symbols
Filter Geometric Tolerances		Filter Notes/Balloons
Filter Datum Features		Filter Weld Symbols
Filter Weld Beads		Filter Datum Targets
Filter Cosmetic Threads		Filter Blocks
Filter Dowel Pin Symbols		Filter Connection Points
Filter Routing Points		

Sheet Metal Toolbar

The Sheet Metal toolbar is used exclusively with sheet metal parts.

Simulation Toolbar

The Simulation toolbar allows for creating motion simulation in assembly documents.

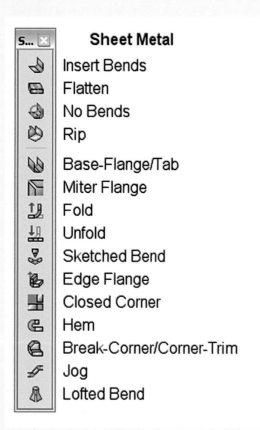

Sheet Metal

	Insert Bends
	Flatten
	No Bends
	Rip
	Base-Flange/Tab
	Miter Flange
	Fold
	Unfold
	Sketched Bend
	Edge Flange
	Closed Corner
	Hem
	Break-Corner/Corner-Trim
	Jog
	Lofted Bend

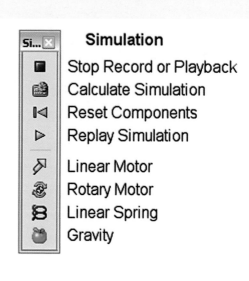

Simulation

	Stop Record or Playback
	Calculate Simulation
	Reset Components
	Replay Simulation
	Linear Motor
	Rotary Motor
	Linear Spring
	Gravity

Sketch Toolbar

The Sketch toolbar contains all commands necessary for creating sketch geometry, and is commonly left turned on at all times.

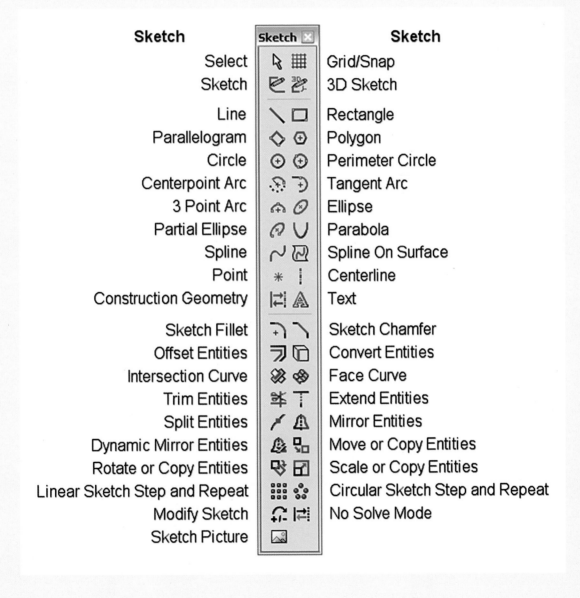

Sketch		Sketch
Select		Grid/Snap
Sketch		3D Sketch
Line		Rectangle
Parallelogram		Polygon
Circle		Perimeter Circle
Centerpoint Arc		Tangent Arc
3 Point Arc		Ellipse
Partial Ellipse		Parabola
Spline		Spline On Surface
Point		Centerline
Construction Geometry		Text
Sketch Fillet		Sketch Chamfer
Offset Entities		Convert Entities
Intersection Curve		Face Curve
Trim Entities		Extend Entities
Split Entities		Mirror Entities
Dynamic Mirror Entities		Move or Copy Entities
Rotate or Copy Entities		Scale or Copy Entities
Linear Sketch Step and Repeat		Circular Sketch Step and Repeat
Modify Sketch		No Solve Mode
Sketch Picture		

SolidWorks Office Toolbar

If SolidWorks Office Professional has been installed on your computer, this toolbar will be available. It is a very convenient method of turning various add-ins (additional software included in the SolidWorks Office Professional suite) on and off.

Spline Tools Toolbar

Turn the Spline Tools toolbar on only if working with spline entity types is a common occurrence in your daily routine.

Standard Toolbar

Leave the Standard toolbar on at all times, as it contains commands for even the most basic SolidWorks functions, such as opening and saving documents.

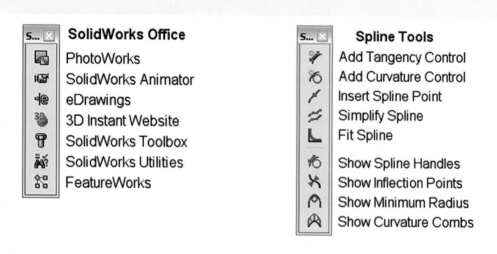

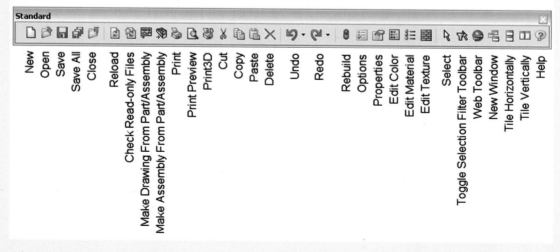

Standard Views Toolbar

The Standard Views toolbar allows for easily changing views, and should always be turned on when working with parts or assemblies. It is not needed when creating drawings.

Surfaces Toolbar

The Surfaces toolbar contains commands associated with surface creation and editing. Typically associated with creating part documents, surfacing functions are usually reserved for complex geometry and are not needed for common mechanical elements.

Table Toolbar

The Table toolbar is most commonly associated with drawing documents, and allows for creating hole or revision tables, to name two.

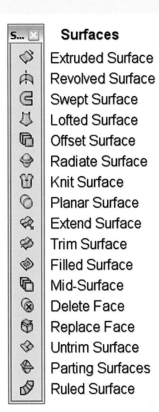

Tools Toolbar

Turn the Tools toolbar on to access commonly used commands such as Measure or Mass Properties.

2D To 3D Toolbar

The 2D To 3D toolbar is a specialized toolbar and is used to create a 3D model from imported 2D geometry. Normally, this toolbar can remain turned off.

View Toolbar

The View toolbar is arguably the second most commonly used toolbar in SolidWorks, and should be left on at all times.

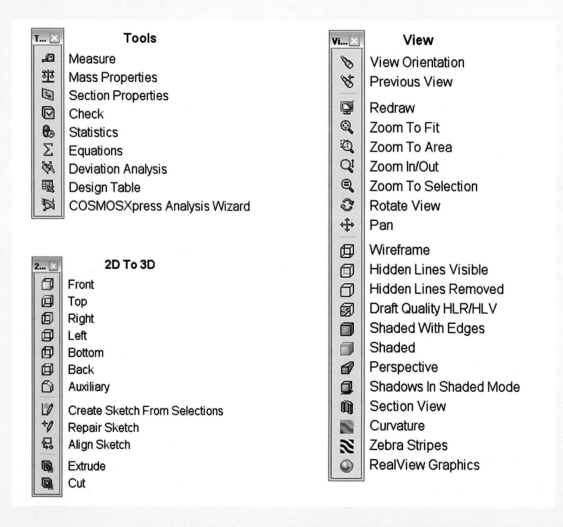